T0211334

Undergraduate Texts in Mathematics

For other titles Published in this series, go to
www.springer.com/series/666

James J. Callahan

Advanced Calculus

A Geometric View

 Springer

James J. Callahan
Department of Mathematics and Statistics
Smith College
Northampton, MA 01063
USA
callahan@math.smith.edu

ISSN 0172-6056
ISBN 978-1-4939-4070-7   ISBN 978-1-4419-7332-0 (eBook)
DOI 10.1007/978-1-4419-7332-0
Springer New York Dordrecht Heidelberg London

Mathematics Subject Classification (2010): 26-01, 26B12, 26B15, 26B10, 26B20, 26A12

Springer is part of Springer Science+Business Media (www.springer.com)

To my teacher, Linus Richard Foy

Preface

A half-century ago, advanced calculus was a well-defined subject at the core of the undergraduate mathematics curriulum. The classic texts of Taylor [19], Buck [1], Widder [21], and Kaplan [9], for example, show some of the ways it was approached. Over time, certain aspects of the course came to be seen as more significant—those seen as giving a rigorous foundation to calculus—and they became the basis for a new course, an introduction to real analysis, that eventually supplanted advanced calculus in the core.

Advanced calculus did not, in the process, become less important, but its role in the curriculum changed. In fact, a bifurcation occurred. In one direction we got calculus on n-manifolds, a course beyond the practical reach of many undergraduates; in the other, we got calculus in two and three dimensions but still with the theorems of Stokes and Gauss as the goal.

The latter course is intended for everyone who has had a year-long introduction to calculus; it often has a name like *Calculus III*. In my experience, though, it does not manage to accomplish what the old advanced calculus course did. Multivariable calculus naturally splits into three parts: (1) several functions of one variable, (2) one function of several variables, and (3) several functions of several variables. The first two are well-developed in Calculus III, but the third is really too large and varied to be treated satisfactorily in the time remaining at the end of a semester. To put it another way: Green's theorem fits comfortably; Stokes' and Gauss' do not.

I believe the common view is that any such limitations of Calculus III are at worst only temporary because a student will eventually progress to the study of general k-forms on n-manifolds, the proper modern setting for advanced calculus. But in the last half-century, undergraduate mathematics has changed in many ways, not just in the flowering of rigor and abstraction. Linear algebra has been brought forward in the curriculum, and with it an introduction to important multivariable functions. Differential equations now have a larger role in the first calculus course, too; students get to see something of their power and necessity. The computer vastly expands the possibilities for computation and visualization.

The premise of this book is that these changes create the opportunity for a new geometric and visual approach to advanced calculus.

*   *   *

More than forty years ago—and long before the curriculum had evolved to its present state—Andrew Gleason outlined a modern geometric approach in a series of lectures, "The Geometric Content of Advanced Calculus" [8]. (In a companion piece [17], Norman Steenrod made a similar assessment of the earlier courses in the calculus sequence.) Because undergraduate analysis bifurcated around the same time, Gleason's insights have not been implemented to the extent that they might have been; nevertheless, they fit naturally into the approach I take in this book.

Let me try to describe this geometric viewpoint and to indicate how it hangs upon recent curricular and technological developments. Geometry has always been bound up with the teaching of calculus, of course. Everyone associates the derivative of a function with the slope of its graph. But when the function becomes a map $\mathbf{f} : \mathbb{R}^n \to \mathbb{R}^p$ with $n, p \geq 2$, we must ask: Where is the graph? What is its slope at a point? Even in the simplest case $n = p = 2$, the graph (a two-dimensional surface) lies in $\mathbb{R}^4$ and thus cannot be visualized directly. Nevertheless, we can get a picture if we turn our attention from the graph to the image, because the image of $\mathbf{f}$ lies in the $\mathbb{R}^2$ target. Computer algebra systems now make such pictures a practical possibility. For example, the *Mathematica* command `ParametricPlot` produces a nonlinear grid that is the image under a given map of a uniform coordinate grid from its source. We can train ourselves to learn as much about a map from its image grid as we learn about a function from its graph.

How do we picture the derivative in this setting? When we dealt with graphs, the derivative of a nonlinear function f at the point a was the linear function whose graph was tangent to the graph of f at a. Tangency implies that, under progressive magnification at the point $(a, f(a))$, the two graphs look more and more alike. At some stage the nonlinear function becomes indistinguishable from the linear one. There are two subtly different concepts at play here, depending on what we mean by "indistinguishable." One is *local linearity* (or *differentiability*): $f(a + \Delta x) - f(a)$ and $f'(a)\Delta x$ are indistinguishable in the technical sense that their difference vanishes to greater than first order in Δx. The other is *looking linear locally*: the graphs themselves are indistinguishable under sufficient magnification. For our function f, there is no difference: f is locally linear precisely where it looks linear locally.

There is a real and important difference, though, when we replace graphs by image grids, as we must do to visualize a map $\mathbf{f} : \mathbb{R}^2 \to \mathbb{R}^2$ and its derivative $d\mathbf{f_a}$. We say $\mathbf{f}$ is *locally linear* (or *differentiable*) at $\mathbf{a}$ if $\mathbf{f}(\mathbf{a} + \Delta \mathbf{x}) - \mathbf{f}(\mathbf{a})$ and $d\mathbf{f_a}(\Delta \mathbf{x})$ are indistinguishable in the sense that their difference vanishes to greater than first order in $\|\Delta \mathbf{x}\|$. By contrast, we say $\mathbf{f}$ *looks linear locally* at $\mathbf{a}$ if the image grid of $\mathbf{f}$ near $\mathbf{a}$ is indistinguishable from the image grid of $d\mathbf{f_a}$ under sufficient magnification. To make the difference clear, consider the quadratic map $\mathbf{q}$ and its derivative at a point $\mathbf{a} = (a, b)$:

$$\mathbf{q} : \begin{cases} u = x^2 - y^2, \\ v = 2xy; \end{cases} \qquad d\mathbf{q_a} = \begin{pmatrix} 2a & -2b \\ 2b & 2a \end{pmatrix}.$$

Because the derivative exists everywhere, $\mathbf{q}$ is locally linear everywhere. Moreover, $\mathbf{q}$ also looks like its derivative under sufficient magnification as long as $\mathbf{a} \neq \mathbf{0}$. But at the origin, $\mathbf{q}$ doubles angles and squares distances, and continues to do so at any magnification. No linear map does this. Thus in no open neighborhood of the origin does $\mathbf{q}$ look like any linear map, and certainly not its derivative, which is the zero map. (There is no contradiction, of course, because the difference between $\mathbf{q}$ and its derivative vanishes to second order at the origin.)

Quite generally, a locally linear map $\mathbf{f} \colon \mathbb{R}^n \to \mathbb{R}^n$ need not look linear at a point; however (as our example suggests), if the derivative is invertible at that point, the map *will* look linear there. In fact, this is the essential geometric content of the inverse function theorem. Here is why. By hypothesis, a linear coordinate change will transform the derivative into the identity map. The local inverse for $\mathbf{f}$ that is provided by the theorem can be viewed as another coordinate change, one that transforms $\mathbf{f}$ itself into the identity map, at least locally. Thus $\mathbf{f}$ must look like its derivative locally because a suitable (composite) coordinate change will transform one into the other. This leads us, in effect, to gather maps into geometric equivalence classes: two maps are equivalent if a coordinate change transforms one into another. In other words, a class consists of different coordinate descriptions of the same geometric action. The invertible maps together make up a single class. (Geometrically, there is only one invertible map!)

For parametrized surfaces $\mathbf{f} \colon \mathbb{R}^2 \to \mathbb{R}^3$, or more generally for maps in which the source and target have different dimensions, invertibility of the derivative is out of the question. The appropriate notion here is maximal rank. Then, at a point where the derivative has maximal rank, the implicit function theorem implies that the map and its derivative once again look alike in a neighborhood of that point. Coordinate changes convert both into the standard form of either a linear injection or a linear projection. For each pair of source–target dimensions, maps whose derivatives have maximal rank at a point make up a single local geometric class.

A nonlinear map can certainly have other local geometric forms; for example, a plane map can fold the plane on itself or it can wrap it doubly on itself (like $\mathbf{q}$, above). The inverse and implicit function theorems imply that all such local geometric forms must therefore occur at points where the derivative fails to have maximal rank. Such points are said to be *singular*. The analysis of the singularities of a differentiable map is an active area of current research that was initiated by Hassler Whitney half a century ago [20] and guided to a mature form by René Thom in the following decades. Although this book is not about map singularities, its geometric approach reflects the way singularities are analyzed. There are further connections. In 1975, I wrote a survey article on singularities of plane maps [2]; one of my aims here is to provide more detailed background for that article.

We do analyze singularities in one familiar setting: a real-valued function f. The target dimension is now 1, so only the zero derivative fails to have maximal rank. This happens precisely at a critical point, where all the linear terms in the Taylor expansion of f vanish. So we turn to the quadratic terms, that is, to the quadratic form Q defined by the Hessian matrix of the function at that point. Taylor's theorem assures us that the Hessian form approximates f near the critical point (up to terms that vanish to third order). We ask: does f also look like its Hessian form near that point?

Some condition is needed; for example, $f(x,y) = x^2 - y^4$ does not look like its quadratic part $Q(x,y) = x^2$ near the origin. Morse's lemma provides the condition: f does look like Q near a critical point if the Hessian matrix has maximal rank. That is to say, a local coordinate change in a neighborhood of the critical point will transform the original function into its Hessian form, in effect, removing all higher-order terms in the Taylor expansion of f.

A nondegenerate Hessian therefore has an invariant geometric meaning, but only at a critical point. At a noncritical point, even concavity, for example, fails to be preserved under all coordinate changes. More generally, if linear terms are present and "robust" in the Taylor expansion of f at a point (i.e., they define a linear map that has maximal rank), quadratic and higher terms have no invariant geometric meaning. This is the implicit function theorem speaking once again.

By asking whether a map looks like the beginning of its Taylor series, we are led to see the underlying geometric character of the inverse and implicit function theorems and Morse's lemma. The question thus provides a way to organize and unify much of our subject and, in so doing, to bring out its simple beauty.

Let me now describe the geometric approach this book takes to another of its central themes: the change of variables formula for integrals.

To fix ideas, suppose we have a double integral, so the change of variables is an invertible map of (a portion of) the plane. Locally, that map looks linear. Each linear map has a characteristic factor by which it magnifies areas. To a nonlinear map we can therefore assign a *local* area magnification factor at each point, the area magnification factor of its local linear approximation at that point. This is the Jacobian.

In the simplest case, the integrand is identically equal to 1, and the value of the integral is just the area of the domain of integration. A change of variables maps that domain to a new one with, in general, a different area. If the map is linear, and has area multiplication factor M, the new area is just M times the old (or the integral of the constant M over the old domain). However, if the map is nonlinear, then we need to proceed in steps. First subdivide the old domain into small regions on each of which the local area magnification factor M (the Jacobian) is essentially constant. The area of the image of one small region is then approximately the product of its own area and the local value of M, and the area of the entire image is approximately the sum of those individual products. To get better approximations, make finer and finer subdivisions; in the limit, we have the area of the new region as the integral of the local area multiplication function M over the original domain. For an arbitrary integrand, transform the integral the same way: multiply the integrand by M. All of this is easily generalized from two to n variables; areas become n-volumes.

A typical proof of the change of variables formula proceeds one dimension at a time; this tends to submerge the geometric force and meaning of the Jacobian M. By contrast, my proof in Chapter 9 follows the geometric argument above. I found it in an article by Jack Schwartz ([16]), who remarks that his proof appears to be new; he could not find a similar argument in any of the standard calculus texts of the time.

*   *   *

One way I have chosen to stress the geometric is by concentrating on what happens in two and three dimensions, where we can construct—with the help of a computer algebra system as needed—illustrations that help us "see" theorems. And this is not a bad thing: the words *theorem* and *theatre* stem from the same Greek root $\theta\varepsilon\overline{\alpha}$, "the act of seeing." In a literal sense, a theorem is "that which is seen." But the eye, and the mind's eye not less, can play tricks. To be certain a theorem is true, we know we must test what we see. Here is where proof comes in: *to prove* means "to test." The cognate form *to probe* makes this more evident; probate tests the validity of a will. Ordinary language supports this meaning, too: yeast is "proofed" before it is used to leaven bread dough, "the proof of the pudding is in the eating," and "the exception proves the rule" because it tests how widely the rule applies.

In much of mathematical exposition, *proving* is given more weight than *seeing*. Jean Dieudonné's seminal *Foundations of Modern Analysis* [4] is a good example. In the preface he argues for the "necessity of a strict adherence to axiomatic methods, with no appeal to 'geometric intuition', at least in the formal proofs: a necessity which we have emphasized by deliberately abstaining from introducing any diagram in the book." As prevalent as it is, the axiomatic tradition is not the only one. René Thom, a contemporary of Dieudonné and Bourbaki, followed a distinctly different geometric tradition in framing the study of map singularities, a study whose outlines have guided the development of this book. Although proof may be given a different weight in the geometric tradition, it still has a crucial role. I believe that a student who sees a theorem more fully has all the more reason to test its validity.

But there is, of course, usually no reason to restrict the proofs themselves to low dimensions. For example, my proof of the inverse function theorem (Chapter 5, p. 169ff.) is for maps on $\mathbb{R}^n$. It elaborates upon Serge Lang's proof for maps on infinite-dimensional Banach spaces [10, 11]. Incidentally, Lang points out that, in finite dimensions, the inverse function theorem is often proven using the implicit function theorem, but that does not work in infinite dimensions. Lang gives the proofs the other way around, and I do the same. Furthermore, because there is so much instructive geometry associated with implicit functions, I provide not just a general proof but a sequence of more gradually complicated ones (Chapter 6) that fold in the growing geometric complexity that additional variables entail. I think the student benefits from seeing all this put together. Other important examples of n-dimensional proofs of theorems that are visualized primarily in $\mathbb{R}^2$ are Taylor's theorem (Chapter 3), the chain rule (Chapter 4), and Morse's lemma (Chapter 7). The definition of the derivative gets the same kind of treatment as the proof of the implicit function theorem, and for the same reason. Unlike the other topics, integral proofs are mainly restricted to two dimensions. One reason is that the many technical details about Jordan content are easiest to see there. Another reason is that the extension to higher dimensions is straightforward and can be carried out by the student.

At a couple of points in the text, I provide brief *Mathematica* commands that generate certain 3D images. Because programs like *Mathematica* are always being updated (and the *Mathematica 5* code I have used in the text has already been superceded), details are bound to change. My aim has simply been to indicate how easy it is to generate useful images. I have also included a simple BASIC program that calculates a Riemann sum for a particular double integral. Again, it is not my aim to advocate for a particular computational tool. Nevertheless, I do think it is important for students to see that programs do have a role—integrals arise out of computations—and that even a simple program can increase our power to estimate the value of an integral.

To help keep the focus on geometry, I have excluded proofs of nearly all the theorems that are associated with introductory real analysis (e.g., those concerning uniform continuity, convergence of sequences of functions, or equality of mixed partial derivatives). I consider real analysis to be a different course, one that is treated thoroughly and well in a variety of texts at different levels, including the classics of Rudin [15] and Protter and Morrey [14]. To be sure, I am recalibrating the balance here between that which is seen and that which is tested.

This book does not attempt to be an exhaustive treatment of advanced calculus. Even so, it has plenty of material for a year-long course, and it can be used for a variety of semester courses. (As I was writing, it occurred to me that a course is like a walk in the woods—a personal excursion—but a text must be like a map of the whole woodland, so that others can take walks of their own choosing.) My own course goes through the basics in Chapters 2–4 and then draws mainly on Chapters 9–11. A rather different one could go from the basics to inverse and implicit functions (Chapters 5 and 6), in preparation for a study of differentiable manifolds. The pace of the book, with its numerous visual examples to introduce new ideas and topics, is particularly suited for independent study. From start to finish, illustrations carry the same weight as text and the two are thoroughly interwoven. The eye has an important role to play.

In addition to the *CUPM Proceedings* [12] that contain the lectures of Gleason and Steenrod, I have been strongly influenced by the content and tone of the beautiful three-volume *Introduction to Calculus and Analysis* [3] by Richard Courant and Fritz John. In particular, I took their approach to integration via Jordan content. At a different level of detail, I adopted their phrase *order of vanishing* as a replacement for the less apt *order of magnitude* for vanishing quantities. For the theorems connecting Riemann and Darboux integrals in Chapter 8, I relied on Protter and Morrey [14]; my own contribution was a number of figures to illustrate their proofs. It was Gleason who argued that the Morse lemma has a place in the undergraduate advanced calculus course. I was fully persuaded after my student Stephanie Jakus (Smith '05) wrote her senior honors thesis on the subject.

The *Feynman Lectures on Physics* [6] have had a pervasive influence on this book. First of all, Feynman's vision of his subject, and his flair for explanation, is awe-inspiring. I felt I could find no better introduction to surface integrals than the context of fluid flux. Because physics works with two-dimensional surfaces in $\mathbb{R}^3$, I also felt justified in concentrating my treatment of surface integrals on this case. I believe students will have learned all they need in order to deal with the integral of a k-form over a k-dimensional parametrized surface patch in $\mathbb{R}^n$, for arbitrary $k < n$. In providing a physical basis for the curl, the *Lectures* prodded me to try to understand it geometrically. The result is a discussion of the curl (in Chapter 11) that—like the discussion of the Morse lemma—has not previously appeared in an advanced calculus text, as far as I am aware.

I thank my students over the last decade for their curiosity, their perseverance, their interest in the subject, and their support. I especially thank Anne Watson (Smith '09), who worked with me to produce and check exercises. My editor at Springer, Kaitlin Leach, makes the rough places smooth; I am most fortunate to have worked with her. I am grateful to Smith College for its generous sabbatical policy; I wrote much of the book while on sabbatical during the 2005–2006 academic year. My deepest debt is to my teacher, Linus Richard Foy, who stimulated my interest in both mathematics and teaching. In his advanced calculus course, I often caught myself trying to follow him along two tracks simultaneously: what he was saying, and how he was saying it.

Amherst, MA
June 2010

James Callahan

Contents

Chapter 1
Starting Points

Abstract Our goal in this book is to understand and work with integrals of functions of several variables. As we show, the integrals we already know from the introductory calculus courses give us a basis for the understanding we need. The key idea for our future work is change of variables. In this chapter, we review how we use a change of variables to compute many one-variable integrals as well as path integrals and certain double integrals that can be evaluated by making a change from Cartesian to polar coordinates.

1.1 Substitution

There are two kinds of integral substitutions. As an example of the first kind, consider the familiar integral

Two kinds of substitutions

$$\int_0^b \frac{dx}{1+x^2}.$$

We know that the substitution $x = \tan s$ is helpful here because $1 + x^2 = 1 + \tan^2 s = \sec^2 s$ and $dx = \sec^2 s \, ds$. Therefore,

$$\int \frac{dx}{1+x^2} = \int \frac{\sec^2 s \, ds}{1+\tan^2 s} = \int ds = s = \arctan x,$$

and we then have

$$\int_0^b \frac{dx}{1+x^2} = \arctan x \Big|_0^b = \arctan b.$$

As an example of the second kind of substitution, take the apparently similar integral

$$\int_0^b \frac{x \, dx}{(1+x^2)^p}, \quad p \neq 1.$$

J.J. Callahan, *Advanced Calculus: A Geometric View*, Undergraduate Texts in Mathematics, DOI 10.1007/978-1-4419-7332-0_1, © Springer Science+Business Media, LLC 2010

The factor x in the numerator suggests the substitution $u = 1 + x^2$. Then $du = 2x\,dx$ and

$$\int \frac{x\,dx}{(1+x^2)^p} = \frac{1}{2}\int \frac{du}{u^p} = \frac{1}{2}\frac{u^{-p+1}}{(-p+1)} = \frac{-1}{2(p-1)(1+x^2)^{p-1}}.$$

Thus,

$$\int_0^b \frac{x\,dx}{(1+x^2)^p} = \frac{1}{2(p-1)}\left[1 - \frac{1}{(1+b^2)^{p-1}}\right].$$

Integral as antiderivative

In these examples, integration is done with the fundamental theorem of calculus. That is, we use the fact that the indefinite integral of a given function f,

$$F = \int f(x)\,dx,$$

is an *antiderivative* of f: $F'(x) = f(x)$. However, the substitutions we used to find the two antiderivatives are different in important ways.

Pullback

We call the first an example of a **pullback** substitution, for reasons that become clear in a moment. In a pullback, we express the variable x itself as some differentiable function $x = \varphi(s)$ of a new variable s. Then $dx = \varphi'(s)\,ds$ and we get

$$\int f(x)\,dx = \int f(\varphi(s))\,\varphi'(s)\,ds = \Phi,$$

where $\Phi(s)$ is an antiderivative of $f(\varphi(s))\,\varphi'(s)$. Here the aim is to choose the function φ so the antiderivative Φ becomes evident. The indefinite integral we want is then $F(x) = \Phi(\varphi^{-1}(x))$, where $s = \varphi^{-1}(x)$ is the *inverse* of the function $x = \varphi(s)$. (In our example, φ is the tangent function and φ^{-1} is the arctangent function; $\Phi(s)$ is just s.) We also use φ^{-1} to get the upper and lower endpoints of the definite integral:

$$\int_a^b f(x)\,dx = \int_{\varphi^{-1}(a)}^{\varphi^{-1}(b)} f(\varphi(s))\,\varphi'(s)\,ds.$$

Push-forward

In the second example, a **push-forward** substitution, we replace some functional expression $g(x)$ involving x with a new variable u. As with $\varphi(s)$, it takes practice and experience to make an effective choice of $g(x)$: the aim is to be able to write

$$f(x) = G'(g(x)) \cdot g'(x) \text{ or } f(x)\,dx = G'(u)\,du$$

for a suitable function $G'(u)$. That is, $du = g'(x)\,dx$ and

$$\int f(x)\,dx = \int G'(g(x))\,g'(x)\,dx = \int G'(u)\,du = G,$$

and the antiderivative is $F(x) = G(g(x))$. In our example,

$$G'(u) = \frac{1}{2u^p} \text{ and } G(u) = \frac{-1}{2(p-1)u^{p-1}}.$$

Note that we use g (and not g^{-1}) to get the endpoints of the transformed definite integral:

$$\int_a^b f(x)\,dx = \int_{g(a)}^{g(b)} G'(u)\,du.$$

To see how the substitutions using φ and g are different, and also to see how they got their names, let us think of them as maps:

Why *pull back* and *push forward*?

$$s \xrightarrow{\ \varphi\ } x \xrightarrow{\ g\ } u$$

Then we can say g pushes forward information about the value of x to the variable u, and φ pulls back that information to s. Note that the pullback needs to be invertible: without a well-defined φ^{-1}, a given value of x may pull back to two or more different values of s or to none at all. This problem does not arise with g, though.

To complete this section, let us review why the differential changes the way it does in a substitution. For example, in the pullback $x = \varphi(s)$, why is $dx = \varphi'(s)\,ds$? The answer might seem obvious: because dx/ds is just another notation for the derivative—that is, $dx/ds = \varphi'(s)$—we simply multiply by ds to get $dx = \varphi'(s)\,ds$. This is a good mnemonic; however, it is not an explanation, because the expressions dx and ds have no independent meaning, at least as far as derivatives are concerned. We must look more carefully at the link between differentials and derivatives.

In a linear function, $x = \varphi(s) = ms + b$, we usually interpret the coefficient m as the slope of the graph: $\Delta x/\Delta s = m$. However, if we rewrite the slope equation in the form $\Delta x = m\,\Delta s$, it becomes natural to interpret m instead as a *multiplier*. That is, our linear map $\varphi : s \mapsto x$ multiplies lengths by the factor m: an interval of length Δs on the s-axis is mapped to an interval of length $\Delta x = m\,\Delta s$ on the x-axis. Furthermore, when $m < 0$, Δs and Δx have opposite orientations, so φ also carries out a "flip." (The role of the coefficient as a multiplier rather than as a slope suggests why it is so commonly represented by the letter "m".)

When $x = \varphi(s)$ is nonlinear, the slope of the graph (or the slope of its tangent line) varies from point to point. Nevertheless, by fixing our attention on a small neighborhood of a particular point $s = s_0$, we still have a way to interpret the derivative as a multiplier. To see how this happens, recall first that we assume φ is differentiable, so

Transformation of differentials

Slope as length multiplier

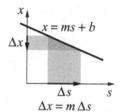

$$\Delta x = m\,\Delta s$$

$$\varphi'(s_0) = \lim_{\Delta s \to 0} \frac{\Delta x}{\Delta s} = \lim_{s \to s_0} \frac{\varphi(s) - \varphi(s_0)}{s - s_0}.$$

According to the meaning of a limit, we can make $\Delta x/\Delta s$ as close to $\varphi'(s_0)$ as we wish by making $\Delta s = s - s_0$ sufficiently small; in other words,

The microscope equation and linear approximations

$$\Delta x \approx \varphi'(s_0)\,\Delta s \ \text{ when } \Delta s \approx 0.$$

To see what this means, focus a microscope at the point (s_0, x_0) and use coordinates $\Delta s = s - s_0$ and $\Delta x = x - x_0$ centered in this window. Then, under sufficient magnification (i.e., with $\Delta s \approx 0$), φ looks like $\Delta x \approx \varphi'(s_0)\,\Delta s$. We call this the **microscope**

equation for $x = \varphi(s)$ at s_0; it is linear, and defines the **linear approximation** of the function φ at s_0.

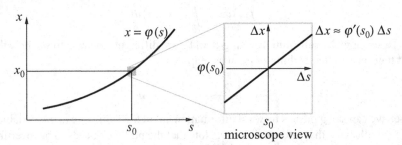

microscope view

φ' is the local
length multiplier

Finally, we can say that φ is *locally linear*, in the sense that $x = \varphi(s)$ comes as close as we wish to its linear approximation $\Delta x \approx \varphi'(s_0) \Delta s$ when s is restricted to a sufficiently small neighborhood of s_0. Thus, because the map $\varphi : s \to x$ is locally linear at s_0, it multiplies lengths (approximately) by $\varphi'(s_0)$ in any sufficiently small neighborhood of $s = s_0$.

Integral as a limit
of Riemann sums

With the microscope equation, we can now see why the differential transforms the way it does when we make a change of variables in an integral. First of all, a definite integral is defined as a limit of Riemann sums. In the simplest case—a left-endpoint Riemann sum with n equal subintervals—we can set $\Delta x = (b-a)/n$ and $x_i = a + (i-1)\Delta x$ and write

$$\int_a^b f(x)\,dx = \lim_{n\to\infty} \sum_{i=1}^{n} f(x_i)\,\Delta x.$$

We think of each term in the sum as the area of a rectangle with height $f(x_i)$ and base Δx, as in the figure at the left, below.

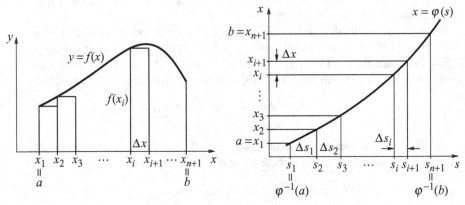

The pullback creates
a new Riemann sum

The figure at the right shows how the substitution $x = \varphi(s)$ pulls back our partition of the interval $a \leq x \leq b$ to a partition of $\varphi^{-1}(a) \leq s \leq \varphi^{-1}(b)$. We set $s_i = \varphi^{-1}(x_i)$ $(i = 1,\ldots,n+1)$ and $\Delta s_i = s_{i+1} - s_i$ $(i = 1,\ldots,n)$. Note that the subintervals Δs_i are generally unequal when φ is nonlinear. In fact, $\Delta s_i \approx \Delta x / \varphi'(s_i)$, by the microscope equation. The pullback allows us to write

$$\sum_{i=1}^{n} f(x_i)\,\Delta x \approx \sum_{i=1}^{n} f(\varphi(s_i))\,\varphi'(s_i)\,\Delta s_i.$$

By choosing n sufficiently large, we can make every Δs_i arbitrarily small and thus can make these two sums arbitrarily close. Notice that the right-hand side is also a Riemann sum, in this case for the function $f(\varphi(s))\,\varphi'(s)$. Therefore, in the limit as $n \to \infty$, the Riemann sums become integrals and we have the equality

$$\int_{a}^{b} f(x)\,dx = \int_{\varphi^{-1}(a)}^{\varphi^{-1}(b)} f(\varphi(s))\,\varphi'(s)\,ds.$$

Thus we see that the justification for the transformation $dx = \varphi'(s)\,ds$ of differentials in integration lies in the transformation $\Delta x \approx \varphi'(s_i)\,\Delta s_i$ that the microscope equation provides for the Riemann sums.

$dx = \varphi'(s)\,ds$

    The microscope equation $\Delta x \approx \varphi'(s_i)\,\Delta s_i$ has one further geometric consequence. In our Riemann sum for the second integral, the standard way to think about each term is as the area of a rectangle with height $f(\varphi(s_i))\,\varphi'(s_i)$ and base Δs_i. However, if we change the proportions and make the height $f(\varphi(s_i))$ and the base $\varphi'(s_i)\,\Delta s_i$, then we have a rectangle that matches (as closely as we wish) the shape of the original rectangle with height $f(x_i)$ and base Δx, because $f(x_i) = f(\varphi(s_i))$ and $\Delta x \approx \varphi'(s_i)\,\Delta s_i$.

Rectangles in the Riemann sums

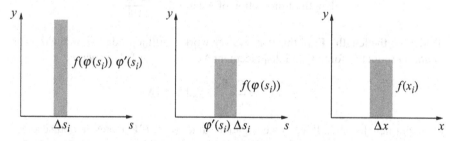

To understand why differentials transform the way they do, we worked with a pullback substitution. We get the same result with a push-forward, though the details are different. Our work has led us to several questions that we ask again when we turn to more general integrals that involve functions of several variables: what different kinds of substitutions occur? What role do inverses play? What is the form of a linear approximation? What is the analogue of the local length multiplier? What are differentials and how do they transform? What is the geometric interpretation of that transformation?

Some questions raised

1.2 Work and path integrals

Path integrals are one of the centerpieces of the first multivariable calculus course, and they are often treated, as we do here, in the context of work.

Force, displacement, and work

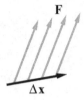

By definition, a force moving a body from one place to another produces work, and the work done is proportional to both the force applied and to the displacement caused. The simplest formula that captures this idea is

$$\text{work} = \text{force} \times \text{displacement}.$$

Although work is a scalar quantity, force and displacement are actually both vectors, and the force is a *field*, that is, a variable function of position. We must elaborate our simple formula to reflect these facts. Consider a straight-line displacement along some vector $\Delta \mathbf{x}$ and a constant force field $\mathbf{F}$ that acts the same way at every point along $\Delta \mathbf{x}$. Only the component of the force that lies in the direction of the displacement does any work; this is the *effective* force $\mathbf{F}_{\text{eff}}$. We can take all this into account in the new formula

$$\text{work} = \|\mathbf{F}_{\text{eff}}\| \|\Delta \mathbf{x}\|.$$

The scalar $\|\mathbf{F}_{\text{eff}}\|$ is the length of the perpendicular projection of $\mathbf{F}$ on $\Delta \mathbf{x}$. Now, in general, for arbitrary vectors $\mathbf{A}$ and $\mathbf{B} \neq \mathbf{0}$,

$$\text{length of projection of } \mathbf{A} \text{ onto } \mathbf{B} = \frac{\mathbf{A} \cdot \mathbf{B}}{\|\mathbf{B}\|}.$$

work $= \mathbf{F} \cdot \Delta \mathbf{x}$

Rewriting the length $\|\mathbf{F}_{\text{eff}}\|$ this way, we see work is still a product; it is the *dot* (or *scalar*) product of force $\mathbf{F}$ and displacement $\Delta \mathbf{x}$, now regarded as *vectors*:

$$\text{work} = W = \frac{\mathbf{F} \cdot \Delta \mathbf{x}}{\|\Delta \mathbf{x}\|} \|\Delta \mathbf{x}\| = \mathbf{F} \cdot \Delta \mathbf{x}.$$

Work is additive

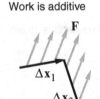

In our new formula, W can take negative values (e.g., if $\mathbf{F}$ makes an obtuse angle with $\Delta \mathbf{x}$). To see why "negative work" must arise, consider a constant force $\mathbf{F}$ that displaces an object along a path consisting of two straight segments $\Delta \mathbf{x}_1$ and $\Delta \mathbf{x}_2$, one immediately followed by the other. We want the total work to be the sum of the work done on the separate segments:

$$\text{total work} = \mathbf{F} \cdot \Delta \mathbf{x}_1 + \mathbf{F} \cdot \Delta \mathbf{x}_2.$$

Orientation matters

We say that work is additive on displacements. In particular, if $\Delta \mathbf{x}_2 = -\Delta \mathbf{x}_1$, then the total work done is 0. Consequently, the work done by $\mathbf{F}$ along $-\Delta \mathbf{x}$ must be the negative of the work done by the same $\mathbf{F}$ along $+\Delta \mathbf{x}$. Orientation matters: reversing the displacement reverses the sign of the work done.

Components of work:
$W = P \Delta x + Q \Delta y$

Let us introduce coordinates into the plane containing the vectors $\mathbf{F}$ and $\Delta \mathbf{x}$ and write $\mathbf{F} = (P, Q)$ and $\Delta \mathbf{x} = (\Delta x, \Delta y)$. Then

$$W = \mathbf{F} \cdot \Delta\mathbf{x} = P\Delta x + Q\Delta y.$$

This formula gives the **coordinate components** of work. It says that, in the x-direction, there is a force of size P acting along a displacement of size Δx, doing work $W_x = P\Delta x$. Similarly, in the y-direction the work done is $W_y = Q\Delta y$. We call W_x and W_y the *components of W in the x- and y-directions*. The following definition summarizes our observations to this point.

Definition 1.1 *The **work** done by the constant force $\mathbf{F} = (P,Q)$ in displacing an object along the line segment $\Delta\mathbf{x} = (\Delta x, \Delta y)$ is*

$$W = \mathbf{F} \cdot \Delta\mathbf{x} = P\Delta x + Q\Delta y = W_x + W_y.$$

Ultimately, we need to deal with variable forces and displacements along curved paths. The prototype is a smooth simple curve C in the plane. We say C is **smooth** if it is the image of a map (an example of a **vector-valued function**)

Displacement along a curved path

$$\mathbf{x} : [a,b] \to \mathbb{R}^2 : t \mapsto (x(t), y(t))$$

(a **parametrization**) whose coordinate functions $x(t)$ and $y(t)$ have continuous derivatives on $a \leq t \leq b$. We call t the **parameter**. In addition, C is **simple** if it has no self-intersections, that is, if $\mathbf{x}$ is 1–1. The parametrization orders the points on C in the following sense: $\mathbf{x}(t_1)$ *precedes* $\mathbf{x}(t_2)$ if $t_1 < t_2$ (i.e., t_1 precedes t_2 in $[a,b]$). The ordering gives C an **orientation**; we write $\vec{C}$ to indicate C is oriented. At any point on $\vec{C}$ where the tangent vector $\mathbf{x}'(t)$ is nonzero, it points in the direction of increasing t, and thus also indicates the orientation of $\vec{C}$. We can immediately extend these ideas to paths in $\mathbb{R}^n$.

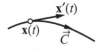

Definition 1.2 *A **smooth, simple, oriented curve** $\vec{C}$ in $\mathbb{R}^n$ is the image of a smooth 1–1 map,*

Parametrizing a smooth simple curve

$$\mathbf{x} : [a,b] \to \mathbb{R}^n : t \mapsto \mathbf{x}(t),$$

*where $\mathbf{x}'(t) \neq 0$ for all $a < t < b$. The point $\mathbf{x}(a)$ is the **start** of $\vec{C}$ and $\mathbf{x}(b)$ is its **end**.*

The simple formula $W = \mathbf{F} \cdot \Delta\mathbf{x}$ for work assumes that the force $\mathbf{F}$ is constant, so the location of the base point $\mathbf{a}$ of the displacement $\Delta\mathbf{x}$ is irrelevant. However, if $\mathbf{F}$ varies, then the work done will depend on $\mathbf{a}$. We must, in fact, treat a linear displacement as we would any displacement, and provide it with a parametrization. A natural one is

Linear displacements as oriented curves

$$\mathbf{x}(t) = \mathbf{a} + t \cdot \Delta\mathbf{x}, \quad 0 \leq t \leq 1.$$

We are now in a position to estimate the work done by a *variable* force as it displaces an object along a smooth, simple, oriented curve $\vec{C}$ in $\mathbb{R}^3$. Force is now a (continuous) **vector field**—that is, a vector-valued function $\mathbf{F}(\mathbf{x})$ that varies (continuously) with position $\mathbf{x}$. To estimate the work done, chop the curve into small pieces. When a piece is small enough, it is essentially straight and the force is essentially constant along it. On this piece, the linear formula for work (Definition 1.1) gives a good approximation. By additivity, the sum of these contributions will approximate

Work done by a variable force

the total work done along the whole curve. To get a better estimate, chop the curve into even smaller pieces.

Partition the curve

In more detail, let $\mathbf{x}_1$, $\mathbf{x}_2$, ..., $\mathbf{x}_{k+1}$ be an ordered sequence of points on $\vec{C}$, with $\mathbf{x}_1$ at the start of $\vec{C}$ and $\mathbf{x}_{k+1}$ at the end. We say $\{\mathbf{x}_i\}$ is a **partition that respects the orientation of $\vec{C}$**. Let the oriented curve $\vec{C}_i$ be the portion of $\vec{C}$ from $\mathbf{x}_i$ to $\mathbf{x}_{i+1}$, and let W_i be the work done by $\mathbf{F}$ along $\vec{C}_i$; then, by the additivity of work,

$$\text{total work done by } \mathbf{F} = \sum_{i=1}^{k} W_i.$$

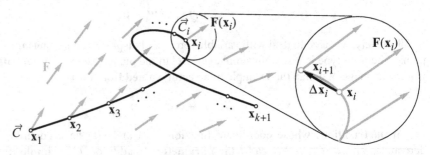

Approximate the work along each segment

Let $\Delta \mathbf{x}_i = \mathbf{x}_{i+1} - \mathbf{x}_i$ be the linear displacement with base point $\mathbf{x}_i$. When $\|\Delta \mathbf{x}_i\|$ is sufficiently small, $\Delta \mathbf{x}_i$ will be as close to the curved segment $\vec{C}_i$ as we wish, because $\vec{C}_i$ is smooth. Moreover, $\mathbf{F}$ will be nearly constant along $\Delta \mathbf{x}_i$, because $\mathbf{F}$ is continuous. In particular, $\mathbf{F}(\mathbf{x})$ will differ by an arbitrarily small amount from its value $\mathbf{F}(\mathbf{x}_i)$ at the base point of $\Delta \mathbf{x}_i$. Therefore, $W_i \approx \mathbf{F}(\mathbf{x}_i) \cdot \Delta \mathbf{x}_i$. If we choose k large enough and make each $\|\Delta \mathbf{x}_i\|$ sufficiently small, then the sum

$$\sum_{i=1}^{k} \mathbf{F}(\mathbf{x}_i) \cdot \Delta \mathbf{x}_i$$

will approximate the total work as closely as we wish. In fact, this last expression is a Riemann sum for a new kind of integral, called a *path*, or *line*, integral that we now define quite generally for smooth, simple, oriented paths in any dimension. Note that the definition does not depend on the parametrization of the path.

Smooth path integral

Definition 1.3 (Smooth path integral) *The integral of the continuous vector-valued function* $\mathbf{F}(\mathbf{x})$ *over the smooth, simple, oriented curve* $\vec{C}$ *in* $\mathbb{R}^n$ *is*

$$\int_{\vec{C}} \mathbf{F} \cdot d\mathbf{x} = \lim_{\substack{k \to \infty \\ mesh \to 0}} \sum_{i=1}^{k} \mathbf{F}(\mathbf{x}_i) \cdot \Delta \mathbf{x}_i,$$

if the limit exists when taken over all ordered partitions $\mathbf{x}_1$, $\mathbf{x}_2$, ..., $\mathbf{x}_{k+1}$ *of* $\vec{C}$ *with* ***mesh*** $= \max_i \|\Delta \mathbf{x}_i\|$ *and* $\Delta \mathbf{x}_i = \mathbf{x}_{i+1} - \mathbf{x}_i$, $i = 1, \dots, k$.

More general paths

We can now define a more general collection of integration paths. If we allow the start and end of $\vec{C}$ to coincide (the tangent directions need not agree) and there are no other self-intersections, we say that $\vec{C}$ is a **simple closed curve**. The definition of the path integral of **F** over $\vec{C}$ is unchanged. A **piecewise-smooth, oriented path** $\vec{C}$ is the union of smooth oriented paths $\vec{C}_1, \vec{C}_2, \ldots, \vec{C}_m$, each of which is either simple or a simple closed curve. We write $\vec{C} = \vec{C}_1 + \cdots + \vec{C}_m$ and define

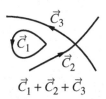

$\vec{C}_1 + \vec{C}_2 + \vec{C}_3$

$$\int_{\vec{C}} \mathbf{F} \cdot d\mathbf{x} = \int_{\vec{C}_1 + \cdots + \vec{C}_m} \mathbf{F} \cdot d\mathbf{x} = \int_{\vec{C}_1} \mathbf{F} \cdot d\mathbf{x} + \cdots + \int_{\vec{C}_m} \mathbf{F} \cdot d\mathbf{x}.$$

The combined path $\vec{C}$ may be neither simple, smooth, nor even connected. A special case arises when the pieces $\vec{C}_i$ fit together, with the end of $\vec{C}_j$ coinciding with the start of $\vec{C}_{j+1}$, for every $j = 1, \ldots, m-1$. Then $\vec{C}$ is a single connected curve that is smooth everywhere except possibly at the points where the pieces join.

$\vec{C}_1 + \vec{C}_2$

Work is a path integral

Because the work done by the force **F** in displacing an object along a smooth, simple, oriented path $\vec{C}_i$ is the limit of the same sums that define the integral of **F** over $\vec{C}_i$, and because we want work to be additive over the sum $\vec{C}$ of such paths in $\mathbb{R}^3$, we make the following definition.

Definition 1.4 *The **work** done by the force **F** along the smooth oriented path $\vec{C}$ is*

$$W = \int_{\vec{C}} \mathbf{F} \cdot d\mathbf{x}.$$

How shall we compute the value of a path integral? Integrals—both ordinary integrals and path integrals—are limits of Riemann sums, but we rarely calculate them that way. To evaluate an ordinary integral, the common practice is to invoke the fundamental theorem of calculus to treat the integral as an antiderivative, and then use various techniques to find the antiderivative. To evaluate a path integral, our practice is to pull it back to an ordinary integral using the parametrization of the path. In the process, we demonstrate that the path integral exists; that is, the Riemann sums defining it have a limit.

Computing a path integral

Let $\vec{C}$ be a smooth, simple, oriented curve, and suppose it is parametrized by $\mathbf{x}(t)$, $a \leq t \leq b$. Let $\mathbf{F}(\mathbf{x})$ be a continuous vector function defined on $\vec{C}$. To decide whether the integral of **F** over the path $\vec{C}$ exists and has a finite value (Definition 1.3), we choose a partition $\mathbf{x}_1, \mathbf{x}_2, \ldots, \mathbf{x}_{k+1}$ that respects the orientation of $\vec{C}$ and form the Riemann sum

Path integral to ordinary integral

$$\sum_{i=1}^{k} \mathbf{F}(\mathbf{x}_i) \cdot \Delta\mathbf{x}_i, \quad \Delta\mathbf{x}_i = \mathbf{x}_{i+1} - \mathbf{x}_i$$

for the path integral. Because $\vec{C}$ is *simple*, the parametrization $\mathbf{x}(t)$ is 1–1, so there is a unique partition

$$a = t_1 < t_2 < \cdots < t_{k+1} = b$$

of $[a, b]$ with $\mathbf{x}_i = \mathbf{x}(t_i)$, $i = 1, \ldots, k+1$. The microscope equation implies that

$$\Delta\mathbf{x}_i = \mathbf{x}(t_{i+1}) - \mathbf{x}(t_i) \approx \mathbf{x}'(t_i)(t_{i+1} - t_i) = \mathbf{x}'(t_i)\Delta t_i$$

when $\Delta t_i = t_{i+1} - t_i \approx 0$. Therefore, if every $\Delta t_i \approx 0$,

$$\sum_{i=1}^{k} \mathbf{F}(\mathbf{x}_i) \cdot \Delta \mathbf{x}_i \approx \sum_{i=1}^{k} \mathbf{F}(\mathbf{x}(t_i)) \cdot \mathbf{x}'(t_i) \Delta t_i.$$

The expression on the right is a Riemann sum for the ordinary integral

$$\int_{a}^{b} \mathbf{F}(\mathbf{x}(t)) \cdot \mathbf{x}'(t) \, dt.$$

Because $\mathbf{F}(\mathbf{x}(t)) \cdot \mathbf{x}'(t)$ is continuous on $a \leq t \leq b$, this ordinary integral exists and is the limit of its Riemann sums. The Riemann sums for the path integral—the sums on the left—must likewise converge, and to the same limit. This establishes the following theorem.

Theorem 1.1. *Suppose $\vec{C}$ is a smooth, simple, oriented curve in $\mathbb{R}^n$ that is parametrized by $\mathbf{x}(t)$, $a \leq t \leq b$. If $\mathbf{F}(\mathbf{x})$ is a continuous vector function defined on $\vec{C}$, then the integral of $\mathbf{F}$ over the path $\vec{C}$ exists, and*

$$\int_{\vec{C}} \mathbf{F} \cdot d\mathbf{x} = \int_{a}^{b} \mathbf{F}(\mathbf{x}(t)) \cdot \mathbf{x}'(t) \, dt. \qquad \square$$

Corollary 1.2 *Suppose $\vec{C} = \vec{C}_1 + \cdots + \vec{C}_m$, where $\vec{C}_j$ is a smooth, simple, oriented curve in $\mathbb{R}^n$ parametrized by $\mathbf{x}_j(t)$, $a_j \leq t \leq b_j$, $j = 1, \ldots, m$; then*

$$\int_{\vec{C}} \mathbf{F} \cdot d\mathbf{x} = \int_{a_1}^{b_1} \mathbf{F}(\mathbf{x}_1(t)) \cdot \mathbf{x}_1'(t) \, dt + \cdots + \int_{a_m}^{b_m} \mathbf{F}(\mathbf{x}_m(t)) \cdot \mathbf{x}_m'(t) \, dt. \qquad \square$$

Pullback substitution

Note that the conversion from path integral to ordinary integral is a pullback: by means of the map $\mathbf{x} : [a,b] \to \mathbb{R}^n$, the integrand $\mathbf{F}(\mathbf{x})$ on $\vec{C}$ is pulled back to $\mathbf{F}(\mathbf{x}(t))$ on $[a,b]$, and the differential $d\mathbf{x}$ is pulled back to $\mathbf{x}'(t) \, dt$.

$$[a,b] \xrightarrow{\ \mathbf{x}\ } \vec{C}$$

$$\int_{a}^{b} \mathbf{F}(\mathbf{x}(t)) \cdot \mathbf{x}'(t) \, dt = \int_{\vec{C}} \mathbf{F} \cdot d\mathbf{x}$$

The equality $d\mathbf{x} = \mathbf{x}' \, dt$ of differentials comes from the microscope equation, $\Delta \mathbf{x} \approx \mathbf{x}' \Delta t$, just as it did in the pullback of ordinary integrals.

Example 1

Here is a simple example in $\mathbb{R}^2$ to illustrate how the pullback substitution works to evaluate a path integral. We take $\mathbf{F} = (0, x)$; this represents a vertical force whose magnitude at any point is proportional to the distance from that point to the y-axis. We take $\vec{C}$ to be the right third of the circle $x^2 + y^2 = 4$, oriented counterclockwise. The map

$$\mathbf{x}(t) = (2\cos t, 2\sin t), \quad -\pi/3 \leq t \leq \pi/3,$$

parametrizes $\vec{C}$, provides the correct orientation, and gives

$$\mathbf{F} = \mathbf{F}(\mathbf{x}(t)) = (0, 2\cos t), \quad d\mathbf{x} = \mathbf{x}' \, dt = (-2\sin t, 2\cos t) \, dt.$$

The work W done by $\mathbf{F}$ along $\vec{C}$ is

$$\int_{\vec{C}} \mathbf{F} \cdot d\mathbf{x} = \int_{-\pi/3}^{\pi/3} (0, 2\cos t) \cdot (-2\sin t, 2\cos t)\, dt = \int_{-\pi/3}^{\pi/3} 4\cos^2 t\, dt = \tfrac{4}{3}\pi + \sqrt{3}.$$

The work done by the same force along the curve $-\vec{C}$ with the opposite orientation should be -2π. To verify this independently, we can parametrize $-\vec{C}$ as

$$\mathbf{x} = (2\sin t, 2\cos t), \quad \pi/6 \le t \le 5\pi/6;$$

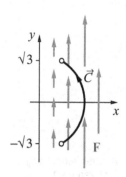

then

$$\mathbf{F} = (0, 2\sin t), \quad d\mathbf{x} = (-2\cos t, -2\sin t)\, dt, \quad \mathbf{F} \cdot d\mathbf{x} = -4\sin^2 t\, dt,$$

and

$$\text{work done along } -\vec{C} = \int_{-\vec{C}} \mathbf{F} \cdot d\mathbf{x} = \int_{\pi/6}^{5\pi/6} -4\sin^2 t\, dt = -\tfrac{4}{3}\pi - \sqrt{3}.$$

A curve can be parametrized in more than one way. Will that change the value of W? Notice that the (unoriented) curve C in our example is also the graph of the function $x = \sqrt{4-y^2}$, $-\sqrt{3} \le y \le \sqrt{3}$. We can therefore use y itself as the parameter. The map

Reparametrizing $\vec{C}$

$$\mathbf{r}(y) = \left(\sqrt{4-y^2}, y \right), \quad -\sqrt{3} \le y \le \sqrt{3},$$

parametrizes $\vec{C}$ with the proper orientation, and

$$\mathbf{F} = \mathbf{F}(\mathbf{r}(y)) = \left(0, \sqrt{4-y^2} \right), \quad d\mathbf{r} = \mathbf{r}'(y)\, dy = \left(\frac{-y}{\sqrt{4-y^2}}, 1 \right) dy.$$

Using the new parametrization to provide the pullback, we find the work done is

$$W = \int_{\vec{C}} \mathbf{F} \cdot d\mathbf{r} = \int_{-\sqrt{3}}^{\sqrt{3}} \sqrt{4-y^2}\, dy = \sqrt{3} + \frac{4\pi}{3}.$$

Thus W is unchanged. The exercises provide a third parametrization of $\vec{C}$; you are asked to verify that it also gives $W = \sqrt{3} + 4\pi/3$.

The example prompts us to consider different parametrizations of the same smooth, simple, oriented curve $\vec{C}$. One way to generate a new parametrization is by a smooth parameter change $t = h(u)$:

Alternate parametrizations

$$\mathbf{r}(u) = \mathbf{x}(h(u)), \quad c \le u \le d.$$

Here $h : [c, d] \to [a, b]$ is continuously differentiable, 1–1, onto, and $h'(u) > 0$ for all $c < u < d$. Note that $\mathbf{r}$ has the same image as $\mathbf{x}$. Furthermore,

$$\mathbf{r}'(u) = \mathbf{x}'(h(u))\,h'(u);$$

because $h'(u) > 0$, the tangent vectors $\mathbf{r}'(u)$ and $\mathbf{x}'(h(u))$ point in the same direction for all $c < u < d$. Thus $\mathbf{r}$ induces the same orientation as $\mathbf{x}$, and provides an alternate parametrization of $\vec{C}$.

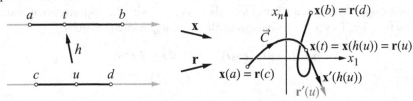

A parameter change for Example 1

The two parametrizations of the counterclockwise-oriented circular arc of radius 2 (Example 1, above) are connected by a parameter change; it is $t = h(y) = \arcsin(y/2)$. To see that this is the formula for h, first note that

$$(2\cos t, 2\sin t) = \mathbf{x}(t) = \mathbf{r}(y) = \left(\sqrt{4 - y^2}, y\right).$$

Equality of the y-components gives $2\sin t = y$, so $t = h(y) = \arcsin(y/2)$. It remains to check that the x-components transform properly under the same parameter change. But because $\cos(\arcsin w) = \sqrt{1 - w^2}$, we have

$$x = 2\cos t = 2\cos h(y) = 2\cos(\arcsin(y/2)) = 2\sqrt{1 - (y/2)^2} = \sqrt{4 - y^2},$$

as required. According to Theorem 1.4, below, any two parametrizations of a smooth simple curve are connected by a smooth change of parameter. This is easiest to see after we have introduced the special arc-length parametrization (p. 15ff.).

Component notation

We saw earlier that, in the simple case of a constant force acting along a straight line, we could break down the work into coordinate components. For example, in $\mathbb{R}^2$, if $\mathbf{F} = (P, Q)$ and $\Delta\mathbf{x} = (\Delta x, \Delta y)$, then W has components $P\Delta x$ and $Q\Delta y$. Let us see how we can do the same for a variable force or, more generally, for any path integral in $\mathbb{R}^n$.

For simplicity, we first take $\mathbf{F}$ and $\vec{C}$ in the (x, y)-plane. Let $\mathbf{x}_i = (x_i, y_i)$ be a fine partition of $\vec{C}$ into oriented segments $\vec{C}_i$, and let $\mathbf{x}_{i+1} - \mathbf{x}_i = \Delta\mathbf{x}_i = (\Delta x_i, \Delta y_i)$. Let $P(x, y)$ and $Q(x, y)$ be the components of $\mathbf{F}$ that act at the point $\mathbf{x} = (x, y)$:

$$\mathbf{F}(x, y) = (P(x, y), Q(x, y)).$$

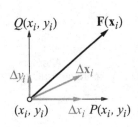

Then we can estimate the work done by $\mathbf{F}$ along the segment $\vec{C}_i$ as

$$\Delta W_i = \mathbf{F}(\mathbf{x}_i) \cdot \Delta\mathbf{x}_i = P(x_i, y_i)\,\Delta x_i + Q(x_i, y_i)\,\Delta y_i.$$

The figure indicates that the work estimate splits into two parts: a horizontal force $P(x_i, y_i)$ acting through the horizontal displacement Δx_i, and a vertical force $Q(x_i, y_i)$ acting through a vertical displacement Δy_i. Our estimate for the total work along $\vec{C}$

is

$$\sum_{i=1}^{k} \Delta W_i = \sum_{i=1}^{k} \mathbf{F}(\mathbf{x}_i) \cdot \Delta \mathbf{x}_i = \sum_{i=1}^{k} P(x_i, y_i) \Delta x_i + Q(x_i, y_i) \Delta y_i.$$

These estimates converge to the path integral—and hence to the work done—as the mesh size of the partition tends to zero. We can regard each expression as a Riemann sum for the limiting integral; the three different ways of writing the Riemann sum lead us to three different ways to write the integral:

$$W = \int_{\vec{C}} dW = \int_{\vec{C}} \mathbf{F} \cdot d\mathbf{x} = \int_{\vec{C}} P \, dx + Q \, dy.$$

On the right is the component form of the work integral.

If we adopt the informal practice of regarding an integral as an infinite sum of "infinitesimal" terms, then the integrand in the work integral is the infinitesimal amount of work dW done along an infinitesimal segment $d\mathbf{x} = (dx, dy)$:

"Infinitesimal" work

$$dW = \mathbf{F} \cdot d\mathbf{x} = (P, Q) \cdot (dx, dy) = P \, dx + Q \, dy.$$

From this point of view, the expressions $P \, dx$ and $Q \, dy$ are the horizontal and vertical components of the infinitesimal work dW.

Moving to $\mathbb{R}^n$, we set $\mathbf{x} = (x_1, \dots, x_n)$, $d\mathbf{x} = (dx_1, \dots, dx_n)$, and take P_i to be the ith component of $\mathbf{F}$, so

Component form of a path integral

$$\mathbf{F}(\mathbf{x}) = (P_1(x_1, \dots, x_n), \dots, P_n(x_1, \dots, x_n)),$$
$$dW = \mathbf{F} \cdot d\mathbf{x} = P_1 \, dx_1 + \dots + P_n \, dx_n.$$

Suppose $\mathbf{x}(t) = (\varphi_1(t), \dots, \varphi_n(t))$, $a \le t \le b$, parametrizes the oriented curve $\vec{C}$; then $d\mathbf{x} = (\varphi_1'(t), \dots, \varphi_n'(t)) \, dt$ and

$$W = \int_{\vec{C}} dW = \int_{\vec{C}} \mathbf{F} \cdot d\mathbf{x} = \int_{\vec{C}} P_1 \, dx_1 + \dots + P_n \, dx_n$$
$$= \int_a^b [P_1(\varphi_1(t), \dots, \varphi_n(t)) \, \varphi_1'(t) + \dots + P_n(\varphi_1(t), \dots, \varphi_n(t)) \, \varphi_n'(t)] \, dt.$$

The final expression is an ordinary integral; it gives us a way to compute the path integral by means of the n pullback substitutions $x_1 = \varphi_1(t), \dots, x_n = \varphi_n(t)$.

As an example of a path integral in $\mathbb{R}^2$ given in terms of its two components, let us determine

Example 2

$$\int_{\vec{C}} xy \, dx + \frac{x^2}{y} \, dy,$$

where $\vec{C}$ is the piecewise-smooth path consisting of the horizontal segment $\vec{C}_1$ from $(1, 7)$ to $(5, 7)$ followed by the vertical segment $\vec{C}_2$ from $(5, 7)$ to $(5, 2)$, traversed in that order. On $\vec{C}_1$, $y = 7$ and $dy = 0$; we can simply take x as the parameter. On $\vec{C}_2$, $x = 5$, $dx = 0$; we take y as the parameter and note that we must integrate "backwards" from $y = 7$ to $y = 2$. The path integral equals

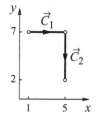

$$\int_1^5 7x\,dx + \int_7^2 \frac{5^2}{y}\,dy = \left.\frac{7x^2}{2}\right|_1^5 + 25\ln y\,\Big|_7^2 = 84 + 25\ln\left(\tfrac{2}{7}\right).$$

Arc length

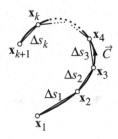

We now define the length of a smooth, simple, oriented curve $\vec{C}$ in $\mathbb{R}^n$. Let $\mathbf{x}_1$, $\mathbf{x}_2, \ldots, \mathbf{x}_{k+1}$ be an ordered partition that respects the orientation of $\vec{C}$. The vectors $\Delta\mathbf{x}_i = \mathbf{x}_{i+1} - \mathbf{x}_i$, $i = 1, \ldots, k$, are straight-line segments that make up what is called a **polygonal approximation** to $\vec{C}$. If we let $\Delta s_i = \|\Delta\mathbf{x}_i\|$ denote the length of the ith segment, then we can express the length of the whole polygonal approximation as the sum

$$L_{\{\mathbf{x}_i\}} = \sum_{i=1}^n \Delta s_i.$$

This expression is also a Riemann sum for a new kind of path integral. If the Riemann sums have a limit, so that the new path integral is defined, then that path integral can serve to define the length of $\vec{C}$.

Definition 1.5 *The **arc length** of the smooth, simple, oriented curve $\vec{C}$ is*

$$L = \lim_{\substack{n\to\infty\\ mesh\to 0}} \sum_{i=1}^n \Delta s_i = \int_C ds, \quad mesh = \max_i \Delta s_i = \max_i \|\Delta\mathbf{x}_i\|,$$

*if this limit exists when taken over all ordered partitions $\{\mathbf{x}_i\}$ of $\vec{C}$; we call **ds** the **element of arc length for $\vec{C}$**.*

Computing arc length

The definition is not a practical computational tool. However, we can compute arc length by following the same approach we took to compute the path integral for work: use the parametrization of the curve to pull back the integration to the real line. The justification is essentially the same as the one for Theorem 1.1, page 10. Let $\mathbf{x}(t)$, $a \le t \le b$ parametrize $\vec{C}$. Because $\vec{C}$ is simple, there is a unique partition

$$a = t_1 < t_2 < \cdots < t_{k+1} = b$$

of $[a, b]$ with $\mathbf{x}_i = \mathbf{x}(t_i)$. If the partition is sufficiently fine, then the microscope equation implies $\Delta\mathbf{x}_i \approx \mathbf{x}'(t_i)\Delta t_i$ (where $\Delta t_i = t_{i+1} - t_i$) as closely as we wish. Therefore

$$L_{\{\mathbf{x}_i\}} = \sum_{i=1}^n \Delta s_i = \sum_{i=1}^n \|\Delta\mathbf{x}_i\| \approx \sum_{i=1}^n \|\mathbf{x}'(t_i)\|\Delta t_i;$$

the last is a Riemann sum for the ordinary integral

$$\int_a^b \|\mathbf{x}'(t)\|\,dt.$$

Because $\|\mathbf{x}'(t)\|$ is continuous, the Riemann sums converge to the integral. The numbers $L_{\{\mathbf{x}_i\}}$ must converge as well, giving the arc length of $\vec{C}$ and proving the following theorem.

Theorem 1.3. *If $\vec{C}$ is a smooth, simple, oriented curve parametrized as $\mathbf{x}(t)$, with $a \le t \le b$, then*

$$\int_C ds = arc\ length\ of\ \vec{C} = \int_a^b \|\mathbf{x}'(t)\|\,dt. \qquad \square$$

Let $\vec{C}_t$ denote the segment of $\vec{C}$ parametrized by $\mathbf{x}(v)$ with $a \le v \le t$. Then the arc length of $\vec{C}_t$ is the function

The arc-length parametrization

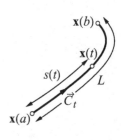

$$s(t) = \int_a^t \|\mathbf{x}'(v)\|\,dv.$$

Note that $0 \le s(t) \le L =$ arc length of $\vec{C}$. By the fundamental theorem of calculus, $s'(t) = \|\mathbf{x}'(t)\|$. Because $\|\mathbf{x}'(t)\| > 0$ on $a < t < b$, the function $s = s(t)$ is invertible on this interval. Let $t = \sigma(s)$ be the inverse (extended to all of $0 \le s \le L$ by setting $a = \sigma(0)$, $b = \sigma(L)$). Then $\mathbf{y}(s) = \mathbf{x}(\sigma(s))$ is a new parametrization of $\vec{C}$, called the **parametrization by arc length**, or the **arc-length parametrization**. The variable s itself is called the **arc-length parameter**. Because $s'(t) = \|\mathbf{x}'(t)\|$, our mnemonic for differentials becomes $ds = s'(t)\,dt = \|\mathbf{x}'(t)\|\,dt$, supporting the equality (Theorem 1.3)

$$\int_C ds = \int_a^b \|\mathbf{x}'(t)\|\,dt.$$

Let us determine the arc length and the arc-length parametrization of the curve $\vec{C} : \mathbf{x}(t) = (t^3/3, t^2/2)$, $0 \le t \le 2$. The arc-length function is the integral (i.e., antiderivative) of the speed of the parameter point as it moves along $\vec{C}$. The velocity of the moving point is the vector $\mathbf{x}' = (t^2, t)$; therefore its speed is $\sqrt{t^4 + t^2} = t\sqrt{t^2 + 1}$, and the arc-length function is

Example 3

$$s(t) = \int_0^t v\sqrt{v^2 + 1}\,dv = \left.\frac{(v^2 + 1)^{3/2}}{3}\right|_0^t = \frac{(t^2 + 1)^{3/2} - 1}{3}.$$

The length of $\vec{C}$ is therefore $s(2) = (5^{3/2} - 1)/3 \approx 3.39$. To find the inverse $t = \sigma(s)$ of the arc-length function, we solve $s = s(t)$ for t:

$$s = \frac{(t^2 + 1)^{3/2} - 1}{3},$$
$$3s + 1 = (t^2 + 1)^{3/2},$$
$$(3s + 1)^{2/3} = t^2 + 1,$$
$$\sqrt{(3s + 1)^{2/3} - 1} = t.$$

Hence the arc-length parametrization of $\vec{C}$ is

$$\mathbf{y}(s) = \mathbf{x}(\sigma(s)) = \left(\frac{((3s + 1)^{2/3} - 1)^{3/2}}{3}, \frac{(3s + 1)^{2/3} - 1}{2} \right).$$

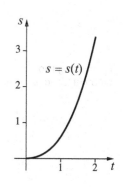

To show how s measures arc length, let us locate the points that are $s = 1, 2,$ and 3 units along $\vec{C}$. We have:

s	t	$x = t^3/3$	$y = t^2/2$
1	$\sqrt{4^{2/3} - 1} = 1.23282$	0.625	0.760
2	$\sqrt{7^{2/3} - 1} = 1.63074$	1.446	1.330
3	$\sqrt{10^{2/3} - 1} = 1.90829$	2.316	1.821

The three points are plotted on the graph below. Their spacing from the origin (where $s = 0$) along $\vec{C}$ appears to match the spacing of the unit segments along the x-axis. Compare this to the uneven spacing of the parameter points $t = 0, 1,$ and 2.

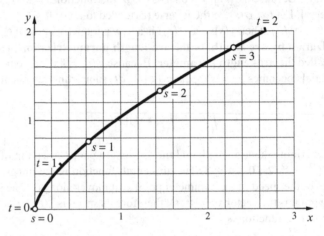

Connecting two parametrizations

Theorem 1.4. *Suppose* $\mathbf{x}(t)$, $a \leq t \leq b$, *and* $\mathbf{r}(u)$, $c \leq u \leq d$, *both parametrize the smooth, simple, oriented curve* $\vec{C}$; *then there is a continuously differentiable parameter change* $h : [c, d] \to [a, b]$ *for which* $\mathbf{r}(u) = \mathbf{x}(h(u))$, $c \leq u \leq d$.

Proof. Suppose $\mathbf{y}(s)$ is the arc-length parametrization of $\vec{C}$, and

$$s(t) = \int_a^t \|\mathbf{x}'(v)\| \, dv, \quad \bar{s}(u) = \int_c^u \|\mathbf{r}'(w)\| \, dw.$$

These functions are continuously differentiable on their domains, and so is the inverse $t = \sigma(s)$ of $s = s(t)$. Therefore,

$$t = h(u) = \sigma(\bar{s}(u))$$

is continuously differentiable on $c \leq u \leq d$, and inasmuch as

$$\mathbf{x}(\sigma(s)) = \mathbf{y}(s), \quad \mathbf{r}(u) = \mathbf{y}(\bar{s}(u)),$$

we have $\mathbf{x}(h(u)) = \mathbf{x}(\sigma(\bar{s}(u))) = \mathbf{y}(\bar{s}(u)) = \mathbf{r}(u)$. $\square$

If we think of the parameter t as measuring time, then $\mathbf{x}'(t)$ gives the velocity of the point $\mathbf{x}(t)$ as it moves along $\vec{C}$, and $\|\mathbf{x}'(t)\|$ gives its speed. For the arc-length parametrization $\mathbf{y}(s) = \mathbf{x}(\sigma(s))$, we therefore have

$$\mathbf{y}'(s) = \frac{d\mathbf{y}}{ds} = \frac{d\mathbf{x}}{dt}\frac{d\sigma}{ds} = \mathbf{x}'(\sigma(s))\frac{1}{ds/dt} = \frac{\mathbf{x}'(\sigma(s))}{\|\mathbf{x}'(\sigma(s))\|} \quad \Longrightarrow \quad \|\mathbf{y}'(s)\| \equiv 1.$$

For this reason, $\mathbf{y}$ is also called the **unit-speed parametrization of** $\vec{C}$. Furthermore, $\mathbf{t}(s) = \mathbf{y}'(s)$ is a **unit tangent vector** for $\vec{C}$ and points in the direction in which $\vec{C}$ is oriented (cf. p. 12).

There is a simple way to reverse a path's orientation with a parameter change. Suppose $\mathbf{x}(t)$, $a \le t \le b$ parametrizes $\vec{C}$. Let $u = a + b - t$; then, as t goes from a to b, u goes from b to a. Therefore, although the map

$$\mathbf{r}(u) = \mathbf{x}(a + b - u), \quad a \le u \le b,$$

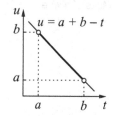

has the same image as $\mathbf{x}$, it induces the opposite orientation on that image. We denote the oppositely-oriented path as $-\vec{C}$. Note that the tangent vectors at corresponding points of $-\vec{C}$ and $\vec{C}$ point in opposite directions:

$$\mathbf{r}'(u) = -\mathbf{x}'(t) \quad \text{when } t = a + b - u.$$

It follows from Theorem 1.4 that any parametrization of $-\vec{C}$ can be obtained from $\mathbf{x}(t)$ by a parameter change $t = h(u)$ for which $h'(u) \le 0$.

The oriented paths $-\vec{C}$ and $+\vec{C}$ have distinct arc-length parameters: they run along their common path C in opposite directions. By contrast, the paths have the same arc length. Although this is evident from the definitions, it is worth deducing the result anew by calculating the integrals using parametrizations. Suppose $\mathbf{x}(t)$, $a \le t \le b$ parametrizes $+\vec{C}$, and $\mathbf{r}(u)$, $c \le u \le d$ parametrizes $-\vec{C}$. Then there is a parameter change $t = h(u)$ with $h' \le 0$, $h(c) = b$, $h(d) = a$. We have

$$\text{arc length of } +\vec{C} = \int_a^b \|\mathbf{x}'(t)\| \, dt = \int_d^c \|\mathbf{x}'(h(u))\| \, h'(u) \, du$$

$$= \int_c^d \|\mathbf{x}'(h(u))\| \, |h'(u)| \, du = \int_c^d \|\mathbf{x}'(h(u)) \, h'(u)\| \, du,$$

reversing the limits of integration by taking $h'(u) = -|h'(u)|$ into account. But because $\mathbf{r}(u) = \mathbf{x}(h(u))$, the chain rule gives

$$\mathbf{r}'(u) = \mathbf{x}'(h(u)) \, h'(u),$$

and hence

$$\int_c^d \|\mathbf{x}'(h(u))\, h'(u)\|\, du = \int_c^d \|\mathbf{r}'(u)\|\, du = \text{arc length of } -\vec{C}.$$

Thus, we can assign to the underlying unoriented path C an arc length equal to the common arc lengths of $+\vec{C}$ and $-\vec{C}$.

The path integral of a scalar function

Note that the integrand of every path integral we have considered—with the exception of the integral for arc length—has been a vector function. With such an integrand, it is essential to pay attention to the relation between the direction of the vector and the direction (i.e., the orientation) of the integration path. With a scalar function, there is no similar concern and, as we have seen, arc length is meaningful for an unoriented path. We now define the integral of a general scalar function over an unoriented path, illustrating the ideas by using mass density.

Mass of a wire

Consider a thin wire in the shape of a space curve C. Suppose that the *mass density* of the wire varies continuously and has the value $\rho(\mathbf{x})$ grams per centimeter at the point $\mathbf{x}$ on C. To estimate the total mass of the wire, partition C by choosing points $\mathbf{x}_1, \ldots, \mathbf{x}_{k+1}$ that run from one end of C to the other. With a sufficiently fine partition, the length of the segment from $\mathbf{x}_i$ to $\mathbf{x}_{i+1}$ is approximately $\Delta s_i = \|\mathbf{x}_{i+1} - \mathbf{x}_i\|$ centimeters, its mass is approximately $\rho(\mathbf{x}_i)\Delta s_i$ grams, and thus

$$\text{total mass of } C \approx \sum_{i=1}^{k} \rho(\mathbf{x}_i)\Delta s_i \text{ grams.}$$

This is a Riemann sum for a new kind of path integral that generalizes the arc-length integral (where $\rho(\mathbf{x}) \equiv 1$). With this path integral, we are then able to write

$$\text{total mass of } C = \int_C \rho(\mathbf{x})\, ds.$$

Scalar path integral

Definition 1.6 *Suppose $f(\mathbf{x})$ is a scalar function defined for all $\mathbf{x}$ on an unoriented simple curve C in $\mathbb{R}^n$; the **path integral of f over C** is*

$$\int_C f(\mathbf{x})\, ds = \lim_{\substack{n \to \infty \\ mesh \to 0}} \sum_{i=1}^{n} f(\mathbf{x})\Delta s_i \quad mesh = \max_i \Delta s_i = \max_i \|\Delta \mathbf{x}_i\|,$$

if the limit exists when taken over all partitions $\{\mathbf{x}_i\}$ of C.

Theorem 1.5. *Suppose f is continuous and C has the smooth parametrization $\mathbf{x}(t)$, $a \le t \le b$; then the path integral of f over C exists, and*

$$\int_C f(\mathbf{x})\, ds = \int_a^b f(\mathbf{x}(t)) \|\mathbf{x}'(t)\|\, dt.$$

Proof. Adapt the argument that was used to prove Theorem 1.3. $\qquad\qquad\square$

Example 4

The theorem implies that the value of the scalar path integral is independent of the parametrization of C; even oppositely oriented parametrizations give the same value. For example, let us evaluate

$$\int_C yz\,ds$$

when C is the helix parametrized as $\mathbf{x}(t) = (\cos(t), \sin(t), t)$, $0 \le t \le 2\pi$. Because $\|\mathbf{x}'(t)\| = \sqrt{2}$ and $yz = t\sin t$, we have

$$\int_C yz\,ds = \int_0^{2\pi} t\sin(t)\,\sqrt{2}\,dt = -2\pi\sqrt{2}.$$

By setting $u = 2\pi - t$, and thus $t = 2\pi - u$, we get the opposite parametrization $\mathbf{r}(u) = \mathbf{x}(2\pi - u)$:

$$\mathbf{r}(u) = (\cos(2\pi - u), \sin(2\pi - u), 2\pi - u) = (\cos u, -\sin u, 2\pi - u).$$

Then $\|\mathbf{r}'(u)\| = \sqrt{2}$ and $yz = (u - 2\pi)\sin u$, so

$$\int_C yz\,ds = \int_0^{2\pi} (u - 2\pi)\sin(u)\,\sqrt{2}\,du$$

$$= \sqrt{2}\int_0^{2\pi} u\sin(u)\,du - 2\pi\sqrt{2}\int_0^{2\pi} \sin(u)\,du.$$

The second integral is zero, so the two evaluations of $f(x,y,z) = yz$ over the helix agree.

We can even express a *vector* integral over an oriented path (e.g., the work done by a force) as the new kind of *scalar* integral over that path without its orientation. Keeping in mind that the scalar path integral uses the "element of arc length" ds (Definition 1.5), let us parametrize the oriented curve $\vec{C}$ by arc length: $\mathbf{x} = \mathbf{y}(s)$, $0 \le s \le L$. Then, for the vector function $\mathbf{F}(\mathbf{x})$,

Rewriting $\int_{\vec{C}} \mathbf{F} \cdot d\mathbf{x}$

$$\int_{\vec{C}} \mathbf{F} \cdot d\mathbf{x} = \int_a^b \mathbf{F}(\mathbf{y}(s)) \cdot \mathbf{y}'(s)\,ds = \int_0^L \mathbf{F}(\mathbf{y}(s)) \cdot \mathbf{t}(s)\,ds,$$

where $\mathbf{t}(s) = \mathbf{y}'(s)$ is the unit tangent vector at the point $\mathbf{y}(s)$ on $\vec{C}$; its direction indicates the orientation of $\vec{C}$. But the last integral is a way to evaluate the scalar function $\mathbf{F} \cdot \mathbf{t}$ over the *unoriented* curve C; in other words, we have

$$\int_{\vec{C}} \mathbf{F} \cdot d\mathbf{x} = \int_C \mathbf{F} \cdot \mathbf{t}\,ds.$$

The path on the right is unoriented; the information about the orientation of $\vec{C}$ has been transferred to the integrand, into the vector $\mathbf{t}$. To confirm this, let $\mathbf{t}_+$ denote the unit tangent for $+\vec{C}$; then $\mathbf{t}_- = -\mathbf{t}_+$ is the unit tangent for $-\vec{C}$, and we have

$$\int_{-\vec{C}} \mathbf{F} \cdot d\mathbf{x} = \int_C \mathbf{F} \cdot \mathbf{t}_-\,ds = \int_C \mathbf{F} \cdot -\mathbf{t}_+\,ds = -\int_{+\vec{C}} \mathbf{F} \cdot d\mathbf{x},$$

as we should.

The scalar $\mathbf{F} \cdot \mathbf{t}$ gives the value of the (signed) projection of $\mathbf{F}$ along $\mathbf{t}$; we call it the *tangential component* of $\mathbf{F}$ along the (oriented) curve $\vec{C}$. Thus, *the work done by a force displacing an object along an oriented curve is the scalar path integral of the tangential component of that force.*

1.3 Polar coordinates

Random variables

Statistics deals with random variables that take on values in a given range with a certain probability. For example, the weights of a 5-ounce bar of soap coming off a production line may be thought of as values of the random variable X because the manufacturing process introduces small fluctuations in weight from bar to bar. Few bars will weigh exactly 5 ounces, but most will have weights close to 5 ounces, some a little higher, some a little lower. The manufacturer can expect that the weights X will be dispersed around the *central value* (here, 5 ounces) in a certain predictable way.

Normal distribution

For many random variables like X, the dispersion follows what is called a *normal distribution*. If $X_{\mu,\sigma}$ is a random variable that follows a normal distribution with mean μ (its central value) and standard deviation σ (its measure of dispersion), then the probability that the value of $X_{\mu,\sigma}$ lies between a and b is equal to the fraction of the area under the entire graph of

$$y = g_{\mu,\sigma}(x) = e^{-(x-\mu)^2/2\sigma^2}$$

that lies between the vertical lines $x = a$ and $x = b$. In other words,

$$\text{Prob}(a \leq X_{\mu,\sigma} \leq b) = \frac{\text{area under graph of } g_{\mu,\sigma} \text{ between } a \text{ and } b}{\text{area under entire graph of } g_{\mu,\sigma}}.$$

These areas are integrals, of course, but the antiderivative of $g_{\mu,\sigma}(x)$ is not one of the elementary functions of the introductory calculus course, so the integrals cannot be found by the usual techniques. Nevertheless, there is a way to find the exact value of the entire area when $\mu = 0$, $\sigma = 1$; it is

$$I = \int_{-\infty}^{\infty} g_{0,1}(x)\,dx = \int_{-\infty}^{\infty} e^{-x^2/2}\,dx = \sqrt{2\pi},$$

as we now show by an ingenious use of polar coordinates.

Showing $I = \sqrt{2\pi}$

The idea is to work with two copies of I and compute I^2 instead of I, using a new "dummy" variable of integration in the second copy of I. With the rule $e^A\,e^B = e^{A+B}$, we then combine the two integrals into one double (iterated) integral:

$$I^2 = \left(\int_{-\infty}^{\infty} e^{-x^2/2} \, dx \right) \left(\int_{-\infty}^{\infty} e^{-y^2/2} \, dy \right)$$
$$= \int_{-\infty}^{\infty} \int_{-\infty}^{\infty} e^{-x^2/2} \, e^{-y^2/2} \, dx \, dy = \int_{-\infty}^{\infty} \int_{-\infty}^{\infty} e^{-(x^2+y^2)/2} \, dx \, dy.$$

There is still no convenient antiderivative, but now make a change to polar coordinates, $x = r\cos\theta$, $y = r\sin\theta$. The new limits of integration are then $0 \le \theta \le 2\pi$, $0 \le r < \infty$, and $dx \, dy$ becomes $r \, dr \, d\theta$. This is the key because it introduces a new factor r into the integrand, and with this new factor, the integrand $r e^{-r^2/2}$ *does* have a simple antiderivative, namely $-e^{-r^2/2}$:

$$I^2 = \int_0^{2\pi} \int_0^{\infty} e^{-r^2/2} r \, dr \, d\theta = \int_0^{2\pi} \left. -e^{-r^2/2} \right|_0^{\infty} d\theta = \int_0^{2\pi} d\theta = 2\pi.$$

Hence $I = \sqrt{2\pi}$.

In the exercises you are asked to show that

$$I_{\mu,\sigma} = \int_{-\infty}^{\infty} e^{-(x-\mu)^2/2\sigma^2} \, dx = \sigma\sqrt{2\pi},$$

which implies

$$\text{Prob}(a \le X_{\mu,\sigma} \le b) = \frac{1}{\sigma\sqrt{2\pi}} \int_a^b e^{-(x-\mu)^2/2\sigma^2} \, dx.$$

If we now combine the factor outside the integral with the integrand function $g_{\mu,\sigma}(x)$ to form the new function

Normal density function

$$f_{\mu,\sigma} = \frac{e^{-(x-\mu)^2/2\sigma^2}}{\sigma\sqrt{2\pi}},$$

we get the more usual formula for the *density function* of the normal distribution with mean μ and standard deviation σ; that is, using $f_{\mu,\sigma}$ we have simply

$$\text{Prob}(a \le X_{\mu,\sigma} \le b) = \text{area under } f_{\mu,\sigma} \text{ from } a \text{ to } b.$$

To what extent is the change to polar coordinates like the coordinate changes we have seen in single-variable integrals? For example, in the transformation from $dx \, dy$ to $r \, dr \, d\theta$, does the factor r play the same role as the factor $\varphi'(s)$ in the pullback from dx to $\varphi'(s) \, ds$? In which case, does r represent a *multiplier*, as $\varphi'(s)$ does in the microscope equation $\Delta x \approx \varphi'(s) \Delta s$. Furthermore, is there a microscope equation (linear approximation) for the polar coordinate change map $M : (r,\theta) \mapsto (x,y)$? If so, is this microscope equation the source of the transformation of differentials, as it is in the one-variable case? And does the multiplier in this new microscope equation involve the derivatives of x and y with respect to r and θ? We explore these questions in the coming chapters.

Comparing coordinate changes

Exercises

1.1. Evaluate $\int_0^\infty \dfrac{dx}{1+x^2}$ and $\int_{-\infty}^1 \dfrac{dx}{1+x^2}$.

1.2. Determine $\int \dfrac{x\,dx}{1+x^2}$. Which type of substitution did you use?

1.3. Carry out a change of variables to evalulate the integral

$$\int_{-R}^R \sqrt{R^2-x^2}\,dx.$$

(This is the area of a semicircle of radius R, and therefore has the value $\pi R^2/2$.) Which type of substitution did you use, pullback or push-forward?

1.4. Determine $\int \dfrac{\arctan x\,dx}{1+x^2}$ and show $\int_0^\infty \dfrac{\arctan x\,dx}{1+x^2} = \dfrac{\pi^2}{8}$.

1.5. a. Determine $\int \dfrac{dw}{w(\ln w)^p}$. Which type of substitution did you use?

   b. Evaluate $I = \int_2^\infty \dfrac{dw}{w(\ln w)^p}$; for which values of p is I finite?

1.6. State a condition that guarantees a function $x = \varphi(s)$ has an inverse. Then use your condition to decide whether each of the following functions is invertible. When possible, find a formula for the inverse of each function that is invertible.

   a. $x = 1/s$.

   b. $x = s + s^3$.

   c. $x = \dfrac{s}{1+s^2}$.

   d. $x = \sinh s = \dfrac{e^s - e^{-s}}{2}$.

   e. $x = \dfrac{s}{\sqrt{1-s^2}}$.

   f. $x = ms + b$.

   g. $x = \cosh s = \dfrac{e^s + e^{-s}}{2}$.

   h. $x = s - s^3$.

   i. $x = \tanh s = \dfrac{\sinh s}{\cosh s}$.

   j. $x = \dfrac{1-s}{1+s}$.

1.7. a. Obtain formulas for $f(s) = \cos(\arcsin s)$ and $g(s) = \tan(\arcsin s)$ directly as functions of s that involve neither trigonometric nor inverse trigonometric functions. Your answers will involve the square root function and polynomial expressions in s.

   b. Compute the derivative of $\cos(\arcsin s)$ using the chain rule and the derivatives of $\cos u$ and $\arcsin s$. Then compute the derivative of $f(s)$ using your expression in part (a). Compare the two derivatives. Do the same for $\tan(\arcsin s)$ and $g(s)$.

1.8. Use $x = \arcsin s$ to show $\int \cos^3 x\,dx = \sin x - \dfrac{\sin^3 x}{3}$.

1.9. a. Write the microscope equation (i.e., the linear approximation) for $\varphi(s) = \sqrt{s}$ at $s = 100$.

   b. Use the microscope equation from part (a) to estimate $\sqrt{102}$ and $\sqrt{99.4}$.

   c. How far are your estimates from those given by a calculator?

   d. Your estimates should be greater than the calculator values; use the graph of $x = \varphi(s)$ to explain why this is so.

1.10. a. Write the microscope equation for $\varphi(s) = 1/s$ at $s = 2$ and use it to estimate $1/2.03$ and $1/1.98$.

   b. How far are your estimates from the values given by a calculator?

   c. Your estimates should be less than the calculator values; use the graph of $x = \varphi(s)$ to explain why this is so.

1.11. Show that $\sqrt{1 + 2h} \approx 1 + h$ when $h \approx 0$.

1.12. a. Determine the microscope equation for $x = \tan s$ at $s = \pi/4$.

   b. Show that $\tan(h + \pi/4) \approx 1 + 2h$ when $h \approx 0$. Is this estimate larger or smaller than the true value? Explain why.

1.13. Determine the local length multiplier for $x = \sin s$ at each of the points $s = 0$, $s = \pi/4$, $s = \pi/2$, $s = 2\pi/3$, and $s = \pi$.

1.14. What is true about the map $\varphi : s \to x$ at a point s_0 where the local length multiplier is negative?

1.15. Consider the hyperbolic sine and hyperbolic cosine functions, $\sinh s$ and $\cosh s$, as defined in Exercise 1.6. Show each is the derivative of the other, and show

$$\cosh^2 s - \sinh^2 s = 1 \quad \text{for all } s.$$

1.16. Use the substitution $x = \sinh s$ to determine $\displaystyle\int \frac{dx}{\sqrt{1 + x^2}}$.

1.17. Determine the work done by the constant force $\mathbf{F} = (2, -3)$ in displacing an object along (a) $\Delta \mathbf{x} = (1, 2)$; (b) $\Delta \mathbf{x} = (1, -2)$; (c) $\Delta \mathbf{x} = (-1, 0)$.

1.18. Determine the work done by the constant force $\mathbf{F} = (7, -1, 2)$ in displacing an object along (a) $\Delta \mathbf{x} = (0, 1, 1)$; (b) $\Delta \mathbf{x} = (1, -2, 0)$; (c) $\Delta \mathbf{x} = (0, 0, 1)$.

1.19. Suppose a constant force $\mathbf{F}$ in the plane does 7 units of work in displacing an object along $\Delta \mathbf{x} = (2, -1)$ and -3 units of work along $\Delta \mathbf{x} = (4, 1)$. How much work does $\mathbf{F}$ do in displacing an object along $\Delta \mathbf{x} = (1, 0)$? Along $\Delta \mathbf{x} = (0, 1)$? Find a nonzero displacement $\Delta \mathbf{x}$ along which $\mathbf{F}$ does no work.

1.20. Let $W(\mathbf{F}, \Delta \mathbf{x})$ be the work done by the constant force $\mathbf{F}$ along the linear displacement $\Delta \mathbf{x}$. Show that W is a linear function of the vectors $\mathbf{F}$ and $\Delta \mathbf{x}$.

1.21. Suppose $\mathbf{F} = (P, Q)$. Determine the unit displacements $\Delta \mathbf{u}$ (i.e., $\|\Delta \mathbf{u}\| = 1$) that yield the maximum and the minimum values of W.

1.22. Suppose the constant force $\mathbf{F} = (P,Q)$ does the work A along the displacement (a,c) and the work B along the displacement (b,d). Determine P and Q. What condition (on a, b, c, and d) must be satisfied for P and Q to be found?

1.23. a. Sketch the curve in the (x,y)-plane given parametrically as

$$x = \frac{2t}{1+t^2}, \quad y = \frac{1-t^2}{1+t^2}.$$

In particular, label the points where $t = -2,-1,0,+1,+2$.

b. Each of the following limits exists; determine the location of each as a point in the (x,y)-plane:

$$\lim_{t \to +\infty} (x(t),y(t)) \quad \lim_{t \to -\infty} (x(t),y(t))$$

c. Compute $\alpha = x(t)^2 + y(t)^2$; your result should be a constant (i.e., independent of t) that is consistent with the sketch of the curve you made in part (a). What is the curve and how does α relate to it?

1.24. Determine the work done by the force field $\mathbf{F}$ in moving a particle along the oriented curve $\vec{C}$, where:

a. $\mathbf{F} = (x,3y)$, $\quad \vec{C}:(\tau^2,\tau^3)$, $\quad 1 \le \tau \le 2$.

b. $\mathbf{F} = (-y,x)$, $\quad \vec{C}$: semicircle of radius 2 at origin, counterclockwise from $(2,0)$ to $(-2,0)$.

c. $\mathbf{F} = (y,x)$, $\quad \vec{C}$: any path from $(5,2)$ to $(7,11)$.

d. $\mathbf{F} = (0,0,-mg)$, $\quad \vec{C}:(2t,t,4-t^2)$, $\quad 0 \le t \le 1$.

e. $\mathbf{F} = (-y,x,1)$, $\quad \vec{C}:(\cos\theta, \sin\theta, 3\theta)$, $\quad 0 \le \theta \le A$.

1.25. Determine $\int_{\vec{C}} \mathbf{F} \cdot d\mathbf{x}$ when

a. $\mathbf{F} = (x+2y,x-y)$, $\quad \vec{C}$: straight line from $(-2,3)$ to $(1,7)$.

b. $\mathbf{F} = (xy,z,x)$, $\quad \vec{C}:(t^2,t,1-t)$, $\quad 0 \le t \le 1$.

c. $\mathbf{F} = \left(\dfrac{-y}{x^2+y^2}, \dfrac{x}{x^2+y^2}\right)$, $\quad \vec{C}:(R\cos t, R\sin t)$, $\quad 0 \le t \le 8\pi$.

1.26. Let $\vec{C}$ be the semicircle of radius 2 centered at the origin, oriented counterclockwise from $(0,-\sqrt{3})$ to $(0,\sqrt{3})$.

a. Show that $\mathbf{r}(u) = \left(\dfrac{4u}{u^2+1}, \dfrac{2u^2-2}{u^2+1}\right)$, $2-\sqrt{3} \le u \le 2+\sqrt{3}$, parametrizes $\vec{C}$. (Cf. Example 1, p. 10, and Exercise 1.23, above.)

b. Using $\mathbf{r}$ to parametrize $\vec{C}$, determine $\int_{\vec{C}} x\,dy$.

We recall here some ideas introduced in the first course in multivariable calculus. Suppose that either

(a) $\oint_{\vec{C}} \mathbf{F} \cdot d\mathbf{x} = 0$ for all *closed* paths $\vec{C}$, or (b) $\int_{\vec{C}_1} \mathbf{F} \cdot d\mathbf{x} = \int_{\vec{C}_2} \mathbf{F} \cdot d\mathbf{x}$

for every pair of oriented paths $\vec{C}_1$ and $\vec{C}_2$ that start at the same point A and end at the same point B. Then the vector field $\mathbf{F}$ is said to be **conservative** if (a) is true, and **path-independent** if (b) is true. It can be shown that a continuous and everywhere-defined vector field is conservative if and only if it is path-independent. The function $\Phi(\mathbf{x})$ is a **potential** for the vector field $\mathbf{F}(\mathbf{x})$ if $\mathbf{F}(\mathbf{x}) = \operatorname{grad} \Phi(\mathbf{x})$ for all $\mathbf{x}$. If $\mathbf{F}$ has a potential, then it is path-independent and (see Exercise 4.36.b, p. 149)

$$\int_{\vec{C}} \mathbf{F} \cdot d\mathbf{x} = \Phi(\mathbf{x}) \Big|_{\text{start of } \vec{C}}^{\text{end of } \vec{C}} .$$

1.27. a. In a coordinate system (x, y, z) where the z-axis is vertical, the gravitational force field at the surface of the earth can be written as $\mathbf{F} = (0, 0, -gm)$, where g is the acceleration due to gravity and m is the mass of a falling object. (Note that g and m are both constant.) Show that $\Phi(x, y, z) = -gmz$ is a potential function for $\mathbf{F}$, demonstrating that $\mathbf{F}$ is a *conservative* field.

b. What is the work done by gravity in moving an object of mass m from the point (a, b, c) to (α, β, γ)? Is this negative if $c < \gamma$? What is the meaning of "negative" work?

c. What is the net work done by gravity in moving an object of mass m from the point (a, b, c) to another point (α, β, c) at the same vertical height as the first? (What does it mean to say that the earth's gravitational field is *conservative* at the surface of the earth?)

1.28. If $\mathbf{x} = (x, y, z)$ is the position of a planet in terms of a coordinate system centered at the sun, then the force of the sun's gravity on the planet is given by $\mathbf{F}(\mathbf{x}) = -\mu \mathbf{x}/r^3$, where μ is a constant (depending on the mass of the sun and of the planet), and $r = \|\mathbf{x}\|$.

a. Write $\mathbf{F}$ explicitly in terms of the space variables x, y, and z.

b. Show that the gravitational force obeys the "inverse square" law: $\|\mathbf{F}\| = \mu/r^2$.

c. Write $\Phi(\mathbf{x}) = \mu/r$ explicitly in terms of the space variable, and show that Φ is a potential for $\mathbf{F}$: $\operatorname{grad} \Phi = \mathbf{F}$. This demonstrates that the gravitational field is conservative.

d. Suppose we choose a unit for distance in such a way that $r = 10$ when our planet is farthest from the sun (called *aphelion*) and $r = 3$ when it is closest to the sun (*perihelion*). How much work does the sun's gravitational field do in moving the planet from aphelion to perihelion?

e. What is the net work done on the planet by the sun's gravity when the planet traverses one complete orbit, from aphelion back to aphelion? (What does it mean to say that the sun's gravitational field is *conservative*?)

1.29. Determine the arc length of each of the following curves.

a. $\mathbf{x}(t) = (3t^2, 4t^2)$, $0 \leq t \leq 1$.

b. $\mathbf{x}(t) = (t^2, t^3)$, $1 \leq t \leq 3$.

c. $\mathbf{x}(t) = (e^t \cos t, e^t \sin t)$, $a \leq t \leq b$.

d. $\mathbf{x}(t) = (\cos t, \sin t, kt)$, $0 \leq t \leq 2\pi$.

e. $\mathbf{x}(t) = \left(\dfrac{1-t^2}{1+t^2}, \dfrac{2t}{1+t^2} \right)$, $-1 \leq t \leq 1$.

f. The ellipse $16x^2 + 9y^2 = 144$. (Suggestion: Use numerical integration.)

1.30. a. Determine the arc-length function $s(t)$ for the circle C of radius R parametrized as $\mathbf{x}(t) = (R\cos t, R\sin t)$.

b. Determine the inverse $t = \sigma(s)$ of the arc-length function and then the corresponding arc-length parametrization $\mathbf{y}(s) = \mathbf{x}(\sigma(s))$.

1.31. Determine the arc-length function $s(t)$ (with $s(0) = 0$) and the arc-length parametrization $\mathbf{y}(s)$ of the curve parametrized as

$$\mathbf{x}(t) = \left(\frac{1-t^2}{1+t^2}, \frac{2t}{1+t^2} \right), \quad -\infty < t < \infty.$$

1.32. Let C be a thin wire formed into the circle of radius R cm centered at the origin. Suppose the mass density of the wire at the point (x, y) is $1 + x^2$ gm/cm. Determine the total mass of the wire.

1.33. Let C be the helix $(x, y, z) = (\cos t, \sin t, t)$, $0 \leq t \leq 4\pi$, and let s be arc length on C. Determine

$$\int_C z\, ds \quad \text{and} \quad \int_C z^2\, ds.$$

1.34. Let C be the circle of radius 5 centered at the point $(4, -3)$, and let s be the arc-length parameter along C, as measured counterclockwise from the origin. Propose a definition for the path integrals

$$\oint_C \cos s\, ds \quad \text{and} \quad \oint_C \cos^2 s\, ds,$$

and then determine their values.

1.35. Is the change from Cartesian to polar coordinates either a pullback or a pushforward substitution, or is it some new type?

1.36. a. Sketch the region D that lies in the first quadrant in the (x, y)-plane between the circles $x^2 + y^2 = 1$ and $x^2 + y^2 = 10$.

b. Describe D in polar coordinates.

c. Change to polar coordinates to evaluate the double integral

$$\iint_D \sin\left(x^2 + y^2\right) dx\,dy.$$

1.37. Let $g_{\mu,\sigma}(x) = e^{-(x-\mu)^2/2\sigma^2}$, as in the text.

a. Show $g_{\mu,\sigma}$ takes its maximum at $x = \mu$ and the graph of $g_{\mu,\sigma}$ has inflection points at $x = \mu \pm \sigma$. Sketch the graph of $z = g_{\mu,\sigma}(x)$ for $\mu - 3\sigma \leq x \leq \mu + 3\sigma$. Do this first with $\mu = 5$ and $\sigma = 2$ and then symbolically with general values for μ and σ.

b. Without repeating the argument in the text that showed $I = \sqrt{2\pi}$, show that

$$\int_{-\infty}^{\infty} g_{\mu,\sigma}(x)\,dx = \sigma\sqrt{2\pi}.$$

You can do this by making an appropriate change of variable that converts this integral to one you can evaluate knowing only that $I = \sqrt{2\pi}$.

1.38. a. Sketch together in the same coordinate plane the graphs of $y = f_{0,\sigma}(x)$ with $\sigma = \frac{1}{2}$, $\sigma = 1$, and $\sigma = 3$. (Use $\mu = 0$ for each.) How do the maximum height and the "width" of the graph vary with σ?

b. Sketch together the graphs of $y = f_{\mu,1}(x)$ with $\mu = -2$, $\mu = 1$, and $\mu = 10$. How do these graphs vary with changing μ?

The various probabilities,

$$\text{Prob}(a \leq X_{\mu,\sigma} \leq b) = \text{area under } \frac{e^{-(x-\mu)^2/2\sigma^2}}{\sigma\sqrt{2\pi}} \text{ from } a \text{ to } b,$$

associated with a normal random variable cannot be computed in terms of the antiderivatives of standard functions. However, because it is important to have these values, strategies have been devised to get access to them. Here is the first: Assume (only for the sake of simplicity) that $0 < a$; then

$$\text{Prob}(a \leq X_{\mu,\sigma} \leq b) = \text{Prob}(0 \leq X_{\mu,\sigma} \leq b) - \text{Prob}(0 \leq X_{\mu,\sigma} \leq a).$$

In other words, it is sufficient to calculate only $\text{Prob}(0 \leq X_{\mu,\sigma} \leq b)$ for various values of b. The following is the second strategy.

1.39. Suppose $Z_{0,1}$ is a normal random variables with mean 0 and standard deviation 1. Continue to assume $X_{\mu,\sigma}$ is a normal random variables with mean μ and standard deviations σ. Show that

$$\cdot \quad \text{Prob}(0 \leq X_{\mu,\sigma} \leq b) = \text{Prob}(0 \leq Z_{0,1} \leq (b - \mu)/\sigma).$$

Suggestion: Consider the push-forward substitution $z = (x - \mu)/\sigma$ and use it to show that

$$\frac{1}{\sigma\sqrt{2\pi}} \int_0^b e^{-(x-\mu)^2/2\sigma^2}\, dx = \frac{1}{\sqrt{2\pi}} \int_0^{(b-\mu)/\sigma} e^{-z^2/2}\, dz.$$

The last result implies that it is sufficient to calculate (e.g., by numerical integration) the values

$$P(z_0) = \mathrm{Prob}(0 \leq Z_{0,1} \leq z_0)$$

for various numbers $z_0 > 0$. In other words, we need only know the distribution of one very special normal random variable, $Z_{0,1}$; all others can be calculated from it. The values $P(z_0)$ are some times called "z-scores"; the probability that a given normal random variable lies in a given range reduces to knowing certain z-scores.

1.40. For simplicity, we assumed that $a > 0$ when we reduced probabilities for $X_{\mu,\sigma}$ to certain z-scores. This assumption is not necessary; describe how to remove it.

Chapter 2
Geometry of Linear Maps

Abstract The geometric meaning of a linear function $x \mapsto y = mx$ is simple and clear: it maps $\mathbb{R}^1$ to itself, multiplying lengths by the factor m. As we show, linear maps $M : \mathbb{R}^n \to \mathbb{R}^n$ also have their multiplication factors of various sorts, for any $n > 1$. In later chapters, these factors play a role in transforming the differentials in multiple integrals that is exactly like the role played by the multiplier $\varphi'(s)$ in the transformation $dx = \varphi'(s)\,ds$ in single-variable integrals. With this in mind, we take up the geometry of linear maps in the simplest case of two variables.

2.1 Maps from $\mathbb{R}^2$ to $\mathbb{R}^2$

Some examples $M : (u,v) \mapsto (x,y)$ illustrate the possibilities that we face.

$$M_1 = \begin{pmatrix} 2 & 0 \\ 0 & \frac{3}{5} \end{pmatrix}$$

$$M_1 : \begin{cases} x = 2u, \\ y = \frac{3}{5}v, \end{cases} \quad \begin{pmatrix} x \\ y \end{pmatrix} = \begin{pmatrix} 2 & 0 \\ 0 & \frac{3}{5} \end{pmatrix} \begin{pmatrix} u \\ v \end{pmatrix}.$$

This map carries horizontal lines to horizontal lines and multiplies horizontal lengths by 2. It carries vertical lines to vertical lines and multiplies vertical lengths by $\frac{3}{5}$. These lines are special: they are the only ones whose directions are left unchanged by the map. (For example, the image of a line with slope $\Delta v/\Delta u = 1$ has the different slope $\Delta y/\Delta x = \frac{3}{5}\Delta v/2\Delta u = 3/10$. See the exercises.) A grid of unit

Horizontal and vertical directions are invariant

J.J. Callahan, *Advanced Calculus: A Geometric View*, Undergraduate Texts in Mathematics, DOI 10.1007/978-1-4419-7332-0_2, © Springer Science+Business Media, LLC 2010

squares in the (u,v)-plane is mapped to a grid of rectangles in the (x,y)-plane, and the sides of the rectangles are parallel to the sides of the squares. Finally, orientation is preserved: a counterclockwise circuit around the unit square in the (u,v)-plane maps to a counterclockwise circuit of its image rectangle in the (x,y)-plane.

$$M_2 = \begin{pmatrix} 1 & 0 \\ 0 & -1 \end{pmatrix}$$

Our second example is also quite simple in form; it is a pure reflection across the horizontal axis:

$$M_2 : \begin{pmatrix} x \\ y \end{pmatrix} = \begin{pmatrix} 1 & 0 \\ 0 & -1 \end{pmatrix} \begin{pmatrix} u \\ v \end{pmatrix}.$$

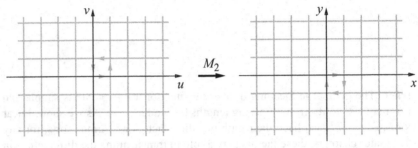

Orientation is reversed

The horizontal and vertical lines are still the invariant ones, and this time even lengths on them are unchanged. Vertical lines are reversed in direction, though, because the vertical multiplier is -1. Orientation of the whole plane is therefore reversed: the counterclockwise circuit in the (u,v)-plane has a clockwise image in the (x,y)-plane. Note that M_1 and M_2 are both diagonal matrices, and the multipliers are their diagonal elements.

$$M_3 = \begin{pmatrix} 0 & -2 \\ 2 & 0 \end{pmatrix}$$

Our third example, although still simple in form, introduces a new action: rotation,

$$M_3 : \begin{pmatrix} x \\ y \end{pmatrix} = \begin{pmatrix} 0 & -2 \\ 2 & 0 \end{pmatrix} \begin{pmatrix} u \\ v \end{pmatrix}.$$

Consider the effect M_3 has on a unit vector $\mathbf{u}$ that makes an angle θ with the positive horizontal axis:

$$M_3(\mathbf{u}) = \begin{pmatrix} 0 & -2 \\ 2 & 0 \end{pmatrix} \begin{pmatrix} \cos\theta \\ \sin\theta \end{pmatrix} = \begin{pmatrix} -2\sin\theta \\ 2\cos\theta \end{pmatrix} = 2 \begin{pmatrix} \cos(\theta + \pi/2) \\ \sin(\theta + \pi/2) \end{pmatrix}.$$

Thus, $M_3(\mathbf{u})$ is two units long and makes an angle $\theta + \pi/2$ with the horizontal axis. (You should check that $-\sin\theta = \cos(\theta + \pi/2)$ and $\cos\theta = \sin(\theta + \pi/2)$.) Every unit vector, and therefore every nonzero vector, is rotated by $\pi/2$. For this linear map, no line is special in the sense that it is preserved with at most a change in length, so there are no length multipliers. Nevertheless, M_3 doubles the length of every vector and it preserves orientation. It is the combination of a rotation (by $90°$) and a uniform dilation (by a factor of 2), as the following figure shows. Any combination of a rotation with a uniform dilation is a linear map of the plane to itself.

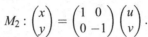

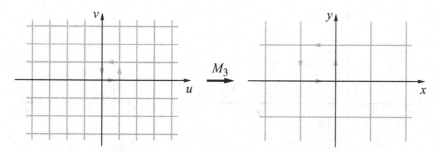

The maps M_1 and M_2 have similarities not shared with M_3; M_1 and M_2 are what we call **strains**. The next two matrices provide us further examples of strains.

Strains

Example 4 has a more complicated formula than the previous ones, but we ultimately show that it is as simple geometrically as the first two.

$$M_4 = \begin{pmatrix} 1 & 2 \\ 2 & 1 \end{pmatrix}$$

$$M_4 : \begin{pmatrix} x \\ y \end{pmatrix} = \begin{pmatrix} 1 & 2 \\ 2 & 1 \end{pmatrix} \begin{pmatrix} u \\ v \end{pmatrix}.$$

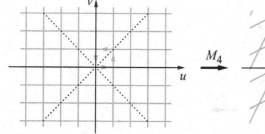

Neither horizontal nor vertical lines are preserved. The image of the grid of unit squares is a grid of congruent parallelograms, but there is apparently little to connect the two grids geometrically. Notice, however, that the diagonals of the square grid are invariant; they are shown dotted in the figure. The image of a vector that lies on the diagonals is just a multiple of itself:

Diagonals are invariant

$$\begin{pmatrix} 1 & 2 \\ 2 & 1 \end{pmatrix} \begin{pmatrix} 1 \\ 1 \end{pmatrix} = \begin{pmatrix} 3 \\ 3 \end{pmatrix} = 3 \begin{pmatrix} 1 \\ 1 \end{pmatrix}, \quad \begin{pmatrix} 1 & 2 \\ 2 & 1 \end{pmatrix} \begin{pmatrix} -1 \\ 1 \end{pmatrix} = \begin{pmatrix} 1 \\ -1 \end{pmatrix} = -1 \begin{pmatrix} -1 \\ 1 \end{pmatrix}.$$

Specifically, the diagonal in the first and third quadrants is stretched by the factor 3, whereas the diagonal in the second and fourth is simply flipped, with no change in length. The presence of a negative multiplier suggests that orientation is reversed, and that is confirmed by the clockwise circuit in the image.

Multipliers are 3 and -1 on diagonals

In the figure below, we have switched to a new grid that is parallel to the invariant diagonals in order to see the geometric action of M_4 more clearly. The basis vectors for the new grid (in both source and target) are

Geometric clarity with a new basis. . .

$$\overline{\mathbf{u}} = \overline{\mathbf{x}} = \begin{pmatrix} 1 \\ 1 \end{pmatrix}, \quad \overline{\mathbf{v}} = \overline{\mathbf{y}} = \begin{pmatrix} -1 \\ 1 \end{pmatrix}.$$

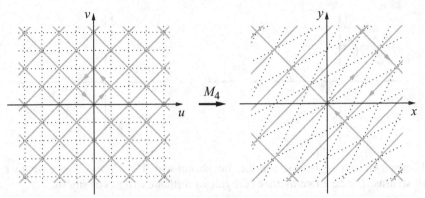

... and new coordinates   Every point (or vector) in the source now has new coordinates $(\overline{u},\overline{v})$ as well as the original coordinates (u,v), so we must be able to change from one to the other. To see how, suppose vector $\mathbf{p}$ has new coordinates $(\overline{u},\overline{v})$; then

$$\mathbf{p} = \overline{u}\begin{pmatrix}1\\1\end{pmatrix} + \overline{v}\begin{pmatrix}-1\\1\end{pmatrix} = \begin{pmatrix}\overline{u}-\overline{v}\\\overline{u}+\overline{v}\end{pmatrix} = (\overline{u}-\overline{v})\begin{pmatrix}1\\0\end{pmatrix} + (\overline{u}+\overline{v})\begin{pmatrix}0\\1\end{pmatrix}.$$

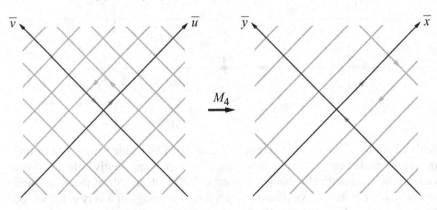

$\overline{M_4} = \begin{pmatrix}3 & 0\\0 & -1\end{pmatrix}$ The original coordinates of $\mathbf{p}$ are therefore $(\overline{u}-\overline{v},\overline{u}+\overline{v})$, so the formulas for the coordinate change and its inverse are

$$u = \overline{u}-\overline{v} \qquad \overline{u} = \tfrac{1}{2}(u+v)$$

$$v = \overline{u}+\overline{v} \qquad \overline{v} = \tfrac{1}{2}(-u+v)$$

The coordinates $(\overline{x},\overline{y})$ and (x,y) change the same way, of course. Using the coordinate changes we can transform the original formulas for the linear map M_4 into formulas that use the new coordinates:

$$\overline{x} = \tfrac{1}{2}(x+y) = \tfrac{1}{2}(u+2v+2u+v) = \tfrac{1}{2}(3u+3v) = 3\overline{u},$$

$$\overline{y} = \tfrac{1}{2}(-x+y) = \tfrac{1}{2}(-u-2v+2u+v) = \tfrac{1}{2}(u-v) = -\overline{v}.$$

Thus, in the new coordinates, our linear map is described by a new matrix:

$$\begin{pmatrix} \overline{x} \\ \overline{y} \end{pmatrix} = \begin{pmatrix} 3 & 0 \\ 0 & -1 \end{pmatrix} \begin{pmatrix} \overline{u} \\ \overline{v} \end{pmatrix}, \quad \overline{M_4} = \begin{pmatrix} 3 & 0 \\ 0 & -1 \end{pmatrix}.$$

With the new matrix, there is no doubt that $\overline{M_4}$ has the same kind of geometric action as M_1 and M_2 (but not M_3!); as a combination of stretches in two different directions, it is a *strain*. Thus, a coordinate change can bring clarity and simplicity to the study of linear maps, just as it can for the study of integration.

It is worth seeing the connection between M_4 and $\overline{M_4}$ directly in terms of matrices. The coordinate change itself is a matrix multiplication, $U = G\overline{U}, \overline{U} = G^{-1}U$, where

$$U = \begin{pmatrix} u \\ v \end{pmatrix}, \quad \overline{U} = \begin{pmatrix} \overline{u} \\ \overline{v} \end{pmatrix}, \quad G = \begin{pmatrix} 1 & -1 \\ 1 & 1 \end{pmatrix}, \quad G^{-1} = \begin{pmatrix} \frac{1}{2} & \frac{1}{2} \\ -\frac{1}{2} & \frac{1}{2} \end{pmatrix}.$$

Equivalence
of matrices

(Notice that the columns of G are the coordinates of the new basis with respect to the old.) The same change $X = G\overline{X}$ and $\overline{X} = G^{-1}X$ happens in the target. In the new coordinates, the map $X = M_4 U$ takes the form

$$\overline{X} = G^{-1}X = G^{-1}M_4 U = G^{-1}M_4 G\overline{U} = \overline{M_4}\,\overline{U}.$$

For us, the object of this string of equalities is the conclusion

$$\overline{M_4} = G^{-1}M_4 G,$$

which leads, finally, to the following definition.

Definition 2.1 *Suppose A and B are $n \times n$ matrices; then we say that B is equivalent to A if there is an invertible matrix G for which $B = G^{-1}AG$.*

What we have just shown about $\overline{M_4}$ and M_4 implies that *if B is equivalent to A, then there is a basis of $\mathbb{R}^n$ on which A acts in the same way that B acts on the standard basis of $\mathbb{R}^n$. Alternately, A and B represent the same linear map in different coordinates.* The matrix G, in $B = G^{-1}AG$, represents the coordinate change.

Note: if $B = G^{-1}AG$, then $A = H^{-1}BH$, where $H = G^{-1}$, so A is equivalent to B when B is equivalent to A. This allows us to say, more symmetrically, that "A and B are equivalent." In the exercises you are asked to show that if C is equivalent to B and B is equivalent to A, then C is also equivalent to A. In other words, equivalent matrices are always *mutually* equivalent. We define an **equivalence class** of $n \times n$ matrices to be the set of all matrices equivalent to some given one. (An example of an equivalence class in a more familiar context is a *rational number*: a rational number is a set of mutually equivalent integer fractions, where two such fractions a/b and c/d are defined to be equivalent if $ad = bc$.) Our aim, which is to identify the different geometric actions of a linear map $M : \mathbb{R}^2 \to \mathbb{R}^2$, is accomplished by determining the equivalence classes of 2×2 matrices.

Equivalence classes
of matrices

$$M_5 = \begin{pmatrix} 0 & 3 \\ 1 & 2 \end{pmatrix}$$

In all our examples where there were invariant lines, those lines were mutually perpendicular. Our next example shows us we cannot expect this to happen in general.

$$M_5 : \begin{pmatrix} x \\ y \end{pmatrix} = \begin{pmatrix} 0 & 3 \\ 1 & 2 \end{pmatrix} \begin{pmatrix} u \\ v \end{pmatrix}.$$

In the figure below, it may appear that M_5 leaves vertical lines invariant. But this is not true: a vertical line in the target is the image of a horizontal line in the source, not a vertical one. In fact, the directions of the invariant lines are indicated by the heavy vectors and the dotted lines in that figure, because

$$\begin{pmatrix} 0 & 3 \\ 1 & 2 \end{pmatrix} \begin{pmatrix} 1 \\ 1 \end{pmatrix} = \begin{pmatrix} 3 \\ 3 \end{pmatrix} = 3 \begin{pmatrix} 1 \\ 1 \end{pmatrix}, \quad \begin{pmatrix} 0 & 3 \\ 1 & 2 \end{pmatrix} \begin{pmatrix} -3 \\ 1 \end{pmatrix} = \begin{pmatrix} 3 \\ -1 \end{pmatrix} = - \begin{pmatrix} -3 \\ 1 \end{pmatrix}.$$

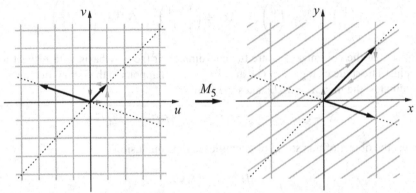

Multipliers are 3 and −1 again

We show how to find invariant lines for an arbitrary linear map immediately below. For the moment, we just observe that M_5 has the same length multipliers as M_4: 3 and −1. The two maps have different invariant lines, though; in particular, the ones for M_5 are not mutually perpendicular. Nevertheless, it is reasonable to call M_5 a strain. With a new coordinate system and grid that is based on vectors along the invariant lines (cf. Exercise 2.2), M_5 has the following form.

$$\begin{pmatrix} \bar{x} \\ \bar{y} \end{pmatrix} = \begin{pmatrix} 3 & 0 \\ 0 & -1 \end{pmatrix} \begin{pmatrix} \bar{u} \\ \bar{v} \end{pmatrix}, \quad \overline{M_5} = \begin{pmatrix} 3 & 0 \\ 0 & -1 \end{pmatrix}.$$

We discover that M_5 is in the same equivalence class as M_4 (because they are both equivalent to $\overline{M_5} = \overline{M_4}$). Indeed, they have the same geometric description: they map the plane to itself by stretching it by a factor of 3 in one direction and simply flipping it—without a stretch—in another. Yet M_5 and M_4 are not the same, because they perform their stretches and flips along different lines (their own invariant lines). It is evident, though, that the invariant lines and the associated multipliers, taken together, characterize each linear map geometrically: two maps with the same multipliers acting on the same lines must be identical.

Multipliers and invariant lines therefore give us an important way to characterize a linear map. They are introduced in the following definition with their usual names. Rather curiously, historical accident has cast those names—*eigen*value and *eigen*vector—half in German and half in English. *Eigen* means "one's own", or "characteristic", and the alternatives *characteristic value* and *characteristic vector* are also used. Furthermore, *eigen* can be translated into French as *propre*, and the terms *proper value* and *proper vector* are likewise in use, but less frequently.

Definition 2.2 *Let $M : \mathbb{R}^n \to \mathbb{R}^n : U \mapsto X$ be a linear map defined by matrix multiplication: $X = MU$. A vector $U \neq 0$ is an **eigenvector of M with eigenvalue λ** if $MU = \lambda U$.*

Note that an eigenvector is nonzero by definition because it has to determine an invariant line. An eigenvalue can be 0, though; it just means M has a nonzero *kernel* consisting of the eigenvectors with eigenvalue zero.

Let us rewrite the "eigen" condition $MU = \lambda U$ first as $MU = \lambda IU$ (where I is the identity matrix), and then as $(M - \lambda I)U = 0$. This says that U is in the kernel of the newly defined matrix $M - \lambda I$. But $U \neq 0$, so $M - \lambda I$ must be noninvertible, implying $\det(M - \lambda I) = 0$. Because the determinant of a matrix is a polynomial function of the elements of the matrix, the expression $p(\lambda) = \det(M - \lambda I)$ is is a polynomial in λ, called the **characteristic polynomial** of M. The equation $p(\lambda) = 0$ is the **characteristic equation** of M.

Theorem 2.1. *Each eigenvalue of M is a root of its characteristic equation.* $\quad\square$

But real polynomials can have complex roots, too. For example, our rotation matrix M_3 has the characteristic polynomial $p(\lambda) = \lambda^2 + 4$ whose roots are $\lambda = \pm 2i$. Furthermore,

$$\begin{pmatrix} 0 & -2 \\ 2 & 0 \end{pmatrix}\begin{pmatrix} 1 \\ -i \end{pmatrix} = \begin{pmatrix} 2i \\ 2 \end{pmatrix} = 2i\begin{pmatrix} 1 \\ -i \end{pmatrix}; \quad \begin{pmatrix} 0 & -2 \\ 2 & 0 \end{pmatrix}\begin{pmatrix} 1 \\ i \end{pmatrix} = \begin{pmatrix} -2i \\ 2 \end{pmatrix} = -2i\begin{pmatrix} 1 \\ i \end{pmatrix}.$$

In other words, when we allow M_3 to act on ordered pairs of complex numbers, we find that M_3 does have invariant directions. A real polynomial always has complex roots, but need not have any real roots. Thus this example suggests that, for the purpose of getting the simple view of the action of matrix multiplication, we use complex n-tuples instead of real ones ($M : \mathbb{C}^n \to \mathbb{C}^n$) to define eigenvectors and eigenvalues (Definition 2.2). With this understanding, every root of the characteristic equation of M becomes an eigenvalue of M.

Characteristics of a linear map

Eigenvectors and eigenvalues

$\lambda = 0$ and the kernel of M

Characteristic equation

Complex roots

Trace and determinant

If M is an 2×2 matrix, we have

$$p(\lambda) = \det\left[\begin{pmatrix} a & b \\ c & d \end{pmatrix} - \lambda \begin{pmatrix} 1 & 0 \\ 0 & 1 \end{pmatrix}\right] = \det\begin{pmatrix} a-\lambda & b \\ c & d-\lambda \end{pmatrix}$$
$$= (a-\lambda)(d-\lambda) - bc = \lambda^2 - (a+d)\lambda + ad - bc$$
$$= \lambda^2 - \operatorname{tr}(M)\lambda + \det(M),$$

where $\operatorname{tr}(M) = a+d$ is the **trace** of M and $\det(M) = ad - bc$ is, of course, its **determinant**. Thus, the eigenvalues of M are the roots of a quadratic equation that involves the trace and determinant of M. If λ_1 and λ_2 are these roots, then

$$(\lambda - \lambda_1)(\lambda - \lambda_2) = \lambda^2 - (\lambda_1 + \lambda_2)\lambda + \lambda_1\lambda_2$$

is also the characteristic polynomial, so we have the following proposition.

Theorem 2.2. *The sum of the eigenvalues of a 2×2 matrix is equal to its trace and their product is equal to its determinant.* ☐

If we write the equation for the eigenvalues of the 2×2 matrix M in the form

$$\lambda_{1,2} = \frac{\operatorname{tr} M \pm \sqrt{\operatorname{tr}^2 M - 4\det M}}{2},$$

we see these roots will be complex when the *discriminant* is negative:

$$\operatorname{tr}^2 M - 4\det M = (a+d)^2 - 4(ad - bc) = (a-d)^2 + 4bc < 0.$$

Because $(a-d)^2 \geq 0$, b and c must be of opposite sign and have $bc < -(a-d)^2/4$ for M to have complex eigenvalues.

As we have seen, equivalent matrices describe the same linear map but in terms of different bases. We would expect, then, that such matrices have the same eigenvalues, and their eigenvectors would be mapped to one another by the coordinate change that connects the matrices.

Eigenvalues of equivalent matrices

Theorem 2.3. *Suppose A and $B = G^{-1}AG$ are equivalent matrices, and $\overline{U}$ is an eigenvector of B with eigenvalue λ. Then $U = G\overline{U}$ is an eigenvector of A with the same eigenvalue λ.*

Proof. Suppose $\overline{U}$ is an eigenvector of B with eigenvalue λ: $B\overline{U} = \lambda\overline{U}$. Then

$$G^{-1}AG\overline{U} = \lambda\overline{U}, \quad \text{so } A(G\overline{U}) = G\lambda\overline{U} = \lambda(G\overline{U}). \qquad \square$$

Corollary 2.4 *Equivalent matrices have the same eigenvalues and therefore the same trace, determinant, and characteristic polynomial.*

Proof. According to the theorem, every eigenvalue of $B = G^{-1}AG$ is an eigenvalue of A. But equivalence is symmetric ($A = H^{-1}BH$ with $H = G^{-1}$), so every eigenvalue of A is an eigenvalue of B. ☐

Even when the eigenvalues and eigenvectors of M are complex, they can provide crucial information about the geometric action of M on the real plane $\mathbb{R}^2$. Consider the map

$$M_6 : \begin{pmatrix} x \\ y \end{pmatrix} = \begin{pmatrix} 1 & -2 \\ 1 & -1 \end{pmatrix} \begin{pmatrix} u \\ v \end{pmatrix}.$$

$$M_6 = \begin{pmatrix} 1 & -2 \\ 1 & -1 \end{pmatrix}$$

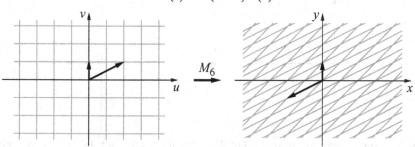

Here $\operatorname{tr} M_6 = 0$ and $\det M_6 = 1$, so the characteristic polynomial is $\lambda^2 + 1$ and the eigenvalues are $\pm i$. However, M_6 is not a rotation: it turns the coordinate axes by different amounts (so their images are not perpendicular, as they would be under a rotation).

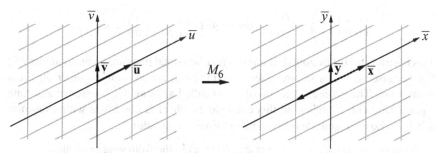

A clearer picture emerges, however, when we consider the action of M_6 on the heavy vectors; note that

$$M_6(\overline{\mathbf{u}}) = \overline{\mathbf{y}}, \quad M_6(\overline{\mathbf{v}}) = -\overline{\mathbf{x}}.$$

Consequently, in terms of the new grid and coordinates (which use $\{\overline{\mathbf{u}}, \overline{\mathbf{v}}\}$ and $\{\overline{\mathbf{x}}, \overline{\mathbf{y}}\}$ as bases in the source and target), our linear map is now described by

$$\overline{M}_6 : \begin{pmatrix} \overline{x} \\ \overline{y} \end{pmatrix} = \begin{pmatrix} 0 & -1 \\ 1 & 0 \end{pmatrix} \begin{pmatrix} \overline{u} \\ \overline{v} \end{pmatrix}.$$

You can check directly that the new matrix $\overline{M}_6$ has the same trace, determinant, characteristic polynomial, and eigenvalues as M_6. But the geometric action of $\overline{M}_6$ is simpler to describe: applied to the standard basis (instead of $\{\overline{\mathbf{u}}, \overline{\mathbf{v}}\}$), $\overline{M}_6$ is just rotation by $90°$. Note, however, that $\overline{M}_6$, applied to the basis $\{\overline{\mathbf{u}}, \overline{\mathbf{v}}\}$, is not a rotation, because M_6 is not a rotation. Thus, although M_6 is not a rotation, it is equivalent to a $90°$ rotation.

M_6 is equivalent to a $90°$ rotation

$$M_7 = \begin{pmatrix} 2 & 2 \\ 0 & 2 \end{pmatrix}$$

There is essentially only one more type of linear map for us to analyze. Here is an example:

$$M_7 : \begin{pmatrix} x \\ y \end{pmatrix} = \begin{pmatrix} 2 & 2 \\ 0 & 2 \end{pmatrix} \begin{pmatrix} u \\ v \end{pmatrix} = 2 \begin{pmatrix} 1 & 1 \\ 0 & 1 \end{pmatrix} \begin{pmatrix} u \\ v \end{pmatrix}.$$

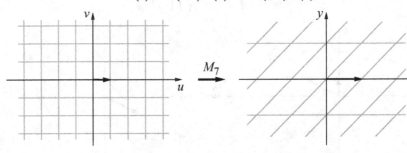

M_7 has only one invariant direction

The horizontal lines are invariant, but the vertical ones are not. In fact, there is no second set of invariant lines. We can trace this shortcoming to the fact that M_7 has only one eigenvalue, $\lambda = 2$; it is a *repeated* root of the characteristic polynomial $\lambda^2 - 4\lambda + 4 = (\lambda - 2)^2$. All eigenvectors are therefore the solutions of the single pair of equations

$$\begin{pmatrix} 0 & 2 \\ 0 & 0 \end{pmatrix} \begin{pmatrix} u \\ v \end{pmatrix} = \begin{pmatrix} 0 \\ 0 \end{pmatrix}, \text{ giving just } \begin{pmatrix} u \\ v \end{pmatrix} = \begin{pmatrix} u \\ 0 \end{pmatrix} = u \begin{pmatrix} 1 \\ 0 \end{pmatrix}.$$

Shears

This eigenvector is horizontal; it implies that horizonal lines are invariant under M_7. But because no vector in any other direction is an eigenvector, no other direction is invariant. The geometric action of M_7 is called a **shear**. Of course we associate eigenvalues with stretches; in this example the shear is combined with a uniform dilation whose magnitude is given by the single eigenvalue 2.

$$M_8 = \begin{pmatrix} 4 & -2 \\ 2 & 0 \end{pmatrix}$$

A shear can take a less recognizable form, as in the following example.

$$M_8 : \begin{cases} x = 4u - 2v, \\ y = 2u, \end{cases} \quad M_8 = \begin{pmatrix} 4 & -2 \\ 2 & 0 \end{pmatrix}.$$

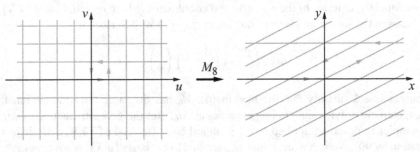

Because $\operatorname{tr} M_8 = \det M_8 = 4$, M_8 has the single eigenvalue $\lambda = 2$. There is an eigenvector in only one direction. In the exercises you are asked to find that eigenvector and then to verify that the coordinate change

$$G: \begin{cases} u = \bar{u} + \bar{v}, \\ v = \bar{u}, \end{cases}$$

converts M_8 into M_7: $G^{-1} M_8 G = M_7$. This implies M_8 is equivalent to a shear combined with a uniform dilation by the factor 2.

Before we proceed to a description of all the different geometric actions of a linear map $M : \mathbb{R}^2 \to \mathbb{R}^2$, it is helpful to comment on a few specific matrices. The matrix

Rotations

$$R_\theta = \begin{pmatrix} \cos\theta & -\sin\theta \\ \sin\theta & \cos\theta \end{pmatrix}$$

rotates the plane by θ radians; it has complex eigenvalues $\lambda_\pm = \cos\theta \pm i\sin\theta$. The matrix

$$C_{a,b} = \begin{pmatrix} a & -b \\ b & a \end{pmatrix}$$

has a similar form, and has the complex eigenvalues, $\lambda_\pm = a \pm ib$, but is not a simple rotation if $a^2 + b^2 \neq 1$. However, the following theorem connects it to a rotation.

Theorem 2.5. *If $(a,b) \neq (0,0)$, then the matrix $C_{a,b}$ rotates the plane by the angle $\theta = \arctan(b/a)$ and then performs a uniform dilation by the factor $\sqrt{a^2 + b^2}$.*

Proof. By hypothesis, $\sqrt{a^2 + b^2} \neq 0$, so we can factor this term out of each component of $C_{a,b}$:

$$C_{a,b} = \sqrt{a^2 + b^2} \begin{pmatrix} \dfrac{a}{\sqrt{a^2+b^2}} & \dfrac{-b}{\sqrt{a^2+b^2}} \\ \dfrac{b}{\sqrt{a^2+b^2}} & \dfrac{a}{\sqrt{a^2+b^2}} \end{pmatrix} = \sqrt{a^2+b^2} \begin{pmatrix} \cos\theta & -\sin\theta \\ \sin\theta & \cos\theta \end{pmatrix}.$$

In the matrix on the right, we have made the replacements

$$\frac{a}{\sqrt{a^2+b^2}} = \cos\theta \quad \text{and} \quad \frac{b}{\sqrt{a^2+b^2}} = \sin\theta$$

by using the angle $\theta = \arctan(b/a)$, as the figure shows. In fact, we can extend $\theta = \arctan(b/a)$ as a function of two variables a and b (cf. Exercise 2.10) to define a unique value of θ in the interval $-\pi < \theta \leq \pi$ for every $(a,b) \neq (0,0)$. That is, we need not require that a and b be positive. Therefore,

$$C_{a,b} = \left(\sqrt{a^2+b^2} \right) R_{\arctan(b/a)}. \qquad \square$$

Suppose M has an eigenvector U with eigenvalue 0; then M collapses $\mathbb{R}^2$ along the direction of U. For this reason, we describe any matrix with a zero eigenvalue as a **collapse**. A rather special example is

Collapse and shear–collapse

$$K = \begin{pmatrix} 0 & 1 \\ 0 & 0 \end{pmatrix} \quad \text{with eigenvector } U = \begin{pmatrix} 1 \\ 0 \end{pmatrix}.$$

Because $\operatorname{tr} K = \det K = 0$, the characteristic equation of K is $\lambda^2 = 0$. This has only a single root, the repeated eigenvalue $\lambda = 0$. Because there is only a single eigendirection, given by the eigenvector U, above, K behaves like a *shear*. Because its sole eigenvalue is 0, it is also a *collapse*; we call it a **shear–collapse**.

Theorem 2.6. *Every linear map $M : \mathbb{R}^2 \to \mathbb{R}^2$ is equivalent to precisely one of the types listed in the following table; M lies in the equivalence class of matrices that have the same eigenvalues and the same number of eigendirections.*

Equivalence Classes of 2×2 Matrices
and Their Representatives

Name	Matrix	Eigenvalues*	Eigendirections
Zero	$\begin{pmatrix} 0 & 0 \\ 0 & 0 \end{pmatrix}$	$0,0$	all
Shear–collapse	$\begin{pmatrix} 0 & 1 \\ 0 & 0 \end{pmatrix}$	$0,0$	one
Strain–collapse	$\begin{pmatrix} \lambda & 0 \\ 0 & 0 \end{pmatrix}$	$0,\lambda$	two
Pure dilation	$\begin{pmatrix} \lambda & 0 \\ 0 & \lambda \end{pmatrix}$	λ,λ	all
Shear–dilation	$\begin{pmatrix} \lambda & \lambda \\ 0 & \lambda \end{pmatrix}$	λ,λ	one
Strain	$\begin{pmatrix} \lambda_1 & 0 \\ 0 & \lambda_2 \end{pmatrix}$	$\lambda_1 \neq \lambda_2$	two
Rotation–dilation	$\begin{pmatrix} a & -b \\ b & a \end{pmatrix}$	$a \pm ib$	none

*An eigenvalue not written as 0 is understood to be nonzero.

Proof. You carry out parts of the proof in the exercises. The basic classification is by the eigenvalues of M:

- *Real and equal*: zero, shear–collapse, pure dilation, shear–dilation
- *Real and unequal*: strain–collapse, strain
- *Complex conjugates*: rotation–dilation

Types are then further separated by the number of eigendirections that M has:

- *None*: rotation–dilation

- *One*: all shears

- *Two*: all strains

- *All*: pure dilations, including zero ☐

If a matrix of a linear map has real eigenvalues and eigenvectors, the eigenvectors determine the map's invariant lines and the eigenvalues give the length multiplication factors along those lines. But even more is true: the product of those factors then tells us how much the map magnifies *areas*; the sign of the product even indicates how the map affects orientation. Furthermore, the area multiplier is just the determinant of the matrix (because the product of the eigenvalues is the determinant), so the area multiplier can be determined directly from the matrix itself, without first calculating the eigenvalues. (This is particularly useful when the eigenvalues and eigenvectors are complex because then the matrix has no usable length multipliers.)

We need a notation that indicates the orientation of a parallelogram, and this is easily obtained. We use $\mathbf{v} \wedge \mathbf{w}$ to denote the parallelogram spanned by the vectors $\mathbf{v}$ and $\mathbf{w}$, *in that order*. Call this the **wedge product** of $\mathbf{v}$ and $\mathbf{w}$. The order determines the "sense of rotation"—either clockwise or counterclockwise—that carries the first-named vector, $\mathbf{v}$, to the second, $\mathbf{w}$. Reversing the order, to $\mathbf{w} \wedge \mathbf{v}$, reverses the sense of rotation; we write $\mathbf{w} \wedge \mathbf{v} = -\mathbf{v} \wedge \mathbf{w}$. A parallelogram has **positive orientation** if it has the same sense of rotation as the positive coordinate axes, and **negative orientation** if it has the opposite sense. (If $\mathbf{v}$ and $\mathbf{w}$ are linearly dependent, $\mathbf{v} \wedge \mathbf{w}$ collapses to a line segment and has no orientation.) As a rule, we take the positive sense of rotation to be counterclockwise. Thus, in the adjacent figure, $\mathbf{v} \wedge \mathbf{w}$ is negatively oriented and $\mathbf{w} \wedge \mathbf{v}$ is positively oriented.

The signed area, area $\mathbf{v} \wedge \mathbf{w}$, will then be determined by the following two stipulations. First, the signed area of the unit square $\mathbf{e}_1 \wedge \mathbf{e}_2$ should be $+1$ (rather than -1). Second, area$(\mathbf{w} \wedge \mathbf{v}) = -$ area$(\mathbf{v} \wedge \mathbf{w})$ for all $\mathbf{v}, \mathbf{w}$.

Theorem 2.7. area $\begin{pmatrix} v_1 \\ v_2 \end{pmatrix} \wedge \begin{pmatrix} w_1 \\ w_2 \end{pmatrix} = \det \begin{pmatrix} v_1 & w_1 \\ v_2 & w_2 \end{pmatrix}$.

Proof. See Exercise 2.15. ☐

The signed area is the determinant of the matrix V whose columns are the coordinates of $\mathbf{v}$ and $\mathbf{w}$, in that order. The matrix represents a linear map

$$\mathbf{x} = V(s,t) = s\mathbf{v} + t\mathbf{w}$$

that maps the unit square $\mathbf{e}_1 \wedge \mathbf{e}_2$ to $\mathbf{v} \wedge \mathbf{w}$. Thus, the orientation and area of $\mathbf{v} \wedge \mathbf{w}$ are determined by a parametrization.

Theorem 2.8. *If $M : \mathbb{R}^2 \to \mathbb{R}^2$ is a linear map and $\mathbf{v} \wedge \mathbf{w}$ is an oriented parallelogram in the source, then $M(\mathbf{v} \wedge \mathbf{w}) = M(\mathbf{v}) \wedge M(\mathbf{w})$ is an oriented parallelogram in the target and*

$$\text{area}\, M(\mathbf{v} \wedge \mathbf{w}) = \det M \times \text{area}(\mathbf{v} \wedge \mathbf{w}).$$

Proof. In Exercise 2.16 you are asked to prove this directly by analyzing the function $\text{area}\, M(\mathbf{v}) \wedge M(\mathbf{w})$. □

Corollary 2.9 *The **area multiplier** of the linear map $M : \mathbb{R}^2 \to \mathbb{R}^2$ is $\det M$. The map M reverses orientation precisely when $\det M < 0$.* □

2.2 Maps from $\mathbb{R}^n$ to $\mathbb{R}^n$

Because we found the area multiplier to be the most salient geometric feature of a linear map of the plane, we can expect that the volume multiplier, and its higher-dimensional analogues, will play a similar role here.

$\mathbf{x} \wedge \mathbf{y}$ in $\mathbb{R}^n$, $n \geq 3$

In $\mathbb{R}^n$, $n \geq 3$, we continue to use $\mathbf{x} \wedge \mathbf{y}$ to denote the oriented parallelogram spanned by the vectors $\mathbf{x}$ and $\mathbf{y}$. When $n = 2$, the orientation of $\mathbf{x} \wedge \mathbf{y}$ is fixed in relation to the orientation of the two coordinate axes. However, when $n \geq 3$, this is not true: an orientation-preserving linear map of $\mathbb{R}^n$ can reverse the orientation of $\mathbf{x} \wedge \mathbf{y}$ (see below, p. 44). Moreover, because the coordinates of $\mathbf{x}$ and $\mathbf{y}$ now make up an $n \times 2$ matrix V—for which the determinant is not even defined—we cannot express $\text{area}(\mathbf{x} \wedge \mathbf{y})$ as the determinant of V. (Let $V^\dagger$ be the *transpose* of the matrix of V; it is a $2 \times n$ matrix. The product $V^\dagger V$ does give a square 2×2 matrix, and $\text{area}^2(\mathbf{x} \wedge \mathbf{y}) = \det V^\dagger V$. See the exercises.)

$\mathbf{x} \times \mathbf{y}$ in $\mathbb{R}^3$

In $\mathbb{R}^3$, the cross-product of two vectors is defined: $\mathbf{p} = \mathbf{x} \times \mathbf{y}$ is the unique vector with length $|\text{area}(\mathbf{x} \wedge \mathbf{y})|$ and with direction orthogonal to both $\mathbf{x}$ and $\mathbf{y}$ so that the three vectors $\mathbf{x}$, $\mathbf{y}$, $\mathbf{p}$—in that order—have the same orientation as the three coordinate axes. We call this the **positive orientation**, and always take it to be *right-handed*, meaning that the thumb, index finger, and middle finger of the right hand can be lined up with the first, second, and third coordinate axes, respectively.

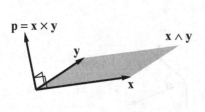
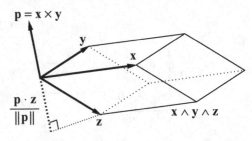

$\mathbf{x} \wedge \mathbf{y} \wedge \mathbf{z}$ in $\mathbb{R}^3$

For vectors $\mathbf{x}$, $\mathbf{y}$, and $\mathbf{z}$ in $\mathbb{R}^3$, we define $\mathbf{x} \wedge \mathbf{y} \wedge \mathbf{z}$ to be the **oriented parallelepiped** spanned by $\mathbf{x}$, $\mathbf{y}$, and $\mathbf{z}$, in that order. Notice that the parallelepiped shown in the figure on the right, above, has *left*-handed, or *negative*, orientation. To calculate its volume, we take $\mathbf{x} \wedge \mathbf{y}$ as base and measure its height as the length of the projection of $\mathbf{z}$ on $\mathbf{p} = \mathbf{x} \times \mathbf{y}$:

$$\text{vol}(\mathbf{x} \wedge \mathbf{y} \wedge \mathbf{z}) = \text{area of base} \cdot \text{height}$$

$$= \|\mathbf{x} \times \mathbf{y}\| \frac{(\mathbf{x} \times \mathbf{y}) \cdot \mathbf{z}}{\|\mathbf{x} \times \mathbf{y}\|} = (\mathbf{x} \times \mathbf{y}) \cdot \mathbf{z}.$$

The quantity $(\mathbf{x} \times \mathbf{y}) \cdot \mathbf{z}$ is called the **scalar triple product** of $\mathbf{x}$, $\mathbf{y}$, and $\mathbf{z}$. Note that the parentheses can be removed, because $\mathbf{x} \times (\mathbf{y} \cdot \mathbf{z})$ is meaningless. The *order* of the three vectors in $\mathbf{x} \times \mathbf{y} \cdot \mathbf{z}$ is still important, though.

Scalar triple product

Theorem 2.10. *The signed volume of the oriented parallelepiped* $\mathbf{x} \wedge \mathbf{y} \wedge \mathbf{z}$ *is the scalar triple product* $\mathbf{x} \times \mathbf{y} \cdot \mathbf{z}$. *The volume is negative precisely when the parallelepiped has negative orientation.*

Proof. The first statement has been proven above. To prove the second, note that the parallelepiped $\mathbf{x} \wedge \mathbf{y} \wedge (\mathbf{x} \times \mathbf{y})$ has positive orientation by definition, at least when $\mathbf{x} \times \mathbf{y} \neq 0$. Therefore, $\mathbf{x} \wedge \mathbf{y} \wedge \mathbf{z}$ has negative orientation precisely when $\mathbf{z}$ and $\mathbf{x} \times \mathbf{y}$ lie on opposite sides of the plane determined by $\mathbf{x} \wedge \mathbf{y}$. But $\mathbf{x} \times \mathbf{y}$ is perpendicular to $\mathbf{x} \wedge \mathbf{y}$, so $\mathbf{z}$ and $\mathbf{x} \times \mathbf{y}$ are on opposite sides of $\mathbf{x} \wedge \mathbf{y}$ when $\mathbf{z}$ makes an obtuse angle with $\mathbf{x} \times \mathbf{y}$, and that is precisely the condition that

$$(\mathbf{x} \times \mathbf{y}) \cdot \mathbf{z} = \text{vol}(\mathbf{x} \wedge \mathbf{y} \wedge \mathbf{z}) < 0. \qquad \square$$

Theorem 2.11. *Let* V *be the matrix whose columns are the coordinates of* $\mathbf{x}$, $\mathbf{y}$, *and* $\mathbf{z}$, *in that order. Then* $\text{vol}(\mathbf{x} \wedge \mathbf{y} \wedge \mathbf{z}) = \det V$.

Volumes and determinants

Proof. Let $\mathbf{x} = (x_1, x_2, x_3)^\dagger$, $\mathbf{y} = (y_1, y_2, y_3)^\dagger$, $\mathbf{z} = (z_1, z_2, z_3)^\dagger$. Then

$$V = \begin{pmatrix} x_1 & y_1 & z_1 \\ x_2 & y_2 & z_2 \\ x_3 & y_3 & z_3 \end{pmatrix},$$

and if we calculate the determinant of V along the third column, we get

$$\det V = \begin{vmatrix} x_2 & y_2 \\ x_3 & y_3 \end{vmatrix} z_1 + \begin{vmatrix} x_3 & y_3 \\ x_1 & y_1 \end{vmatrix} z_2 + \begin{vmatrix} x_1 & y_1 \\ x_2 & y_2 \end{vmatrix} z_3;$$

note the order of the rows in the second determinant. On the other hand,

$$\mathbf{x} \times \mathbf{y} = \left(\begin{vmatrix} x_2 & y_2 \\ x_3 & y_3 \end{vmatrix}, \begin{vmatrix} x_3 & y_3 \\ x_1 & y_1 \end{vmatrix}, \begin{vmatrix} x_1 & y_1 \\ x_2 & y_2 \end{vmatrix} \right)^\dagger,$$

so

$$\text{vol}(\mathbf{x} \wedge \mathbf{y} \wedge \mathbf{z}) = \mathbf{x} \times \mathbf{y} \cdot \mathbf{z} = \begin{vmatrix} x_2 & y_2 \\ x_3 & y_3 \end{vmatrix} z_1 + \begin{vmatrix} x_3 & y_3 \\ x_1 & y_1 \end{vmatrix} z_2 + \begin{vmatrix} x_1 & y_1 \\ x_2 & y_2 \end{vmatrix} z_3 = \det V. \qquad \square$$

Corollary 2.12 *The parallelepiped* $\mathbf{x} \wedge \mathbf{y} \wedge \mathbf{z}$ *has positive orientation if and only if the linear map* $V : \mathbb{R}^3 \to \mathbb{R}^3$ *that maps the standard basis* $\mathbf{e}_1, \mathbf{e}_2, \mathbf{e}_3$ *to* $\mathbf{x}, \mathbf{y}, \mathbf{z}$, *in that order, has* $\det V > 0$.

Proof. We use two fundamental results of linear algebra: (a) a linear map is uniquely defined by its action on a basis; and (b) the matrix V whose columns are the coordinates of $\mathbf{x}, \mathbf{y}, \mathbf{z}$, in that order, has $V(\mathbf{e}_1) = \mathbf{x}$, $V(\mathbf{e}_2) = \mathbf{y}$, $V(\mathbf{e}_3) = \mathbf{z}$. By the preceding theorems, $\det V > 0$ if and only if $\mathbf{x} \wedge \mathbf{y} \wedge \mathbf{z}$ has positive orientation. $\square$

Corollary 2.13 *An ordered set of vectors* $\{\mathbf{x}, \mathbf{y}, \mathbf{z}\}$ *has positive orientation if and only if it is the image of the standard basis* $\{\mathbf{e}_1, \mathbf{e}_2, \mathbf{e}_3\}$ *under a linear map* M *with* $\det M > 0$. $\square$

Corollary 2.14 *If* $M : \mathbb{R}^3 \to \mathbb{R}^3$ *is a linear map and* $\mathbf{x} \wedge \mathbf{y} \wedge \mathbf{z}$ *is an oriented parallelepiped in the source, then* $M(\mathbf{x} \wedge \mathbf{y} \wedge \mathbf{z}) = M(\mathbf{x}) \wedge M(\mathbf{y}) \wedge M(\mathbf{z})$ *is an oriented parallelepiped in the target and*

$$\operatorname{vol} M(\mathbf{x} \wedge \mathbf{y} \wedge \mathbf{z}) = \det M \times \operatorname{vol}(\mathbf{x} \wedge \mathbf{y} \wedge \mathbf{z}).$$

Proof. This is analogous to Theorem 2.8 (p. 42) and is proven the same way. $\square$

Corollary 2.15 *The* **volume multiplier** *for the linear map* $M : \mathbb{R}^3 \to \mathbb{R}^3$ *is* $\det M$. *The map* M *reverses orientation precisely when* $\det M < 0$. $\square$

Suppose $\mathbf{x} \times \mathbf{y} \neq \mathbf{0}$. Then there is a linear map $L : \mathbb{R}^3 \to \mathbb{R}^3$ with positive determinant for which $L(\mathbf{x}) = \mathbf{y}$, $L(\mathbf{y}) = \mathbf{x}$ (see Exercise 2.21). Consequently, orientation-preserving linear maps of $\mathbb{R}^3$ need not preserve the orientation of 2-dimensional parallelograms that lie in $\mathbb{R}^3$. (We still orient such objects; see pp. 388ff.) The sign of $\operatorname{area}(\mathbf{x} \wedge \mathbf{y})$ has no intrinsic geometric significance in $\mathbb{R}^3$; thus we always take $\operatorname{area}(\mathbf{x} \wedge \mathbf{y}) \geq 0$.

There is a remarkable connection between $\mathbf{x} \wedge \mathbf{y}$ and its projections onto the three coordinate planes. First of all, if $\mathbf{x} = (x_1, x_2, x_3)^{\dagger}$, $\mathbf{y} = (y_1, y_2, y_3)^{\dagger}$, then

$$\operatorname{area}(\mathbf{x} \wedge \mathbf{y}) = \|\mathbf{x} \times \mathbf{y}\| = \left\| \left(\begin{vmatrix} x_2 & y_2 \\ x_3 & y_3 \end{vmatrix}, \begin{vmatrix} x_3 & y_3 \\ x_1 & y_1 \end{vmatrix}, \begin{vmatrix} x_1 & y_1 \\ x_2 & y_2 \end{vmatrix} \right)^{\dagger} \right\|$$

$$= \sqrt{\begin{vmatrix} x_2 & y_2 \\ x_3 & y_3 \end{vmatrix}^2 + \begin{vmatrix} x_3 & y_3 \\ x_1 & y_1 \end{vmatrix}^2 + \begin{vmatrix} x_1 & y_1 \\ x_2 & y_2 \end{vmatrix}^2}.$$

Let $\mathbf{x}_i$ denote the projection of $\mathbf{x}$ onto the coordinate plane $u_i = 0$, $i = 1, 2, 3$, and similarly for $\mathbf{y}_i$. Then $\mathbf{x}_i \wedge \mathbf{y}_i$ is a parallelogram in a 2-dimensional plane whose area is therefore a simple 2×2 determinant:

$$\operatorname{area}(\mathbf{x}_1 \wedge \mathbf{y}_1) = \begin{vmatrix} x_2 & y_2 \\ x_3 & y_3 \end{vmatrix}, \quad \operatorname{area}(\mathbf{x}_2 \wedge \mathbf{y}_2) = \begin{vmatrix} x_3 & y_3 \\ x_1 & y_1 \end{vmatrix}, \quad \operatorname{area}(\mathbf{x}_3 \wedge \mathbf{y}_3) = \begin{vmatrix} x_1 & y_1 \\ x_2 & y_2 \end{vmatrix}.$$

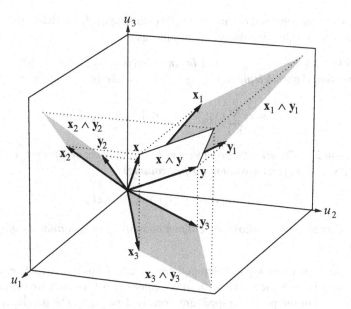

We can therefore rewrite our expression for area$(\mathbf{x} \wedge \mathbf{y})$ as

A "Pythagorean"
theorem

$$\text{area}(\mathbf{x} \wedge \mathbf{y}) = \sqrt{\text{area}^2(\mathbf{x}_1 \wedge \mathbf{y}_1) + \text{area}^2(\mathbf{x}_2 \wedge \mathbf{y}_2) + \text{area}^2(\mathbf{x}_3 \wedge \mathbf{y}_3)};$$

in other words, the square of the area of a parallelogram is equal to the sum of the squares of the areas of its projections onto the three coordinate planes. We can think of this as a "Pythagorean" theorem whose more usual form deals with lengths rather than areas, but relates, in the same way, the length of a vector to the lengths of its projections to the three coordinate axes.

Although we cannot visualize $\mathbb{R}^n$ directly when $n > 3$, we do carry over geometric concepts by analogy. For example, we continue to say the vector $\mathbf{x}$ has length $\|\mathbf{x}\| = \sqrt{\mathbf{x} \cdot \mathbf{x}}$ and the angle θ between the vectors $\mathbf{x}$ and $\mathbf{y}$ (assuming $\mathbf{x} \neq 0 \neq \mathbf{y}$) is

Geometry in $\mathbb{R}^n$, $n > 3$

$$\theta = \arccos\left(\frac{\mathbf{x} \cdot \mathbf{y}}{\|\mathbf{x}\| \|\mathbf{y}\|}\right).$$

Of course, $\arccos q$ is defined only when $|q| \leq 1$, so our definition of θ makes sense only if $|\mathbf{x} \cdot \mathbf{y}| \leq \|\mathbf{x}\| \|\mathbf{y}\|$ for all vectors $\mathbf{x}, \mathbf{y}$ in $\mathbb{R}^n$. This fact is established in the exercises. In what follows, $\{\mathbf{e}_1, \mathbf{e}_2, \ldots, \mathbf{e}_n\}$ is the standard basis for $\mathbb{R}^n$. As you can see, the definitions relate the orientation and volume of an n-dimensional parallelepiped to the determinant of a certain $n \times n$ matrix. We review the definition of an $n \times n$ determinant in the exercises.

Definition 2.3 *An ordered set* $\{\mathbf{v}_1, \mathbf{v}_2, \ldots, \mathbf{v}_n\}$ *has* **positive orientation** *if* $\det V > 0$, *where* $V : \mathbb{R}^n \to \mathbb{R}^n$ *is the linear map defined by the conditions* $V(\mathbf{e}_i) = \mathbf{v}_i$ *and* $i = 1, 2, \ldots, n$. *The set has* **negative orientation** *if* $\det V < 0$.

We can now construct the analogue of a parallelepiped, and define its volume and orientation, by extending the wedge product as follows.

n-parallelepipeds

Definition 2.4 *Let* $\{\mathbf{v}_1, \mathbf{v}_2, \ldots, \mathbf{v}_n\}$ *be an ordered set of vectors in* $\mathbb{R}^n$; *the **oriented** **n-dimensional parallelepiped*** $\mathbf{v}_1 \wedge \mathbf{v}_2 \wedge \cdots \wedge \mathbf{v}_n$ *is the set of vectors*

$$\mathbf{w} = \sum_{i=1}^{n} t_i \mathbf{v}_i, \quad 0 \le t_i \le 1, \quad i = 1, \ldots, n.$$

Orientation and volume

Definition 2.5 *The **orientation** of* $\mathbf{v}_1 \wedge \mathbf{v}_2 \wedge \cdots \wedge \mathbf{v}_n$ *is the orientation of the ordered set* $\{\mathbf{v}_1, \mathbf{v}_2, \ldots, \mathbf{v}_n\}$; *its **n-volume** (or just **volume**) is*

$$\mathrm{vol}(\mathbf{v}_1 \wedge \mathbf{v}_2 \wedge \cdots \wedge \mathbf{v}_n) = \det V,$$

where V *is the matrix whose jth column consists of the coordinates of* $\mathbf{v}_j$; *that is,* $V(\mathbf{e}_j) = \mathbf{v}_j$.

$n \times n$ determinants

The volume of an *n*-parallelepiped can be either positive, negative, or zero. If the volume is zero, then $\det V = 0$, so the columns of V (which are the coordinates of the edges of the parallelepiped) are linearly dependent. The parallelepiped does not fill out an *n*-dimensional region in $\mathbb{R}^n$. Our final statement about volumes is the analogue of similar results in $\mathbb{R}^2$ and $\mathbb{R}^3$, and is proven the same way.

Theorem 2.16. *The **volume multiplier** of the linear map* $M : \mathbb{R}^n \to \mathbb{R}^n$ *is* $\det M$; *that is,*

$$\mathrm{vol}\, M(\mathbf{v}_1 \wedge \cdots \wedge \mathbf{v}_n) = \mathrm{vol}\, M(\mathbf{v}_1) \wedge \cdots \wedge M(\mathbf{v}_n) = \det M \times \mathrm{vol}(\mathbf{v}_1 \wedge \cdots \wedge \mathbf{v}_n)$$

for every oriented n-parallelepiped $\mathbf{v}_1 \wedge \cdots \wedge \mathbf{v}_n$. *The map* M *reverses orientation precisely when* $\det M < 0$.																	□

2.3 Maps from $\mathbb{R}^n$ to $\mathbb{R}^p$, $n \ne p$

Image and kernel subspaces are graphs

A good example of a map between spaces of the same dimension is a coordinate change. Of course, a coordinate change has to work both ways; that is, the map must be invertible. When the source and target have different dimensions, invertibility is out of the question, but the geometric action of such linear maps still has a simple description. When the source is larger, the map cannot be one-to-one: the kernel of the map must be a linear subspace of positive dimension in the source. When the target is larger, the map cannot be onto: the image must be a linear subspace of strictly smaller dimension than the target. As we show, each of these subspaces is the graph of a new linear map that is defined implicitly by the original one.

Rank–nullity theorem

When $L : \mathbb{R}^n \to \mathbb{R}^p$ is a linear map, the **kernel**, or **null space**, of L is the linear subspace $\ker L$ of the source $\mathbb{R}^n$ that consists of all vectors $\mathbf{v}$ for which $L(\mathbf{v}) = 0$. The **image** of L is the linear subspace $\mathrm{im}\, L$ of the target consisting of all vectors $\mathbf{x}$

of the form $\mathbf{x} = L(\mathbf{v})$, for some $\mathbf{v}$ in the source. We call $r = \dim \operatorname{im} L$ the **rank** of L and $k = \dim \ker L$ its **nullity**. The **rank–nullity theorem** of linear algebra says that

$$r + k = \text{rank of } L + \text{nullity of } L = \dim \text{source of } L = n.$$

To analyze linear maps $L : \mathbb{R}^n \to \mathbb{R}^p$ for which $n \neq p$, let us first assume $n > p$. Because the image is a linear subspace of the target, we always have $r \leq p$. Because $n - p > 0$, the rank–nullity theorem implies $k = n - r \geq n - p > 0$. In other words, the kernel of L has positive dimension, at least as large as $n - p$. Let us now look more closely at $\ker L$.

Kernel L has positive dimension when $n > p$

We begin with an example in which $n = 3$ and $p = 1$, so L has the general form $x = L(u,v,w) = au + bv + cw$. How can we describe $\ker L$? To illustrate, suppose $L(u,v,w) = u - 2v - 3w$. The kernel of L is the locus of points in (u,v,w)-space that satisfy the equation

Example: one equation in three variables

$$u - 2v - 3w = 0.$$

The figure shows this locus is a (2-dimensional) plane through the origin. We can solve the equation for w, for example, and get

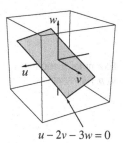

$u - 2v - 3w = 0$

$$w = \frac{u - 2v}{3}.$$

The original equation $u - 2v - 3w = 0$ therefore implies that w is a (linear) function of u and v. Thus, we can view the plane in the figure as either a *locus* (i.e., the locus of zeros of $L(u,v,w)$) or a *graph* (e.g., the graph of $w = (u - 2v)/3$).

Of course, this is not the only functional relation that is implicit in the equation $u - 2v - 3w = 0$; we get others by solving for u or v:

Implicit functions

$$u = 2v + 3w \quad \text{or} \quad v = \frac{u - 3w}{2}.$$

In each case, however, precisely one of the variables is expressed in terms of the other two.

We can say the same about an arbitrary linear function of three variables:

$\ker L = \operatorname{graph} M$

$$x = L(u,v,w) = au + bv + cw.$$

We have already seen that $\dim \ker L$ is at least $n - p = 2$. If $a = b = c = 0$, then every point satisfies the kernel equation $au + bv + cw = 0$, and $\dim \ker L = 3$. Otherwise, at least one of the coefficients is different from zero. Suppose $c \neq 0$; then we can solve the kernel equation for w in terms of u and v:

$$w = -\frac{a}{c}u - \frac{b}{c}v = M(u,v).$$

The linear function $M : \mathbb{R}^2 \to \mathbb{R}^1$ that expresses w in terms of u and v is implicitly defined by the equation $L(u,v,w) = 0$. In fact, the graph of M is the locus of the equation $L(u,v,w) = 0$. The dimension of a graph is equal to the dimension of its source (see below); therefore $\dim \ker L = \dim \operatorname{graph} M = 2$.

Graphs in general

Because we have found that the kernel of one linear map can be the graph of another, we pause to state some facts about graphs generally.

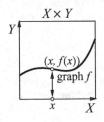

Definition 2.6 *The **graph** of an arbitrary map $f : X \to Y$ is the subset of the product $X \times Y$ that is defined by*

$$\operatorname{graph} f = \{(x, f(x)) \mid x \in X\}.$$

The definition makes it clear that there is a 1–1 correspondence between the source X and $\operatorname{graph} f$. If the map is linear, we can say more.

Theorem 2.17. *If $L : \mathbb{R}^n \to \mathbb{R}^p$ is linear, then $\operatorname{graph} L$ is a linear subspace of the product $\mathbb{R}^n \times \mathbb{R}^p = \mathbb{R}^{n+p}$ and $\dim \operatorname{graph} L = n$.*

Proof. Do this as an exercise.   □

Example: p equations in $p + k$ variables

Now consider a general linear map $L : \mathbb{R}^n \to \mathbb{R}^p : \mathbf{v} \mapsto \mathbf{x}$ with $n > p$. To begin, assume that L has maximal rank, so $r = p$ and $k = \dim \ker L = n - p$. If $A = (a_{ij})$ is the $p \times n$ matrix representing L, then the vector kernel equation $L(\mathbf{v}) = \mathbf{0}$ translates into a system of p ordinary equations in the $n = p + k$ unknowns $\mathbf{v} = (v_1, v_2, \ldots, v_n)$ and the coefficients a_{ij}:

$$a_{11}v_1 + a_{12}v_2 + \cdots + a_{1n}v_n = 0,$$
$$a_{21}v_1 + a_{22}v_2 + \cdots + a_{2n}v_n = 0,$$
$$\vdots$$
$$a_{p1}v_1 + a_{p2}v_2 + \cdots + a_{pn}v_n = 0.$$

Our previous example suggests that we should be able to solve these equations so as to express p of the unknowns as linear functions of the remaining $k = n - p$.

Solving the equations

To solve the equations, note first that the rank of L is the number of linearly independent columns of A. By our assumption, A must have p linearly independent columns. By rearranging them (and with them the variables v_j), if necessary, we can assume that the final p columns of A are linearly independent. They form an invertible $p \times p$ submatrix, C. The initial $n - p = k$ columns form a $p \times k$ submatrix B. With these identifications, the kernel equations take the form

$$
\begin{aligned}
\mathbf{0} &= A\mathbf{v} \\
&= (B\,C)\begin{pmatrix} \mathbf{u} \\ \mathbf{w} \end{pmatrix} \\
&= B\mathbf{u} + C\mathbf{w}
\end{aligned}
\qquad
\begin{pmatrix} 0 \\ 0 \\ \vdots \\ 0 \end{pmatrix}
=
\overbrace{
\begin{pmatrix}
a_{11} & \cdots & a_{1k} \\
a_{21} & \cdots & a_{2k} \\
\vdots & \ddots & \vdots \\
a_{p1} & \cdots & a_{pk}
\end{pmatrix}}^{B_{p \times k}}
\overbrace{
\begin{pmatrix}
a_{1,k+1} & \cdots & a_{1,k+p} \\
a_{2,k+1} & \cdots & a_{2,k+p} \\
\vdots & \ddots & \vdots \\
a_{p,k+1} & \cdots & a_{p,k+p}
\end{pmatrix}}^{C_{p \times p}}
\left.\begin{pmatrix} u_1 \\ \vdots \\ u_k \\ w_1 \\ \vdots \\ w_p \end{pmatrix}\right\}
\begin{matrix} \\[-2pt] \mathbf{u} \\[10pt] \mathbf{w} \end{matrix}
= (B\,C)\begin{pmatrix} \mathbf{u} \\ \mathbf{w} \end{pmatrix},
$$

or just $B\mathbf{u} + C\mathbf{w} = 0$ in terms of matrices. Because C is invertible, we can solve for $\mathbf{w}$:

The implicit functions

$$\mathbf{w} = -C^{-1}B\mathbf{u} = M(\mathbf{u}),$$
$$w_1 = \beta_{11}u_1 + \cdots + \beta_{1k}u_k,$$
$$w_2 = \beta_{21}u_1 + \cdots + \beta_{2k}u_k,$$
$$\vdots$$
$$w_p = \beta_{p1}u_1 + \cdots + \beta_{pk}u_k.$$

This is what we want: the equation $\mathbf{w} = M(\mathbf{u})$ expresses p of the variables as linear functions of the remaining k. The $p \times k$ matrix $-C^{-1}B = (\beta_{ij})$ that represents M is constructed from certain submatrices of the matrix A that represents L. Finally, the argument shows that the graph of M is precisely the kernel of L, because

$\ker L = \operatorname{graph} M$

$$\mathbf{w} = M(\mathbf{u}) \text{ if and only if } L(\mathbf{u},\mathbf{w}) = B\mathbf{u} + C\mathbf{w} = \mathbf{0}.$$

That is, a point $(\mathbf{u},\mathbf{w})$ is in the graph of M (so $\mathbf{w} = M(\mathbf{u})$) if and only if it is in the kernel of L.

Notice, incidentally, how our general result echoes what we found in the first example, in which $L(u,v,w) = au + bv + cw$. We had $A = (a\ b\ c)$ (a 1×3 matrix), $B = (a\ b)$, and $C = (c)$, implying

$$C^{-1} = \left(\frac{1}{c}\right) \text{ and } -C^{-1}B = \left(-\frac{a}{c} \quad -\frac{b}{c}\right).$$

The following theorem summarizes our result in both an algebraic form involving equations, and a geometric form involving maps. The condition that the original equations are linearly independent implies that the rank of the associated linear map is p.

Theorem 2.18. Algebraically: *A set of p linearly independent linear equations in $k + p$ variables implicitly defines p of the variables as linear functions of the remaining k variables.* Geometrically: *The kernel of a linear map $L : \mathbb{R}^{k+p} \to \mathbb{R}^p$ of maximal rank p is the graph of another linear map $M : \mathbb{R}^k \to \mathbb{R}^p$.* $\qquad\square$

One such map $L : \mathbb{R}^{k+p} \to \mathbb{R}^p$ of maximal rank p is just the identity on the last p variables:

$$w_1 = v_{k+1} = x_1,$$
$$w_2 = v_{k+2} = x_2,$$
$$\vdots \qquad (O_{p\times k}\ I_{p\times p}) \begin{pmatrix} \mathbf{u} \\ \mathbf{w} \end{pmatrix} = \mathbf{x}.$$
$$w_p = v_{k+p} = x_p,$$

The $p \times k$ zero matrix $O_{p\times k}$ eliminates the $\mathbf{u}$ variables $u_1 = v_1, \ldots, u_k = v_k$ from the formulas. Geometrically, L projects $\mathbb{R}^{k+p} = \mathbb{R}^k \times \mathbb{R}^p$ onto its second factor. It projects the first factor $\mathbb{R}^k$ (the kernel of L) to $\mathbf{0}$, and it projects the parallel translate of $\mathbb{R}^k$ by the vector $(\mathbf{u}, \mathbf{w})$ (for an arbitrary $\mathbf{u}$) to the point $\mathbf{w} = \mathbf{x}$ in the target. Although this example may seem special, the following theorem shows that, in a sense, it is the only possibility. Because the theorem is geometric, it is helpful to write the linear map in its original form $L : \mathbb{R}^n \to \mathbb{R}^p$, keeping in mind it is an onto map with $n > p$.

Every *onto* map is a projection

Theorem 2.19. *If the linear map $L : \mathbb{R}^n \to \mathbb{R}^p$ is onto, then there is a linear coordinate change H in the source $\mathbb{R}^n$ that transforms L into the projection that is the identity on the final p variables.*

Proof. We assume, as in the proof of Theorem 2.18, that variables in the source have been permuted so that L has the form

$$L = \left(B_{p \times k} \; C_{p \times p} \right), \quad \mathbf{x} = L(\mathbf{u}, \mathbf{w}) = \left(B \; C \right) \begin{pmatrix} \mathbf{u} \\ \mathbf{w} \end{pmatrix} = B\mathbf{u} + C\mathbf{w},$$

where $k = n - p \geq 0$ and the square submatrix C is invertible. Define $H : \mathbb{R}^n \to \mathbb{R}^n :$ $(\overline{\mathbf{u}}, \overline{\mathbf{w}}) \mapsto (\mathbf{u}, \mathbf{w})$ as the pullback

$$H : \begin{cases} \mathbf{u} = \overline{\mathbf{u}}, \\ \mathbf{w} = -C^{-1}B\overline{\mathbf{u}} + C^{-1}\overline{\mathbf{w}}, \end{cases} \qquad H = \begin{pmatrix} I_{k \times k} & O_{k \times p} \\ -(C^{-1}B)_{p \times k} & C^{-1}_{p \times p} \end{pmatrix}.$$

Now H is invertible,

$$H^{-1} = \begin{pmatrix} I & O \\ B & C \end{pmatrix},$$

so H is a valid coordinate change. Applying H to $\mathbf{x} = B\mathbf{u} + C\mathbf{w}$ gives

$$\mathbf{x} = B\overline{\mathbf{u}} + C\left[-C^{-1}B\overline{\mathbf{u}} + C^{-1}\overline{\mathbf{w}} \right] = B\overline{\mathbf{u}} - B\overline{\mathbf{u}} + \overline{\mathbf{w}} = \overline{\mathbf{w}}$$

(i.e., $\mathbf{x} = \overline{\mathbf{w}}$). The matrix for L in the new coordinates is

$$\overline{L} = LH = \left(B \; C \right) \begin{pmatrix} I & O \\ -C^{-1}B & C^{-1} \end{pmatrix} = \left(O \; I \right).$$

Thus $\overline{L}$ is the identity on the second component of $(\overline{\mathbf{u}}, \overline{\mathbf{w}})$; it projects $(\overline{\mathbf{u}}, \overline{\mathbf{w}})$ to $\overline{\mathbf{w}}$. $\square$

Corollary 2.20 *If the linear map $L : \mathbb{R}^n \to \mathbb{R}^p$ is onto, and Y is a linear subspace of the target of dimension q, then its preimage*

$$L^{-1}(Y) = \{\mathbf{v} \text{ in } \mathbb{R}^n : L(\mathbf{v}) \text{ is in } Y\}$$

is a linear subspace of dimension $q + k$, where $k = n - p \geq 0$.

Proof. The theorem provides new coordinates $(\overline{\mathbf{u}}, \overline{\mathbf{w}})$ in $\mathbb{R}^k \times \mathbb{R}^p = \mathbb{R}^n$ in which L becomes a projection onto the second factor. Then

$$L^{-1}(Y) = \{(\overline{\mathbf{u}}, \overline{\mathbf{w}}) : \overline{\mathbf{w}} \text{ is in } Y\} = \mathbb{R}^k \times Y,$$

because $\bar{\mathbf{u}}$ is an arbitrary point in $\mathbb{R}^k$. Hence $\dim L^{-1}(Y) = k + \dim Y$. Standard arguments in linear algebra show that $L^{-1}(Y)$ is a linear subspace. □

It is intuitively clear that, if L projects a larger space onto a smaller one, the pullback $L^{-1}(Y)$ will always be larger than the original Y. The corollary says that the difference is equal to the difference in dimension of the spaces themselves:

Codimension

$$\dim L^{-1}(Y) - \dim Y = k = \dim \mathbb{R}^n - \dim \mathbb{R}^p.$$

But this means

$$\dim \mathbb{R}^n - \dim L^{-1}(Y) = \dim \mathbb{R}^p - \dim Y.$$

Definition 2.7 *The **codimension** of a linear subspace Y of a vector space V is*

$$\operatorname{codim} Y = \dim V - \dim Y.$$

Think of *codimension* as "dimension of the complement" or "complement's dimension." A complement of Y in V is a linear subspace Z for which $V = Y \times Z$. Obviously, $\dim Z = \dim V - \dim Y = \operatorname{codim} Y$. In the previous corollary, it is thus more useful to work with the *codimension* of a linear subspace than its *dimension*, because pullback alters dimension but preserves codimension.

Corollary 2.21 *If the linear map $L : \mathbb{R}^n \to \mathbb{R}^p$ is onto, and Y is a linear subspace of $\mathbb{R}^p$, then $\operatorname{codim} L^{-1}(Y) = \operatorname{codim} Y$.* □

Theorem 2.19 provides a classification of onto linear maps that is similar to the classification of linear maps of the plane by Theorem 2.6 (p. 40). There we found several classes, each with a typical representative (that shares eigenvalues with all members of the class). Here there is only a single class, and projection is chosen as the typical representative of that class.

Classifying linear maps that are *onto*

It remains to consider the kernel of $L : \mathbb{R}^n \to \mathbb{R}^p$ when the rank of L is no longer maximal. In this case it turns out not to matter that $n > p$. We can illustrate this with a simple example.

Assume rank is not maximal

$$
\begin{aligned}
u \quad\;\; - \; w &= 0, \\
v - 2w &= 0, \\
u + v - 3w &= 0, \\
u - v + \; w &= 0.
\end{aligned}
$$

These four equations in three variables describe the kernel of a particular linear map $L : \mathbb{R}^3 \to \mathbb{R}^4$. The maximum possible rank of L is 3. However, the actual rank is only $r = 2$: only two of the four equations are linearly independent; the other two are linear combinations of those. By the rank–nullity theorem, the dimension of the kernel of L is $k = n - r = 1$. We expect, therefore, that the kernel of L is the graph of some other linear map $M : \mathbb{R}^1 \to \mathbb{R}^2$. Indeed, the kernel equations imply

$$M : \begin{cases} u = w, \\ v = 2w. \end{cases}$$

This is the linear map we seek. The following theorem generalizes this example; it makes no assumption about the relative sizes of n and p.

ker$L$ is a graph
when $r < n$

Theorem 2.22 (Linear implicit function theorem). *If $L : \mathbb{R}^n \to \mathbb{R}^p$ has rank $r < n$, then ker$L$ in $\mathbb{R}^n$ is the graph of a linear map $M : \mathbb{R}^{n-r} \to \mathbb{R}^r$.*

Proof. We consider cases; in every case, $r \leq p$. If $r = p$, Theorem 2.18 applies. If $r = 0$, then L is identically zero and ker$L = \mathbb{R}^n$. Therefore, the "zero" linear map $M : \mathbb{R}^n \to \mathbb{R}^0 : \mathbf{v} \mapsto 0$ has graph$M = $ ker$L$.

The only remaining possibility is $0 < r < p$. Then $q = p - r > 0$ of the equations that determine ker$L$ depend linearly on the remaining r equations. Select q dependent equations and discard them. Then use the remaining $p - q = r > 0$ equations to define a new linear map $L^* : \mathbb{R}^n \to \mathbb{R}^r$. Because the discarded kernel equations add no information, ker$L^* = $ ker$L$.

By construction, L^* does have maximal rank r, so Theorem 2.18 applies again: the kernel equations for L^* implicitly define r of the variables $v_1, \dots, v_n$ as linear functions of the remaining $n - r$ variables. In other words, ker$L^* = $ ker$L$ is the graph of a linear map $M : \mathbb{R}^{n-r} \to \mathbb{R}^r$. $\square$

One linear map $L : (u_1, \dots, u_k, w_1, \dots, w_r) \mapsto (x_1, \dots, x_q, y_1, \dots, y_r)$ of rank r is given by

$\mathbb{R}^r$
$(\mathbf{u}, \mathbf{w})$

$\mathbb{R}^r$
$(\mathbf{0}, \mathbf{w})$

L

$\mathbb{R}^k$

$\mathbb{R}^q$

$$L : \begin{cases} x_1 = 0, \\ \quad \vdots \\ x_q = 0, \\ y_1 = w_1, \\ \quad \vdots \\ y_r = w_r, \end{cases} \qquad L = \begin{pmatrix} O_{q \times k} & O_{q \times r} \\ O_{r \times k} & I_{r \times r} \end{pmatrix}.$$

To make this different from the previous example (p. 49), it is sufficient to require $q > 0$. The kernel of L consists of the points $(\mathbf{u}, \mathbf{0})$, and the image of L consists of the points $(\mathbf{0}, \mathbf{y})$. If we write the source as $\mathbb{R}^k \times \mathbb{R}^r$ and the target as $\mathbb{R}^q \times \mathbb{R}^r$, then L is the projection that is the identity on the second components:

$$L : (\mathbf{u}, \mathbf{w}) \mapsto (\mathbf{0}, \mathbf{w}).$$

The next theorem shows that this example is essentially the only possibility.

Theorem 2.23. *Let $L : \mathbb{R}^n \to \mathbb{R}^p$ be a linear map of rank r, and let $k = n - r$, $q = p - r$. Then there are coordinates in the source and target for which the matrix representing L is*

$$\Pi = \begin{pmatrix} O_{q \times k} & O_{q \times r} \\ O_{r \times k} & I_{r \times r} \end{pmatrix}.$$

Proof. We obtain coordinates for the source and for the target in which the map L is represented by multiplication by the given matrix Π. According to the rank–nullity theorem, $k = n - r$ is the dimension of the kernel of L. Let $\{U_1, \ldots, U_k\}$ be any basis for the kernel; add additional vectors $W_1, \ldots, W_r$ so that

$$\{U_1, \ldots, U_k, W_1, \ldots, W_r\}$$

is a basis for the entire source $\mathbb{R}^{k+r}$. Then any vector $\mathbf{v}$ in $\mathbb{R}^{k+R}$ can be written as

$$\mathbf{v} = u_1 U_1 + \cdots + u_k U_k + w_1 W_1 + \cdots + w_r W_r,$$

and (inasmuch as every $L(U_i) = 0$)

$$L(\mathbf{v}) = w_1 L(W_1) + \cdots + w_r L(W_r).$$

Because $\mathbf{v}$ is an arbitrary vector in the source, the vectors $L(\mathbf{v})$ constitute the entire image of L, and the equation for $L(\mathbf{v})$ shows that the vectors $L(W_j)$ span the image. Because the image has dimension r, the vectors $Y_j = L(W_j)$, $j = 1, \ldots, r$ must, in fact, form a basis for the image. Add additional vectors $X_1, \ldots, X_q$ so that

$$\{X_1, \ldots, X_q, Y_1, \ldots, Y_r\}$$

is a basis for the entire target $\mathbb{R}^{q+r}$. Then, in terms of these two bases, the coordinates of $\mathbf{v}$ and $L(\mathbf{v})$ are

$$\mathbf{v} \leftrightarrow (u_1, \ldots, u_k, w_1, \ldots, w_r) = (\mathbf{u}, \mathbf{w}),$$
$$L(\mathbf{v}) \leftrightarrow (0, \ldots, 0, w_1, \ldots, w_r) = (\mathbf{0}, \mathbf{w}).$$

Multiplication by Π gives

$$\begin{pmatrix} \mathbf{x} \\ \mathbf{y} \end{pmatrix} = \begin{pmatrix} O & O \\ O & I \end{pmatrix} \begin{pmatrix} \mathbf{u} \\ \mathbf{w} \end{pmatrix} = \begin{pmatrix} \mathbf{0} \\ \mathbf{w} \end{pmatrix};$$

thus Π does indeed represent L in terms of these coordinates. □

We now switch our attention from kernels to images. We show that when $\mathrm{im}\,L$ is a proper subspace of the target of L, it too can be thought of as the graph of a linear map implicitly defined by L. As we did for kernels, we assume first that L has maximal rank.

When is $\mathrm{im}\,L$ a graph?

We begin with an example. Consider the linear map $L : \mathbb{R}^2 \to \mathbb{R}^3 : (u,v) \mapsto (x,y,z)$ given by

Example: three equations in two variables

$$au + bv = x,$$
$$cu + dv = y,$$
$$eu + fv = z.$$

If L has maximum rank, namely 2, then precisely two of the equations are linearly independent. By rearranging them, if necessary, we may assume the first two equations are. Then we can solve these two equations for u and v in terms of x and y. It is perhaps easiest to see this if we work with matrices:

$$\begin{pmatrix} a & b \\ c & d \end{pmatrix} \begin{pmatrix} u \\ v \end{pmatrix} = \begin{pmatrix} x \\ y \end{pmatrix}, \qquad \begin{pmatrix} u \\ v \end{pmatrix} = \begin{pmatrix} a & b \\ c & d \end{pmatrix}^{-1} \begin{pmatrix} x \\ y \end{pmatrix} = \frac{1}{D} \begin{pmatrix} d & -b \\ -c & a \end{pmatrix} \begin{pmatrix} x \\ y \end{pmatrix}.$$

The implicit function

Here $D = ad - bc$, the determinant; linear independence of the first two equations implies $D \neq 0$. We can now express z directly in terms of x and y:

$$z = e\frac{dx - by}{D} + f\frac{-cx + ay}{D} = \frac{de - cf}{D}x + \frac{af - be}{D}y = M(x,y).$$

In geometric terms, the image of L is a plane in the target, $\mathbb{R}^3$. We have shown that if (x,y,z) are the coordinates of a point in this plane, then z is not independent of x and y, but depends linearly on them. The points of the image plane are of the form $(x,y,M(x,y))$; in other words, the image of L is the graph of M.

Example: $n+q$ equations in n variables

To generalize our example, take a linear map $L : \mathbb{R}^n \to \mathbb{R}^{n+q}$, $q > 0$, with maximal rank $r = n$. Let $A = (a_{ij})$ be the matrix representing L. Assume, by rearranging the rows of A, if neccesary, that the *first n rows* are linearly independent. Then, if we write the equation $L(\mathbf{v}) = A\mathbf{x} = \mathbf{x}$ in the form,

$$\begin{matrix} B \left\{ \begin{pmatrix} a_{11} & \cdots & a_{1n} \\ \vdots & \ddots & \vdots \\ a_{n1} & \cdots & a_{nn} \\ a_{n+1,1} & \cdots & a_{n+1,n} \\ \vdots & \ddots & \vdots \\ a_{n+q,1} & \cdots & a_{n+q,n} \end{pmatrix} \right. \\ C \left\{ \right. \end{matrix} \begin{pmatrix} v_1 \\ \vdots \\ v_n \end{pmatrix} = \begin{matrix} \left. \begin{pmatrix} y_1 \\ \vdots \\ y_n \\ z_{n+1} \\ \vdots \\ z_{n+q} \end{pmatrix} \right\} \mathbf{y} \\ \\ \left. \right\} \mathbf{z} \end{matrix} \qquad \begin{pmatrix} B \\ C \end{pmatrix} \mathbf{v} = \begin{pmatrix} \mathbf{y} \\ \mathbf{z} \end{pmatrix},$$

we can express this in terms of the submatrices B and C as a pair of matrix equations:

$$B\mathbf{v} = \mathbf{y}, \quad C\mathbf{v} = \mathbf{z}.$$

The implicit functions

We rearranged the rows of A to guarantee that the $n \times n$ submatrix B is invertible. Hence we can solve for $\mathbf{v}$ in terms of $\mathbf{y}$: $\mathbf{v} = B^{-1}\mathbf{y}$ and we can then express $\mathbf{z}$ in terms of $\mathbf{y}$:

$$\mathbf{z} = CB^{-1}\mathbf{y}.$$

In terms of the components $\mathbf{y}, \mathbf{z}$ of $\mathbf{x}$, the equation $\mathbf{z} = CB^{-1}\mathbf{y}$ takes the form

$$z_{n+1} = \gamma_{11}y_1 + \cdots + \gamma_{1n}y_n,$$
$$z_{n+2} = \gamma_{21}y_1 + \cdots + \gamma_{2n}y_n,$$
$$\vdots$$
$$z_{n+q} = \gamma_{q1}y_1 + \cdots + \gamma_{qn}y_n.$$

In this way we have expressed the image of L as a system of q linear equations in n variables. The coefficients γ_{ij} in these equations are the components of the $q \times n$ matrix CB^{-1} that represents a certain linear map $M : \mathbb{R}^n \to \mathbb{R}^q$. We see that a point $(\mathbf{y}, \mathbf{z})$ is in the image of L if and only if $\mathbf{z} = M(\mathbf{y})$, that is, if and only if it is of the form $(\mathbf{y}, M(\mathbf{y}))$. These are precisely the points of the graph of M.

Theorem 2.24. *Suppose the linear map $L : \mathbb{R}^n \to \mathbb{R}^{n+q}$ has maximal rank n. Then the image of L in $\mathbb{R}^{n+q}$ is the graph of a linear map $M : \mathbb{R}^n \to \mathbb{R}^q$.*   □

In these circumstances, the image $L(\mathbb{R}^n)$ is n-dimensional. Therefore, if P is a parallelepiped in the source and $L(P)$ is its image, both are n-dimensional, and both have n-volume. It is natural, then, to ask what is the volume multiplier for L. Of course, because the target dimension is larger than n, the image parallelepiped $L(P)$ cannot be oriented and the sign of its n-volume will have no meaning. Thus, we always understand the volume multiplier for L to be nonnegative. Let us consider first the special case $L : \mathbb{R}^2 \to \mathbb{R}^3$ when the image is a 2-dimensional plane.

Volume multiplier

Theorem 2.25. *Suppose $L : \mathbb{R}^2 \to \mathbb{R}^3$ has maximal rank 2 and is represented by the 3×2 matrix (a_{ij}), $i = 1, 2, 3$, $j = 1, 2$. Then*

$$L = \begin{pmatrix} a_{11} & a_{12} \\ a_{21} & a_{22} \\ a_{31} & a_{32} \end{pmatrix}$$

$$\textit{area multiplier of } L = \sqrt{\begin{vmatrix} a_{21} & a_{22} \\ a_{31} & a_{32} \end{vmatrix}^2 + \begin{vmatrix} a_{11} & a_{12} \\ a_{31} & a_{32} \end{vmatrix}^2 + \begin{vmatrix} a_{11} & a_{12} \\ a_{21} & a_{22} \end{vmatrix}^2}.$$

Proof. By linearity, the area multiplier will equal the area of the image of a parallelogram of unit area. In particular, take P to be the unit square

$$P = \begin{pmatrix} 1 \\ 0 \end{pmatrix} \wedge \begin{pmatrix} 0 \\ 1 \end{pmatrix}; \quad \text{then } L(P) = \begin{pmatrix} a_{11} \\ a_{21} \\ a_{31} \end{pmatrix} \wedge \begin{pmatrix} a_{12} \\ a_{22} \\ a_{32} \end{pmatrix}.$$

It follows from page 44 that the area of the parallelogram $L(P)$ is

$$\sqrt{\begin{vmatrix} a_{21} & a_{22} \\ a_{31} & a_{32} \end{vmatrix}^2 + \begin{vmatrix} a_{11} & a_{12} \\ a_{31} & a_{32} \end{vmatrix}^2 + \begin{vmatrix} a_{11} & a_{12} \\ a_{21} & a_{22} \end{vmatrix}^2}. \qquad □$$

According to Exercise 2.26.c. (p. 64), the area of the same parallelogram $L(P)$ can also be written as $\sqrt{\det L^\dagger L}$. More generally, then, Exercise 2.36 and the discussion leading up to it establish the following ("Pythagorean") theorem.

Theorem 2.26. *Suppose $L : \mathbb{R}^n \to \mathbb{R}^{n+q}$ has maximal rank n; then the n-volume multiplier for L is $\sqrt{\det L^\dagger L}$.*   □

Because the rank of L equals the dimension of the source, the kernel of L is zero so L is a 1–1 map. The next theorem says we can split the target into two factors so that L is just the identity mapping onto the first factor. This is a special case of Theorem 2.23, but it is worth having another proof that follows different lines.

Splitting the target of a 1–1 map

Theorem 2.27. *Suppose $L : \mathbb{R}^n \to \mathbb{R}^{n+q}$ is 1–1; then a coordinate change H in the target transforms L into the matrix*

$$\overline{L} = \begin{pmatrix} I_{n \times n} \\ O_{q \times n} \end{pmatrix}.$$

Proof. To begin, we assume (as in the proof of Theorem 2.24) that L has the matrix form

$$\begin{pmatrix} B \\ C \end{pmatrix} \mathbf{v} = \begin{pmatrix} \mathbf{y} \\ \mathbf{z} \end{pmatrix}, \quad \text{that is,} \quad \begin{array}{l} B\mathbf{v} = \mathbf{y}, \\ C\mathbf{v} = \mathbf{z}, \end{array}$$

with B an $n \times n$ invertible matrix. Define $H : \mathbb{R}^{n+q} \to \mathbb{R}^{n+q}$ by

$$H : \begin{cases} \overline{\mathbf{y}} = B^{-1}\mathbf{y}, \\ \overline{\mathbf{z}} = -CB^{-1}\mathbf{y} + \mathbf{z}, \end{cases} \qquad H = \begin{pmatrix} B^{-1} & O \\ -CB^{-1} & I \end{pmatrix}$$

Because $\det H = \det(B^{-1}) \neq 0$, H is a valid coordinate change. Moreover,

$$\overline{L} = HL = \begin{pmatrix} B^{-1} & O \\ -CB^{-1} & I \end{pmatrix} \begin{pmatrix} B \\ C \end{pmatrix} = \begin{pmatrix} I \\ O \end{pmatrix}, \qquad \begin{array}{l} \overline{\mathbf{y}} = B^{-1}(B\mathbf{v}) = \mathbf{v}, \\ \overline{\mathbf{z}} = -CB^{-1}(B\mathbf{v}) + C\mathbf{v} = \mathbf{0}. \end{array} \qquad \square$$

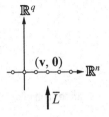

Injections

By analogy with projections, a linear map with the special form of $\overline{L}$ is sometimes called an **injection**. The theorem thus says that every 1–1 linear map is (equivalent to) an injection. The following corollary stands in the same relation to Theorem 2.27 that Corollaries 2.20 and 2.21 do to Theorem 2.19. It says that a 1–1 linear map preserves dimension under push-forward.

Corollary 2.28 *If the linear map $L : \mathbb{R}^n \to \mathbb{R}^p$ is 1–1, and X is a linear subspace of dimension k in the source, then $L(X)$ is a linear subspace of the same dimension k in the target.*

Proof. In terms of the $(\overline{\mathbf{y}}, \overline{\mathbf{z}})$ coordinates in the proof of Theorem 2.27, the linear map L becomes the injection $(\overline{\mathbf{y}}, \overline{\mathbf{z}}) = \overline{L}(\mathbf{v}) = (\mathbf{v}, \mathbf{0})$. Thus, for any subspace X, we have $\overline{L}(X) = X \times \mathbf{0}$, implying $\dim L(X) = \dim \overline{L}(X) = \dim X$. $\square$

Assume rank is not maximal

The final possibility to consider is the image of a linear map $L : \mathbb{R}^n \to \mathbb{R}^p$ whose rank may not be maximal. In this case we need make no assumption about the relative sizes of n and p.

Theorem 2.29. *If $L : \mathbb{R}^n \to \mathbb{R}^p$ has rank $r < p$, then $\operatorname{im} L$ in $\mathbb{R}^p$ is the graph of a linear map $M : \mathbb{R}^r \to \mathbb{R}^{p-r}$.*

Proof. As a preliminary step, write $L : \mathbb{R}^n \to \mathbb{R}^{r+q}$, where $q = p - r$ and $q > 0$ by hypothesis. The rest of the proof now proceeds by analogy with the proof of Theorem 2.22.

In every case, $r \leq n$. If $r = n$, the previous theorem applies. If $r = 0$, then L is identically 0 and $\operatorname{im} L = \mathbb{R}^0$. In this case, $\operatorname{im} L$ is the graph of the linear map $M : \mathbb{R}^0 \to \mathbb{R}^p : 0 \mapsto (0,0,\ldots,0)$.

The only remaining possibility is $0 < r < n$. Let A be the $p \times n$ matrix that represents L. The n columns of A are elements of $\mathbb{R}^p$ that span $\operatorname{im} L$ in $\mathbb{R}^p$. Our assumption implies that only r of the n columns are linearly independent; the remaining $n - r > 0$ columns depend linearly upon these. Delete the dependent columns from A to create a new matrix A^* of size $p \times r = (r + q) \times r$, and let $L^* : \mathbb{R}^r \to \mathbb{R}^{r+q}$ be the linear map defined by A^*. Because the columns of A^* and A span the same r-dimensional linear subspace of $\mathbb{R}^p = \mathbb{R}^{r+q}$, we have $\operatorname{im} L^* = \operatorname{im} L$.

By construction, L^* has maximal rank r, so Theorem 2.24 applies: $\operatorname{im} L^* = \operatorname{im} L$ is the graph of a linear map $M : \mathbb{R}^r \to \mathbb{R}^q = \mathbb{R}^{p-r}$. $\qquad\square$

Exercises

2.1. Show that the linear map M_1 (p. 29) alters the slope of any line that is neither horizontal nor vertical. Specifically, show that if a line has slope $\Delta v / \Delta u = m$, its image has slope $\Delta y / \Delta x = 3m/10 \neq m$ if $m \neq 0, \infty$.

2.2. Determine the coordinate change from (u, v) to $(\overline{u}, \overline{v})$ (and from (x, y) to $(\overline{x}, \overline{y})$) that converts the matrix M_5 into $\overline{M_5}$ (cf. p. 34)).

2.3. Let A, B, and C be $n \times n$ matrices. Suppose C is equivalent to B (cf. Definition 2.1, p. 33) and B is equivalent to A; show that C is equivalent to A.

2.4. This question concerns the map $M : \mathbb{R}^2 \to \mathbb{R}^2 : (u, v) \mapsto (x, y)$ defined by the matrix

$$M = \begin{pmatrix} 6 & 2 \\ 2 & 3 \end{pmatrix}.$$

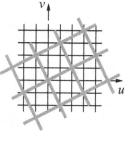

a. Determine the area multiplier for M.

b. Sketch in the (x, y)-plane the image of the standard unit grid from the (u, v)-plane.

c. Show that the *image* of the line in the direction of the vector $(u, v) = (2, 1)$ is the line in the direction of the vector $(x, y) = (2, 1)$ (in the (x, y)-plane). In other words, show that this line is invariant under M.

d. Show that the same is true for the line in the direction of the vector $(u, v) = (-1, 2)$.

e. Sketch in the (x, y)-plane the image of the solid gray grid shown in the (u, v)-plane. (Notice that the lines in this grid are parallel to the lines from parts (c) and (d).)

f. What is the shape of the image of a single square from the new grid? What are the dimensions of that shape if you use that solid gray grid to define new unit lengths? What is the area of that shape in terms of these new units? What is the area multiplier in terms of these new units?

2.5. Determine the eigenvalues and eigenvectors of each of the following matrices/maps.

a. $M = \begin{pmatrix} 1 & 2 \\ 2 & -2 \end{pmatrix}$ b. $M = \begin{pmatrix} 2 & 1 \\ 3 & 0 \end{pmatrix}$ c. $M = \begin{pmatrix} 0 & \sqrt{6} \\ \sqrt{6} & -1 \end{pmatrix}$

2.6. Carry out an analysis similar to what you did in Exercise 2.4 for the linear map defined by each of the the following matrices.

a. $M = \begin{pmatrix} 1 & 2 \\ 2 & -2 \end{pmatrix}$ b. $M = \begin{pmatrix} 1 & 1 \\ 2 & 0 \end{pmatrix}$

In particular, construct a grid whose image is "parallel to itself." Note that, in the second case, the grid consists of parallelograms rather than rectangles (or squares). Determine the **linear multipliers** for the map and show that the sides of the grid are stretched by these factors. Determine the **area multiplier** for the map and indicate how your diagram confirms that value. Comment on how the map affects orientation.

2.7. Consider the following from Example 8 (p. 38):

$$M_8 = \begin{pmatrix} 4 & -2 \\ 2 & 0 \end{pmatrix}, \quad \overline{u} = \overline{x} = \begin{pmatrix} 1 \\ 1 \end{pmatrix}, \quad \overline{v} = \overline{y} = \begin{pmatrix} 1 \\ 0 \end{pmatrix}.$$

a. You know M_8 has the repeated eigenvalue 2, and that therefore the kernel of the matrix $M_8 - 2I$ contains all the eigenvectors of M_8. Show that the dimension of the image of $M_8 - 2I$ is 1; by the rank–nullity theorem, the dimension of the set of eigenvectors of M_8 is only 1, not 2.

b. Show that $M_8(\overline{u}) = 2\overline{x}$, $M_8(\overline{v}) = 2\overline{x} + 2\overline{y}$. In particular, identify an eigenvector of M_8.

c. Let $(\overline{u}, \overline{v})$ be coordinates based on $\overline{u}$ and $\overline{v}$, and let $(\overline{x}, \overline{y})$ be coordinates based, in the same way, on $\overline{x}$ and $\overline{y}$. Show that the coordinate change from the standard basis to the new one is

$$G : \begin{cases} u = \overline{u} + \overline{v}, & x = \overline{x} + \overline{y}, \\ v = \overline{u}, & y = \overline{x}. \end{cases}$$

d. Show that, in terms of the new coordinates, M_8 becomes

$$\overline{M_8} : \begin{cases} \overline{x} = 2\overline{u} + 2\overline{v}, \\ \overline{y} = 2\overline{v}, \end{cases} \qquad \overline{M_8} = \begin{pmatrix} 2 & 2 \\ 0 & 2 \end{pmatrix}.$$

2.8. a. Show that the complex numbers $e^{\pm i\theta} = \cos\theta \pm i\sin\theta$ are the eigenvalues of the matrix

$$R_\theta = \begin{pmatrix} \cos\theta & -\sin\theta \\ \sin\theta & \cos\theta \end{pmatrix}.$$

b. Show that R_θ rotates the plane by θ radians by showing that

$$\begin{pmatrix} \cos\theta & -\sin\theta \\ \sin\theta & \cos\theta \end{pmatrix} \begin{pmatrix} r\cos\alpha \\ r\sin\alpha \end{pmatrix} = \begin{pmatrix} r\cos(\alpha+\theta) \\ r\sin(\alpha+\theta) \end{pmatrix}.$$

In other words, R_θ maps the point with polar coordinates (r,α) to the point with polar coordinates $(r,\alpha+\theta)$. Explain why this implies R_θ has no real eigenvectors when $\theta \neq n\pi$, n integer.

  c. Find a (complex) eigenvector for each of the eigenvalues of R_θ, $\theta \neq n\pi$.

2.9. Show that the only matrix equivalent (cf. Definition 2.1, p. 33) to the uniform dilation $D = \lambda I$ is D itself.

2.10. Let $\arctan(y/x)$, viewed as a function of two variables, be defined in terms of the usual arctangent function for all $(x,y) \neq (0,0)$ as follows:

$$\arctan(y/x) = \begin{cases} \arctan(y/x), & 0 < x, \\ \pi/2, & x = 0,\ 0 < y, \\ -\pi/2, & x = 0,\ y < 0, \\ \arctan(y/x) - \pi, & x < 0,\ y < 0, \\ \arctan(y/x) + \pi, & x < 0,\ 0 \le y. \end{cases}$$

  a. Show that $\arctan(y/x)$ is continuous across the y-axis, and is thus continuous on $\mathbb{R}^2 \setminus \{(x,0) | x \le 0\}$; this is the plane with the origin and the negative x-axis deleted.

  b. The graph $z = \arctan(y/x)$ is a *spiral ramp*; sketch it.

2.11. Suppose the 2×2 matrix M has real unequal eigenvalues λ_1 and λ_2, with corresponding eigenvectors $\mathbf{u}_1$ and $\mathbf{u}_2$.

  a. Explain why $\mathbf{u}_1$ and $\mathbf{u}_2$ are linearly independent.

  b. Let G be the matrix whose columns are $\mathbf{u}_1$ and $\mathbf{u}_2$, in that order. Explain why G is invertible, and then show

$$G^{-1}MG = \begin{pmatrix} \lambda_1 & 0 \\ 0 & \lambda_2 \end{pmatrix}.$$

2.12. Suppose the 2×2 matrix M has the repeated eigenvalue $\lambda \neq 0$, but has only a single eigendirection, along the eigenvector $\mathbf{u}$. The purpose of this exercise is to show that M is equivalent to the standard shear–dilation matrix

$$S_\lambda = \begin{pmatrix} \lambda & \lambda \\ 0 & \lambda \end{pmatrix}.$$

  a. Let $\mathbf{e}_1$ and $\mathbf{e}_2$ be the standard basis vectors. Show by direct computation that the vectors $(M - \lambda I)\mathbf{e}_1$ and $(M - \lambda I)\mathbf{e}_2$ are both eigenvectors of M. Conclude that $(M - \lambda I)\mathbf{w}$ is an eigenvector of M for every $\mathbf{w}$ in $\mathbb{R}^2$. Suggestion: The vectors $(M - \lambda I)\mathbf{e}_1$ and $(M - \lambda I)\mathbf{e}_2$ are the columns of the matrix

$$M - \lambda I = \begin{pmatrix} a - \lambda & b \\ c & d - \lambda \end{pmatrix}.$$

    b. The image of $M - \lambda I$ is 1-dimensional and contains $\lambda \mathbf{u}$; why? Conclude that there is a vector $\mathbf{v}$ for which $(M - \lambda I)\mathbf{v} = \lambda \mathbf{u}$. Explain why $\mathbf{u}$ and $\mathbf{v}$ are linearly independent.

    c. Let G be the matrix whose columns are $\mathbf{u}$ and $\mathbf{v}$. Explain why G is invertible, and show that $G^{-1}MG = S_\lambda$.

2.13. Suppose the real 2×2 matrix M has complex eigenvalues $a \pm bi$, $b \neq 0$, and the real vectors $\mathbf{u}$ and $\mathbf{v}$ form the complex eigenvector $\mathbf{u} + i\mathbf{v}$ for M with eigenvalue $a - bi$ (note the difference in signs). The purpose of this exercise is to show that M is equivalent to the standard rotation–dilation matrix $C_{a,b}$ (cf. p. 39).

    a. Show that the following *real* matrix equations are true:

$$M\mathbf{u} = a\mathbf{u} + b\mathbf{v}, \quad M\mathbf{v} = -b\mathbf{u} + a\mathbf{v}.$$

    b. Let G be the matrix whose columns are $\mathbf{u}$ and $\mathbf{v}$, in that order. Show that $MG = GC_{a,b}$.

    c. Show that the real vectors $\mathbf{u}$ and $\mathbf{v}$ are linearly independent in $\mathbb{R}^2$. Suggestion: first show $\mathbf{u} \neq \mathbf{0}$, $\mathbf{v} \neq \mathbf{0}$. Then suppose there are real numbers r, s for which $r\mathbf{u} + s\mathbf{v} = \mathbf{0}$. Show that $\mathbf{0} = M(r\mathbf{u} + s\mathbf{v})$ implies that $-s\mathbf{u} + r\mathbf{v} = \mathbf{0}$, and hence that $r = s = 0$.

    d. Conclude that G is invertible and $G^{-1}MG = C_{a,b}$.

2.14. Notice that, in Exercise 2.6, the map whose invariant grid was rectangular (and hence whose eigenvectors were orthogonal) was the one whose matrix was *symmetric*. A matrix is symmetric if it is equal to its own transpose; the 2×2 symmetric matrices are

$$M = \begin{pmatrix} a & b \\ b & c \end{pmatrix} \quad a, b, \text{ and } c \text{ real}.$$

The purpose of this exercise is to show that a symmetric matrix always has linearly independent orthogonal eigenvectors that define an invariant rectangular grid.

    a. Show that the eigenvalues of a 2×2 symmetric matrix M are real. (Suggestion: Look at the formula for the roots of the quadratic equation that gives the eigenvalues and focus on the part of the formula that leads to complex values.)

    b. Suppose a 2×2 symmetric matrix M has unequal eigenvalues. Show that the eigenvectors that correspond to these eigenvalues are orthogonal (e.g., their dot product is zero).

c. Suppose a 2×2 symmetric matrix M has equal eigenvalues. Show that this can happen only if $b = 0$ and $a = c$. (Again, look at the formula for the roots.) This implies M must reduce to a multiple of the identity matrix.

In the last case (c), every nonzero vector is an eigenvector. Therefore, for every symmetric 2×2 matrix M, $\mathbb{R}^2$ has a basis consisting of orthogonal eigenvectors of M. Thus if a linear map $M : \mathbb{R}^2 \to \mathbb{R}^2$ is represented by a symmetric matrix, there is a grid of squares that is stretched into a parallel grid of rectangles (as happened in Exercises 2.4 and 2.6a).

2.15. The purpose of this exercise is to show that the area of $\mathbf{v} \wedge \mathbf{w}$ is the determinant of the matrix whose columns are the coordinates of $\mathbf{v}$ and $\mathbf{w}$.

a. Let θ be the angle between $\mathbf{v}$ and $\mathbf{w}$, taken from $\mathbf{v}$ to $\mathbf{w}$ and chosen so that $-\pi < \theta \le \pi$. Thus, θ is negative precisely when $\mathbf{v} \wedge \mathbf{w}$ has negative orientation. Show that $\operatorname{area}(\mathbf{v} \wedge \mathbf{w}) = \|\mathbf{v}\|\|\mathbf{w}\| \sin\theta$. (Notice that this has a negative value when $\mathbf{v} \wedge \mathbf{w}$ has negative orientation.)

b. Show that $\operatorname{area}^2(\mathbf{v} \wedge \mathbf{w}) = \|\mathbf{v}\|^2\|\mathbf{w}\|^2 - (\mathbf{v} \cdot \mathbf{w})^2$.

c. Let $\mathbf{v} = (v_1, v_2)$, $\mathbf{w} = (w_1, w_2)$. Show that

$$\|\mathbf{v}\|^2\|\mathbf{w}\|^2 - (\mathbf{v} \cdot \mathbf{w})^2 = (v_1 w_2 - v_2 w_1)^2,$$

and hence that $\operatorname{area}(\mathbf{v} \wedge \mathbf{w}) = \pm(v_1 w_2 - v_2 w_1)$.

d. Show that requiring the area of the unit square $(1,0) \wedge (0,1)$ be positive implies we should choose the positive root in part (c):

$$\operatorname{area}(\mathbf{v} \wedge \mathbf{w}) = v_1 w_2 - v_2 w_1 = \det\begin{pmatrix} v_1 & w_1 \\ v_2 & w_2 \end{pmatrix}.$$

2.16. Suppose $M = \begin{pmatrix} a & b \\ c & d \end{pmatrix}$, $\mathbf{v} = \begin{pmatrix} v_1 \\ v_2 \end{pmatrix}$, $\mathbf{w} = \begin{pmatrix} w_1 \\ w_2 \end{pmatrix}$.

a. Let V be the matrix whose columns are $\mathbf{v}$ and $\mathbf{w}$, and let $\overline{V}$ be the matrix whose columns are $M(\mathbf{v})$ and $M(\mathbf{w})$. Show that $\overline{V} = MV$ and hence that $\det \overline{V} = \det M \times \det V$.

b. Conclude that

$$\operatorname{area} M(\mathbf{v} \wedge \mathbf{w}) = \operatorname{area} M(\mathbf{v}) \wedge M(\mathbf{w}) = \det M \times \operatorname{area}(\mathbf{v} \wedge \mathbf{w}).$$

2.17. The aim of this exercise is to show that the determinant of a 2×2 matrix can be viewed as a certain function $D(\mathbf{v}, \mathbf{w})$ of its columns $\mathbf{v}$ and $\mathbf{w}$ that is uniquely characterized by the following three properties.

Defining properties of the determinant

- $D(\mathbf{e}_1, \mathbf{e}_2) = 1$, where $\mathbf{e}_i$ is the ith column of the identity matrix.
- $D(\mathbf{v}, \mathbf{w}) = 0$ if $\mathbf{v} = \mathbf{w}$.
- $D(\mathbf{v}, \mathbf{w})$ is a linear function of each of its arguments $\mathbf{v}$ and $\mathbf{w}$. That is,

$$D(t\mathbf{v}, \mathbf{w}) = tD(\mathbf{v}, \mathbf{w}),$$
$$D(\mathbf{v}_1 + \mathbf{v}_2, \mathbf{w}) = D(\mathbf{v}_1, \mathbf{w}) + D(\mathbf{v}_2, \mathbf{w}),$$

and similarly for the second argument. We say D is **bilinear**.

a. Show that D is **antisymmetric**; that is, $D(\mathbf{y}, \mathbf{x}) = -D(\mathbf{x}, \mathbf{y})$ for all $\mathbf{x}$ and $\mathbf{y}$. Suggestion: First expand $D(\mathbf{x} + \mathbf{y}, \mathbf{x} + \mathbf{y})$ to four terms using the bilinearity of D (this is a kind of "FOIL"). The second property will guarantee two of those terms are zero; the remaining two terms then give the result.

b. Show that $D(\mathbf{e}_2, \mathbf{e}_1) = -1$.

c. Let $\mathbf{v} = v_1\mathbf{e}_1 + v_2\mathbf{e}_2$ and $\mathbf{w} = w_1\mathbf{e}_1 + w_2\mathbf{e}_2$. Using the bilinearity of D ("FOIL" again!), show that

$$D(\mathbf{v}, \mathbf{w}) = v_1 w_2 D(\mathbf{e}_1, \mathbf{e}_2) + v_2 w_1 D(\mathbf{e}_2, \mathbf{e}_1) = v_1 w_2 - v_2 w_1,$$

proving that $D(\mathbf{v}, \mathbf{w}) = \det \begin{pmatrix} v_1 & w_1 \\ v_2 & w_2 \end{pmatrix}$.

The determinant in higher dimensions

The reason we have paused here to characterize the determinant of a 2×2 matrix by these three properties is that we later use (suitable modifications of) the same three properties to define the determinant of an $n \times n$ matrix. (See the exercises below). The properties of $D(\mathbf{v}, \mathbf{w})$ are the properties of $\text{area}(\mathbf{v} \wedge \mathbf{w})$; thus the determinant of an $n \times n$ matrix will be connected with the volume of an n-dimensional parallelepiped, the analogue of a 2-dimensional parallelogram.

2.18. Determine $(5, 2, -1) \times (3, 4, 2)$ and $(1, 1, 1) \times (1, 1, -1)$.

2.19. Determine the volume of the parallelepiped:

a. $(5, 2, -1) \wedge (3, 4, 2) \wedge (1, 0, -1)$.

b. $(1, 1, 1) \wedge (1, 1, -1) \wedge (1, -1, -1)$.

2.20. Consider the linear map $M : \mathbb{R}^3 \to \mathbb{R}^3$ and the parallelepiped P defined by

$$M = \begin{pmatrix} 0 & 1 & 0 \\ 0 & 0 & 2 \\ -1 & 0 & 0 \end{pmatrix}, \quad P = \begin{pmatrix} 1 \\ 0 \\ 0 \end{pmatrix} \wedge \begin{pmatrix} 1 \\ 1 \\ 0 \end{pmatrix} \wedge \begin{pmatrix} 1 \\ 1 \\ 1 \end{pmatrix}.$$

a. What is the orientation of P? What is the volume of P?

b. Describe the parallelepiped $M(P)$ by listing its edges, in proper order.

c. Determine directly from your answer in (b) the orientation and volume of $M(P)$.

d. What is the volume multiplier of M? Does this value account for the orientation and volume of $M(P)$ you found in part (c)?

2.21. Suppose $\mathbf{z} = \mathbf{x} \times \mathbf{y} \neq \mathbf{0}$. Show that the linear map $L : \mathbb{R}^3 \to \mathbb{R}^3$ defined by

$$L(\mathbf{x}) = \mathbf{y}, \quad L(\mathbf{y}) = \mathbf{x}, \quad L(\mathbf{z}) = -\mathbf{z},$$

has positive determinant and maps $\mathbf{x} \wedge \mathbf{y}$ to $\mathbf{y} \wedge \mathbf{x}$.

2.22. Show that $\mathrm{vol}(\mathbf{x} \wedge \mathbf{y} \wedge \bar{\mathbf{z}}) = \mathrm{vol}(\mathbf{x} \wedge \mathbf{y} \wedge \mathbf{z})$, where $\bar{\mathbf{z}} = \mathbf{z} + \alpha \mathbf{x} + \beta \mathbf{y}$ and α and β are arbitrary. One way to do this is to note that the two parallelepipeds $\mathbf{x} \wedge \mathbf{y} \wedge \mathbf{z}$ and $\mathbf{x} \wedge \mathbf{y} \wedge \bar{\mathbf{z}}$ have the same base $\mathbf{x} \wedge \mathbf{y}$, and their third edges $\mathbf{z}$ and $\bar{\mathbf{z}}$ lie the same distance above that base. Draw a picture.

2.23. The parallelpipeds obtained from $\mathbf{x} \wedge \mathbf{y} \wedge \mathbf{z}$ by an arbitrary permutation of $\mathbf{x}$, $\mathbf{y}$, and $\mathbf{z}$ are all equal as sets. However, they differ in orientation. Show that the cyclic permutations $\mathbf{y} \wedge \mathbf{z} \wedge \mathbf{x}$ and $\mathbf{z} \wedge \mathbf{x} \wedge \mathbf{y}$ have the same orientation as $\mathbf{x} \wedge \mathbf{y} \wedge \mathbf{z}$, but the permutations that simply transpose a pair of edges, namely $\mathbf{y} \wedge \mathbf{x} \wedge \mathbf{z}$, $\mathbf{x} \wedge \mathbf{z} \wedge \mathbf{y}$, and $\mathbf{z} \wedge \mathbf{y} \wedge \mathbf{x}$, all have the opposite orientation. We can express this in the following way.

$$\mathbf{y} \wedge \mathbf{z} \wedge \mathbf{x} = \mathbf{z} \wedge \mathbf{x} \wedge \mathbf{y} = \mathbf{x} \wedge \mathbf{y} \wedge \mathbf{z},$$
$$\mathbf{x} \wedge \mathbf{z} \wedge \mathbf{y} = \mathbf{y} \wedge \mathbf{x} \wedge \mathbf{z} = \mathbf{z} \wedge \mathbf{y} \wedge \mathbf{x} = -\mathbf{x} \wedge \mathbf{y} \wedge \mathbf{z}.$$

2.24. The purpose of this exercise is to show that the determinant of a 3×3 matrix is a certain function D of its columns that is uniquely defined by the following three properties. This is exactly analogous to the 2×2 case as dealt with in Exercise 2.17 (p. 61), and is preparation for the $n \times n$ case addressed below (Exercise 2.28).

- $D(\mathbf{e}_1, \mathbf{e}_2, \mathbf{e}_3) = 1$, where $\mathbf{e}_i$ is the ith column of the identity matrix.
- $D(\mathbf{x}, \mathbf{y}, \mathbf{z}) = 0$ if any two of the columns $\mathbf{x}$, $\mathbf{y}$, $\mathbf{z}$ are equal.
- $D(\mathbf{x}, \mathbf{y}, \mathbf{z})$ is a linear function of each of its arguments $\mathbf{x}$, $\mathbf{y}$, and $\mathbf{z}$. That is,

$$D(t\mathbf{x}, \mathbf{y}, \mathbf{z}) = t D(\mathbf{x}, \mathbf{y}, \mathbf{z}),$$
$$D(\mathbf{x}_1 + \mathbf{x}_2, \mathbf{y}, \mathbf{z}) = D(\mathbf{x}_1, \mathbf{y}, \mathbf{z}) + D(\mathbf{x}_2, \mathbf{y}, \mathbf{z}),$$

and similarly for the second and third arguments.) We say D is **multilinear**.

a. Show that D is **antisymmetric**; that is, D changes sign when any two columns are interchanged:

$$D(\mathbf{x}, \mathbf{z}, \mathbf{y}) = D(\mathbf{z}, \mathbf{y}, \mathbf{x}) = D(\mathbf{y}, \mathbf{x}, \mathbf{z}) = -D(\mathbf{x}, \mathbf{y}, \mathbf{z}).$$

(Compare this with the previous exercise.)

b. Show that

$$D(\mathbf{e}_1, \mathbf{e}_3, \mathbf{e}_2) = D(\mathbf{e}_3, \mathbf{e}_2, \mathbf{e}_1) = D(\mathbf{e}_2, \mathbf{e}_1, \mathbf{e}_3) = -1,$$
$$D(\mathbf{e}_2, \mathbf{e}_3, \mathbf{e}_1) = D(\mathbf{e}_3, \mathbf{e}_1, \mathbf{e}_2) = D(\mathbf{e}_1, \mathbf{e}_2, \mathbf{e}_3) = +1.$$

c. Write $\mathbf{x}$, $\mathbf{y}$, $\mathbf{z}$ in terms of their coordinates with respect to the standard basis:

$$\mathbf{x} = \begin{pmatrix} x_1 \\ x_2 \\ x_3 \end{pmatrix} = x_1 \begin{pmatrix} 1 \\ 0 \\ 0 \end{pmatrix} + x_2 \begin{pmatrix} 0 \\ 1 \\ 0 \end{pmatrix} + x_3 \begin{pmatrix} 0 \\ 0 \\ 1 \end{pmatrix} = x_1 \mathbf{e}_1 + x_2 \mathbf{e}_2 + x_3 \mathbf{e}_3;$$

similarly, $\mathbf{y} = y_1\mathbf{e}_1 + y_2\mathbf{e}_2 + y_3\mathbf{e}_3$; $\mathbf{z} = z_1\mathbf{e}_1 + z_2\mathbf{e}_2 + z_3\mathbf{e}_3$. Using the multilinearity of D, expand $D(\mathbf{x},\mathbf{y},\mathbf{z})$ as a sum of 27 terms of the form $x_i y_j z_k D(\mathbf{e}_i, \mathbf{e}_j, \mathbf{e}_k)$. Note that, in a given one of these expressions, the indices i, j, k need not be distinct.

d. Precisely 21 of the 27 terms you just obtained are automatically zero. Which ones, and why?

e. Show that the remaining six terms yield

$$D(\mathbf{x},\mathbf{y},\mathbf{z}) = \det \begin{pmatrix} x_1 & y_1 & z_1 \\ x_2 & y_2 & z_2 \\ x_3 & y_3 & z_3 \end{pmatrix},$$

the familiar 3×3 determinant. Hence, the 3×3 determinant is uniquely determined by the three properties defining D.

2.25. Show that $D(\mathbf{x},\mathbf{y},\bar{\mathbf{z}}) = D(\mathbf{x},\mathbf{y},\mathbf{z})$, where $\bar{\mathbf{z}} = \mathbf{z} + \alpha\mathbf{x} + \beta\mathbf{y}$ and α and β are arbitrary. This is obviously the same result as Exercise 2.22 above; however, you should prove it here using only the properties of D defined and deduced in the previous exercise.

2.26. Let $P = \mathbf{v} \wedge \mathbf{w}$ be the parallelogram in $\mathbb{R}^3 : (u_1, u_2, u_3)$ spanned by $\mathbf{v} = (1,1,2)$ and $\mathbf{w} = (1,0,-1)$.

a. For $i = 1,2,3$, describe in the form $P_i = \mathbf{v}_i \wedge \mathbf{w}_i$ the projection of P onto the coordinate plane $u_i = 0$. For clarity, write $\mathbf{v}_i$ and $\mathbf{w}_i$ as elements of $\mathbb{R}^2$ rather than $\mathbb{R}^3$.

b. Determine the areas of the three projections P_i; then use the "Pythagorean" theorem to calculate

$$\text{area}\,P = \sqrt{\text{area}^2 P_1 + \text{area}^2 P_2 + \text{area}^2 P_3}.$$

c. Let V be the 3×2 matrix whose columns are the components of $\mathbf{v}$ and $\mathbf{w}$, in that order, and let $V^\dagger$ be its transpose. Show that

$$V^\dagger V = \begin{pmatrix} \mathbf{v}\cdot\mathbf{v} & \mathbf{v}\cdot\mathbf{w} \\ \mathbf{w}\cdot\mathbf{v} & \mathbf{w}\cdot\mathbf{w} \end{pmatrix},$$

implying $\det V^\dagger V = \|\mathbf{v}\|^2\|\mathbf{w}\|^2 - (\mathbf{v}\cdot\mathbf{w})^2$. Show that this is $\text{area}^2 P$ (cf. Exercise 2.15, p. 61). Confirm that this value of $\text{area}\,P$ agrees with the value you found in part (b).

2.27. Show that $(\mathbf{x}\cdot\mathbf{y})^2 \leq \|\mathbf{x}\|^2\|\mathbf{y}\|^2$ for all vectors $\mathbf{x}$ and $\mathbf{y}$ in $\mathbb{R}^n$. (Suggestion: both sides are zero if $\mathbf{y} = 0$, so assume $\mathbf{y} \neq 0$ and let

$$\mathbf{z} = \mathbf{x} - \frac{\mathbf{x}\cdot\mathbf{y}}{\|\mathbf{y}\|^2}\mathbf{y}.$$

Now consider the implications of $0 \leq \mathbf{z}\cdot\mathbf{z}$.)

The next exercise defines the determinant of an $n \times n$ matrix V as a function D of its columns that satisfies certain properties. We have already seen how this approach works with a 2×2 matrix (Exercise 2.17, p. 61) and with a 3×3 matrix (Exercise 2.24).

As we saw in the earlier exercises, and see again in the $n \times n$ case, an essential property of D is antisymmetry. A description of this property involves rearranging, or permuting, the columns. When there were only two or three columns, this was simple to follow. However, it is useful here to pause and introduce some facts about permutations of an arbitrary number, n, of objects.

A **permutation on n elements** is an invertible map π of the set $\{1, 2, \ldots, n\}$ to itself. A **transposition** is a permutation $\tau_{i,j}$ that interchanges the elements i and j and leaves all the others unchanged: $\tau_{i,j}(i) = j$; $\tau_{i,j}(j) = i$; $\tau_{i,j}(k) = k, k \neq i, j$. The **product** of two permutations is the permutation that results from their composition: $(\pi_1 \cdot \pi_2)(i) = (\pi_1 \pi_2)(i) = \pi_1(\pi_2(i))$. The identity map is a permutation and it is the identity element in the product. (With this product, the set S_n of permutations on n elements is a **group** with $n!$ elements.)

Permutations and transpositions

Every permutation can be written as a product of transpositions. The number of transpositions in such a product is not unique, but its *parity* is. Therefore, we say a permutation π is **even**, and write $\operatorname{sgn} \pi = +1$, if π is always the product of an *even* number of transpositions; we say it is **odd**, and write $\operatorname{sgn} \pi = -1$, if it is always the product of an *odd* number of transpositions.

Even and odd permutations

If $v_1, v_2, \ldots v_n$, are the columns of the matrix V, then $D(v_1, v_2, \ldots v_n)$ is a function that satisfies the following conditions:

The defining properties of D

- $D(e_1, e_2, \ldots, e_n) = 1$, where e_i is the ith column of the $n \times n$ identity matrix.

- $D(v_1, v_2, \ldots, v_n) = 0$ if any two of the columns v_i are equal.

- D is multilinear; that is, $D(v_1, v_2, \ldots, v_n)$ is a linear function of each of its arguments $v_1, v_2, \ldots, v_n$.

Note: It is not yet evident that there is such a function D; the exercise shows that D exists.

2.28. a. Show that D is **antisymmetric**; that is, D changes signs when any two columns are interchanged. In terms of permutations: if π is a transposition, then

$$D(v_{\pi(1)}, v_{\pi(2)}, \ldots, v_{\pi(n)}) = -D(v_1, v_2, \ldots, e_n).$$

   b. Suppose π is a permutation of $\{1, 2, \ldots, n\}$. Show that

$$D(e_{\pi(1)}, e_{\pi(2)}, \ldots, e_{\pi(n)}) = \operatorname{sgn} \pi = \pm 1.$$

   c. Suppose that the map $\pi : \{1, 2, \ldots, n\} \to \{1, 2, \ldots, n\}$ is not a permutation, that is, π is not onto or one-to-one. Explain why

$$D(e_{\pi(1)}, e_{\pi(2)}, \ldots, e_{\pi(n)}) = 0.$$

d. First write each column $\mathbf{v}_i$ as a linear combination of the standard basis elements $\mathbf{e}_k$:

$$\mathbf{v}_1 = v_{11}\mathbf{e}_1 + v_{21}\mathbf{e}_2 + \cdots + v_{n1}\mathbf{e}_n,$$

$$\mathbf{v}_2 = v_{12}\mathbf{e}_1 + v_{22}\mathbf{e}_2 + \cdots + v_{n2}\mathbf{e}_n,$$

$$\vdots$$

$$\mathbf{v}_n = v_{1n}\mathbf{e}_1 + v_{2n}\mathbf{e}_2 + \cdots + v_{nn}\mathbf{e}_n.$$

Then, using the multilinearity of D, expand $D(\mathbf{v}_1, \mathbf{v}_2, \ldots, \mathbf{v_n})$ as a sum of n^n terms of the form

$$v_{\pi(1),1} v_{\pi(2),2} \cdots v_{\pi(n),n} D(\mathbf{e}_{\pi(1)}, \mathbf{e}_{\pi(2)}, \ldots, \mathbf{e}_{\pi(n)}).$$

e. Most of the n^n terms you just obtained in the previous part are automatically zero; why? Which ones are not?

f. Conclude that

$$D(\mathbf{v}_1, \mathbf{v}_2, \ldots, \mathbf{v}_n) = \sum_{\pi \text{ in } S_n} (\operatorname{sgn} \pi)\, v_{\pi(1),1}\, v_{\pi(2),2} \cdots v_{\pi(n),n}.$$

This formula for D shows that there is one and only one function D that satisfies the three given properties.

This exercise gives us a way to define the determinant of an $n \times n$ matrix.

Definition 2.8 *Suppose $V = (v_{ij})$ is the $n \times n$ matrix that has the element v_{ij} in the ith row and jth column. Let $\mathbf{v}_j = (v_{ij})$, $i = 1, \ldots, n$, denote the jth column of V, so $V = (\mathbf{v}_1, \mathbf{v}_2, \ldots, \mathbf{v}_n)$. The **determinant of V** is*

$$\det V = D(\mathbf{v}_1, \mathbf{v}_2, \ldots, \mathbf{v}_n) = \sum_{\pi \text{ in } S_n} (\operatorname{sgn} \pi)\, v_{\pi(1),1}\, v_{\pi(2),2} \cdots v_{\pi(n),n}.$$

Thus, the determinant of V is the sum of all possible products of n elements of V, one taken from each row and each column, switching the sign of a particular product if the row indices represent an odd permutation of the columns indices.

2.29. Write out the 24 terms of the determinant of a 4×4 matrix.

2.30. Suppose that $V = (v_{ij})$ and $v_{ij} = 0$ if $i > j$. This is called an *upper triangular* matrix, because all entries below the main diagonal are zero. Show that $\det V = v_{11} v_{22} \cdots v_{nn}$.

2.31. Let A and B be 2×2 matrices, and let O denote the 2×2 matrix whose elements are all 0. Find the determinant of each of the following 4×4 matrices in terms of $\det A$ and $\det B$.

$$M_1 = \begin{pmatrix} A & O \\ O & B \end{pmatrix}, \quad M_2 = \begin{pmatrix} O & A \\ B & O \end{pmatrix}, \quad M_3 = \begin{pmatrix} A & B \\ O & O \end{pmatrix}.$$

2.32. Show that a square matrix A with a row or a column of zeros has $\det A = 0$.

2.33. Show that A and $A^\dagger$ have the same determinant.

2.34. The *minor M_{ij}* of A is the $(n-1) \times (n-1)$ matrix obtained by deleting the ith row and jth column of A. Show that we can "expand $\det A$ by minors along the ith row" in the following way.

$$\det A = (-1)^{i+1} a_{i1} \det M_{i1} + (-1)^{i+2} a_{i2} \det M_{i2} + \cdots + (-1)^{i+n} a_{in} \det M_{in}.$$

Write the analogous formula to "expand by minors along the jth column."

k-parallelepipeds

The definition of a parallelogram in $\mathbb{R}^3$ suggests we can define similar objects in $\mathbb{R}^n$, $n > 3$; in fact, we can generalize Definition 2.4 to define a parallelepiped of any dimension $k \leq n$ in $\mathbb{R}^n$.

The **k-dimensional parallelepiped** $\mathbf{v}_1 \wedge \mathbf{v}_2 \wedge \cdots \wedge \mathbf{v}_k$ is the set of vectors

$$\mathbf{w} = \sum_{i=1}^{k} t_i \mathbf{v}_i,$$

where $\mathbf{v}_1, \ldots, \mathbf{v}_k$ are vectors in $\mathbb{R}^n$ and $0 \leq t_i \leq 1$, $i = 1, \ldots, k$. We take $k \leq n$, but if $k \neq n$, $\mathbf{v}_1 \wedge \cdots \wedge \mathbf{v}_k$ is not oriented. We continue to call a 2-parallelepiped $\mathbf{v} \wedge \mathbf{w}$ a *parallelogram*.

2.35. Let $\mathbf{v} \wedge \mathbf{w}$ be a parallelogram in $\mathbb{R}^n$, and let V be the $n \times 2$ matrix whose columns are the coordinates of the vectors $\mathbf{v}$ and $\mathbf{w}$, in that order. Show that $\det V^\dagger V = \|\mathbf{v}\|^2 \|\mathbf{w}\|^2 - (\mathbf{v} \cdot \mathbf{w})^2$. By Exercise 2.26, we can take this to be $\text{area}^2(\mathbf{v} \wedge \mathbf{w})$.

3-volume

Now that we have the area of a parallelogram, we can define the k-volume of a k-parallelepiped inductively on k. We start with a 3-parallelepiped $\mathbf{v}_1 \wedge \mathbf{v}_2 \wedge \mathbf{v}_3$. Think of this as having base $\mathbf{v}_1 \wedge \mathbf{v}_2$; then we want the "3-volume" to be the area of the base times the perpendicular height:

$$\text{vol}(\mathbf{v}_1 \wedge \mathbf{v}_2 \wedge \mathbf{v}_3) = \text{area}(\mathbf{v}_1 \wedge \mathbf{v}_2) \|\mathbf{h}\|.$$

Here $\mathbf{h}$ is the vector that is orthogonal to the base $\mathbf{v}_1 \wedge \mathbf{v}_2$ and in the plane that contains $\mathbf{v}_3$ and is parallel to the base. A vector in that parallel plane is of the form

$$\mathbf{h} = \mathbf{v}_3 - a_1 \mathbf{v}_1 - a_2 \mathbf{v}_2,$$

for some real numbers a_1 and a_2. The orthogonality condition on $\mathbf{h}$ gives us two equations,

$$0 = \mathbf{v}_1 \cdot \mathbf{h} = \mathbf{v}_1 \cdot \mathbf{v}_3 - a_1 \mathbf{v}_1 \cdot \mathbf{v}_1 - a_2 \mathbf{v}_1 \cdot \mathbf{v}_2,$$
$$0 = \mathbf{v}_2 \cdot \mathbf{h} = \mathbf{v}_2 \cdot \mathbf{v}_3 - a_1 \mathbf{v}_2 \cdot \mathbf{v}_1 - a_2 \mathbf{v}_2 \cdot \mathbf{v}_2,$$

that we can convert into a matrix equation for the unknowns a_1 and a_2:

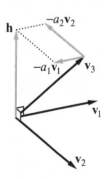

$$\begin{pmatrix} \mathbf{v}_1 \cdot \mathbf{v}_1 & \mathbf{v}_1 \cdot \mathbf{v}_2 \\ \mathbf{v}_2 \cdot \mathbf{v}_1 & \mathbf{v}_2 \cdot \mathbf{v}_2 \end{pmatrix} \begin{pmatrix} a_1 \\ a_2 \end{pmatrix} = \begin{pmatrix} \mathbf{v}_1 \cdot \mathbf{v}_3 \\ \mathbf{v}_2 \cdot \mathbf{v}_3 \end{pmatrix}.$$

There are unique values for a_1 and a_2—that is, $\mathbf{h}$ is uniquely defined—precisely when the matrix on the left is invertible. But notice that the determinant of that matrix is, by Exercise 2.26, the square of the area of the base parallelogram $\mathbf{v}_1 \wedge \mathbf{v}_2$. If $\text{area}(\mathbf{v}_1 \wedge \mathbf{v}_2) = 0$, the 3-volume of $\mathbf{v}_1 \wedge \mathbf{v}_2 \wedge \mathbf{v}_3$ is zero; otherwise, we can find $\mathbf{h}$, as above, and obtain $\text{vol}(\mathbf{v}_1 \wedge \mathbf{v}_2 \wedge \mathbf{v}_3)$. In fact, the volume is a determinant:

$$\begin{vmatrix} \mathbf{v}_1 \cdot \mathbf{v}_1 & \mathbf{v}_1 \cdot \mathbf{v}_2 & 0 \\ \mathbf{v}_2 \cdot \mathbf{v}_1 & \mathbf{v}_2 \cdot \mathbf{v}_2 & 0 \\ 0 & 0 & \mathbf{h} \cdot \mathbf{h} \end{vmatrix} = \begin{vmatrix} \mathbf{v}_1 \cdot \mathbf{v}_1 & \mathbf{v}_1 \cdot \mathbf{v}_2 \\ \mathbf{v}_2 \cdot \mathbf{v}_1 & \mathbf{v}_2 \cdot \mathbf{v}_2 \end{vmatrix} \mathbf{h} \cdot \mathbf{h}$$

$$= \text{area}^2(\mathbf{v}_1 \wedge \mathbf{v}_2) \, \|\mathbf{h}\|^2 = \text{vol}^2(\mathbf{v}_1 \wedge \mathbf{v}_2 \wedge \mathbf{v}_3).$$

Now replace the zeros in the 3×3 determinant with $\mathbf{v}_1 \cdot \mathbf{h}$ and $\mathbf{v}_2 \cdot \mathbf{h}$, as appropriate. Then substitute for the right-hand factor $\mathbf{h}$ in each entry in the third column its expression as a linear combination of $\mathbf{v}_1$, $\mathbf{v}_2$, and $\mathbf{v}_3$:

$$\begin{vmatrix} \mathbf{v}_1 \cdot \mathbf{v}_1 & \mathbf{v}_1 \cdot \mathbf{v}_2 & \mathbf{v}_1 \cdot \mathbf{h} \\ \mathbf{v}_2 \cdot \mathbf{v}_1 & \mathbf{v}_2 \cdot \mathbf{v}_2 & \mathbf{v}_2 \cdot \mathbf{h} \\ \mathbf{h} \cdot \mathbf{v}_1 & \mathbf{h} \cdot \mathbf{v}_2 & \mathbf{h} \cdot \mathbf{h} \end{vmatrix} = \begin{vmatrix} \mathbf{v}_1 \cdot \mathbf{v}_1 & \mathbf{v}_1 \cdot \mathbf{v}_2 & \mathbf{v}_1 \cdot \mathbf{v}_3 - a_1 \mathbf{v}_1 \cdot \mathbf{v}_1 - a_2 \mathbf{v}_1 \cdot \mathbf{v}_2 \\ \mathbf{v}_2 \cdot \mathbf{v}_1 & \mathbf{v}_2 \cdot \mathbf{v}_2 & \mathbf{v}_2 \cdot \mathbf{v}_3 - a_1 \mathbf{v}_2 \cdot \mathbf{v}_1 - a_2 \mathbf{v}_2 \cdot \mathbf{v}_2 \\ \mathbf{h} \cdot \mathbf{v}_1 & \mathbf{h} \cdot \mathbf{v}_2 & \mathbf{h} \cdot \mathbf{v}_3 - a_1 \mathbf{h} \cdot \mathbf{v}_1 - a_2 \mathbf{h} \cdot \mathbf{v}_2 \end{vmatrix}$$

$$= \begin{vmatrix} \mathbf{v}_1 \cdot \mathbf{v}_1 & \mathbf{v}_1 \cdot \mathbf{v}_2 & \mathbf{v}_1 \cdot \mathbf{v}_3 \\ \mathbf{v}_2 \cdot \mathbf{v}_1 & \mathbf{v}_2 \cdot \mathbf{v}_2 & \mathbf{v}_2 \cdot \mathbf{v}_3 \\ \mathbf{h} \cdot \mathbf{v}_1 & \mathbf{h} \cdot \mathbf{v}_2 & \mathbf{h} \cdot \mathbf{v}_3 \end{vmatrix}.$$

To get the simpler expression in the last step, above, we have used the multilinearity of the determinant and the fact that the determinant of a matrix with two equal columns is zero; cf. Exercise 2.25. The net result is that $\mathbf{h}$ is replaced by $\mathbf{v}_3$ in the third column. The same substitution for the factor $\mathbf{h}$ in the third row, followed by a similar addition of rows, will leave us $\mathbf{h}$ replaced by $\mathbf{v}_3$ in the third row. We discover that the volume is a determinant involving the vectors $\mathbf{v}_i$ in a symmetric way.

$$\begin{vmatrix} \mathbf{v}_1 \cdot \mathbf{v}_1 & \mathbf{v}_1 \cdot \mathbf{v}_2 & \mathbf{v}_1 \cdot \mathbf{v}_3 \\ \mathbf{v}_2 \cdot \mathbf{v}_1 & \mathbf{v}_2 \cdot \mathbf{v}_2 & \mathbf{v}_2 \cdot \mathbf{v}_3 \\ \mathbf{v}_3 \cdot \mathbf{v}_1 & \mathbf{v}_3 \cdot \mathbf{v}_2 & \mathbf{v}_3 \cdot \mathbf{v}_3 \end{vmatrix} = \text{vol}^2(\mathbf{v}_1 \wedge \mathbf{v}_2 \wedge \mathbf{v}_3)$$

$\det V^\dagger V = \text{vol}^2$ Finally, if V is the $n \times 3$ matrix whose columns are $\mathbf{v}_1$, $\mathbf{v}_2$, $\mathbf{v}_3$, in that order, then

$$V^\dagger V = \begin{pmatrix} \mathbf{v}_1 \cdot \mathbf{v}_1 & \mathbf{v}_1 \cdot \mathbf{v}_2 & \mathbf{v}_1 \cdot \mathbf{v}_3 \\ \mathbf{v}_2 \cdot \mathbf{v}_1 & \mathbf{v}_2 \cdot \mathbf{v}_2 & \mathbf{v}_2 \cdot \mathbf{v}_3 \\ \mathbf{v}_3 \cdot \mathbf{v}_1 & \mathbf{v}_3 \cdot \mathbf{v}_2 & \mathbf{v}_3 \cdot \mathbf{v}_3 \end{pmatrix},$$

so $\det V^\dagger V = \text{vol}^2(\mathbf{v}_1 \wedge \mathbf{v}_2 \wedge \mathbf{v}_3)$.

We consider next the 4-parallelepiped $\mathbf{v}_1 \wedge \mathbf{v}_2 \wedge \mathbf{v}_3 \wedge \mathbf{v}_4$; it has base, a 3-parallelepiped : $\mathbf{v}_1 \wedge \mathbf{v}_2 \wedge \mathbf{v}_3$,

height vector, orthogonal to base : $\mathbf{h} = \mathbf{v}_4 - a_1\mathbf{v}_1 - a_2\mathbf{v}_2 - a_3\mathbf{v}_3$,

and we can define its 4-volume as the product of the 3-volume of its base with the length of its height vector. The same, rather lengthy, argument we have just carried out allows us to determine the square of the 4-volume as $\det V^\dagger V$, where V is the matrix whose columns are the vectors $\mathbf{v}_i$. This process of establishing the volume of a k-parallelepiped from the volume of a $(k-1)$-parallelepiped is an example of *mathematical induction*, and shows for every $k \le n$ that the squared k-volume is $\det V^\dagger V$.

2.36. Show that the square of the n-volume of an n-parallelepiped, as defined in the text, equals $\det V^\dagger V$, as derived in these exercises.

2.37. Let V be the $n \times 3$ matrix whose columns are $\mathbf{v}_1$, $\mathbf{v}_2$, $\mathbf{v}_3$; verify that

$$V^\dagger V = \begin{pmatrix} \mathbf{v}_1 \cdot \mathbf{v}_1 & \mathbf{v}_1 \cdot \mathbf{v}_2 & \mathbf{v}_1 \cdot \mathbf{v}_3 \\ \mathbf{v}_2 \cdot \mathbf{v}_1 & \mathbf{v}_2 \cdot \mathbf{v}_2 & \mathbf{v}_2 \cdot \mathbf{v}_3 \\ \mathbf{v}_3 \cdot \mathbf{v}_1 & \mathbf{v}_3 \cdot \mathbf{v}_2 & \mathbf{v}_3 \cdot \mathbf{v}_3 \end{pmatrix}.$$

2.38. Let $\mathbf{h} = \mathbf{v}_k - a_1\mathbf{v}_1 - \cdots - a_{k-1}\mathbf{v}_{k-1}$, where $a_1, \ldots, a_{k-1}$ are arbitrary real numbers. Show that $\mathbf{h}$ is in the (hyper)plane that contains $\mathbf{v}_k$ and is parallel to the linear subspace of $\mathbb{R}^n$ spanned by $\mathbf{v}_1, \ldots, \mathbf{v}_{k-1}$.

2.39. Find a vector $\mathbf{h}$ in the plane that contains $\mathbf{v}_3$ and is parallel to $\mathbf{v}_1 \wedge \mathbf{v}_2$, when

a. $\mathbf{v}_1 = (1,0,1,0)$, $\mathbf{v}_2 = (2,1,1,1)$, $\mathbf{v}_3 = (0,1,2,0)$.
b. $\mathbf{v}_1 = (1,0,-1,0)$, $\mathbf{v}_2 = (2,1,1,1)$, $\mathbf{v}_3 = (0,1,2,0)$.

2.40. Determine the rank and nullity of each of the following matrices, viewing each as a linear map.

a. $\begin{pmatrix} 1 & 2 & 3 \\ 4 & 5 & 6 \end{pmatrix}$
b. $\begin{pmatrix} 0 & 1 & 0 & 0 \\ 0 & 0 & 1 & 0 \\ 0 & 0 & 0 & 1 \\ 1 & 1 & 1 & 1 \end{pmatrix}$
c. $\begin{pmatrix} 1 & -2 \\ -2 & 4 \\ 3 & -6 \end{pmatrix}$
d. $\begin{pmatrix} 1 & 1 & 0 \\ 0 & 1 & 1 \\ 1 & 0 & 1 \end{pmatrix}$

2.41. a. Solve the equations

$$5u + 3v - 3w + x = 0,$$
$$3u + 2v + 6w - 2x = 0,$$

for u and v in terms of w and x.

b. Can you solve for w and x in terms of u and v? What happens?
c. Can you solve for u and x in terms of v and w? What is the result?

2.42. This question concerns the linear map $L : \mathbb{R}^3 \to \mathbb{R}^3$ defined by the equations

$$L : \begin{cases} x = u + v, \\ y = v + w, \\ z = u - w. \end{cases}$$

a. What is the dimension of $\ker L$? Give a basis for $\ker L$.

b. Find the matrix for a linear map $M : \mathbb{R}^p \to \mathbb{R}^q$ whose graph is the kernel of L. What are the values of p and q?

c. Write M as a set of q equations in p variables.

d. What is the dimension of $\operatorname{im} L$? Give a basis for $\operatorname{im} L$.

e. Find the matrix for a linear map $A : \mathbb{R}^j \to \mathbb{R}^k$ whose graph is the image of L. What are the values of j and k?

f. Write A as a set of k equations in j variables.

2.43. a. Find all solutions (u, v) to the equations

$$u - 2v = 5,$$
$$4v - 2u = -10,$$

and sketch the solution set in the (u, v)-plane.

b. Describe the solution set in (a) as the graph of a suitable function. Is your sketch in (a) the graph of that function?

c. Describe the relation between the solution set in part (a) to the set of solutions to the equations

$$u - 2v = 0,$$
$$4v - 2u = 0.$$

2.44. Let $L : \mathbb{R}^n \to \mathbb{R}^p$ be an arbitrary linear map, and let

$$V = \{(\mathbf{u}, L(\mathbf{u})) \mid \mathbf{u} \in \mathbb{R}^n\} \subset \mathbb{R}^n \times \mathbb{R}^p = \mathbb{R}^{n+p}$$

be the graph of L. The purpose of this exercise is to show that V is a linear subspace of $\mathbb{R}^{n+p}$ of dimension n.

a. Show that the sum of two vectors in V is also in V. That is, given $\mathbf{v}_1 = (\mathbf{u}_1, L(\mathbf{u}_1))$ and $\mathbf{v}_2 = (\mathbf{u}_2, L(\mathbf{u}_2))$ with $\mathbf{u}_1$ and $\mathbf{u}_2$ in $\mathbb{R}^n$, show that $\mathbf{v}_1 + \mathbf{v}_2$ also has the form $(\mathbf{w}, L(\mathbf{w}))$ for some suitable $\mathbf{w}$ in $\mathbb{R}^n$.

b. Show that any scalar multiple of a vector in V is also in V.

c. Suppose $\mathcal{B} = \{\mathbf{u}_1, \mathbf{u}_2, \dots, \mathbf{u}_n\}$ is a basis for $\mathbb{R}^n$. Let $\mathbf{v}_j = (\mathbf{u}_j, L(\mathbf{u}_j))$ for $j = 1, 2, \dots, n$. Show that
   i. $\mathcal{G} = \{\mathbf{v}_1, \mathbf{v}_2, \dots, \mathbf{v}_n\}$ is a linearly independent set of vectors in $\mathbb{R}^{n+p}$;
   ii. $\mathcal{G}$ spans V; that is, any vector in V can be written as a linear combination of the vectors $\{\mathbf{v}_1, \mathbf{v}_2, \dots, \mathbf{v}_n\}$ that span V.

d. Explain why $\dim V = \dim \operatorname{graph} L = n$.

Chapter 3
Approximations

Abstract Approximations are at the heart of calculus. In Chapter 1 we saw that the transformation of differentials $dx = \varphi'(s)\,ds$ can be traced back to the linear approximation $\Delta x \approx \varphi'(s)\,\Delta s$ (the microscope equation), and that the factor $\varphi'(s)$ represented a local length multiplier. We also suggested there that the transformation $dx\,dy = r\,dr\,d\theta$ of differentials from Cartesian to polar coordinates has the same explanation: the polar coordinate change map has a linear approximation (a two-variable "microscope" equation) and the factor r is the local area multiplier for that map. In this chapter we construct a variety of useful approximations to nonlinear functions of one or more variables. However, we save for the following chapter a discussion of the most important approximation, the *derivative* of a map.

3.1 Mean-value theorems

The derivative indicates how much a function changes. It does this in the microscope equation, for example, and also in a similar equation called the law of the mean. First, consider the microscope equation, in this form:

$$f(x) - f(a) \approx f'(a)(x-a) \quad \text{for } x \approx a.$$

Measuring change with the derivative

This says each change $x - a$ in input produces a change $f(x) - f(a)$ in output that is approximately $f'(a)$ times as large. Although this equation is only approximate, note that the multiplier $f'(a)$ that links the changes is the same for all x in some sufficiently small neighborhood of a.

The **law of the mean** makes a similar statement, but there are important differences. It says that for each fixed x, say $x = b$, we can find a "mean value" c somewhere between a and b for which

Law of the mean

$$f(b) - f(a) = f'(c)(b-a).$$

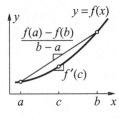

The link between changes in input and output is now exact rather than approximate, and b need not be near a, as in the microscope equation. However, these benefits come at a cost: the location of the c varies with b and, in fact, is not usually explicitly known, even when a and b are. Also, f' has to be continuous from a to b. As the figure indicates, c is to be chosen so that the slope $f'(c)$ is equal to the slope of the line segment from $(a, f(a))$ to $(b, f(b))$. We can put the law this way:

When x changes from a to b, then $f(x)$ undergoes a change that is exactly $f'(c)$ times as large.

The law of the mean is the first of several mean-value theorems we consider, and it is the most basic: all others follow from it.

Bounding the magnitude of change

Our ignorance of the location of c makes it difficult to use the law of the mean in some circumstances. However, it is natural to assume that we do have information about f and its derivative. In particular, we can make use of a bound on the size of $|f'(x)|$.

Theorem 3.1 (Mean-value theorem). *Suppose $f(x)$ is continuously differentiable for all x between a and b; then*

$$|f(b) - f(a)| \leq \max_{a \leq x \leq b} |f'(x)||b - a|. \qquad \Box$$

Although the law of the mean states explicitly—by means of an equality—how much the function grows (i.e., how $f(b) - f(a)$ depends on $b - a$), the mean-value theorem just provides a bound on that growth in terms of a bound on the derivative. As we show below, there is no direct extension of the law of the mean to *vector-valued* functions, that is, to functions $\mathbf{x} = \mathbf{f}(\mathbf{v})$ where the value $\mathbf{x}$ is a vector in $\mathbb{R}^p$, with $p \geq 2$. For a vector-valued function, we are only able to bound its growth by the size of its derivative, as in Theorem 3.1.

The law of the mean is frequently called the *mean-value theorem*; however, we reserve the latter term for the general theorem that extends to vector-valued functions.

Integral mean-value theorem

Continuing on, we formulate an integral law of the mean.

Theorem 3.2 (Law of the mean for integrals). *If $f(x)$ is a continuous function on the interval from a to b, then*

$$\int_a^b f(x)\,dx = f(c)(b - a)$$

for at least one point c in that interval.

Proof. Let $F(x)$ be an antiderivative of $f(x)$ (i.e., $F'(x) = f(x)$). The fundamental theorem of calculus guarantees that F exists because f is continuous. If we now apply the law of the mean to F, we find

$$\int_a^b f(x)\,dx = F(b) - F(a) = F'(c)(b - a) = f(c)(b - a)$$

for some c between a and b. $\qquad \Box$

We can connect this with the *mean value* of f. By definition, the **mean value** of a function $y = f(x)$ on the interval $a \leq x \leq b$ is the constant $\overline{f}$ whose integral over $[a,b]$ (the hatched region in the figure) is equal to the integral of $f(x)$ over $[a,b]$ (the shaded region):

$$\int_a^b f(x)\,dx = \int_a^b \overline{f}\,dx = \overline{f} \times (b-a).$$

This leads to the more familiar form of the definition:

$$\text{mean value of } f \text{ on } [a,b] = \overline{f} = \frac{1}{b-a} \int_a^b f(x)\,dx.$$

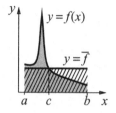

Because $\overline{f}$ lies between the maximum and minimum values of $f(x)$ on $[a,b]$, and because f is continuous, there is at least one c between a and b for which $f(c) = \overline{f}$. In other words, in the equation

$$\int_a^b f(x)\,dx = f(c)(b-a)$$

provided by the integral law of the mean, $f(c)$ is in fact the mean value of f on the interval $[a,b]$.

Because we know $f(c) = \overline{f}$ in the integral law of the mean, we can sometimes determine c explicitly. For example, let $f(x) = \sqrt{r^2 - x^2}$. The graph of $y = f(x)$ is a semicircle of radius r on the interval $[-r,r]$, so the area under it is $\pi r^2/2$ and

$$\frac{\pi r^2}{2} = \int_{-r}^{r} \sqrt{r^2 - x^2}\,dx = f(c)\,2r = 2r\sqrt{r^2 - c^2}.$$

We can solve this for c and get $c = \pm r\sqrt{1 - \pi^2/16}$; see the exercises.

The integral law of the mean has us compute an integral by extracting the mean value of its integrand. The following theorem makes a more general assertion: there are circumstances where we can compute an integral by extracting the mean value of just a part of its integrand.

Theorem 3.3 (Generalized law of the mean for integrals). *If $f(x)$ and $g(x)$ are continuous on the interval $[a,b]$ and $g(x) \geq 0$ there, then*

$$\int_a^b f(x)g(x)\,dx = f(c) \int_a^b g(x)\,dx$$

for at least one point c in that interval.

Proof. Our proof here echoes the previous argument, which said that $\overline{f} = f(c)$ for some c because $\overline{f}$ lies between the minimum and maximum values of the continuous function f.

Note first that if $g(x) = 0$ for all x in the interval $[a,b]$, then both integrals equal 0 and there is nothing to prove. So we assume $g(x) > 0$ for at least some x in the interval; because g is continuous and $g(x) \geq 0$, it also follows that

$$\int_a^b g(x)\,dx > 0.$$

Now let m and M be the minimum and maximum values of $f(x)$ on $[a,b]$; thus $m \le f(x) \le M$. Because $0 \le g(x)$, we also have

$$mg(x) \le f(x)g(x) \le Mg(x)$$

for all x in the interval, and therefore

$$m \int_a^b g(x)\,dx \le \int_a^b f(x)g(x)\,dx \le M \int_a^b g(x)\,dx.$$

We have already noted that the integral of g is nonzero, so we can divide these inequalities by it and conclude that the expression

$$\frac{\displaystyle\int_a^b f(x)g(x)\,dx}{\displaystyle\int_a^b g(x)\,dx}$$

lies between the minimum and maximum values of $f(x)$ on the interval. Therefore, because f is continuous, this expression must equal $f(c)$ for at least one value c in the interval. $\qquad\square$

The condition $g(x) \ge 0$ can be changed to $g(x) \le 0$ without affecting the truth of the theorem; however, the theorem does fail if $g(x)$ changes sign on the interval. You can explore these points in the exercises. The proof of the generalized law of the mean can be adapted to have the result stated as an inequality, as with the mean-value theorem for functions.

Corollary 3.4 *If $f(x)$ and $g(x) \ge 0$ are continuous on $[a,b]$, then*

$$\left| \int_a^b f(x)g(x)\,dx \right| \le \max_{a \le x \le b} |f(x)| \left| \int_a^b g(x)\,dx \right|. \qquad\qquad\square$$

Law of the mean for multivariable functions

Until now we have considered only functions of a single variable, but there are analogous mean-value theorems for functions of several variables. To extend the law of the mean to such functions, it is convenient for us to recast the law in a slightly different form. Let $\Delta x = b - a$; then the point c that lies between a and b can be written as $c = a + \theta\Delta x$ for some $0 \le \theta \le 1$, and the law itself can be written as

$$f(a + \Delta x) = f(a) + f'(a + \theta\Delta x)\Delta x.$$

Now consider a function of several variables $F(\mathbf{x}) = F(x_1, \ldots, x_n)$. The analogue of the ordinary derivative $f'(x)$ is the gradient of F (constructed with the differential operator ∇, "nabla"):

$$\nabla F(\mathbf{x}) = \operatorname{grad} F(\mathbf{x}) = \left(\frac{\partial F}{\partial x_1}(\mathbf{x}), \ldots, \frac{\partial F}{\partial x_n}(\mathbf{x}) \right).$$

Theorem 3.5 (Law of the mean). *Suppose $F(\mathbf{x}) = F(x_1, \ldots, x_n)$ is a continuously differentiable function of $\mathbf{x}$; then*

$$F(\mathbf{a} + \Delta\mathbf{x}) - F(\mathbf{a}) = \nabla F(\mathbf{a} + \theta\Delta\mathbf{x}) \cdot \Delta\mathbf{x},$$

for some $0 \le \theta \le 1$.

Proof. There is a simple way to reduce this to a single-variable question; let $f(t) = F(\mathbf{a} + t\Delta\mathbf{x}) = F(a_1 + t\Delta x_1, \ldots, a_n + t\Delta x_n)$. Then the chain rule gives us the derivative of f:

$$f'(t) = \frac{\partial F}{\partial x_1}(\mathbf{a} + t\Delta\mathbf{x})\Delta x_1 + \cdots + \frac{\partial F}{\partial x_n}(\mathbf{a} + t\Delta\mathbf{x})\Delta x_n = \nabla F(\mathbf{a} + t\Delta\mathbf{x}) \cdot \Delta\mathbf{x}.$$

This is the scalar (or dot) product of the gradient with the vector $\Delta\mathbf{x}$. By the law of the mean for $f(t)$, we know there is a $0 \le \theta \le 1$ for which the following is true:

$$\begin{aligned} F(\mathbf{a} + \Delta\mathbf{x}) - F(\mathbf{a}) = f(1) - f(0) &= f'(\theta)(1 - 0) \\ &= \nabla F(\mathbf{a} + \theta\Delta\mathbf{x}) \cdot \Delta\mathbf{x}. \end{aligned} \qquad \square$$

Thus, even for a function $F(\mathbf{x})$ of several variables, if we use the dot product for multiplication, we can express the law of the mean in the following way.

When $\mathbf{x}$ changes from $\mathbf{a}$ to $\mathbf{b} = \mathbf{a} + \Delta\mathbf{x}$, the change in $F(\mathbf{x})$ is exactly ∇F times as large, where the gradient ∇F is evaluated at some intermediate (or "mean") point along the line from $\mathbf{a}$ to $\mathbf{b}$.

We can rephrase the multivariable law of the mean as an inequality that bounds the growth of F in terms of a bound on its derivative (i.e., its gradient):

Corollary 3.6 (Mean-value theorem) *If $z = F(\mathbf{x})$ is continuously differentiable on the line that connects $\mathbf{a}$ and $\mathbf{b}$, then*

$$|F(\mathbf{b}) - F(\mathbf{a})| \le \max_{\mathbf{x}} \|\nabla F(\mathbf{x})\| \|\mathbf{b} - \mathbf{a}\|,$$

where the maximum is taken over all points $\mathbf{x}$ on the line from $\mathbf{a}$ to $\mathbf{b}$. $\qquad \square$

For functions of two variables, we have a natural extension of the integral law of the mean. The proof follows the pattern of all previous proofs; see the exercises. We deal with functions of three or more variables in a later chapter, after we discuss their integrals.

Double integrals

Definition 3.1 *The **mean value** $\overline{F}$ of the function $F(x,y)$ on the domain D in $\mathbb{R}^2$ is*

$$\overline{F} = \frac{1}{\operatorname{area} D} \iint_D F(x,y)\,dx\,dy.$$

Theorem 3.7 (Law of the mean for double integrals). *Let $F(x,y)$ be a continuous function on a connected domain D in $\mathbb{R}^2$. Then there is at least one point (c,d) in D where F takes on its mean value $\overline{F}$; thus*

$$\iint\limits_{D} F(x,y)\,dx\,dy = F(c,d) \times \operatorname{area} D. \qquad \square$$

The problem with vector-valued functions

To complete the present study of mean-value theorems, let us consider *vector-valued* functions $\mathbf{x} = \mathbf{f}(\mathbf{v})$, where $\mathbf{x}$ is a vector in $\mathbb{R}^p$, $p \geq 2$. For even in the simplest such case—a vector-valued function $\mathbf{x} = \mathbf{f}(t)$ of a single variable, which defines a curve in $\mathbb{R}^p$—we now show there can be no direct extension of the law of the mean as an equality.

To see why, consider the helix $\mathbf{x} = \mathbf{f}(t) = (\cos t, \sin t, t)$ in $\mathbb{R}^3$. Let us try to express the change $\mathbf{f}(2\pi) - \mathbf{f}(0)$ in the form $\mathbf{f}'(c)(2\pi - 0)$ for some suitable mean value c between 0 and 2π. The vector $\Delta\mathbf{f} = \mathbf{f}(2\pi) - \mathbf{f}(0) = (0,0,2\pi)$ is vertical, but the derivative

$$\mathbf{f}'(c) = (-\sin c, \cos c, 1)$$

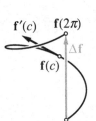

is never vertical, because its first two components are never simultaneously zero. Therefore, no scalar multiple of $\mathbf{f}'(c)$ will ever equal $\Delta\mathbf{f}$, even approximately. In particular, there is no number c for which

$$\mathbf{f}(2\pi) - \mathbf{f}(0) = \mathbf{f}'(c)(2\pi - 0).$$

Even though the law of the mean itself fails to hold, we can still get a *bound* on the size of the change in $\mathbf{f}$ that is exactly analogous to the bound provided by Theorem 3.1; in fact, we have

$$\|\mathbf{f}(2\pi) - \mathbf{f}(0)\| \leq \max \|\mathbf{f}'(t)\| |2\pi - 0|,$$

because the left-hand side equals 2π and the right-hand side equals $2\sqrt{2}\pi$.

Theorem 3.8 (Mean-value theorem). *If $\mathbf{f} : I \to \mathbb{R}^p$ has a continuous derivative on an interval I that contains a and b, then*

$$\|\mathbf{f}(b) - \mathbf{f}(a)\| \leq \max_{a \leq t \leq b} \|\mathbf{f}'(t)\| |b - a|.$$

Proof. Because $\mathbf{f}(b) - \mathbf{f}(a) = \displaystyle\int_a^b \mathbf{f}'(t)\,dt$, we have

$$\|\mathbf{f}(b) - \mathbf{f}(a)\| = \left\| \int_a^b \mathbf{f}'(t)\,dt \right\| \leq \max_{a \leq t \leq b} \|\mathbf{f}'(t)\| \left| \int_a^b dt \right| \leq \max_{a \leq t \leq b} \|\mathbf{f}'(t)\| |b - a|. \qquad \square$$

Extension to multivariable inputs

This theorem relies on the fact that we can regard the derivative $\mathbf{f}'(t)$ as a vector of the same sort as $\mathbf{f}(t)$ itself; this, in turn, is a consequence of the fact that the input is just a single variable t. If, instead, $\mathbf{x} = \mathbf{f}(\mathbf{v})$, where $\mathbf{v}$ is in $\mathbb{R}^p$, $p \geq 2$, then the *derivative of* $\mathbf{f}$ is something new. We define this new derivative below (Defini-

tion 3.16, p. 99) and analyze it in the following two chapters. At that time, we state
and prove a natural extension of Theorem 3.8 for maps (Theorem 4.15, p. 140).

3.2 Taylor polynomials in one variable

You are probably familiar with Taylor polynomials for $f(x)$ written in the form   New approximations

$$f(a) + f'(a)(x-a) + \frac{f''(a)}{2!}(x-a)^2 + \cdots + \frac{f^{(n)}(a)}{n!}(x-a)^n.$$

These expressions give us new ways to approximate $f(x)$. One obvious benefit of
any polynomial approximation is that it can be computed using only the four basic
operations of arithmetic; most functions are not computable in this sense. Taylor
polynomials become better approximations as n increases, and as x gets closer to a.
We also see less obvious reasons that make them valuable.

Consider successive terms in the Taylor polynomial to be separate contributions   Properties of the
Taylor approximation
to the approximation. Then we find that the lowest-order term makes the largest
contribution (at least when x is close to a) whereas the succeeeding terms, involving
higher and higher powers of $x - a$, contribute less and less to the total. Also, because
a Taylor polynomial gives a better approximation to $f(x)$ the closer x is to a, we
write the polynomial instead in terms of the variable $\Delta x = x - a$ that indicates how
close x is to a. We took the same approach when we reformulated the law of the
mean (pp. 74ff.)

Definition 3.2 *Suppose* $y = f(x)$ *has derivatives up to order n at $x = a$; then the*   Taylor polynomial;
$\Delta x = x - a$
Taylor polynomial of degree n for f at $x = a$ is

$$P_{n,a}(\Delta x) = f(a) + f'(a)\Delta x + \frac{f''(a)}{2!}(\Delta x)^2 + \cdots + \frac{f^{(n)}(a)}{n!}(\Delta x)^n.$$

Notice that this expression for $P_{n,a}$ includes each of the Taylor polynomials $P_{0,a}$,
$P_{1,a}, P_{2,a}, \ldots, P_{n-1,a}$ as an initial part.

In many cases, it is easy to calculate these polynomials. For example, if $f(x) =$ Estimates and errors
$\sqrt{x}$, $n = 3$, and $a = 100$, then

$$P_{3,100}(\Delta x) = 10 + \frac{\Delta x}{20} - \frac{(\Delta x)^2}{8\,000} + \frac{(\Delta x)^3}{1\,600\,000}.$$

Let us see how this gives us approximate values of $\sqrt{x}$ when $x \approx 100$. First we
build a sequence of estimates of $\sqrt{102}$ (so $\Delta x = 2$) by adding in the terms of the
polynomial, one at a time. This allows us to see how the approximation improves as
the degree increases. Second, we then do the same for $\sqrt{120}$ ($\Delta x = 20$). Comparing
the two sets of estimates allows us to see how the approximation improves as Δx
decreases. In all cases our focus is on the **error**: that is, on the difference between
the true value and the approximation.

Contribution made by
each term in $P_{3,100}(2)$

First, we consider how each term in the cubic Taylor polynomial

$$P_{3,100}(2) = 10 + \frac{2}{20} - \frac{4}{8\,000} + \frac{8}{1\,600\,000} = 10.099\,505$$

contributes to the estimate of $\sqrt{102} = 10.099\,504\,938\,362\ldots$. Here are the results in a table:

Degree	Term	Sum	Error $= \sqrt{102} -$ Sum	
0	10	10	$0.0995\ldots$	$\approx\ 1 \times 10^{-1}$
1	0.1	10.1	$-0.000495\ldots$	$\approx -5 \times 10^{-4}$
2	-0.0005	10.0995	$0.000004938\ldots$	$\approx\ 5 \times 10^{-6}$
3	0.000005	10.099505	$-0.0000000616\ldots$	$\approx -6 \times 10^{-8}$

Thus we see that higher terms contribute less and less to the sum, but they effectively "fine-tune" the estimate. In fact, the contributions drive the error down *exponentially*; that is, the error at each stage made by the intermediate sum $P_{k,100}(2)$ is roughly of size 10^{-ak-b}, for some $a > 0$.

Comparative errors for
$\sqrt{102}$ and $\sqrt{120}$

Of course the terms get smaller because their coefficients do, and this is clearly the result of the choice of the original function $f(x) = \sqrt{x}$. For a different function, the coefficients may not be so obliging. Nevertheless, by comparing the errors that $P_{3,100}(\Delta x)$ makes for different values of Δx, we largely eliminate the effect of the coefficients. At the same time, we see how the error is connected to the size of Δx, our second objective. The following table gives comparative information for $\sqrt{120} = 10.954\,451\,15\ldots (\Delta x = 20)$.

Degree	Term	Sum	Error $= \sqrt{120} -$ Sum	
0	10	10	$0.954\ldots$	$\approx\ 1 \times 10^{0}$
1	1	11	$-0.0455\ldots$	$\approx -5 \times 10^{-2}$
2	-0.05	10.95	$0.00445\ldots$	$\approx\ 4 \times 10^{-3}$
3	0.005	10.955	$-0.000548\ldots$	$\approx -5 \times 10^{-4}$

Compare the rightmost columns of the two tables for $k = 2$ or 3: $x = 102$ is only $1/10$ as far from 100 as $x = 120$, and the error that $P_{k,100}$ makes in estimating $\sqrt{102}$ is, roughly speaking, only about $1/10^{k+1}$ times as large as the error for $\sqrt{120}$. Later (p. 83), we confirm this is true even for $k > 3$.

Errors and
Taylor's theorem

Our experiments with $\sqrt{x}$ suggest that that we should study how

$$\text{error} = f(a + \Delta x) - P_{n,a}(\Delta x)$$

depends on Δx and on n in general. The result is contained in Taylor's theorem, which we state and prove below. It spells out the error, and with it we are able to see that $P_{n,a}(\Delta x)$ makes a smaller error than any other polynomial of the same degree in approximating $f(x)$ near a. Before we state the theorem, let us look first at the simple case when $n = 0$.

The Taylor polynomial for $n = 0$ is just the constant function $P_{0,a}(\Delta x) = f(a)$, so we have the estimate

The error for $P_{0,a}(\Delta x)$

$$\text{error} = f(a + \Delta x) - f(a) \approx f'(a)\Delta x$$

by using the microscope equation. Because $f'(a)$ is fixed, this expression already tells us the error is roughly proportional to the size of Δx itself. So if x_1 is 1/10 as far from a as x_2 (and x_2 is still sufficiently close to a), then the error that $P_{0,a}$ makes in estimating $f(x_1)$ will be about 1/10 as large as the error in estimating $f(x_2)$.

By contrast, the law of the mean (p. 74) gives us the exact error, but in terms of a quantity $0 \leq t \leq 1$ whose value we may not be able to determine effectively:

$$f(a + \Delta x) - f(a) = f'(a + t\Delta x)\Delta x.$$

Because the derivative of $f(a + t\Delta x)$ with respect to t is $f'(a + t\Delta x)\Delta x$ (chain rule), we can also express the exact error as an integral:

$$\int_0^1 f'(a + t\Delta x)\Delta x\, dt = f(a + t\Delta x)\Big|_0^1 = f(a + \Delta x) - f(a).$$

Although this integral is more complicated-looking than the other ways of writing the error, it turns out to be the most useful. Let us rewrite the last equation as

$$f(a + \Delta x) = f(a) + \int_0^1 f'(a + t\Delta x)\Delta x\, dt = P_{0,a}(\Delta x) + R_{0,a}(\Delta x);$$

we call

$$R_{0,a}(\Delta x) = \Delta x \int_0^1 f'(a + t\Delta x)\, dt$$

the **remainder** because it is what is left after we subtract the Taylor polynomial from the function. Of course it is also the error we make in replacing the function value by the polynomial value. We are now ready to state Taylor's theorem; it expresses the remainder for a general $P_{n,a}$ as an integral.

Theorem 3.9 (Taylor). *If $f(x)$ has continuous derivatives up to order $n + 1$ on an interval containing a and $a + \Delta x$, then*

Taylor's formula with remainder

$$f(a + \Delta x) = f(a) + f'(a)\Delta x + \frac{f''(a)}{2!}(\Delta x)^2 + \cdots + \frac{f^{(n)}(a)}{n!}(\Delta x)^n + R_{n,a}(\Delta x),$$

where $R_{n,a}(\Delta x) = \dfrac{(\Delta x)^{n+1}}{n!}\displaystyle\int_0^1 f^{(n+1)}(a + t\Delta x)(1 - t)^n\, dt.$

Proof by induction

Proof. Because the theorem is essentially a sequence of formulas—one for each value of n—we prove them one at a time, "by induction on n." That is, we first prove the formula involving $P_{0,a}$, then use it to prove the one involving $P_{1,a}$, then use that to prove the one involving $P_{2,a}$, and so on, generating each new remainder as we go. To prove Taylor's formula for $P_{0,a}$,

$$f(a+\Delta x) = f(a) + \Delta x \int_0^1 f'(a+t\Delta x)\,dt,$$

just set $\varphi(t) = f(a+t\Delta x)$ and use the fundamental theorem of calculus in the form

$$\varphi(1) = \varphi(0) + \int_0^1 \varphi'(t)\,dt.$$

First induction step   The induction that takes us from one formula to the next is just an integration by parts carried out on the remainder integral. The integration by parts formula we use is

$$\int_\alpha^\beta u\,dv = uv\Big|_\alpha^\beta - \int_\alpha^\beta v\,du.$$

To begin, we set $u = f'(a+t\Delta x)$ and $dv = dt$ in the formula for $P_{0,a}$. Then $du = f''(a+t\Delta x)\Delta x\,dt$ and $v = t+C$, where C is a constant of integration that we determine in a moment. We have

$$R_{0,a}(\Delta x) = \Delta x \int_0^1 f'(a+t\Delta x)\,dt$$

$$= \Delta x\,f'(a+t\Delta x)(t+C)\Big|_0^1 - \Delta x \int_0^1 (t+C)f''(a+t\Delta x)\,\Delta x\,dt$$

$$= \Delta x\,f'(a+\Delta x)(1+C) - \Delta x\,f'(a)C - (\Delta x)^2 \int_0^1 (t+C)f''(a+t\Delta x)\,dt.$$

If we now set $C = -1$ the first term on the right disappears and the second becomes $+f'(a)\Delta x$, which is exactly the linear term in $P_{1,a}(x)$. Thus the formula with $P_{0,a}$ becomes (note the sign changes)

$$f(a+\Delta x) = f(a) + R_{0,a}(\Delta x) = \underbrace{f(a) + f'(a)\Delta x}_{P_{1,a}(\Delta x)} + \underbrace{(\Delta x)^2 \int_0^1 f''(a+t\Delta x)(1-t)\,dt}_{R_{1,a}(\Delta x)}.$$

The error for $P_{1,a}(x)$ The error in replacing $f(a+\Delta x)$ by $P_{1,a}(\Delta x)$ is thus the new remainder,

$$R_{1,a}(\Delta x) = (\Delta x)^2 \int_0^1 f''(a+t\Delta x)(1-t)\,dt.$$

Second induction step   The second induction step starts with Taylor's formula when $n = 1$:

$$f(a+\Delta x) = P_{1,a}(\Delta x) + R_{1,a}(\Delta x)$$

$$= f(a) + f'(a)\Delta x + (\Delta x)^2 \int_0^1 f''(a+t\Delta x)(1-t)\,dt.$$

To integrate by parts here, use $u = f''(a+t\Delta x)$, $dv = (1-t)\,dt$, and $v = -(1-t)^2/2$; then

$$R_{1,a}(\Delta x) = (\Delta x)^2 \int_0^1 f''(a + t\Delta x)(1 - t)\, dt$$

$$= -(\Delta x)^2 f''(a + t\Delta x)\frac{(1 - t)^2}{2}\bigg|_0^1 + (\Delta x)^2 \int_0^1 f^{(3)}(a + t\Delta x)\Delta x \frac{(1 - t)^2}{2}\, dt$$

$$= \frac{f''(a)}{2}(\Delta x)^2 + \frac{(\Delta x)^3}{2}\int_0^1 f^{(3)}(a + t\Delta x)(1 - t)^2\, dt.$$

With $R_{1,a}(\Delta x)$ replaced by the two terms in the last line, our previous equation for $f(a + \Delta x)$ (i.e., Taylor's formula when $n = 1$) now reads

$$f(a + \Delta x) = \underbrace{f(a) + f'(a)\Delta x + \frac{f''(a)}{2}(\Delta x)^2}_{P_{2,a}(\Delta x)} + \underbrace{\frac{(\Delta x)^3}{2}\int_0^1 f^{(3)}(a + t\Delta x)(1 - t)^2\, dt}_{R_{2,a}(\Delta x)}.$$

As we see, this has become Taylor's formula when $n = 2$.

In the next step, we are able to see how the factorial expressions arise. Our starting point is $f(a + \Delta x) = P_{2,a}(\Delta x) + R_{2,a}(\Delta x)$, and we must integrate $R_{2,a}(\Delta x)$ by parts. If we use $u = f^{(3)}(a + t\Delta x)$ and $v = -(1 - t)^3/3$, then

Third induction step

$$R_{2,a}(\Delta x) = -\frac{(\Delta x)^3}{3 \cdot 2}f^{(3)}(a + t\Delta x)(1 - t)^3\bigg|_0^1 + \frac{(\Delta x)^3}{3 \cdot 2}\int_0^1 f^{(4)}(a + t\Delta x)\Delta x\,(1 - t)^3\, dt$$

$$= \frac{f^{(3)}(a)}{3!}(\Delta x)^3 + \frac{(\Delta x)^4}{3!}\int_0^1 f^{(4)}(a + t\Delta x)(1 - t)^3\, dt.$$

Consequently, Taylor's formula when $n = 2$ becomes

$$f(a + \Delta x) = P_{2,a}(\Delta x) + R_{2,a}(\Delta x)$$

$$= \underbrace{P_{2,a}(\Delta x) + \frac{f^{(3)}(a)}{3!}(\Delta x)^3}_{P_{3,a}(\Delta x)} + \underbrace{\frac{(\Delta x)^4}{3!}\int_0^1 f^{(4)}(a + t\Delta x)(1 - t)^3\, dt}_{R_{3,a}(\Delta x)},$$

which is just Taylor's formula when $n = 3$.

To complete the induction, we must transform Taylor's formula when $n = k$,

General induction step

$$f(a + \Delta x) = P_{k,a}(\Delta x) + R_{k,a}(\Delta x),$$

(where k is any nonnegative integer) into the corresponding formula when $n = k + 1$,

$$f(a + \Delta x) = P_{k+1,a}(\Delta x) + R_{k+1,a}(\Delta x).$$

This is another integration by parts (see the exercises):

$$R_{k,a}(\Delta x) = \frac{(\Delta x)^{k+1}}{k!} \int_0^1 f^{(k+1)}(a + t\Delta x)(1-t)^k dt$$

$$= \frac{f^{(k+1)}(a)}{(k+1)!}(\Delta x)^{k+1} + \frac{(\Delta x)^{k+2}}{(k+1)!} \int_0^1 f^{(k+2)}(a + t\Delta x)(1-t)^{k+1} dt.$$

It implies $f(a + \Delta x)$

$$= \underbrace{P_{k,a}(\Delta x) + \frac{f^{(k+1)}(a)}{(k+1)!}(\Delta x)^{k+1}}_{P_{k+1,a}(\Delta x)} + \underbrace{\frac{(\Delta x)^{k+2}}{(k+1)!} \int_0^1 f^{(k+2)}(a + t\Delta x)(1-t)^{k+1} dt}_{R_{k+1,a}(\Delta x)},$$

completing the general induction step.   □

Forms of the remainder

Our approximations of $\sqrt{102}$ and $\sqrt{120}$ suggested that the error $R_{n,a}(\Delta x)$ gets small as Δx does. In fact, in those examples we saw that the error got small faster than Δx; $R_{n,a}(\Delta x)$ vanished like $(\Delta x)^{n+1}$. This is true in general; to see why, we first write $R_{n,a}(\Delta x)$ in some alternate forms.

Corollary 3.10 (Lagrange's form of the remainder) *For each $\Delta x \approx 0$, there is a* $\theta = \theta(\Delta x)$ *with* $0 \le \theta \le 1$ *for which*

$$R_{n,a}(\Delta x) = \frac{f^{(n+1)}(a + \theta\Delta x)}{(n+1)!}(\Delta x)^{n+1}.$$

Proof. With the generalized integral law of the mean (Theorem 3.3), we can extract $f^{(n+1)}(a + t\Delta x)$ from the integral defining $R_{n,a}(\Delta x)$, and then compute the integral of the remaining function, $(1-t)^n$, exactly. Thus, for a given Δx, there is a point θ in the interval $[0, 1]$ for which

$$R_{n,a}(\Delta x) = \frac{(\Delta x)^{n+1}}{n!} \int_0^1 f^{(n+1)}(a + t\Delta x)(1-t)^n dt$$

$$= (\Delta x)^{n+1} \frac{f^{(n+1)}(a + \theta\Delta x)}{n!} \int_0^1 (1-t)^n dt$$

$$= (\Delta x)^{n+1} \frac{f^{(n+1)}(a + \theta\Delta x)}{n!} \left(-\frac{(1-t)^{n+1}}{n+1} \right)\Big|_0^1$$

$$= \frac{f^{(n+1)}(a + \theta\Delta x)}{(n+1)!}(\Delta x)^{n+1}.   \qquad\qquad□$$

Taylor's formula and the law of the mean

If we write Taylor's formula for $n = 0$ using the Lagrange form of the remainder, we get

$$f(a + \Delta x) = f(a) + f'(a + \theta\Delta x)\Delta x,$$

which is just the law of the mean. Thus we can see Taylor's formula with Lagrange's remainder as an extension of the law of the mean that incorporates higher powers of the displacement Δx.

With the remainder in Lagrange's form we can see why we said, on page 78, that the error $P_{k,100}$ makes in estimating $\sqrt{102}$ would be only about $1/10^{k+1}$ times the error in estimating $\sqrt{120}$. Consider first $k = 3$; because $f(x) = \sqrt{x}$, Comparative estimates of $\sqrt{102}$ and $\sqrt{120}$

$$R_{3,100}(\Delta x) = \frac{f^{(4)}(100 + \theta \Delta x)}{4!}(\Delta x)^4 = \frac{-15}{16 \times 24\,(100 + \theta \Delta x)^{7/2}}(\Delta x)^4.$$

Because the number $100 + \theta \Delta x$ certainly lies between 100 and 120, the coefficient $-15/(16 \times 24\,(100 + \theta \Delta x)^{7/2})$ will lie in the narrow range from -4×10^{-9} to -2×10^{-9} for both estimates. So the main cause of the difference between the two errors must be the factor $(\Delta x)^4$: in the two cases, its values are $2^4 = 16$ and $20^4 = 16 \times 10^4$. This is why $R_{3,100}(2)$ is only about $1/10^4$ times as large as $R_{3,100}(20)$. Furthermore, because $100 \leq 100 + \theta \Delta x \leq 120$, we see that the errors themselves must lie in the following ranges:

$$-6.4 \times 10^{-8} < R_{3,100}(2) < -3.2 \times 10^{-8},$$
$$-6.4 \times 10^{-4} < R_{3,100}(20) < -3.2 \times 10^{-4}.$$

In fact, we already obtained by direct calculation the values $R_{3,100}(2) \approx -6 \times 10^{-8}$ and $R_{3,100}(20) \approx -5 \times 10^{-4}$; they fit into these ranges, as they should.

Now take an arbitrary $k \geq 2$; then (see the exercises) How the comparative error depends on k

$$R_{k,100}(\Delta x) = \frac{\pm 1 \cdot 3 \cdots (2k-1)}{2^{k+1}(k+1)!(100 + \theta \Delta x)^{k+1/2}}(\Delta x)^{k+1}.$$

The term $1/(100 + \theta \Delta x)^{k+1/2}$ again varies over a small range of values when we have $0 \leq \Delta x \leq 20$; the main cause of the variation of $R_{k,100}(\Delta x)$ with Δx is the factor $(\Delta x)^{k+1}$. Therefore,

$$\frac{R_{k,100}(2)}{R_{k,100}(20)} \approx \frac{2^{k+1}}{20^{k+1}} = \frac{1}{10^{k+1}},$$

so the error $P_{k,100}$ makes in estimating $\sqrt{102}$ is only about $1/10^{k+1}$ times as large as the error estimating $\sqrt{120}$.

Because $f^{(n+1)}$ is a continuous function, the factor $f^{(n+1)}(a + \theta \Delta x)$ in the Lagrange remainder is as close as we wish to $f^{(n+1)}(a)$ if Δx is sufficiently close to 0. This gives us another form of the remainder. Taylor's formula and the microscope equation

Corollary 3.11 (Generalized microscope equation) *When* $\Delta x \approx 0$,

$$R_{n,a}(\Delta x) \approx \frac{f^{(n+1)}(a)}{(n+1)!}(\Delta x)^{n+1}. \qquad \square$$

Notice that when $n = 0$, $R_{0,a}(\Delta x) \approx f'(a)\Delta x$, and Taylor's formula becomes the microscope equation,

$$f(a + \Delta x) \approx f(a) + f'(a)\Delta x.$$

This is why we call the statement in the corollary the generalized microscope equation. Taylor's formula thus generalizes both the microscope equation and the law of the mean (Lagrange's form).

The next term is most of the remainder

The generalized microscope equation is a remarkable result. It says that the remainder looks more and more like the next term in the Taylor expansion of f, the more we magnify the graph of the remainder in a microscope window centered at $\Delta x = 0$. Thus, because most of the error at the nth stage is equal to the term of degree $n + 1$, we can eliminate most of the error by adding that term to the nth stage (i.e., to $P_{n,a}$). The result is, of course, the next Taylor polynomial, $P_{n+1,a}$. This is perhaps the simplest and most intuitive way of seeing how the Taylor polynomials arise.

An example

Here is a visual example to help make it clear how much the remainder looks like the next term in the Taylor expansion. Let us again use the function $f(x) = \sqrt{x}$ at $a = 100$, but this time just the second-degree approximation instead of the third. According to the generalized microscope equation, the graph of the remainder,

$$R_{2,100}(\Delta x) = f(100 + \Delta x) - P_{2,100}(\Delta x) = \sqrt{100 + \Delta x} - \left(10 + \frac{\Delta x}{20} - \frac{(\Delta x)^2}{8\,000}\right),$$

Macroscopic versus microscopic

should look like the cubic term in $P_{3,100}$, at least if the domain is suitably restricted to a small interval around $\Delta x = 0$. Below we see two views of this graph.

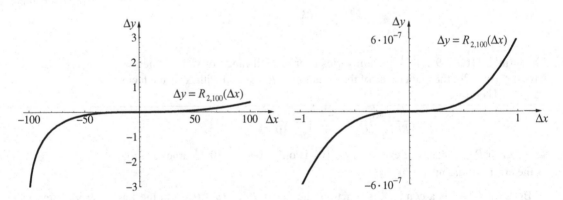

On the left, where it is plotted over a large domain, $-100 < \Delta x < 100$, the graph looks only vaguely like a cubic. It fails to have the necessary symmetry, for example. But on the right, where its domain has been shrunk to $-1 \leq \Delta x \leq 1$ (and the vertical scale has been exaggerated), the graph is now indistinguishable from the graph of the cubic

$$\Delta y = \frac{(\Delta x)^3}{1\,600\,000} = 6.25 \times 10^{-7} \times (\Delta x)^3.$$

So $R_{2,100}(\Delta x)$ does indeed look like the cubic term $(\Delta x)^3/1\,600\,000$ in a sufficiently small window centered at $\Delta x = 0$. Note that we needed to exaggerate the vertical scale; the graph would otherwise have appeared to be just a horizontal line! The vertical scale is linked to the small coefficient $(1/1\,600\,000 \approx 6 \times 10^{-7})$ of $(\Delta x)^3$.

The following corollary says that $|R_{n,a}(\Delta x)|$ is bounded by (a multiple of) $|\Delta x|^{n+1}$. It relies on the continuity of the $(n+1)$st derivative of f, which was one of the original hypotheses of Taylor's theorem. The case $n = 0$ is the ordinary mean-value theorem for f (Theorem 3.1, p. 72).

A bound on $|R_{n,a}(\Delta x)|$

Corollary 3.12 *Let $f^{(n+1)}(a + \Delta x)$ be continuous for $|\Delta x| \leq r$; then*

$$|R_{n,a}(\Delta x)| \leq \max_x |f^{(n+1)}(x)| \frac{|\Delta x|^{n+1}}{(n+1)!},$$

where the maximum is taken over all x between a and $a + \Delta x$, inclusive.

Proof. When $0 \leq t \leq 1$, then $x = a + t\Delta x$ lies between a and $a + \Delta x$, inclusive. The continuous function $f^{(n+1)}(a + t\Delta x)$ has a finite maximum on this closed interval. Therefore we have

$$|R_{n,a}(\Delta x)| = \left| \frac{(\Delta x)^{n+1}}{n!} \int_0^1 f^{(n+1)}(a + t\Delta x)(1 - t)^n\, dt \right|$$

$$\leq \frac{|\Delta x|^{n+1}}{n!} \max_{0 \leq t \leq 1} |f^{(n+1)}(a + t\Delta x)| \int_0^1 (1 - t)^n\, dt$$

$$= \max_x |f^{(n+1)}(x)| \frac{|\Delta x|^{n+1}}{(n+1)!},$$

because the value of the last integral is $1/(n+1)$.   □

With this corollary we finally have a simple and useful way to describe the size of the error and, in particular, how rapidly it vanishes as $\Delta x \to 0$. Here is the basic idea (expressed in terms of a variable t): although any positive power of a variable t *vanishes* (i.e., tends to 0) as $t \to 0$, a higher power vanish more rapidly, or as we say, *to a higher order*. For example, t^3 vanishes to a higher order than t^2 because the quotient t^3/t^2 also vanishes as $t \to 0$. We say that t **vanishes to the first order**, and t^p **vanishes to order p** (for any positive power $p > 0$).

The order to which
a power vanishes

To describe the order of vanishing of an *arbitrary* function $\varphi(t)$ as $t \to 0$, we define what it means for $\varphi(t)$ to vanish to higher order than a given power of t, mimicking the way we compared t^3 and t^2. For the moment, we are concerned with the order of vanishing of $\varphi(t)$ only as $t \to 0$; later in the section we generalize to the case $t \to a$ for a arbitrary.

Vanishing to order
greater than p:
little oh notation

Definition 3.3 *We say $\boldsymbol{\varphi(t)}$ vanishes to order greater than p (at the origin), and write $\boldsymbol{\varphi(t) = o(p)}$, if*

$$\lim_{t \to 0} \frac{\varphi(t)}{t^p} = 0.$$

The symbol "o" is called *little oh* and is meant to suggest the word *order*. Read "$o(p)$" as either "of order greater than p" or just "*little oh* of p."

Using ratios
to compare
orders of vanishing
The condition $\varphi(t) = o(p)$ is imprecise: when $\varphi(t)$ vanishes to higher order than t^p, we do not know how much higher: if $\varphi(t) = o(p)$, then $\varphi(t) = o(q)$ for every $q < p$. To get a more precise condition, let us look more closely at the ratio $\varphi(t)/t^p$ in the case when $\varphi(t) = Ct^m$ $(C \neq 0)$; then, for $0 < p < \infty$, we have

$$\lim_{t \to 0} \frac{\varphi(t)}{t^p} = \begin{cases} 0 & p < m, \\ C & p = m, \\ \infty & p > m. \end{cases}$$

It is evident that we get the most information about $\varphi(t)$ not when the limit is zero but when it takes a finite nonzero value: that is, when $p = m$. Our example suggests that, to gain additional precision about the order of vanishing of an arbitrary $\varphi(t)$, we should focus on the value of p for which the ratio $\varphi(t)/t^p$ has a finite nonzero limit. This is certainly the right idea, but there are two technical stumbling blocks.

Functions can vanish
unlike any power
First, consider the example of $\varphi(t) = 1/\ln|t|$. This vanishes at 0, but there is no $p > 0$ for which the limit $\varphi(t)/t^p$ is finite and nonzero (see the exercises). There is no way around this problem; some functions that do vanish still fail to vanish like any positive power of t.

With several variables,
the ratio usually has
no limit
Second, consider the two-variable function $\varphi(x,y) = x^2 + 2y^2$. (In the next section we extend Taylor's theorem to functions of several variables; the remainder is likewise a function of several variables, and we have to consider its order of vanishing.) It is reasonable to say $\varphi(x,y)$ vanishes to the same order as $x^2 + y^2$; they are both homogeneous quadratic polynomials. However,

$$\lim_{(x,y) \to (0,0)} \frac{x^2 + 2y^2}{x^2 + y^2}$$

does not exist. One way to see this is to note that, on the radial line $y = mx$, $\varphi(x, mx) = (1 + 2m^2)/(1 + m^2)$, a quantity that takes values between 1 and 2 as m varies. Nevertheless, we do have

$$1 \leq \frac{x^2 + 2y^2}{x^2 + y^2} \leq 2 \text{ for all } (x,y) \neq (0,0),$$

and this is enough to guarantee that $x^2 + 2y^2$ and $x^2 + y^2$ each vanishes as rapidly as the other. In fact, it is sufficient if such upper and lower bounds exist for all (x,y) sufficiently close to $(0,0)$.

Vanishing to
the same order
Definition 3.4 *The functions $\varphi(t)$ and $\psi(t)$ vanish to the same order (at the origin) if there are positive constants δ, C_1, and C_2 for which*

$$C_1 \leq \left| \frac{\varphi(t)}{\psi(t)} \right| \leq C_2 \text{ for all } 0 < |t| < \delta.$$

We can rewrite the inequalities one at a time so as to indicate that each function "dominates" the other in a completely symmetric way:

$$|\psi(t)| \leq \frac{1}{C_1} |\varphi(t)|; \quad |\varphi(t)| \leq C_2 |\psi(t)|.$$

According to the first, $\psi(t)$ vanishes at least to the same order as $\varphi(t)$; according to the second, $\varphi(t)$ vanishes at least to the same order as $\psi(t)$. In the following definition, we have only one of these two inequalities, and the comparison is being made with a power.

Definition 3.5 *We say* $\varphi(t)$ *vanishes at least to order p (at the origin), and write* $\varphi(t) = O(p)$, *if there are positive constants* δ, C *for which* $|\varphi(t)| \leq C|t|^p$ *when* $|t| < \delta$. *Otherwise, we say* $\varphi(t)$ *fails to vanish to order p, and write* $\varphi(t) \neq O(p)$.

Vanishing at least to order p: big oh notation

The symbol "O" is called *big oh* and like "o" it is meant to suggest the word *order*. Read "$O(p)$" as either "of order at least p" or as "*big oh of* p." Note the following.

- $\varphi(t) = O(p)$ implies $\varphi(t) = O(\alpha)$ for all $0 < \alpha < p$.

- $\varphi(t) \neq O(p)$ implies $\varphi(t) \neq O(\beta)$ for all $\beta > p$.

- $\varphi(t) = o(p)$ implies $\varphi(t) = O(p)$, but the converse is not true: if $\varphi(t)$ vanishes at least to order p, there is no reason to think $\varphi(t)$ vanishes to higher order than p (e.g., $t^p = O(p)$ but $t^p \neq o(p)$). See also Exercise 3.15.

(We use O and o to indicate "order of vanishing;" however, in other settings they are used to indicate "order of magnitude." We avoid this phrase, though, because *magnitude* implies, etymologically at least, "largeness," not the "smallness" with which we are dealing.)

Big oh notation gives us the right level of precision to describe the order of vanishing of an approximation error. With it we get a convenient and vivid way to rewrite Taylor's formula. The first step is to restate Corollary 3.12 (p. 85) in the new language.

Corollary 3.13 $R_{n,a}(\Delta x) = O(n+1)$. □

Next, we enlarge the meaning of $O(p)$ to allow it to stand for an otherwise unspecified function that vanishes at least to order p (or even allow it to stand for the set of such functions). Then, with this in mind, we can rewrite Taylor's formula in the following simple form that indicates just the order of the remainder:

$O(p)$ as a function; Taylor's formula

$$f(a + \Delta x) = f(a) + f'(a)\Delta x + \frac{f''(a)}{2!}(\Delta x)^2 + \cdots + \frac{f^{(n)}(a)}{n!}(\Delta x)^n + O(n+1).$$

In words: $f(a + \Delta x)$ equals the Taylor polynomial of degree n plus *some unspecified function that vanishes to order* $n+1$ *in* Δx. Often this level of precision is all we need. Consider, for example, the infinite Taylor series for $f(x) = \ln x$ at $a = 1$:

$$\ln(1 + \Delta x)$$

$$= \Delta x - \frac{(\Delta x)^2}{2} + \cdots + (-1)^{n-1} \frac{(\Delta x)^n}{n} + (-1)^n \frac{(\Delta x)^{n+1}}{n+1} + (-1)^{n+1} \frac{(\Delta x)^{n+2}}{n+2} + \cdots.$$

$$\underbrace{\phantom{= \Delta x - \frac{(\Delta x)^2}{2} + \cdots + (-1)^{n-1} \frac{(\Delta x)^n}{n}}}_{P_{n,1}(\Delta x)} \quad \underbrace{\phantom{(-1)^n \frac{(\Delta x)^{n+1}}{n+1} + (-1)^{n+1} \frac{(\Delta x)^{n+2}}{n+2} + \cdots}}_{O(n+1)}$$

The first n terms constitute the Taylor polynomial $P_{n,1}(\Delta x)$; the rest are the remainder $O(n+1)$ in an explicit form. This shows how apt it is to think of $O(n+1)$ as a shorthand for "the terms that vanish at least to order $n+1$."

Functions that agree at least to order p

Definition 3.6 *We say $\varphi(t)$ and $\psi(t)$ agree at least to order p in t, and write $\varphi(t) = \psi(t) + O(p)$, if $\varphi(t) - \psi(t) = O(p)$.*

With this definition, we can put Taylor's formula,

$$f(a + \Delta x) = P_{n,a}(\Delta x) + O(n+1),$$

into these words: "$f(a + \Delta x)$ and $P_{n,a}(\Delta x)$ agree (or are equal) at least to order $n+1$ in Δx when Δx is near 0."

The "best fitting" approximation

Taylor's theorem tells us just half the story about the Taylor polynomial, namely, how well it approximates a given function. The other half of the story is that the Taylor polynomial is unique: no other polynomial of the same degree approximates the function as well. Theorem 3.14, below, explains just what this means.

Lemma 3.1. *If $\varphi(t)/t^p \to \infty$ as $t \to 0$, then $\varphi(t) \neq O(p)$.*

Proof. (By contradiction.) Suppose that $\varphi(t) = O(p)$; then $|\varphi(t)/t^p|$ would be bounded when $t \approx 0$. However, this contradicts the hypothesis that $\varphi(t)/t^p \to \infty$ as $t \to 0$.   □

Theorem 3.14. *Suppose $Q(\Delta x)$ is a polynomial of degree n that differs from the Taylor polynomial $P_{n,a}(\Delta x)$ at least in the term of degree k; then $f(a + \Delta x) - Q(\Delta x)$ fails to vanish to order $k+1$, and hence*

$$f(a + \Delta x) - Q(\Delta x) \neq O(n+1).$$

Proof. The difference $S(\Delta x) = Q(\Delta x) - P_{n,a}(\Delta x)$ is also a polynomial of degree n:

$$S(\Delta x) = a_0 + a_1 \Delta x + \cdots + a_k (\Delta x)^k + a_{k+1} (\Delta x)^{k+1} + \cdots + a_n (\Delta x)^n,$$

and $a_k \neq 0$ by hypothesis. Therefore

$$\frac{S(\Delta x)}{(\Delta x)^{k+1}} = \frac{a_0}{(\Delta x)^{k+1}} + \frac{a_1}{(\Delta x)^k} + \cdots + \frac{a_k}{\Delta x} + a_{k+1} + \cdots + a_n (\Delta x)^{n-k-1},$$

and this becomes infinite as $\Delta x \to 0$ (even if $a_0 = a_1 = \cdots = a_{k-1} = 0$), because $a_k \neq 0$. The error made by using $Q(\Delta x)$ to approximate $f(a + \Delta x)$ is

$$f(a+\Delta x) - Q(\Delta x) = f(a+\Delta x) - P_{n,a}(\Delta x) - S(\Delta x) = R_{n,a}(\Delta x) - S(\Delta x).$$

Therefore

$$\frac{f(a+\Delta x) - Q(\Delta x)}{(\Delta x)^{k+1}} = \frac{R_{n,a}(\Delta x)}{(\Delta x)^{k+1}} - \frac{S(\Delta x)}{(\Delta x)^{k+1}}$$

and, as we have seen, the second term becomes infinite as $t \to 0$. However, the first term remains bounded, because $R_{n,a}(\Delta x) = O(k+1)$ for all $k \le n$. Therefore, the two terms together become infinite. It follows from the lemma that

$$f(a+\Delta x) - Q(\Delta x) \ne O(k+1),$$

and hence $f(a+\Delta x) - Q(\Delta x) \ne O(n+1)$.   □

Corollary 3.15 *If $K \le n$ is the degree of the lowest order term where Q and $P_{n,a}$ differ, then*

$$f(a+\Delta x) - Q(\Delta x) = O(K), \quad f(a+\Delta x) - Q(\Delta x) \ne O(K+1).$$

Proof. Exercise 3.21.   □

Finally, we see how a Taylor polynomial becomes a better approximation as its degree increases. There is one case where this fails to happen, namely when an increase in the degree leaves the polynomial unchanged. For example, with $f(x) = \sin x$, the first- and second-degree Taylor polynomials at the origin are

Comparing $P_{n-1,a}$ and $P_{n,a}$

$$P_1(\Delta x) = \Delta x \text{ and } P_2(\Delta x) = \Delta x,$$

so $P_2(\Delta x)$ will be no better than $P_1(\Delta x)$ in approximating $f(0+\Delta x) = \sin \Delta x$. The problem is that P_2 lacks a quadratic term, because $f''(0) = 0$. The following corollary avoids this case by requiring $f^{(n)}(a) \ne 0$, guaranteeing that the two polynomials are indeed different.

Corollary 3.16 *Suppose $f^{(n)}(a) \ne 0$; then $R_{n,a}(\Delta x)$ vanishes to a higher order than $R_{n-1,a}(\Delta x)$:*

$$R_{n,a}(\Delta x) = O(n+1) \text{ but } R_{n-1,a}(\Delta x) \ne O(n+1).$$

Proof. Take $Q(\Delta x) = P_{n-1,a}(\Delta x)$ in the previous corollary; then $K = n$, because the term of degree n in $Q = P_{n-1,a}$ is 0, but in $R_{n,a}$ it is $f^{(n)}(a)(\Delta x)^n/n! \ne 0$. Therefore $R_{n-1,a}(\Delta x) \ne O(n+1)$.   □

We end with a summary of definitions and results about the order of vanishing of functions at an arbitrary point.

Order of vanishing at an arbitrary point

Definition 3.7 *We say $\varphi(t)$ **vanishes to order greater than p at $t = a$**, and write $\varphi(t) = o(p)$ as $t \to a$, if*

$$\lim_{t \to a} \frac{\varphi(t)}{(t-a)^p} = 0.$$

Thus, $\varphi(t) = o(p)$ is an abbreviation for $\varphi(t) = o(p)$ as $t \to 0$. We continue to use the briefer form unless clarity requires the longer one.

Definition 3.8 *The functions* $\boldsymbol{\varphi(t)}$ *and* $\boldsymbol{\psi(t)}$ *vanish to the same order at $t = a$ if there are positive constants δ, C_1, and C_2 for which*

$$C_1 \le \left| \frac{\varphi(t)}{\psi(t)} \right| \le C_2 \text{ for all } 0 < |t - a| < \delta.$$

Definition 3.9 *We say* $\boldsymbol{\varphi(t)}$ *vanishes at least to order p at $t = a$, and write* $\boldsymbol{\varphi(t) = O(p)}$ *as $t \to a$, if there are positive constants δ, C for which $|\varphi(t)| \le C|t - a|^p$ when $|t - a| < \delta$. Otherwise, we say* $\boldsymbol{\varphi(t)}$ *fails to vanish to order p at $t = a$, and write* $\boldsymbol{\varphi(t) \ne O(p)}$ *as $t \to a$.*

Thus $\varphi(t) = O(p)$ is an abbreviation for $\varphi(t) = O(p)$ as $t \to 0$.

Definition 3.10 *We say* $\boldsymbol{\varphi(t)}$ *and* $\boldsymbol{\psi(t)}$ *agree at least to order p at $t = a$, and write* $\boldsymbol{\varphi(t) = \psi(t) + O(p)}$ *as $t \to a$, if $\varphi(t) - \psi(t) = O(p)$ as $t \to a$.*

3.3 Taylor polynomials in several variables

Taylor polynomials
in two variables

We obtain Taylor polynomials for a function of two variables in a natural way from the one-variable version. However, the formulas are messy and therefore harder to interpret. For example, a polynomial in two or more variables can have several terms of the same degree. The collection of terms of a given degree forms a *homogeneous* polynomial. A homogeneous polynomial of degree k in two variables has the general form

$$Q(x,y) = A_0 x^k + A_1 x^{k-1} y + A_2 x^{k-2} y^2 + \cdots + A_{k-1} x y^{k-1} + A_k y^k$$
$$= \sum_{i+j=k} A_j x^i y^j.$$

The Taylor polynomial of degree n for a function $z = f(x,y)$ at $(x,y) = (a,b)$ consists of terms that are homogeneous polynomials in $\Delta x = x - a$ and $\Delta y = y - b$; there is a homogeneous polynomial of every degree between 0 and n. The terms involve the binomial coefficients

$$\binom{k}{j} = \frac{k!}{j!(k-j)!} = \binom{k}{k-j}$$

and partial derivatives of f. For the sake of visual clarity, we use subscripts to write the partial derivatives (e.g., $\partial^3 f / \partial x^2 \, \partial y = f_{x^2 y}$).

Definition 3.11 *Suppose all partial derivatives of $f(x,y)$ up to order n exist at $(x,y) = (a,b)$; then the **Taylor polynomial of degree n for f at (a,b)** is*

$$P_{n,(a,b)}(\Delta x, \Delta y) = f(a,b) + f_x(a,b)\,\Delta x + f_y(a,b)\,\Delta y$$
$$+ \frac{1}{2!}\left(f_{xx}(a,b)\,(\Delta x)^2 + 2f_{xy}(a,b)\,\Delta x\,\Delta y + f_{yy}(a,b)\,(\Delta y)^2\right)$$
$$+ \cdots + \frac{1}{n!}\sum_{i+j=n}\binom{n}{j}f_{x^i y^j}(a,b)\,(\Delta x)^i\,(\Delta y)^j$$

Theorem 3.17 (Taylor). *If $f(x,y)$ has continuous partial derivatives up to order $n+1$ on an open set that contains the line segment from (a,b) to $(a+\Delta x, b+\Delta y)$, then*

Taylor's formula for functions of two variables

$$f(a+\Delta x, b+\Delta y) = P_{n,(a,b)}(\Delta x, \Delta y) + R_{n,(a,b)}(\Delta x, \Delta y),$$

where $R_{n,(a,b)}(\Delta x, \Delta y)$

$$= \frac{1}{n!}\sum_{i+j=n+1}\binom{n+1}{j}(\Delta x)^i (\Delta y)^j \int_0^1 f_{x^i y^j}(a+t\Delta x, b+t\Delta y)(1-t)^n\, dt.$$

Proof. The idea is to have the two-variable formula emerge from Taylor's formula for a suitably chosen function of one variable. We can assume $(\Delta x, \Delta y) \neq (0,0)$, for otherwise there is nothing to prove. In that case, there is a unique unit vector (α, β) for which $(\Delta x, \Delta y) = s(\alpha, \beta)$ with $s > 0$. Let

$$F(s) = f(a+s\alpha, b+s\beta) = f(a+\Delta x, b+\Delta y).$$

Taylor's formula for $F(s)$ at $s = 0$ is

$$F(s) = F(0) + F'(0)s + \cdots + \frac{F^{(n)}(0)}{n!}s^n + \frac{s^{n+1}}{n!}\int_0^1 F^{(n+1)}(ts)(1-t)^n\, dt.$$

We claim this will turn into Taylor's formula for $f(a+\Delta x, b+\Delta y)$ when we express each derivative of F in terms of α and β and partial derivatives of f. One application of the chain rule gives

Derivatives of $F(s)$

$$F'(s) = f_x(a+s\alpha, b+s\beta)\,\alpha + f_y(a+s\alpha, b+s\beta)\,\beta.$$

A second application gives

$$F''(s) = f_{xx}(a+s\alpha, b+s\beta)\,\alpha^2 + f_{xy}(a+s\alpha, b+s\beta)\,\alpha\beta$$
$$+ f_{yx}(a+s\alpha, b+s\beta)\,\beta\alpha + f_{yy}(a+s\alpha, b+s\beta)\,\beta^2,$$
$$= f_{xx}(a+s\alpha, b+s\beta)\,\alpha^2 + 2f_{xy}(a+s\alpha, b+s\beta)\,\alpha\beta$$
$$+ f_{yy}(a+s\alpha, b+s\beta)\,\beta^2.$$

To get a clearer idea of the patterns being generated here, we calculate one more derivative. Applying the chain rule to each of the functions $f_{xx}(a+s\alpha, b+s\beta)$, $f_{xy}(a+s\alpha, b+s\beta)$, and $f_{yy}(a+s\alpha, b+s\beta)$, we get

$$F'''(s) = f_{xxx}(a+s\alpha, b+s\beta)\,\alpha^3 + f_{xxy}(a+s\alpha, b+s\beta)\,\alpha^2\beta$$
$$+ 2f_{xyx}(a+s\alpha, b+s\beta)\,\alpha^2\beta + 2f_{xyy}(a+s\alpha, b+s\beta)\,\alpha\beta^2$$
$$+ f_{yyx}(a+s\alpha, b+s\beta)\,\alpha\beta^2 + f_{yyy}(a+s\alpha, b+s\beta)\,\beta^3$$
$$= f_{xxx}(a+s\alpha, b+s\beta)\,\alpha^3 + 3f_{xxy}(a+s\alpha, b+s\beta)\,\alpha^2\beta$$
$$+ 3f_{xyy}(a+s\alpha, b+s\beta)\,\alpha^2\beta + f_{yyy}(a+s\alpha, b+s\beta)\,\beta^3.$$

We have used $f_{xyx} = f_{xxy}$, and so forth, to combine terms.

A binomial expansion   For $k = 1$, 2, and 3, the formula for $F^{(k)}(s)$ is a sum of partial derivatives of f in which the numerical coefficients are the binomial coefficients. For an arbitrary k, the formula is

$$F^{(k)}(s) = \sum_{i+j=k} \binom{k}{j} f_{x^i y^j}(a+s\alpha, b+s\beta)\,\alpha^i\beta^j.$$

The next step is to determine the factor $F^{(k)}(0)\,s^k$ that appears in the kth term of the Taylor polynomial for $F(s)$. We have

$$F^{(k)}(0)\,s^k = \sum_{i+j=k} \binom{k}{j} f_{x^i y^j}(a,b)\,s^k\alpha^i\beta^j = \sum_{i+j=k} \binom{k}{j} f_{x^i y^j}(a,b)\,(s\alpha)^i(s\beta)^j$$

$$= \sum_{i+j=k} \binom{k}{j} f_{x^i y^j}(a,b)\,(\Delta x)^i(\Delta y)^j.$$

These expresssions are equal because $s^k\alpha^q\beta^p = s^{i+j}\alpha^i\beta^j = (s^i\alpha^i)(s^j\beta^j)$ and, furthermore, $s\alpha = \Delta x$, $s\beta = \Delta y$.

Determining   At this point we have found all the terms in the Taylor polynomial $P_{n,(a,b)}$. The
$R_{n,(a,b)}(\Delta x, \Delta y)$ final step is to see how the remainder $R_{n,(a,b)}$ emerges from the remainder for $F(s)$. That remainder is

$$\frac{s^{n+1}}{n!} \int_0^1 F^{(n+1)}(ts)(1-t)^n\,dt$$

$$= \frac{1}{n!} \int_0^1 \left(\sum_{i+j=n+1} \binom{n+1}{j} f_{x^i y^j}(a+ts\alpha, b+ts\beta)\,s^{n+1}\alpha^i\beta^j \right)(1-t)^n\,dt$$

$$= \frac{1}{n!} \sum_{i+j=n+1} \binom{n+1}{j}(s\alpha)^i(s\beta)^j \int_0^1 f_{x^i y^j}(a+t\Delta x, b+t\Delta y)(1-t)^n\,dt$$

$$= R_{n,(a,b)}(\Delta x, \Delta y)$$

when we set $(s\alpha)^i = (\Delta x)^i$, $(s\beta)^j = (\Delta y)^j$.   □

Simplified notation   The large and unwieldy expression for the Taylor polynomial of a function of two variables gets even worse when there are more input variables. Before moving on to this, we introduce a simplifying notation for the two-variable polynomial that makes the r-variable case clearer.

The first step is to use vector notation. Thus, we write $\mathbf{x} = (x, y)$, $\mathbf{a} = (a, b)$, $\Delta\mathbf{x} = (\Delta x, \Delta y) = \mathbf{x} - \mathbf{a}$. The second step is to express the various partial derivatives in vector fashion, as well. A familiar example is the vector differential operator "nabla" $\nabla = (\partial/\partial x, \partial/\partial y)$ that is used for the gradient: $\operatorname{grad} f = \nabla f$. The operator we need is the dot product of $\Delta\mathbf{x}$ and ∇:

$$\Delta\mathbf{x} \cdot \nabla = \Delta x \frac{\partial}{\partial x} + \Delta y \frac{\partial}{\partial y}$$

The differential operator $\Delta\mathbf{x} \cdot \nabla$

This operator produces a certain "mixture" of the partial derivatives of any function it operates on:

$$(\Delta\mathbf{x} \cdot \nabla) f(\mathbf{x}) = \Delta x \frac{\partial f}{\partial x}(x, y) + \Delta y \frac{\partial f}{\partial y}(x, y).$$

In particular, $(\Delta\mathbf{x} \cdot \nabla) f(\mathbf{a})$ is just the linear homogeneous (i.e., first-degree) part of the Taylor polynomial for f at $\mathbf{a}$.

To get the homogeneous parts of higher degree, just apply the same differential operator to its previous output. In other words, compose $\Delta\mathbf{x} \cdot \nabla$ with itself, treating Δx and Δy as constants with respect to the partial differential operators $\partial/\partial x$ and $\partial/\partial y$. The resulting operator involves second derivatives:

Composing $\Delta\mathbf{x} \cdot \nabla$ with itself

$$\begin{aligned}
(\Delta\mathbf{x} \cdot \nabla)^2 &= (\Delta\mathbf{x} \cdot \nabla) \circ (\Delta\mathbf{x} \cdot \nabla) \\
&= \left(\Delta x \frac{\partial}{\partial x} + \Delta y \frac{\partial}{\partial y} \right) \circ \left(\Delta x \frac{\partial}{\partial x} + \Delta y \frac{\partial}{\partial y} \right) \\
&= (\Delta x)^2 \frac{\partial^2}{\partial x^2} + 2\Delta x \Delta y \frac{\partial^2}{\partial x \partial y} + (\Delta y)^2 \frac{\partial^2}{\partial y^2},
\end{aligned}$$

You should check that $(\Delta\mathbf{x} \cdot \nabla)^2 f(\mathbf{a})/2!$ is the homogeneous quadratic part of the Taylor polynomial of f at $\mathbf{a}$.

Repeated composition produces operators $(\Delta\mathbf{x} \cdot \nabla)^k$ involving derivatives of order k for any positive integer k. Because $\Delta\mathbf{x} \cdot \nabla$ is a binomial expression, each such power of $\Delta\mathbf{x} \cdot \nabla$ can be expanded as if it were an ordinary binomial:

A binomial expansion

$$(\Delta\mathbf{x} \cdot \nabla)^k = \left(\Delta x \frac{\partial}{\partial x} + \Delta y \frac{\partial}{\partial y} \right)^k = \sum_{i+j=k} \binom{k}{j} (\Delta x)^i (\Delta y)^j \frac{\partial^k}{\partial x^i \partial y^j}.$$

Notice that this is indeed a homogeneous polynomial of degree k in the variables Δx and Δy.

In terms of $\Delta\mathbf{x} \cdot \nabla$, the Taylor polynomial for $f(\mathbf{x})$ at $\mathbf{a}$ is just

Taylor's formula in terms of $\Delta\mathbf{x} \cdot \nabla$

$$P_{n,\mathbf{a}}(\Delta\mathbf{x}) = f(\mathbf{a}) + (\Delta\mathbf{x} \cdot \nabla) f(\mathbf{a}) + \frac{(\Delta\mathbf{x} \cdot \nabla)^2 f(\mathbf{a})}{2!} + \cdots + \frac{(\Delta\mathbf{x} \cdot \nabla)^n f(\mathbf{a})}{n!},$$

a much simpler expression than in the original definition (Definition 3.11)! The formula for the remainder is simplified in the same way:

$$R_{n,\mathbf{a}}(\Delta\mathbf{x}) = \frac{1}{n!}\int_0^1 (\Delta\mathbf{x}\cdot\nabla)^{n+1} f(\mathbf{a}+t\Delta\mathbf{x})(1-t)^n\, dt.$$

Functions of r variables

Let us move, finally, to the case of a function of r variables. As we have seen, the differential operator $\Delta\mathbf{x}\cdot\nabla$ plays the crucial role in the new notation. When there are r variables instead of two, so that

$$\mathbf{x} = (x_1, x_2, \ldots, x_r), \quad \mathbf{a} = (a_1, a_2, \ldots, a_r), \quad \Delta\mathbf{x} = \mathbf{x} - \mathbf{a},$$

our differential operator becomes a multinomial,

$$\Delta\mathbf{x}\cdot\nabla = \Delta x_1 \frac{\partial}{\partial x_1} + \cdots + \Delta x_r \frac{\partial}{\partial x_r},$$

instead of a binomial. Consequently, we can no longer represent the higher powers $(\Delta\mathbf{x}\cdot\nabla)^k$ using the binomial expansion.

A multinomial expansion

However, there is a way to expand multinomials that is exactly analogous to the binomial expansion. It uses the **multinomial coefficients**

$$\binom{k}{p_1 \ p_2 \ \cdots \ p_r} = \frac{k!}{p_1!p_2!\cdots p_r!}, \quad p_1 + p_2 + \cdots + p_r = k;$$

the **multinomial expansion** is

$$(\Delta\mathbf{x}\cdot\nabla)^k = \sum_{p_1+\cdots+p_r=k} \binom{k}{p_1 \ \cdots \ p_r}(\Delta x_1)^{p_1}\cdots(\Delta x_r)^{p_r} \frac{\partial^k}{\partial x_1^{p_1}\cdots\partial x_r^{p_r}}.$$

Taylor's formula for functions of r variables

Theorem 3.18 (Taylor). *If $f(\mathbf{x})$ has $n+1$ continuous derivatives on an open set containing the line segment from $\mathbf{a}$ to $\mathbf{a}+\Delta\mathbf{x}$, then*

$$f(\mathbf{a}+\Delta\mathbf{x}) = \sum_{k=0}^{n} \frac{1}{k!}(\Delta\mathbf{x}\cdot\nabla)^k f(\mathbf{a}) + R_{n,\mathbf{a}}(\Delta\mathbf{x}),$$

where $R_{n,\mathbf{a}}(\Delta\mathbf{x}) = \dfrac{1}{n!}\displaystyle\int_0^1 (\Delta\mathbf{x}\cdot\nabla)^{n+1} f(\mathbf{a}+t\Delta\mathbf{x})(1-t)^n\, dt.$ $\qquad\square$

Forms of the remainder

The remainder $R_{n,\mathbf{a}}(\Delta\mathbf{x})$ can be rewritten in different forms, just as in the one-variable case. The proofs are the same as the one-variable versions.

Corollary 3.19 (Lagrange's form of the remainder) *For each $\Delta\mathbf{x}\approx 0$, there is a $\theta = \theta(\Delta\mathbf{x})$ with $0 \le \theta \le 1$ for which*

$$R_{n,\mathbf{a}}(\Delta\mathbf{x}) = \frac{1}{(n+1)!}(\Delta\mathbf{x}\cdot\nabla)^{n+1} f(\mathbf{a}+\theta\Delta\mathbf{x}). \qquad\square$$

The next corollary asserts that the remainder for the Taylor polynomial of degree n is approximately the highest-degree homogeneous part of the Taylor polynomial of degree $n+1$.

Corollary 3.20 (Generalized microscope equation) *When* $\Delta\mathbf{x} \approx \mathbf{0}$,

$$R_{n,\mathbf{a}}(\Delta\mathbf{x}) \approx \frac{1}{(n+1)!}(\Delta\mathbf{x} \cdot \nabla)^{n+1} f(\mathbf{a}).$$ □

Also as in the one-variable case, the Taylor polynomial provides the "best fit" to a function near a given point, among all polynomials of the same degree. To see this, we use the same device we employed in the one-variable case, namely, the order of vanishing of the remainder. The definitions are analogous. Let

$$\mathbf{t} = (t_1,\ldots,t_r), \quad \|\mathbf{t}\| = \sqrt{t_1^2 + \cdots + t_r^2},$$

and suppose $z = \varphi(\mathbf{t})$ is a real-valued function that vanishes at the origin: $\varphi(\mathbf{0}) = 0$. You can extend these definitions to the case where φ vanishes at an arbitrary point $\mathbf{a}$, as on pages 89–90.

Definition 3.12 *We say the function* $\boldsymbol{\varphi}(\mathbf{t})$ *vanishes to order greater than* $\boldsymbol{p}$, *and write* $\boldsymbol{\varphi}(\mathbf{t}) = o(p)$, *if*

$$\lim_{\mathbf{t}\to 0} \frac{\varphi(\mathbf{t})}{\|\mathbf{t}\|^p} = 0.$$

Order of vanishing of a multivariable function

Definition 3.13 *We say* $\boldsymbol{\varphi}(\mathbf{t})$ *vanishes to order at least* $\boldsymbol{p}$, *and write* $\boldsymbol{\varphi}(\mathbf{t}) = O(p)$, *if there are positive constants* δ, C *for which* $|\varphi(\mathbf{t})| \leq C\|\mathbf{t}\|^p$ *when* $\|\mathbf{t}\| < \delta$. *Otherwise, we say* $\boldsymbol{\varphi}(\mathbf{t})$ *fails to vanish to order* $\boldsymbol{p}$, *and write* $\boldsymbol{\varphi}(\mathbf{t}) \neq O(p)$.

For example, any linear function $z = L(\mathbf{t}) = m_1 t_1 + \cdots + m_r t_r$ vanishes at least to order 1: $|L(\mathbf{t})| \leq C\|\mathbf{t}\|$, for some C. The graph of $z = C\|\mathbf{t}\|$ is a cone, whereas the graph of $z = |L(\mathbf{t})|$ resembles a (hyper)plane that has been folded upward along the set where $L(\mathbf{t}) = 0$. The cone can always be elongated enough (by increasing C) to make the folded hyperplane lie below it.

A linear function vanishes at least to first order

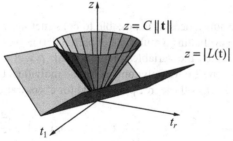

Corollary 3.21 $R_{n,\mathbf{a}}(\Delta\mathbf{x}) = O(n+1)$.

Proof. The proof of Taylor's theorem (Theorem 3.17) used the one-variable function

$$F(s) = f(\mathbf{a} + s\mathbf{u}) = f(\mathbf{a} + \Delta\mathbf{x}),$$

where $\mathbf{u}$ is a unit vector, $s > 0$, and $s\mathbf{u} = \Delta\mathbf{x}$. (The theorem was stated and proved when $\Delta\mathbf{x}$ was 2-dimensional, but nothing needs to be changed in higher dimensions.)

In the proof, Taylor's formula for $F(s)$ at $s = 0$ became Taylor's formula for $f(\mathbf{x})$ at $\mathbf{x} = \mathbf{a}$. In particular, the remainder $R_{n,\mathbf{a}}(\Delta\mathbf{x})$ was just the remainder $R_{n,0}(s)$ for $F(s)$. We know $R_{n,0}(s)$ vanishes at least to order $n+1$ in s, so there are positive constants δ and C for which $|R_{n,0}(s)| \leq C|s|^{n+1}$ when $|s| < \delta$. But $s = \|\Delta\mathbf{x}\|$, so

$$|R_{n,\mathbf{a}}(\Delta\mathbf{x})| = |R_{n,0}(s)| \leq C|s|^{n+1} = C\|\Delta\mathbf{x}\|^{n+1}$$

when $\|\Delta\mathbf{x}\| < \delta$.   □

The Taylor polynomial provides the "best fit"

Thus, $f(\mathbf{a} + \Delta\mathbf{x})$ agrees with its Taylor polynomial $P_{n,\mathbf{a}}(\Delta\mathbf{x})$ at least up to order $n+1$ in $\Delta\mathbf{x}$. There is no other polynomial of degree n for which this is true; according to the following theorem (which mimics the one-variable case), the agreement is always of lower order.

Theorem 3.22. *Suppose $Q(\Delta\mathbf{x})$ is a polynomial of degree n that differs from the Taylor polynomial $P_{n,\mathbf{a}}(\Delta\mathbf{x})$ at least in some term of degree $k \leq n$; then*

$$f(\mathbf{a} + \Delta\mathbf{x}) - Q(\Delta\mathbf{x}) \neq O(k+1).$$

Proof. We can use the idea of the proof of the last corollary. Write $\Delta\mathbf{x} = s\mathbf{u}$ for a suitable $s > 0$ and unit vector $\mathbf{u}$. Then the one-variable function $q(s) = Q(\Delta\mathbf{x}) = Q(s\mathbf{u})$ is a polynomial of degree n.

Let $p_{n,0}(s) = P_{n,\mathbf{a}}(s\mathbf{u})$; then $p_{n,0}(s)$ is the Taylor polynomial of degree n for $F(s) = f(\mathbf{a} + s\mathbf{u})$. Therefore, $p_{n,0}(s)$ and $q(s)$ differ at least in the term of degree k, implying

$$F(s) - q(s) \neq O(k+1)$$

(as functions of s) by Theorem 3.14. Because $F(s) = f(\mathbf{a} + \Delta\mathbf{x})$ and $q(s) = Q(\Delta\mathbf{x})$, we have

$$f(\mathbf{a} + \Delta\mathbf{x}) - Q(\Delta\mathbf{x}) \neq O(k+1)$$

as functions of $\Delta\mathbf{x}$.   □

The Taylor polynomial of certain products

In certain circumstances, it is possible to construct the Taylor polynomial more directly, without evaluating a multitude of partial derviatives. For example, if $f(x,y)$ is a product in which the variables are separated, $f(x,y) = g(x)h(y)$, we can just multiply together the Taylor polynomials for the individual factors g and h. To illustrate, we construct the 4th-degree polynomial for $e^x \cos y$ at $(x,y) = (0,0)$ from

$$e^x = 1 + x + \frac{x^2}{2} + \frac{x^3}{3!} + \frac{x^4}{4!} + O(5), \quad \cos y = 1 - \frac{y^2}{2} + \frac{y^4}{4!} + O(6).$$

Now just distribute the terms of $\cos y$ over the terms of e^x:

$$e^x \cos y = \left(1 + x + \frac{x^2}{2} + \frac{x^3}{6} + \frac{x^4}{24} + O(5)\right) \times 1$$

$$- \left(1 + x + \frac{x^2}{2} + \frac{x^3}{6} + \frac{x^4}{24} + O(5)\right) \times \frac{y^2}{2}$$

$$+ \left(1 + x + \frac{x^2}{2} + \frac{x^3}{6} + \frac{x^4}{24} + O(5)\right) \times \frac{y^4}{24}$$

$$+ \left(1 + x + \frac{x^2}{2} + \frac{x^3}{6} + \frac{x^4}{24} + O(5)\right) \times O(6)$$

$$= 1 + x + \frac{x^2}{2} + \frac{x^3}{6} + \frac{x^4}{24} - \frac{y^2}{2} - \frac{xy^2}{2} - \frac{x^2 y^2}{4} + \frac{y^4}{24} + O(5)$$

$$= 1 + x + \frac{x^2 - y^2}{2!} + \frac{x^3 - 3xy^2}{3!} + \frac{x^4 - 6x^2 y^2 + y^4}{4!} + O(5).$$

All of the individual products (e.g., $x^3 y^2 / 12$) that do not appear explicitly in the last two lines have been absorbed into the symbol $O(5)$ because they vanish at least to order 5. You should check that this agrees with the definition of $P_{4,(0,0)}(x,y)$ for $e^x \cos y$; see Exercise 3.25.

The last possibility we need to consider is a nonlinear map $\mathbf{f} : V^p \to \mathbb{R}^q$, where V^p is an open set in $\mathbb{R}^p$,

Taylor polynomials for vector-valued functions

$$\mathbf{f} : \begin{cases} x_1 = f_1(v_1, \ldots, v_p), \\ x_2 = f_2(v_1, \ldots, v_p), \\ \vdots \\ x_q = f_q(v_1, \ldots, v_p). \end{cases}$$

This is a vector-valued function, $\mathbf{x} = \mathbf{f}(\mathbf{v})$, and the Taylor polynomial of degree n at $\mathbf{v} = \mathbf{a}$ for $\mathbf{f}$ is just the vector of Taylor polynomials for the individual component functions $f_i(\mathbf{v})$. That is, let $P_{i;n,\mathbf{a}}(\Delta \mathbf{v})$ be the Taylor polynomial of degree n for $f_i(\mathbf{v})$ at $\mathbf{v} = \mathbf{a}$. Then the polynomial map $\mathbf{P}_{n,\mathbf{a}} : \mathbb{R}^p \to \mathbb{R}^q$,

$$\mathbf{P}_{n,\mathbf{a}} : \begin{cases} x_1 = P_{1;n,\mathbf{a}}(\Delta v_1, \ldots, \Delta v_p), \\ x_2 = P_{2;n,\mathbf{a}}(\Delta v_1, \ldots, \Delta v_p), \\ \vdots \\ x_q = P_{q;n,\mathbf{a}}(\Delta v_1, \ldots, \Delta v_p), \end{cases}$$

is the Taylor polynomial for the map $\mathbf{f}$ at the point $\mathbf{v} = \mathbf{a}$. Likewise, the remainder is the vector of the corresponding remainder functions $x_i = R_{i;n,\mathbf{a}}(\Delta \mathbf{v})$. It is the map $\mathbf{R}_{n,\mathbf{a}} : V_0^p \to \mathbb{R}^q$, where V_0^p is a suitable open neighborhood of $\mathbf{0}$:

$$\mathbf{R}_{n,\mathbf{a}} : \begin{cases} x_1 = R_{1;n,\mathbf{a}}(\Delta v_1, \dots, \Delta v_p), \\ x_2 = R_{2;n,\mathbf{a}}(\Delta v_1, \dots, \Delta v_p), \\ \quad \vdots \\ x_q = R_{q;n,\mathbf{a}}(\Delta v_1, \dots, \Delta v_p). \end{cases}$$

Taylor's formula with remainder

Taylor's formula then holds for the maps themselves:

$$\mathbf{f}(\mathbf{a} + \Delta\mathbf{v}) = \mathbf{P}_{n,\mathbf{a}}(\Delta\mathbf{v}) + \mathbf{R}_{n,\mathbf{a}}(\Delta\mathbf{v}).$$

We can even describe the order of vanishing of the remainder. Suppose $\mathbf{\Phi} : T^p \to \mathbb{R}^q$ is a vector-valued function that vanishes at the origin: $\mathbf{\Phi}(\mathbf{0}) = \mathbf{0}$.

Order of vanishing of a vector-valued function

Definition 3.14 *We say the function $\mathbf{\Phi}(\mathbf{t})$ **vanishes to order greater than** p, and write $\mathbf{\Phi}(\mathbf{t}) = o(p)$, if*

$$\lim_{\mathbf{t}\to 0} \frac{\|\mathbf{\Phi}(\mathbf{t})\|}{\|\mathbf{t}\|^p} = 0.$$

Definition 3.15 *We say $\mathbf{\Phi}(\mathbf{t})$ **vanishes at least to order** p, and write $\mathbf{\Phi}(\mathbf{t}) = O(p)$, if there are positive constants δ, C for which $\|\mathbf{\Phi}(\mathbf{t})\| \leq C\|\mathbf{t}\|^p$ when $\|\mathbf{t}\| < \delta$. Otherwise, we say $\mathbf{\Phi}(\mathbf{t})$ **fails to vanish to order** p, and write $\mathbf{\Phi}(\mathbf{t}) \neq O(p)$.*

Theorem 3.23. $\mathbf{R}_{n,\mathbf{a}}(\Delta\mathbf{v}) = O(n+1)$.

Proof. We must show there are positive numbers δ, C for which

$$\|\mathbf{R}_{n,\mathbf{a}}(\Delta\mathbf{v})\| \leq C\|\Delta\mathbf{v}\|^{n+1} \text{ when } \|\Delta\mathbf{v}\| < \delta.$$

Each component of $\mathbf{R}_{n,\mathbf{a}}(\Delta\mathbf{v})$ is just a real-valued function, so we know it vanishes at least to order $n+1$ (Taylor's theorem for real-valued functions, Theorem 3.18, and Corollary 3.21). Hence, for each $i = 1, \dots, q$, there are positive numbers δ_i, C_i for which

$$|R_{i;n,\mathbf{a}}(\Delta\mathbf{v})| \leq C_i\|\Delta\mathbf{v}\|^{n+1} \text{ when } \|\Delta\mathbf{v}\| < \delta_i.$$

All the inequalities remain true when we take $\|\Delta\mathbf{v}\| < \delta$, where δ is the smallest of $\delta_1, \dots, \delta_q$.

For the magnitude of the vector-valued function $\mathbf{R}_{n,\mathbf{a}}(\Delta\mathbf{v})$ we have

$$\|\mathbf{R}_{n,\mathbf{a}}(\Delta\mathbf{v})\|^2 = |R_{1;n,\mathbf{a}}(\Delta\mathbf{v})|^2 + \dots + |R_{q;n,\mathbf{a}}(\Delta\mathbf{v})|^2$$
$$\leq C_1^2\|\Delta\mathbf{v}\|^{2(n+1)} + \dots + C_q^2\|\Delta\mathbf{v}\|^{2(n+1)}$$

when $\|\Delta\mathbf{v}\| < \delta$. Therefore, if we set $C = \sqrt{C_1^2 + \dots C_q^2}$, then

$$\|\mathbf{R}_{n,\mathbf{a}}(\Delta\mathbf{v})\| \leq C\|\Delta\mathbf{v}\|^{n+1}. \qquad \square$$

The Taylor polynomial map of degree 1

For our future work, the Taylor polynomial map of degree 1 is the most important. In terms of components, it is

$$\mathbf{P}_{1,\mathbf{a}} : \begin{cases} x_1 = f_1(\mathbf{a}) + \dfrac{\partial f_1}{\partial v_1}(\mathbf{a})\Delta v_1 + \cdots + \dfrac{\partial f_1}{\partial v_p}(\mathbf{a})\Delta v_p, \\[2ex] x_2 = f_2(\mathbf{a}) + \dfrac{\partial f_2}{\partial v_1}(\mathbf{a})\Delta v_1 + \cdots + \dfrac{\partial f_2}{\partial v_p}(\mathbf{a})\Delta v_p, \\[2ex] \quad\vdots \\[1ex] x_q = f_q(\mathbf{a}) + \dfrac{\partial f_q}{\partial v_1}(\mathbf{a})\Delta v_1 + \cdots + \dfrac{\partial f_q}{\partial v_p}(\mathbf{a})\Delta v_p. \end{cases}$$

The initial constant terms are the components of the vector $\mathbf{f}(\mathbf{a})$. The remaining terms are linear in $\Delta v_1, \ldots, \Delta v_p$; they are naturally represented by a linear map acting on the vector $\Delta \mathbf{v}$:

The derivative of $\mathbf{f}$

$$\mathrm{df}_{\mathbf{a}}(\Delta\mathbf{v}) = \begin{pmatrix} \dfrac{\partial f_1}{\partial v_1}(\mathbf{a}) & \cdots & \dfrac{\partial f_1}{\partial v_p}(\mathbf{a}) \\[2ex] \vdots & & \vdots \\[2ex] \dfrac{\partial f_q}{\partial v_1}(\mathbf{a}) & \cdots & \dfrac{\partial f_q}{\partial v_p}(\mathbf{a}) \end{pmatrix} \begin{pmatrix} \Delta v_1 \\ \vdots \\ \Delta v_p \end{pmatrix}.$$

Definition 3.16 *The **derivative** of the map* $\mathbf{f} : V^p \to \mathbb{R}^q$ *at* $\mathbf{a}$ *is the linear map* $\mathrm{df}_{\mathbf{a}} : \mathbb{R}^p \to \mathbb{R}^q$ *that is represented by the* $q \times p$ *matrix with components* $\partial f_i / \partial v_j(\mathbf{a})$.

In terms of the derivative, the Taylor polynomial map of degree 1 for $\mathbf{f}$ at $\mathbf{a}$ is

$$\mathbf{P}_{1,\mathbf{a}}(\Delta\mathbf{v}) = \mathbf{f}(\mathbf{a}) + \mathrm{df}_{\mathbf{a}}(\Delta\mathbf{v}),$$

and Taylor's formula is

$$\mathbf{f}(\mathbf{a} + \Delta\mathbf{v}) = \mathbf{f}(\mathbf{a}) + \mathrm{df}_{\mathbf{a}}(\Delta\mathbf{v}) + O(2).$$

In the next chapter we study in detail how $\mathrm{df}_{\mathbf{a}}$ approximates $\mathbf{f}$ near $\mathbf{a}$.

Here are two examples of Taylor approximations to maps. The first map is the polar coordinate change

Examples

$$\mathbf{f} : \begin{cases} x = r\cos\theta, \\ y = r\sin\theta. \end{cases}$$

At the point $(r, \theta) = (r_0, 0)$ (so $\Delta r = r - r_0$, $\Delta\theta = \theta$), the Taylor polynomial map of degree 3 is

$$\mathbf{f} : \begin{cases} x = r_0 + \Delta r - \dfrac{r_0}{2}(\Delta\theta)^2 - \dfrac{1}{2}(\Delta r)(\Delta\theta)^2 + O(4), \\[2ex] y = r_0\Delta\theta + (\Delta r)(\Delta\theta) - \dfrac{r_0}{6}(\Delta\theta)^3 + O(4). \end{cases}$$

Notice that the polynomial terms are just the products of $r = r_0 + \Delta r$ with the familiar Taylor polynomials for the cosine and sine functions.

The second map is

$$\mathbf{g}: \begin{cases} x = u^3 - 3uv^2, \\ y = 3u^2v - v^3. \end{cases}$$

The derivative of $\mathbf{g}$ at the point $\mathbf{a} = (a,b)$ is given by the matrix

$$d\mathbf{g}_{(a,b)} = \begin{pmatrix} 3u^2 - 3v^2 & -6uv \\ 6uv & 3u^2 - 3v^2 \end{pmatrix}\Bigg|_{(u,v)=(a,b)} = \begin{pmatrix} 3a^2 - 3b^2 & -6ab \\ 6ab & 3a^2 - 3b^2 \end{pmatrix}.$$

The determinant of $d\mathbf{g}_\mathbf{a}$ has the simple form

$$\det(d\mathbf{g}_\mathbf{a}) = (3a^2 - 3b^2)^2 + (6ab)^2 = 9a^4 - 18a^2b^2 + 9b^4 + 36a^2b^2 = 9(a^2 + b^2)^2;$$

thus $d\mathbf{g}_\mathbf{a}$ is invertible for all $\mathbf{a} \neq \mathbf{0}$.

Exercises

3.1. Determine the mean value of each of the following functions on the given domain.

    a. $f(x) = x^n$ on $[0,1]$.

    b. $f(x) = \sin x$ on $[0,\pi]$.

    c. $f(x,y) = x^2 + y^2$ on $x^2 + y^2 \le 1$. (Suggestion: Use polar coordinates.)

3.2. a. Let R be the rectangle $[a,b] \times [c,d]$ in the (x,y)-plane. Determine the coordinates (ξ, η) of the point at the center of R.

    b. Show that the mean value of $f(x,y) = \alpha x + \beta y + \gamma$ on R is the value $f(\xi, \eta) = \alpha\xi + \beta\eta + \gamma$ of f at the center of the rectangle.

3.3. Show that the mean value of a linear function on a circular disk in the (x,y)-plane is its value at the center of the disk.

3.4. Find c in $[0,1]$ so that $\displaystyle\int_0^1 x^n\, dx = c^n$.

3.5. Find c in $[0,\pi]$ so that $\displaystyle\int_0^\pi \sin x\, dx = \pi \sin c$.

3.6. Assume $0 < a < b$. Find c for which

$$\int_a^b x^n\, dx = c^n(b-a),$$

and show that c lies between a and b.

3.7. Find a point (α, β) in the unit disk for which

$$\iint\limits_{x^2+y^2\leq1} (x^2+y^2)\,dx\,dy = \pi(\alpha^2+\beta^2).$$

3.8. a. Let $f(x) = \sqrt{r^2-x^2}$, $-r \leq x \leq r$. Show that

$$\int_{-r}^{r} f(x)\,dx = 2r f(c),$$

where $c = \pm r\sqrt{1-\pi^2/16}$. (Suggestion: Evaluate the integral and use that value to find c.)

b. Sketch the graph $y = f(x)$ on the interval $-r \leq x \leq r$. In the sketch, mark one of the points c and sketch the horizontal graph $y = f(c)$ to make a rectangle over the interval $[-r,r]$. Does this rectangle appear to contain the same area as the semicircular graph of $f(x)$ over $[-r,r]$? Is this what you expect?

3.9. a. Prove the generalized integral law of the mean when the condition $g(x) \geq 0$ on $[a,b]$ has been changed to $g(x) \leq 0$.

b. Suppose $f(x) = g(x) = x$. Show that there is no c in the interval $[-1,1]$ for which

$$\int_{-1}^{1} f(x)g(x)\,dx = f(c)\int_{-1}^{1} g(x)\,dx.$$

Why does this not contradict the generalized integral law of the mean?

3.10. Prove the law of the mean for double integrals (Theorem 3.7). Why does the domain D have to be connected?

3.11. a. Obtain estimates for the numerical values of $\sqrt{102}$ and $\sqrt{120}$ using the Taylor polynomials $P_{n,100}(\Delta x)$ of degree $n = 2, 3, 4$, and 5 for $f(x) = \sqrt{x}$ at $a = 100$. Use these values to verify that, for each n, the error estimating $\sqrt{102}$ is only about $1/10^{n+1}$ times the size of the error estimating $\sqrt{120}$.

b. For each $n = 2, 3, 4$, and 5, sketch the graph of the remainder function $y = R_{n,100}(\Delta x) = \sqrt{100+\Delta x} - P_{n,100}(\Delta x)$ on the interval $-1 \leq \Delta x \leq 1$. How does your graph indicate that $R_{n,100}(\Delta x) = O(n+1)$?

c. For each $n = 2, 3, 4$, and 5, there is a C_n for which

$$R_{n,100}(\Delta x) = \sqrt{100+\Delta x} - P_{n,100}(\Delta x) \approx C_n(\Delta x)^{n+1}.$$

Determine C_n and sketch $R_{n,100}(\Delta x)$ and $C_n(\Delta x)^{n+1}$ together on the same axes to indicate that $R_{n,100}(\Delta x) = O(n+1)$. In each case, take $-1 \leq \Delta x \leq 1$.

3.12. a. Construct the Taylor polynomials $P_{n,1}(\Delta x)$ centered at $a = 1$ for $f(x) = \ln x$; take $n = 1, 2, 3$, and 4.

b. Obtain estimates for $\ln 1.02$ and $\ln 1.2$ using the four different Taylor polynomials you found in part (a), and determine the error in each of these estimates. Is the error that $P_{n,1}$ makes for $\ln 1.02$ only about $1/10^{n+1}$ times the size of the error the same polynomial makes for $\ln 1.2$? Explain.

c. For each $n = 1, 2, 3$, and 4, sketch the graph of the function $y = R_{n,1}(\Delta x) =$ $\ln(1 + \Delta x) - P_{n,1}(\Delta x)$ on the interval $-0.3 \leq \Delta x \leq 0.3$. Does your graph demonstrate that $R_{n,1}(\Delta x) = O(n+1)$? How, or why not?

3.13. Prove the induction step

$$R_{k,a}(\Delta x) = \frac{(\Delta x)^{k+1}}{k!} \int_0^1 f^{(k+1)}(a + t\Delta x)(1-t)^k \, dt$$

$$= \frac{f^{(k+1)}(a)}{(k+1)!} (\Delta x)^{k+1} + \frac{(\Delta x)^{k+2}}{(k+1)!} \int_0^1 f^{(k+2)}(a + t\Delta x)(1-t)^{k+1} \, dt$$

$$= \frac{f^{(k+1)}(a)}{(k+1)!} (\Delta x)^{k+1} + R_{k+1,a}(\Delta x)$$

in the proof of Taylor's theorem.

3.14. Prove l'Hôpital's rule in the following form. Suppose $f(a) = f'(a) = \cdots = f^{(n-1)}(a) = 0$, $g(a) = g'(a) = \cdots = g^{(n-1)}(a) = 0$, and either $f^{(n)}(a) \neq 0$ or $g^{(n)}(a) \neq 0$ (or both); then

$$\lim_{x \to a} \frac{f(x)}{g(x)} = \begin{cases} \infty & \text{if } g^{(n)}(a) = 0, \\ \dfrac{f^{(n)}(a)}{g^{(n)}(a)} & \text{otherwise.} \end{cases}$$

(Suggestion: Use Taylor's formula with Lagrange's form of the remainder, for both $f(a + \Delta x)$ and $g(a + \Delta x)$.)

3.15. Let $\varphi(t) = t^\alpha$, $1 < \alpha < 2$. Show that $\varphi(t) = o(1)$ but $\varphi(t) \neq O(2)$.

3.16. Use the fact that e^x grows faster than any positive power of x (i.e., $x^p/e^x \to 0$ as $x \to +\infty$ for any $p > 0$) to show that $\psi(u) = \exp(-1/|u|)$ vanishes to order greater than p for any $p > 0$. It follows that we can define $\psi(0) = 0$. Sketch the graph of $t = \psi(u)$ and determine the image of ψ on the t-axis.

3.17. Show that the condition $\varphi(t) = o(p)$ can be expressed in the following way. Given any $\varepsilon > 0$, there is a $\delta > 0$ so that

$$|t| < \delta \implies |\varphi(t)| \leq \varepsilon |t|^p.$$

(Suggestion: The fact that $\varphi(t)/t^p \to 0$ as $t \to 0$ means that $|\varphi(t)/t^p|$ can be made smaller than any preassigned $\varepsilon > 0$ by making $|t|$ sufficiently small, i.e., by making $|t| < \delta$ for some suitable δ.)

Note: Although this formulation of "little oh" may seem more cumbersome, it has the advantage of avoiding using a quotient, a useful feature in some of our later work (cf. p. 133). This formulation is also more like our definition of "big oh."

3.18. Let $\varphi(t) = -1/\ln t$, $0 < t < 1$, $\varphi(0) = 0$; the graph of φ appears in the margin. The goal of this exercise is to show that $\varphi(|t|) = -1/\ln|t|$ vanishes to order less than p for any $p > 0$. This is true if, for a given $p > 0$ and for every θ sufficiently small, $t^p < \varphi(t)$ for every $0 < t < \theta$ (the contrapositive of Exercise 3.17).

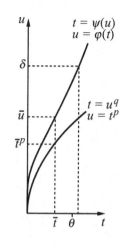

a. Show that $t = \psi(u)$ (Exercise 3.16) is invertible on $u \geq 0$, and show that $u = \varphi(t)$ is the inverse.

b. Fix $p > 0$, set $q = 1/p$, and choose $\delta > 0$ so that $|u| < \delta \implies \psi(u) < |u|^q$ (as provided by Exercise 3.17 with any $\varepsilon < 1$). Now take any $0 < \theta < \psi(\delta)$ and let $0 < \bar{t} < \theta$ be arbitrary. Set $\bar{u} = \varphi(\bar{t})$ and show that $\bar{t}^p < \bar{u} = \varphi(\bar{t})$.

3.19. Let $f(x) = \sqrt{x}$; show that

$$f^{(k+1)}(x) = \frac{\pm 1 \cdot 3 \cdots (2k-1)}{2^{k+1} x^{k+1/2}}.$$

3.20. In the microscope equation, $\Delta y \approx f'(a) \Delta x$, the nature of the approximation is unclear. Show that the microscope equation has the more explicit form $\Delta y = f'(a) \Delta x + O(2)$; in words, "$\Delta y$ agrees with $f'(a) \Delta x$ at least up to order 2 in Δx."

3.21. Adapt the proof of Theorem 3.14 to prove Corollary 3.15.

3.22. The purpose of this exercise is to show that the extent to which a polynomial approximates a given function near a given point depends on the extent to which it matches the Taylor polynomial constructed at that point (cf. Theorem 3.14 and Corollary 3.15).

a. Show that $P(x) = 1 + x + \frac{1}{2}x^2 + \frac{1}{6}x^3$ is the Taylor polynomial of degree 3 at $x = 0$ for the function e^x.

b. Sketch the graph of $y = R(x) = e^x - P(x)$ in a small neighborhood of $x = 0$ to demonstrate that $R(x) = O(4)$, as required by Taylor's theorem.

c. Sketch the graph of $y = V_1(x) = e^x - (1 + x + \frac{1}{2}x^2 + \frac{1}{3}x^3)$ in a small neighborhood of $x = 0$. Determine the value of p for which $V_1(x) = O(p)$ and $V_1(x) \neq O(p+1)$.

d. Sketch the graph of $y = V_2(x) = e^x - (1 + x + \frac{1}{3}x^2 + \frac{1}{6}x^3)$ in a small neighborhood of $x = 0$. Determine the value of p for which $V_2(x) = O(p)$ and $V_2(x) \neq O(p+1)$.

e. Sketch the graph of $y = V_3(x) = e^x - (1 + 1.1x + \frac{1}{2}x^2 + \frac{1}{6}x^3)$ in a small neighborhood of $x = 0$. Determine the value of p for which $V_3(x) = O(p)$ and $V_3(x) \neq O(p+1)$.

f. Sketch the graph of $y = V_4(x) = e^x - (1.1 + x + \frac{1}{2}x^2 + \frac{1}{6}x^3)$ in a small neighborhood of $x = 0$. Determine the value of p for which $V_4(x) = O(p)$ and $V_4(x) \neq O(p+1)$.

3.23. Write the Taylor polynomial of degree 2 for the given function at the given point.
    a. $e^x \sin y$ at $(0,0)$ d. $\ln(x^2 + y^2)$ at $(1,0)$
    b. $\cos x \cos y$ at $(0, \pi/2)$ e. xyz at $(1, -2, 4)$
    c. $x^3 - 3x + y^2$ at $(-1,0)$ f. $1 - \cos\theta + \frac{1}{2}v^2$ at $(\pi, 0)$

3.24. Write the Taylor polynomial of degree 4 for $(x^2 + y^2)^2 - (x^2 + y^2)$ at the point $(x,y) = (1/2, 1/2)$.

3.25. Show that the Taylor polynomial of degree 4 for $e^x \cos y$ at $(x,y) = (0,0)$, as obtained from the definition, agrees with the computation done on page 96.

3.26. Write out in words what "$O(p) \cdot O(q) = O(p+q)$" means, and prove it.

3.27. Construct the Taylor polynomial of degree 2 centered at the point $(\rho, \theta, \varphi) = (\rho_0, \pi/2, 0)$ for the *spherical coordinate change*

$$\mathbf{s}: \begin{cases} x = \rho \cos\theta \cos\varphi, \\ y = \rho \sin\theta \cos\varphi, \\ z = \rho \sin\varphi; \end{cases} \quad \begin{array}{l} -\pi \leq \theta \leq \pi, \\ -\pi/2 \leq \varphi \leq \pi/2. \end{array}$$

3.28. a. Suppose $\mathbf{L}: \mathbb{R}^p \to \mathbb{R}^q$ is linear; show $\mathbf{L}(\Delta\mathbf{u})$ vanishes at least to first order in $\Delta\mathbf{u}$. In fact, show there is a positive number C for which $\|\mathbf{L}(\Delta\mathbf{u})\| \leq C\|\Delta\mathbf{u}\|$ for all $\Delta\mathbf{u}$.

    b. The smallest number C for which this inequality holds is called the **norm** of the linear map $\mathbf{L}$, written $\|\|\mathbf{L}\|\|$. It follows that $\|\mathbf{L}(\Delta\mathbf{u})\| \leq \|\|\mathbf{L}\|\| \, \|\Delta\mathbf{u}\|$ for all $\Delta\mathbf{u}$. Show that

$$\|\|\mathbf{L}\|\| = \max_{\|\Delta\mathbf{u}\|=1} \|\mathbf{L}(\Delta\mathbf{u})\|.$$

    c. Suppose the linear map $\mathbf{L}: \mathbb{R}^p \to \mathbb{R}^p : \Delta\mathbf{u} \mapsto \Delta\mathbf{x}$ is invertible. Show that $\mathbf{L}$ and $\mathbf{L}^{-1}$ vanish exactly to order 1 in the sense that there are bounding constants $0 < A_1 \leq A_2$, $0 < B_1 \leq B_2$ for which

$$A_1 \leq \frac{\|\mathbf{L}(\Delta\mathbf{u})\|}{\|\Delta\mathbf{u}\|} \leq A_2 \text{ and } B_1 \leq \frac{\|\mathbf{L}^{-1}(\Delta\mathbf{x})\|}{\|\Delta\mathbf{x}\|} \leq B_2$$

    for all $\Delta\mathbf{u}, \Delta\mathbf{x} \neq \mathbf{0}$. (This is an adaptation of Definition 3.4 to multivariable functions.)

    d. Show that we can take $B_1 = 1/A_2$, $B_2 = 1/A_1$ in part (b).

Chapter 4
The Derivative

Abstract The derivative of a map is the linear term in its Taylor approximation; it is a map itself. Because linear approximations are simpler than those of higher order, and because linear maps are easier to visualize than nonlinear ones, the derivative is an especially important part of the study of maps. It gives us valuable local information. We study the derivative in this chapter, beginning with the familiar connection to tangents.

4.1 Differentiability

Analytically, a function $y = f(x)$ is differentiable at a point if a certain limit exists; geometrically, the graph of the function must have a tangent at that point. When there are several input variables, $y = f(x_1, \ldots, x_p)$, the geometric characterization is the same—the graph must have a tangent—but the analytic one becomes uncertain: Is it enough for the partial derivatives to exist, or must the directional derivatives exist in all directions, or is even more necessary? In this section, we introduce the derivative map to settle the question and to make a clear connection between the analytic and geometric aspects of differentiability.

Differentiability of $y = f(x_1, \ldots, x_n)$

According to the usual definition, $y = f(x)$ is differentiable at $x = a$ if

Differentiability in terms of "little oh"

$$\lim_{\Delta x \to 0} \frac{f(a + \Delta x) - f(a)}{\Delta x} = f'(a),$$

for some finite number $f'(a)$ that we then call the derivative of f at a. In that case, we can rewrite the limit expression in the form

$$\lim_{\Delta x \to 0} \frac{f(a + \Delta x) - f(a) - f'(a)\Delta x}{\Delta x} = 0.$$

This says that the numerator, as a function of Δx, vanishes to order greater than 1 (cf. Definition 3.3, p. 85). In other words, the usual definition of differentiability is

J.J. Callahan, *Advanced Calculus: A Geometric View*, Undergraduate Texts in Mathematics,   105
DOI 10.1007/978-1-4419-7332-0_4, © Springer Science+Business Media, LLC 2010

equivalent to the following equality involving "little oh":

$$f(a+\Delta x) = \underbrace{f(a)+f'(a)\Delta x}_{\substack{\text{values along} \\ \text{tangent line at } a}} + o(1).$$

$$\underbrace{}_{\substack{\text{values of } f \\ \text{near } a}}$$

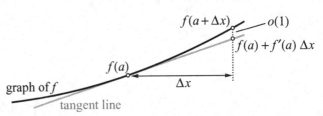

Differentiability and local linearity

We recognize $y = f(a) + f'(a)\Delta x$ as the formula for the tangent line at $x = a$, so the equation tells us what it means for the graph of f to have a tangent: the gap between the graph of f and its tangent line vanishes more rapidly than the horizontal displacement from the point of tangency. We can take this as the *geometric* definition of differentiability. Another name for differentiability, understood geometrically, is *local linearity*: under sufficiently high magnification (i.e., for Δx sufficiently small), the graph of f at $x = a$ is indistinguishable from the linear graph of the tangent there.

Comparison with Taylor's theorem

Notice that our new geometric formula for differentiability is similar to Taylor's formula,

$$f(a+\Delta x) = f(a) + f'(a)\Delta x + O(2).$$

The difference lies solely in the order of vanishing of the remainder; Taylor's formula has the stronger condition $R_{1,a}(\Delta x) = O(2)$. (For example, $t^{4/3} = o(1)$ but $t^{4/3} \neq O(2)$; see also Exercise 3.15, p. 102.) But the hypothesis that Taylor's formula rests upon is stronger, too: Taylor's theorem requires that f have a continuous second derivative on an open interval that contains a and $a + \Delta x$. However, as we have seen, the limit defining the derivative leads us to the formula

$$f(a+\Delta x) = f(a) + f'(a)\Delta x + o(1)$$

that involves "little oh" rather than "big oh."

Differentiability for $z = f(x,y)$

Let us move on to the differentiability of $z = f(x,y)$ at $(x,y) = (a,b)$, and approach it from the geometric point of view. In terms of coordinates $\Delta x = x - a$, $\Delta y = y - b$ centered at (a,b), an arbitrary plane has a formula that we can write as

$$z = c + p\Delta x + q\Delta y.$$

We require the gap $f(a+\Delta x, b+\Delta y) - (c + p\Delta x + q\Delta y)$ to vanish more rapidly than the horizontal displacement $\sqrt{(\Delta x)^2 + (\Delta y)^2}$ to the point of tangency (a,b).

Definition 4.1 *The function $z = f(x,y)$ is **differentiable**, or **locally linear**, at $(x,y) = (a,b)$ if there are constants c, p, and q for which*

$$f(a+\Delta x, b+\Delta y) - (c + p\Delta x + q\Delta y) = o(1).$$

*In that case, the graph of $z = c + p\Delta x + q\Delta y$ is the **tangent plane** to the graph of f at the point (a,b).*

Theorem 4.1. *If $z = f(x,y)$ is differentiable at (a,b), then both partial derivatives exist at (a,b) and the equation of the tangent plane there is*

$$z = f(a,b) + f_x(a,b)\Delta x + f_y(a,b)\Delta y.$$

Proof. In terms of the definition, we must show

$$c = f(a,b), \quad p = \frac{\partial f}{\partial x}(a,b), \quad q = \frac{\partial f}{\partial y}(a,b);$$

in particular, we must show that the two partial derivatives exist. For a start, the expression

$$f(a+\Delta x, b+\Delta y) - (c + p\Delta x + q\Delta y)$$

must vanish when $\Delta x = \Delta y = 0$. This implies $c = f(a,b)$. Now keep $\Delta y = 0$ but let Δx vary. The hypothesis then becomes

$$f(a+\Delta x, b) - (f(a,b) + p\Delta x) = o(1),$$

and it means (cf. Definition 3.3, p. 85)

$$0 = \lim_{\Delta x \to 0} \frac{f(a+\Delta x, b) - f(a,b) - p\Delta x}{\Delta x} = \lim_{\Delta x \to 0} \frac{f(a+\Delta x, b) - f(a,b)}{\Delta x} - p.$$

Therefore

$$p = \lim_{\Delta x \to 0} \frac{f(a+\Delta x, b) - f(a,b)}{\Delta x} = \frac{\partial f}{\partial x}(a,b);$$

that is, the partial derivative exists and has the value p. The value of q is determined in a similar way, by fixing $\Delta x = 0$ and letting Δy vary.   □

The partial derivatives that define the tangent plane are, at the same time, the components of the 1×2 matrix that defines the derivative of f at (a,b) (Definition 3.16, p. 99). The following corollary makes explicit this (natural!) connection between differentiability and the derivative.

Derivative of f

Corollary 4.2 *If $z = f(x,y)$ is differentiable at (a,b), then the derivative $\mathrm{d}f_{(a,b)} : \mathbb{R}^2 \to \mathbb{R}$ exists and*

$$f(a+\Delta x, b+\Delta y) = f(a,b) + \mathrm{d}f_{(a,b)}(\Delta x, \Delta y) + o(1).$$ □

A reasonable question to ask at this point is: if both partial derivatives of $f(x,y)$ exist at (a,b), is f then differentiable at (a,b)? In other words, if the plane

Do partial derivatives imply differentiability?

$$z = f(a,b) + f_x(a,b)\Delta x + f_y(a,b)\Delta y$$

is defined, is it not automatically the tangent plane to the graph of f at (a,b)? Is it not guaranteed that the gap between this plane and the graph of f must vanish to higher order than the horizontal displacement $(\Delta x, \Delta y)$ from the point (a,b)?

Counterexample: the "manta ray"

It fact, the answer is *no*. Here is an example that illustrates the contrary (a *counterexample*):

$$f(x,y) = \begin{cases} 0 & \text{if } (x,y) = (0,0), \\ \dfrac{x^2 y}{x^2 + y^2} & \text{otherwise.} \end{cases}$$

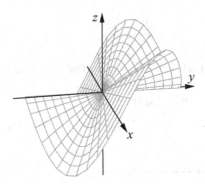 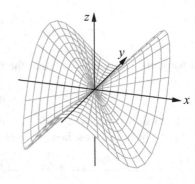

A bundle of lines through the origin

The two figures are different views of the graph of $z = f(x,y)$; it looks vaguely like a manta ray swimming along the y-axis. The essential thing to note is that the graph is made up of a bundle of straight lines through the origin. An easy way to confirm this is to put a polar coordinate overlay on the graph. That is, let $x = r\cos\theta$, $y = r\sin\theta$; then, away from the origin,

$$z = f(x,y) = \frac{r^3 \cos^2 \theta \sin \theta}{r^2} = r\cos^2 \theta \sin \theta.$$

When θ is fixed (as it is along a radial line), this is the straight line $z = mr$ of slope $m = \cos^2 \theta \sin \theta$. The *Mathematica 5* code that produces the figures makes use of this overlay:

```
ParametricPlot3D[{r Cos[t], r Sin[t], r Cos[t]^2 Sin[t]},
    {r, 0, 1}, {t, 0, 2 Pi}, PlotPoints -> {10, 49},
    BoxRatios -> {1, 1, .8}, Axes -> False, Boxed -> False,
    ViewPoint -> {3.500, -0.680, 1.221}]
```

No plane is tangent to the graph at the origin

Let us now see why no plane is tangent to the graph at the origin. Along the radial line $\theta = \theta_0$, the graph is the straight line of slope $m = \cos^2 \theta_0 \sin \theta_0$. If θ_0 is an integer multiple of $\pi/2$, then $m = 0$ and the radial line lies along an axis. So this slope is a partial derivative; we find $f_x(0,0) = f_y(0,0) = 0$. Therefore, if the graph were to have a tangent at the origin, Theorem 4.1 would force it to be the (x,y)-plane itself: $z = 0$. In that case, the slope of the graph in any direction at the origin would be 0. But the figures make it clear (and the formula $m = \cos^2 \theta_0 \sin \theta_0$ confirms) that the slope of the radial line in the direction $\theta_0 \neq k\pi/2$ is nonzero. The manta ray graph has no tangent plane at the origin; the function f is not differentiable there.

So the mere existence of the partial derivatives of $z = f(x,y)$ at a point is not enough to guarantee that f is differentiable at that point, i.e., that its graph has a tangent plane there. But suppose we impose a stronger condition, one requiring that all directional derivatives exist. We recall the definition of a directional derivative for a function of p variables (where p need not equal 2).

Directional derivatives

Definition 4.2 *Let $\mathbf{u}$ be a unit vector; then the **directional derivative** of $z = f(\mathbf{x})$ at the point $\mathbf{x} = \mathbf{a}$ in the direction $\mathbf{u}$ is*

$$D_{\mathbf{u}}f(\mathbf{a}) = \frac{d}{dt}f(\mathbf{a}+t\mathbf{u})\Big|_{t=0},$$

when the expression on the right exists.

Let $\mathbf{u} = \mathbf{e}_i = (0,\ldots,0,1,0,\ldots,0)$ (i.e., 1 in the ith place, 0 elsewhere); the derivative in the direction $\mathbf{e}_i$ is just the usual **partial derivative**:

$$D_{\mathbf{e}_i}f(\mathbf{a}) = \frac{\partial f}{\partial x_i}(\mathbf{a}).$$

Suppose we now require that the directional derivatives of $z = f(x,y)$ in all directions (and not just the axis directions) exist at a point. Will this stronger condition guarantee that the graph of f has a tangent at that point? In fact, the manta ray is still a counterexample. In the direction $\mathbf{u} = (\cos\theta, \sin\theta)$, the directional derivative of f at the origin is

Do directional derivatives imply differentiability?

$$D_{(\cos\theta,\sin\theta)}f(0,0) = \cos^2\theta\sin\theta.$$

(In the given direction, the graph of f is a straight line of slope $\cos^2\theta\sin\theta$.) Thus, even though all the directional derivatives of f exist at the origin, there is (still) no tangent plane. The existence of all directional derivatives of f at a point is not enough to guarantee that f is differentiable there.

Although the existence of directional derivatives does not guarantee differentiability, the converse is true, according to the following theorem.

Directional derivatives from the derivative

Theorem 4.3. *If $z = f(x,y)$ is differentiable at (a,b), then all directional derivatives exist at (a,b). In fact, $D_{(\alpha,\beta)}f(a,b) = df_{(a,b)}(\alpha,\beta)$.*

Proof. The proof is probably easier to follow in vector notation. We set $(\alpha,\beta) = \mathbf{u}$; then, by definition, the directional derivative is

$$D_{\mathbf{u}}f(\mathbf{a}) = \lim_{t\to 0}\frac{f(\mathbf{a}+t\mathbf{u})-f(\mathbf{a})}{t} = \lim_{t\to 0}\frac{df_{\mathbf{a}}(t\mathbf{u})+o(1)}{t}.$$

We have $f(\mathbf{a}+t\mathbf{u})-f(\mathbf{a}) = df_{\mathbf{a}}(t\mathbf{u})+o(1)$ because f is differentiable at $\mathbf{a}$. But then

$$\lim_{t\to 0}\frac{df_{\mathbf{a}}(t\mathbf{u})+o(1)}{t} = \lim_{t\to 0}\frac{t\,df_{\mathbf{a}}(\mathbf{u})}{t} + \lim_{t\to 0}\frac{o(1)}{t} = df_{\mathbf{a}}(\mathbf{u})+0 = df_{\mathbf{a}}(\mathbf{u}).$$

We have used the linearity of $df_{\mathbf{a}}$ to write $df_{\mathbf{a}}(t\mathbf{u}) = t\,df_{\mathbf{a}}(\mathbf{u})$. Also, $o(1)/t \to 0$ as $t \to 0$, by definition of $o(1)$. $\qquad\qquad\qquad\qquad\qquad\qquad\qquad\qquad\qquad\qquad\square$

The gradient vector

The gradient of $z = f(x,y)$ at (a,b) is the *vector* $\operatorname{grad} f(a,b) = \nabla f(a,b)$ in $\mathbb{R}^2$ whose components are $f_x(a,b)$ and $f_y(a,b)$. Of course these are, at the same time, the components of the matrix that represents the derivative $df_{(a,b)} : \mathbb{R}^2 \to \mathbb{R}^1$. Conceptually, $\nabla f(a,b)$ and $df_{(a,b)}$ are different; the first is a vector, the second is a linear map. However, because matrix multiplication involves the scalar (dot) product, the two are connected:

$$df_{(a,b)}(\alpha,\beta) = \begin{pmatrix} f_x(a,b) & f_y(a,b) \end{pmatrix} \begin{pmatrix} \alpha \\ \beta \end{pmatrix} = \nabla f(a,b) \cdot (\alpha,\beta).$$

This connection allows us to express the previous theorem in a way that is probably more familiar.

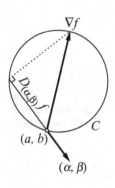

Corollary 4.4 *If* $z = f(x,y)$ *is differentiable at* (a,b), *then*

$$D_{(\alpha,\beta)}f(a,b) = \nabla f(a,b) \cdot (\alpha,\beta).$$
$\qquad\qquad\qquad\qquad\qquad\qquad\qquad\qquad\qquad\qquad\qquad\qquad\qquad\qquad\square$

Let C be the circle that has the vector $\nabla f(a,b)$ as diameter. Then, because the scalar $\nabla f(a,b) \cdot (\alpha,\beta)$ is the perpendicular projection of $\nabla f(a,b)$ on the line in the direction of (α,β), we can realize $D_{(\alpha,\beta)}f$ as the length of the chord of C that lies in the direction (α,β). If (α,β) makes an obtuse angle with ∇f, extend $-(\alpha,\beta)$ and note that then $D_{(\alpha,\beta)}f \le 0$.

The hypothesis that f is differentiable is crucial in the corollary. If we merely know that all the directional derivatives exist (including the partial derivatives), we cannot conclude that $D_{(\alpha,\beta)}f(a,b) = \nabla f(a,b) \cdot (\alpha,\beta)$. The manta ray at the origin is once again a counterexample: we have

$$D_{(\alpha,\beta)}f(0,0) = \alpha^2\beta \quad \text{but} \quad \nabla f(0,0) \cdot (\alpha,\beta) = (0,0) \cdot (\alpha,\beta) = 0.$$

Local linearity with level curves

Because we used tangents to define differentiability, it is natural to use tangents to illustrate local linearity. For example, consider the function $f(x,y) = x^2 - y^2$ at the point $(a,b) = (2,-1)$. The graph of $z = f(x,y)$ is a curved surface in $\mathbb{R}^3$ and the graph of the derivative

$$df_{(2,-1)}(\Delta x, \Delta y) = f_x(2,-1)\Delta x + f_y(2,-1)\Delta y = 4\Delta x + 2\Delta y$$

is a plane. We expect that, under sufficient magnification at the point $(x,y,z) = (2,-1,f(2,-1))$ in $\mathbb{R}^3$, the graph of f will become indistinguishable from the tangent plane; cf. Exercise 4.2.

We, however, take a different approach, comparing instead the level sets of f and $df_{(2,-1)}$ in windows centered at $(2,-1)$ in the (x,y)-plane. The window on the left below shows level curves

$$f(2+\Delta x, -1+\Delta y) - f(2,-1) = \Delta z, \quad \Delta z = -5, -3.75, \ldots, 5,$$

in the square $1 \leq x \leq 3$, $-2 \leq y \leq 0$. By design, the level curve $\Delta z = 0$ passes through the origin of the $(\Delta x, \Delta y)$-window. Obviously f is nonlinear: its level sets are curved, and they are unequally spaced.

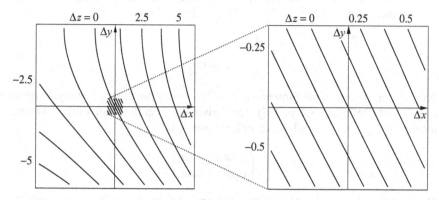

Under tenfold magnification (see the right window; spacing between levels has been cut by the same factor of 10), the local linearity of f begins to emerge. The level curves of f are now essentially straight, parallel, and evenly spaced: the hallmarks of a linear function. Thus, at this magnification, f looks linear. We must now just check that the apparently linear function we see in this window agrees with the derivative,

Local linearity emerges

$$df_{(2,-1)}(\Delta x, \Delta y) = 4\Delta x + 2\Delta y.$$

The level curves $4\Delta x + 2\Delta y = C$ of the derivative are parallel straight lines with common slope $\Delta y / \Delta x = -2$, just as in the microscope window on the right. But this is not yet enough; we need to show that a given line represents the same level for f and for $df_{(2,-1)}$. This is made easier by the fact that f itself has a simple formula:

Comparing level curves of f and $df_{(2,-1)}$

$$
\begin{aligned}
z &= f(2 + \Delta x, -1 + \Delta y) \\
&= (2 + \Delta x)^2 - (-1 + \Delta y)^2 = 4 + 4\Delta x + (\Delta x)^2 - 1 + 2\Delta y - (\Delta y)^2 \\
&= 3 + df_{(2,-1)}(\Delta x, \Delta y) + (\Delta x)^2 - (\Delta y)^2.
\end{aligned}
$$

Along the diagonals of the window, $(\Delta x)^2 = (\Delta y)^2$, so

$$\Delta z = f(2 + \Delta x, -1 + \Delta y) - 3 = df_{(2,-1)}(\Delta x, \Delta y);$$

in other words, f and its derivative agree exactly on the diagonals. It follows that, everywhere in the right window, a given level curve for f is indistinguishable from the level curve for $df_{(2,-1)}$ at the same level.

4.2 Maps of the plane

Our goal here is to understand what differentiability means geometrically for a map of the plane. Suppose $\mathbf{f}\colon U^2 \to \mathbb{R}^2$ has the coordinate form

$$\mathbf{f}\colon \begin{cases} x = f(u,v), \\ y = g(u,v). \end{cases}$$

Here U^2 is a *window* of the form $|u-a| < p$, $|v-b| < q$ (or, more generally, an *open set* in $\mathbb{R}^2$; cf. Definition 8.4, p. 277). The derivative of $\mathbf{f}$ at the point $\mathbf{a} = (a,b)$ is the linear map $d\mathbf{f_a}\colon \mathbb{R}^2 \to \mathbb{R}^2$ whose coordinate matrix is

$$d\mathbf{f}_{(a,b)} = \begin{pmatrix} f_u(a,b) & f_v(a,b) \\ g_u(a,b) & g_v(a,b) \end{pmatrix},$$

(Definition 3.16, p. 99). The graph of $\mathbf{f}$ and the graph of $d\mathbf{f}_{(a,b)}$ are both 2-dimensional surfaces, but they lie in the 4-dimensional (u,v,x,y)-space, so we cannot visualize them directly. In particular, we cannot see how—or whether—the graph of the derivative is tangent to the graph of the map.

We faced this dimension problem with the graph of a linear map of the plane in Chapter 2. There, we solved the problem by looking at images instead of graphs; we do the same here. Differentiability is then manifested as local linearity: we compare the image of $\mathbf{f}$ in a microscope window centered at $\mathbf{a}$ to the image of the linear map $d\mathbf{f_a}$ in that window.

The polar coordinate change is the map that pulls back Cartesian coordinates to polar coordinates:

$$\mathbf{f}\colon \begin{cases} x = r\cos\theta, \\ y = r\sin\theta. \end{cases}$$

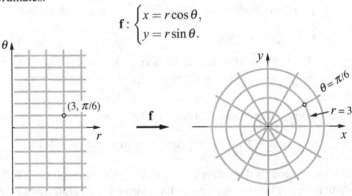

The map $\mathbf{f}$ puts a grid of rays and concentric circles on the (x,y)-plane, corresponding to the rectangular grid $\theta = $ constant and $r = $ constant in the (r,θ)-plane itself. By convention, only the positive half of each ray is used; that is, we assume $r > 0$. (In other words, the domain U^2 for $\mathbf{f}$ is the open right half-plane.) Sometimes it is useful to allow $r = 0$, as well. This is the θ-axis; the map $\mathbf{f}$ collapses it to a single point, the origin in the target.

According to Taylor's theorem, any smooth map is approximately a polynomial Local behavior of **f** near $(r, \theta) = (3, \pi/6)$
when its domain is restricted to a small enough region. The closeness of the ap-
proximation is directly related to the degree of the polynomial, but even a linear
polynomial can provide an impressive approximation. Let us focus on the point
$(r, \theta) = (3, \pi/6)$ and see how **f** becomes approximately linear as it acts on smaller
and smaller regions centered at this point. In the process we also see that the approximation is precisely the linear term in the Taylor polynomial of **f**, the map that
we call the *derivative* of **f** at $(r, \theta) = (3, \pi/6)$ and denote with the symbol $d\mathbf{f}_{(3,\pi/6)}$.

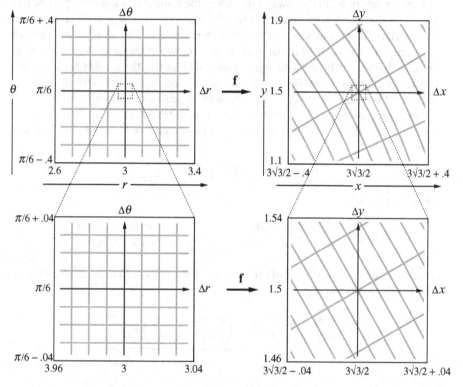

The figure above shows what **f** does to a a grid of squares in a small window
centered at the point $(r, \theta) = (3, \pi/6)$. The image is a grid of radial lines and circular
arcs in another small window centered at the image point $(x, y) = (3\sqrt{3}/2, 3/2)$. To
describe the action of **f** in these windows it is natural to use coordinates Δr, $\Delta \theta$, Δx,
and Δy that measure displacements from the center of each window:

$$\Delta r = r - 3, \qquad \Delta x = x - 3\sqrt{3}/2,$$
$$\Delta \theta = \theta - \pi/6, \qquad \Delta y = y - 3/2.$$

In the figure, there are two pairs of windows at different levels of magnification. In Views of **f** in two "microscope" windows
the upper pair the grid spacing is 0.1 units and the windows themselves measure
0.8 units on a side. The radial lines we see in the image window therefore have a

separation of $\Delta\theta = 0.1$ radians and the concentric circular arcs have a separation of $\Delta r = 0.1$ units. At this level of magnification the arcs are still noticeably curved.

Each lower window is a tenfold magnification of the center of the window above it. The radial lines in the image are now only $\Delta\theta = 0.01$ radians apart; they look nearly parallel. The concentric circular arcs are spaced $\Delta r = 0.01$ units apart and likewise appear to be straight and parallel. In this microscopic view, **f** looks like a linear map, because it maps a grid of congruent squares to a grid of (nearly) congruent parallelograms, rectangles, in fact.

Near $(3, \pi/6)$, f is a stretch and rotation

Can we describe **f** in the lower windows in the fashion of a linear map? The line $\Delta\theta = 0$ in the source (the Δr-axis) is just the horizontal line $\theta = \pi/6$, so its image is the radial line that makes an angle of $\pi/6$ radians, or $30°$, as it passes through the origin $(\Delta x, \Delta y) = (0,0)$ of the microscope window in the target. In other words, **f** rotates the Δr-axis by $30°$; the figure makes it clear that the whole $(\Delta r, \Delta\theta)$-plane undergoes essentially the same rotation. In addition, before **f** rotates the plane it stretches it vertically; by eye, the stretch factor appears to be about 3.

The derivative $\mathrm{df}_{(r,\theta)}$

Now compare this action with the action of the derivative $\mathrm{df}_{(3,\pi/6)}$. At an arbitrary point $(r, \theta) = (r_0, \theta_0)$, the derivative $\mathrm{df}_{(r_0,\theta_0)}$ is defined (see p. 112) to be the linear map

$$\mathrm{df}_{(r_0,\theta_0)} : \mathbb{R}^2 \to \mathbb{R}^2 : \begin{pmatrix} \Delta r \\ \Delta\theta \end{pmatrix} \mapsto \begin{pmatrix} \Delta x \\ \Delta y \end{pmatrix}$$

whose matrix is

$$\mathrm{df}_{(r_0,\theta_0)} = \begin{pmatrix} \partial x/\partial r & \partial x/\partial\theta \\ \partial y/\partial r & \partial y/\partial\theta \end{pmatrix}\Bigg|_{(r,\theta)=(r_0,\theta_0)} = \begin{pmatrix} \cos\theta_0 & -r_0\sin\theta_0 \\ \sin\theta_0 & r_0\cos\theta_0 \end{pmatrix}.$$

The derivative splits into two factors

Notice that $\mathrm{df}_{(r_0,\theta_0)}$ factors neatly into a pair of matrices, a stretch (or *strain*) S_{1,r_0}, followed by a rotation R_{θ_0}:

$$\begin{pmatrix} \cos\theta_0 & -r_0\sin\theta_0 \\ \sin\theta_0 & r_0\cos\theta_0 \end{pmatrix} = \underbrace{\begin{pmatrix} \cos\theta_0 & -\sin\theta_0 \\ \sin\theta_0 & \cos\theta_0 \end{pmatrix}}_{R_{\theta_0}} \underbrace{\begin{pmatrix} 1 & 0 \\ 0 & r_0 \end{pmatrix}}_{S_{1,r_0}}.$$

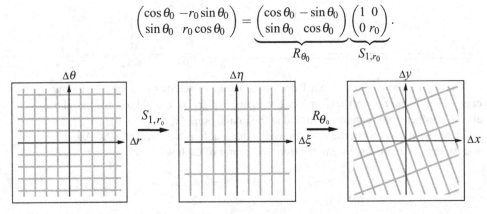

Factoring $\mathrm{df}_{(r_0,\theta_0)}$ means that we can describe its effect in two stages. First, S_{1,r_0} stretches the plane vertically (i.e., in the direction of the $\Delta\theta$-axis) by the factor r_0; then R_{θ_0} rotates the result by θ_0 radians. (The coordinate names $\Delta\xi$, $\Delta\eta$ in the intermediate window are just arbitrary choices.)

In particular, we can now conclude that $df_{(3,\pi/6)}$ is a threefold vertical stretch followed by a 30° rotation. But this is exactly how **f** itself behaves in a microscope window centered at $(r, \theta) = (3, \pi/6)$. This suggests we say that **f** is *locally linear* in a small neighborhood of $(3, \pi/6)$.

$\mathbf{f} \approx df$ near $(3, \pi/6)$

The example is leading us to say that a map $\mathbf{f} : U^2 \to \mathbb{R}^2 : \mathbf{u} \mapsto \mathbf{x}$ will be locally linear—or differentiable—at a point $\mathbf{u} = \mathbf{a}$ if

$$\Delta \mathbf{x} = \mathbf{f}(\mathbf{a} + \Delta \mathbf{u}) - \mathbf{f}(\mathbf{a})$$

differs from a linear function of $\Delta \mathbf{u}$ by an amount that vanishes faster than $\Delta \mathbf{u}$. Here is a precise definition.

Definition 4.3 *The map* $\mathbf{x} = \mathbf{f}(\mathbf{u})$ *is **differentiable**, or **locally linear**, at* $\mathbf{u} = \mathbf{a}$ *if there is a linear map* $\mathbf{L} : \mathbb{R}^2 \to \mathbb{R}^2$, *called the **derivative** of* **f** *at* **a**, *for which*

Differentiability and local linearity

$$\mathbf{f}(\mathbf{a} + \Delta \mathbf{u}) = \mathbf{f}(\mathbf{a}) + \mathbf{L}(\Delta \mathbf{u}) + o(1).$$

Theorem 4.5. *If* $\mathbf{f} : U^2 \to \mathbb{R}^2$ *is differentiable at* $\mathbf{u} = \mathbf{a}$, *then* $\mathbf{L} = df_{\mathbf{a}}$. *In particular, all the partial derivatives appearing in the matrix* $df_{\mathbf{a}}$ *exist.*

Proof. See Chapter 4.4, where the theorem is restated (as Theorem 4.6) for the general case $\mathbf{f} : U^p \to \mathbb{R}^q$. $\square$

The theorem makes it clear that if **f** is locally linear at **a**, then its linear approximation is its derivative $df_{\mathbf{a}}$. Note that if we rewrite the **window equation**

The microscope equation

$$\Delta \mathbf{x} = \mathbf{f}(\mathbf{a} + \Delta \mathbf{u}) - \mathbf{f}(\mathbf{a}) = df_{\mathbf{a}}(\Delta \mathbf{u}) + o(1)$$

without the remainder term $o(1)$, we get an approximation that is, in effect, a new form of the **microscope equation**:

$$\Delta \mathbf{x} \approx df_{\mathbf{u}_0}(\Delta \mathbf{u}).$$

In other words, the microscope equation emerges as a (rather condensed) way of expressing the differentiability or local linearity of a map. We have already noted the connection between the microscope equation and Taylor's theorem in Chapter 3 (p. 83; p. 95).

Definition 4.4 *If* $\mathbf{f} : U^2 \to \mathbb{R}^2$ *is differentiable at* **a**, *its **local area multiplier** at* **a** *is the area multiplier of its derivative* $df_{\mathbf{a}}$.

Local area multiplier

For the polar coordinate map $\mathbf{x} = \mathbf{f}(\mathbf{r})$, we find that the area multiplier of $df_{\mathbf{r}_0}$ is r_0:

$$\det df_{\mathbf{r}_0} = \det \begin{pmatrix} \cos \theta_0 & -r_0 \sin \theta_0 \\ \sin \theta_0 & r_0 \cos \theta_0 \end{pmatrix} = r_0 \cos^2 \theta_0 + r_0 \sin^2 \theta_0 = r_0.$$

Thus we say that r_0 is the **local area multiplier** for **f** itself at the point $\mathbf{r}_0 = (r_0, \theta_0)$. It is evident in the figure below that the local area multiplier of **f** varies from point to point and increases with the radius r; our calculations show that the local multiplier is exactly r.

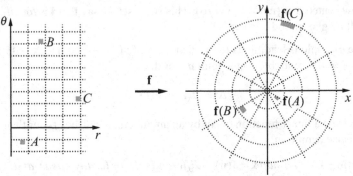

Local linearity *versus* "looking linear locally"

For plane maps **f** like the polar coordinate map, we have used the notions

$$\mathbf{f} \text{ is arbitrarily close to } d\mathbf{f_a} \text{ near } \mathbf{a} \quad \text{and} \quad \mathbf{f} \text{ "looks like" } d\mathbf{f_a} \text{ near } \mathbf{a}$$

more or less interchangeably. However, the two are subtly different. The first, of course, is what we now call *local linearity*. The second, however, may not be true if $d\mathbf{f_a}$ fails to be invertible; it is a stronger condition. To help bring into sharper focus the distinction between these two notions—and to see how the second condition can fail—we analyze a second map.

Second example: a quadratic map

Consider the quadratic map $\mathbf{f} : \mathbb{R}^2 \to \mathbb{R}^2$, defined by the equations

$$\mathbf{f} : \begin{cases} x = u^2 - v^2, \\ y = 2uv. \end{cases}$$

Although the action of the polar coordinate change map was immediately evident on a global level (i.e., on the entire right half-plane $r > 0$), the same is not true for the quadratic map **f**. However, the action is not hard to describe; we now show **f** squares the distance of any point from the origin and doubles the angle that point makes with the positive horizontal axis.

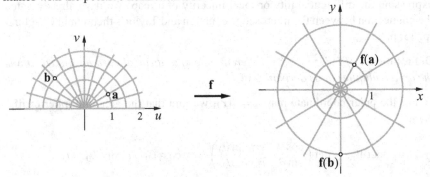

Polar coordinate overlays

We can show that **f** acts this way by translating our formulas for **f** into polar coordinates, because they provide the angles and distances we wish to measure.

That is, think of polar coordinates (r, θ) as an "overlay" on the (x, y)-plane, and introduce the same overlay on the (u, v)-plane using new polar coordinates (ρ, φ): $u = \rho \cos \varphi$, $v = \rho \sin \varphi$. Then the formulas for $\mathbf{f}$ that define x and y in terms of u and v translate into expressions for r and θ in terms of ρ and φ:

$$r \cos \theta = x = u^2 - v^2 = \rho^2 \cos^2 \varphi - \rho^2 \sin^2 \varphi = \rho^2 \cos 2\varphi,$$
$$r \sin \theta = y = 2uv = 2\rho \cos \varphi \cdot \rho \sin \varphi = \rho^2 \sin 2\varphi.$$

In other words, $r \cos \theta = \rho^2 \cos 2\varphi$ and $r \sin \theta = \rho^2 \sin 2\varphi$, so

$$r = \rho^2 \quad \text{and} \quad \theta = 2\varphi.$$

Thus, in terms of the polar coordinate overlays on the source and target, $\mathbf{f}$ squares the distance of a point from the origin ($r = \rho^2$) and it doubles the angle that that point makes with the horizontal ($\theta = 2\varphi$).

f doubles angles and squares distances from the origin

The angle-doubling means $\mathbf{f}$ "fans out" the upper half-plane $v \geq 0$ in the source to cover the entire target (x, y)-plane. The lower half-plane $v \leq 0$ also covers the entire (x, y)-plane, so the source covers the target twice, except for the origin. More precisely, let V^2 be the plane minus the origin: $V^2 = \mathbb{R}^2 \setminus (0, 0)$. Then $\mathbf{f} : V^2 \to V^2$ is a 2–1 map. Every point in the target V^2 is the image of exactly two points in the source V^2 (that lie 180° apart at the same distance from the origin). The unit circle maps to itself. A concentric circle inside the unit circle maps to another one even closer to the origin; one outside the unit circle is mapped to another farther from the origin.

f is a "double cover"

With this clear picture of the global behavior of $\mathbf{f}$, it is easy to analyze its local behavior near a given point. For example, take the point $(u, v) = (\sqrt{3}/2, 1/2)$. Its image is $(x, y) = (1/2, \sqrt{3}/2)$. These points are 1 unit from the origin ($\rho = r = 1$) and make angles of $\varphi = 30° = \pi/6$ radians and $\theta = 60° = \pi/3$ radians, respectively.

Behavior of f near $(u, v) = (\sqrt{3}/2, 1/2)$

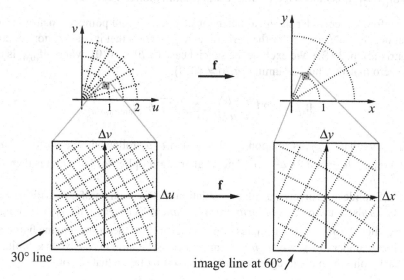

30° line   image line at 60°

The figure above uses the polar coordinate overlays to show how **f** acts in microscope windows centered at these two points. In the polar grid in the source window, the spacing between adjacent circular arcs is $\Delta\rho = 1/36 \approx 0.028$ units; the spacing between radial line segments is $\Delta\varphi = 1.5° \approx 0.026$ radians. Because these numbers are nearly equal and relatively small, the grid looks approximately square.

Rotation and linear expansion

The target window shows the image of the source grid under the map **f**. The $30°$-line from the source gives us our bearings. At the macroscopic level, **f** maps it to the $60°$-line, so in the microscope window **f** rotates it by $30°$. The entire grid is carried along by this action, so **f** rotates all points in the source by $30°$, more or less. Obviously, there is also linear expansion we must take into account. The image grid is still approximately square; thus **f** must be close to a uniform dilation. Moreover, a single square in the image grid is about the size of a 2×2 square in the source grid, so the linear expansion factor is about 2 (and the area expansion factor is about 4). Therefore, in the microscope windows it appears that **f** approximately doubles all lengths and rotates points by about $30°$, or $\pi/6$ radians. In other words, **f** approximates the linear map $2R_{\pi/6}$ in a small neighborhood of $(u, v) = (\sqrt{3}/2, 1/2)$.

The derivative of f at $(\sqrt{3}/2, 1/2)$

We expect, therefore, that the derivative of **f** at $(\sqrt{3}/2, 1/2)$ must equal $2R_{\pi/6}$. Can we confirm this? First of all, at an arbitrary point $(u, v) = (a, b)$, the derivative of **f** is given by the matrix

$$\mathbf{df}_{(a,b)} = \begin{pmatrix} \partial x/\partial u & \partial x/\partial v \\ \partial y/\partial u & \partial y/\partial v \end{pmatrix}\bigg|_{(u,v)=(a,b)} = \begin{pmatrix} 2a & -2b \\ 2b & 2a \end{pmatrix} = 2\begin{pmatrix} a & -b \\ b & a \end{pmatrix}.$$

Therefore,

$$\mathbf{df}_{(\sqrt{3}/2,1/2)} = 2\begin{pmatrix} \sqrt{3}/2 & -1/2 \\ 1/2 & \sqrt{3}/2 \end{pmatrix} = 2\begin{pmatrix} \cos\pi/6 & -\sin\pi/6 \\ \sin\pi/6 & \cos\pi/6 \end{pmatrix} = 2R_{\pi/6},$$

so $\mathbf{df}_{(\sqrt{3}/21/2)}$ is indeed the local linear approximation to **f** at $(\sqrt{3}/2, 1/2)$.

$\mathbf{df}_{(a,b)}$ is a rotation–dilation matrix

Before we consider the local behavior of **f** at a second point, we pause to note that our formula, above, for the derivative $\mathbf{df}_{(a,b)}$ shows that it is a rotation–dilation matrix (cf. p. 39 ff.). We exclude the special case $(a, b) = (0, 0)$, where $\mathbf{df}_{(0,0)}$ is just the zero matrix. Thus, assuming $(a, b) \neq (0, 0)$,

$$\mathbf{df}_{(a,b)} = 2\begin{pmatrix} a & -b \\ b & a \end{pmatrix} = 2\sqrt{a^2 + b^2}\, R_{\arctan(b/a)}.$$

This says that $\mathbf{df}_{(a,b)}$ is rotation by $\theta = \arctan(b/a)$ followed by a uniform linear dilation by the factor $2\sqrt{a^2 + b^2}$. The local area multiplier for **f** at (a, b) is therefore $4(a^2 + b^2)$.

Conformal maps

In Euclidean geometry, a rotation–dilation matrix such as $\mathbf{df}_{(a,b)}$, with $(a, b) \neq (0, 0)$, is also known as a *similarity transformation*. Even when $\mathbf{df}_{(a,b)}$ alters lengths (i.e., when $2\sqrt{a^2 + b^2} \neq 1$), angles remain unchanged.; therefore, a plane figure and its image under $\mathbf{df}_{(a,b)}$ are *similar*. A map such as **f** whose derivative is a similarity at each point in an open region is said to be **conformal** on that region.

For our second illustration, let us study the local behavior of **f** at the point $(u,v) = (-3\sqrt{2}/4, 3\sqrt{2}/4)$ (so $(\rho, \varphi) = (3/2, 3\pi/4)$). The derivative is

$$\mathbf{df}_{(-3\sqrt{2}/4, 3\sqrt{2}/4)} = \begin{pmatrix} -3\sqrt{2}/2 & -3\sqrt{2}/2 \\ 3\sqrt{2}/2 & -3\sqrt{2}/2 \end{pmatrix} = 3R_{3\pi/4}.$$

Therefore, if **f** is locally linear at $(-3\sqrt{2}/4, 3\sqrt{2}/4)$, we expect **f** will approximately triple all lengths and rotate all points by $3\pi/4$ radians (i.e., $135°$) in a microscope window centered at $(-3\sqrt{2}/4, 3\sqrt{2}/4)$.

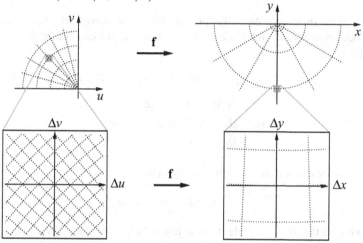

The figure shows the action of **f**. To describe it, we again use a polar grid overlay in the source. At the macroscopic level, **f** "fans out" the second quadrant to cover the third and fourth quadrants. At the microscopic level, the spacing between concentric circular arcs in the source grid is $\Delta\rho = 1/36 \approx 0.028$ units, just as it was in our first illustration. However, we have reduced the spacing between radial lines in the grid to $\Delta\varphi = 1° = \pi/180$ radians, but because $\rho \approx 1.5$ in the window, the width between adjacent rays is about $1.5 \times \pi/180 \approx 0.026$ units. (By the definition of radian measure, an angle of φ radians at the center of a circle of radius ρ cuts off an arc of length $\rho \cdot \varphi$ on the circle.) This adjustment keeps the source grid roughly square.

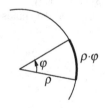

It is evident from the image in the target window that **f** is, once again, approximately a rotation coupled with a uniform dilation. This time the $135°$-line is our landmark; **f** maps it to the $270°$-line in the target, rotating all points in the window therefore by about $135°$. A square in the image grid is about the size of a 3×3 square in the source grid, so the linear dilation factor is about 3. Hence $d\mathbf{f}$ is indeed the local linear approximation to **f** at $(-3\sqrt{2}/4, 3\sqrt{2}/4)$.

Our third illustration analyzes the action of **f** at the origin. Here we finally get to see the difference between *looking linear* and *local linearity*. At the origin, the local action of **f** is the same as its global action: in any microscope window, no matter how small, **f** doubles angles and squares lengths. But no linear map does this, so **f** near the origin does not "look like" any linear map. In particular, **f** does not "look like" its derivative $d\mathbf{f}_{(0,0)}$ there. Nevertheless, **f** *is* locally linear at the origin; we show this in order to confirm that **f** is well-approximated by $d\mathbf{f}_{(0,0)}$ there.

Local linearity for
the quadratic map

How can this be? How can **f** be *locally linear* at a point and not *look linear* at that point? How can a map be "well-approximated" by a linear map and not "look like" that linear map? The explanation lies in the definition of local linearity. A map **f** is **locally linear** at $\mathbf{u} = \mathbf{a}$ (p. 115) if

$$\Delta \mathbf{x} = \mathbf{f}(\mathbf{a} + \Delta \mathbf{u}) - \mathbf{f}(\mathbf{a}) = d\mathbf{f}_{\mathbf{a}}(\Delta \mathbf{u}) + \mathbf{o}(1);$$

that is, if the difference $\Delta \mathbf{x} - d\mathbf{f}_{\mathbf{a}}(\Delta \mathbf{u})$ vanishes more rapidly than $\Delta \mathbf{u}$. To see that this is indeed true for our quadratic map, note first that we can compute $\Delta \mathbf{x} = (\Delta x, \Delta y)$ exactly:

$$\Delta x = (a + \Delta u)^2 - (b + \Delta v)^2 - (a^2 - b^2)$$
$$= 2a\,\Delta u - 2b\,\Delta v + (\Delta u)^2 - (\Delta v)^2,$$
$$\Delta y = 2(a + \Delta x)(b + \Delta v) - 2ab$$
$$= 2b\,\Delta u + 2a\,\Delta v + 2\,\Delta x\,\Delta y.$$

These window equations take the vector form

$$\Delta \mathbf{x} = d\mathbf{f}_{\mathbf{a}}(\Delta \mathbf{u}) + \mathbf{f}(\Delta \mathbf{u}),$$

showing us that the remainder term is just $\mathbf{f}(\Delta \mathbf{u})$. But $\mathbf{f}(\Delta \mathbf{u}) = \mathbf{O}(2)$ because $\mathbf{f}(\Delta \mathbf{u})$ is quadratic, and this, in turn, implies the weaker condition $\mathbf{f}(\Delta \mathbf{u}) = \mathbf{o}(1)$ for local linearity. Thus, **f** is "well-approximated" by the linear map $d\mathbf{f}_{\mathbf{a}}$ near **a**, for every point **a**, *including the origin*. But whether **f** "looks like" its linear approximation $d\mathbf{f}_{\mathbf{a}}$ in a microscope window centered at **a** will depend on the relative sizes of the two terms in the formula for $\Delta \mathbf{x}$.

At $(\sqrt{3}/2, 1/2)$,
$2R_{\pi/6}(\Delta \mathbf{u})$ dominates

For example, in the window centered at $\mathbf{a} = (\sqrt{3}/2, 1/2)$ (i.e., the first window), the window equation for **f** is

$$\Delta \mathbf{x} = 2R_{\pi/6}(\Delta \mathbf{u}) + \mathbf{f}(\Delta \mathbf{u}).$$

The linear term is the rotation and uniform dilation $2R_{\pi/6}(\Delta \mathbf{u})$. This linear map is invertible, so it vanishes exactly to order 1 (Exercise 3.28, p.104). By contrast, the second term is the remainder $\mathbf{f}(\Delta \mathbf{u})$ and, as such, vanishes at least to order 2. Thus, when $\Delta \mathbf{u} \approx \mathbf{0}$ (in other words, in the microscope window), the linear term *dominates*, precisely because it vanishes to a lower order in $\Delta \mathbf{u}$. This is why the map **f** looks like its linear approximation $2R_{\pi/6}$ near $(\sqrt{3}/2, 1/2)$.

The behavior of **f** in the second window (where $\mathbf{a} = (-3\sqrt{2}/4, 3\sqrt{2}/4)$) is entirely similar: the linear term $3R_{3\pi/4}(\Delta \mathbf{u})$ is invertible so it again dominates the quadratic one $\mathbf{f}(\Delta \mathbf{u})$. Thus in the second window **f** looks like its linear approximation $3R_{3\pi/4}(\Delta \mathbf{u})$.

At the origin, the window equation still expresses the local linearity of **f**, but it has the fundamentally different form

$$\Delta\mathbf{x} = d\mathbf{f}_0(\Delta\mathbf{u}) + \boldsymbol{O}(2) = \mathbf{0} + \mathbf{f}(\Delta\mathbf{u}).$$

The linear term, which had been dominant in the other windows, here contributes nothing. It vanishes to *infinite* order, meaning it vanishes at least to order p for every $p > 0$. The value of $\Delta\mathbf{x}$ is determined solely by the quadratic term $\mathbf{f}(\Delta\mathbf{u})$. This is what accounts for the angle-doubling and distance-squaring. By default, $\mathbf{f}(\Delta\mathbf{u})$ is the dominant term; in fact, it vanishes exactly to order 2 (see Exercise 4.12), so we are justified in saying it dominates any map (such as $d\mathbf{f}_0$) that vanishes to higher order. Thus, in a microscope window centered at the origin, **f** does not look like its linear approximation, because the linear approximation is not the dominant term in $\Delta\mathbf{x}$. Instead, **f** looks like (indeed, is equal to) the quadratic term $\mathbf{f}(\Delta\mathbf{u})$.

In summary, **f** will look like its linear approximation when that linear approximation is invertible, but need not otherwise. At the moment, we are relying only on an intuitive understanting of what it means for one map to "look like" another. We make the idea precise in the chapter on inverse maps (Chapter 5), where we say that two maps look alike if we can transform one into the other by a coordinate change. This is the same approach we took in Chapter 2 for linear maps. In that case, the coordinate change also involved finding a certain inverse map.

4.3 Parametrized surfaces

For another useful set of examples to illustrate the role of the derivative, we turn to surfaces in $\mathbb{R}^3$ given parametrically. Such a surface is the image of a map $\mathbf{f}: U^2 \to \mathbb{R}^3$ of a 2-dimensional region U^2 (in the same way that a parametrized curve is the image of a 1-dimensional interval). Our aim is to see how the derivative $d\mathbf{f}_\mathbf{u}$ is related to the map **f** near **u**.

Our first example is the unit sphere in $\mathbb{R}^3$, given parametrically as

$$\mathbf{f}: \begin{cases} x = \cos\theta \cos\varphi, \\ y = \sin\theta \cos\varphi, \\ z = \sin\varphi. \end{cases}$$

The image is indeed the unit sphere centered at the origin because every image point is exactly 1 unit from the origin:

$$x^2 + y^2 + z^2 = \cos^2\theta \cos^2\varphi + \sin^2\theta \cos^2\varphi + \sin^2\varphi = \cos^2\varphi + \sin^2\varphi = 1.$$

Because $\cos\theta$ and $\sin\theta$ have period 2π, it is sufficient to take $-\pi \le \theta \le \pi$. When $\varphi = 0$, we have

$$x = \cos\theta, \quad y = \sin\theta, \quad z = 0;$$

thus $\mathbf{f}(\theta,0)$ traces out the unit circle in the (x,y)-plane, the *equator* of the sphere. When $\varphi = \pm\pi/2$, we have $x = y = 0$ and $z = \pm 1$, the north and south poles of the sphere. It follows that $\mathbf{f}$ already covers the entire image even if we restrict θ and φ to the rectangular domain U^2:

$$-\pi \leq \theta \leq \pi, \quad -\frac{\pi}{2} \leq \varphi \leq \frac{\pi}{2}.$$

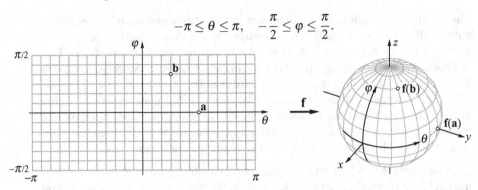

θ = longitude;
φ = latitude

Note how the images of the θ- and φ-axes appear on the sphere, as the equator and the prime meridian, respectively. The parameters θ and φ are evidently just the familiar *longitude* and *latitude*. The points **a** and **b** that are marked on the (θ,φ)-plane, and their images $\mathbf{f}(\mathbf{a})$ and $\mathbf{f}(\mathbf{b})$ on the sphere, are the two sites where we compare $\mathbf{f}$ with its derivative.

We first view the action of $\mathbf{f}$ itself in a microscope window centered at each point, and then compare that with the action of the derivative at that point. The source is 2-dimensional, so a window will still be a small square. The target, however, is 3-dimensional, so each target window will be a small cube.

Action of $\mathbf{f}$ at $(\pi/2, 0)$

The first point $\mathbf{a} = (\theta, \varphi) = (\pi/2, 0)$ has its image $\mathbf{f}(\mathbf{a})$ on the equator at 90° east longitude, a point that has target coordinates $(x,y,z) = (0,1,0)$. In the figure below, the microscope window centered at $(\theta, \varphi) = (\pi/2, 0)$ is a square 0.2 units on a side; the target window is a cube of the same dimensions centered at $(x,y,z) = (0,1,0)$. Following the figure is *Mathematica* 5 code that produces the image of the $(\Delta\theta, \Delta\varphi)$-plane in the target window.

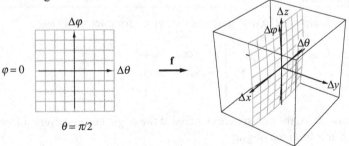

```
ParametricPlot3D[{Cos[u] Sin[v], Sin[u] Sin[v], Cos[v]},
    {u, Pi/2 - 0.1, Pi/2 + 0.1}, {v, -0.1, 0.1},
    PlotRange->{{-0.1,0.1}, {0.9,1.1}, {-0.1,0.1}},
    ViewPoint->{3.103, 2.109, 2.299}, PlotPoints->9]
```

The image is the portion of the sphere that lies in this window; it is nearly flat because the window is small. It approximates the plane $\Delta y = 0$, that is, the $(\Delta x, \Delta z)$-plane. It appears that $\mathbf{f}$ preserves lengths and angles: the image grid has the same size and shape as the source grid. The image of the $\Delta \varphi$-axis does not quite coincide with the Δz-axis, but is tangent to it. Likewise, the image of the $\Delta \theta$-axis is tangent to the Δx-axis, but has the opposite orientation.

The target window shows a small piece of the sphere

Let us now determine the action of the derivative $\mathbf{df}_{(\pi, 2, 0)}$ in the same microscope window. At an arbitrary point (θ, φ), the derivative map $\mathbf{df}_{(\theta, \varphi)} : \mathbb{R}^2 \to \mathbb{R}^3 : \Delta \boldsymbol{\theta} \mapsto \Delta \mathbf{x}$ given by the 3×2 matrix

Action of $\mathbf{df}_{(\pi/2, 0)}$

$$\mathbf{df}_{(\theta, \varphi)} = \begin{pmatrix} -\sin\theta \cos\varphi & -\cos\theta \sin\varphi \\ \cos\theta \cos\varphi & -\sin\theta \sin\varphi \\ 0 & \cos\varphi \end{pmatrix}.$$

For each (θ, φ) (except when $\cos \varphi = 0$), the image will therefore be a plane in $\mathbb{R}^3$. When $(\theta, \varphi) = (\pi/2, 0)$, the map $\Delta \mathbf{x} = \mathbf{df}_{(\pi/2, 0)}(\Delta \boldsymbol{\theta})$ is

$$\begin{pmatrix} \Delta x \\ \Delta y \\ \Delta z \end{pmatrix} = \begin{pmatrix} -1 & 0 \\ 0 & 0 \\ 0 & 1 \end{pmatrix} \begin{pmatrix} \Delta \theta \\ \Delta \varphi \end{pmatrix}, \quad \text{or just} \quad \begin{aligned} \Delta x &= -\Delta \theta, \\ \Delta y &= 0, \\ \Delta z &= \Delta \varphi. \end{aligned}$$

This is relatively easy to interpret. The equation $\Delta y = 0$ tells us the image is the $(\Delta x, \Delta z)$-plane. The $\Delta \varphi$-axis is mapped to the Δz-axis without stretching, and the $\Delta \theta$-axis is mapped to the Δx-axis, reversing direction but without stretching.

Our visual evidence indicates that $\mathbf{df}_{(\pi/2, 0)}$ is just the "flattening-out" of $\mathbf{f}$ in a microscope window centered at $(\pi/2, 0)$. It seems reasonable to say that $\mathbf{f}$ is locally linear at $(\pi/2, 0)$ and "looks like" its derivative there.

$\mathbf{f}$ looks like $\mathbf{df}_{(\pi/2, 0)}$ near $(\pi/2, 0)$

At the second point $\mathbf{b} = (\theta, \varphi) = (\pi/4, \pi/3)$, both $\mathbf{f}$ and its derivative are a bit more complicated to describe. The image $\mathbf{f}(\mathbf{b})$ lies in the northern hemisphere, at 60° north latitude and 45° east longitude; its target coordinates are $(x, y, z) = (\sqrt{2}/4, \sqrt{2}/4, \sqrt{3}/2)$.

Action of $\mathbf{f}$ at $(\pi/4, \pi/3)$

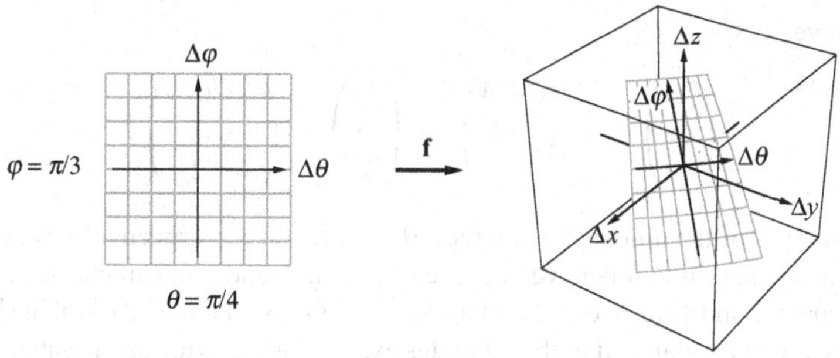

In the figure above, the microscope windows (both the square and the cube) are again 0.2 units on a side. The image is nearly flat, and is only about half as wide as it is tall. The image of the $\Delta \theta$-axis is horizontal; that is, it lies in the $(\Delta x, \Delta y)$-plane. The image of the $\Delta \varphi$-axis lies in the vertical plane where $\Delta x = \Delta y$. It would seem that we cannot tell by eye the angle between this image and the Δz-axis. But remember that we are just viewing a small portion of the sphere at 60° north latitude. That makes it obvious that the image of the $\Delta\varphi$-axis makes an angle of 60° with the vertical at the center of the window.

f does not *yet* look linear

The sides of the image seem pinched together, more so at the top than at the bottom; the grid of latitude and longitude lines is relatively far from rectangular. This is only to be expected, though. Away from the equator, longitude lines do pinch together toward the poles. In a linear map, parallel lines always have parallel images, so we must conclude that **f** does not look linear, at least at this scale.

But according to Taylor's theorem, **f** is indeed locally linear (everywhere away from the poles). We see that better when we magnify the view; the figure below is a tenfold magnification over the previous one.

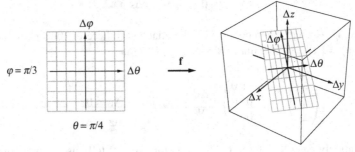

f under further magnification

At this magnification, the quadrilaterals in the image grid now look like congruent rectangles, so **f** now looks like a linear map. In the $\Delta\varphi$-direction, lengths are unaltered (the image rectangles are as tall as the original squares in the source), but in the $\Delta\theta$-direction, lengths are halved.

Action of $df_{(\pi/4,\pi/3)}$

The map $\Delta\mathbf{x} = df_{(\pi/4,\pi/3)}(\Delta\boldsymbol{\theta})$ defined by the derivative is

$$\begin{pmatrix} \Delta x \\ \Delta y \\ \Delta z \end{pmatrix} = \frac{1}{4}\begin{pmatrix} -\sqrt{2} & -\sqrt{6} \\ \sqrt{2} & -\sqrt{6} \\ 0 & 2 \end{pmatrix}\begin{pmatrix} \Delta\theta \\ \Delta\varphi \end{pmatrix}.$$

The two vectors

$$\begin{pmatrix} \Delta x \\ \Delta y \\ \Delta z \end{pmatrix} = \begin{pmatrix} -\sqrt{2}/4 \\ \sqrt{2}/4 \\ 0 \end{pmatrix}, \quad \begin{pmatrix} \Delta x \\ \Delta y \\ \Delta z \end{pmatrix} = \begin{pmatrix} -\sqrt{6}/4 \\ -\sqrt{6}/4 \\ 1/2 \end{pmatrix},$$

are the images of the unit vectors on the $\Delta\theta$- and $\Delta\varphi$-axes, respectively. We can see immediately that these image vectors are orthogonal, and you can check that their lengths are $1/2$ and 1, respectively. Thus, a square grid in the $(\Delta\theta, \Delta\varphi)$-plane has for its image a rectangular grid with rectangles exactly half as wide as the squares. The image of the $\Delta\theta$-axis lies in the $(\Delta x, \Delta y)$-plane because the image vector has $\Delta z = 0$. For a similar reason, the image of the $\Delta\varphi$-axis lies in the vertical plane $\Delta x = \Delta y$. Moreover, because the dot product of the image vector with the unit vector in the Δz-direction is $2/4 = 1/2$, the image makes an angle of 60° with the Δz-axis.

Once again we have compelling visual evidence. This time it indicates that $\mathrm{df}_{(\pi/4,\pi/3)}$ matches $\mathbf{f}$ in a sufficiently small microscope window centered at the point $(\pi/4,\pi/3)$. It seems reasonable to say that $\mathbf{f}$ is locally linear at $(\pi/4,\pi/3)$ and "looks like" its derivative there.

$\mathbf{f}$ looks like $\mathrm{df}_{(\pi/4,\pi/3)}$ near $(\pi/4,\pi/3)$

We noted that unit vectors on the $\Delta\theta$- and $\Delta\varphi$-axes are mapped to orthogonal vectors in the target that have lengths $1/2$ and 1, respectively. Thus, a unit square maps to a rectangle with area $1/2$; the local area multiplier at $(\theta,\varphi) = (\pi/4,\pi/3)$ appears to be $1/2$. In fact, the area multiplier for the linear map $\mathrm{df}_{(\pi/4,\pi/3)} : \mathbb{R}^2 \to \mathbb{R}^3$ is (Theorem 2.25, p. 55)

Area magnification at $(\pi/4,\pi/3)$

$$\sqrt{\left|\begin{matrix} \frac{\sqrt{2}}{4} & -\frac{\sqrt{6}}{4} \\ 0 & \frac{1}{2} \end{matrix}\right|^2 + \left|\begin{matrix} 0 & \frac{1}{2} \\ -\frac{\sqrt{2}}{4} & -\frac{\sqrt{6}}{4} \end{matrix}\right|^2 + \left|\begin{matrix} -\frac{\sqrt{2}}{4} & -\frac{\sqrt{6}}{4} \\ \frac{\sqrt{2}}{4} & -\frac{\sqrt{6}}{4} \end{matrix}\right|^2} = \sqrt{\tfrac{1}{32} + \tfrac{1}{32} + \tfrac{6}{32}} = \tfrac{1}{2}.$$

A similar calculation at $(\theta,\varphi) = (\pi/2,0)$ (using the matrix for $\mathrm{df}_{(\pi/2,0)}$) gives a local area magnification factor of

$$\sqrt{\left|\begin{matrix} 0 & 0 \\ 0 & 1 \end{matrix}\right|^2 + \left|\begin{matrix} 0 & 1 \\ -1 & 0 \end{matrix}\right|^2 + \left|\begin{matrix} -1 & 0 \\ 0 & 0 \end{matrix}\right|^2} = \sqrt{0+1+0} = 1;$$

this agrees with our discussion of the local action of $\mathbf{f}$ in the microscope window at $(\pi/2,0)$.

These examples lead us to the following definition.

Definition 4.5 *If the surface parametrization* $\mathbf{f}: U^2 \to \mathbb{R}^3$ *is differentiable at* $\mathbf{a}$, *its* ***local area multiplier*** *is the area multiplier of its derivative* $\mathrm{df}_{\mathbf{a}} : \mathbb{R}^2 \to \mathbb{R}^3$.

Local area multiplier

At an arbitrary point (θ,φ) in the domain of the sphere parametrization, the local area multiplier is $\cos\varphi$; see the exercises.

The next example is called a *crosscap*. It has a simple parametrization $\mathbf{f}: \mathbb{R}^2 \to \mathbb{R}^3$ in terms of polynomials defined on the entire plane.

The crosscap

$$\mathbf{f}: \begin{cases} x = u, \\ y = uv, \\ z = -v^2. \end{cases} \qquad \begin{aligned} \mathbf{a} &= (1,0) & \mathbf{f}(\mathbf{a}) &= (1,0,0) \\ \mathbf{b} &= (-1,1) & \mathbf{f}(\mathbf{b}) &= (-1,-1,-1) \\ \mathbf{c} &= (0,0) & \mathbf{f}(\mathbf{c}) &= (0,0,0) \end{aligned}$$

The image is a kind of parabolic arch that "crosses through" itself in the way shown in the figure. The u-axis is mapped to the x-axis along the ridge of the arch. The image of the v-axis folds back on itself along the line of self-intersection; both halves map to the negative z-axis. We do a local analysis at three different points. This time, though, we first compute the derivative and then compare it to the map itself.

Action of df at a = $(1,0)$

At an arbitrary point $\mathbf{p} = (p,q)$, the derivative $d\mathbf{f_p} : \mathbb{R}^2 \to \mathbb{R}^3$ is given by the 3×2 matrix

$$\begin{pmatrix} 1 & 0 \\ q & p \\ 0 & -2q \end{pmatrix}.$$

At $(p,q) = \mathbf{a} = (1,0)$, the derivative $d\mathbf{f_a}$ is the map

$$\begin{pmatrix} \Delta x \\ \Delta y \\ \Delta z \end{pmatrix} = \begin{pmatrix} 1 & 0 \\ 0 & 1 \\ 0 & 0 \end{pmatrix} \begin{pmatrix} \Delta u \\ \Delta v \end{pmatrix} \quad \text{or} \quad \begin{aligned} \Delta x &= \Delta u, \\ \Delta y &= \Delta v, \\ \Delta z &= 0. \end{aligned}$$

This is just the identity map of the $(\Delta u, \Delta v)$-plane to the $(\Delta x, \Delta y)$-plane in the target. All lengths and angles are preserved; the local area magnification factor is therefore equal to 1.

Action of f near a = $(1,0)$

Compare this with the action of $\mathbf{f}$ itself in a square microscope window, 0.2 units on a side, centered at $(p,q) = (1,0)$. Its image is the portion of the crosscap that appears in the small cubical window of the same dimensions, centered at $\mathbf{f(a)} = (1,0,0)$.

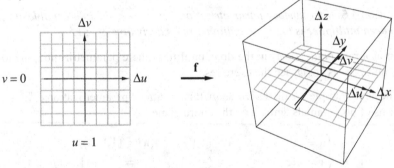

Apart from the slight curving (which would become even less noticeable if we increased the magnification), the image is the same size and shape as the source: $\mathbf{f}$ essentially preserves all lengths and angles in mapping the microscope window to the target, so $\mathbf{f}$ "looks like" $d\mathbf{f}_{(1,0)}$ near $(1,0)$.

Action of df at b = $(-1,1)$

At $(p,q) = \mathbf{b} = (-1,1)$, the derivative $d\mathbf{f_b}$ is the map

$$\begin{pmatrix} \Delta x \\ \Delta y \\ \Delta z \end{pmatrix} = \begin{pmatrix} 1 & 0 \\ 1 & -1 \\ 0 & -2 \end{pmatrix} \begin{pmatrix} \Delta u \\ \Delta v \end{pmatrix} \quad \text{or} \quad \begin{aligned} \Delta x &= \Delta u, \\ \Delta y &= \Delta u - \Delta v, \\ \Delta z &= -2\Delta v. \end{aligned}$$

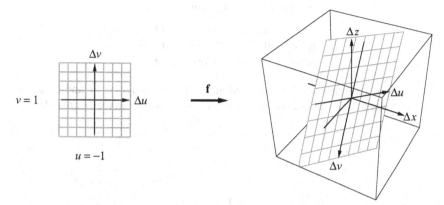

As always, the image is spanned by the images of the unit vectors in the Δu- and Δv-directions. In this case, the spanning vectors are $(1,1,0)$ and $(0,-1,-2)$ in $(\Delta x, \Delta y, \Delta z)$-space. The first lies in the $(\Delta x, \Delta y)$-plane, and the second lies in the $(\Delta y, \Delta z)$-plane, pointing downward (i.e., in the negative Δz-direction). The image of a coordinate grid of unit squares in the source consists of congruent parallelograms whose sides have lengths $\sqrt{2}$ and $\sqrt{5}$. Locally, areas are tripled (see the exercises).

The figure above shows the action of $\mathbf{f}$ itself in a microscope window centered at $(p, q) = \mathbf{b} = (-1, 1)$. The source window is a square 0.2 units on a side; the image cube is larger, about 0.4 units on a side, so that it can contain the entire image. As we see, the image of the Δu-axis lies in the $(\Delta x, \Delta y)$-plane, whereas the image of the Δv-axis lies in the $(\Delta y, \Delta z)$-plane, oriented so that the positive Δv-axis points down. The image coordinate grid appears to consist of congruent parallelograms that are taller than they are wide. The same figure serves to represent both $d\mathbf{f_b}$ and $\mathbf{f}$ near $\mathbf{b}$; it shows that $\mathbf{f}$ "looks like" its linear approximation near $\mathbf{b} = (-1, 1)$.

Action of f near
$\mathbf{b} = (-1, 1)$

At the origin the situation is not so simple. Perhaps this is to be expected, because it is the place where the crosscap "crosses" itself. The derivative $d\mathbf{f_c}$ is the linear map

Action of df
at $\mathbf{c} = (0, 0)$

$$\begin{pmatrix} \Delta x \\ \Delta y \\ \Delta z \end{pmatrix} = \begin{pmatrix} 1 & 0 \\ 0 & 0 \\ 0 & 0 \end{pmatrix} \begin{pmatrix} \Delta u \\ \Delta v \end{pmatrix} \quad \text{or} \quad \begin{aligned} \Delta x &= \Delta u, \\ \Delta y &= 0, \\ \Delta z &= 0. \end{aligned}$$

The rank of the matrix has dropped to 1. This means the image has dimension 1 instead of 2; instead of a plane, it has collapsed to a line. Indeed, the equations indicate the image is just the Δx-axis. Furthermore, the local area magnification factor is equal to 0.

In the exercises you compute the window equation

Action of f near
$\mathbf{c} = (0, 0)$

$$\Delta x = \mathbf{f}(\mathbf{p} + \Delta \mathbf{u}) - \mathbf{f}(\mathbf{p}) = d\mathbf{f_p}(\Delta \mathbf{u}) + \mathbf{R}_{1,\mathbf{p}}(\Delta \mathbf{u})$$

for the crosscap map $\mathbf{f}$ at an arbitrary point $\mathbf{p}$. Because $\mathbf{f}$ is a simple quadratic map, you get an explicit (quadratic) formula for the remainder $\mathbf{R}_{1,\mathbf{p}}$. At the origin, $\mathbf{p} = \mathbf{c} = \mathbf{0}$, the formula for $\Delta \mathbf{x}$ reduces to

$$\Delta \mathbf{x} = d\mathbf{f}_0(\Delta \mathbf{u}) + \mathbf{R}_{1,0}(\Delta \mathbf{u}),$$

$$\begin{aligned}
\Delta x &= & \Delta u & + & 0 & = O(1), \\
\Delta y &= & 0 & + & \Delta u \Delta v & = O(2), \\
\Delta z &= & 0 & + & -(\Delta v)^2 & = O(2).
\end{aligned}$$

Thus, in a microscope window centered at $\mathbf{x} = \mathbf{0}$, the values of Δy and Δz are an order of magnitude smaller than Δx. And that is what we see in the figure below.

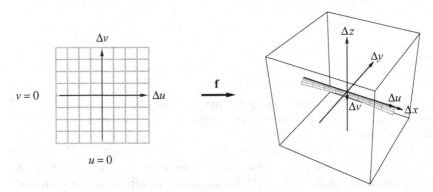

The image has been squeezed in the Δv-direction so that it fits into a narrow tube along the Δx-axis. The source and target windows are both 0.2 units on a side, but the tube's dimensions are only 0.01×0.01 in the Δy- and Δz-directions; they are an order of magnitude smaller than the long dimension.

How $\mathbf{f}$ differs from $d\mathbf{f}_0$

Does $\mathbf{f}$ "look like" $d\mathbf{f}_0$ in this window? Not quite. In the Δx-direction, the derivative $d\mathbf{f}_0$ vanishes only to order 1 but the remainder $\mathbf{R}_{1,0}$ vanishes to infinite order. The derivative dominates, and $\mathbf{f}$ does indeed look like $d\mathbf{f}_0$ in that direction. But in the Δy- and Δz-directions, $d\mathbf{f}_0$ now vanishes to infinite order, but $\mathbf{R}_{1,0}$—and thus $\mathbf{f}$ itself—vanishes only to order 2. The remainder dominates; $\mathbf{f}$ therefore looks like the remainder in those directions, and not like its derivative. So, even though $\mathbf{f}$ is "well-approximated" by its derivative at $\mathbf{u} = \mathbf{0}$ (i.e., even though $\mathbf{f}$ is differentiable), it does not look like its derivative in a microscope window centered there.

$\mathbf{f}$ "looks like" $d\mathbf{f}$ if $d\mathbf{f}$ has maximal rank

In our study of the quadratic map in the previous section we noted that the quadratic map (pp. 116–121) failed to look like its derivative locally at a point where the derivative itself failed to be invertible. Invertibility is out of the question here, because the source and target have different dimensions (the derivative is not a square matrix). The proper analogue is *maximal rank*:

If $d\mathbf{f}_\mathbf{a}$ has maximal rank, $\mathbf{f}$ will *look like* $d\mathbf{f}_\mathbf{a}$ near $\mathbf{a}$.

We investigate this point further in the chapter on implicit functions (Chapter 6).

4.4 The chain rule

In elementary calculus, every formula is assembled from a few simple types of functions—think of them as "atoms"—by using arithmetic and composition. We calculate the derivative of any such formula by knowing the derivatives of the atoms and the rules for differentiating arbitrary sums, products, quotients, and chains (or compositions). In this section we develop rules for derivatives of maps that generalize the sum, product, and chain rules. As in the one-variable case, the most important is the chain rule; with it we show how the derivatives of a map and its inverse are related.

Before considering the differentiation rules, we must say what it means for a map *Differentiability* between spaces of arbitrary dimension to be differentiable. Our definition is just the *of a map* generalization of the ones we have used in special cases (Definitions 4.1, p. 106 and 4.3, p. 115); U^p is a *window* of the form $|u_i - a_i| < q_i$, $i = 1, \cdots, p$.

Definition 4.6 *The map* $\mathbf{f} : U^p \to \mathbb{R}^q$ *is* **differentiable**, *or* **locally linear**, *at* $\mathbf{u} = \mathbf{a}$ *if there is a linear map* $\mathbf{L} : \mathbb{R}^p \to \mathbb{R}^q$, *called the* **derivative** *of* $\mathbf{f}$ *at* $\mathbf{a}$, *for which*

$$\mathbf{f}(\mathbf{a} + \Delta\mathbf{u}) = \mathbf{f}(\mathbf{a}) + \mathbf{L}(\Delta\mathbf{u}) + \mathbf{o}(1).$$

Theorem 4.6. *Suppose* $\mathbf{f} : U^p \to \mathbb{R}^q$ *is differentiable at* $\mathbf{u} = \mathbf{a}$; *then* $\mathbf{L} = d\mathbf{f_a}$. *In particular, if the component functions of* $\mathbf{f}$ *are* $f_i(\mathbf{u})$, *then all the partial derivatives* $\partial f_i / \partial u_j(\mathbf{a})$ *exist.*

Proof. Suppose the element in the *i*th row and *j*th column of the matrix representing $\mathbf{L}$ is ℓ_{ij}. We show that the partial derivative $\partial f_i / \partial u_j(\mathbf{a})$ exists and is equal to ℓ_{ij}. Let $\mathbf{e}_j = (0, \ldots, 0, 1, 0, \ldots, 0)$, the vector in $\mathbb{R}^p$ with 1 in the *j*th coordinate and 0 elsewhere. Then, by definition,

$$\frac{\partial f_i}{\partial u_j}(\mathbf{a}) = \lim_{h \to 0} \frac{f_i(\mathbf{a} + h\mathbf{e}_j) - f_i(\mathbf{a})}{h}.$$

By hypothesis, $f_i(\mathbf{a} + h\mathbf{e}_j) - f_i(\mathbf{a}) = L_i(h\mathbf{e}_j) + R_i(h\mathbf{u}) = hL_i(\mathbf{e}_j) + R_i(h\mathbf{u})$, where L_i is the *i*th component of $\mathbf{L}$ and R_i is the *i*th component of the remainder map that is represented by the symbol $\mathbf{o}(1)$. Therefore,

$$\frac{\partial f_i}{\partial u_j}(\mathbf{a}) = \lim_{h \to 0} \frac{f_i(\mathbf{a} + h\mathbf{e}_j) - f_i(\mathbf{a})}{h} = \lim_{h \to 0} \frac{hL_i(\mathbf{e}_j) + R_i(h\mathbf{e}_j)}{h}$$

$$= L_i(\mathbf{e}_j) + \lim_{h \to 0} \frac{R_i(h\mathbf{e}_j)}{h} = \ell_{ij}.$$

The final equation holds because $L_i(\mathbf{e}_j) = \ell_{ij}$, and $R_i(h\mathbf{e}_j)/h \to 0$ is simply a consequence of $R_i(\Delta\mathbf{u}) = o(1)$.   □

The theorem shows $\mathbf{L}$ is unique. We continue to equate differentiability and local *Differentiability is* linearity: $\mathbf{f}$ is differentiable at $\mathbf{a}$ if and only if $\Delta\mathbf{x} = \mathbf{f}(\mathbf{a} + \Delta\mathbf{u}) - \mathbf{f}(\mathbf{a})$ agrees with the *local linearity* linear map $df_a(\Delta u)$ to order greater than 1 as $\Delta u \to 0$. We use this characterization repeatedly in the rest of the section.

Derivative of a sum

Given two maps $f, g : U^p \to \mathbb{R}^q$, their sum and difference,

$$(f \pm g)(u) = f(u) \pm g(u),$$

are themselves maps of the same type: $f \pm g : U^p \to \mathbb{R}^q$. The following theorem extends to maps the familiar rule: *the derivative of a sum is the sum of the derivatives.*

Theorem 4.7. *If f and g are differentiable on U^p, then so are $f \pm g$, and*

$$d(f \pm g)_a(\Delta u) = df_a(\Delta u) \pm dg_a(\Delta u)$$

at any point a in U^p.

Proof. The statements are probably intuitively clear, but we prove them to illustrate our characterization of a differentiable map. Because f and g are differentiable at a, we can write

$$\begin{aligned}
(f \pm g)(a + \Delta u) &= f(a + \Delta u) \pm g(a + \Delta u) \\
&= f(a) + df_a(\Delta u) + o(1) \pm (g(a) + dg_a(\Delta u) + o(1)) \\
&= (f \pm g)(a) + (df_a \pm dg_a)(\Delta u) + o(1).
\end{aligned}$$

Thus $(f \pm g)(a + \Delta u) - (f \pm g)(a)$ agrees with the linear map $(df_a \pm dg_a)(\Delta u)$ to order greater than 1; by Theorem 4.6, the maps $f \pm g$ are differentiable at a, and $d(f \pm g)_a = df_a \pm dg_a$. $\square$

Derivative of a scalar product

There are product rules, too, at least when the products themselves are defined in meaningful ways. For example, we can compute the ordinary cross-product (or vector product) of maps whose target is $\mathbb{R}^3$, and we can compute the dot product (or scalar product) of any maps whose targets have the same dimension.

Theorem 4.8. *If $f, g : U^p \to \mathbb{R}^q$ are differentiable on U^p, then so is the scalar product $f \cdot g : U^p \to \mathbb{R}$, and*

$$d(f \cdot g)_a(\Delta u) = f(a) \cdot dg_a(\Delta u) + df_a(\Delta u) \cdot g(a).$$

Proof. By definition of the scalar product function, and then by the differentiability of f and g, we have

$$\begin{aligned}
(f \cdot g)(a + \Delta u &= f(a + \Delta u) \cdot g(a + \Delta u) \\
&= \big(f(a) + df_a(\Delta u) + o(1)\big) \cdot \big(g(a) + dg_a(\Delta u) + o(1)\big).
\end{aligned}$$

When we expand the right-hand side, we get nine individual scalar terms, five of which have $o(1)$ as a factor. Those five therefore all vanish to order greater than 1, so we combine them into a single (scalar) symbol $o(1)$:

$$(\mathbf{f} \cdot \mathbf{g})(\mathbf{a} + \Delta\mathbf{u}) = \mathbf{f}(\mathbf{a}) \cdot \mathbf{g}(\mathbf{a}) + \mathbf{f}(\mathbf{a}) \cdot \mathrm{dg}_{\mathbf{a}}(\Delta\mathbf{u}) + \mathrm{df}_{\mathbf{a}}(\Delta\mathbf{u}) \cdot \mathbf{g}(\mathbf{a})$$
$$+ \mathrm{df}_{\mathbf{a}}(\Delta\mathbf{u}) \cdot \mathrm{dg}_{\mathbf{a}}(\Delta\mathbf{u}) + o(1).$$

To decide whether $\mathbf{f} \cdot \mathbf{g}$ is differentiable, Theorem 4.6 suggests we rewrite the last equation (after setting $\mathbf{f}(\mathbf{a}) \cdot \mathbf{g}(\mathbf{a}) = (\mathbf{f} \cdot \mathbf{g})(\mathbf{a})$) in the form

$$(\mathbf{f} \cdot \mathbf{g})(\mathbf{a} + \Delta\mathbf{u}) - (\mathbf{f} \cdot \mathbf{g})(\mathbf{a}) = \mathbf{f}(\mathbf{a}) \cdot \mathrm{dg}_{\mathbf{a}}(\Delta\mathbf{u}) + \mathrm{df}_{\mathbf{a}}(\Delta\mathbf{u}) \cdot \mathbf{g}(\mathbf{a})$$
$$+ \mathrm{df}_{\mathbf{a}}(\Delta\mathbf{u}) \cdot \mathrm{dg}_{\mathbf{a}}(\Delta\mathbf{u}) + o(1).$$

Let us now consider, in turn, the first three terms on the right. We know $\mathrm{dg}_{\mathbf{a}}(\Delta\mathbf{u})$ is a linear function of $\Delta\mathbf{u}$, and so is its dot product with the scalar $\mathbf{f}(\mathbf{a})$. The second term is likewise a linear function of $\Delta\mathbf{u}$. In the third term, each of the two factors is linear; by Exercise 3.28 (p. 104), each factor vanishes at least to order 1. That is, there are constants $C_{\mathbf{f}}$ and $C_{\mathbf{g}}$ such that

$$\|\mathrm{df}_{\mathbf{a}}(\Delta\mathbf{u})\| \leq C_{\mathbf{f}}\|\Delta\mathbf{u}\|, \quad \|\mathrm{dg}_{\mathbf{a}}(\Delta\mathbf{u})\| \leq C_{\mathbf{g}}\|\Delta\mathbf{u}\|.$$

Therefore, because $|A \cdot B| \leq \|A\|\|B\|$ for any two vectors in $\mathbb{R}^q$,

$$|\mathrm{df}_{\mathbf{a}}(\Delta\mathbf{u}) \cdot \mathrm{dg}_{\mathbf{a}}(\Delta\mathbf{u})| \leq \|\mathrm{df}_{\mathbf{a}}(\Delta\mathbf{u})\|\|\mathrm{dg}_{\mathbf{a}}(\Delta\mathbf{u})\| \leq C_{\mathbf{f}}C_{\mathbf{g}}\|\Delta\mathbf{u}\|^2,$$

implying that $\mathrm{df}_{\mathbf{a}}(\Delta\mathbf{u}) \cdot \mathrm{dg}_{\mathbf{a}}(\Delta\mathbf{u})$ is $O(2)$ and hence can be absorbed into the term denoted $o(1)$. Thus we see $(\mathbf{f} \cdot \mathbf{g})(\mathbf{a} + \Delta\mathbf{u}) - (\mathbf{f} \cdot \mathbf{g})(\mathbf{a})$ agrees with the linear function $\mathbf{f}(\mathbf{a}) \cdot \mathrm{dg}_{\mathbf{a}}(\Delta\mathbf{u}) + \mathrm{df}_{\mathbf{a}}(\Delta\mathbf{u}) \cdot \mathbf{g}(\mathbf{a})$ to order greater than 1, so it follows that $\mathrm{d}(\mathbf{f} \cdot \mathbf{g})_{\mathbf{a}}(\Delta\mathbf{u}) = \mathbf{f}(\mathbf{a}) \cdot \mathrm{dg}_{\mathbf{a}}(\Delta\mathbf{u}) + \mathrm{df}_{\mathbf{a}}(\Delta\mathbf{u}) \cdot \mathbf{g}(\mathbf{a})$. □

We turn now to the chain rule. For functions of a single variable, the chain rule is commonly written two different ways, corresponding to the two ways we write derivatives. Suppose $s = f(u)$ and $x = \varphi(s)$; then $x = \varphi(f(u))$ and x therefore depends on u through the action of a new function composed of f and φ. We write the composed function as $x = (\varphi \circ f)(u)$, and write its derivative in terms of the derivatives of the components f and φ as either

The chain rule for one-variable functions

$$\frac{dx}{du} = \frac{dx}{ds}\frac{ds}{du} \quad \text{or} \quad (\varphi \circ f)'(a) = \varphi'(f(a))f'(a).$$

These are two formulations of the chain rule. The first uses the *Leibniz notation* for derivatives; its appeal is that it looks like an ordinary rule for multiplying fractions. The second calls attention to the fact that the derivatives of the individual functions φ and f must be evaluated at different points. It also reminds us that x depends on u one way (namely, through $\varphi \circ f$) but on s a different way (namely, through φ alone). The Leibniz notations dx/du and dx/ds suggest the same thing, though somewhat more obliquely.

Now suppose $\mathbf{f} : U^p \to \mathbb{R}^q$ and $\boldsymbol{\varphi} : S^q \to \mathbb{R}^r$ are differentiable maps with the image of $\mathbf{f}$ contained in the domain of $\boldsymbol{\varphi}$: $\mathbf{f}(U^p) \subseteq S^q$. Then the composite $\boldsymbol{\varphi} \circ \mathbf{f} : U^p \to \mathbb{R}^r$ is defined for all $\mathbf{u}$ in U^p: $(\boldsymbol{\varphi} \circ \mathbf{f})(\mathbf{u}) = \boldsymbol{\varphi}(\mathbf{f}(\mathbf{u}))$. Visually, we can think of $\mathbf{f}$ and $\boldsymbol{\varphi}$ as maps coming one after another in a linear "chain" as on the left, below.

Diagrams of maps

However, to show how they are related to the composite map $\boldsymbol{\varphi} \circ \mathbf{f}$, it is natural to put the three maps in a triangle:

$$U^p \xrightarrow{\ \mathbf{f}\ } S^q \xrightarrow{\ \boldsymbol{\varphi}\ } \mathbb{R}^r$$

The chain rule for maps says that the derivative of a composite is the composite of its derivatives. We state and prove the theorem below. You should check that the proof is just a rigorous version of the following "plausibility argument" based on the microscope equations for $\mathbf{f}$, $\boldsymbol{\varphi}$, and $\boldsymbol{\varphi} \circ \mathbf{f}$. Starting with

$$\Delta \mathbf{s} \approx \mathrm{df}_{\mathbf{a}}(\Delta \mathbf{u}) \ \text{ and } \ \Delta \mathbf{x} \approx \mathrm{d}\boldsymbol{\varphi}_{\mathbf{f}(\mathbf{a})}(\Delta \mathbf{s}),$$

it follows that

$$\Delta \mathbf{x} \approx \mathrm{d}\boldsymbol{\varphi}_{\mathbf{f}(\mathbf{a})}\big(\mathrm{df}_{\mathbf{a}}(\Delta \mathbf{u})\big) = (\mathrm{d}\boldsymbol{\varphi}_{\mathbf{f}(\mathbf{a})} \circ \mathrm{df}_{\mathbf{a}})(\Delta \mathbf{u}).$$

But, by definition, $\Delta \mathbf{x} \approx \mathrm{d}(\boldsymbol{\varphi} \circ \mathbf{f})_{\mathbf{a}}(\Delta \mathbf{u})$; because the linear map in the microscope equation is unique, $\mathrm{d}(\boldsymbol{\varphi} \circ \mathbf{f})_{\mathbf{a}} = \mathrm{d}\boldsymbol{\varphi}_{\mathbf{f}(\mathbf{a})} \circ \mathrm{df}_{\mathbf{a}}$.

Theorem 4.9 (Chain rule). *If* $\mathbf{f} : U^p \to S^q$ *is differentiable at* $\mathbf{a}$, *and* $\boldsymbol{\varphi} : S^q \to \mathbb{R}^r$ *is differentiable at* $\mathbf{f}(\mathbf{a})$, *then the composite map* $\boldsymbol{\varphi} \circ \mathbf{f} : U^p \to \mathbb{R}^r$ *is differentiable at* $\mathbf{a}$ *and*

$$\mathrm{d}(\boldsymbol{\varphi} \circ \mathbf{f})_{\mathbf{a}} = \mathrm{d}\boldsymbol{\varphi}_{\mathbf{f}(\mathbf{a})} \circ \mathrm{df}_{\mathbf{a}}.$$

Proof. It is possible to prove this result in terms of the component functions of the maps. However, we work directly with the maps themselves. According to our characterization of differentiability (and taking into account that $\mathrm{d}\boldsymbol{\varphi}_{\mathbf{f}(\mathbf{a})}\big(\mathrm{df}_{\mathbf{a}}(\Delta \mathbf{u})\big) = \mathrm{d}\boldsymbol{\varphi}_{\mathbf{f}(\mathbf{a})} \circ \mathrm{df}_{\mathbf{a}}(\Delta \mathbf{u})$ by definition), we must therefore show

$$(\boldsymbol{\varphi} \circ \mathbf{f})(\mathbf{a} + \Delta \mathbf{u}) - (\boldsymbol{\varphi} \circ \mathbf{f})(\mathbf{a}) = \mathrm{d}\boldsymbol{\varphi}_{\mathbf{f}(\mathbf{a})}\big(\mathrm{df}_{\mathbf{a}}(\Delta \mathbf{u})\big) + o(1).$$

This will prove that the derivative of $\boldsymbol{\varphi} \circ \mathbf{f}$ at $\mathbf{a}$ is $\mathrm{d}\boldsymbol{\varphi}_{\mathbf{f}(\mathbf{a})} \circ \mathrm{df}_{\mathbf{a}}$.

To begin, the differentiability of $\mathbf{f}$ at $\mathbf{a}$ allows us to write

$$(\boldsymbol{\varphi} \circ \mathbf{f})(\mathbf{a} + \Delta \mathbf{u}) = \boldsymbol{\varphi}\big(\mathbf{f}(\mathbf{a} + \Delta \mathbf{u})\big) = \boldsymbol{\varphi}\big(\underbrace{\mathbf{f}(\mathbf{a})}_{\mathbf{b}} + \underbrace{\mathrm{df}_{\mathbf{a}}(\Delta \mathbf{u}) + o(\Delta \mathbf{u})}_{\Delta \mathbf{s}}\big).$$

We write $o(\Delta \mathbf{u})$ here, instead of just $o(1)$, to stress that this particular remainder vanishes (to order greater than 1) with $\Delta \mathbf{u}$. Now use the differentiability of $\boldsymbol{\varphi}$ at $\mathbf{f}(\mathbf{a}) = \mathbf{b}$ to expand the right-hand side. This yields a second remainder $o(\Delta \mathbf{s})$ that vanishes with $\Delta \mathbf{s}$ and is thus distinct from $o(\Delta \mathbf{u})$:

$$\boldsymbol{\varphi}(\mathbf{b}) + d\boldsymbol{\varphi}_\mathbf{b}(\Delta\mathbf{s}) + o(\Delta\mathbf{s}) = \boldsymbol{\varphi}(\mathbf{f}(\mathbf{a})) + d\boldsymbol{\varphi}_{\mathbf{f}(\mathbf{a})}(d\mathbf{f}_\mathbf{a}(\Delta\mathbf{u}) + o(\Delta\mathbf{u})) + o(\Delta\mathbf{s})$$
$$= (\boldsymbol{\varphi} \circ \mathbf{f})(\mathbf{a}) + d\boldsymbol{\varphi}_{\mathbf{f}(\mathbf{a})}(d\mathbf{f}_\mathbf{a}(\Delta\mathbf{u})) + d\boldsymbol{\varphi}_{\mathbf{f}(\mathbf{a})}(o(\Delta\mathbf{u})) + o(\Delta\mathbf{s}).$$

We used the linearity of $d\boldsymbol{\varphi}_{\mathbf{f}(\mathbf{a})}$ to split the second term into two. We write the new remainder as $o(\Delta\mathbf{s})$ to indicate that it is a function of $\Delta\mathbf{s}$ rather than $\Delta\mathbf{u}$ and that, as such, it vanishes to order greater than 1 with $\Delta\mathbf{s}$. At this stage we have

$$(\boldsymbol{\varphi} \circ \mathbf{f})(\mathbf{a} + \Delta\mathbf{u}) - (\boldsymbol{\varphi} \circ \mathbf{f})(\mathbf{a}) = d\boldsymbol{\varphi}_{\mathbf{f}(\mathbf{a})}(d\mathbf{f}_\mathbf{a}(\Delta\mathbf{u})) + d\boldsymbol{\varphi}_{\mathbf{f}(\mathbf{a})}(o(\Delta\mathbf{u})) + o(\Delta\mathbf{s}).$$

It remains only to show that the last two terms on the right vanish to order greater than 1 in $\Delta\mathbf{u}$.

Lemma 4.1. $d\boldsymbol{\varphi}_{\mathbf{f}(\mathbf{a})}(o(\Delta\mathbf{u})) = o(\Delta\mathbf{u})$.

Proof. Because $d\boldsymbol{\varphi}_{\mathbf{f}(\mathbf{a})}$ is a linear map, we know (Exercise 3.28, p. 104) there is a positive constant C for which $\|d\boldsymbol{\varphi}_{\mathbf{f}(\mathbf{a})}(o(\Delta\mathbf{u}))\| \leq C\|o(\Delta\mathbf{u}))\|$. Therefore,

$$\lim_{\Delta\mathbf{u} \to 0} \frac{\|d\boldsymbol{\varphi}_{\mathbf{f}(\mathbf{a})}(o(\Delta\mathbf{u}))\|}{\|\Delta\mathbf{u}\|} \leq \lim_{\Delta\mathbf{u} \to 0} \frac{C\|o(\Delta\mathbf{u})\|}{\|\Delta\mathbf{u}\|} = 0,$$

by the definition of $o(\Delta\mathbf{u})$. Thus $d\boldsymbol{\varphi}_{\mathbf{f}(\mathbf{a})}(o(\Delta\mathbf{u})) = o(\Delta\mathbf{u})$. □

Lemma 4.2. *Let* $\Delta\mathbf{s} = d\mathbf{f}_\mathbf{a}(\Delta\mathbf{u}) + o(\Delta\mathbf{u})$; *then* $\Delta\mathbf{s} = O(\Delta\mathbf{u})$.

Proof. The first term, $d\mathbf{f}_\mathbf{a}(\Delta\mathbf{u})$, is linear, so by Exercise 3.28.a, it vanishes at least to order 1 in $\Delta\mathbf{u}$. The second term certainly vanishes at least to order 1 in $\Delta\mathbf{u}$, so $\Delta\mathbf{s} = O(\Delta\mathbf{u})$. □

Lemma 4.3. *If* $\Delta\mathbf{s} = O(\Delta\mathbf{u})$, *then* $o(\Delta\mathbf{s}) = o(\Delta\mathbf{u})$.

Proof. We must show $\|o(\Delta\mathbf{s})\|/\|\Delta\mathbf{u}\| \to 0$ as $\Delta\mathbf{u} \to 0$; note that there are different variables in the numerator and the denominator. The two variables are linked, however: $\Delta\mathbf{s} = O(\Delta\mathbf{u})$. In fact, this hypothesis means $\Delta\mathbf{s} \to 0$ as $\Delta\mathbf{u} \to 0$, suggesting that we write

$$\frac{\|o(\Delta\mathbf{s})\|}{\|\Delta\mathbf{u}\|} = \frac{\|o(\Delta\mathbf{s})\|}{\|\Delta\mathbf{s}\|} \cdot \frac{\|\Delta\mathbf{s}\|}{\|\Delta\mathbf{u}\|}.$$

Now the second factor on the right is bounded as $\Delta\mathbf{u} \to 0$, because $\Delta\mathbf{s} = O(\Delta\mathbf{u})$. The first factor tends to zero as $\Delta\mathbf{s} \to 0$, by definition of $o(\Delta\mathbf{s})$. Because $\Delta\mathbf{s} \to 0$ as $\Delta\mathbf{u} \to 0$, it appears we have shown that $\|o(\Delta\mathbf{s})\|/\|\Delta\mathbf{u}\|$ does indeed tend to 0 as $\Delta\mathbf{u} \to 0$.

But $\Delta\mathbf{s}$ may be zero for some $\Delta\mathbf{u} \neq 0$, so the first factor $\|o(\Delta\mathbf{s})\|/\|\Delta\mathbf{s}\|$ is undefined and the argument fails. We need to avoid quotients here. Fortunately, Exercise 3.17 (p. 102) provides an alternate formulation of "little oh" without quotients. The alternate formulation of the condition $o(\Delta\mathbf{s}) = o(\Delta\mathbf{u})$ that we seek to prove is as follows. For any given $\varepsilon > 0$, we must be able to find a $\delta > 0$ so that

$$\|o(\Delta\mathbf{s})\| \leq \varepsilon\|\Delta\mathbf{u}\| \quad \text{when} \quad \|\Delta\mathbf{u}\| < \delta.$$

To find δ, first note that $\Delta \mathbf{s} = \mathbf{O}(\Delta \mathbf{u})$ means, by definition, that there are positive constants δ_1 and C for which $\|\Delta \mathbf{s}\| \leq C\|\Delta \mathbf{u}\|$ when $\|\Delta \mathbf{u}\| < \delta_1$. The alternate formulation of "little oh" implies that, for the ε already given, we can choose δ_2 so that

$$\|\mathbf{o}(\Delta \mathbf{s})\| \leq \frac{\varepsilon}{C}\|\Delta \mathbf{s}\| \quad \text{when} \quad \|\Delta \mathbf{s}\| < \delta_2.$$

Finally, if we let δ be the smaller of δ_1 and δ_2/C, then $\|\Delta \mathbf{u}\| < \delta$ implies first that $\|\Delta \mathbf{s}\| \leq C\|\Delta \mathbf{u}\| < C\delta \leq \delta_2$ and consequently that

$$\|\mathbf{o}(\Delta \mathbf{s})\| \leq \frac{\varepsilon}{C}\|\Delta \mathbf{s}\| \leq \varepsilon\|\Delta \mathbf{u}\|. \qquad \square$$

To complete the proof of the theorem, we just need to combine the results of Lemmas 4.2 and 4.3 to conclude that if $\Delta \mathbf{s} = d\mathbf{f_a}(\Delta \mathbf{u}) + \mathbf{o}(\Delta \mathbf{u})$, then $\mathbf{o}(\Delta \mathbf{s}) = \mathbf{o}(\Delta \mathbf{u})$.
$\square$

Example: a chain of maps of the plane

Here is an example that shows how the chain rule works for a pair of maps of the plane. The first, $\mathbf{f}$, is the polar coordinate map and the second, $\boldsymbol{\varphi}$, is the conformal quadratic map; these are Examples 1 and 2 in Chapter 4.2.

$$\boldsymbol{\varphi} : \begin{cases} x = u^2 - v^2, \\ y = 2uv. \end{cases} \qquad \mathbf{f} : \begin{cases} u = \rho \cos \varphi, \\ v = \rho \sin \varphi, \end{cases}$$

Their composite is

$$\boldsymbol{\varphi} \circ \mathbf{f} : \begin{cases} x = \rho^2 \cos^2 \varphi - \rho^2 \sin^2 \varphi = \rho^2 \cos 2\varphi, \\ y = 2\rho \cos \varphi \cdot \rho \sin \varphi = \rho^2 \sin 2\varphi. \end{cases}$$

With these formulas for the component functions of $\boldsymbol{\varphi} \circ \mathbf{f}$, we can compute the derivative directly, without using the chain rule. At an arbitrary point (ρ, φ), the derivative is

$$d(\boldsymbol{\varphi} \circ \mathbf{f})_{(\rho, \varphi)} = \begin{pmatrix} 2\rho \cos 2\varphi & -2\rho^2 \sin 2\varphi \\ 2\rho \sin 2\varphi & 2\rho^2 \cos 2\varphi \end{pmatrix}.$$

Let us compare this with the derivative obtained with the chain rule. We start with the derivatives of the individual maps $\boldsymbol{\varphi}$ and $\mathbf{f}$:

$$d\boldsymbol{\varphi}_{(u,v)} = \begin{pmatrix} 2u & -2v \\ 2v & 2u \end{pmatrix}, \quad d\mathbf{f}_{(\rho,\varphi)} = \begin{pmatrix} \cos \varphi & -\rho \sin \varphi \\ \sin \varphi & \rho \cos \varphi \end{pmatrix}.$$

These have been evaluated at arbitrary points in the domains of the maps. But, in the chain rule, $d\boldsymbol{\varphi}$ must be evaluated at $\mathbf{f}(\rho, \varphi) = (\rho \cos \varphi, \rho \sin \varphi)$. Thus,

$$d\boldsymbol{\varphi}_{\mathbf{f}(\rho,\varphi)} = \begin{pmatrix} u & -v \\ v & u \end{pmatrix}\Bigg|_{\substack{u = \rho \cos \varphi \\ v = \rho \sin \varphi}} = \begin{pmatrix} 2\rho \cos \varphi & -2\rho \sin \varphi \\ 2\rho \sin \varphi & 2\rho \cos \varphi \end{pmatrix},$$

and the matrix product we seek is

$$d\boldsymbol{\varphi}_{f(\rho,\varphi)} \circ d\mathbf{f}_{(\rho,\varphi)} = \begin{pmatrix} 2\rho\cos\varphi & -2\rho\sin\varphi \\ 2\rho\sin\varphi & 2\rho\cos\varphi \end{pmatrix} \begin{pmatrix} \cos\varphi & -\rho\sin\varphi \\ \sin\varphi & \rho\cos\varphi \end{pmatrix}$$

$$= \begin{pmatrix} 2\rho\cos^2\varphi - 2\rho\sin^2\varphi & -2\rho^2\cos\varphi\sin\varphi - 2\rho^2\sin\varphi\cos\varphi \\ 2\rho\sin\varphi\cos\varphi + 2\rho\cos\varphi\sin\varphi & -2\rho^2\sin^2\varphi + 2\rho^2\cos^2\varphi \end{pmatrix}$$

$$= \begin{pmatrix} 2\rho\cos 2\varphi & -2\rho^2\sin 2\varphi \\ 2\rho\sin 2\varphi & 2\rho^2\cos 2\varphi \end{pmatrix} = d(\boldsymbol{\varphi}\circ\mathbf{f})_{(\rho,\varphi)}.$$

The chain rule evidently holds in this case.

For a second example, we consider two maps of the plane defined by arbitrary component functions:

Example: arbitrary maps of the plane

$$\boldsymbol{\varphi}: \begin{cases} x = \varphi(s,t), \\ y = \psi(s,t), \end{cases} \qquad \mathbf{f}: \begin{cases} s = f(u,v), \\ t = g(u,v). \end{cases}$$

The derivatives of these maps are the matrices

$$d\boldsymbol{\varphi}_{(s,t)} = \begin{pmatrix} \dfrac{\partial\varphi}{\partial s} & \dfrac{\partial\varphi}{\partial t} \\ \dfrac{\partial\psi}{\partial s} & \dfrac{\partial\psi}{\partial t} \end{pmatrix}, \quad d\mathbf{f}_{(u,v)} = \begin{pmatrix} \dfrac{\partial f}{\partial u} & \dfrac{\partial f}{\partial v} \\ \dfrac{\partial g}{\partial u} & \dfrac{\partial g}{\partial v} \end{pmatrix},$$

whose product is the derivative of the composite map $\boldsymbol{\varphi}\circ\mathbf{f}$:

$$d(\boldsymbol{\varphi}\circ\mathbf{f})_{(u,v)} = \begin{pmatrix} \dfrac{\partial\varphi}{\partial s}\dfrac{\partial f}{\partial u} + \dfrac{\partial\varphi}{\partial t}\dfrac{\partial g}{\partial u} & \dfrac{\partial\varphi}{\partial s}\dfrac{\partial f}{\partial v} + \dfrac{\partial\varphi}{\partial t}\dfrac{\partial g}{\partial v} \\ \dfrac{\partial\psi}{\partial s}\dfrac{\partial f}{\partial u} + \dfrac{\partial\psi}{\partial t}\dfrac{\partial g}{\partial u} & \dfrac{\partial\psi}{\partial s}\dfrac{\partial f}{\partial v} + \dfrac{\partial\psi}{\partial t}\dfrac{\partial g}{\partial v} \end{pmatrix}.$$

When we write out the components of the composite map,

$$\boldsymbol{\varphi}\circ\mathbf{f}: \begin{cases} x = \varphi(f(u,v),g(u,v)), \\ y = \psi(f(u,v),g(u,v)), \end{cases}$$

we express x and y directly as functions of u and v; if we use $\partial x/\partial u$, and so forth, to denote the partial derivatives of these functions, then

$$d(\boldsymbol{\varphi}\circ\mathbf{f})_{(u,v)} = \begin{pmatrix} \dfrac{\partial x}{\partial u} & \dfrac{\partial x}{\partial v} \\ \dfrac{\partial y}{\partial u} & \dfrac{\partial y}{\partial v} \end{pmatrix}.$$

We now have two formulas for the derivative $d(\boldsymbol{\varphi}\circ\mathbf{f})_{(u,v)}$. Together they give us the chain rule for the individual component functions:

$$\frac{\partial x}{\partial u} = \frac{\partial \varphi}{\partial s}\frac{\partial f}{\partial u} + \frac{\partial \varphi}{\partial t}\frac{\partial g}{\partial u}, \qquad \frac{\partial x}{\partial v} = \frac{\partial \varphi}{\partial s}\frac{\partial f}{\partial v} + \frac{\partial \varphi}{\partial t}\frac{\partial g}{\partial v},$$

$$\frac{\partial y}{\partial u} = \frac{\partial \psi}{\partial s}\frac{\partial f}{\partial u} + \frac{\partial \psi}{\partial t}\frac{\partial g}{\partial u}, \qquad \frac{\partial y}{\partial v} = \frac{\partial \psi}{\partial s}\frac{\partial f}{\partial v} + \frac{\partial \psi}{\partial t}\frac{\partial g}{\partial v}.$$

Chain rule for component functions

There are clear patterns in these four equations that allow us to see what form the component derivatives will take in the general case. We start with two variables (u and v) that first determine two others (s and t) directly and then two more (x and y) indirectly. The partial derivative of x with respect to u, for example, must have terms that take into account how x varies with u via s (viz. $\partial \varphi/\partial s \cdot \partial f/\partial u$) and via t ($\partial \varphi/\partial t \cdot \partial g/\partial u$)).

In the general case, p variables ($u_1, \ldots, u_p$) first determine the values of q new variables ($s_1, \ldots, s_q$) directly, and then r additional variables ($x_1, \ldots, x_r$) indirectly:

$$x_k = \varphi_k(s_1,\ldots,s_q) \quad \text{and} \quad s_j = f_j(u_1,\ldots,u_p),$$

for $k = 1,\ldots,r$ and $j = 1,\ldots,q$. Thus, a partial derivative of x_k, for example, must take into account how x_k varies with each u_i via each of the q intermediate variables s_j:

$$\frac{\partial x_k}{\partial u_i} = \frac{\partial \varphi_k}{\partial s_1}\frac{\partial f_1}{\partial u_i} + \cdots + \frac{\partial \varphi_k}{\partial s_q}\frac{\partial f_q}{\partial u_i}, \quad i = 1,\ldots,p.$$

The following theorem summarizes this discussion, with the single variable y replacing the various $x_1, \ldots, x_r$.

Theorem 4.10. *Suppose the functions* $y = \varphi(s_1,\ldots,s_q)$ *and* $s_j = f_j(u_1,\ldots,u_p)$, *with* $j = 1,\ldots,q$, *are all differentiable; then*

$$\frac{\partial y}{\partial u_i} = \sum_{j=1}^{q} \frac{\partial \varphi}{\partial s_j}\frac{\partial f_j}{\partial u_i}, \quad i = 1,\ldots,p. \qquad \Box$$

Derivative of the inverse

The following corollary of the chain rule says that the derivative of the inverse (of a given map) is the inverse of the derivative (of that map).

Corollary 4.11 *Suppose* $\mathbf{f}: U^n \to S^n$ *is invertible, and* $\mathbf{f}^{-1}: S^n \to U^n$ *is its inverse. Suppose that both* $\mathbf{f}$ *and* $\mathbf{f}^{-1}$ *are differentiable; then*

$$(d\mathbf{f_u})^{-1} = d(\mathbf{f}^{-1})_{\mathbf{f(u)}}.$$

Proof. Let $\mathbf{I}: U^n \to U^n$ be the identity map, $\mathbf{I(u)} = \mathbf{u}$. Then $\mathbf{f}^{-1} \circ \mathbf{f} = \mathbf{I}$; by the chain rule

$$I = d\mathbf{I_u} = d(\mathbf{f}^{-1})_{\mathbf{f(u)}} \circ d\mathbf{f_u},$$

where I is the linear map represented by the $n \times n$ identity matrix. The equation implies that $d(\mathbf{f}^{-1})_{\mathbf{f(u)}}$ is the inverse of $d\mathbf{f_u}$. $\qquad \Box$

Local orientation and volume magnification

The corollary focuses our attention on maps $\mathbf{f}: U^n \to S^n$ whose source and target have the same dimension. We have already studied some examples when $n = 2$ in the second section of this chapter; in particular, we saw we could use the area multiplier of the derivative to assign a local area multiplier (p. 115) to the map itself at each point. The following definition carries these ideas over to higher dimensions.

Definition 4.7 *Suppose* $\mathbf{f} : U^n \to \mathbb{R}^n$ *is differentiable; the* **local volume multiplier** *of* $\mathbf{f}$ *at* $\mathbf{a}$, *written* $J_{\mathbf{f}}(\mathbf{a})$, *is* $\det d\mathbf{f}_{\mathbf{a}}$, *the volume multiplier of the derivative of* $\mathbf{f}$ *at* $\mathbf{a}$. *Also,* $\mathbf{f}$ *is* **orientation-preserving** *or* **reversing** *at* $\mathbf{a}$ *according as its derivative* $d\mathbf{f}_{\mathbf{a}}$ *is orientation-preserving or reversing.*

The chain rule implies that the local volume multiplier of a composite is the product of their individual multipliers, as we would expect.

Volume magnification in a chain

Corollary 4.12 *If* $\mathbf{f} : U^n \to S^n$ *and* $\boldsymbol{\varphi} : S^n \to \mathbb{R}^n$ *are differentiable, then*

$$J_{\boldsymbol{\varphi} \circ \mathbf{f}}(\mathbf{a}) = J_{\boldsymbol{\varphi}}(\mathbf{f}(\mathbf{a}))\, J_{\mathbf{f}}(\mathbf{a}).$$

Proof. The proof is just a consequence of the fact that the determinant of a product is the product of the determinants:

$$J_{\boldsymbol{\varphi} \circ \mathbf{f}}(\mathbf{a}) = \det d(\boldsymbol{\varphi} \circ \mathbf{f})_{\mathbf{a}} = \det(d\boldsymbol{\varphi}_{\mathbf{f}(\mathbf{a})} \circ d\mathbf{f}_{\mathbf{a}})$$

$$= \det d\boldsymbol{\varphi}_{\mathbf{f}(\mathbf{a})}\, \det d\mathbf{f}_{\mathbf{a}} = J_{\boldsymbol{\varphi}}(\mathbf{f}(\mathbf{a}))\, J_{\mathbf{f}}(\mathbf{a}). \qquad \square$$

The traditional name for $J_{\mathbf{f}}(\mathbf{a})$ is the **Jacobian**; hence the letter "J". In this context, the matrix $d\mathbf{f}_{\mathbf{a}}$ is the **Jacobian matrix** (the Jacobian itself is always the *determinant*). The Jacobian plays a central role in multiple integrals. We show it is the analogue of the factor $\varphi'(s)$ that appears in the transformation $dx = \varphi'(s)\,ds$ of differentials (pp. 3–5). For that reason, we write it another way (also traditional) that suggests the connection with derivatives more directly. To illustrate, let

The Jacobian

$$\mathbf{f} : \begin{cases} x = f(u,v), \\ y = g(u,v); \end{cases}$$

then our alternate notation for the Jacobian of $\mathbf{f}$ is

$$J(u,v) = \frac{\partial(x,y)}{\partial(u,v)} = \frac{\partial(f,g)}{\partial(u,v)}.$$

Here we write $J(u,v)$ without the subscript for the map $\mathbf{f}$. This is frequently done, and it directs attention to the Jacobian as a function of the input variables. The second and third expressions are the more common ones; they remind us that the Jacobian involves partial derivatives. The Jacobian of the polar coordinate map is

Jacobian notation

$$\frac{\partial(x,y)}{\partial(r,\theta)} = \begin{vmatrix} \dfrac{\partial}{\partial r} r\cos\theta & \dfrac{\partial}{\partial \theta} r\cos\theta \\[2mm] \dfrac{\partial}{\partial r} r\sin\theta & \dfrac{\partial}{\partial \theta} r\sin\theta \end{vmatrix} = \begin{vmatrix} \cos\theta & -r\sin\theta \\ \sin\theta & r\cos\theta \end{vmatrix} = r.$$

This agrees with our original determination of the local area multiplier for polar coordinates (p. 115).

Definition 4.8 *Suppose* $\mathbf{f}: U^n \to \mathbb{R}^n$ *is differentiable and has component functions*

$$x_i = f_i(u_1,\ldots,u_n), \quad i = 1,\ldots,n.$$

Then the **Jacobian** *of* $\mathbf{f}$ *is the determinant*

$$J(u_1,\ldots,u_n) = \frac{\partial(x_1,\ldots,x_n)}{\partial(u_1,\ldots,u_n)} = \frac{\partial(f_1,\ldots,f_n)}{\partial(u_1,\ldots,u_n)} = \det\left(\frac{\partial f_i}{\partial u_j}\right) = \det d\mathbf{f}_{\mathbf{u}}.$$

The following are restatements of Corollaries 4.11 and 4.12 using Jacobian notation. Because they deal with inverses and with Jacobians, it becomes practical to replace function names by names of output variables (e.g., to replace $s_i = f_i(u_1,\ldots,u_n)$ by $s_i = s_i(u_1,\ldots,u_n)$).

Corollary 4.13 *If a map and its inverse are both differentiable,*

$$\mathbf{f}: \begin{cases} s_1 = s_1(u_1,\ldots,u_n), \\ \vdots \\ s_n = s_n(u_1,\ldots,u_n), \end{cases} \qquad \mathbf{f}^{-1}: \begin{cases} u_1 = u_1(s_1,\ldots,s_n), \\ \vdots \\ u_n = u_n(s_1,\ldots,s_n), \end{cases}$$

then

$$\frac{\partial(u_1,\ldots,u_n)}{\partial(s_1,\ldots,s_n)} = \frac{1}{\dfrac{\partial(s_1,\ldots,s_n)}{\partial(u_1,\ldots,u_n)}}. \qquad \square$$

Chain rule for Jacobians

Corollary 4.14 *If the following are differentiable,*

$$\boldsymbol{\varphi}: \begin{cases} x_1 = x_1(s_1,\ldots,s_n), \\ \vdots \\ x_n = x_n(s_1,\ldots,s_n), \end{cases} \qquad \mathbf{f}: \begin{cases} s_1 = s_1(u_1,\ldots,u_n), \\ \vdots \\ s_n = s_n(u_1,\ldots,u_n), \end{cases}$$

then

$$\frac{\partial(x_1,\ldots,x_n)}{\partial(u_1,\ldots,u_n)} = \frac{\partial(x_1,\ldots,x_n)}{\partial(s_1,\ldots,s_n)} \frac{\partial(s_1,\ldots,s_n)}{\partial(u_1,\ldots,u_n)}. \qquad \square$$

These results obviously remind us of the one-variable cases: if $u = u(s)$ is the inverse of $s = s(u)$, and $x = x(s)$, then

$$\frac{du}{ds} = \frac{1}{ds/du} \quad \text{and} \quad \frac{dx}{du} = \frac{dx}{ds}\frac{ds}{du}.$$

Local area multiplier on a surface patch

Although Jacobians are determinants and consequently involve an equal number of input and output variables, they can appear in other circumstances. For example, the parametrization of a surface patch, $\mathbf{f} : U^2 \to \mathbb{R}^3$, involves three functions of two real variables:

$$\mathbf{f} : \begin{cases} x = f(u,v), \\ y = g(u,v), \\ z = h(u,v). \end{cases}$$

The linear map $d\mathbf{f_a} : \mathbb{R}^2 \to \mathbb{R}^3$ has an area magnification factor that is a kind of "Pythagorean formula" (Theorem 2.25, p. 55). The formula involves the 2×2 minors of $d\mathbf{f_a}$, which can be written in a simple and direct way using Jacobians; the result is given in the following definition.

Definition 4.9 *The **local area multiplier** for the parametrized surface patch* $\mathbf{f} : U^2 \to \mathbb{R}^3 : (u,v) \mapsto (x,y,z)$ *is*

$$M(u,v) = \sqrt{\left[\frac{\partial(y,z)}{\partial(u,v)}\right]^2 + \left[\frac{\partial(z,x)}{\partial(u,v)}\right]^2 + \left[\frac{\partial(x,y)}{\partial(u,v)}\right]^2}.$$

For example, the crosscap we analyzed earlier (pp. 126–128) has the parametrization

Example: the crosscap

$$\mathbf{f} : \begin{cases} x = u, \\ y = uv, \\ z = -v^2, \end{cases} \qquad d\mathbf{f_u} = \begin{pmatrix} 1 & 0 \\ v & u \\ 0 & -2v \end{pmatrix},$$

so the local area multiplier is

$$\sqrt{\left|\begin{matrix} v & u \\ 0 & -2v \end{matrix}\right|^2 + \left|\begin{matrix} 0 & -2v \\ 1 & 0 \end{matrix}\right|^2 + \left|\begin{matrix} 1 & 0 \\ v & u \end{matrix}\right|^2} = \sqrt{4v^4 + 4v^2 + u^2}.$$

At $(u,v) = (1,0)$, the multiplier is 1. This agrees with what we saw earlier: near $(1,0)$, $\mathbf{f}$ preserves areas. At $(u,v) = (-1,1)$, the multiplier is 3. This too agrees with our earlier analysis: near $(-1,1)$, $\mathbf{f}$ triples areas.

The chain rule gives us a way to prove the mean-value theorem for maps of the form $\mathbf{f} : U^p \to \mathbb{R}^q$. The mean-value theorem says that, for any two points $\mathbf{a}$ and $\mathbf{b}$ in U^p,

A mean-value theorem
for maps

$$\|\mathbf{f}(\mathbf{b}) - \mathbf{f}(\mathbf{a})\| \le M\|\mathbf{b} - \mathbf{a}\|,$$

where M is a bound on the size of the derivative of $\mathbf{f}$ at points along the line from $\mathbf{a}$ to $\mathbf{b}$. We need to establish what the "size" of the derivative is.

In Theorem 3.8 and the discussion preceding it (pp. 76–77), we had $q = 1$. Therefore we were able to identify the derivative with a vector and the size of the derivative with the magnitude of that vector. When $q \ge 2$, the derivative is a more general linear map; to measure its size, we use its *norm* (see Exercise 3.28.b, p. 104). The norm of $d\mathbf{f_u}$ is

The *norm* of
a derivative

$$\||d\mathbf{f_u}\|| = \max_{\|\Delta\mathbf{v}\|=1} \|d\mathbf{f_u}(\Delta\mathbf{v})\|;$$

we have $\|\mathbf{df_u}(\Delta\mathbf{v})\| \leq \||\mathbf{df_u}\|\| \|\Delta\mathbf{v}\|$ for all $\Delta\mathbf{v}$ in $\mathbb{R}^p$. The norm of a linear map is the largest amount by which it stretches any vector.

Theorem 4.15 (Mean-value theorem). *Suppose the map* $\mathbf{f}: U^p \to \mathbb{R}^q$ *is continuously differentiable and* U^p *contains every point on the line segment from* $\mathbf{a}$ *to* $\mathbf{b}$. *Then*

$$\|\mathbf{f}(\mathbf{b}) - \mathbf{f}(\mathbf{a})\| \leq \max_{\mathbf{u}} \||\mathbf{df_u}\|\| \|\mathbf{b} - \mathbf{a}\|,$$

where the maximum is taken over all points $\mathbf{u}$ *on the line from* $\mathbf{a}$ *to* $\mathbf{b}$.

Write the difference as an integral

Proof. We begin by constructing the analogue of the error formula

$$\int_0^1 f'(a + t\Delta x)\Delta x\, dt = f(a + \Delta x) - f(a)$$

with which we began the discussion of Taylor's theorem (cf. pp. 78–79). Set $\Delta\mathbf{u} = \mathbf{b} - \mathbf{a}$; then $\mathbf{u}(t) = \mathbf{a} + t\Delta\mathbf{u}$, $0 \leq t \leq 1$, is the line segment from $\mathbf{a}$ to $\mathbf{b}$. All the points on this segment are also in U^p; therefore the map

$$\boldsymbol{\varphi}(t) = \mathbf{f}(\mathbf{u}(t)) = \mathbf{f}(\mathbf{a} + t\Delta\mathbf{u})$$

is continuously differentiable on $[0, 1]$. By the chain rule,

$$\boldsymbol{\varphi}'(t) = \mathbf{d}\boldsymbol{\varphi}_t = \mathbf{df}_{\mathbf{u}(t)}(\mathbf{du}_t) = \mathbf{df}_{\mathbf{a}+t\Delta\mathbf{u}}(\Delta\mathbf{u});$$

we have used $\mathbf{du}_t = \mathbf{u}'(t) = \Delta\mathbf{u}$. Thus

$$\int_0^1 \mathbf{df}_{\mathbf{a}+t\Delta\mathbf{u}}(\Delta\mathbf{u})\, dt = \int_0^1 \boldsymbol{\varphi}'(t)\, dt = \boldsymbol{\varphi}(1) - \boldsymbol{\varphi}(0)$$
$$= \mathbf{f}(\mathbf{a} + \Delta\mathbf{u}) - \mathbf{f}(\mathbf{a}) = \mathbf{f}(\mathbf{b}) - \mathbf{f}(\mathbf{a}).$$

This is, in fact, just Taylor's formula with remainder in degree 0 for the map $\mathbf{f}$. It implies

$$\|\mathbf{f}(\mathbf{b}) - \mathbf{f}(\mathbf{a})\| = \left\|\int_0^1 \mathbf{df}_{\mathbf{a}+t\Delta\mathbf{u}}(\Delta\mathbf{u})\, dt\right\| \leq \int_0^1 \|\mathbf{df}_{\mathbf{a}+t\Delta\mathbf{u}}(\Delta\mathbf{u})\|\, dt.$$

Because $\Delta\mathbf{u} = \mathbf{b} - \mathbf{a}$ is fixed, we have

$$\|\mathbf{df}_{\mathbf{a}+t\Delta\mathbf{u}}(\Delta\mathbf{u})\| \leq \max_{0 \leq t \leq 1} \|\mathbf{df}_{\mathbf{a}+t\Delta\mathbf{u}}(\Delta\mathbf{u})\| \leq \max_{0 \leq t \leq 1} \||\mathbf{df}_{\mathbf{a}+t\Delta\mathbf{u}}\|\| \|\Delta\mathbf{u}\|.$$

The right-hand side is independent of t, so (with $\mathbf{u} = \mathbf{a} + t\Delta\mathbf{u}$)

$$\|\mathbf{f}(\mathbf{b}) - \mathbf{f}(\mathbf{a})\| \leq \max_{0 \leq t \leq 1} \||\mathbf{df}_{\mathbf{a}+t\Delta\mathbf{u}}\|\| \|\Delta\mathbf{u}\| \leq \max_{\mathbf{u}} \||\mathbf{df_u}\|\| \|\mathbf{b} - \mathbf{a}\|. \qquad \square$$

Exercises

4.1. Determine the derivative of the given function $z = f(x,y)$ at the given point $(x,y) = (a,b)$. Write the derivative as a linear function of the variables $\Delta x = x - a$, $\Delta y = y - b$.

   a. $f(x,y) = 7x - 3y + 9$, $(a,b) = (4,5)$.
   b. $f(x,y) = 1 - \cos x + y^2/2$, $(a,b) = (0,1)$.
   c. $f(x,y) = \arctan(y/x)$, $(a,b) = (4,-3)$.
   d. $f(x,y) = \alpha x^2 + 2\beta xy + \gamma y^2 + \delta x + \varepsilon y + \kappa$, (a,b) arbitrary (and α, β, γ, δ, ε, κ are all constants).

4.2. a. Let $f(x,y) = x^2 - y^2$. Plot the graph of $z = f(x,y)$ in a window centered at $(x,y) = (2,-1)$, making the window small enough for the graph to appear to be a flat plane.
   b. Confirm that the plane appearing in (a) is the graph of the derivative $df_{(2,-1)}$. Because the derivative is expressed in terms of the displacement variables $\Delta x = x - 2$, $\Delta y = y + 1$, it is necessary to include the constant term $f(2,-1) = 3$ in the equation for the second graph.
   c. Construct a contour plot of $f(x,y)$ in two windows centered at $(x,y) = (2,-1)$. Make the first window 2.0 units on a side, and make the second 0.2. Confirm that f changes its appearance from nonlinear to linear from the first to the second window, and confirm that the level curves of f are indistinguishable from the level curves of its derivative there.

4.3. a. Determine the equation of the tangent plane to the graph of $z = f(x,y) = \sin x \sin y$ at the point $(x,y) = (\pi/3, -\pi/2)$.
   b. Sketch, together, the graph of f and this tangent plane over the square $0 \le x \le \pi$, $-\pi \le y \le 0$.
   c. Select a smaller square centered at $(\pi/3, -\pi/2)$ on which the graph of f and this tangent plane become indistinguishable; sketch the two surfaces over that square.
   d. Sketch together contour plots of f and the derivative of f at $(\pi/3, -\pi/2)$ on the square $0 \le x \le \pi$, $-\pi \le y \le 0$. Sketch them again in the smaller square you selected in part (c).

4.4. a. Explain what $\cos t - 1 = o(1)$ means, and then show that it is true. (Suggestion: Use l'Hôpital's rule.)
   b. Is $\sin t = o(1)$ true? Explain.
   c. Explain what $\sin t - t = o(2)$ means, and then show that it is true. Is it true that $\sin t - t = o(3)$? Explain. Is $\sin t - t = O(3)$ true? Explain.

4.5. Show $f(x) = x^2 \sin(1/x)$ is differentiable at $x = 0$, that $f'(0) = 0$, and even that $f(\Delta x) = f(0) + f'(0)\Delta x + O(2)$. Show that, nevertheless, $f''(0)$ does not exist. Moreover, show that $f'(x)$ is not a continuous function: If $x_n = 1/(2\pi n)$, then $f'(x_n) = 1$; however, $f'(x_n) \nrightarrow f'(0)$ even though $x_n \to 0$.

4.6. Let $z = f(x,y)$ be the "manta ray" function (pp. 108–109). Show analytically that $z = 0$ is not the tangent plane to f at the origin by showing the gap $f(\Delta x, \Delta y) - f(0,0)$ does not vanish to order greater than 1; that is, show directly (cf. Definition 3.14, p. 98) that the ratio

$$\frac{f(\Delta x, \Delta y) - f(0,0)}{\|(\Delta x, \Delta y)\|} = \frac{(\Delta x)^2 \Delta y / ((\Delta x)^2 + (\Delta y)^2)}{\sqrt{(\Delta x)^2 + (\Delta y)^2}}$$

does not have the limit 0 as $(\Delta x, \Delta y) \to (0,0)$.

4.7. Let $f(0,0) = 0$, $f(x,y) = \dfrac{2xy}{\sqrt{x^2 + y^2}}$ for $(x,y) \neq (0,0)$.

a. Sketch the graph of $z = f(x,y)$ near the origin. Use a polar coordinate overlay to clarify the picture.

b. Show that the partial derivatives $f_x(0.0)$ and $f_y(0,0)$ exist, and determine their values.

c. Show that the directional derivative $D_{\mathbf{u}}f(0,0)$ does not exist if $\mathbf{u}$ is not an axis direction. Explain this result in terms of the graph of f.

d. Conclude that f, like the "manta ray" counterexample, fails to be differentiable at the origin. (In fact, for both this function and the "manta ray," $f(x,y) = O(1)$ is true but $f(x,y) = o(1)$ is false.)

4.8. Let $f(0,0) = 0$, $f(x,y) = \dfrac{x^3 y}{x^4 + y^2}$ for $(x,y) \neq (0,0)$.

a. Sketch the graph of $z = f(x,y)$ near the origin. (A polar coordinate overlay is not as helpful here; it does not simplify our view of the graph.)

b. Show that the directional derivative $D_{\mathbf{u}}f(0,0)$ exists and equals 0 in every direction $\mathbf{u}$. (In particular, $f_x(0,0) = f_y(0,0) = 0$; therefore, if f were differentiable at $(0,0)$, the tangent plane to its graph at the origin would be the (x,y)-plane.)

c. Compute the partial derivatives $f_x(x,y)$ and $f_y(x,y)$ at any arbitrary point, and show they are not continuous at $(x,y) = (0,0)$.

d. Add to your sketch in part (a) the curve $z = f(x,x^2)$ in the graph of f that lies over the parabola $y = x^2$. Show that $z = x/2$ along the parabola, implying that f vanishes exactly to order 1 on the parabola. Conclude that f cannot be differentiable at the origin.

4.9. Let $\mathbf{f}: \mathbb{R}^2 \to \mathbb{R}^2$ and $\mathbf{g}: \mathbb{R}^2 \to \mathbb{R}^2$ be given by

$$\mathbf{f}: \begin{cases} x = au + bv + k, \\ y = cu + dv + l, \end{cases} \qquad \mathbf{g}: \begin{cases} r = \alpha x + \beta y + \kappa, \\ s = \gamma x + \delta y + \lambda. \end{cases}$$

a. Determine the derivative $d\mathbf{f_u}$ at an arbitrary point $\mathbf{u} = (u, v)$. Does the derivative depend on (u, v)? How is $d\mathbf{f_u}$ related to $\mathbf{f}$?

b. Determine $d\mathbf{g_x}$ and then use the (components of the) substitution $\mathbf{x} = \mathbf{f}(\mathbf{u})$ to express the derivative in terms of $\mathbf{u}$: $d\mathbf{g_x} = d\mathbf{g}_{\mathbf{f(u)}}$.

c. Compute the components of the composite map $\mathbf{h} = \mathbf{g} \circ \mathbf{f}$. That is, determine r and s in terms of u and v; $\mathbf{h}(\mathbf{u}) = \mathbf{g}(\mathbf{f}(\mathbf{u}))$.

d. Determine $d\mathbf{h_u}$ and verify that $d\mathbf{h_u} = d\mathbf{g}_{\mathbf{f(u)}} \cdot d\mathbf{f_u}$. (This is the chain rule; see Theorem 4.9, p. 132.)

4.10. Let $\mathbf{f} : U^2 \to \mathbb{R}^2$ be the polar coordinate map, and let $\mathbf{f}^{-1} : \mathbb{R}^2 \to U^2$ be its inverse:

$$\mathbf{f} : \begin{cases} x = r\cos\theta, \\ y = r\sin\theta, \end{cases} \qquad \mathbf{f}^{-1} : \begin{cases} r = \sqrt{x^2 + y^2}, \\ \theta = \arctan(y/x). \end{cases}$$

a. Compute the matrix of the derivative $d\mathbf{f}_{\mathbf{x}}^{-1}$ at an arbitrary point $\mathbf{x} = (x, y)$.

b. Compute the inverse matrix $(d\mathbf{f_r})^{-1}$ at an arbitrary point $\mathbf{r} = (r, \theta)$.

c. Use the coordinate change $\mathbf{x} = \mathbf{f}(\mathbf{r})$ to express $d\mathbf{f}_{\mathbf{x}}^{-1}$ in terms of $\mathbf{r}$: $d\mathbf{f}_{\mathbf{x}}^{-1} = d\mathbf{f}_{\mathbf{f(r)}}^{-1}$. Then verify that $(d\mathbf{f_r})^{-1} = d\mathbf{f}_{\mathbf{f(r)}}^{-1}$. (The derivatives of inverse maps are themselves inverses of each other. To see this it is first necessary, however, to express them in terms of the same variables. See Theorem 4.11, p. 136.)

d. Compute the determinants

$$\det d\mathbf{f}_{\mathbf{x}}^{-1} \quad \text{and} \quad \det d\mathbf{f_r},$$

and show that they are reciprocals. Use an appropriate change of variables on one of the expressions to compare the two.

4.11. Let $\mathbf{x} = \mathbf{f}(\mathbf{r})$ be the polar coordinate map of the previous exercise, and let $\mathbf{u} = \mathbf{g}(\mathbf{x})$ be the map

$$\mathbf{g} : \begin{cases} u = x^2 + y^2, \\ v = x^2 - y^2. \end{cases}$$

a. Determine $d\mathbf{g_x}$ at an arbitrary point $\mathbf{x} = (x, y)$; then use the coordinate change $\mathbf{x} = \mathbf{f}(\mathbf{r})$ to express the derivative in terms of $\mathbf{r} = (r, \theta)$: $d\mathbf{g}_{\mathbf{f(r)}}$.

b. Compute the components of the composite map $\mathbf{h} = \mathbf{g} \circ \mathbf{f}$. That is, determine u and v in terms of r and θ; $\mathbf{u} = \mathbf{h}(\mathbf{r}) = \mathbf{g}(\mathbf{f}(\mathbf{r}))$.

c. Verify that

$$d\mathbf{h_r} = \begin{pmatrix} 2r & 0 \\ 2r\cos 2\theta & -r^2 \sin 2\theta \end{pmatrix}$$

and also verify that $d\mathbf{h_r} = d\mathbf{g}_{\mathbf{f(r)}} \cdot d\mathbf{f_r}$. (This is another instance of the chain rule.)

In Exercises 4.12–4.20, $\mathbf{f} : \mathbb{R}^2 \to \mathbb{R}^2$ is the quadratic map discussed in the text:

$$\mathbf{f} : \begin{cases} x = u^2 - v^2, \\ y = 2uv. \end{cases}$$

4.12. Compute $\|\mathbf{f}(\Delta \mathbf{u})\|^2$ to show that $\|\mathbf{f}(\Delta \mathbf{u})\| = \|\Delta \mathbf{u}\|^2$. Note: this means that $\mathbf{f}(\Delta \mathbf{u})$ vanishes exactly to order 2 (by the extension of Definition 3.4 used in Exercise 3.28). That is, there are positive constants C_1, C_2 for which

$$C_1 \leq \frac{\|\mathbf{f}(\Delta \mathbf{u})\|}{\|\Delta \mathbf{u}\|^2} \leq C_2$$

for all $\Delta \mathbf{u} \neq \mathbf{0}$. (It is evident we can take $C_1 = C_2 = 1$.)

4.13. Let $\mathcal{U}^2 = \{(x,y) \mid y > 0\}$ be the upper half-plane, and let $\mathbf{g}_+ : \mathcal{U}^2 \to \mathbb{R}^2$ be the map

$$\mathbf{g}_+ : \begin{cases} u = \sqrt{\dfrac{\sqrt{x^2+y^2}+x}{2}}, \\[4mm] v = \sqrt{\dfrac{\sqrt{x^2+y^2}-x}{2}}. \end{cases}$$

a. Show that $\mathbf{f}(\mathbf{g}_+(\mathbf{x})) = \mathbf{x}$ for all $\mathbf{x}$ in $\mathcal{U}^2$. In other words, $\mathbf{g}_+$ is a (partial) inverse for $\mathbf{f}$.

b. Describe the action of $\mathbf{g}_+$ in terms of polar coordinate overlays on the source and target (cf. pp. 116–121). In other words, describe what happens to the angle a point makes with the positive horizontal axis change, and what happens to its distance from the origin change.

c. Describe the image of $\mathbf{g}_+(\mathcal{U}^2)$

4.14. Show that $\mathbf{g}_+(\mathbf{x})$ (Exercise 4.13) can be extended to the two sides of the x-axis ($y = 0$) as follows:

$$\begin{matrix} u = \sqrt{x}, \\ v = 0, \end{matrix} \quad \text{if } x \geq 0; \qquad \begin{matrix} u = 0, \\ v = \sqrt{|x|}, \end{matrix} \quad \text{if } x \leq 0.$$

What, therefore, is the image of the x-axis under this extension of $\mathbf{g}_+$?

4.15. a. Compute $d(\mathbf{g}_+)_{\mathbf{x}}$, where $\mathbf{g}_+$ is the map of Exercise 4.13.

b. Use the coordinate change $x = u^2 - v^2$, $y = 2uv$ (provided by the inverse map $\mathbf{f}$) to express $d(\mathbf{g}_+)_{\mathbf{x}}$ in terms of u and v, giving $d(\mathbf{g}_+)_{\mathbf{f}(\mathbf{u})}$.

c. Verify that $d(\mathbf{g}_+)_{\mathbf{f}(\mathbf{u})}$ is the inverse of $d\mathbf{f}_{\mathbf{u}}$.

4.16. The object of this exercise is to study the action of $\mathbf{g}_+$ (Exercise 4.13) in a microscope window centered at $\mathbf{x}_0 = (1/2, \sqrt{3}/2)$.

a. Determine the center $\mathbf{g}_+(\mathbf{x}_0)$ of the target window.

   b. Explain why g_+ maps the 60°-line in the window at x_0 to the 30°-line in the target window. (Suggestion: Use results from Exercise 4.13.)

   c. Show that $d(g_+)_{x_0} = \lambda R_\theta$, for certain $\lambda > 0$ and $\theta < 0$; R_θ is rotation by θ radians. What are the values of λ and θ? Does g_+ "look like" its derivative $d(g_+)_{x_0}$ in this microscope window? Explain, in terms of what you know about the action of g_+.

4.17. Let $\mathcal{L}^2 = \{(x,y) \mid y < 0\}$ be the lower half-plane, and let $g_- : \mathcal{L}^2 \to \mathbb{R}^2$ be the map

$$g_- : \begin{cases} u = -\sqrt{\dfrac{\sqrt{x^2+y^2}+x}{2}}, \\[2ex] v = \sqrt{\dfrac{\sqrt{x^2+y^2}-x}{2}}. \end{cases}$$

Note that g_- differs from g_+ only in the sign of u.

   a. Show that $f(g_-(x)) = x$ for all x in $\mathcal{L}^2$. In other words, g_- is also a partial inverse for f.

   b. Describe the action of g_- in terms of polar coordinate overlays on the source and target (cf. Exercise 4.13).

   c. Determine the image $g_-(\mathcal{L}^2)$.

4.18. a. Show that $g_-(x)$ can be extended to the two sides of the x-axis ($y = 0$) as follows:

$$\begin{aligned} u = -\sqrt{x}, & \\ v = 0, & \end{aligned} \quad \text{if } x \geq 0; \qquad \begin{aligned} u = 0, & \\ v = \sqrt{|x|}, & \end{aligned} \quad \text{if } x \leq 0.$$

What, therefore, is the image of the x-axis under this extension of g_-?

   b. Show that g_+ and g_- agree on the negative x-axis but disagree on the positive x-axis. Excluding, therefore, the positive x-axis from the domain of g_-, explain how g_+ and g_- together define a single map on the whole plane $\mathbb{R}^2$ that serves as an inverse for f. Show that this combined map is not continuous across the positive x-axis.

4.19. a. Compute $d(g_-)_x$, where g_- is given in Exercise 4.17.

   b. Use the coordinate change $x = u^2 - v^2$, $y = 2uv$ (provided by the inverse map $x = f(u)$) to express $d(g_-)_x$ in terms of $u = (u,v)$, giving $d(g_-)_{f(u)}$.

   c. Verify that $d(g_-)_{f(u)}$ is the inverse of df_u.

4.20. The object of this exercise is to study the action of g_- (Exercise 4.17) in a microscope window centered at $x_0 = (0, -9/4)$.

   a. Determine the center $g_-(x_0)$ of the target window.

   b. Explain why g_- maps the 270°-line in the window at x_0 to the 135°-line in the target window.

c. Show that $d(\mathbf{g}_-)_{\mathbf{x}_0} = \lambda R_\theta$, for certain $\lambda > 0$ and $\theta < 0$. What are the values of λ and θ? Does $\mathbf{g}_-$ "look like" its derivative $d(\mathbf{g}_-)_{\mathbf{x}_0}$ in this microscope window? Explain, in terms of what you know about the action of $\mathbf{g}_-$.

4.21. Let $\mathbf{f}: \mathbb{R}^2 \to \mathbb{R}^2 : (u,v) \mapsto (x,y)$ be the map defined by

$$\mathbf{f}: \begin{cases} x = u, \\ y = v^2. \end{cases}$$

a. Describe the global behavior of $\mathbf{f}$ by showing what happens to an ordinary Cartesian grid in the source. In particular, indicate the effect of the equation $y = v^2$. This map is sometimes called a *fold*; your picture should explain why.

b. Determine the derivative $d\mathbf{f}_{(a,b)}$ at each point $(u,v) = (a,b)$.

c. Show that the local area multiplier of $\mathbf{f}$ at (a,b) is b. Hence the local area multiplier along the horizontal axis (the u-axis) is 0; why? What feature of the map $\mathbf{f}$ does this reflect?

d. Sketch the effect of $\mathbf{f}$ in a microscope window centered at $(u,v) = (3,2)$, and indicate how this corresponds to the effect of the derivative $d\mathbf{f}_{(3,2)}$.

e. Sketch the effect of $\mathbf{f}$ in a microscope window centered at $(u,v) = (3,0)$, and indicate how this corresponds to the effect of the derivative $d\mathbf{f}_{(3,0)}$. (Note: the local area multiplier here is 0, and the derivative $d\mathbf{f}_{(3,0)}$ is non-invertible. Thus we do not expect $\mathbf{f}$ to look like $d\mathbf{f}_{(3,0)}$ in a microscope window centered at $(3,0)$.)

4.22. a. Obtain the derivative $d\mathbf{q}_{\mathbf{a}}$ of the map

$$\mathbf{q}: \begin{cases} x = u^3 - 3uv^2, \\ y = 3u^2v - v^3. \end{cases}$$

b. Show that $d\mathbf{q}_{(a,b)}$ is a similarity transformation (cf. p. 118), that is, a rotation by an angle θ combined with a uniform dilation by a factor λ. Determine θ and λ in terms of a and b. Conclude that $\mathbf{q}$ is conformal on the whole plane minus the origin. Why must the origin be excluded?

c. Use a polar coordinate overlay to create a description of the action of $\mathbf{q}$ that is analogous to the description of the quadratic map (as one that doubles angles and squares distances from the origin).

4.23. Repeat all the steps of the previous exercise for the map

$$\mathbf{s}: \begin{cases} x = u^4 - 6u^2v^2 + v^4, \\ y = 4u^3v - 4uv^3. \end{cases}$$

In particular, find $\lambda = 4(a^2 + b^2)^{3/2}$ and $\theta = \arctan\left(\dfrac{3a^2b - b^3}{a^3 - a3b^2}\right)$.

4.24. a. Show that the local area magnification factor at the parameter point (θ, φ) on the unit sphere is $\cos \varphi$.

   b. Show that the arc length of the "parallel of latitude" at latitude φ_0 is $2\pi \cos \varphi_0$. Show that the arc length of a "meridian of longitude" at longitude θ_0 is π, independent of θ_0.

4.25. a. Compute the matrix of the derivative $df_{(\pi/3, \pi/6)}$ of the unit sphere map f, and describe its image in $\mathbb{R}^3$.

   b. Sketch the image of the unit sphere map f in a microscope window at the point $(\pi/3, \pi/6)$. Show that the window can be made small enough so the image is indistinguishable from the image of the derivative $df_{(\pi/3, \pi/6)}$.

4.26. Show that the local area magnification factor at the parameter point (p, q) on the crosscap is $\sqrt{p^2 + 4q^2 + 4q^4}$. Confirm that the local area magnification factors at the points $\mathbf{a}$, $\mathbf{b}$, and $\mathbf{c}$ discussed in the text are, respectively, 1, 3, and 0.

4.27. Show that the window equation

$$\Delta x = f(p + \Delta u) - f(p) = df_p(\Delta u) + R_{1,p}(\Delta u)$$

for the crosscap parametrization f at an arbitrary point $p = (p, q)$ can be written with the remainder $R_{1,p}$ explicitly as

$$\begin{aligned}
\Delta x &= & df_p(\Delta u) &+ R_{1,p}(\Delta u), \\
\Delta x &= & \Delta u &+ & 0, \\
\Delta y &= q\,\Delta u + p\,\Delta v + & & \Delta u\,\Delta v, \\
\Delta z &= & -2q\,\Delta v &+ -(\Delta v)^2.
\end{aligned}$$

Note that $R_{1,p}(\Delta u)$ is purely quadratic in Δu and is independent of p.

In exercises 4.28–4.30, the map $t_{R,a} : U^2 \to \mathbb{R}^3$ (where $0 < a < R$) parametrizes a torus:

$$t_{R,a} : \begin{cases} x = (R + a\cos\varphi)\cos\theta, \\ y = (R + a\cos\varphi)\sin\theta, \\ z = a\sin\varphi, \end{cases} \qquad \begin{aligned} 0 &\le \theta \le 2\pi, \\ -\pi &\le \varphi \le \pi. \end{aligned}$$

4.28. Make a sketch of the entire image of $t_{3,1}$. From this, describe what R and a measure on the torus. What happens to the shape of the torus if $R < a$?

4.29. a. Compute the (matrix of the) derivative $d(t_{R,a})_{(\theta, \varphi)}$ at an arbitrary point (θ, φ) and for arbitrary R and a.

   b. Determine the local area magnification factor of $t_{R,a}$ at an arbitrary point (θ, φ), and confirm that it is independent of θ. For which value of φ is the factor largest, and for which is it smallest? Is this consistent with your sketch of the action of $t_{3,1}$?

4.30. a. Sketch the image of $\mathbf{t}_{3,1}$ in a microscope window at $(\theta, \varphi) = (\pi/4, \pi/3)$, and compare it to the image of the derivative of $\mathbf{t}_{3,1}$ at that point.

   b. Do the same at the point $(\theta, \varphi) = (0, \pi/2)$.

4.31. Let $x = g(v) > 0$ be a smooth function for $a \le v \le b$. The surface parametrized as

$$\mathbf{r}: \begin{cases} x = g(v) \cos \theta, \\ y = g(v) \sin \theta, \\ z = v, \end{cases} \qquad \begin{matrix} 0 \le \theta \le 2\pi, \\ a \le v \le b, \end{matrix}$$

is a **surface of revolution**. The curve $x = g(z)$ is called its **generator**.

   a. Determine the derivative $d\mathbf{r}_{(\theta,v)}$ at an arbitrary point, and determine the local area magnification factor.

   b. Confirm that the local area magnification factor is independent of θ, and is smallest where $g(v)$ has its minimum.

   c. Show that when $g(v) = 2 + \sin kv$, and k is properly chosen, the largest local area magnification factor does not occur where $g(v)$ has its maximum.

4.32. Suppose $y = f(x)$ is differentiable at $x = a$; show that $df_a(\Delta x) = f'(a) \cdot \Delta x$.

4.33. Show that a linear map is differentiable everywhere, and is its own derivative: if $\mathbf{L}: \mathbb{R}^p \to \mathbb{R}^q$ is linear then $d\mathbf{L}_\mathbf{a} = \mathbf{L}$, for every $\mathbf{a}$ in $\mathbb{R}^p$.

4.34. Let $\mathbf{f}: U^2 \to \mathbb{R}^2: (\rho, \varphi) \mapsto (u, v)$ be the polar coordinate map (p. 134), and let

$$\mathbf{f}^{-1}: \begin{cases} \rho = \sqrt{u^2 + v^2}, \\ \varphi = \arctan(y/x), \end{cases}$$

be its inverse. Show that their derivatives are inverses; that is, show that

$$d(\mathbf{f}^{-1})_{\mathbf{f}(\rho,\varphi)} = (d\mathbf{f}_{(\rho,\varphi)})^{-1}.$$

Note that, for the equality to hold, the two sides must be expressed in terms of the same variables; thus, the derivative $d\mathbf{f}^{-1}_{(u,v)}$ must be determined at the point $(u, v) = (\rho, \varphi) = (\rho \cos \varphi, \rho \sin \varphi)$.

4.35. Determine the Jacobians $\partial(x,y)/\partial(u,v)$ and $\partial(u,v)/\partial(x,y)$ when

$$x = u^3 - 3uv^2,$$
$$y = 3u^2v - v^3.$$

4.36. Let $\varphi(t) = \Phi(x(t), y(t))$, where Φ is differentiable and $x = x(t)$ and $y = y(t)$ be differentiable functions of t.

   a. Verify that $\varphi'(t) = \Phi_x(x(t),y(t))x'(t) + \Phi_y(x(t),y(t))y'(t)$, or, more briefly, $\varphi' = \text{grad}\,\Phi \cdot (x', y')$.

b. Suppose Φ is a *potential* for $\mathbf{F}$: $\mathbf{F}(\mathbf{x}) = \text{grad}\,\Phi(\mathbf{x})$ (cf. p. 25), and $\vec{C}$ is an oriented curve. Fill in the details to show that

$$\int_{\vec{C}} \mathbf{F} \cdot d\mathbf{x} = \int_a^b d\varphi = \Phi(\mathbf{x})\Big|_{\text{start of } \vec{C}}^{\text{end of } \vec{C}}.$$

4.37. Suppose $\mathbf{f}: U^2 \to \mathbb{R}^2 : (u,v) \mapsto (x(u,v), y(u,v))$ is a continuously differentiable map, not necesarily invertible. Let $\vec{C}$ be a piecewise-smooth oriented curve in U^2 for which $\mathbf{f}(\vec{C})$ is also piecewise smooth and oriented (cf. p. 9). Assume $\vec{C}$ and $\mathbf{f}(\vec{C})$ have a common decomposition into smooth oriented curves:

$$\vec{C} = \vec{C}_1 + \cdots + \vec{C}_m, \quad \mathbf{f}(\vec{C}) = \mathbf{f}(\vec{C}_1) + \cdots + \mathbf{f}(\vec{C}_m);$$

each $\vec{C}_i$ and $\mathbf{f}(\vec{C}_i)$ is either a simple closed curve or a simple curve (i.e., no self-intersections). If $\mathbf{u} = \mathbf{u}_i(t)$, $a_i \le t \le b_i$ is a continuously differentiable parametrization of $\vec{C}_i$, $i = 1, \ldots, m$, then $\mathbf{x} = \mathbf{f}(\mathbf{u}_i(t))$ is a continuously differentiable parametrization of $\mathbf{f}(\vec{C}_i)$.

a. Let $P(x,y)$ and $Q(x,y)$ be continuously differentiable functions defined on $\mathbf{f}(U^2)$; show that

$$\int_{\mathbf{f}(\vec{C}_i)} P\,dx + Q\,dy = \int_{\vec{C}_i} (P^*x_u + Q^*y_u)\,du + (P^*x_v + Q^*y_v)\,dv.$$

Here, $P^* = P^*(u,v) = P(x(u,v), y(u,x))$, $x_u = \partial x / \partial u$, and so forth. The equation describes how the path integral on the left is transformed into the one on the right by the change of variables $(x,y) = \mathbf{f}(u,v)$.

b. Deduce that

$$\int_{\mathbf{f}(\vec{C})} P\,dx + Q\,dy = \int_{\vec{C}} (P^*x_u + Q^*y_u)\,du + (P^*x_v + Q^*y_v)\,dv.$$

4.38. Let $\mathbf{f}: (r, \theta) \mapsto (x, y)$ be the polar coordinate map, and let $\vec{C}$ be any continuously differentiable oriented curve in the (r, θ)-plane with $r > 0$. Determine how the path integral

$$I = \int_{\mathbf{f}(\vec{C})} \frac{-y}{x^2 + y^2}\,dx + \frac{x}{x^2 + y^2}\,dy$$

is transformed by polar coordinates. Use the transformed integral to show that

$$I = \Delta\theta = \theta\Big|_{\text{start of } \mathbf{f}(\vec{C})}^{\text{end of } \mathbf{f}(\vec{C})}.$$

4.39. Let $\mathbf{f}: (x,y) = (u^2 - v^2, 2uv)$ be the quadratic map, and let $\vec{C}$ be any continuously differentiable oriented curve in the (u,v)-plane that avoids the origin. Show that

$$\int_{\mathbf{f}(\vec{C})} \frac{-y}{x^2+y^2}\, dx + \frac{x}{x^2+y^2}\, dy = 2 \int_{\vec{C}} \frac{-v}{u^2+v^2}\, du + \frac{u}{u^2+v^2}\, dv$$

and conclude that

$$\int_{\mathbf{f}(\vec{C})} \frac{-y}{x^2+y^2}\, dx + \frac{x}{x^2+y^2}\, dy = 2\arctan\left(\frac{v}{u}\right)\Bigg|_{\text{start of }\vec{C}}^{\text{end of }\vec{C}}.$$

Chapter 5
Inverses

Abstract Inverses help us solve equations: if $5 = x^3$, then $x = \sqrt[3]{5}$. Equations also imply relations between their variables. For example, if $x^2 + y^2 - 1 = 0$, then we can "solve for y" to get either $y = +\sqrt{1-x^2}$ or $y = -\sqrt{1-x^2}$. We soon learn that a formula for an inverse or for an implicitly defined function is seldom available. Usually, the most we can expect to know is that such a function exists. As we show, even this apparently limited knowledge can simplify and clarify our view of a problem, the same way that changing coordinates can simplify an integration. In this chapter, we look only briefly at explicit formulas. We give the bulk of our attention to the way inverses give us a powerful tool for understanding maps, and to the conditions that guarantee their existence. The next chapter does the same for implicitly defined functions.

5.1 Solving equations

The first inverse operations we learn are subtraction and division; after all, $x = y/m$ is the inverse of $y = mx$. And division is the first place where we see that an inverse may not exist: "You cannot divide by zero" is the way we say that $y = 0 \times x$ has no inverse. We use subtraction and division to solve equations, at the start, just linear equations of the form $y = mx + b$. After this come polynomial equations and the square root $x = \sqrt{y}$, introduced as the inverse of $y = x^2$. The square root function shows us that an inverse may have a restricted domain of definition ($y \geq 0$ in this case) and a restricted range (we need $x = -\sqrt{y}$ along with $x = +\sqrt{y}$).

Early examples

For each new function in calculus, an inverse is introduced with it; the exponential and logarithm functions provide a good example. The immediate use of inverses is in solving equations, including even those that give alternate formulas for inverses themselves. For example, the hyperbolic cosine function $y = \cosh x$ has an inverse that is written simply $x = \operatorname{arccosh} y$ (or $x = \cosh^{-1} y$). But we can get a different— and possibly more useful—expression for the inverse by solving the defining equation

Inverse of the hyperbolic cosine

J.J. Callahan, *Advanced Calculus: A Geometric View*, Undergraduate Texts in Mathematics, 151
DOI 10.1007/978-1-4419-7332-0_5, © Springer Science+Business Media, LLC 2010

$$y = \cosh x = \frac{e^x + e^{-x}}{2}$$

for x algebraically. Some simple computations give

$$2ye^x = e^{2x} + 1 \quad \text{and then} \quad e^{2x} - 2ye^x + 1 = 0.$$

Notice that this is an ordinary quadratic equation in e^x; the quadratic formula (an inverse!) gives

$$e^x = \frac{2y \pm \sqrt{4y^2 - 4}}{2} = y \pm \sqrt{y^2 - 1}.$$

Branches of the inverse We can finally solve for x itself by using the logarithm (yet another inverse):

$$x = \ln\left(y \pm \sqrt{y^2 - 1}\right) = \operatorname{arccosh} y.$$

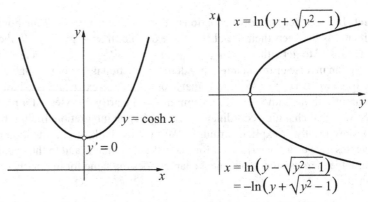

The "$\pm$" in the formula for x means that the inverse splits into two parts, or **branches**, with a separate formula for each. The graph of $x = \operatorname{arccosh} y$ on the right, above, helps us see why. It is the reflection of the graph of $y = \cosh x$ across the line $y = x$. It splits into two halves at the point ($x = 0$) where $y' = 0$:

$$\text{upper,} \, x \geq 0 : \quad x = \ln\left(y + \sqrt{y^2 - 1}\right),$$
$$\text{lower,} \, x \leq 0 : \quad x = \ln\left(y - \sqrt{y^2 - 1}\right).$$

The two branches imply that we should think of the inverse as a 1–2 map: for each $y > 1$, the inverse gives two x-values.

There is more to say here: the graphs of those two branches are symmetric across the y-axis, implying that the two corresponding x-values must be negatives of each other. In other words, the equation of the lower half should be

$$x = -\ln\left(y + \sqrt{y^2 - 1}\right).$$

There is no conflict, however. Note that

$$\left(y - \sqrt{y^2 - 1}\right)\left(y + \sqrt{y^2 - 1}\right) = y^2 - (y^2 - 1) = 1,$$

so

$$\ln\left(y - \sqrt{y^2 - 1}\right) = \ln\left(\frac{1}{y + \sqrt{y^2 - 1}}\right) = -\ln\left(y + \sqrt{y^2 - 1}\right).$$

Finally, notice that the term $y^2 - 1$ under the radical implies the inverse is defined only for $y \geq 1$, a fact borne out by the graph.

In a similar way, you can show that $\operatorname{arcsinh} y = \ln\left(y + \sqrt{y^2 + 1}\right)$ (there is no "$\pm$" ambiguity here) and use this formula with the pullback substitution $y = \sinh x$ to show that

Inverses of other hyperbolic functions

$$\int \frac{dy}{\sqrt{1 + y^2}} = \ln\left(y + \sqrt{y^2 + 1}\right).$$

See the exercises for this and other questions involving the hyperbolic functions and their inverses.

Inverses play a crucial role in solving problems even when there is no formula or explicit expression for the inverse in terms of elementary functions. For example, consider the differential equation

Inverses without formulas

$$\frac{dy}{dx} = \frac{y}{y - 1}.$$

This equation, as written, indicates that y changes with x, so x is the independent variable. Thus, we are looking for a function $y = f(x)$ for which the equation

$$f'(x) = \frac{f(x)}{f(x) - 1}$$

is an identity in x, at least for all x in some interval.

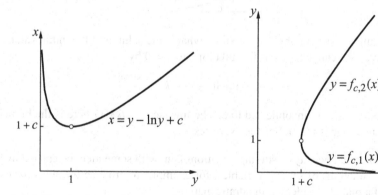

A solution is shown above, at the right. We can obtain this solution and others by using the method of separation of variables. The method begins by rewriting the original differential equation as

Separating variables

$$dx = \frac{y - 1}{y}\, dy = \left(1 - \frac{1}{y}\right) dy$$

(a *differential equation* was originally "an equation involving differentials"). Integrating this, we get an expression involving an arbitrary constant c,

$$x = y - \ln(y) + c, \quad y > 0,$$

whose graph is shown on the left, above. Of course, the function $y = f(x)$ we seek is the inverse, whose graph is shown on the right.

Branches of the solution

We put a hole in the graph of $x = y - \ln y + c$ at $(x,y) = (1 + c, 1)$ because the original differential equation is undefined when $y = 1$. The inverse hence has two branches; we call them $y = f_{c,1}(x)$ and $y = f_{c,2}(x)$. The branches have different ranges, $0 < f_{c,1}(x) < 1$, $1 < f_{c,2}(x)$, but the same domain, $x > 1 + c$. We can describe the two functions by their graphs or by the words "the two branches of the inverse of $x = y - \ln y + c$." They have no formulas.

Initial-value problems

Separate branches here are welcomed, because they provide the flexibility needed to solve different initial-value problems. For example, sketched below are the two particular solutions f_α and f_β to the differential equation that satisfy the different initial conditions

$$f_\alpha(1) = \tfrac{1}{2}, \qquad f_\beta(0) = 2.$$

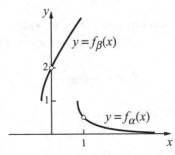

These examples raise an obvious question: what is the solution if the initial value is not positive? We can extend our formula for x to $y < 0$ by

$$x = y - \ln(-y) + k,$$

where k is a constant unconnected to c. The inverse $y = f_{k,3}(x)$ here is the branch we need; its range is $y < 0$. See the exercises.

Solving two equations in two unknowns

Our rather ad hoc way of solving equations can, with some luck, be carried over to several functions of several variables, for example, to produce formulas for the inverse of a map. Consider the quadratic map

$$\mathbf{f} : \begin{cases} x = u^2 - v^2, \\ y = 2uv, \end{cases}$$

from Chapter 4. The inverse of $\mathbf{f}$ expresses u and v in terms of x and y. We do this— that is, we solve for u and v—by isolating each of these variables in its own separate equation. The key is to notice that

$$x^2 + y^2 = u^4 - 2u^2v^2 + v^4 + 4u^2v^2 = (u^2 + v^2)^2.$$

We are then able to isolate u and v by adding and then by subtracting the pair of equations

$$u^2 + v^2 = \sqrt{x^2 + y^2},$$
$$u^2 - v^2 = x;$$

this gives us the components of $\mathbf{f}^{-1}$:

$$u = \pm\sqrt{\frac{\sqrt{x^2+y^2}+x}{2}}, \quad v = \pm\sqrt{\frac{\sqrt{x^2+y^2}-x}{2}}.$$

(The expressions for u and v are real because $\sqrt{x^2+y^2} = u^2 + v^2 \geq 0$.)

The "$\pm$" signs put the image point in the four different quadrants of the (u,v)-plane. To decide which signs to use, recall that the original map $\mathbf{f}: (u,v) \mapsto (x,y)$ "doubled angles." In particular, it mapped the first quadrant of the (u,v)-plane to the upper half-plane $y \geq 0$ and the second quadrant to the lower half-plane $y \leq 0$. Thus $\mathbf{f}^{-1}$ maps $y \geq 0$ to the first quadrant,

Choosing signs for $\mathbf{f}^{-1}$

$$u = +\sqrt{\frac{\sqrt{x^2+y^2}+x}{2}}, \quad v = +\sqrt{\frac{\sqrt{x^2+y^2}-x}{2}}, \quad \text{if } y \geq 0,$$

and $y \leq 0$ to the second quadrant,

$$u = -\sqrt{\frac{\sqrt{x^2+y^2}+x}{2}}, \quad v = +\sqrt{\frac{\sqrt{x^2+y^2}-x}{2}}, \quad \text{if } y \leq 0.$$

These are the formulas for $\mathbf{g}_+$ (Exercise 4.13) and $\mathbf{g}_-$ (Exercise 4.17) on pages 144ff. What happens on the overlap $y = 0$? If $x < 0$ and $y = 0$, then

Do the formulas agree on the overlap?

$$v = \sqrt{\frac{\sqrt{x^2}-x}{2}} = \sqrt{\frac{|x|-x}{2}} = \sqrt{\frac{-x-x}{2}} = \sqrt{-x} = \sqrt{|x|} > 0,$$

and

$$u = \pm\sqrt{\frac{|x|+x}{2}} = \pm\sqrt{\frac{-x+x}{2}} = 0.$$

The two pairs of formulas agree: for both, the image of the negative x-axis is the positive v-axis. On the other hand, if $x > 0$ and $y = 0$, then $|x| = x$, so

$$v = \sqrt{\frac{|x|-x}{2}} = 0, \quad \text{and} \quad u = \pm\sqrt{\frac{|x|+x}{2}} = \pm\sqrt{|x|}.$$

Here there is a conflict: the first pair of formulas (where the sign of u is "+") maps the positive x-axis to the positive u-axis, but the second pair maps it to the negative u-axis.

$\mathbf{f}^{-1}$ is discontinuous

One way to eliminate the conflict is to remove the positive x-axis from the domain of the second pair of formulas. Then $\mathbf{f}^{-1}$ is well defined on the whole plane $\mathbb{R}^2$. However, there is a cost: along the positive x-axis, $\mathbf{f}^{-1}$ is *discontinuous*. For example, as the points $\mathbf{p}$ and $\mathbf{q}$, below, become arbitrarily close, their images do not.

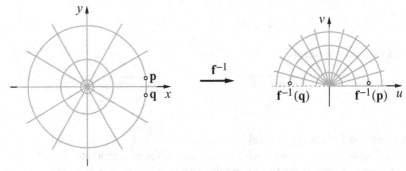

Give $\mathbf{f}^{-1}$
a second branch

There is a radical way to solve this problem that builds on the fact $\mathbf{f}$ is a 2–1 map (because diametrically opposite points in the (u,v)-plane have the same image under $\mathbf{f}$). This suggests that $\mathbf{f}^{-1}$ is, more properly, a 1–2 map and therefore has a second branch. (Consider, for a moment, the 1-dimensional analogue of $\mathbf{f}$: $x = f(u) = u^2$; f^{-1} has the two familiar branches $u = +\sqrt{x}$ and $u = -\sqrt{x}$. Not coincidentally, the two u-values are opposites.) If $\mathbf{f}^{-1}$ already assigns to the point $\mathbf{x}$ the point $\mathbf{u}$ in the upper half-plane, then we can easily get the second branch by having $\mathbf{f}^{-1}$ assign to $\mathbf{x}$ also the diametrically opposite point $-\mathbf{u}$ in the lower half-plane:

$$\mathbf{f}^{-1}(\mathbf{x}) = \pm\mathbf{u}.$$

Here is a set of formulas that expresses both branches in terms of components. The two branches are distinguished from each other by the "$\pm$"signs:

$$(u,v) = \mathbf{f}^{-1}(x,y) = \begin{cases} \pm\left(\sqrt{\dfrac{\sqrt{x^2+y^2}+x}{2}},\sqrt{\dfrac{\sqrt{x^2+y^2}-x}{2}}\right), & y \geq 0, \\[3em] \pm\left(-\sqrt{\dfrac{\sqrt{x^2+y^2}+x}{2}},\sqrt{\dfrac{\sqrt{x^2+y^2}-x}{2}}\right), & y \leq 0. \end{cases}$$

The second branch eliminates the discontinuity along the positive x-axis. For example, $\mathbf{f}^{-1}(\mathbf{q})$ in the figure above now becomes the pair of points $\pm\mathbf{f}^{-1}(\mathbf{q})$, and similarly $\mathbf{f}^{-1}(\mathbf{p})$ branches into $\pm\mathbf{f}^{-1}(\mathbf{p})$. Then, although $\pm\mathbf{f}^{-1}(\mathbf{q})$ is not close to $\pm\mathbf{f}^{-1}(\mathbf{p})$, it *is* close to $\mp\mathbf{f}^{-1}(\mathbf{p})$.

Solving equations
by finding fixed points

There is another important method we can use to solve equations: find the fixed points of a suitably chosen map by iterating the map. We show immediately below how this gives us a valuable computational tool; in Chapter 5.3, we show that it provides the theoretical key to the proof of the inverse function theorem.

Definition 5.1 *Suppose $g : X \to X$ is a map of a set X to itself; then $\widehat{x}$ is a **fixed point** of g if $g(\widehat{x}) = \widehat{x}$.*

Suppose, now, we must solve the numerical equation $y = f(x)$ for x when y is given. Let $g(x) = f(x) - y + x$; then the following chain of implications shows that every fixed point of g is a solution of $f(x) = y$, and conversely:

$$g(\widehat{x}) = \widehat{x} \iff f(\widehat{x}) - y + \widehat{x} = \widehat{x} \iff f(\widehat{x}) = y.$$

The formula $g(x) = f(x) - y + x$ is just one way to construct a function whose fixed points are the solutions to $y = f(x)$; there are others. One familiar example is provided by the *Newton–Raphson method* for finding roots of $f(x) = 0$:

$$g(x) = x - \frac{f(x)}{f'(x)}.$$

Another is provided by the ancient *Babylonian algorithm* for finding square roots. We look at this in detail in order to see how well the fixed-point approach lends itself to computation. Given $a > 0$, our goal is to find $\widehat{x} > 0$ so that $\widehat{x}^2 = a$. We have

The Babylonian algorithm

$$\widehat{x} = a/\widehat{x} \text{ so } 2\widehat{x} = \widehat{x} + a/\widehat{x} \text{ and } \widehat{x} = \frac{\widehat{x} + a/\widehat{x}}{2}.$$

In other words, $\widehat{x} = \sqrt{a}$ is a fixed point of

$$g(x) = \frac{x + a/x}{2}.$$

But g is just the function; the algorithm itself tells us how to find $\widehat{x}$: pick x_0 arbitrarily (but reasonably close to $\sqrt{a}$), and then set

$$x_1 = g(x_0), \quad x_2 = g(x_1), \quad x_3 = g(x_2),$$

and so on. The sequence $x_0, x_1, x_2, \ldots$ converges to the fixed point $\widehat{x} = \sqrt{a}$. An example makes it clear how rapid this convergence can be. Take $a = 6$ and let $x_0 = 2$. Then

n	x_n	x_n^2
1	2.5	6.25
2	2.45	6.0025
3	2.449489795918367	6.000000260308205
4	2.449489742783179	6.000000000000004
5	2.449489742783178	5.999999999999999

To fifteen decimal places, $\widehat{x} = 2.449489742783178$. The convergence here is especially rapid: the number of correct digits roughly doubles with each iteration.

Fixed points
by iteration
The Babylonian algorithm suggests the following general procedure for finding a fixed point. Take a point x_0 and construct its *iterates* under g: $x_{n+1} = g(x_n)$, $n = 0, 1, 2, \ldots$. If the sequence has a limit, let $\widehat{x}$ be that limit. Then

$$g(\widehat{x}) = g\left(\lim_{n \to \infty} x_n\right) = \lim_{n \to \infty} g(x_n) = \lim_{n \to \infty} x_{n+1} = \widehat{x}$$

if g is continuous. (Continuity is needed to be sure that the limit can be taken either before or after g is evaluated.) Thus $\widehat{x}$ is a fixed point of g. The Newton–Raphson method is implemented by the same kind of iteration.

Contraction mappings
Of course, in order to use this procedure, we have to make certain that the iterates have a limit, and the map g is continuous. For a *contraction mapping* (Definition 5.3, p. 167), these conditions are satisfied, and the *contraction mapping principle* (Theorem 5.1, p. 167) then guarantees the existence of a unique fixed point.

5.2 Coordinate changes

We already use coordinate changes in integration, to simplify an integrand or to convert it into a more recognizable form. In this section we put coordinate changes to larger use, to simplify the geometry of a map. For instance, we saw in Chapter 4 that a map frequently "looked like" its derivative near a point. The derivative, being linear, was essentially simple; the resemblance between the map and its derivative meant that the map itself was simple, too, at least near that point. Our goal in this section is to explain what it means for one map to *look like* a second; in fact, it means that, when the first map is expressed using appropriate new coordinates, it will be *identical to* the second one. To see how coordinate changes play this vital role, we consider several examples.

Example 1:
$f(x) = \sqrt{x}$ at $x = 1$
At the point $x = 1$, the tangent to the graph of $y = f(x) = \sqrt{x}$ is a straight line of slope $1/2$. Let us analyze f in a window centered at $(x, y) = (1, 1)$, first using coordinates $\Delta x = x - 1$, $\Delta y = y - 1$ based at the center of the window. Then

$$\Delta y = y - 1 = \sqrt{x} - 1 = \sqrt{1 + \Delta x} - 1,$$

so the formula for f in the window coordinates is $\Delta y = -1 + \sqrt{1 + \Delta x}$. The graph is the familar one shown in black on the right, above. With it, in gray, is the graph of the derivative, $\Delta y = df_1(\Delta x) = \frac{1}{2}\Delta x$. The black and gray graphs "share ink" near $\Delta x = 0$, ample evidence that the square root map "looks like" its derivative there. But we can do even more: with the proper coordinate change $\Delta x = \varphi(\Delta s)$, we can make the formula for f become $\Delta y = \frac{1}{2}\Delta s$. In the new $(\Delta s, \Delta y)$ window, the graph of f will be straight.

A pullback to simplify the formula for f

How can we find φ? Because our goal is to simplify the formula for f, and because that formula involves $\sqrt{1 + \Delta x}$, a reasonable approach is to make $1 + \Delta x$ a perfect square. Thus, let

$$1 + \Delta x = 1 + \Delta s + \frac{(\Delta s)^2}{4} = \left(1 + \frac{\Delta s}{2}\right)^2.$$

Then

$$\Delta x = \Delta s + \frac{(\Delta s)^2}{4} = \varphi(\Delta s)$$

is a pullback substitution that does what we want:

$$f : \Delta y = -1 + \sqrt{1 + \Delta x} = -1 + \sqrt{1 + \varphi(\Delta s)}$$

$$= -1 + \sqrt{1 + \Delta s + \frac{(\Delta s)^2}{4}} = -1 + \left(1 + \frac{\Delta s}{2}\right) = \tfrac{1}{2}\Delta s.$$

Thus the formula for f in the $(\Delta s, \Delta y)$ window is identical to the formula for df_1 in the $(\Delta x, \Delta y)$ window.

Let us extend our pullback to a map $\boldsymbol{\varphi} : (\Delta s, \Delta y) \mapsto (\Delta x, \Delta y)$ of one window to the other:

The pullback map

$$\boldsymbol{\varphi} : \begin{cases} \Delta x = \varphi(\Delta s), \\ \Delta y = \Delta y. \end{cases}$$

We see the effect of $\boldsymbol{\varphi}$ in the figure above, on the left. For a start, $\boldsymbol{\varphi}$ pulls back the uniform grid to the nonuniform one shown. The numbers at the bottom of the vertical grid lines are the Δx-values in both cases. Pick a vertical line with the same Δx value in each of the windows; you should check that, at a point where the black graphs cross those lines, the Δy coordinates agree. This means that the black line in the $(\Delta s, \Delta y)$ window is the graph of the same function—namely f—as the black curved line in the $(\Delta x, \Delta y)$ window.

The pullback gradually stretches the grid on the left and compresses it on the right. This is just a geometric manifestation of the nonlinearity of the map $\boldsymbol{\varphi}$. Near the origin, there is virtually no distortion in the grid. In other words, the "coordinate change" does not change anything there. (This is a consequence of $\varphi'(0) = 1$.) The nonlinearity of $\boldsymbol{\varphi}$ makes it possible to straighten the curved graph of f. Of course, the same nonlinearity causes the straight tangent line to bend into a curve, in this case the parabolic curve

Nonlinearity of the pullback

$$\Delta y = \frac{1}{2}\left(\Delta s + \frac{(\Delta s)^2}{4}\right).$$

In terms of its own coordinate, the $(\Delta s, \Delta y)$ window covers the horizontal range $-2 \leq \Delta s \leq 2(-1+\sqrt{2}) \approx 0.8$. The points $\Delta s = -2, -1$, and 0.8 are marked on the Δs-axis; compare them to nearby values of Δx.

Changing y
instead of x

We just converted f into its derivative by changing the source variable x. We can accomplish the same thing in a different way by making an appropriate change in the target variable y. In the exercises you are asked to find an explicit formula for a push-forward substitution $\Delta w = \psi(\Delta y)$ that converts

$$f : \Delta y = -1 + \sqrt{1 + \Delta x}$$

into $\Delta w = \frac{1}{2}\Delta x$. The figure below shows the form that the coordinate change takes: to straighten the graph of f, the bottom of the $(\Delta x, \Delta y)$ window must be compressed (quite severely near $\Delta y = -1$), and the top stretched.

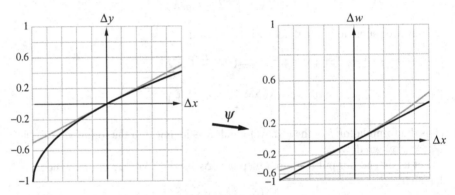

Example 2:
semi-log paper

The two maps φ and ψ suggest a general principle: to convert a curved graph to a straight one, plot it on a suitable nonuniform grid. Perhaps the most familiar example of this is semi-log graph paper, on which an exponential function plots as a straight line. We take this now as our second example of a coordinate change that simplifies the geometry of functions.

To be concrete, consider the function $g(x) = 3 \times 10^{0.1x}$. We use base 10 here because the usual semi-log paper is geared to it (rather than to base e, for example). On the left, below, is the graph of g; it has, of course, the familar shape of an exponential curve. The coordinate change $Y = \log_{10} y$ (a push-forward substitution) gives

$$Y = \log_{10} y = \log_{10}\left(3 \times 10^{0.1x}\right) = 0.1x + \log_{10} 3,$$

making Y a linear function of x. Its graph is the straight line shown in black, on the right. For comparison, the graph of $y = 10^x$ is shown in gray. It is also a straight line, with a slope 10 times steeper than the black graph.

You can verify that the **semi-log** map,

$$\mathbf{sl}: \begin{cases} x = x, \\ Y = \log_{10} y, \end{cases}$$

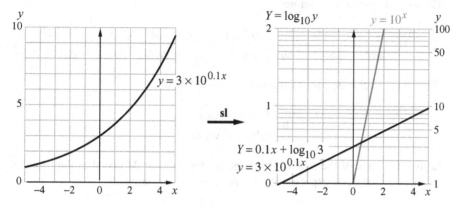

shown above, converts any exponential function $y = Ba^{kx}$ into a linear one:

$$Y = (k\log_{10} a)\, x + \log_{10} B;$$

see the exercises. It compresses the image of a uniform grid more and more in the vertical direction. In particular, notice that the vertical spacing is $\Delta y = 1$ on the lower half of the image grid but $\Delta y = 10$ on the upper. (Although the grid immediately below $Y = 0$ is not shown, you will find it repeats the same nonuniform pattern but with a spacing of $\Delta y = 0.1$.) Semi-log paper has two virtues that are commonly exploited. First, it allows data values that vary over several orders of magnitude to be plotted in a small space. Second, it makes exponential growth or decline easier to discern and to quantify, by plotting it on a straight line. You can explore these features, and the related log–log map, in the exercises.

For our third example we move to a 2-dimensional source and target, and to the quadratic map (Chapter 4.2)

$$\mathbf{f}: \begin{cases} x = u^2 - v^2, \\ y = 2uv, \end{cases} \qquad \mathbf{df}_{(a,b)}: \begin{pmatrix} \Delta x \\ \Delta y \end{pmatrix} = \begin{pmatrix} 2a & -2b \\ 2b & 2a \end{pmatrix} \begin{pmatrix} \Delta u \\ \Delta v \end{pmatrix}.$$

Let us see how a suitable coordinate change near an arbitrary point $(u,v) = (a,b)$ can convert $\mathbf{f}$ into its derivative $\mathbf{df}_{(a,b)}$. As usual, we set

$$\begin{aligned} \Delta u &= u - a, & \Delta x &= x - (a^2 - b^2), \\ \Delta v &= v - b, & \Delta y &= y - 2ab, \end{aligned}$$

to get coordinates $(\Delta u, \Delta v)$ in a window centered at $(u,v) = (a,b)$ and coordinates $(\Delta x, \Delta y)$ in a window centered at the image point $(x,y) = \mathbf{f}(a,b) = (a^2 - b^2, 2ab)$.

In window coordinates, we represent the map **f** by the **window map** $\Delta \mathbf{f}$ (see below, p. 172) defined by

$$(\Delta x, \Delta y) = \Delta \mathbf{f}(\Delta u, \Delta v) = \mathbf{f}(a + \Delta u, b + \Delta v) - \mathbf{f}(a, b).$$

The formula for $\Delta \mathbf{f}$ is therefore

$$\Delta x = (a + \Delta u)^2 - (b + \Delta v)^2 - (a^2 - b^2) \quad \Delta y = 2(a + \Delta u)(b + \Delta v) - 2ab$$
$$= 2a\,\Delta u - 2b\,\Delta v + (\Delta u)^2 - (\Delta v)^2, \qquad = 2b\,\Delta u + 2a\,\Delta v + 2\,\Delta u\,\Delta v.$$

Our goal is to change coordinates in the source window,

$$\mathbf{h} : \begin{cases} \Delta s = h(\Delta u, \Delta v), \\ \Delta t = k(\Delta u, \Delta v), \end{cases}$$

so that, in terms of the new coordinates, the formula for $\Delta \mathbf{f}$ becomes the formula for $d\mathbf{f}_{(a,b)}$. That is, $\Delta \mathbf{f}$ expresses Δx and Δy as the linear functions

$$\Delta \mathbf{f} : \begin{cases} \Delta x = 2a\,\Delta s - 2b\,\Delta t, \\ \Delta y = 2b\,\Delta s + 2a\,\Delta t. \end{cases}$$

Solving for Δs and Δt   Note that we now have two expressions for Δx (and, likewise, for Δy), one involving Δu and Δv, the other Δs and Δt. To find the functions h and k that connect Δs and Δt with Δu and Δv, we can begin by equating those expressions (in matrix form):

$$\begin{pmatrix} 2a & -2b \\ 2b & 2a \end{pmatrix} \begin{pmatrix} \Delta s \\ \Delta t \end{pmatrix} = \begin{pmatrix} 2a & -2b \\ 2b & 2a \end{pmatrix} \begin{pmatrix} \Delta u \\ \Delta v \end{pmatrix} + \begin{pmatrix} (\Delta u)^2 - (\Delta v)^2 \\ 2\,\Delta u\,\Delta v \end{pmatrix}.$$

Then, to solve for Δs and Δt, we need only multiply by the appropriate inverse matrix:

$$\begin{pmatrix} \Delta s \\ \Delta t \end{pmatrix} = \begin{pmatrix} \Delta u \\ \Delta v \end{pmatrix} + \frac{1}{2(a^2 + b^2)} \begin{pmatrix} a & b \\ -b & a \end{pmatrix} \begin{pmatrix} (\Delta u)^2 - (\Delta v)^2 \\ 2\,\Delta u\,\Delta v \end{pmatrix}.$$

This is the coordinate change we seek; in effect, $\mathbf{h} = (d\mathbf{f_a})^{-1} \circ \Delta \mathbf{f}$. The individual components of $\mathbf{h}$ are

$$\mathbf{h} : \begin{cases} \Delta s = h(\Delta u, \Delta v) = \Delta u + \dfrac{a(\Delta u)^2 + 2b\,\Delta u\,\Delta v - a(\Delta v)^2}{2(a^2 + b^2)}, \\[4mm] \Delta t = k(\Delta u, \Delta v) = \Delta v + \dfrac{-b(\Delta u)^2 + 2a\,\Delta u\,\Delta v + b(\Delta v)^2}{2(a^2 + b^2)}. \end{cases}$$

Incidentally, it is not yet evident that $\mathbf{h}$ is a coordinate change: that is, that the map $\mathbf{h}(\Delta u, \Delta v)$ has an inverse defined in some neighborhood W of $(\Delta s, \Delta t) = (0, 0)$. In fact, there is such a local inverse, but rather than go through a proof in this particular case, we simply appeal to the inverse function theorem, proven later in this chapter. (In particular, see Corollary 5.4, page 176. It says $\mathbf{h}$ will have a local inverse at

$(\Delta s, \Delta t) = (0,0)$ if its derivative is continuous near $(0,0)$ and invertible at $(0,0)$. All these conditions are satisfied; in particular, $\mathbf{dh}_{(0,0)} = I$, the identity map.)

By using the vectors   Why **f** "looks like" $\mathbf{df_a}$

$$\Delta\mathbf{u} = (\Delta u, \Delta v), \quad \Delta\mathbf{s} = (\Delta s, \Delta t), \quad \Delta\mathbf{x} = (\Delta x, \Delta y), \quad \mathbf{a} = (a, b),$$

we can write the formula that connects the maps $\mathbf{f}$, $\mathbf{h}$, and $\mathbf{df_a}$ as

$$\mathbf{f}(\mathbf{a} + \Delta\mathbf{u}) - \mathbf{f}(\mathbf{a}) = \Delta\mathbf{f}(\Delta\mathbf{u}) = \Delta\mathbf{x} = \mathbf{df_a}(\mathbf{h}(\Delta\mathbf{u})) = \mathbf{df_a}(\Delta\mathbf{s}).$$

Think of the formula this way. Each point $\mathbf{p}$ in the window centered at $\mathbf{a}$ has two different coordinate labels, $\Delta\mathbf{u}$ and $\Delta\mathbf{s}$. The map $\mathbf{h}$ connects those labels. The image of $\mathbf{p}$ under the action of $\mathbf{f}$ (i.e., $\Delta\mathbf{f}$) has coordinate $\Delta\mathbf{x}_1 = \Delta\mathbf{f}(\Delta\mathbf{u})$. The image of $\mathbf{p}$ under the action of $\mathbf{df_a}$ has coordinate $\Delta\mathbf{x}_2 = \mathbf{df_a}(\Delta\mathbf{s})$. But $\Delta\mathbf{x}_1 = \Delta\mathbf{x}_2$; these are the coordinates of the same point $\mathbf{q}$. Thus, $\mathbf{f}$ (written as $\Delta\mathbf{f}$ in the window W) and $\mathbf{df_a}$ both map $\mathbf{p}$ to $\mathbf{q}$. That is why $\mathbf{f}$ "looks like" $\mathbf{df_a}$; they are just different coordinate descriptions of the same map. All of this is diagrammed on the left, below, and summarized more briefly on the right.

Thus we have $\Delta\mathbf{f} = \mathbf{df_a} \circ \mathbf{h}$. If we think of composition of maps as a kind of prod-   **f** *factors through* **h**
uct, then we can say $\Delta\mathbf{f}$ *factors into* $\mathbf{h}$ and $\mathbf{df_a}$. In effect, we constructed the coordinate change map $\mathbf{h}$ so that, in a small window centered at $\mathbf{a}$, $\Delta\mathbf{f}$ **factors through** $\mathbf{h}$.

We can get a better idea how the coordinate change $\mathbf{h}$ converts $\Delta\mathbf{f}$ (or $\mathbf{f}$) into   Converting $\Delta\mathbf{f}$ to $\mathbf{df}$ at $(\sqrt{3}/2, 1/2)$
$\mathbf{df}_{(a,b)}$ by focusing on a specific point. In the figure below, we have taken $(a,b) =$
$(\sqrt{3}/2, 1/2)$ and used windows that measure 1 unit on a side. Thus, the square grid
in the $(\Delta s, \Delta t)$-window at the top has a spacing of 0.1 unit. The same grid appears
in the lower-left window, "pulled back" by $\mathbf{h}$; it becomes curved there because $\mathbf{h}$ is
nonlinear. The lower-left window therefore demonstrates concretely what we said
above: that each point in the source has two sets of coordinates. The curved grid
provides $(\Delta s, \Delta t)$ whereas the "native" coordinates (whose square grid is not drawn)
are $(\Delta u, \Delta v)$. Thus, for example,

$$(-0.3, -0.1) = (\Delta s, \Delta t) \leftrightarrow \mathbf{p} \leftrightarrow (\Delta u, \Delta v) \approx (-0.3754, -0.0996).$$

The curved grid is the key to visualizing both the action of $\mathbf{f}$ (as $\Delta\mathbf{f}$) and the   Mapping the curved grid
connection between $\mathbf{f}$ and its derivative. First, follow $\mathbf{f}$ as it maps the source on the
left directly to the target on the right; it sends the curved grid to the grid of large
squares, straightening all grid lines in the process. Second, follow $\mathbf{h}$ and $\mathbf{df}_{(\sqrt{3}/2, 1/2)}$
into and out of the upper window. This time $\mathbf{h}$ itself straightens the curved grid,
mapping it to the "native" grid in the $(\Delta s, \Delta t)$-window. The linear map $\mathbf{df}_{(\sqrt{3}/2, 1/2)}$

then carries the $(\Delta s, \Delta t)$ grid to the grid of large squares in the target. Now we already know that

$$\mathbf{df}_{(\sqrt{3}/2,1/2)} = 2R_{\pi/6}$$

(rotation by $\pi/6$ radians with all lengths doubled; see p. 118), so the large squares in the $(\Delta x, \Delta y)$-window are 0.2 units on a side and make an angle of 30° with the horizontal.

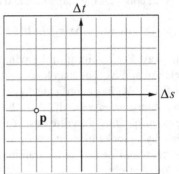

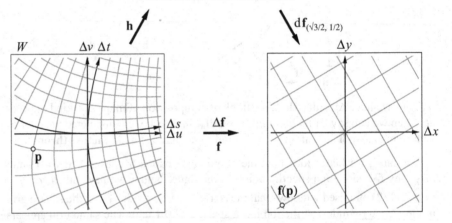

see p. 118

Details of the $(\Delta s, \Delta t)$ **coordinates**

The figure has even more to say about the map **h** that pulls back the $(\Delta s, \Delta t)$ coordinates to the $(\Delta u, \Delta v)$ window. The curves that make up the new grid are, of course, just contour lines of the two functions

$$\Delta s = h(\Delta u, \Delta v), \quad \Delta t = k(\Delta u, \Delta v)$$

defined in the window. Contours of h give the roughly vertical curves; contours of k give the roughly horizontal ones. The following Mathematica 5 code generates the curved grid in the $(\Delta u, \Delta v)$-window.

```
scon = ContourPlot[u + Sqrt[3](u^2 - v^2)/4 + u v/2,
    {u, -.5, .5}, {v, -.5, .5}, Contours ->
    {-.6, -.5, -.4, -.3, -.2, -.1, 0, .1, .2, .3, .4, .5, .6},
    ContourShading -> False, FrameTicks -> None]
```

```
tcon = ContourPlot[v + Sqrt[3]u v/2 + (v^2 - u^2)/4 ,
    {u, -.5, .5}, {v, -.5, .5}, Contours ->
    {-.6, -.5, -.4, -.3, -.2, -.1, 0, .1, .2, .3, .4, .5, .6},
    ContourShading -> False, FrameTicks -> None]
```

```
Show[scon, tcon]
```

The two sets of curves are everywhere orthogonal. This is not automatic. It happens because the map **h** is conformal (p. 118); see Exercise 5.19. Note furthermore how the axes are pulled back: the Δs-axis is tangent to the Δu-axis, and the Δt-axis to the Δv-axis. Moreover, the grid squares around the origin undergo almost no distortion: the pullbacks are nearly the same size and shape as the original. This is a consequence of $d\mathbf{h_0} = I$.

Because the coordinate lines $\Delta s =$ constant and $\Delta t =$ constant become curved when they appear in the $(\Delta u, \Delta v)$-window, we say that Δs and Δt are **curvilinear coordinates** there. As we have just seen, curvilinear coordinates can simplify our view of a map. This is a trade-off, of course: to simplify the map, we complicate the coordinates. But this is a cost we have already accepted when, for example, we plot exponential functions on semi-log paper. We have also accepted it when we use polar coordinates: it was a curved polar coordinate grid that first clarified the action of the quadratic map **f**. Here is our earlier view (p. 117) of the local action of **f** in a window centered at $(\sqrt{3}/2, 1/2)$:

Curvilinear coordinates

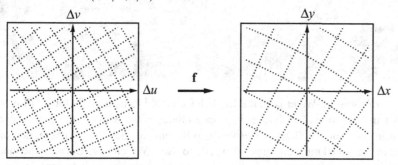

Compare this figure now with the new one (p. 164) that uses the curvilinear $(\Delta s, \Delta t)$ coordinates in the $(\Delta u, \Delta v)$-window.

Summary: Under a suitable coordinate change, a complicated situation can often be made simpler; for example, it may be possible to convert a map locally into its derivative.

5.3 The inverse function theorem

As we have seen, coordinate changes give us a powerful tool to simplify the description of a map. But a coordinate change must be invertible, a condition that is often difficult to verify directly. In this section we state and prove the inverse function theorem. Simply put, the theorem says that if the derivative of a map is continuous and invertible at a point, the map itself is invertible locally near that point. The proof uses several tools, beginning with the *contraction mapping principle*, which can be nicely illustrated by the model village found in Bourton-on-the-Water.

The nested models of Bourton-on-the-Water

Bourton-on-the-Water is in the English Cotswold hills, near Oxford. Filling the back garden of one of its houses is a scale model of the whole village. Now every point in the model village corresponds to a point in the actual village, so some point in the model must correspond exactly to *itself*. Which one? Because the model contains a copy of everything in the village, you would expect it to contain a copy of the model itself. It does; you can see it in the foreground of the photo above. The model of the model is small, of course; it covers only a few square meters. And that smaller model likewise contains a copy of everything, so it has a still smaller copy of itself. In theory, the nested copies could go on forever, getting smaller and smaller and converging, ultimately, to a single point. However, the third iteration was the last that was practical to build. Now return to our question about the point in the model that corresponds to itself. It is in the first model, by definition, but it must be in the second model, too, and the third, and so on. The point that corresponds to itself—the point that is left fixed by the model—must therefore be the limit point of the nested sequence of models.

A more formal view

A little more formally, the model defines a map $\mathbf{m} : V \to V$ of the village, V, to itself, and the point in the model that corresponds to itself is the fixed point of that map (Definition 5.1, p. 157). The model is built at the scale $\sigma = 1/9$. Thus, if $\mathbf{x}$ and

$\mathbf{y}$ are any points in V, and $\mathbf{m}(\mathbf{x})$ and $\mathbf{m}(\mathbf{y})$ are their copies in the model, then

$$\|\mathbf{m}(\mathbf{x}) - \mathbf{m}(\mathbf{y})\| = \sigma \|\mathbf{x} - \mathbf{y}\|.$$

Now $\mathbf{x}$ appears in the model of the model at the point

$$\mathbf{m}(\mathbf{m}(\mathbf{x})) = (\mathbf{m} \circ \mathbf{m})(\mathbf{x}) = \mathbf{m}^2(\mathbf{x});$$

it appears in the model of the model of the model at $\mathbf{m}^3(\mathbf{x})$, and so forth. For the kth iterate of the model, the scale factor would be σ^k:

$$\|\mathbf{m}^k(\mathbf{x}) - \mathbf{m}^k(\mathbf{y})\| = \sigma^k \|\mathbf{x} - \mathbf{y}\|.$$

Because $\sigma < 1$, $\sigma^k \to 0$ as $k \to \infty$; the size of the kth iterate shrinks to zero. Intuitively, this forces the nested models to converge to a single point, say $\mathbf{p}$. Now $\mathbf{p}$ is in every iterate, so it must be the fixed point of the model: $\mathbf{m}(\mathbf{p}) = \mathbf{p}$.

Although the contraction mapping principle can be stated quite generally, we need only the special circumstance of maps defined on a *ball* in $\mathbb{R}^n$.

Contraction mapping principle

Definition 5.2 *The **ball** B_r of radius r centered at the origin in $\mathbb{R}^n$ is the set of all points $\mathbf{x}$ in $\mathbb{R}^n$ for which $\|\mathbf{x}\| \leq r$.*

Definition 5.3 *A **contraction mapping** on B_r is a map $\mathbf{m} : B_r \to B_r$ for which*

$$\|\mathbf{m}(\mathbf{x}) - \mathbf{m}(\mathbf{y})\| \leq \sigma \|\mathbf{x} - \mathbf{y}\|$$

for some $\sigma < 1$ and for all $\mathbf{x}, \mathbf{y}$ in B_r.

A contraction mapping is thus somewhat more general than a "scale model" map, where all distances are contracted by exactly the same factor. Here the factor can vary, as long as it is bounded by a fixed $\sigma < 1$. The additional generality does not weaken the contraction mapping principle, nor does it make the proof more difficult.

Theorem 5.1 (Contraction mapping principle). *A contraction mapping $\mathbf{m}$ on B_r has a unique fixed point $\widehat{\mathbf{x}}$ in B_r. Moreover, for any $\mathbf{x}$ in B_r,*

$$\widehat{\mathbf{x}} = \lim_{k \to \infty} \mathbf{m}^k(\mathbf{x}).$$

Proof. Pick $\mathbf{x}_0$ arbitrarily in B_r, and let $\mathbf{x}_k = \mathbf{m}^k(\mathbf{x}_0)$. For the "telescoping" sum

$$\mathbf{x}_k - \mathbf{x}_{k+l} = \mathbf{x}_k - \mathbf{x}_{k+1} + \mathbf{x}_{k+1} - \mathbf{x}_{k+2} + \cdots + \mathbf{x}_{k+l-1} - \mathbf{x}_{k+l},$$

we have

$$\|\mathbf{x}_k - \mathbf{x}_{k+l}\| \leq \|\mathbf{x}_k - \mathbf{x}_{k+1}\| + \|\mathbf{x}_{k+1} - \mathbf{x}_{k+2}\| + \cdots + \|\mathbf{x}_{k+l-1} - \mathbf{x}_{k+l}\|.$$

Now, for any $i \geq 0$,

$$\|\mathbf{x}_{k+i} - \mathbf{x}_{k+i+1}\| = \|\mathbf{m}^{k+i}(\mathbf{x}_0) - \mathbf{m}^{k+i}(\mathbf{x}_1)\| \leq \sigma^{k+i} \|\mathbf{x}_0 - \mathbf{x}_1\|;$$

therefore the previous inequality implies

$$\|\mathbf{x}_k - \mathbf{x}_{k+l}\| \le \sigma^k \|\mathbf{x}_0 - \mathbf{x}_1\| + \sigma^{k+1}\|\mathbf{x}_0 - \mathbf{x}_1\| + \cdots + \sigma^{k+l-1}\|\mathbf{x}_0 - \mathbf{x}_1\|$$
$$\le \sigma^k \|\mathbf{x}_0 - \mathbf{x}_1\| \left(1 + \sigma + \sigma^2 + \cdots \sigma^{l-1}\right)$$
$$= \sigma^k \|\mathbf{x}_0 - \mathbf{x}_1\| \frac{1 - \sigma^l}{1 - \sigma}.$$

Because $\sigma^k \to 0$ as $k \to \infty$, the last inequality implies that

$$\lim_{k \to \infty} \|\mathbf{x}_k - \mathbf{x}_{k+l}\| = 0 \quad \text{for any integer } l > 0.$$

Let $x_k^{(j)}$ be the jth coordinate of $\mathbf{x}_k$; then

$$\left|x_k^{(j)} - x_{k+l}^{(j)}\right|^2 \le \left|x_k^{(1)} - x_{k+l}^{(1)}\right|^2 + \cdots + \left|x_k^{(n)} - x_{k+l}^{(n)}\right|^2 = \|\mathbf{x}_k - \mathbf{x}_{k+l}\|^2.$$

Thus, for each fixed $l \ge 0$ and for every $j = 1, 2, \ldots, n$,

$$\lim_{k \to \infty} \left|x_k^{(j)} - x_{k+l}^{(j)}\right| = 0.$$

Lemma 5.1. *If y_k is a sequence of real numbers for which $|y_k - y_{k+l}| \to 0$ as $k \to \infty$, (for any positive integer l), then y_k has a limiting value, $\widehat{y}$, as $k \to \infty$.*

Proof. See an analysis text for a proof of this basic fact ("Every Cauchy sequence of real numbers has a limit").   □

The lemma permits us to define

$$\widehat{x}^{(j)} = \lim_{k \to \infty} x_k^{(j)}, \quad j = 1, 2, \ldots, n, \text{ and then } \widehat{\mathbf{x}} = (\widehat{x}^{(1)}, \ldots, \widehat{x}^{(n)}).$$

In other words, $\widehat{\mathbf{x}} = \lim_{k \to \infty} \mathbf{x}_k$.

Lemma 5.2. *The contraction map $\mathbf{m} : B_r \to B_r$ is continuous at every point of B_r.*

Proof. We must show $\mathbf{m}(\mathbf{x}_n) \to \mathbf{m}(\mathbf{x})$ when $\mathbf{x}_n \to \mathbf{x}$. But we have

$$\|\mathbf{m}(\mathbf{x}_n) - \mathbf{m}(\mathbf{x})\| \le \sigma \|\mathbf{x}_n - \mathbf{x}\| \to 0.$$

(In fact, the inequality implies that $\mathbf{m}$ is *uniformly* continuous, even if $\sigma \ge 1$; see an analysis text.)   □

Because $\mathbf{m}$ is continuous (i.e., it commutes with limit processes),

$$\mathbf{m}(\widehat{\mathbf{x}}) = \mathbf{m}\left(\lim_{k \to \infty} \mathbf{x}_k\right) = \lim_{k \to \infty} \mathbf{m}(\mathbf{x}_k) = \lim_{k \to \infty} \mathbf{x}_{k+1} = \widehat{\mathbf{x}},$$

so $\widehat{\mathbf{x}}$ is a fixed point of $\mathbf{m}$. Here is a different argument, which does not depend explicitly on the continuity of $\mathbf{m}$. It begins with the "telescoping" sum

$$\widehat{\mathbf{x}} - \mathbf{m}(\widehat{\mathbf{x}}) = \widehat{\mathbf{x}} - \mathbf{x}_{k+1} + \mathbf{m}(\mathbf{x}_k) - \mathbf{m}(\widehat{\mathbf{x}}),$$

which implies

$$\|\widehat{\mathbf{x}} - \mathbf{m}(\widehat{\mathbf{x}})\| \le \|\widehat{\mathbf{x}} - \mathbf{x}_{k+1}\| + \|\mathbf{m}(\mathbf{x}_k) - \mathbf{m}(\widehat{\mathbf{x}})\| \le \|\widehat{\mathbf{x}} - \mathbf{x}_{k+1}\| + \sigma \|\mathbf{x}_k - \widehat{\mathbf{x}}\|,$$

an inequality that is true for all k. But because $\mathbf{x}_k \to \widehat{\mathbf{x}}$ as $k \to \infty$, the right-hand side vanishes as $k \to \infty$, leaving $\|\widehat{\mathbf{x}} - \mathbf{m}(\widehat{\mathbf{x}})\| = 0$, or $\widehat{\mathbf{x}} = \mathbf{m}(\widehat{\mathbf{x}})$.

It remains to verify that the fixed point is unique. If $\widehat{\mathbf{y}}$ is also a fixed point, then

$$\|\widehat{\mathbf{x}} - \widehat{\mathbf{y}}\| = \|\mathbf{m}(\widehat{\mathbf{x}}) - \mathbf{m}(\widehat{\mathbf{x}})\| \le \sigma \|\widehat{\mathbf{x}} - \widehat{\mathbf{y}}\|.$$

If $\|\widehat{\mathbf{x}} - \widehat{\mathbf{y}}\| \ne 0$, we can divide both sides of this inequality and get $1 \le \sigma$, contradicting our assumption that $\sigma < 1$. This forces $\widehat{\mathbf{x}} = \widehat{\mathbf{y}}$. $\qquad\square$

Theorem 5.2 (Inverse function theorem). *Suppose* $\mathbf{f} : U^n \to \mathbb{R}^n$ *is continuously differentiable on* U^n, *and its derivative is invertible at the point* $\mathbf{a}$ *in* U^n. *Then* $\mathbf{f}$ *itself is invertible on the image* $\mathrm{df}_{\mathbf{a}}(B)$ *in the target of some ball* B *of positive radius centered at the point* $\mathbf{a}$. *The inverse is continuously differentiable on its domain, and* $\mathrm{d}(\mathbf{f}^{-1})_{\mathbf{q}} = (\mathrm{df}_{\mathbf{f}^{-1}(\mathbf{q})})^{-1}$ *for all* $\mathbf{q}$ *in* $\mathrm{df}_{\mathbf{a}}(B)$.

Proof. Our proof expands an argument found in Lang [10, 11].. The proof is long; therefore we split it into a number of steps.

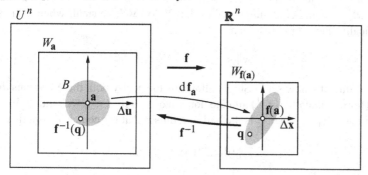

The inverse is a *local* object, that is, one defined essentially in some window centered at the image point $\mathbf{f}(\mathbf{a})$. Therefore, we begin by setting up windows and introducing window coordinates. Let $W_{\mathbf{a}}$ be a window centered at $\mathbf{a}$ in the source, with coordinate $\Delta\mathbf{u} = \mathbf{u} - \mathbf{a}$. Similarly, let $W_{\mathbf{f}(\mathbf{a})}$ be a window in the target centered at $\mathbf{f}(\mathbf{a})$; its coordinate is $\Delta\mathbf{x} = \mathbf{x} - \mathbf{f}(\mathbf{a})$. Finding $\mathbf{f}^{-1}$ means solving the equation

$$\mathbf{f}(\mathbf{a} + \Delta\mathbf{u}) - \mathbf{f}(\mathbf{a}) = \Delta\mathbf{x}$$

for $\Delta\mathbf{u}$ in $W_{\mathbf{a}}$, given $\Delta\mathbf{x}$ in some suitable region in $W_{\mathbf{f}(\mathbf{a})}$. In fact, the $\Delta\mathbf{u}$ we seek is a fixed point of the map

$$\mathbf{g}(\Delta\mathbf{u}) = \Delta\mathbf{u} + (\mathrm{df}_{\mathbf{a}})^{-1}[\Delta\mathbf{x} - (\mathbf{f}(\mathbf{a} + \Delta\mathbf{u}) - \mathbf{f}(\mathbf{a}))].$$

Step 1

The bulk of the proof of invertibility involves showing that $\mathbf{g}$ is a contraction mapping on a suitable ball B_r, implying that $\Delta\mathbf{u}$ exists. The map $(\mathrm{d}\mathbf{f_a})^{-1}$ is needed here to bring the point $\Delta\mathbf{x} - (\mathbf{f}(\mathbf{a}+\Delta\mathbf{u}) - \mathbf{f}(\mathbf{a}))$ from $W_{\mathbf{f}(\mathbf{a})}$ back to $W_\mathbf{a}$, where it can be added to $\Delta\mathbf{u}$.

Step 2

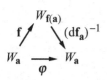

Our analysis of $\mathbf{g}$ begins with the portion

$$\boldsymbol{\varphi}(\Delta\mathbf{u}) = (\mathrm{d}\mathbf{f_a})^{-1}\big(\mathbf{f}(\mathbf{a}+\Delta\mathbf{u}) - \mathbf{f}(\mathbf{a})\big),$$

which is a map $\boldsymbol{\varphi} : W_\mathbf{a} \to W_\mathbf{a}$. By the chain rule,

$$\mathrm{d}\boldsymbol{\varphi}_{\Delta\mathbf{v}} = (\mathrm{d}\mathbf{f_a})^{-1} \circ \mathrm{d}\mathbf{f}_{\mathbf{a}+\Delta\mathbf{v}}$$

for every $\Delta\mathbf{v}$ in $W_\mathbf{a}$. By hypothesis, $\mathrm{d}\mathbf{f}_{\mathbf{a}+\Delta\mathbf{v}}$ depends continuously on $\Delta\mathbf{v}$; therefore, $\mathrm{d}\boldsymbol{\varphi}_{\Delta\mathbf{v}}$ depends continuously on $\Delta\mathbf{v}$, as well. Furthermore, $\mathrm{d}\boldsymbol{\varphi}_0 = (\mathrm{d}\mathbf{f_a})^{-1} \circ \mathrm{d}\mathbf{f_a} = I$.

Step 3

Define $\mathbf{h} : W_\mathbf{a} \to W_\mathbf{a}$ by

$$\mathbf{h}(\Delta\mathbf{u}) = \Delta\mathbf{u} - \boldsymbol{\varphi}(\Delta\mathbf{u});$$

then $\mathrm{d}\mathbf{h}_{\Delta\mathbf{v}} = I - \mathrm{d}\boldsymbol{\varphi}_{\Delta\mathbf{v}}$ is likewise continuous as a function of $\Delta\mathbf{v}$. It follows that the real-valued function

$$N(\Delta\mathbf{v}) = \max_{\|\Delta\mathbf{u}\|=1} \|\mathrm{d}\mathbf{h}_{\Delta\mathbf{v}}(\Delta\mathbf{u})\|$$

is a continuous function of $\Delta\mathbf{v}$, as well. Because $\mathrm{d}\mathbf{h}_0 = \mathbf{0}$, the zero linear map, we have $N(\mathbf{0}) = 0$, and continuity implies that $N(\Delta\mathbf{v})$ will be small when $\Delta\mathbf{v}$ is small. Specifically, choose $r > 0$ so that

$$\|\Delta\mathbf{v}\| < 2r \implies |N(\Delta\mathbf{v})| < \tfrac{1}{2}.$$

Step 4

The value of r now set in Step 3 allows us to specify both the $\Delta\mathbf{x}$-values that we allow (for the domain of $\mathbf{f}^{-1}$) and the domain of $\mathbf{g}$ itself. First, we require that $\Delta\mathbf{x}$ be in the image $\mathrm{d}\mathbf{f_a}(B_{r/2})$ ($B_{r/2}$ is the ball of radius $r/2$ at the center of $W_\mathbf{a}$); this means

$$\|(\mathrm{d}\mathbf{f_a})^{-1}(\Delta\mathbf{x})\| \leq r/2.$$

Second, we require that the domain of $\mathbf{g}$ be restricted to B_r; this means

$$\|\Delta\mathbf{u}\| \leq r.$$

The next three steps show that $\mathbf{g}$ maps B_r to itself and is a contraction mapping there.

Step 5

Because $\mathbf{g} = \mathbf{h} + (\mathrm{d}\mathbf{f_a})^{-1}(\Delta\mathbf{x})$, that is, $\mathbf{g}$ and $\mathbf{h}$ just "differ by a constant," most of what we still need to prove about $\mathbf{g}$ can be done by working with $\mathbf{h}$.

Lemma 5.3. *If $\|\Delta\mathbf{v}\| < 2r$, then $\|\mathrm{d}\mathbf{h}_{\Delta\mathbf{v}}(\Delta\mathbf{w})\| < \tfrac{1}{2}\|\Delta\mathbf{w}\|$ for all $\Delta\mathbf{w}$.*

Proof. We can assume $\Delta\mathbf{w}$ is nonzero; then $\Delta\mathbf{u} = \Delta\mathbf{w}/\|\Delta\mathbf{w}\|$ is a unit vector and $\Delta\mathbf{w} = \|\Delta\mathbf{w}\|\,\Delta\mathbf{u}$. Because $\mathrm{d}\mathbf{h}_{\Delta\mathbf{v}}$ is linear, we have

$$\|\mathrm{d}\mathbf{h}_{\Delta\mathbf{v}}(\Delta\mathbf{w})\| = \|\mathrm{d}\mathbf{h}_{\Delta\mathbf{v}}(\|\Delta\mathbf{w}\|\Delta\mathbf{u})\| = \|\Delta\mathbf{w}\|\,\|\mathrm{d}\mathbf{h}_{\Delta\mathbf{v}}(\Delta\mathbf{u})\| \leq \|\Delta\mathbf{w}\|\,N(\Delta\mathbf{v}) < \tfrac{1}{2}\|\Delta\mathbf{w}\|.$$

We have used the fact that $\|\mathbf{dh}_{\Delta v}(\Delta \mathbf{u})\| \le N(\Delta \mathbf{v})$ for any $\Delta \mathbf{v}$ and for all unit vectors $\Delta \mathbf{u}$, by the definition of N (Step 2).   □

Suppose $\Delta \mathbf{u}_1$ and $\Delta \mathbf{u}_2$ are in the ball of radius r, then the entire line segment $\Delta \mathbf{v} = \Delta \mathbf{u}_1 + t(\Delta \mathbf{u}_2 - \Delta \mathbf{u}_1)$ $(0 \le t \le 1)$ is likewise, and we have

Step 6

$$\|\mathbf{dh}_{\Delta \mathbf{u}_1 + t(\Delta \mathbf{u}_2 - \Delta \mathbf{u}_1)}(\Delta \mathbf{w})\| \le \tfrac{1}{2}\|\Delta \mathbf{w}\|$$

for all $\Delta \mathbf{w}$. This inequality and the continuous differentiability of $\mathbf{h}$ (Step 2) allow us to use the mean value theorem for maps (Theorem 4.15, p. 140) to conclude

$$\|\mathbf{h}(\Delta \mathbf{u}_2) - \mathbf{h}(\Delta \mathbf{u}_1)\| \le \tfrac{1}{2}\|\Delta \mathbf{u}_2 - \Delta \mathbf{u}_1\|.$$

Moreover, if we set $\Delta \mathbf{u}_1 = \mathbf{0}$, then $\|\mathbf{h}(\Delta \mathbf{u}_2)\| \le \tfrac{1}{2}\|\Delta \mathbf{u}_2\| \le r/2$. In other words, $\mathbf{h}$ maps the ball of radius r into the ball of radius $r/2$.

We now move on to $\mathbf{g}$ itself.

Step 7

Lemma 5.4. *For any* $\Delta \mathbf{x}$ *in* $\mathbf{df_a}(B_{r/2})$, $\mathbf{g} : B_r \to B_r$.

Proof. Because $\Delta \mathbf{u}$ is in B_r, by hypothesis, we have $\|\mathbf{h}(\Delta \mathbf{u}\| \le r/2$ (Step 6). Also by hypothesis, $\|(\mathbf{df_a})^{-1}(\Delta \mathbf{x})\| \le r/2$, so

$$\|\mathbf{g}(\Delta \mathbf{u})\| = \|\mathbf{h}(\Delta \mathbf{u}) + (\mathbf{df_a})^{-1}(\Delta \mathbf{x})\| \le \|\mathbf{h}(\Delta \mathbf{u})\| + \|(\mathbf{df_a})^{-1}(\Delta \mathbf{x})\| \le r. \qquad □$$

Lemma 5.5. *For any* $\Delta \mathbf{x}$ *in* $\mathbf{df_a}(B_{r/2})$, $\mathbf{g}$ *is a contraction mapping on* B_r.

Proof. Because $\mathbf{g}(\Delta \mathbf{u}) = \mathbf{h}(\Delta \mathbf{u}) + (\mathbf{df_a})^{-1}(\Delta \mathbf{x})$, it follows that

$$\mathbf{g}(\Delta \mathbf{u}_2) - \mathbf{g}(\Delta \mathbf{u}_1) = \mathbf{h}(\Delta \mathbf{u}_2) - \mathbf{h}(\Delta \mathbf{u}_1);$$

therefore, by Step 6,

$$\|\mathbf{g}(\Delta \mathbf{u}_2) - \mathbf{g}(\Delta \mathbf{u}_1)\| = \|\mathbf{h}(\Delta \mathbf{u}_2) - \mathbf{h}(\Delta \mathbf{u}_1)\| \le \tfrac{1}{2}\|\Delta \mathbf{u}_2 - \Delta \mathbf{u}_1\|. \qquad □$$

Let $\widehat{\Delta \mathbf{x}}$ be an arbitrary point in $\mathbf{df_a}(B_{r/2})$. This choice determines a specific map $\mathbf{g} : B_r \to B_r$, and $\mathbf{g}$ has a unique fixed point $\widehat{\Delta \mathbf{u}}$ in B_r, by the contraction mapping principle. Because $\widehat{\Delta \mathbf{x}}$ determines $\widehat{\Delta \mathbf{u}}$ uniquely, and

Step 8

$$\mathbf{f}(\mathbf{a} + \widehat{\Delta \mathbf{u}}) - \mathbf{f}(\mathbf{a}) = \widehat{\Delta \mathbf{x}},$$

we now have the required inverse map $\mathbf{f}^{-1} : \mathbf{df_a}(B_{r/2}) \to B_r : \widehat{\Delta \mathbf{x}} \mapsto \widehat{\Delta \mathbf{u}}$.

Before showing that $\mathbf{f}^{-1}$ is continuously differentiable, we pause to call attention to the relation between a map and the way we write it within a window. For example, in $W_\mathbf{a}$ and $W_{\mathbf{f}(\mathbf{a})}$, the equation $\mathbf{x} = \mathbf{f}(\mathbf{u})$ has the form

Step 9

$$\mathbf{f}(\mathbf{a}) + \underline{\Delta \mathbf{x}} = \mathbf{x} = \mathbf{f}(\mathbf{u}) = \mathbf{f}(\mathbf{a} + \Delta \mathbf{u}) = \mathbf{f}(\mathbf{a}) + \underline{\mathbf{f}(\mathbf{a} + \Delta \mathbf{u}) - \mathbf{f}(\mathbf{a})}.$$

The underlined elements are equal (this is the "window equation"), and they
define the **window map** $\Delta\mathbf{f}$ for $\mathbf{f}$:

$$\Delta\mathbf{x} = \Delta\mathbf{f}(\Delta\mathbf{u}) = \mathbf{f}(\mathbf{a} + \Delta\mathbf{u}) - \mathbf{f}(\mathbf{a}).$$

Conversely, we can reconstruct the original formula $\mathbf{x} = \mathbf{f}(\mathbf{u})$ from its window equation $\Delta\mathbf{x} = \Delta\mathbf{f}(\Delta\mathbf{u})$. Furthermore, by solving the window equation for $\Delta\mathbf{u}$, we obtain
the window equation of the inverse $\mathbf{u} = \mathbf{f}^{-1}(\mathbf{x})$ (at the point $\mathbf{b} = \mathbf{f}(\mathbf{a})$):

$$\Delta\mathbf{u} = \Delta\mathbf{f}^{-1}(\Delta\mathbf{x}) = \mathbf{f}^{-1}(\mathbf{b} + \Delta\mathbf{x}) - \mathbf{f}^{-1}(\mathbf{b}).$$

Step 10
In preparation for showing $\mathbf{f}^{-1}$ is differentiable, we first show that it is uniformly
continuous, by working with the window map $\Delta\mathbf{f}^{-1}$.

Lemma 5.6. *There is a positive constant K such that, for any two points $\Delta\mathbf{x}_1$, $\Delta\mathbf{x}_2$
and their corresponding images $\Delta\mathbf{u}_1$, $\Delta\mathbf{u}_2$ under $\Delta\mathbf{f}^{-1}$,*

$$\|\Delta\mathbf{u}_2 - \Delta\mathbf{u}_1\| \leq K\|\Delta\mathbf{x}_2 - \Delta\mathbf{x}_1\|.$$

Proof. Recall the definition of the map $\mathbf{h}$ (now written using the window map for $\mathbf{f}$):

$$\mathbf{h}(\Delta\mathbf{u}) = \Delta\mathbf{u} - \boldsymbol{\varphi}(\Delta\mathbf{u}) = \Delta\mathbf{u} - (d\mathbf{f_a})^{-1}(\Delta\mathbf{f}(\Delta\mathbf{u})) = \Delta\mathbf{u} - (d\mathbf{f_a})^{-1}(\Delta\mathbf{x}).$$

If we now evaluate this equation at $\Delta\mathbf{u}_1$ and then at $\Delta\mathbf{u}_2$, subtract the first from the
second, and use the linearity of $(d\mathbf{f_a})^{-1}$, we get

$$\Delta\mathbf{u}_2 - \Delta\mathbf{u}_1 = \mathbf{h}(\Delta\mathbf{u}_2) - \mathbf{h}(\Delta\mathbf{u}_1) + (d\mathbf{f_a})^{-1}(\Delta\mathbf{x_2} - \Delta\mathbf{x_1}).$$

It follows that

$$\|\Delta\mathbf{u}_2 - \Delta\mathbf{u}_1\| \leq \|\mathbf{h}(\Delta\mathbf{u}_2) - \mathbf{h}(\Delta\mathbf{u}_1)\| + \|(d\mathbf{f_a})^{-1}(\Delta\mathbf{x_2} - \Delta\mathbf{x_1})\|$$
$$\leq \tfrac{1}{2}\|\Delta\mathbf{u}_2 - \Delta\mathbf{u}_1\| + C\|\Delta\mathbf{x}_2 - \Delta\mathbf{x}_1\|.$$

for some positive C (see Exercise 3.28, p. 104). The first term on the right side of
the second inequality is a consequence of the contraction property of $\mathbf{h}$ (Step 6). A
final subtraction gives

$$\tfrac{1}{2}\|\Delta\mathbf{u}_2 - \Delta\mathbf{u}_1\| \leq C\|\Delta\mathbf{x}_2 - \Delta\mathbf{x}_1\|,$$

implying we can take $K = 2C$. $\qquad\qquad\square$

The lemma establishes that $\Delta\mathbf{f}^{-1}$ is uniformly continuous (see the comment in
the proof of Lemma 5.2, above).

Step 11
Because we claim the derivative of $\mathbf{f}^{-1}$ at the point $\mathbf{q}$ will be $(d\mathbf{f_p})^{-1}$, where
$\mathbf{p} = \mathbf{f}^{-1}(\mathbf{q})$ is a point in the ball B_r, we must first show $d\mathbf{f_p}$ is invertible.

Lemma 5.7. *Suppose $\mathbf{p} = \mathbf{a} + \Delta\mathbf{v}$ for some $\|\Delta\mathbf{v}\| \leq r$ (i.e., $\mathbf{p}$ is in B_r); then $d\mathbf{f_p}$ is
invertible.*

Proof. From the definition

$$\mathbf{h}(\Delta \mathbf{u}) = \Delta \mathbf{u} + (\mathrm{df_a})^{-1}(f(\mathbf{a} + \Delta \mathbf{u}) - \mathbf{f}(\mathbf{a}))$$

in Step 2, it follows that

$$\mathrm{dh}_{\Delta \mathbf{v}} = I - (\mathrm{df_a})^{-1} \circ \mathrm{df}_{\mathbf{a}+\Delta \mathbf{v}}, \quad \text{or} \quad I = \mathrm{dh}_{\Delta \mathbf{v}} + (\mathrm{df_a})^{-1} \circ \mathrm{df_p}.$$

Therefore, when the maps in the last equation are supplied with input $\Delta \mathbf{u}$, we get

$$\Delta \mathbf{u} = \mathrm{dh}_{\Delta \mathbf{v}}(\Delta \mathbf{u}) + (\mathrm{df_a})^{-1}(\mathrm{df_p}(\Delta \mathbf{u})),$$

and hence

$$\|\Delta \mathbf{u}\| \le \|\mathrm{dh}_{\Delta \mathbf{v}}(\Delta \mathbf{u})\| + \|(\mathrm{df_a})^{-1}(\mathrm{df_p}(\Delta \mathbf{u}))\|$$

$$\le \tfrac{1}{2}\|\Delta \mathbf{u}\| + C\|\mathrm{df_p}(\Delta \mathbf{u})\|$$

for some $C > 0$, exactly as in Lemma 5.6. A bit of algebra now gives

$$\frac{1}{2C}\|\Delta \mathbf{u}\| \le \|\mathrm{df_p}(\Delta \mathbf{u})\|.$$

This inequality implies $\mathrm{df_p}(\Delta \mathbf{u}) \ne \mathbf{0}$ when $\Delta \mathbf{u} \ne \mathbf{0}$. In other words, the kernel of $\mathrm{df_p}$ contains only $\mathbf{0}$, so $\mathrm{df_p}$ is invertible. $\qquad \square$

We now show $\mathbf{f}^{-1}$ is differentiable at an arbitrary point $\mathbf{q}$ in the domain $\mathrm{df_a}(B_{r/2})$, and its derivative is $(\mathrm{df_p})^{-1}$ there ($\mathbf{p} = \mathbf{f}^{-1}(\mathbf{q})$). We work in windows $W_{\mathbf{q}}$ and $W_{\mathbf{p}}$ with local coordinates $\Delta \mathbf{y}$ and $\Delta \mathbf{v}$, respectively, and with the window equations

Step 12

$$\Delta \mathbf{v} = \Delta \mathbf{f}^{-1}(\Delta \mathbf{y}) = \mathbf{f}^{-1}(\mathbf{q} + \Delta \mathbf{y}) - \mathbf{f}^{-1}(\mathbf{q}),$$

$$\Delta \mathbf{y} = \Delta \mathbf{f}(\Delta \mathbf{v}) = \mathbf{f}(\mathbf{p} + \Delta \mathbf{v}) - \mathbf{f}^{-1}(\mathbf{p}).$$

To prove $\mathbf{f}^{-1}$ is differentiable at $\mathbf{q}$, and has derivative $(\mathrm{df_p})^{-1}$ there, it is necessary and sufficient to show $\mathbf{R}(\Delta \mathbf{y}) = \boldsymbol{o}(\Delta \mathbf{y})$, where

$$\Delta \mathbf{v} = \Delta \mathbf{f}^{-1}(\Delta \mathbf{y}) = (\mathrm{df_p})^{-1}(\Delta \mathbf{y}) + \mathbf{R}(\Delta \mathbf{y}).$$

To analyze $\mathbf{R}$, apply $\mathrm{df_p}$ to both sides of this equation,

$$\mathrm{df_p}(\Delta \mathbf{v}) = \Delta \mathbf{y} + \mathrm{df_p}(\mathbf{R}(\Delta \mathbf{y})),$$

and then solve for $\Delta \mathbf{y} = \Delta \mathbf{f}(\Delta \mathbf{v})$ to get

$$\Delta \mathbf{f}(\Delta \mathbf{v}) = \mathrm{df_p}(\Delta \mathbf{v}) - \underbrace{\mathrm{df_p}(\mathbf{R}(\Delta \mathbf{y}))}_{o(\Delta \mathbf{v})}.$$

This equation expresses the differentiability of $\mathbf{f}$ itself at $\mathbf{p}$, so the last term must be $o(\Delta\mathbf{v})$, as indicated. Because $L(o(\mathbf{u})) = o(\mathbf{u})$ for any linear map L (cf. the proof of Lemma 4.1, p. 133),

$$\mathbf{R}(\Delta\mathbf{y}) = (\mathrm{d}\mathbf{f_p})^{-1}(o(\Delta\mathbf{v})) = o(\Delta\mathbf{v}).$$

However, we need $o(\Delta\mathbf{y})$ on the right, not just $o(\Delta\mathbf{v})$. The uniform continuity of $\Delta\mathbf{f}^{-1}$ (Lemma 5.6) in this setting implies $\|\Delta\mathbf{v}\| \leq K\|\Delta\mathbf{y}\|$. Therefore, as $\Delta\mathbf{y} \to \mathbf{0}$, we also have $\Delta\mathbf{v} \to \mathbf{0}$ and

$$\frac{\|\mathbf{R}(\Delta\mathbf{y})\|}{\|\Delta\mathbf{y}\|} \leq K\frac{\|\mathbf{R}(\Delta\mathbf{y})\|}{\|\Delta\mathbf{v}\|} \to 0.$$

(Lemma 4.2, p. 133 makes essentially the same point.) This proves that $\mathbf{R}(\Delta\mathbf{y}) = o(\Delta\mathbf{y})$ and thus that $\mathbf{f}^{-1}$ is differentiable at $\mathbf{q}$.

The last fact to prove is that the derivative $\mathrm{d}(\mathbf{f}^{-1})_\mathbf{q}$ depends continuously on $\mathbf{q}$. But $\mathrm{d}(\mathbf{f}^{-1})_\mathbf{q} = (\mathrm{d}\mathbf{f}_{\mathbf{f}^{-1}(\mathbf{q})})^{-1} = (\mathrm{d}\mathbf{f_p})^{-1}$; therefore we can use the following chain of argument.

- The entries of the $n \times n$ matrix $(\mathrm{d}\mathbf{f_p})^{-1}$ are polynomial functions of the entries of $\mathrm{d}\mathbf{f_p}$, and hence depend continuously on them.

- $\mathrm{d}\mathbf{f_p}$ depends continuously on $\mathbf{p}$.

- $\mathbf{p} = \mathbf{f}^{-1}(\mathbf{q})$ depends continuously on $\mathbf{q}$.

This completes the proof of the inverse function theorem.   □

Corollary 5.3 *Suppose $\mathbf{f}: U^n \to \mathbb{R}^n$ satisfies the conditions of the inverse function theorem at the point $\mathbf{a}$ in U^n. Then the image $\mathbf{f}(U^n)$ contains a ball $\widehat{B}$ of positive radius centered at the point $\mathbf{f}(\mathbf{a})$ in the target of $\mathbf{f}$.*

Proof. The conditions of the inverse function theorem apply to the inverse map $\mathbf{f}^{-1}$ at $\mathbf{f}(\mathbf{a})$. For $\widehat{B}$ take the ball provided by the theorem.   □

The inverse function theorem assumes that the derivative is continuous. This hypothesis is invoked, for example, at the point in the proof where the mean value theorem of maps is used (Steps 6 and 7) to show that $\mathbf{g}$ was a contraction mapping. But is continuity necessary? Our proof needs it, but does the theorem itself? Can we find a better proof that dispenses with that hypothesis?

In fact, the hypothesis is indispensable: there are differentiable functions that have an invertible derivative at a point but are not themselves invertible on any open neighborhood of that point. One such example is

$$f(x) = \frac{x}{2} + \frac{x^2}{2}\sin\frac{\pi}{x} \quad \text{if } x \neq 0, \quad f(0) = 0.$$

The function undergoes infinitely many oscillations near $x = 0$, but because the graph is squeezed between the parabolas $y = (x \pm x^2)/2$, it follows that $f'(0) = \frac{1}{2}$.

The derivative map $df_0(\Delta y) = \frac{1}{2}\Delta x$ is thus invertible. Now consider $f'(x)$ for $x \neq 0$; a direct computation gives

$$f'(x) = \frac{1}{2} + x\sin\frac{\pi}{x} - \frac{\pi}{2}\cos\frac{\pi}{x}.$$

At the points $1/n \to 0$, $f'(1/n)$ is alternately positive and negative. In particular, $f'(1/2n) = (1-\pi)/2$, so

$$\frac{1-\pi}{2} = \lim_{n\to\infty} f'\left(\frac{1}{2n}\right) \neq f'\left(\lim_{n\to\infty}\frac{1}{2n}\right) = f'(0) = \frac{1}{2}.$$

Thus, although f is differentiable everywhere, that derivative is not continuous at the origin.

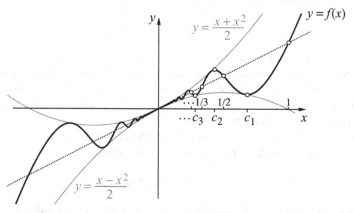

However, f' is continuous away from the origin, so it must change sign at some point c_n between $1/(n+1)$ and $1/n$. That is, $f'(c_n) = 0$; in fact the c_n are alternately local maxima and minima of f. Because $1/n$ is an infinite sequence that converges to 0, the same must be true of the interlaced sequence c_n. Near each local extremum c_n, f is a 2–1 map. Inasmuch as $c_n \to 0$, there is no open interval around $x = 0$ on which f is 1–1. In other words, the oscillations make f noninvertible near the origin.

$f(x)$ is not 1–1
near $x = 0$

In our analysis of several examples of maps of the plane in Chapter 4.2, we found that a map usually "looked like" its derivative locally. When we returned to the quadratic map above (pp. 161–165), we actually converted the map into its derivative within a window by expressing the derivative in terms of appropriate curvilinear coordinates. The curvilinear coordinates were supplied by a map $\mathbf{h}$ for which

When will
$\mathbf{f}$ "look like" $d\mathbf{f_a}$?

$$\Delta\mathbf{f} = d\mathbf{f_a} \circ \mathbf{h}.$$

We claimed that the new variables could indeed serve as coordinates; that is, we claimed, in effect, that $\mathbf{h}$ was invertible. Because $d\mathbf{f_a}$ was obviously invertible, we defined $\mathbf{h}$ as

$$\mathbf{h} = (d\mathbf{f_a})^{-1} \circ \Delta\mathbf{f}.$$

But this, in itself, does not prove $\mathbf{h}$ invertible. Now, however, we can settle the question: If $\mathbf{f}$ is continuously differentiable (and thus invertible in a neighborhood of $\mathbf{a}$, by the inverse function theorem), then $\mathbf{h}$ is invertible and

$$\mathbf{h}^{-1} = \Delta \mathbf{f}^{-1} \circ d\mathbf{f_a}.$$

This discussion leads to the following corollary of the inverse function theorem.

Corollary 5.4 *If $\mathbf{f} : U^n \to \mathbb{R}^n$ is continuously differentiable on an open neighborhood U^n of $\mathbf{a}$, and $d\mathbf{f_a}$ is invertible, then there is a coordinate change $\mathbf{h} : V^n \to S^n$ on some possibly smaller neighborhood V^n of $\mathbf{a}$ for which $\Delta \mathbf{f} = d\mathbf{f_a} \circ \mathbf{h}$.* ◻

Comparison with Taylor's theorem

Before leaving the inverse function theorem, let us compare it with Taylor's theorem as a tool for understanding the geometric action of a map. Suppose we use Taylor's theorem to expand the map $\mathbf{f} : U^n \to \mathbb{R}^n$ at a point $\mathbf{a}$:

$$\Delta \mathbf{x} = \Delta \mathbf{f}(\Delta \mathbf{u}) = d\mathbf{f_a}(\Delta \mathbf{u}) + \boldsymbol{O}(2).$$

This equation was the basis for our frequent observation, in Chapter 4, that the derivative $d\mathbf{f_a}$ approximates $\mathbf{f}$ in a sufficiently small window centered at $\mathbf{a}$. The equation gives only an approximation because the difference $\boldsymbol{O}(2)$ between $\Delta \mathbf{f}$ and $d\mathbf{f_a}$ is nonzero, in general. But the approximation is a good one in the sense that the difference vanishes like $\|\Delta \mathbf{u}\|^2$ as $\Delta \mathbf{u} \to \mathbf{0}$.

By contrast, suppose $d\mathbf{f_a}$ is invertible. The inverse function theorem then says that new coordinates $\Delta \mathbf{s} = \mathbf{h}(\Delta \mathbf{u})$ can be found so that

$$\Delta \mathbf{x} = d\mathbf{f_a}(\Delta \mathbf{s}).$$

In these circumstances, $d\mathbf{f_a}$ equals $\Delta \mathbf{f}$ in a sufficiently small window (at least when $\Delta \mathbf{f}$ is expressed in terms of the proper curvilinear coordinates); the remainder $\boldsymbol{O}(2)$ is dispensed with. There are some minor technical differences, too. For Taylor's theorem, the components of $\mathbf{f}$ must have continuous second derivatives; for the inverse function theorem, only continuous first derivatives are needed.

The Taylor approximation goes a long way toward clarifying the action of $\mathbf{f}$; however, the inverse function theorem provides the ultimate simplification: it shows that $\mathbf{f}$ is essentially linear near $\mathbf{a}$. Perhaps most significantly, the inverse function theorem gives us a new tool to analyze maps: curvilinear coordinates and, more generally, alternative coordinate systems.

Exercises

5.1. Show that $\operatorname{arcsinh} y = \ln \left(y + \sqrt{y^2 + 1} \right)$; use this, the pullback substitution $y = \sinh x$, and other properties of hyperbolic functions to show

$$\int \frac{dy}{\sqrt{1 + y^2}} = \ln \left(y + \sqrt{y^2 + 1} \right).$$

5.2. a. Show that $\operatorname{arctanh} y = \frac{1}{2} \ln \left(\frac{1+y}{1-y} \right)$.

b. Use the pullback $y = \tanh x$ (*not* partial fractions) to determine $\int \frac{dy}{1-y^2}$.

5.3. Use $y = \sinh x$ to show $\int \frac{dy}{(1+y^2)^{3/2}} = \frac{y}{\sqrt{1+y^2}}$.

5.4. Use suitable hyperbolic pullbacks to determine

$$\int \frac{dy}{(y^2-1)^{3/2}} \quad \text{and} \quad \int \frac{dy}{(1-y^2)^{3/2}}.$$

5.5. a. Sketch the graphs of $y = \operatorname{sech} x = 1/\cosh x$ and its inverse $x = \operatorname{arcsech} y$.

b. Show that $\operatorname{arcsech} y = \ln \left(\frac{1 \pm \sqrt{1-y^2}}{y} \right)$.

c. Use both the graph and the formula for $x = \operatorname{arcsech} y$ to explain why its domain is $0 < y \le 1$.

d. Show that the two halves of the graph of $x = \operatorname{arcsech} y$ are negatives of each other by showing

$$-\ln \left(\frac{1 + \sqrt{1-y^2}}{y} \right) = \ln \left(\frac{1 - \sqrt{1-y^2}}{y} \right).$$

5.6. a. Sketch the graph of $x = y - \ln(-y) + k$, $y < 0$; use $k = 2$. Indicate the position of the "landmark" point $(x,y) = (-1+k, -1)$ on the graph. Determine the limiting value of x as $y \to -\infty$, and check that your sketch reflects this fact.

b. Sketch the inverse $y = f_{k,3}(x)$ of the function in part a. What are the domain and the range of $f_{k,3}$?

c. Show that the differential equation $y' = y/(y-1)$ has yet another solution $y = f_4(x) \equiv 0$.

d. Make a sketch of the (x,y)-plane that indicates there is precisely one solution of the differential equation $y' = y/(y-1)$ through each point (x,y) in which $y \ne 1$. This sketch exhibits, visually, the general solution to the differential equation.

5.7. Find the general solution of the differential equation

$$\frac{dy}{dx} = \frac{y}{y^2 + 1}.$$

Describe the solution in words and make a sketch that reflects its salient features.

5.8. a. Solve the following equations for x and y.

$$\theta = \arctan\left(\frac{y}{x-1}\right), \qquad \varphi = \arctan\left(\frac{y}{x+1}\right).$$

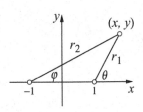

Note: θ and φ are called the **biangular coordinates** for the plane.

  b. Compute the Jacobians

$$\frac{\partial(x,y)}{\partial(\theta,\varphi)} \quad \text{and} \quad \frac{\partial(\theta,\varphi)}{\partial(x,y)},$$

and show by direct computation that they are reciprocals.

5.9. a. Solve the following equations for x and y:

$$r_1 = \sqrt{(x-1)^2 + y^2}, \quad r_2 = \sqrt{(x+1)^2 + y^2}.$$

Note: r_1 and r_2 are called the **two-center bipolar coordinates** for the plane.

  b. Compute the Jacobians

$$\frac{\partial(x,y)}{\partial(r_1,r_2)} \quad \text{and} \quad \frac{\partial(r_1,r_2)}{\partial(x,y)},$$

and verify directly that they are reciprocals.

5.10. The following equations express Cartesian coordinates for space in terms of **spherical coordinates** ρ, θ, and φ (cf. Exercise 3.27, p. 104):

$$x = \rho\cos\theta\cos\varphi,$$
$$y = \rho\sin\theta\cos\varphi,$$
$$z = \rho\sin\varphi.$$

  a. Solve the equations for the spherical coordinates ρ, θ, φ.

  b. Compute the Jacobians

$$\frac{\partial(x,y,z)}{\partial(\rho,\theta,\varphi)} \quad \text{and} \quad \frac{\partial(\rho,\theta,\varphi)}{\partial(x,y,z)},$$

and verify directly that they are reciprocals.

5.11. The following equations express Cartesian coordinates for space in terms of **cylindrical coordinates** r, θ, z:

$$x = r\cos\theta,$$
$$y = r\sin\theta,$$
$$z = z.$$

Show that

$$\frac{\partial(r,\theta,z)}{\partial(x,y,z)} = \frac{1}{\sqrt{x^2+y^2}} = \frac{1}{r},$$

and explain why this is already evident from your knowledge of polar coordinates in the plane.

5.12. a. Show that the formulas that express spherical coordinates (ρ,θ,φ) in terms of cylindrical coordinates *in the new order* (r,z,θ) are identical to the formulas that express cylindrical coordinates *in the same order* in terms of Cartesian coordinates:

$$
\begin{aligned}
\rho &= f(r,z,\theta), & r &= f(x,y,z), \\
\theta &= g(r,z,\theta), & z &= g(x,y,z), \\
\varphi &= h(r,z,\theta), & \theta &= h(x,y,z).
\end{aligned}
$$

It will be sufficient to determine f, g, and h.

b. Determine

$$\frac{\partial(\rho,\theta,\varphi)}{\partial(r,\theta,z)} \quad \text{and} \quad \frac{\partial(r,z,\theta)}{\partial(x,y,z)}$$

and verify directly that

$$\frac{\partial(\rho,\theta,\varphi)}{\partial(x,y,z)} = \frac{\partial(\rho,\theta,\varphi)}{\partial(r,\theta,z)}\frac{\partial(r,\theta,z)}{\partial(x,y,z)}.$$

5.13. Use the Babylonian algorithm to determine $\sqrt{10}$ to 12 decimal places accuracy. Take $x_0 = 3$; how many iterations were required? Take $x_0 = 10$; how many iterations are required now?

5.14. a. To solve $x^2 = a$, the Babylonian algorithm first rewrites the equation as $x = a/x$ and then finds iterates of the average of the left- and right-hand sides: $g(x) = (x+a/x)/2$. This suggests solving $x^3 = a$ by iterating on the average of x and a/x^2: $g_1(x) = (x+a/x^2)/2$. Compute $\sqrt[3]{10}$ by finding the fixed point of g_1 with $x_0 = 2$. Convergence is relatively slow; how many iterates were needed to get 8 decimal places accuracy?

b. Convergence can be sped up by iterating on a weighted average of x and a/x^2: $g_2(x) = (2x+a/x^2)/3$. Compute $\sqrt[3]{10}$ by finding the fixed point of g_2 with $x_0 = 2$. How many iterates were needed to get 8 decimal places accuracy? How does this compare with convergence using g_1?

c. Devise an effective algorithm for solving $x^4 = a$ (that is not just the original Babylonian algorithm applied to the pair $y^2 = a$, $x^2 = y$). Use your algorithm to find $\sqrt[4]{120}$.

5.15. Use the Newton–Raphson method to find (to 6 decimal places) the three real roots of $f(x) = x^3 - 3x + 1$. Sketch the graphs of $y = x^3 - 3x$ and $y = f(x)$ to get the approximate locations of the roots to serve as initial values x_0 for the three iterations.

5.16. Consider the linear map $P : \mathbb{R}^2 \to \mathbb{R}^2$ given by

$$P : \begin{cases} s = \frac{1}{2}x - \frac{\sqrt{3}}{2}y, \\ t = \frac{\sqrt{3}}{2}x + \frac{1}{2}y. \end{cases}$$

a. The map $P : (x,y) \mapsto (s,t)$ is rotation by angle α; what is the value of α?

b. In the (s,t)-plane, sketch the images of the x- and y-axes. According to your sketch, does rotation by α turn the positive s-axis to the image of the positive x-axis?

c. Now let P pull back the variables s and t to provide a second coordinate system in the (x,y)-plane. Describe how the new (s,t) coordinate grid is related to the original (x,y) grid. In particular, describe the position of the positive s-axis in relation to the positive x-axis.

d. If $R_\theta : (x,y) \mapsto (s,t)$ is rotation by the angle θ, how does R_θ pull back the (s,t) coordinate grid to the (x,y)-plane? In particular, describe where the positive s-axis appears in this pullback.

5.17. This exercise uses the semi-log map **sl** (cf. page 161) and the fact that **sl** transforms the exponential function $y = Ba^{kx}$ into the linear function

$$Y = (k \log_{10} a)x + \log_{10} B.$$

a. Plot the US population census data for 1790–1900 on semi-log graph paper and verify that the points lie approximately on a straight line L. Let the horizontal coordinate x be years since 1790.

b. Estimate the slope and Y-intercept of the line L.

c. Use the estimates to obtain an exponential function $y = B10^{kx}$ that approximates the US cenus values, where x denotes years since 1790.

5.18. Define the log–log map (cf. page 161) $\mathbf{L} : (x,y) \to (X,Y)$ by

$$\mathbf{L} : \begin{cases} X = \log_{10} x, \\ Y = \log_{10} y; \end{cases} \quad x, y > 0.$$

a. Show that the image of the graph of $y = ax^p$ under the map $\mathbf{L}$ is a straight line. Determine the equation of this line.

b. Using log–log paper in which the coordinates of the lower left hand corner are $(X,Y) = (0.1, 0.1)$, plot the graphs of $Y = (2/3)X + 4$ and $Y = -2X + 1$. Sketch the pullbacks of these graphs in the (x,y)-plane (using an ordinary uniform coordinate grid).

5.19. This exercise concerns the quadratic map $\mathbf{f}$ of Example 3 (p. 161) and the local coordinate change $\mathbf{h}$ that factors the window map $\Delta\mathbf{f}$ (cf. p. 163). Write $\mathbf{h}$ as $\mathbf{h_a}$ to reflect the fact that this coordinate change depends on the point $\mathbf{a} = (a,b)$ in the (u,v)-plane at which $\Delta\mathbf{f}$ is constructed, and then write the input $(\Delta u, \Delta v)$ of $\mathbf{h_a}$ more simply as $\mathbf{p} = (p,q)$.

   a. Verify that $\mathbf{f(u)} = \mathbf{f(-u)}$ for every $\mathbf{u}$.

   b. Verify that $\mathbf{h_a(0)} = \mathbf{0}$, $\mathbf{h_a(-a)} = -\frac{1}{2}\mathbf{a}$, and $\mathbf{h_a(-2a)} = \mathbf{0}$.

   c. Show that $\mathbf{h_a}$ fails to be 1–1 on any neighborhood of $-\mathbf{a}$ by showing that $\mathbf{h_a}(-\mathbf{a}(1+\varepsilon)) = \mathbf{h_a}(-\mathbf{a}(1-\varepsilon))$ for any ε.

   d. As noted in the text, $\mathbf{h_a}$ is invertible on any sufficiently small square window W centered at $\mathbf{p} = \mathbf{0}$. Give an upper bound on the length of the side of W.

   e. Show that $\mathbf{h_a}$ is conformal everywhere inside W (part d) by showing the derivative $d(\mathbf{h_a})_\mathbf{p}$ is a dilation–rotation matrix (or similarity transformation, p. 118) for each point $\mathbf{p}$ in W.

   f. Show that the dilation factor of $d(\mathbf{h_a})_\mathbf{p}$ is

$$\frac{(a+p)^2 + (b+q)^2}{a^2 + b^2},$$

and conclude that $d(\mathbf{h_a})_{-\mathbf{a}}$ is the zero linear map.

   g. Determine the rotation angle θ of $d(\mathbf{h_a})_\mathbf{p}$ in terms of $\mathbf{a}$ and $\mathbf{p}$, and deduce that $\theta > 0$ when $\mathbf{p} = (p,q)$ is above the line $q = (b/a)p$ and $\theta < 0$ below it. Confirm this fact in the figure on page 164 that illustrates the action of of $\mathbf{h}_{(\sqrt{3}/2, 1/2)}$.

5.20. Let W be the infinite strip $-\pi/2 \leq x \leq \pi/2$ in the (x,y)-plane; let $\mathbf{s} : W \to \mathbb{R}^2$ be the map

$$\mathbf{s} : \begin{cases} u = \sin x \, \cosh y, \\ v = \cos x \, \sinh y. \end{cases}$$

   a. Determine the derivative $d\mathbf{s}_{(x,y)}$. Determine the Jacobian $J(x,y)$ and show that $J > 0$ everywhere except at the two points $(x,y) = (\pm\pi/2, 0)$.

   b. Show that the map $\mathbf{s}$ is conformal (cf. Exercise 5.19) everywhere except at the two points $(x,y) = (\pm\pi/2, 0)$.

   c. Show that the image of the horizontal line segment $y = b$ is the upper half of the ellipse

$$\left(\frac{u}{\cosh b}\right)^2 + \left(\frac{v}{\sinh b}\right)^2 = 1$$

   if $b > 0$, and the lower half if $b < 0$. What happens if $b = 0$?

   d. Show that the image of the vertical line $x = a$ is the right branch of the hyperbola

$$\left(\frac{u}{\sin a}\right)^2 - \left(\frac{v}{\cos a}\right)^2 = 1$$

if $0 < a < \pi/2$ and the left branch if $-\pi/2 < a < 0$. What happens if $a = 0$, $-\pi/2$, or $\pi/2$?

e. Conclude that **s** is invertible on $W \setminus (\pm\pi/2, 0)$ (i.e., W with the two points $(\pm\pi/2, 0)$ removed), and thus defines a coordinate change there.

f. The coordinate change **s** puts curvilinear (x, y) coordinates on the (u, v)-plane. Sketch that coordinate grid in the square $|u| \leq 2$, $|v| \leq 2$. How does this grid manifest the conformality of the map **s**?

g. Sketch the same curvilinear (x, y)-grid on the larger square where $|u| \leq 20$, $|v| \leq 20$. On this square, the grid should look like the polar coordinate grid; does it? Is conformality still evident?

5.21. Let U be the right half-plane $u > 0$, and let $\mathbf{h} : U \to \mathbb{R}^2$ be the map

$$\mathbf{h} : \begin{cases} x = ue^{-v}, \\ y = ue^{v}. \end{cases}$$

a. Show that the image $\mathbf{h}(U)$ is the first quadrant $Q : x > 0, y > 0$.

b. Find the inverse $\mathbf{h}^{-1}$ on Q to show that $\mathbf{h}$ is a coordinate change.

c. Determine the Jacobians $\partial(x, y)/\partial(u, v)$ and $\partial(u, v)/(\partial(x, y)$, and show that they are reciprocals.

d. Sketch the curvilinear (u, v)-coordinate grid on Q, and the curvilinear (x, y)-grid on U. Is the map $\mathbf{h}$ conformal?

5.22. Consider the map $\mathbf{m} : \mathbb{R}^2 \to \mathbb{R}^2$ defined by

$$\mathbf{m} : \begin{cases} p = e^{s} \cosh t, \\ q = e^{s} \sinh t. \end{cases}$$

a. Determine the image $M = \mathbf{m}(\mathbb{R}^2)$ in the (p, q)-plane, and sketch the curvilinear (s, t)-coordinate grid there.

b. Determine the inverse $\mathbf{m}^{-1}$ on M to show that $\mathbf{m}$ is a coordinate change. Sketch the curvilinear (p, q)-grid in the (s, t)-plane.

c. Determine the Jacobians $\partial(p, q)/\partial(s, t)$ and $\partial(s, t)/(\partial(p, q)$, and show that they are reciprocals.

d. You should notice similarities between the maps $\mathbf{h}$ and $\mathbf{m}$ of the previous exercise and this one. Show that the coordinate changes

$$\mathbf{u} : \begin{cases} u = e^{s}, \\ v = t; \end{cases} \qquad \mathbf{p} : \begin{cases} p = (y + x)/2, \\ q = (y - x)/2; \end{cases}$$

convert $\mathbf{h}$ into $\mathbf{m}$. That is, show $\mathbf{m} = \mathbf{p} \circ \mathbf{h} \circ \mathbf{u}$, and sketch all these maps together.

5.23. Let U be the upper half-plane $y > 0$ and let $\mathbf{a} : U \to \mathbb{R}^2 : (x,y) \to (\theta, \varphi)$ be the map defined by the biangular coordinates (cf. Exercise 5.8):

$$\mathbf{a} : \begin{cases} \theta = \arctan y/(x-1), \\ \varphi = \arctan y/(x+1). \end{cases}$$

a. Show that the image $\mathbf{a}(U)$ is the triangular region

$$T : \begin{array}{l} 0 < \theta < \pi, \\ 0 < \varphi < \theta. \end{array}$$

Conclude that $\mathbf{a} : U \to T$ is a coordinate change.

b. Identify, within T, the images of the lines $x = \pm 1$. Also indicate the limit of $\mathbf{a}(x,y)$ as $y \to 0$ and (i) $x < -1$; (ii) $-1 < x < 1$; and (iii) $1 < x$.

c. The curvilinear (x,y)-coordinate grid in T is shown in the margin. Identify the grid lines $x = $ const. and $y = $ const., and indicate how x and y vary through the grid. Indicate how the grid illustrates the limits you determined in the previous part.

d. Referring to the curvilinear (x,y)-coordinate grid, indicate the geometric action of $\mathbf{a}^{-1}$ on T. That is, indicate how $\mathbf{a}^{-1}$ "opens up" T to become the upper half-plane. Does $\mathbf{a}^{-1}$ reverse orientation? How is the answer to this question indicated in the geometric action?

e. Draw the (θ, φ)-coordinate grid in the (x,y)-plane. (This is, in fact, easy to do; do you see why?)

5.24. Let U be the upper half-plane $y > 0$ and let $\mathbf{b} : U \to \mathbb{R}^2 : (x,y) \to (r_1, r_2)$ be the map defined by the two-center bipolar coordinates (cf. Exercise 5.9):

$$\mathbf{b} : \begin{cases} r_1 = \sqrt{(x-1)^2 + y^2}, \\ r_2 = \sqrt{(x+1)^2 + y^2}. \end{cases}$$

a. Show that r_1 and r_2 satisfy the inequalities

$$2 < r_1 + r_2, \quad -2 < r_2 - r_1 < 2.$$

This defines a "half-infinite" strip S in the (r_1, r_2)-plane; $\mathbf{b}(U) = S$.

b. Explain why the map $\mathbf{b} : U \to S$ is a coordinate change. It follows that the (image of the) Cartesian (x,y)-grid defines curvilinear coordinates in S. Sketch this curvilinear grid. Are the curvilinear grid lines perpendicular?

c. Sketch, in U, the curvilinear coordinate grid defined by r_1 and r_2. (This is easy to do.)

d. The map $\mathbf{b}$ is well defined on the x-axis. Sketch the image of the x-axis in the (r_1, r_2)-plane; note in particular the images of the points $(\pm 1, 0)$. How is the image related to S?

5.25. The following map $\sigma : U^4 \to \mathbb{R}^4 : (r,\mathbf{t}) \to \mathbf{x}$ defines the analogue of spherical coordinates on $\mathbb{R}^4$:

$$\sigma : \begin{cases} x_1 = r\cos t_1\, \cos t_2\, \cos t_3, \\ x_2 = r\sin t_1\, \cos t_2\, \cos t_3, \\ x_3 = r\sin t_2\, \cos t_3, \\ x_4 = r\sin t_3; \end{cases} \qquad U^4 : \begin{aligned} & 0 < r, \\ & -\pi < t_1 < \pi, \\ & -\pi/2 < t_2 < \pi/2, \\ & -\pi/2 < t_3 < \pi/2. \end{aligned}$$

a. Describe the image $V^4 = \sigma(U^4)$.

b. Obtain the derivative $d\mathbf{s}_{(r,\mathbf{t})}$ and show that $\det(d\mathbf{s}_{(r,\mathbf{t})}) = r^3 \cos t_2 \cos^2 t_3$.

c. Deduce that σ is locally invertible everywhere in V^4.

d. Find a formula for the (global) inverse of σ on V^4.

Chapter 6
Implicit Functions

Abstract Given a relation between two variables expressed by an equation of the form $f(x,y) = k$, we often want to "solve for y." That is, for each given x in some interval, we expect to find one and only one value $y = \varphi(x)$ that satisfies the relation. The function φ is thus implicit in the relation; geometrically, the locus of the equation $f(x,y) = k$ is a curve in the (x,y)-plane that serves as the graph of the function $y = \varphi(x)$. The problem of implicit functions—and the aim of this chapter—is to determine the function φ from the relation f, or at least to determine that φ exists when its exact form cannot be found. There are analogues of this problem in all dimensions; that is, x and y can be vectors, and the relation $f(x,y) = k$ can expand into a set of equations. However, we begin our analysis with a single equation, because the various impediments to finding the implicit function already occur there.

6.1 A single equation

Perhaps the most familiar example of an implicitly defined function is provided by the equation $f(x,y) = x^2 + y^2$. The locus $f(x,y) = k$ is a circle of radius $\sqrt{k}$ if $k > 0$; we can view it as the graph of two different functions,

The circle as a graph

$$y = \varphi_{\pm}(x) = \pm\sqrt{k - x^2}.$$

But if $k < 0$, the locus is the empty set; there is no implicit function at all. This is the first impediment: there may be no pairs (x,y) whatsoever that satisfy the relation $f(x,y) = k$. We need to know, somehow, that the relation is nonempty; that is, there is at least one point (a,b) for which $f(a,b) = k$, so $\varphi(a) = b$. Think of this point as a kind of "seed" from which the function $y = \varphi(x)$ can "grow."

For example, when $x^2 + y^2 = k > 0$, we can take either $(0, +\sqrt{k})$ or $(0, -\sqrt{k})$ as a seed; then φ_+ grows out of $(0, \sqrt{k})$, and φ_- grows out of $(0, -\sqrt{k})$. This example calls attention to the fact that we must expect the implicit function to be local, that is, to be defined only on part of the locus $f(x,y) = k$. Different parts of the locus may

"Growing φ from a seed" on the locus

therefore be graphs of different implicit functions. Any point of the form $(a,b) = (a, +\sqrt{k-a^2})$, with $-\sqrt{k} < a < \sqrt{k}$ would serve equally well as a seed for φ_+; likewise, any point $(a, -\sqrt{k-a^2})$ would serve for φ_-.

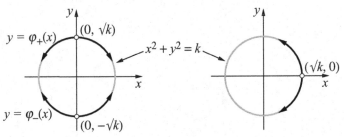

Points on the locus
that are not seeds

This leaves only the points $(\pm\sqrt{k}, 0)$. As the figure on the right shows, $(\sqrt{k}, 0)$ cannot be a seed for a function of x, because the circle has no y-value at all when $x > \sqrt{k}$ and gives two y-values near $y = 0$ when $x < \sqrt{k}$ and x is arbitrarily close to $\sqrt{k}$. There is a similar problem for $(-\sqrt{k}, 0)$. (Of course, $(\sqrt{k}, 0)$ serves perfectly well as a seed for a function $x = \psi(y)$, but we concentrate on x as the independent variable for the moment.) In a different way, there is no seed when $k = 0$. Certainly there is a point on the locus—namely $(a, b) = (0, 0)$—but nothing can grow out of it, because the entire locus $x^2 + y^2 = 0$ is just this single point.

Although there is nothing wrong with having two different parts of the locus be the graph of two different implicit functions, we do require that only one implicit function φ should be able to grow out of a given seed on that locus. This is a significant restriction, and places yet another impediment in the way of obtaining φ. We can illustrate the problem with the quadratic equation $f(x,y) = y^2 - x^2 = 0$. The locus is a pair of lines that cross at the origin. Hence, we find that four different implicit functions grow out of a seed at the origin:

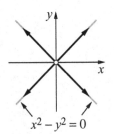

$$\varphi_1(x) = x, \quad \varphi_2(x) = -x, \quad \varphi_3(x) = |x|, \quad \varphi_4(x) = -|x|.$$

Nothing in either the locus or the seed indicates which of these we should choose; therefore we have failed to determine the implicit function we seek.

A "flat" function has
a 2-dimensional locus

The same problem appears in an even more exaggerated form when the locus is not a curve but is a full 2-dimensional region, such as the unit disk $D : x^2 + y^2 \le 1$. This is what happens for the "flat" function that is defined by the formula

$$f(x,y) = \text{the square of the distance from } (x,y) \text{ to } D$$

$$= \begin{cases} 0 & \text{if } x^2 + y^2 \le 1, \\ (\sqrt{x^2 + y^2} - 1)^2 & \text{otherwise.} \end{cases}$$

The graph of f is flat on D, so the locus $f(x,y) = 0$ is D itself; see below. The points at zero distance from D are precisely the points of D. (Because f measures the square of the distance to D, it is differentiable everywhere, including on the boundary of D.) It is clear that the graph of any continuous function $y = \varphi(x)$ that grows out of a seed (a,b) in the interior of D will lie in D, at least for x sufficiently near a.

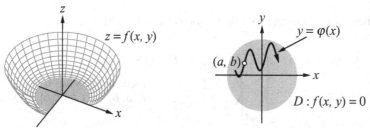

Notice what is true about the tangent to the locus $f(x,y) = k$ at a putative seed point (a,b) in each of our examples thus far. When there was no uniquely defined $\varphi(x)$, either there was no tangent $(x^2 + y^2 = 0)$, there was more than one tangent (both $x^2 - y^2 = 0$ and the function vanishing on the unit disk), or there was a vertical tangent $(x^2 + y^2 = k$ at $(\pm\sqrt{k}, 0))$. We were successful only when the locus had a single nonvertical tangent at the seed point. It seems reasonable, then, to conjecture that this is a sufficient condition for the existence of a unique implicit function of x.

A conjecture

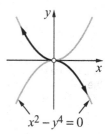

Unfortunately, a single nonvertical tangent is not enough. Consider the locus $y^2 - x^4 = (y - x^2)(y + x^2) = 0$. It is the union of the two parabolas $y = x^2$ and $y = -x^2$, and it has a single nonvertical tangent at every point, including the origin. Nevertheless, there are still four different implicit functions that grow out of the seed at the origin; the figure shows one of them. (Of course, every other point $(a, \pm a^2)$, $a \neq 0$, on the locus is the seed for a unique implicit function.)

To revise the conjecture, let us make use of the fact that an implicit function is a local object. Then we can search for it with basic tools of local analysis, in particular, with Taylor's theorem. Thus, if we suppose that $f(x,y)$ has continuous second derivatives in a neighborhood of (a,b), its first-order Taylor expansion is

Linearizing the locus

$$f(x,y) = f(a,b) + f_x(a,b)\Delta x + f_y(a,b)\Delta y + O(2)$$

in a window centered at (a,b); $\Delta x = x - a$ and $\Delta y = y - b$ are the usual window coordinates. Because $f(a,b) = k$, the equation of the locus $f(x,y) = k$ reduces to

$$f_x(a,b)\Delta x + f_y(a,b)\Delta y + O(2) = 0$$

in the window. That is, the terms after $f(a,b)$ in the Taylor expansion must sum to zero. Within the window, this is the equation of the locus. If we delete the higher-order term $O(2)$, the remaining equation is called the **linearization of the locus at** (a,b):

$$f_x(a,b)\Delta x + f_y(a,b)\Delta y = 0.$$

This is a linear equation; if at least one coefficient $f_x(a,b)$, $f_y(a,b)$ is nonzero, it is the equation of a straight line, the tangent line to the locus at (a,b). Furthermore, if $f_y(a,b) \neq 0$, we can solve the linearized equation for Δy,

$$\Delta y = -\frac{f_x(a,b)}{f_y(a,b)}\,\Delta x,$$

implying that the tangent has finite slope $m = -f_x(a,b)/f_y(a,b)$.

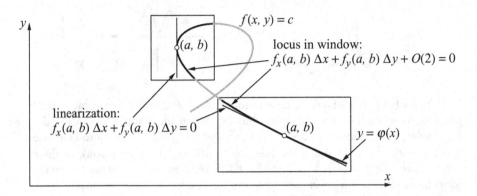

The nature of the linearization

In the figure above, the locus $f(x,y) = k$ has been linearized at two different points (a,b), with fundamentally different results. At the lower right, the linearization has finite slope and implicitly defines the linear function $\Delta y = m\Delta x$. In particular, $f_y(a,b) \neq 0$. Furthermore, it is evident that the locus itself determines a unique implicit function $y = \varphi(x)$ when x is near a and $\varphi(a) = b$. We can even see that φ is differentiable at $x = a$ and

$$\varphi'(a) = m = -\frac{f_x(a,b)}{f_y(a,b)} = -\frac{f_x(a,\varphi(a))}{f_y(a,\varphi(a))}.$$

Now compare this to what happens at the second point, in the upper left. The linearization there is the vertical line $\Delta x = 0$. This means that $f_y(a,b) = 0$, and the linearization is not the graph of any implicit (linear) function of the form $\Delta y = m\Delta x$. Likewise, no implicit function of x can grow out of the seed point (a,b) on the original locus.

The linearization determines the implicit function

According to our evidence, the condition $f_y(a,b) \neq 0$ guarantees that the linearized locus determines a unique implicit function of x, and that is enough to ensure that the original locus does too. The evidence also suggests that we can connect the derivative of φ (where it exists) to the partial derivatives of f, extending the formula for $\varphi'(a)$ we found above. Just differentiate the identity $k = f(x,\varphi(x))$ using the chain rule to get

$$0 = \frac{d}{dx}f(x,\varphi(x)) = f_x(x,\varphi(x)) + f_y(x,\varphi(x))\,\varphi'(x).$$

Because f_y is continuous by hypothesis, the condition $f_y(a,b) \neq 0$ implies $f_y(x,y) \neq 0$ for all (x,y) sufficiently close to (a,b). This allows us to solve for $\varphi'(x)$:

$$\varphi'(x) = -\frac{f_x(x, \varphi(x))}{f_y(x, \varphi(x))}.$$

Theorem 6.1 (Implicit function theorem). *If $f(x,y)$ has continuous first derivatives in a neighborhood of the point (a,b), and $f(a,b) = k$, $f_y(a,b) \neq 0$, then there is a unique function $y = \varphi(x)$ defined and continuously differentiable on an open interval I containing a for which*

- $f(x, \varphi(x)) = k$ *for all x in I.*

- $\varphi(a) = b.$

- $\varphi'(x) = -\dfrac{f_x(x, \varphi(x))}{f_y(x, \varphi(x))}$ *for all x in I.*

Before we prove the implicit function theorem, let us take a closer look at the condition $f_y(a,b) \neq 0$. It expresses our informal conjecture that the locus $f(x,y) = k$ should have a single nonvertical tangent at the seed point (a,b), but it is both more precise and more restrictive. For example, although the locus $f(x,y) = y^2 - x^4 = 0$ (p. 187) appeared to have a single horizontal tangent at the origin, we find $f_x(0,0) = f_y(0,0) = 0$, so the linearized locus is not a horizontal line; it is not a line at all. We call the origin a *critical point* of f.

No seed at
a critical point

Definition 6.1 *We say (a,b) is a **critical point** of the differentible function $f(x,y)$ if $f_x(a,b) = f_y(a,b) = 0$. If either partial derivative is nonzero, we say (a,b) is a **regular point** of f.*

The implicit function theorem rules out any critical point of f as a seed. Indeed, in most of the problematic examples that led to our original conjecture, we were attempting to make a critical point be a seed.

So suppose (a,b) is a regular point of the function $z = f(x,y)$. Either $f_y(a,b) \neq 0$ and the locus $f(x,y) = f(a,b)$ is the graph of a differentiable function of x near (a,b) (by the implicit function theorem), or else $f_x(a,b) \neq 0$ and, switching the roles of x and y, we see the locus is the graph of a differentiable function of y. In either case, the locus is the graph of some differentiable function and is thus a differentiable curve near (a,b).

Near a regular point,
the locus is a curve

Definition 6.2 *If (a,b) is a regular point of the continuously differentiable function $f(x,y)$, and $f(a,b) = k$, then we say (a,b) is a **regular point** of the curve $f(x,y) = k$. If all points on the locus are regular, we say the curve itself is **regular**.*

The locus $f(x,y) = k$ is one of the *level sets*, or *contours*, of f. At a regular point of a contour, at least one of the partial derivatives of f is different from zero. By continuity, that derivative remains nonzero at all sufficiently nearby points. Therefore, near the given regular point, all contours are regular. The following theorem says even more: it says that a suitably chosen coordinate change will "straighten out" those contours. This implies that there is essentially only one way to arrange the contours near a regular point. It also leads to a quick proof of the implicit function theorem.

Theorem 6.2. *Suppose (a,b) is a regular point of a function $z = f(x,y)$ that has continuous first derivatives. Then there is a coordinate change $(u,v) = \mathbf{h}(x,y)$ defined on a window centered at (a,b) that transforms the level curves of f into the coordinate lines $v = \text{constant}$.*

Proof. At least one of $f_x(a,b)$, $f_y(a,b)$ is nonzero; suppose $f_y(a,b) \neq 0$. Define $\mathbf{h}$ by the formulas

$$\mathbf{h} : \begin{cases} u = x, \\ v = f(x,y). \end{cases}$$

(If $f_y(a,b) = 0$, then $f_x(a,b) \neq 0$; set $u = y$ instead; see Exercise 6.10.) Because f has continuous first derivatives, so does $\mathbf{h}$. Moreover,

$$d\mathbf{h}_{(x,y)} = \begin{pmatrix} 1 & 0 \\ f_x(x,y) & f_y(x,y) \end{pmatrix},$$

so $\det d\mathbf{h}_{(a,b)} = f_y(a,b) \neq 0$, implying that $d\mathbf{h}_{(a,b)}$ is invertible. By the inverse function theorem (Theorem 5.2, p. 169), $\mathbf{h}$ has a continuously differentiable inverse defined on a neighborhood of $\mathbf{h}(a,b) = (a, f(a,b)) = (a,k)$. Thus $\mathbf{h}$ is a valid coordinate change near (a,b). Because $v = f(x,y)$, $\mathbf{h}$ transforms each level curve $f(x,y) = \lambda$ into the coordinate line $v = \lambda$. $\square$

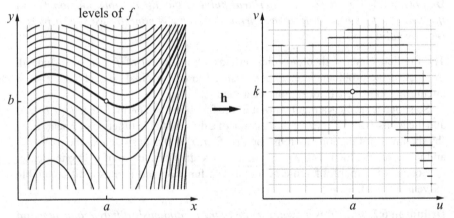

levels of f

Straightening level curves with a nonlinear shear

Thus, near a regular point, the level curves of a real-valued function are part of a curvilinear coordinate system; near that point, those curves are always roughly parallel and evenly spaced. We can see this in the figure above, which shows a coordinate change that straightens the levels of $f(x,y) = y^2 - 4x^2(2x - 1)$ near the point $(a,b) = (0.18, 0.417)$; $k = 0.35$. (Note: The origin is not at the intersection of the coordinate axes in either figure.) The coordinate change $\mathbf{h}$ is a nonlinear shear. It maps each vertical line to itself; this is the geometric meaning of the equation $u = x$ in the definition of $\mathbf{h}$. Each vertical line just slides up or down, stretched by different amounts at different points. At the point (α, β) on $x = \alpha$, the vertical stretch factor is $f_y(\alpha, \beta)$. For example, at points in the lower half of the figure above, we can see that the stretch factor is less than 1 (though still positive), because $\mathbf{h}$ shrinks vertical distances there. You can analyze another example, with a simpler function $f(x,y)$, in Exercise 6.8.

We take up now the proof of the implicit function theorem, Theorem 6.1. Let us write the inverse of $\mathbf{h}$ in terms of components:

$$\mathbf{h}^{-1} : \begin{cases} x = u, \\ y = g(u,v). \end{cases}$$

Proof of the implicit function theorem

The first component is just the first component of $\mathbf{h}$ itself. The second component, g, is a continuously differentiable function of u and v. The inverse relation between $\mathbf{h}^{-1}$ and $\mathbf{h}$ implies

$$(x,y) = \mathbf{h}^{-1}(\mathbf{h}(x,y)) = \mathbf{h}^{-1}(x, f(x,y)) = (x, g(x, f(x,y))),$$

and, in particular, $y = g(x, f(x,y))$. Therefore, if $v = f(x,y) = k$, then $y = g(x,k)$. This is the implicit function we seek: $\varphi(x) = g(x,k)$. $\qquad\qquad\square$

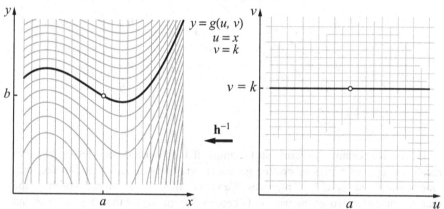

The idea behind the proof of Theorem 6.1 is that it becomes easy to find the implicit function if we use the right coordinate system. We choose coordinates $(x,y) \mapsto (u,v)$ in which the level curves of $f(x,y)$ become coordinate lines $v = \lambda$. Then the implicit function defined by $f(x,y) = k$ is just the constant function $v = k$ in the new coordinates, and the inverse coordinate change $(u,v) \mapsto (x,y)$ converts this line $v = k$ into the graph of a (generally nonlinear) function $y = \varphi(x)$.

Making coordinate choices

Underlying the proof is a basic principle we have used several times: coordinates are just labels for points, and we should choose labels that make the geometry most intelligible. In this setting there are two fundamental objects, both geometric: the first is the plane $\mathcal{P}$ and its constituent points $\mathbf{p}$; the second is a real-valued map $\mathcal{F} : \mathcal{P} \to \mathbb{R} : \mathbf{p} \mapsto \mathcal{F}(\mathbf{p})$ defined on $\mathcal{P}$. To describe that map, we introduce analytic tools: coordinates (x,y) to label the points, and the appropriate expression $f(x,y)$ to represent $\mathcal{F}$:

Geometry underlies analysis

$$\text{analytic level: } \mathbb{R}^2 \xrightarrow{f} \mathbb{R} \qquad (x,y) \xrightarrow{f} z$$
$$\uparrow \qquad \| \qquad\qquad \uparrow \qquad \|$$
$$\text{geometric level: } \mathcal{P} \xrightarrow{\mathcal{F}} \mathbb{R} \qquad \mathbf{p} \xrightarrow{\mathcal{F}} z$$

In practice, we start at the analytic level. Conceptually, though, it helps to let the geometry come first; the analysis is then overlaid as a language with which to describe it. For example, equations such as $f(x,y) = k$ are a way to describe level curves of the more fundamental geometric map $\mathbf{p} \mapsto \mathcal{F}(\mathbf{p})$.

Coordinates as languages

We can take the language analogy further. Just as the world of objects and ideas is described by a variety of human languages, the geometry of points and maps can be described by a variety of coordinate systems and analytic expressions. Two human languages are connected by a pair of translation dictionaries; the geometric analogue is a coordinate change. The following diagram shows how the two coordinate systems we used to analyze the level curves of $\mathcal{F}$ are related by the coordinate change $\mathbf{h}$. In the diagram, "v" stands for a *coordinate function* as well as a *coordinate*; as a function, it assigns to the ordered pair (u,v) the number v.

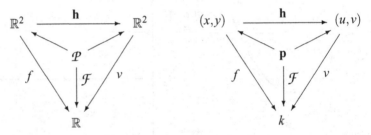

The geometric view

Before we resume our work on the implicit function theorem, let us pause to recall a couple of places where the geometric view has already come to the fore. One was in the study of 2×2 matrices. We viewed them as certain maps of the plane that are characterized geometrically (Theorem 2.6, p. 40) by their eigenvalues and eigenvectors. Like f and v in the diagram above, different matrices can represent the same geometric map. We defined two matrices to be *equivalent* if a linear coordinate change would convert one into the other, and we saw that equivalent matrices had the same geometric action. Another place where we ended up with a geometric view was with the inverse function theorem. According to Corollary 5.4 (p. 176), which was suggested by our work in the example on pages 161–165, a nonlinear map $\mathbf{f}: U^n \to \mathbb{R}^n$ looks like its linear approximation $d\mathbf{f_a}$ near any point $\mathbf{a}$ where the linear approximation is invertible.

Geometry simplifies

So geometry helps us simplify, and it does so by "lumping together" things (such as equivalent matrices) that we would otherwise treat as distinct. In the diagram above, the coordinate change $\mathbf{h}$ that converts f into v allows us to "lump together" those two functions, and therefore to say that f is essentially a coordinate function. The simplification is this: near a regular point, any real-valued function is essentially just a coordinate function. From this observation we then get, first, the structure of the level curves, and second, the implicit function theorem.

The move from two variables to three is straightforward. The meaning of a regular point of a function or of a locus carries over in a natural way, as does the geometric viewpoint generally. The following theorem and corollary are the main results. Their proofs follow the proofs of the two-variable versions (see the exercises), and also follow from the n-variable versions, below.

Regular points with three variables

Theorem 6.3. *Suppose (a,b,c) is a regular point of a function $s = f(x,y,z)$ that has continuous first derivatives. Then there is a coordinate change $(u,v,w) = \mathbf{h}(x,y,z)$ defined on a window centered at (a,b,c) that transforms the level sets of f into the coordinate planes $w = $ constant.* □

Corollary 6.4 (Implicit function theorem) *Suppose the function $s = f(x,y,z)$ has continuous first derivatives in some open neighborhood of a point (a,b,c), and $f(a,b,c) = k$. If $f_z(a,b,c) \neq 0$, then there is a unique function $z = \varphi(x,y)$ defined on an open neighborhood N of (a,b) for which*

- $f(x,y,\varphi(x,y)) = k$ *for all (x,y) in N.*

- $\varphi(a,b) = c$.

- φ *has continuous first derivatives on N, and*

$$\varphi_x(x,y) = -\frac{f_x(x,y,\varphi(x,y))}{f_z(x,y,\varphi(x,y))}, \quad \varphi_y(x,y) = -\frac{f_y(x,y,\varphi(x,y))}{f_z(x,y,\varphi(x,y))}. \qquad \square$$

As stated, the corollary connects the condition that the partial derivative of f with respect to z is nonzero to the conclusion that z depends on x and y near (a,b,c). However, if instead it is the partial derivative with respect to either y or x that is nonzero, we get the same conclusion *mutatis mutandis* ("the necessary changes being made"). Corollary 6.4 thus stands for three different statements; for example, if $f_y(a,b,c) \neq 0$, then $y = \psi(x,z)$ for some ψ and

Different implications of the corollary

$$\psi_x(x,z) = -\frac{f_x(x,\psi(x,z),z)}{f_y(x,\psi(x,z),z)}, \quad \psi_z(x,z) = -\frac{f_z(x,\psi(x,z),z)}{f_y(x,\psi(x,z),z)}.$$

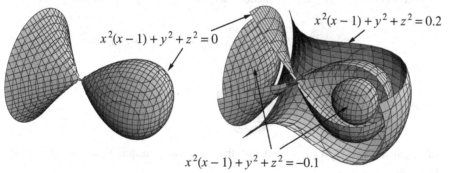

$$x^2(x-1) + y^2 + z^2 = 0$$
$$x^2(x-1) + y^2 + z^2 = 0.2$$
$$x^2(x-1) + y^2 + z^2 = -0.1$$

Even though the theorem and corollary above settle our questions about a function of three variables, it is still valuable to look at an individual level set and its

An example

linearization at a point, as we did above with a function of two variables. We see above, left, the zero level of

$$f(x,y,z) = x^2(x-1) + y^2 + z^2;$$

on the right, we get some idea how the zero level is nested within the collection of nearby level sets. (Parts of some levels have been "peeled away" to help see inside.)

We expect the locus $f(x,y,z) = k$ to be 2-dimensional, although it may fail to be a proper surface at one or more of its points. That is, a point may fail to be a regular point of the locus. This is what happens to $f(x,y,z) = 0$ at the origin, where it has the shape of the vertex of a cone: no coordinate change can convert the vertex into a simple plane. (In this example, however, every nearby level set contains only regular points of f, at which Theorem 6.3 applies.) In general, a locus $f(x,y,z) = k$ can exhibit all the irregularities that $f(x,y) = k$ did, and many more besides (see Exercise 6.13).

Linearization and
the tangent planeBy considering the linearization of $s = f(x,y,z)$, we can see once again what prompts the partial derivative condition that leads us to an implicit function. We begin by assuming, as before, that f has continuous partial derivatives. Then the first-order Taylor expansion of f at a seed point (i.e., $f(a,b,c) = k$) is

$$f(x,y,z) = f(a,b,c) + f_x(a,b,c)\Delta x + f_y(a,b,c)\Delta y + f_z(a,b,c)\Delta z + O(2)$$

in a window centered at (a,b,c). We take

$$f_x(a,b,c)\Delta x + f_y(a,b,c)\Delta y + f_z(a,b,c)\Delta z + O(2) = 0$$

to be the equation of the locus $f(x,y,z) = k$ in that window, and

$$f_x(a,b,c)\Delta x + f_y(a,b,c)\Delta y + f_z(a,b,c)\Delta z = 0$$

to be the linearization of that locus. If (a,b,c) is a regular point of f, then at least one of the coefficients is nonzero, and the equation describes a plane. Because the difference between the locus and its linearization at (a,b,c) vanishes at least to second order in $(\Delta x, \Delta y, \Delta z)$, the plane is tangent to the original locus at (a,b,c).

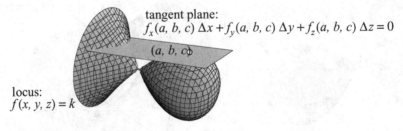

tangent plane:
$$f_x(a, b, c)\,\Delta x + f_y(a, b, c)\,\Delta y + f_z(a, b, c)\,\Delta z = 0$$

(a, b, c)

locus:
$f(x, y, z) = k$

More particularly, if $f_z(a,b,c) \neq 0$, then the linearization implicitly defines the linear function

$$\Delta z = -\frac{f_x(a,b,c)}{f_z(a,b,c)}\Delta x - \frac{f_y(a,b,c)}{f_z(a,b,c)}\Delta y,$$

and this is another version of the equation of the tangent plane at (a,b,c). The analogy with the two-variable case means that the implicit function $z = \varphi(x,y)$ has this equation as its linear approximation near $(x,y) = (a,b)$, and

$$\varphi_x(a,b) = -\frac{f_x(a,b,c)}{f_z(a,b,c)}, \quad \varphi_y(a,b) = -\frac{f_y(a,b,c)}{f_z(a,b,c)}.$$

Finally, let us suppose the number of variables $x_1, x_2, \ldots, x_n$ is arbitrary, with a single relation $f(x_1, \ldots, x_n) = k$ holding between them. Then we expect one of the variables, say x_n, to depend on the others, implying there is a function $x_n = \varphi(x_1, \ldots, x_{n-1})$. The graph of φ is an $(n-1)$-dimensional hypersurface in $\mathbb{R}^n$. Nearby level sets $f(x_1, \ldots, x_n) = \lambda$, for λ near k, should be nested hypersurfaces that together fill a portion of $\mathbb{R}^n$. These expectations are borne out at a regular point of $f(x)$, that is, at a point $(a_1, \ldots, a_n)$ where at least one partial derivative $\partial f/\partial x_i(a_1, \ldots, a_n)$ is nonzero.

Regular points
with n variables

Theorem 6.5. *Suppose the function $f : X^n \to \mathbb{R} : \mathbf{x} \mapsto f(\mathbf{x})$ has continuous first derivatives on X^n, and $\mathbf{x} = \mathbf{a}$ is a regular point of $s = f(\mathbf{x})$. Then there is a coordinate change $\mathbf{u} = \mathbf{h}(\mathbf{x})$ defined on a window W^n centered at $\mathbf{x} = \mathbf{a}$ that transforms the level sets of f into the coordinate hyperplanes $u_n = $ constant.*

Proof. Because $\mathbf{a}$ is a regular point of f, at least one of the partial derivatives $f_i(\mathbf{a})$ is nonzero. (We define f_i to be the partial derivative of f with respect to the i-th variable, x_i.) By permuting the variables x_i, if necessary, we may suppose that $f_n(\mathbf{a}) \neq 0$. Define $\mathbf{h} : X^n \to \mathbb{R}^n : \mathbf{x} \mapsto \mathbf{u}$ by

$$\mathbf{h} : \begin{cases} u_1 = x_1, \\ \quad \vdots \\ u_{n-1} = x_{n-1}, \\ \quad u_n = f(x_1, \ldots, x_n). \end{cases}$$

Then $\mathbf{h}$ is continuously differentiable on X^n because f is, and

$$d\mathbf{h}_\mathbf{x} = \begin{pmatrix} 1 & \cdots & 0 & 0 \\ \vdots & \ddots & \vdots & \vdots \\ 0 & \cdots & 1 & 0 \\ f_1(\mathbf{x}) & \cdots & f_{n-1}(\mathbf{x}) & f_n(\mathbf{x}) \end{pmatrix}.$$

Therefore $\det d\mathbf{h}_\mathbf{a} = f_n(\mathbf{a}) \neq 0$, and the inverse function theorem implies $\mathbf{h}$ is invertible on some neighborhood W^n of $\mathbf{a}$. The coordinate change $\mathbf{h}$ transforms the level set $f(\mathbf{x}) = \lambda$ into the coordinate hyperplane $u_n = \lambda$. $\quad\square$

Corollary 6.6 (Implicit function theorem) *Suppose the function $s = f(x_1, \ldots, x_n)$ has continuous first derivatives on some neighborhood of a point $(a_1, \ldots, a_n)$, and $f(a_1, \ldots, a_n) = k$. If $f_n(a_1, \ldots, a_n) \neq 0$, then there is a unique function $x_n = \varphi(x_1, \ldots, x_{n-1})$ defined on an open neighborhood N^{n-1} of $(a_1, \ldots, a_{n-1})$ for which*

- $f(x_1,\ldots,x_{n-1},\varphi(x_1,\ldots,x_{n-1})) = k$ *for all* $(x_1,\ldots,x_{n-1})$ *in* N^{n-1}.

- $\varphi(a_1,\ldots,a_{n-1}) = a_n$.

- φ *has continuous first derivatives on* N^{n-1}, *and for* $i = 1,\ldots,n-1$,

$$\varphi_i(x_1,\ldots,x_{n-1}) = -\frac{f_i(x_1,\ldots,x_{n-1},\varphi(x_1,\ldots,x_{n-1}))}{f_n(x_1,\ldots,x_{n-1},\varphi(x_1,\ldots,x_{n-1}))}.$$

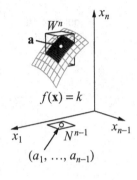

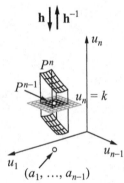

Proof. Let $\mathbf{h}$ be the coordinate change in Theorem 6.5; because it is the identity on the first $n-1$ coordinates, it is a nonlinear shear that maps each vertical line $(x_1,\ldots,x_{n-1}) = (c_1,\ldots,c_{n-1})$ to itself. Its inverse must do the same, and thus has the form

$$\mathbf{h}^{-1}:\begin{cases} x_1 = u_1, \\ \quad\vdots \\ x_{n-1} = u_{n-1}, \\ x_n = g(u_1,\ldots,u_n). \end{cases}$$

Here g is a real-valued function with continuous derivatives on a neighborhood P^n of the image point $\mathbf{h}(\mathbf{a})$. Let P^{n-1} be the intersection of P^n and the horizontal plane $u_n = k$. Because $\mathbf{h}$ and $\mathbf{h}^{-1}$ preserve vertical lines, it is convenient to put the target space $(u_1,\ldots,u_n)$ directly below the source. Also, for visual clarity, P^n is shown as the sheared image of a box W^n centered at $\mathbf{a}$.

Because $\mathbf{h}^{-1} \circ \mathbf{h}$ is the identity, we can write

$$\begin{aligned} (x_1,\ldots,x_{n-1},x_n) &= \mathbf{h}^{-1}(\mathbf{h}(x_1,\ldots,x_{n-1},x_n)) \\ &= \mathbf{h}^{-1}(u_1,\ldots,u_{n-1},f(x_1,\ldots,x_{n-1},x_n)). \end{aligned}$$

Now assume that the point $(u_1,\ldots,u_{n-1},f(x_1,\ldots,x_{n-1},x_n))$ is in P^{n-1}. In particular, this means $f(x_1,\ldots,x_n) = k$, and we can write

$$\begin{aligned} (x_1,\ldots,x_{n-1},x_n) &= \mathbf{h}^{-1}(u_1,\ldots,u_{n-1},k) \\ &= (x_1,\ldots,x_{n-1},g(x_1,\ldots,x_{n-1},k)). \end{aligned}$$

The figure makes it clear that N^{n-1} (which we must still define) and P^{n-1} are in the same vertical column. Thus we make N^{n-1} the projection of P^{n-1} to the coordinate plane $x_n = 0$; that is, $(x_1,\ldots,x_{n-1})$ is in N^{n-1} if and only if $(x_1,\ldots,x_{n-1},k)$ is in P^{n-1}. Finally, if we set $\varphi(x_1,\ldots,x_{n-1}) = g(x_1,\ldots,x_{n-1},k)$ when $(x_1,\ldots,x_{n-1})$ is in N^{n-1}, then

$$f(x_1,\ldots,x_n) = k \quad\Longleftrightarrow\quad x_n = \varphi(x_1,\ldots,x_{n-1})$$

and $\varphi(a_1,\ldots,a_{n-1}) = g(a_1,\ldots,a_{n-1},k) = a_n$. The expressions for the partial derivatives of φ follow from the chain rule applied to

$$k = f(x_1,\ldots,x_{n-1},\varphi(x_1,\ldots,x_{n-1}));$$

we find $0 = f_i + f_n \cdot \varphi_i$, from which it follows that $\varphi_i = -f_i/f_n$. $\qquad\square$

"Symmetrizing" the implicit function theorem

As it is written, the implicit function theorem assumes that $f_n(\mathbf{a}) \neq 0$ at the seed point $\mathbf{a}$. But suppose $f_n(\mathbf{a}) = 0$; the theorem still holds, *mutatis mutandis*, if some other partial derivative is nonzero there. For example, if $f_j(\mathbf{a}) \neq 0$, then we can solve for x_j in terms of the other variables to get a function

$$x_j = \psi(x_1, \ldots, \widehat{x}_j, \ldots, x_n).$$

Here the circumflex is used to indicate that the variable x_j is missing from the list. The theorem implies $\psi(a_1, \ldots, \widehat{a}_j, \ldots, a_n) = a_j$, and

$$\psi_i(x_1, \ldots, \widehat{x}_j, \ldots, x_n) = -\frac{f_i(x_1, \ldots, \psi(x_1, \ldots, \widehat{x}_j, \ldots, x_n), \ldots, x_n)}{f_j(x_1, \ldots, \psi(x_1, \ldots, \widehat{x}_j, \ldots, x_n), \ldots, x_n)}$$

for every $i = 1, \ldots, \widehat{j}, \ldots, n$.

Near a regular point $\mathbf{a}$, f looks like $\mathrm{d}f_{\mathbf{a}}$

Suppose $z = f(\mathbf{x})$ has continuous second partial derivatives at $\mathbf{a}$, allowing us to write the first-order Taylor expansion of f (in terms of window coordinates) at $\mathbf{a}$:

$$\Delta z = f(\mathbf{a} + \Delta \mathbf{x}) - f(\mathbf{a}) = f_1(\mathbf{a})\Delta x_1 + \cdots + f_n(\mathbf{a})\Delta x_n + O(2)$$
$$= \mathrm{d}f_{\mathbf{a}}(\Delta \mathbf{x}) + O(2).$$

If $\mathbf{a}$ is a regular point of f, then it follows from Theorem 6.5 that there are new curvilinear coordinates in which the higher-order terms $O(2)$ disappear: f is transformed into precisely its linear approximation $\mathrm{d}f_{\mathbf{a}}$ near $\mathbf{a}$. The details are in the following corollary, which incidentally is stronger than Taylor's theorem because it requires only continuous first partial derivatives for f.

Corollary 6.7 *Suppose* $\mathbf{x} = \mathbf{a}$ *is a regular point of the continuously differentiable function* $z = f(\mathbf{x})$. *Then there is a coordinate change* $\Delta \mathbf{v} = \mathbf{g}(\Delta \mathbf{x})$ *in a window centered at* $\mathbf{a}$ *for which*

$$\Delta z = f(\mathbf{a} + \mathbf{g}^{-1}(\Delta \mathbf{v})) - \mathbf{f}(\mathbf{a}) = \mathrm{d}f_{\mathbf{a}}(\Delta \mathbf{v}) = f_1(\mathbf{a})\Delta v_1 + \cdots + f_n(\mathbf{a})\Delta v_n.$$

Proof. Express the coordinate change $\mathbf{h}$ that is provided by Theorem 6.5 in window coordinates $\Delta \mathbf{x}$ centered at $\mathbf{a}$ and $\Delta \mathbf{u}$ centered at $(a_1, \ldots, a_{n-1}, f(\mathbf{a}))$ (so that $\Delta \mathbf{u} = \mathbf{h}(\Delta \mathbf{x})$ and $\mathbf{h}(\mathbf{0}) = \mathbf{0}$):

$$\mathbf{h} : \begin{cases} \Delta u_1 = \Delta x_1, \\ \quad \vdots \\ \Delta u_{n-1} = \Delta x_{n-1}, \\ \quad \Delta u_n = f(\mathbf{a} + \Delta \mathbf{x}) - f(\mathbf{a}); \end{cases} \qquad \mathrm{d}\mathbf{h}_0 = \begin{pmatrix} 1 & \cdots & 0 & 0 \\ \vdots & \ddots & \vdots & \vdots \\ 0 & \cdots & 1 & 0 \\ f_1(\mathbf{a}) & \cdots & f_{n-1}(\mathbf{a}) & f_n(\mathbf{a}) \end{pmatrix}.$$

In terms of these coordinates, f is already transformed into the simple linear function $\Delta z = \Delta u_n$. That is,

$$\Delta z = f(\mathbf{a} + \mathbf{h}^{-1}(\Delta \mathbf{u})) - f(\mathbf{a}) = \Delta u_n.$$

Now consider the linear map $\Delta\mathbf{u} = d\mathbf{h}_0(\Delta\mathbf{v})$, whose nth component function is

$$\Delta u_n = f_1(\mathbf{a})\Delta v_1 + \cdots + f_n(\mathbf{a})\Delta v_n.$$

Thus, if we set $\mathbf{g} = d\mathbf{h}_0^{-1} \circ \mathbf{h}$ (so that $\mathbf{g}^{-1} = \mathbf{h}^{-1} \circ d\mathbf{h}_0$), then $\mathbf{g}$ is a valid change of window coordinates in a window centered at $\mathbf{a}$, and we get

$$\begin{aligned}
\Delta z &= f(\mathbf{a} + \mathbf{g}^{-1}(\Delta\mathbf{v})) - f(\mathbf{a}) \\
&= f(\mathbf{a} + \mathbf{h}^{-1}(\Delta\mathbf{u})) - f(\mathbf{a}) \\
&= f_1(\mathbf{a})\Delta v_1 + \cdots + f_n(\mathbf{a})\Delta v_n. \qquad\qquad \square
\end{aligned}$$

Near a regular point, f becomes a coordinate function

Although there is a certain "fitness" to showing that a function can be transformed exactly into its derivative at a regular point, it is useful to know that it can also be transformed into a simple coordinate function. In other words, there is a curvilinear coordinate system in which one of the coordinates is just the value of f there. This result, stated in the next corollary, has already been demonstrated in the last proof.

Corollary 6.8 *Suppose* $\mathbf{x} = \mathbf{a}$ *is a regular point of the continuously differentiable function* $z = f(\mathbf{x})$. *Then there is a coordinate change* $\Delta\mathbf{u} = \mathbf{h}(\Delta\mathbf{x})$ *in a window centered at* $\mathbf{a}$ *for which*

$$\Delta z = f(\mathbf{a} + \mathbf{h}^{-1}(\Delta\mathbf{u})) - \mathbf{f}(\mathbf{a}) = \Delta u_n. \qquad\qquad \square$$

6.2 A pair of equations

We now suppose that two separate conditions have been imposed on our variables:

$$f(x_1,\ldots,x_n) = k, \quad g(x_1,\ldots,x_n) = l.$$

Because we expect these conditions will allow us to solve for two of the variables in terms of the remaining ones, it is natural to assume that $n > 2$, that is, that there are more variables than conditions. However, if we set aside the matter of implicit functions for the moment, and just consider the geometric implications of the two conditions, there is no reason to exclude $n = 1$ or 2. We take up this possibility in the last section (cf. pp. 214ff.). To begin, however, we assume $n = 3$; this is the most complicated case that we can visualize fully.

In general, the locus is a space curve

Thus, we are dealing with two conditions $f(x,y,z) = k$ and $g(x,y,z) = l$ on three variables. We expect the locus $f = k$ to be a 2-dimensional surface $\mathcal{S}_f$ in $\mathbb{R}^3$, and $g = l$ to be another such surface, $\mathcal{S}_g$. Of course, either of these could fail to look like an ordinary surface at one or more points; for the moment, though, let us assume they are completely regular. The locus determined by two conditions together is the intersection $\mathcal{S}_f \cap \mathcal{S}_g$. We expect the intersection of two surfaces to be a curve in space, but it may not be, even when the surfaces themselves are regular; see the counterexamples below.

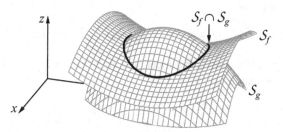

For simplicity, let us assume the intersection locus is, indeed, an ordinary space curve. We should then be able to describe it parametrically:

Solving for two
of the variables

$$\mathcal{S}_f \cap \mathcal{S}_g : (x(t), y(t), z(t)), \quad t_1 \leq t \leq t_2.$$

However, we should not be introducing a new variable t. In the spirit of implicit functions, we should, instead, try to express two of x, y, z in terms of the remaining one, and involve no new variable at all. To recover these implicit functions from the parametrization, let us suppose that $x = x(t)$ is invertible, perhaps for t in some smaller interval. If $t = \tau(x)$ is the inverse (on $x_1 \leq x \leq x_2$), then

$$(x(t), y(t), z(t)) = (x, y(\tau(x)), z(\tau(x))) = (x, \varphi(x), \psi(x)).$$

If $\mathcal{S}_f \cap \mathcal{S}_g$ has more than one point with a given x-value (as in the figure above), we cannot parametrize all of it by x. Restricting the values of x as necessary, we can solve for y and z in terms of x:

$$\begin{array}{ll} f(x,y,z) = k, \\ g(x,y,z) = l, \end{array} \quad \Longleftrightarrow \quad \begin{array}{l} y = \varphi(x), \\ z = \psi(x), \end{array} \quad x_1 \leq x \leq x_2.$$

Although we can expect $x(t)$ to be invertible on only a part of the locus $\mathcal{S}_f \cap \mathcal{S}_g$, on a different part we may find that $y(t)$ is invertible. On that subset we can solve for x and z in terms of y; on a subset where $z(t)$ is invertible, we can solve for x and y in terms of z. To summarize: two conditions on three variables implicitly define two of the variables in terms of the third.

Before we discuss precise conditions that allow us to determine those implicit functions, let us see how two ordinary surfaces can fail to intersect in a curve. Thus, we suppose (a, b, c) is a regular point on each of surfaces,

Faulty intersections

$$\mathcal{S}_f : f(a,b,c) = k, \quad \mathcal{S}_g : g(a,b,c) = l.$$

This means each locus has a well-defined tangent plane at (a, b, c); moreover (Theorem 6.3, p. 193), each locus can be locally transformed into a flat plane by a suitable coordinate change. For these reasons we can regard the locus as a regular surface near (a, b, c).

In each of the following three examples, (a,b,c) is the origin. Moreover, each surface is the graph of a continuously differentiable function, so every point is regular. The first example is

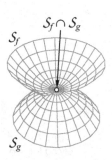

$$S_f : x^2 + y^2 - z = 0, \quad S_g : -x^2 - y^2 - z = 0.$$

The surfaces are parabolic bowls that meet only at the origin; $S_f \cap S_g$ is a point, not a curve. The second example is

$$S_f : x^2 + y^2 - z = 0, \quad S_g : z = 0.$$

This time the intersection is a pair of crossed lines, so, once again, it is not a curve. For the third example, let S_f be the graph of the "flat" function defined on page 186, and let S_g again be the horizontal plane $z = 0$. The intersection is the whole unit disk in the (x,y)-plane.

Transversality and general position

The problem with each example is that, even though the two surfaces meet at the origin, they do not cut cleanly across each other at that point. We say that the surfaces fail to be *transverse* there. Surfaces whose intersections are all transverse are also said to be *in general position* with respect to each other. We prove that the intersection of two regular surfaces in general position is a regular curve. To do this, we need to make our informal definition of transversality precise. The key is to note that whenever two surfaces are transverse in the informal sense, so are their tangent planes. But it is much easier to check transversality for the planes than for the surfaces: two planes passing through the same point are either different or identical.

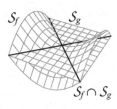

Definition 6.3 *We say that two surfaces in $\mathbb{R}^3$ are **transverse** at a regular point of intersection if they have different tangent planes at that point.*

For surfaces that are given as the loci of equations, the following theorem gives us a simple and convenient analytic criterion for transversality.

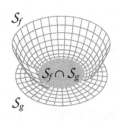

Theorem 6.9. *Suppose $\mathbf{a} = (a,b,c)$ lies on both surfaces $S_f : f(x,y,z) = k$ and $S_g : g(x,y,z) = l$, and is a regular point of both f and g. Then S_f and S_g are transverse at $\mathbf{a}$ if and only if the matrix*

$$M = \begin{pmatrix} f_x(\mathbf{a}) & f_y(\mathbf{a}) & f_z(\mathbf{a}) \\ g_x(\mathbf{a}) & g_y(\mathbf{a}) & g_z(\mathbf{a}) \end{pmatrix}$$

has rank 2.

Proof. The equations of the tangent planes of S_f and S_g at $\mathbf{a}$ are, respectively,

$$f_x(\mathbf{a})\,\Delta x + f_y(\mathbf{a})\,\Delta y + f_z(\mathbf{a})\,\Delta z = 0,$$
$$g_x(\mathbf{a})\,\Delta x + g_y(\mathbf{a})\,\Delta y + g_z(\mathbf{a})\,\Delta z = 0.$$

Because $\mathbf{a}$ is a regular point of both functions, each of these equations has at least one nonzero coefficient, and thus determines a well-defined plane. The two planes are different if and only if their coefficent vectors

$$(f_x(\mathbf{a}), f_y(\mathbf{a}), f_z(\mathbf{a})) \quad \text{and} \quad (g_x(\mathbf{a}), g_y(\mathbf{a}), g_z(\mathbf{a}))$$

are not scalar multiples of each other, and this is true if and only if the matrix M has rank 2. $\square$

We now return to the question of implicit functions. As in the past, the answer is a consequence of a coordinate change that makes the loci straight. Thus, instead of just $S_f : f(x,y,z) = k$, we look at the whole family of nearby surfaces $f(x,y,z) = \kappa$, for $\kappa \approx k$. These are nested surfaces that fill a region containing the seed point (a,b,c). Likewise, we look at the family of surfaces $g(x,y,z) = \lambda$ ($\lambda \approx l$) that are nested around S_g; they too fill a region around the seed point. If S_f and S_g are transverse at (a,b,c), then (as we prove in a moment), all surfaces in the first family are in general position with respect to all those in the second. After they are straightened by the coordinate change, members of the two families look like the spacers in a case of wine bottles; they intersect in parallel straight lines.

Straightening
the surfaces

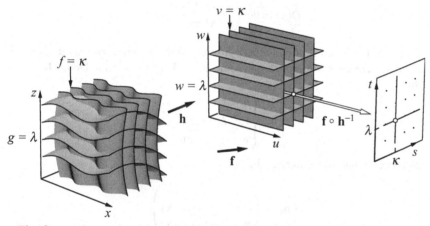

The figure above suggests we should consolidate the functions f and g into a map $\mathbf{f} : X^3 \to \mathbb{R}^2$,

The map defined by
f and g

$$\mathbf{f} : \begin{cases} s = f(x,y,z), \\ t = g(x,y,z). \end{cases}$$

Then the surface S_f is the pullback of the (vertical) coordinate line $s = k$ by $\mathbf{f}$:

$$\mathbf{f}^{-1}(k,t) = \{(x,y,z) : f(x,y,z) = k\}.$$

The other surfaces in the same family are the pullbacks $\mathbf{f}^{-1}(\kappa, t)$ of the other vertical coordinate lines. The pullbacks $\mathbf{f}^{-1}(s, \lambda)$ of the horizontal coordinate lines are the second family of surfaces, that is, the ones nested with S_g. The intersection curve $S_f \cap S_g$ is the pullback of the single point (k,l) in the (s,t)-plane. (The terms *locus* and *pullback* are roughly equivalent. The first is older and commonly used with real-valued functions; the second is used more generally with maps to an arbitrary target.)

The map $\mathbf{f}$ also gives us a convenient way to indicate when $\mathcal{S}_f$ and $\mathcal{S}_g$ are transverse at the seed point $\mathbf{a} = (a, b, c)$, because the matrix M of the previous theorem is the derivative of $\mathbf{f}$ at $\mathbf{a}$. Here is the theorem that "straightens" the surfaces $f = \kappa$ and $g = \lambda$ simultaneously.

Theorem 6.10. *Let* $\mathbf{f} : X^3 \to \mathbb{R}^2$ *be continuously differentiable in a neighborhood* X^3 *of a point* $\mathbf{a} = (a, b, c)$, *let* $\mathbf{f}(a, b, c) = (k, l)$, *and assume the derivative* $d\mathbf{f}_\mathbf{a} : \mathbb{R}^3 \to \mathbb{R}^2$ *has maximal rank. Then, on a smaller neighborhood* N^3 *of* (a, b, c), *there is a coordinate change* $\mathbf{h} : N^3 \to \mathbb{R}^3$ *that maps each pullback* $\mathbf{f}^{-1}(\kappa, \lambda)$ *to the coordinate line* $v = \kappa$, $w = \lambda$ *in* $\mathbf{h}(N^3)$.

Proof. The 2×3 matrix $d\mathbf{f}_\mathbf{a}$ has rank 2; therefore it has two linearly independent columns. By permuting the variables x, y, z, if necessary, we may assume the second and third columns are linearly independent. Thus,

$$D(\mathbf{x}) = \det \begin{pmatrix} f_y(\mathbf{x}) & f_z(\mathbf{x}) \\ g_y(\mathbf{x}) & g_z(\mathbf{x}) \end{pmatrix} \neq 0$$

when $\mathbf{x} = \mathbf{a}$. Because $\mathbf{f}$ is continuously differentiable, $D(\mathbf{x})$ is a continuous function of $\mathbf{x}$ and therefore remains nonzero in some neighbborhood Y^3 of $\mathbf{a}$. Define $\mathbf{h} : Y^3 \to \mathbb{R}^3$ by

$$\mathbf{h} : \begin{cases} u = x, \\ v = f(x, y, z), \\ w = g(x, y, z). \end{cases}$$

Then $\mathbf{h}$ is continuously differentiable, and

$$d\mathbf{h}_\mathbf{x} = \begin{pmatrix} 1 & 0 & 0 \\ f_x(\mathbf{x}) & f_y(\mathbf{x}) & f_z(\mathbf{x}) \\ g_x(\mathbf{x}) & g_y(\mathbf{x}) & g_z(\mathbf{x}) \end{pmatrix}.$$

By construction, $\det d\mathbf{h}_\mathbf{x} = D(\mathbf{x}) \neq 0$ for all $\mathbf{x}$ in Y^3. According to the inverse function theorem, $\mathbf{h}$ (on a possibly small neighborhood N^3) has a continuously differentiable inverse $\mathbf{h}^{-1}$. By the definition of $\mathbf{f}$,

$$\begin{aligned} f(x, y, z) = \kappa &\iff v = \kappa, \\ g(x, y, z) = \lambda &\iff w = \lambda, \end{aligned}$$

whenever (x, y, z) is in N^3. $\square$

Extensions of the theorem

Note that the proof goes beyond what the theorem states: it shows that the coordinate change transforms the individual surfaces $f(x, y, z) = \kappa$ into the coordinate planes $v = \kappa$, and the surfaces $g(x, y, z) = \lambda$ into the coordinate planes $w = \lambda$ of a second family. Moreover, because the two families of coordinate planes are obviously in general position with respect to each other, the same must be true of the original curved families.

The figure also makes it clear that the map

$$\mathbf{f} \circ \mathbf{h}^{-1} : (u,v,w) \rightarrow (s,t)$$

is just projection along the first component. That is, each coordinate line $(u,v,w) = (u,\kappa,\lambda)$ parallel to the u-axis projects to the single point (κ,λ). Putting it another way: the pullback (by $\mathbf{f} \circ \mathbf{h}^{-1}$) of any point in the (s,t)-plane is the line parallel to the u-axis that projects to that point.

Corollary 6.11 (Implicit function theorem) *Let $f(x,y,z)$ and $g(x,y,z)$ have continuous first derivatives in some neighborhood of a point $\mathbf{a} = (a,b,c)$, and let $f(a,b,c) = k$, $g(a,b,c) = l$. If the determinant*

$$\begin{vmatrix} f_y(\mathbf{a}) & f_z(\mathbf{a}) \\ g_y(\mathbf{a}) & g_z(\mathbf{a}) \end{vmatrix}$$

is nonzero, then there are unique functions $y = \varphi(x)$, $z = \psi(x)$ defined on an open interval I containing $x = a$ for which

- $f(x,\varphi(x),\psi(x)) = k$ *and* $g(x,\varphi(x),\psi(x)) = l$ *for all x in I.*

- $\varphi(a) = b$, $\psi(a) = c$.

- φ *and* ψ *are continuously differentiable on I, and*

$$\varphi'(x) = -\frac{\begin{vmatrix} f_x(x,\varphi(x),\psi(x)) & f_z(x,\varphi(x),\psi(x)) \\ g_x(x,\varphi(x),\psi(x)) & g_z(x,\varphi(x),\psi(x)) \end{vmatrix}}{\begin{vmatrix} f_y(x,\varphi(x),\psi(x)) & f_z(x,\varphi(x),\psi(x)) \\ g_y(x,\varphi(x),\psi(x)) & g_z(x,\varphi(x),\psi(x)) \end{vmatrix}},$$

$$\psi'(x) = -\frac{\begin{vmatrix} f_y(x,\varphi(x),\psi(x)) & f_x(x,\varphi(x),\psi(x)) \\ g_y(x,\varphi(x),\psi(x)) & g_x(x,\varphi(x),\psi(x)) \end{vmatrix}}{\begin{vmatrix} f_y(x,\varphi(x),\psi(x)) & f_z(x,\varphi(x),\psi(x)) \\ g_y(x,\varphi(x),\psi(x)) & g_z(x,\varphi(x),\psi(x)) \end{vmatrix}}.$$

Proof. Let $\mathbf{h}$ be the coordinate change in Theorem 6.10. Because $\mathbf{h}$ is the identity on the first coordinate, the same must be true of its inverse:

$$\mathbf{h}^{-1} : \begin{cases} x = u, \\ y = p(u,v,w), \\ z = q(u,v,w). \end{cases}$$

Because $\mathbf{h}^{-1} \circ \mathbf{h}$ is the identity where it is defined,

$$(x,y,z) = \mathbf{h}^{-1} \circ \mathbf{h}(x,y,z)$$
$$= \mathbf{h}^{-1}(x,f(x,y,z),g(x,y,z))$$
$$= (x,p(x,f(x,y,z),g(x,y,z)),q(x,f(x,y,z),g(x,y,z))),$$

implying

$$y = p(x, f(x,y,z), g(x,y,z)), \quad z = q(x, f(x,y,z), g(x,y,z)).$$

These equations reduce to

$$y = p(x,k,l), \quad z = q(x,k,l),$$

when $f(x,y,z) = k$ and $g(x,y,z) = l$, that is, when (x,y,z) lies in $\mathcal{S}_f \cap \mathcal{S}_g$. Let $p(x,k,l) = \varphi(x)$ and $q(x,k,l) = \psi(x)$; as components of the coordinate change $\mathbf{h}^{-1}$, these functions are continuously differentiable in an open neighborhood of $x = a$. By construction,

$$k = f(x,y,z) = f(x, \varphi(x), \psi(x)),$$
$$l = g(x,y,z) = g(x, \varphi(x), \psi(x)),$$

verifying the first condition on φ and ψ.

Because $\mathbf{h}(a,b,c) = (a,k,l)$, it follows that $\mathbf{h}^{-1}(a,k,l) = (a,b,c)$. In terms of components,

$$(a,b,c) = \mathbf{h}^{-1}(a,k,l) = (a, p(a,k,l), q(a,k,l)) = (a, \varphi(a), \psi(a)),$$

so $b = \varphi(a)$ and $c = \psi(a)$, thus verifying the second condition.

We obtain the derivatives of φ and ψ by applying the chain rule to the equations

$$k = f(x, \varphi(x), \psi(x)), \quad l = g(x, \varphi(x), \psi(x)).$$

Suppressing the arguments of the functions for clarity, we find

$$0 = \frac{dk}{dx} = \frac{d}{dx} f(x, \varphi(x), \psi(x)) = f_x + f_y \cdot \varphi' + f_z \cdot \psi',$$
$$0 = \frac{dl}{dx} = \frac{d}{dx} g(x, \varphi(x), \psi(x)) = g_x + g_y \cdot \varphi' + g_z \cdot \psi'.$$

If we write these equations in the matrix form

$$\begin{pmatrix} f_y & f_z \\ g_y & g_z \end{pmatrix} \begin{pmatrix} \varphi' \\ \psi' \end{pmatrix} = \begin{pmatrix} -f_x \\ -g_x \end{pmatrix},$$

we can solve them using Cramer's rule to get

$$\varphi' = \frac{\begin{vmatrix} -f_x & f_z \\ -g_x & g_z \end{vmatrix}}{\begin{vmatrix} f_y & f_z \\ g_y & g_z \end{vmatrix}} = -\frac{\begin{vmatrix} f_x & f_z \\ g_x & g_z \end{vmatrix}}{\begin{vmatrix} f_y & f_z \\ g_y & g_z \end{vmatrix}}, \quad \psi' = \frac{\begin{vmatrix} f_y & -f_x \\ g_y & -g_x \end{vmatrix}}{\begin{vmatrix} f_y & f_z \\ g_y & g_z \end{vmatrix}} = -\frac{\begin{vmatrix} f_y & f_x \\ g_y & g_x \end{vmatrix}}{\begin{vmatrix} f_y & f_z \\ g_y & g_z \end{vmatrix}}. \qquad \square$$

We can also express the hypothesis and conclusion of the implicit function theorem in terms of Jacobians (cf. p. 137). The hypothesis is

Jacobians

$$\left.\frac{\partial(f,g)}{\partial(y,z)}\right|_{\mathbf{x}=\mathbf{a}} \neq 0,$$

and the implicit functions $y = \varphi(x)$, $z = \psi(x)$ have derivatives given by

$$\frac{dy}{dx} = -\frac{\partial(f,g)}{\partial(x,z)}\Big/\frac{\partial(f,g)}{\partial(y,z)}, \qquad \frac{dz}{dx} = -\frac{\partial(f,g)}{\partial(y,x)}\Big/\frac{\partial(f,g)}{\partial(y,z)}$$

Expressed this way, the derivatives are strikingly similar in form to the derivative of the function $y = \varphi(x)$ that is implicitly defined by the single equation $f(x,y) = k$ (Theorem 6.1, p. 189):

$$\frac{dy}{dx} = -\frac{\partial f}{\partial x}\Big/\frac{\partial f}{\partial y}.$$

Let us return to the question we have already addressed several times before (Chapters 4.2, 4.3, 5.2, 5.3, especially pp. 119–121, 128–128, 163–165, 175–176): to what extent—and in what way—does a map look like its linear approximation near a given point? Theorem 6.10 deals with a map $\mathbf{f} : U^3 \to \mathbb{R}^2$. Under that assumption that $d\mathbf{f_a}$ has maximal rank (namely 2), it shows that a suitable coordinate change will make $\mathbf{f}$ look like the linear projection $\Pi : \mathbb{R}^{1+2} \to \mathbb{R}^2 : (x,y,z) \mapsto (y,z)$. But according to Theorem 2.19 (p. 50), a coordinate change will likewise make $d\mathbf{f_a}$ into the same projection Π. Coordinate changes thus make $\mathbf{f}$ look like $d\mathbf{f_a}$ near $\mathbf{a}$.

When does $\mathbf{f}$ "look like" $d\mathbf{f_a}$?

Maximal rank is essential. To see this, consider the map $\mathbf{f} : \mathbb{R}^3 \to \mathbb{R}^2 : (x,y,z) \to (s,t)$:

Maximal rank is essential

$$\mathbf{f} : \begin{cases} s = x, \\ t = (y - z)^3. \end{cases}$$

We have

$$d\mathbf{f_x} = \begin{pmatrix} 1 & 0 & 0 \\ 0 & 3(y-z)^2 & -3(y-z)^2 \end{pmatrix},$$

so the rank of $d\mathbf{f_x}$ is only 1, not 2, at all points in the plane $z = y$ (thus including the origin). Near the origin, $\mathbf{f}$ is geometrically different from $d\mathbf{f_0}$; see Exercise 6.16.

6.3 The general case

In the general case, p equations constrain the values of $k + p$ variables. We expect to find that p of the variables are implicitly determined by the remaining k. Under what conditions can we guarantee that happens, and which variables will be functions of which? Here is the same question, in geometric terms: given a map from (an open set in) $\mathbb{R}^{k+p}$ to $\mathbb{R}^p$, what does the pullback of a point look like? As we have already seen in the low-dimension cases, an answer to the first will follow readily from an answer to the second.

Because the source of the map is split into two factors, with k real variables in one and p in the other, it is useful to split the derivative of the map into the two parts—its *"partial" derivatives*—that act separately on these two factors. To define them, we assume $\mathbf{f} : X^{k+p} \to \mathbb{R}^n : (\mathbf{x}, \mathbf{y}) \to \mathbf{z}$ is differentiable and $\mathbf{x} = (x_1, \ldots, x_k)$, $\mathbf{y} = (y_1, \ldots, y_p)$. If

$$\mathbf{f} : \begin{cases} z_1 = f_1(x_1, \ldots, x_k, y_1, \ldots, y_p), \\ \quad\vdots \\ z_n = f_n(x_1, \ldots, x_k, y_1, \ldots, y_p), \end{cases}$$

then the derivative of $\mathbf{f}$ is given by the $n \times (k+p)$ matrix

$$d\mathbf{f}_{(\mathbf{x},\mathbf{y})} = \begin{pmatrix} f_{11} & \cdots & f_{1k} & f_{1,k+1} & \cdots & f_{1,k+p} \\ \vdots & \ddots & \vdots & \vdots & \ddots & \vdots \\ f_{n1} & \cdots & f_{nk} & f_{n,k+1} & \cdots & f_{n,k+p} \end{pmatrix},$$

where

$$f_{ij} = \begin{cases} \dfrac{\partial f_i}{\partial x_j}(\mathbf{x}, \mathbf{y}) & \text{if } j = 1, \ldots, k, \\[2ex] \dfrac{\partial f_i}{\partial y_q}(\mathbf{x}, \mathbf{y}) & \text{if } j = k+q \text{ and } q = 1, \ldots, p, \end{cases}$$

and $i = 1, \ldots, n$.

Definition 6.4 *The **partial derivatives** of $\mathbf{f} : X^{k+p} \to \mathbb{R}^n$ are the linear maps $\partial_1 \mathbf{f}_{(\mathbf{x},\mathbf{y})} = \partial_{\mathbf{x}} \mathbf{f}_{(\mathbf{x},\mathbf{y})} : \mathbb{R}^k \to \mathbb{R}^n$ and $\partial_2 \mathbf{f}_{(\mathbf{x},\mathbf{y})} = \partial_{\mathbf{y}} \mathbf{f}_{(\mathbf{x},\mathbf{y})} : \mathbb{R}^p \to \mathbb{R}^n$ given by the matrices*

$$\partial_1 \mathbf{f}_{(\mathbf{x},\mathbf{y})} = \begin{pmatrix} f_{11} & \cdots & f_{1k} \\ \vdots & \ddots & \vdots \\ f_{n1} & \cdots & f_{nk} \end{pmatrix}, \quad \partial_2 \mathbf{f}_{(\mathbf{x},\mathbf{y})} = \begin{pmatrix} f_{1,k+1} & \cdots & f_{1,k+p} \\ \vdots & \ddots & \vdots \\ f_{n,k+1} & \cdots & f_{n,k+p} \end{pmatrix}.$$

If the derivative of $\mathbf{f}$ is continuous, then so are its partial derivatives. The notation "∂_1" signifies *the partial derivative with respect to the first factor*, and the alternate notation "$\partial_{\mathbf{x}}$" signifies *the partial derivative with respect to the $\mathbf{x}$ factor*. As we have done for functions of two real variables (e.g., as with $f_1(x,y) = f_x(x,y)$), we use these notations interchangeably.

Theorem 6.12. *Suppose the map $\mathbf{f} : X^{k+p} \to \mathbb{R}^p$ is continuously differentiable, and the derivative $d\mathbf{f}_{(\mathbf{a},\mathbf{b})} : \mathbb{R}^{k+p} \to \mathbb{R}^p$ has maximal rank p. Then there is a coordinate change $\mathbf{h} : (\mathbf{x}, \mathbf{y}) \to (\mathbf{u}, \mathbf{v})$ defined in a neighborhood N^{k+p} of $(\mathbf{a}, \mathbf{b})$ that transforms $\mathbf{f}$ into the projection $\Pi : (\mathbf{u}, \mathbf{v}) \mapsto \mathbf{v}$; that is, $\mathbf{f} \circ \mathbf{h}^{-1} = \Pi$.*

Proof. We know p columns of $d\mathbf{f}_{\mathbf{a}}$ are linearly independent. By permuting the variables, if necessary, we may assume that the final p columns are, so the partial derivative $\partial_{\mathbf{y}}\mathbf{f}_{(\mathbf{a},\mathbf{b})} : \mathbb{R}^p \to \mathbb{R}^p$ is invertible. Now use the component functions of $\mathbf{f}$ to define

$$
\mathbf{h}: \begin{cases} u_1 = x_1, \\ \quad \vdots \\ u_k = x_k, \\ v_1 = f_1(x_1,\ldots,x_k,y_1,\ldots,y_p), \\ \quad \vdots \\ v_p = f_p(x_1,\ldots,x_k,y_1,\ldots,y_p); \end{cases} \qquad \text{schematically,} \quad \mathbf{h}: \begin{cases} \mathbf{u} = \mathbf{x}, \\ \mathbf{v} = \mathbf{f}(\mathbf{x},\mathbf{y}). \end{cases}
$$

Then $\mathbf{h}$ is continuously differentiable on X^{k+p} (because $\mathbf{f}$ is), and

$$
d\mathbf{h}_{(\mathbf{x},\mathbf{y})} = \begin{pmatrix} I & O \\ \partial_{\mathbf{x}}\mathbf{f}_{(\mathbf{x},\mathbf{y})} & \partial_{\mathbf{y}}\mathbf{f}_{(\mathbf{x},\mathbf{y})} \end{pmatrix}, \quad \det d\mathbf{h}_{(\mathbf{x},\mathbf{y})} = \det \partial_{\mathbf{y}}\mathbf{f}_{(\mathbf{x},\mathbf{y})},
$$

implying that $\det d\mathbf{h}_{(\mathbf{a},\mathbf{b})} \neq 0$. By the inverse function theorem (Theorem 5.2, p. 169), $\mathbf{h}$ is invertible on some smaller neighborhood N^{k+p} of $(\mathbf{a},\mathbf{b})$.

To show that $\mathbf{f} \circ \mathbf{h}^{-1} = \boldsymbol{\Pi}$, first write $\mathbf{h}^{-1}$ schematically as

$$
\mathbf{h}^{-1}: \begin{cases} \mathbf{x} = \mathbf{u}, \\ \mathbf{y} = \mathbf{g}(\mathbf{u},\mathbf{v}), \end{cases}
$$

for a suitable map $\mathbf{g} : \mathbf{h}(N^{k+p}) \to \mathbb{R}^p$. By the definition of an inverse,

$$
(\mathbf{u},\mathbf{v}) = \mathbf{h} \circ \mathbf{h}^{-1}(\mathbf{u},\mathbf{v}) = \mathbf{h}(\mathbf{u},\mathbf{g}(\mathbf{u},\mathbf{v})) = (\mathbf{u},\mathbf{f}(\mathbf{u},\mathbf{g}(\mathbf{u},\mathbf{v})))
$$

implying

$$
\mathbf{v} = \mathbf{f}(\mathbf{u},\mathbf{g}(\mathbf{u},\mathbf{v})) = \mathbf{f}(\mathbf{h}^{-1}(\mathbf{u},\mathbf{v})),
$$

as desired. More simply, we know $\mathbf{f} \circ \mathbf{h}^{-1}(\mathbf{u},\mathbf{v}) = \mathbf{v}$ because $\mathbf{f}$ is the second component of $\mathbf{h}$, and $\mathbf{h} \circ \mathbf{h}^{-1}(\mathbf{u},\mathbf{v}) = (\mathbf{u},\mathbf{v})$. $\qquad\square$

Corollary 6.13 (Implicit function theorem) *Suppose* $\mathbf{f} : X^{k+p} \to \mathbb{R}^p : (\mathbf{x},\mathbf{y}) \to \mathbf{z}$ *is continuously differentable and* $\mathbf{f}(\mathbf{a},\mathbf{b}) = \mathbf{k}$. *If the partial derivative map* $\partial_{\mathbf{y}}\mathbf{f}_{(\mathbf{a},\mathbf{b})} : \mathbb{R}^p \to \mathbb{R}^p$ *is invertible, then there is a unique map* $\mathbf{y} = \boldsymbol{\varphi}(\mathbf{x})$ *defined on a neighborhood* N^k *of* $\mathbf{x} = \mathbf{a}$ *in* $\mathbb{R}^k$ *for which*

- $\mathbf{f}(\mathbf{x},\boldsymbol{\varphi}(\mathbf{x})) = \mathbf{k}$ *for all* $\mathbf{x}$ *in* N^k.

- $\boldsymbol{\varphi}(\mathbf{a}) = \mathbf{b}$.

- $\boldsymbol{\varphi}$ *is continuously differentiable on* N^k, *and*

$$
d\boldsymbol{\varphi}_{\mathbf{x}} = -\left(\partial_{\mathbf{y}}\mathbf{f}_{(\mathbf{x},\boldsymbol{\varphi}(\mathbf{x}))}\right)^{-1} \circ \partial_{\mathbf{x}}\mathbf{f}_{(\mathbf{x},\boldsymbol{\varphi}(\mathbf{x}))} : \mathbb{R}^k \to \mathbb{R}^p.
$$

Proof. Let $\mathbf{h}$ be the coordinate change defined on the neighborhood N^{k+p} of $(\mathbf{a},\mathbf{b})$ in $\mathbb{R}^{k+p}$, as provided by Theorem 6.12; let $\mathbf{h}^{-1}$ be its inverse. We wrote

$$\mathbf{h}^{-1}(\mathbf{u},\mathbf{v}) = (\mathbf{u},\mathbf{g}(\mathbf{u},\mathbf{v}))$$

for a suitably defined continuously differentiable map $\mathbf{g}$ on $P^{k+p} = \mathbf{h}(N^{k+p})$, and we saw that

$$\mathbf{v} = \mathbf{f}(\mathbf{u},\mathbf{g}(\mathbf{u},\mathbf{v}))$$

for every $(\mathbf{u},\mathbf{v})$ in P^{k+p}. In particular, $\mathbf{k} = \mathbf{f}(\mathbf{u},\mathbf{g}(\mathbf{u},\mathbf{k}))$.

Now define N^k by the condition

$$\mathbf{u} \text{ is in } N^k \quad \Longleftrightarrow \quad (\mathbf{u},\mathbf{k}) \text{ is in } P^{k+p},$$

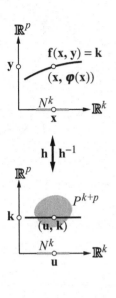

and set $\boldsymbol{\varphi}(\mathbf{x}) = \mathbf{g}(\mathbf{x},\mathbf{k})$. Then $\boldsymbol{\varphi}$ is defined on all of N^k, and the equation $\mathbf{k} = \mathbf{f}(\mathbf{u},\mathbf{g}(\mathbf{u},\mathbf{k}))$ translates into

$$\mathbf{f}(\mathbf{x},\boldsymbol{\varphi}(\mathbf{x})) = \mathbf{k} \text{ for every } \mathbf{x} \text{ in } N^k.$$

This verifies the first condition. Also, because $\mathbf{h}(\mathbf{a},\mathbf{b}) = (\mathbf{a},\mathbf{f}(\mathbf{a},\mathbf{b})) = (\mathbf{a},\mathbf{k})$,

$$(\mathbf{a},\mathbf{b}) = \mathbf{h}^{-1}(\mathbf{a},\mathbf{k}) = (\mathbf{a},\mathbf{g}(\mathbf{a},\mathbf{k})) = (\mathbf{a},\boldsymbol{\varphi}(\mathbf{a})),$$

it follows that $\boldsymbol{\varphi}(\mathbf{a}) = \mathbf{b}$, verifying the second condition.

The third condition follows from the chain rule applied to the equation $\mathbf{k} = \mathbf{f}(\mathbf{x},\boldsymbol{\varphi}(\mathbf{x}))$. To carry out the differentiation, it will be helpful to define the map $\boldsymbol{\Phi} : N^k \to \mathbb{R}^{k+p} : \mathbf{x} \mapsto (\mathbf{x},\boldsymbol{\varphi}(\mathbf{x}))$. Then $\mathbf{k} = \mathbf{f} \circ \boldsymbol{\Phi}(\mathbf{x})$, so

$$O = \underset{p \times k}{\mathrm{d}\mathbf{f}_{\boldsymbol{\Phi}(\mathbf{x})}} \circ \underset{p \times (k+p)}{\mathrm{d}\boldsymbol{\Phi}_{\mathbf{x}}} = \left(\partial_{\mathbf{x}}\mathbf{f}_{(\mathbf{x},\boldsymbol{\varphi}(\mathbf{x}))} \ \partial_{\mathbf{y}}\mathbf{f}_{(\mathbf{x},\boldsymbol{\varphi}(\mathbf{x}))}\right) \underset{(k+p) \times k}{\begin{pmatrix} I \\ \mathrm{d}\boldsymbol{\varphi}_{\mathbf{x}} \end{pmatrix}}$$

$$= \underset{p \times k}{\partial_{\mathbf{x}}\mathbf{f}_{(\mathbf{x},\boldsymbol{\varphi}(\mathbf{x}))}} \circ \underset{k \times k}{I} + \underset{p \times p}{\partial_{\mathbf{y}}\mathbf{f}_{(\mathbf{x},\boldsymbol{\varphi}(\mathbf{x}))}} \circ \underset{p \times k}{\mathrm{d}\boldsymbol{\varphi}_{\mathbf{x}}}.$$

Using the invertibility of $\partial_{\mathbf{y}}\mathbf{f}_{(\mathbf{x},\boldsymbol{\varphi}(\mathbf{x}))}$, we can solve for $\mathrm{d}\boldsymbol{\varphi}_{\mathbf{x}}$ to get

$$\mathrm{d}\boldsymbol{\varphi}_{\mathbf{x}} = -\left(\partial_{\mathbf{y}}\mathbf{f}_{(\mathbf{x},\boldsymbol{\varphi}(\mathbf{x}))}\right)^{-1}\partial_{\mathbf{x}}\mathbf{f}_{(\mathbf{x},\boldsymbol{\varphi}(\mathbf{x}))},$$

verifying the third condition.   □

Summary

This final version of the implicit function theorem echoes the first one (Theorem 6.1, p. 189). In broad outline, it tells us that the locus $\mathbf{f}(\mathbf{x},\mathbf{y}) = \mathbf{k}$ is the graph of a map $\mathbf{y} = \boldsymbol{\varphi}(\mathbf{x})$ for which $\mathrm{d}\boldsymbol{\varphi}_{\mathbf{x}} = \left(\partial_{\mathbf{y}}\mathbf{f}_{(\mathbf{x},\boldsymbol{\varphi}(\mathbf{x}))}\right)^{-1} \circ \partial_{\mathbf{x}}\mathbf{f}_{(\mathbf{x},\boldsymbol{\varphi}(\mathbf{x}))}$, assuming only that $\partial_{\mathbf{y}}\mathbf{f}_{(\mathbf{x},\mathbf{y})}$ is invertible at a seed point $(\mathbf{x},\mathbf{y}) = (\mathbf{a},\mathbf{b})$, where $\mathbf{f}(\mathbf{a},\mathbf{b}) = \mathbf{k}$. The key to the proof is that $\mathbf{f}$ is equivalent to a projection near $(\mathbf{a},\mathbf{b})$; in turn, this follows (Theorem 2.19) from the fact that $\mathrm{d}\mathbf{f}_{(\mathbf{a},\mathbf{b})}$ is onto.

Submersions

A map whose derivative is onto is called a *submersion*. Ultimately, the proof of the implicit function theorem can be traced back to the simple fact that $\mathbf{f}$ is a submersion. Submersions have useful behavior with important consequences (beyond the implicit function theorem) that we now pause to explore.

Definition 6.5 *A continuously differentiable map* $\mathbf{f}: X^n \to \mathbb{R}^p$ *is a* **submersion at** $\mathbf{c}$ *if* $d\mathbf{f_c}: \mathbb{R}^n \to \mathbb{R}^p$ *is onto.*

Theorem 6.14. *A map* $\mathbf{f}: X^n \to \mathbb{R}^p$ *is a submersion at* $\mathbf{c}$ *if and only if there is a coordinate change* $\mathbf{h}: N^n \to \mathbb{R}^n$ *defined on a neighborhood* N^n *of* $\mathbf{c}$ *for which* $\mathbf{f} \circ \mathbf{h}^{-1}$ *is a projection.*

Proof. Notice that the "only if" part of the theorem is just a restatement of Theorem 6.12. To prove the converse (the "if" part), let $\mathbf{f} \circ \mathbf{h}^{-1} = \boldsymbol{\Pi}$, a projection. Then

$$d\mathbf{f_x} = d\boldsymbol{\Pi}_{\mathbf{h(x)}} \circ d\mathbf{h_x} = \boldsymbol{\Pi} \circ d\mathbf{h_x},$$

so $\mathbf{f}$ is continuously differentiable on N^n and $d\mathbf{f_x}$ is onto for every $\mathbf{x}$ in N^n because $\boldsymbol{\Pi}$ and $d\mathbf{h_x}$ are both onto. $\square$

Thus, submersions are precisely the maps that are locally equivalent to projections. Moreover, because $\mathbf{f} \circ \mathbf{h}^{-1} = \boldsymbol{\Pi} = d\mathbf{f_x} \circ d\mathbf{h_x^{-1}}$, the local coordinate change $\mathbf{h}^{-1} \circ d\mathbf{h_x}$ transforms $\mathbf{f}$ into its linear approximation $d\mathbf{f_x}$. This is the generalization of Corollary 6.7, page 197. The next result is a generalization of Corollary 6.8. The result following that is a consequence of the fact that a submersion is equivalent to a local projection.

> $\mathbf{f}$ "looks like" $d\mathbf{f}$

Corollary 6.15 *If* $\mathbf{f}: X^n \to \mathbb{R}^p$ *is a submersion at* $\mathbf{c}$, *then there are curvilinear coordinates defined near* $\mathbf{c}$ *in which* p *of the* n *coordinate functions are the component functions of* $\mathbf{f}$.

Proof. This follows immediately from the definition of the coordinate change $\mathbf{h}$ in the proof of Theorem 6.12. $\square$

Corollary 6.16 *If* $\mathbf{f}: X^n \to \mathbb{R}^p$ *is a submersion at* $\mathbf{c}$, *then* $\mathbf{f}$ *maps* X^n *onto a neighborhood of* $\mathbf{f(c)}$. $\square$

Submersions give us a valuable way to describe and deal with curved surfaces. To see how this happens, consider first the locus $\mathcal{S}_\mathbf{f}: \mathbf{f(x,y)} = \mathbf{k}$ defined by the submersion $\mathbf{f}$. In general, $\mathcal{S}_\mathbf{f}$ is a curved subset of $\mathbb{R}^{k+p}$, but of a special kind. For suppose $\mathbf{c} = (\mathbf{a}, \mathbf{b})$ is a seed point of $\mathbf{f}$; that is, $\mathbf{f(a,b)} = \mathbf{k}$. Then the proof of the implicit function theorem provides a coordinate change $\mathbf{h}: (\mathbf{x,y}) \to (\mathbf{u,v})$ that "straightens" $\mathcal{S}_\mathbf{f}$ locally and makes it a flat k-dimensional plane near $\mathbf{c}$. In effect, $(u_1, \ldots, u_k, v_1, \ldots, v_p)$ provides new curvilinear coordinates in $(\mathbf{x,y})$-space in which equations of the form $v_1 = \kappa_1, \ldots, v_p = \kappa_p$ specify $\mathcal{S}_\mathbf{f}$ and the variables $(u_1, \ldots, u_k)$ provide a system of curvilinear coordinates on $\mathcal{S}_\mathbf{f}$ itself. The k coordinates $u_1, \ldots, u_k$, imply $\mathcal{S}_\mathbf{f}$ is k-dimensional. We now use this characterization of $\mathcal{S}_\mathbf{f}$ as the basis of the definition of an embedded surface patch.

> Embedded surface patches

Definition 6.6 *A set* $\mathcal{S}$ *in* $\mathbb{R}^n$ *is an* **embedded surface patch of dimension** k **at the point** $\mathbf{c}$ *if there are coordinates* $(u_1, \ldots, u_k, v_1, \ldots, v_{n-k})$ *in a window* W^n *centered at* $\mathbf{c}$ *so that* $\mathcal{S}$ *is given by the conditions* $v_1 = \kappa_1, \ldots, v_{n-k} = \kappa_{n-k}$ *there. The variables* $\mathbf{u} = (u_1, \ldots, u_k)$ *provide coordinates on* $\mathcal{S}$ *in* W^n.

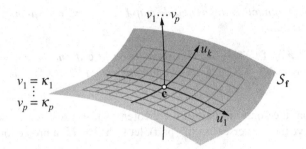

**Surface patches of
dimension 0 or n**

We may abbreviate this term to *surface patch*, *embedding* or just *patch*. We can extend the definition to allow $k = 0$ and $k = n$. An embedded surface patch of dimension 0 at $\mathbf{c}$ is just the point $\mathbf{c}$ itself; it is specified by a full set of n equalities $v_1 = \kappa_1, \ldots, v_n = \kappa_n$. An embedded surface patch of dimension n at $\mathbf{c}$ is just an open set containing $\mathbf{c}$. It is specified by an *empty* set of equalities.

Theorem 6.17. *Suppose* $\mathbf{f} : X^n \to \mathbb{R}^p$ *is a submersion at a point* $\mathbf{c}$ *in* X^n *and* $\mathbf{f}(\mathbf{c}) = \mathbf{k}$. *Then the pullback* $\mathbf{f}^{-1}(\mathbf{k})$ *is an embedded surface patch of dimension* $n - p$ *at the point* $\mathbf{c}$. $\qquad\square$

**Surface patches
and pullbacks**

This is just Theorem 6.12 restated using surface patches. The following theorem is its converse; the two taken together imply surface patches are precisely the pullbacks of points by submersions.

Theorem 6.18. *Suppose* S *is an embedded surface patch of dimension* k *at a point* $\mathbf{c}$ *in* $\mathbb{R}^n$. *Then there is a submersion* $\mathbf{g} : X^n \to \mathbb{R}^{n-k}$ *at* $\mathbf{c}$ *for which* $S = \mathbf{g}^{-1}(\mathbf{g}(\mathbf{c}))$.

Proof. By hypothesis, there is a window X^n centered at $\mathbf{c}$ and a coordinate change $\mathbf{h} : X^n \to \mathbb{R}^{k+(n-k)} : \mathbf{x} \to (\mathbf{u}, \mathbf{v})$ in terms of which S is given by the equations $v_1 = v_1(\mathbf{c}), \ldots, v_{n-k} = v_{n-k}(\mathbf{c})$, where the constants $v_j(\mathbf{c})$ are the $\mathbf{v}$-coordinates of the point $\mathbf{c}$. Let us write the components of $\mathbf{h}$ as

$$\mathbf{h} : \begin{cases} u_1 = h_1(x_1, \ldots, x_n), \\ \quad\vdots \\ u_k = h_k(x_1, \ldots, x_n), \\ v_1 = g_1(x_1, \ldots, x_n), \\ \quad\vdots \\ v_{n-k} = g_{n-k}(x_1, \ldots, x_n), \end{cases}$$

and let $\mathbf{g} : X^n \to \mathbb{R}^{n-k}$ be defined by

$$\mathbf{g} : \begin{cases} v_1 = g_1(x_1, \ldots, x_n), \\ \quad\vdots \\ v_{n-k} = g_{n-k}(x_1, \ldots, x_n). \end{cases}$$

Then $\mathbf{g}$ is continuously differentiable because $\mathbf{h}$ is. Moreover, for every $\mathbf{x}$ in W^n, the matrix $d\mathbf{g_x}$ has maximal rank $n-k$ because it makes up the last $n-k$ rows of the invertible matrix $d\mathbf{h_x}$. In particular, $d\mathbf{g_c}$ is onto, and

$$S = \mathbf{g}^{-1}(v_1(\mathbf{c}), \dots, v_{n-k}(\mathbf{c})) = \mathbf{g}^{-1}(\mathbf{g}(\mathbf{c})). \qquad \square$$

Dimension and codimension

Although the point $\mathbf{g}(\mathbf{c})$ has dimension 0, its pullback $\mathbf{g}^{-1}(\mathbf{g}(\mathbf{c}))$ has dimension k. The pullback does not preserve dimension. However, the difference in the dimensions of the point and its containing space, namely $n-k$, is the same as the difference in the dimensions of the pullback and its containing space. This suggests that, in discussing pullbacks, we focus on this difference, called the *codimension*. We have already done this for vector spaces and pullbacks of onto linear maps. According to Definition 2.7 (p. 51), the *codimension* of a vector subspace W in a vector space $\mathcal{V}$ is

$$\operatorname{codim} W = \dim \mathcal{V} - \dim W.$$

By Corollary 2.21, page 51, any *onto* linear map $L : \mathbb{R}^n \to \mathbb{R}^p$ preserves codimension under pullback; that is, if W is a subspace of codimension m in the target $\mathbb{R}^p$, then the subspace $L^{-1}(W)$ has the same codimension m in the source $\mathbb{R}^n$.

Definition 6.7 *We say an embedded surface patch S of dimension k in $\mathbb{R}^n$ has **codimension $m = n - k$**.*

Note that the codimension of a surface patch is the number of equations (including $m = 0$ for an open set) that define the patch in Definition 6.6. Furthermore, the codimension of the surface patch in Theorem 6.18 equals the dimension of the target of the map $\mathbf{g}$ that defines the patch.

Theorem 6.19. *Suppose $\mathbf{f} : X^n \to \mathbb{R}^p$ is a submersion at $\mathbf{c}$, and S is an embedded surface patch of codimension m at the point $\mathbf{k} = \mathbf{f}(\mathbf{c})$ in $\mathbb{R}^p$. Then $\mathbf{f}^{-1}(S)$ is an embedded surface patch of codimension m at $\mathbf{c}$ in X^n.*

Submersions preserve embeddings under pullback

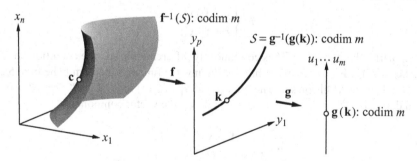

Proof. According to Theorem 6.18, there is a submersion $\mathbf{g} : W^p \to \mathbb{R}^m$ at $\mathbf{k}$ for which $S = \mathbf{g}^{-1}(\mathbf{g}(\mathbf{k}))$. Because

$$\mathbf{f}^{-1}(S) = \mathbf{f}^{-1}(\mathbf{g}^{-1}(\mathbf{g}(\mathbf{k})) = (\mathbf{g} \circ \mathbf{f})^{-1}(\mathbf{g}(\mathbf{k})) = (\mathbf{g} \circ \mathbf{f})^{-1}((\mathbf{g} \circ \mathbf{f})(\mathbf{c})),$$

it is sufficient to show that $\mathbf{g} \circ \mathbf{f} : X^n \to \mathbb{R}^m$ is a submersion at $\mathbf{c}$. But by the chain rule,

$$d(\mathbf{g} \circ \mathbf{f})_{\mathbf{c}} = d\mathbf{g}_{\mathbf{k}} \circ d\mathbf{f}_{\mathbf{c}},$$

and the composite is onto because the individual maps are. $\square$

Immersions and injections

Submersions handle pullbacks properly; however, they do not behave well with push-forwards. That is, if S is a surface patch at $\mathbf{c}$, and $\mathbf{f}$ is a submersion at $\mathbf{c}$, the image $\mathbf{f}(S)$ is not, in general, a surface patch; see the exercises. We have faced this dilemma already with linear maps in Chapter 2.3, and we resolved it there (Corollary 2.28, p. 56) by switching from onto to 1–1 linear maps. To handle push-forwards properly, we use an immersion, that is, a map whose derivative is 1–1.

Definition 6.8 *A continuously differentiable map* $\mathbf{f} : X^n \to \mathbb{R}^p$ *is an **immersion at** $\mathbf{c}$ if* $d\mathbf{f}_{\mathbf{c}} : \mathbb{R}^n \to \mathbb{R}^p$ *is 1–1 (implying $n \leq p$).*

Recall (Theorem 2.27, p. 56) that every 1–1 linear map can be transformed in a simple form, called an *injection* $J : \mathbb{R}^n \to \mathbb{R}^{n+q}$, that is analogous to a projection $\Pi : \mathbb{R}^{p+k} \to \mathbb{R}^p$. The analogy is easily seen by looking at their matrix representatives:

$$J = \begin{pmatrix} I_{n \times n} \\ O_{q \times n} \end{pmatrix}, \quad \Pi = \begin{pmatrix} O_{p \times k} & I_{p \times p} \end{pmatrix}.$$

Theorem 6.20. *A map* $\mathbf{f} : X^n \to \mathbb{R}^{n+q}$ *is an immersion at* $\mathbf{c}$ *if and only if there is a coordinate change* $\mathbf{h} : N^{n+q} \to \mathbb{R}^{n+q}$ *defined on a neighborhood* N^{n+q} *of* $\mathbf{f}(\mathbf{c})$ *for which* $\bar{\mathbf{f}} = \mathbf{h} \circ \mathbf{f}$ *is an injection.*

Proof. To prove the "only if" part, we assume $\mathbf{f}$ is an immersion at $\mathbf{c}$. Hence the $(n + q) \times n$-matrix $d\mathbf{f}_{\mathbf{c}} : \mathbb{R}^n \to \mathbb{R}^{n+q}$ is 1–1 and consequently has n linearly independent rows. By rearranging the rows (and the corresponding target variables), if necessary, we assume that the first n rows are linearly independent. Write the target variables as $(\mathbf{y}, \mathbf{z})$, where $\mathbf{y} = (y_1, \ldots, y_n)$, $\mathbf{z} = (z_1, \ldots, z_q)$, and write $\mathbf{f}$ in terms of (vector) components as

$$\mathbf{f} : \begin{cases} \mathbf{y} = \mathbf{f}_1(\mathbf{x}), \\ \mathbf{z} = \mathbf{f}_2(\mathbf{x}). \end{cases}$$

In particular, $\mathbf{f}_1 : X^n \to \mathbb{R}^n$ is continuously differentiable and the condition on $d\mathbf{f}_{\mathbf{c}}$ makes $d(\mathbf{f}_1)_{\mathbf{c}} : \mathbb{R}^n \to \mathbb{R}^n$ invertible. By the inverse function theorem (Theorem 5.2, p. 169) $\mathbf{f}_1$ is invertible on some neighborhood N_1^n of $\mathbf{f}_1(\mathbf{c})$ in $\mathbb{R}^n$. Let $N^{n+q} = N_1^n \times \mathbb{R}^q$, and define $\mathbf{h} : N^{n+q} \to \mathbb{R}^{n+q} : (\mathbf{y}, \mathbf{z}) \to (\bar{\mathbf{y}}, \bar{\mathbf{z}})$ by the vector components

$$\mathbf{h} : \begin{cases} \bar{\mathbf{y}} = \mathbf{f}_1^{-1}(\mathbf{y}), \\ \bar{\mathbf{z}} = -\mathbf{f}_2(\mathbf{f}_1^{-1}(\mathbf{y})) + \mathbf{z}. \end{cases}$$

Because its components are continuously differentiable on N^{n+q}, $\mathbf{h}$ is a valid coordinate change, and it transforms $\mathbf{f}$ into

$$\begin{aligned}
\bar{\mathbf{f}}(\mathbf{x}) = \mathbf{h} \circ \mathbf{f}(\mathbf{x}) &= \mathbf{h}(\mathbf{f}_1(\mathbf{x}), \mathbf{f}_2(\mathbf{x})) \\
&= (\mathbf{f}_1^{-1}(\mathbf{f}_1(\mathbf{x})), -\mathbf{f}_2(\mathbf{f}_1^{-1}(\mathbf{f}_1(\mathbf{x}))) + \mathbf{f}_2(\mathbf{x})) \\
&= (\mathbf{x}, -\mathbf{f}_2(\mathbf{x}) + \mathbf{f}_2(\mathbf{x})) \\
&= (\mathbf{x}, \mathbf{0}).
\end{aligned}$$

This is an injection.

To prove the converse (the "if" part), assume $J = \mathbf{h} \circ \mathbf{f}$ is an injection. Rearrange the variables, if necessary, so that $J(\mathbf{x}) = (\mathbf{x}, \mathbf{0})$ in $\mathbb{R}^{n+q}$. Then

$$\mathbf{f}(\mathbf{x}) = \mathbf{h}^{-1}(\mathbf{x}, \mathbf{0}),$$

implying that $\mathbf{f}$ is continuously differentiable wherever it is defined. We must show it is defined on some open neighborhood of $\mathbf{c}$.

Because $\mathbf{h}(N^{n+q})$ is an open set containing $\mathbf{h}(\mathbf{f}(\mathbf{c}))$ (by the inverse function theorem), it contains an open "rectangle" $X^n \times Y^q$ centered at $\mathbf{h}(\mathbf{f}(\mathbf{c})) = (\mathbf{c}, \mathbf{0})$. Therefore, for any $\mathbf{x}$ in X^n, $\mathbf{h}^{-1}(\mathbf{x}, \mathbf{0}) = \mathbf{f}(\mathbf{x})$ is defined. Finally,

$$d\mathbf{f}_{\mathbf{x}} = d\mathbf{h}_{J(\mathbf{x})}^{-1} \circ dJ_{\mathbf{x}} = d\mathbf{h}_{(\mathbf{x},\mathbf{0})}^{-1} \circ J$$

so $d\mathbf{f}_{\mathbf{x}}$ is 1–1 because injections and invertible maps are 1–1. $\qquad \square$

Thus, immersions are precisely the maps that are locally equivalent to injections. Moreover, because $\mathbf{h} \circ \mathbf{f} = J = d\mathbf{h}_{\mathbf{f}(\mathbf{x})} \circ d\mathbf{f}_{\mathbf{x}}$, we see that coordinate changes locally transform $\mathbf{f}$ into its linear approximation, so $\mathbf{f}$ "looks like" $d\mathbf{f}$. The following corollary is an immediate consequence of the fact that an immersion is a local injection.

> **f "looks like" df**

Corollary 6.21 *If* $\mathbf{f} : X^n \to \mathbb{R}^{n+q}$ *is an immersion at* $\mathbf{c}$, *then* $\mathbf{f}$ *is 1–1 on a neighborhood of* $\mathbf{c}$. $\qquad \square$

The next theorem, which says that the image of a surface patch under an immersion is still a surface patch of the same dimension, has a more complicated proof than its analogue for submersions because surface patches are naturally determined by submersions (under pullbacks).

Theorem 6.22. *Suppose* $\mathbf{f} : X^n \to \mathbb{R}^{n+q}$ *is an immersion at* $\mathbf{c}$, *and* S *is an embedded surface patch of dimension* k *at* $\mathbf{c}$ *in* X^n. *Then the image* $\mathbf{f}(S)$ *is an embedded surface patch of the same dimension* k *at* $\mathbf{f}(\mathbf{c})$ *in* $\mathbb{R}^{n+q}$.

> **Immersions preserve embeddings under push-forward**

Proof. By the definition of an embedded surface patch (Definition 6.6), there is a coordinate change $\mathbf{g} : X^n \to \mathbb{R}^n : \mathbf{x}_{(n)} \to (\mathbf{y}_{(k)}, \mathbf{z}_{(n-k)})$ that "straightens" S near $\mathbf{c}$. Let us suppose that $\mathbf{g}(S)$ is given by equations $z_1 = \kappa_1, \ldots, z_{n-k} = \kappa_{n-k}$ (i.e., $\mathbf{z} = \boldsymbol{\kappa}$) in the new coordinates. To prove that $\mathbf{f}(S)$ is an embedded surface patch of dimension k at $\mathbf{f}(\mathbf{c})$, it is sufficient to find new coordinates $(\mathbf{u}_{(k)}, \mathbf{v}_{(n-k)}, \mathbf{w}_{(q)})$ in a neighborhood N^{n+q} of $\mathbf{f}(\mathbf{c})$ in which $\mathbf{f}(S)$ is specified by the $n - k + q$ equations $\mathbf{v} = \boldsymbol{\kappa}$, $\mathbf{w} = \mathbf{0}$.

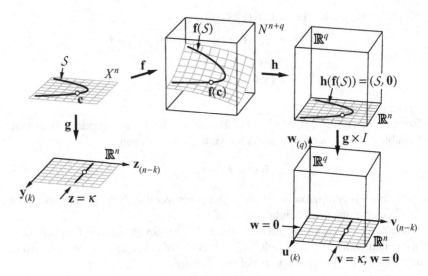

The figure shows how to build the new coordinates. First, use the hypothesis that
f is an immersion at **c** to get (from Theorem 6.20) a coordinate change **h** : $N^{n+q} \to$
$\mathbb{R}^n \times \mathbb{R}^q : \mathbf{r}_{(n+q)} \to (\mathbf{s}_{(n)}, \mathbf{t}_{(q)})$ on a neighboorhood N^{n+q} of **f**(**c**) that transforms **f** into
an injection to the first n coordinates:

$$(\mathbf{s},\mathbf{t}) = (\mathbf{h} \circ \mathbf{f})(\mathbf{x}) = (\mathbf{x},\mathbf{0}); \quad (\mathbf{h} \circ \mathbf{f})(S) = (S,\mathbf{0}).$$

Next, use the coordinate change **g** already introduced to define

$$\mathbf{g} \times I : \mathbb{R}^n \times \mathbb{R}^q \to \mathbb{R}^k \times \mathbb{R}^{n-k} \times \mathbb{R}^q : (\mathbf{s},\mathbf{t}) \to (\mathbf{u}_{(k)}, \mathbf{v}_{(n-k)}, \mathbf{w}_{(q)})$$

on a neighborhood of $(\mathbf{c},\mathbf{0})$ in $\mathbb{R}^p$. Then the composite coordinate change $(\mathbf{g} \times I) \circ \mathbf{h}$
"straightens" **f**(S) near **f**(**c**): **f**(S) is given by the $n - k + q$ equations $\mathbf{v} = \boldsymbol{\kappa}, \mathbf{w} = \mathbf{0}$.   □

Corollary 6.23 *If* **f** : $X^n \to \mathbb{R}^{n+q}$ *is an immersion at a point* **c**, *then* **f**(X^n) *is an*
embedded surface patch of dimension n at the point **f**(**c**) *in* $\mathbb{R}^{n+q}$.

Proof. Because X^n is an open set in $\mathbb{R}^n$, we can view it as a surface patch of dimension n (i.e., of codimension 0) at **c** in $\mathbb{R}^n$. It follows from the theorem that its image
f(X^n) is a surface patch of dimension n at **f**(**c**) in $\mathbb{R}^{n+q}$.   □

Parametrized curves
and surfaces

vWe can think of the previous corollary as the basis for our study of curves
(in Chapter 1.2) and surfaces in space (Chapter 4.3). To see the connection, let **f** :
$X^n \to \mathbb{R}^p$ be an arbitrary continuously differentiable map (i.e., not necessarily an
immersion); it is given by p component functions of n real variables:

$$\mathbf{f} : \begin{cases} v_1 = f_1(x_1,\ldots,x_n), \\ \quad \vdots \\ v_p = f_p(x_1,\ldots,x_n). \end{cases}$$

If $n = 1$ then $\mathbf{f}$ defines the parametrization of a curve in $\mathbb{R}^p$. Our earlier definition (Definition 1.2, p. 7) is more restrictive: $\mathbf{f}$ must be smooth (i.e., have continuous derivatives of all orders) and $d\mathbf{f}_t = \mathbf{f}'(t)$ must have a nonzero derivative at each interior point t. Smoothness is mainly just a technical convenience, but the requirement that $d\mathbf{f}_t \neq \mathbf{0}$ means $\mathbf{f}$ must be an immersion at each interior point. Consequently, a curve as given by Definition 1.2 is actually embedded at each point: there are curvilinear coordinates $(x_1, \ldots, x_p)$ in $\mathbb{R}^p$ in which the curve is specified by the conditions $x_2 = \cdots x_n = 0$, and x_1 serves as a parameter along the curve.

If $n = 2$ and $p = 3$, then $\mathbf{f}$ parametrizes an ordinary surface in space. Such a map is an immersion at a point $\mathbf{c}$ only if the derivative $d\mathbf{f}_\mathbf{c}$ has rank 2, its maximal rank. In the examples in Chapter 4.3, $\mathbf{f}$ was indeed an immersion at most points, so the image surface was embedded there. That is, (by Corollary 6.23) we could introduce coordinates (p, q, r) in a neighborhood of such a point so that the surface was given locally by the equation $r = 0$ and (p, q) could serve as coordinates on the surface near the point.

The notable example of a nonimmersion is the crosscap. The crosscap map

$$\mathbf{f} : \begin{cases} x = u, \\ y = uv, \\ z = -v^2, \end{cases} \qquad d\mathbf{f}_\mathbf{u} = \begin{pmatrix} 1 & 0 \\ v & u \\ 0 & -2v \end{pmatrix}$$

fails to be an immersion at the origin (cf. pp. 127–128).

Exercises

6.1. a. Determine the location of the three saddle points and one relative maximum of $f(x,y) = (3y^2 - x^2)(x - 1)$.

    b. Plot together the graphs of $z = f(x,y)$ and $z = 0$ on the square for which $-0.3 \leq x \leq 1.3$, $-0.8 \leq y \leq 0.8$. Determine the locus $f(x,y) = 0$ from the intersection of the two graphs, and note the location of the four critical points of f in relation to this intersection.

6.2. Solve the equation $e^{xy} = 1$ for y near the point $(2,0)$. What is dy/dx at that point? Sketch the locus $e^{xy} = 1$.

6.3. Solve the equation $e^{xy} = e$ for y near the point $(2, 1/2)$. What is dy/dx at that point? Sketch the locus $e^{xy} = e$.

6.4. Solve the equation $y^2 - 2y\cos x - \sin^2 x = 0$ for y, and determine dy/dx. Sketch the locus $y^2 - 2y\cos x - \sin^2 x = 0$.

6.5. Solve the equation $y^2 - 2y\cos x + \sin^2 x = 0$ for y; for which values of x is y undefined (as a function of x)? Determine dy/dx; where is $dy/dx = \infty$? Sketch the locus $y^2 - 2y\cos x + \sin^2 x = 0$.

6.6. a. Solve the equation $x^2 + 3xy + 4y^2 = 14$ for x in terms of y. Determine dx/dy when $y = 1$.

  b. Sketch the locus $f(x,y) = x^2 + 3xy + 4y^2 = 14$.

  c. Determine the implicit function $y = \varphi(x)$ for which $f(x, \varphi(x)) = 14$ and $\varphi(2) = 1$. Determine $\varphi'(2)$ and relate it to the value of dx/dy that you found in part (a).

6.7. Determine the linearization of the locus $f(x,y) = 0$ at the given point (a,b). Indicate whether the linearization is the tangent line to the locus at that point.

  a. $f(x,y) = y^2 + x^2(x+1)$; $(a,b) = (-1,0)$.
  b. $f(x,y) = y^2 + x^2(x+1)$; $(a,b) = (0,0)$.
  c. $f(x,y) = (3y^2 - x^2)(x-1)$; $(a,b) = (1,0)$.
  d. $f(x,y) = (3y^2 - x^2)(x-1)$; $(a,b) = (1,1/\sqrt{3})$.
  e. $f(x,y) = (3y^2 - x^2)(x-1)$; $(a,b) = (0,0)$.
  f. $f(x,y) = x^3 + y^3$; $(a,b) = (0,0)$.
  g. $f(x,y) = x^3 + y^3$; $(a,b) = (1,-1)$.

6.8. a. Sketch representative level curves of $f(x,y) = xy^2$ in the window W for which $1 \le x \le 2$, $1 \le y \le 2$. Verify that every point of W is a regular point of f.

  b. Obtain the map $\mathbf{h} : W \to \mathbb{R}^2$ that straightens the level curves of f, using the construction in the proof of Theorem 6.2. Then, using a suitable parametrization of a level curve, verify that it does indeed have a horizontal image under $\mathbf{h}$.

  c. Show that the image of W is the set

$$1 \le u \le 2, \quad u \le v \le 4u.$$

  Sketch level curves of f that meet either the top or the bottom of W, and then sketch their images under $\mathbf{h}$. Where do those images meet the boundary of $\mathbf{h}(W)$?

  d. Obtain the formula for the inverse of $\mathbf{h}$ on $\mathbf{h}(W)$.

6.9. a. Sketch representative level curves of $f(x,y) = x^2 + y^2$ in window W for which $1 \le x \le 2$, $1 \le y \le 2$. Verify that every point of W is a regular point of f.

  b. Obtain the map $\mathbf{h} : W \to \mathbb{R}^2$ that straightens the level curves of f, using the construction in the proof of Theorem 6.2. Then, using a suitable parametrization of a level curve, verify that it does indeed have a horizontal image under $\mathbf{h}$.

  c. Show that the image of W is the set

$$0 \le u \le 2, \quad u^2 + 1 \le v \le u^2 + 4,$$

and sketch the image, including imges of the level curves of f.

6.10. a. Let $f(x,y) = x^2 + y^2$ and let Z be the window $0 \leq x \leq 2$, $-1 \leq y \leq 1$. Verify that every point of Z except the origin is a regular point of f. Sketch the level curves of f in Z. Note that $f_y = 0$ at the center of Z.

b. Show that the map $\mathbf{h} : Z \to \mathbb{R}^2$,

$$\mathbf{h} : \begin{cases} u = y, \\ v = f(x,y), \end{cases}$$

is a valid coordinate change near $(1,0)$; that is, show $\mathbf{h}$ is continuously differentiable with a continuously differentiable inverse in a neighborhood of $(1,0)$.

c. Show that $\mathbf{h}$ "straightens out" the level curves of f. Describe the salient geometric features of the action of $\mathbf{h}$; in particular, indicate what happens to a horizontal line in Z.

6.11. Consider the function

$$f(u,v,w) = \frac{1+w}{1-w} \frac{1-u}{1+u} \frac{1-v}{1+v},$$

for $-1 < u,v,w < 1$. Think of u, v, and w as speeds expressed as fractions of the speed of light. Note that $f(0,0,0) = 1$. This exercise studies the implicit function $w = \varphi(u,v)$ defined near $(u,v) = (0,0)$ by the equation $f(u,v,w) = 1$.

a. Use f to compute the partial derivatives $\partial\varphi/\partial u$ and $\partial\varphi/\partial v$, and deduce that $\varphi(u,v) = u + v + O(2)$.

b. Show that

$$w = \varphi(u,v) = \frac{u+v}{1+uv} = u \oplus v.$$

This defines a binary operation called the *law of addition of velocities* in special relativity. That is, if observer A is moving away from observer B with velocity u (as a fraction of the speed of light), and B is moving away from C along the same straight line with velocity v, then A will be moving away from C with velocity $w = u \oplus v$. According to part (a), if u and v are small, then $u \oplus v \approx u + v$, but not otherwise.

c. Show that $u \oplus v$ is defined for all $|u| < 1$ and $|v| < 1$, and that $|u \oplus v| < 1$.

d. Show that $\lim_{u \to 1} u \oplus v = 1$, allowing us to extend φ so that $1 \oplus v = 1$ (and, by symmetry, $u \oplus 1 = 1$).

Thus, if A now represents a photon (a light particle), it moves away from B and from C at the same speed, even though B is moving in relation to C. Special relativity is built on the premise that the speed of light is an invariant for all observers moving uniformly in relation to each other.

6.12. Prove Theorem 6.3 and Corollary 6.4. (Suggestion: Adapt the proofs of their 2-dimensional analogues.)

6.13. Show that any set of points in the (x,y)-plane that can occur as the zero-locus of a function of x and y can occur as the zero-locus of a suitably chosen function of x, y, and z in the $(x,y,0)$-plane. (Suggestion: consider $f(x,y,z) = [g(x,y)]^2 + z^2$.)

6.14. Sketch the intersection of the surfaces $S_f : x^2 + y^3 - z = 0$, $S_g : z = 0$. Is the intersection the graph of a continuously differentiable function $y = \varphi(x)$ within the plane $z = 0$? Explain. Address the same question using a function of the form $x = \psi(y)$.

6.15. Show that the surfaces defined by $(x+y)^3 - z = 0$ and $z = 0$ intersect in a straight line. Verify that the surfaces are not transverse at any intersection point.

6.16. Let $\mathbf{f} : \mathbb{R}^3 \to \mathbb{R}^2$ be defined by $(s,t) = \mathbf{f}(x,y,z) = (x, (y-z)^3)$.

a. Determine the image of $d\mathbf{f}_0$ and show thereby that it is 1-dimensional.

b. Show that $\mathbf{f}$ maps any window centered at $\mathbf{x} = (0,0,0)$ onto a small window centered at $\mathbf{s} = (0,0)$. In particular, show that, for any a, b near 0, the equation $\mathbf{f}(x,y,z) = (a,b)$ has a one-parameter family (i.e., a curve) of solutions. Determine that curve.

c. Conclude that $\mathbf{f}$ does not "look like" $d\mathbf{f}_0$ near the origin.

6.17. In (x,y,z)-space, $x^2 + y^2 = r^2$ is a cylinder of radius $r > 0$ whose axis is the z-axis, and $x^2 + z^2 = 1$ is a cylinder of radius 1 whose axis is the y-axis.

a. Sketch the intersection of the two cylinders when $r^2 = 3/4$. Now let r be arbitrary, assuming only that $r < 1$. Find implicit functions $y = \varphi(x)$ and $z = \psi(x)$ determined by the equations $x^2 + y^2 = r^2$ and $x^2 + z^2 = 1$. Do this for each of the four seed points $(0, \pm r, \pm 1)$. What are the domains of definition of φ and ψ?

b. Sketch the intersection of the two cylinders when $r^2 = 4$. Now let r be arbitrary, assuming only that $r > 1$. Find implicit functions $y = \varphi(x)$ and $z = \psi(x)$ determined by the equations $x^2 + y^2 = r^2$ and $x^2 + z^2 = 1$ and the four seed points $(0, \pm r, \pm 1)$. Now what are the domains of definition of φ and ψ?

c. The implicit functions take simple forms when $r = 1$. What are those forms, and what is the shape of the intersection?

Chapter 7
Critical Points

Abstract At a regular point, the linear terms of a function determine its local behavior, and there is a local coordinate change that transforms the function into one of the new coordinates. At a critical point, the linear terms vanish, but there is still an analogous result for the quadratic terms, called *Morse's lemma*. However, the quadratic terms may not determine the local behavior, but when they do (the critical point is then said to be *nondegenerate*), Morse's lemma provides a local coordinate change that transforms the function into a sum of positive and negative squares of the new coordinates. In this chapter we analyze Morse's lemma and use it to characterize critical points.

7.1 Functions of one variable

Let us see how a coordinate change can transform $y = f(x)$ into a pure square near a critical point $x = a$. As happens so often in local analysis, the key tool is Taylor's theorem. We need the first-order expansion; it helps us to write the remainder using the explicit integral formula that is given in the original formulation of the theorem (Theorem 3.9, p. 79). In fact, because $f'(a) = 0$, the only nonconstant term in the expansion is the remainder:

$$f(a + \Delta x) = f(a) + h(\Delta x)(\Delta x)^2.$$

The variable coefficient $h(\Delta x)$ in the remainder term is the integral

$$h(\Delta x) = \int_0^1 f''(a + t\Delta x)(1 - t)\,dt.$$

Because we need $h(\Delta x)$ to be continuously differentiable for all Δx near 0, we require f to have a continuous third derivative. Then

J.J. Callahan, *Advanced Calculus: A Geometric View*, Undergraduate Texts in Mathematics, DOI 10.1007/978-1-4419-7332-0_7, © Springer Science+Business Media, LLC 2010

$$h'(\Delta x) = \int_0^1 f'''(a + t\Delta x)\, t(1 - t)\, dt.$$

Substituting $\Delta x = 0$ gives

$$h(0) = f''(a) \int_0^1 (1 - t)\, dt = \frac{f''(a)}{2}, \quad h'(0) = f'''(a) \int_0^1 t(1 - t)\, dt = \frac{f'''(a)}{6}.$$

If a coordinate change $\Delta x \to \Delta u$ is to transform $\Delta y = f(a + \Delta x) - f(a)$ into a pure square, $\Delta y = \pm(\Delta u)^2$, then Δu must be

$$\Delta u = p(\Delta x) = \Delta x \sqrt{|h(\Delta x)|}.$$

It remains to see whether p is a valid coordinate change. Before we do this, note how the "$\pm$" comes into play in the formula for Δy. Because

$$(\Delta u)^2 = (\Delta x)^2 |h(\Delta x)| = \begin{cases} (\Delta x)^2 h(\Delta x) = \Delta y & \text{if } h(\Delta x) \geq 0, \\ -(\Delta x)^2 h(\Delta x) = -\Delta y & \text{if } h(\Delta x) < 0, \end{cases}$$

it follows that

$$\Delta y = \begin{cases} +(\Delta u)^2 & \text{if } h(\Delta x) \geq 0, \\ -(\Delta u)^2 & \text{if } h(\Delta x) < 0. \end{cases}$$

Now consider the function p. Formal differentiation gives

$$p'(\Delta x) = \sqrt{|h(\Delta x)|} \pm \frac{h'(\Delta x)}{2\sqrt{|h(\Delta x)|}} \Delta x,$$

implying p has a continuous derivative on any interval where $h(\Delta x) \neq 0$. (The sign in the formula for p' is chosen to be the sign of $h(\Delta x)$.) Moreover,

$$p'(0) = \sqrt{|h(0)|} = \sqrt{|f''(a)|/2}.$$

Thus, if $f''(a) \neq 0$, the inverse function theorem implies that p is a valid coordinate change on an open interval containing $\Delta x = 0$, and we then have

$$y = \begin{cases} f(a) + (\Delta u)^2 & \text{if } f''(a) > 0, \\ f(a) - (\Delta u)^2 & \text{if } f''(a) < 0. \end{cases}$$

If $f''(a) = 0$, our argument fails to obtain the coordinate change p, but it is natural to ask if a better argument would repair the problem. The answer is no; that is, if $f''(a) = 0$, there may be no new coordinate Δu for which $y = f(a) \pm (\Delta u)^2$. We can see this geometrically, because the equation $y = f(a) + (\Delta u)^2$ necessarily implies f has a minimum at a, and $y = f(a) - (\Delta u)^2$ implies f has a maximum there. But the function $f(x) = x^3$ has a critical point at the origin for which $f''(0) = 0$, and the origin is neither a minimum nor a maximum.

For functions of a single variable, the preceding discussion establishes two results: Morse's lemma and the more familiar second derivative test.

Theorem 7.1 (Morse's lemma). *Suppose $y = f(x)$ has a continuous third derivative on an open interval that includes a critical point $x = a$ where $f''(a) \neq 0$. Then in a sufficiently small window centered at $x = a$ there is a coordinate change $\Delta u = p(\Delta x)$ for which*

$$\Delta y = \pm(\Delta u)^2,$$

where the sign of $(\Delta u)^2$ is chosen to be the sign of $f''(a)$. ☐

Theorem 7.2 (Second derivative test). *Suppose $y = f(x)$ has a continuous third derivative on an open interval containing a critical point $x = a$; then the critical point is*

- *A local minimum of f if $f''(a) > 0$*
- *A local maximum of f if $f''(a) < 0$*

If $f''(a) = 0$, the test is inconclusive. ☐

Thus, a function "looks like" its quadratic approximation near a point where the linear approximation breaks down (i.e., at a critical point), assuming the quadratic approximation does not itself break down. We already have names to distinguish between points where the linear approximation to a function breaks down and where it does not (*critical* and *regular* points, respectively). Morse's lemma suggests we make a similar distinction for critical points. Thus we say a critical point is *degenerate* if the quadratic approximation "breaks down," or "degenerates," in the sense that it fails to determine the local behavior of the function. Otherwise, we say the critical point is *nondegenerate*. For a function of one variable, the situation is clearcut: a critical point is degenerate if and only if the second derivative vanishes. For functions of more than one variable, there are several second partial derivatives; as we show in the following sections, a critical point may be degenerate even though all of those second derivatives are nonzero. The relation between degeneracy and the second derivatives is more subtle.

Degeneracy of a critical point

To see how Morse's lemma works, let us apply it to $f(x) = x - x^3/3$ at the critical point $x = 1$. Because $f''(1) = -2$, the point is a local maximum. In terms of window coordinates $(\Delta x, \Delta y)$ centered at $(x, y) = (1, 2/3)$, the formula for f becomes

$$\begin{aligned}
\Delta y &= f(1 + \Delta x) - f(1) \\
&= 1 + \Delta x - \left(1 + 3\Delta x + 3(\Delta x)^2 + (\Delta x)^3\right)/3 - (1 - 1/3) \\
&= -(\Delta x)^2 - (\Delta x)^3/3 = (\Delta x)^2\left(-1 - \Delta x/3\right).
\end{aligned}$$

Thus $h(\Delta x) = -1 - \Delta x/3$, and so $\Delta y = -(\Delta u)^2$ when we set

$$\Delta u = p(\Delta x) = \Delta x\sqrt{1 + \Delta x/3}.$$

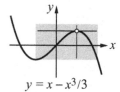

$y = x - x^3/3$

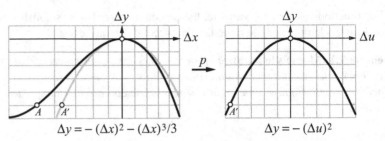

$$\Delta y = -(\Delta x)^2 - (\Delta x)^3/3 \qquad\qquad \Delta y = -(\Delta u)^2$$

The coordinate change p maps the nonuniform grid on the left, above, to the uniform grid on the right, transforming the original cubic curve into a simple parabola. Notice that p pushes points to the left of the origin closer together horizontally; this happens because $0 < p'(\Delta x) < 1$ when $-2 < \Delta x < 0$. To the right of the origin, where $1 < p'(\Delta x)$, points are pushed apart. Finally, because $p'(0) = 1$, the two grids have essentially the same spacing near $\Delta x = 0$. That implies the cubic and the parabola "share ink" near the origin, as the gray copy of the parabola on the left makes clear.

A coordinate change can reverse concavity

The figure also shows that the coordinate change reverses the concavity of part of the graph. For example, at the point A the original cubic is concave up, but at its image A' the parabola is concave down. We associate concavity with the sign of the second derivative, so a coordinate change can reverse the sign of the second derivative. If this were to happen at a critical point (A is not a critical point), the second derivative test would be completely meaningless.

How derivatives depend on coordinates

Let us see how a coordinate change can alter the sign of the second derivative at a noncritical point. Assume $y = f(x)$ is a differentiable function with $f(0) = 0$, and let $x = h(u)$ be a coordinate change with $h(0) = 0$. Then

$$y = f(x) = f(h(u)) = g(u)$$

defines the transformed function g, and we compare $g''(0)$ with $f''(0)$. We have

$$g'(u) = f'(x) \cdot h'(u) \text{ and } g''(u) = f''(x) \cdot \left(h'(u)\right)^2 + f'(x) \cdot h''(u),$$

and our assumptions about the values of h and f at the origin give us

$$g'(0) = f'(0) \cdot h'(0) \text{ and } g''(0) = f''(0) \cdot \left(h'(0)\right)^2 + f'(0) \cdot h''(0).$$

Because h is a coordinate change near the origin, $h'(0) \neq 0$ and the first equation implies

$$g'(0) = 0 \quad \Longleftrightarrow \quad f'(0) = 0.$$

In other words, the origin is a critical point in one coordinate system if and only if it is a critical point in the other. A critical point is thus a geometric property of a function; its presence does not depend upon the coordinates used to descibe the function.

The second derivative at critical and noncritical points

Now suppose the origin is a critical point. Then the equation for $g''(0)$ reduces to

$$g''(0) = f''(0) \cdot \left(h'(0)\right)^2,$$

implying that the second derivatives of f and g have the same sign at the origin, and confirming what is implicit in the second derivative test. If, on the contrary, the origin is not a critical point, then $f'(0) \neq 0$ and the equation for $g''(0)$ now includes the term $f'(0) \cdot h''(0)$. When this additional term is taken into account, $g''(0)$ may well differ in sign from $f''(0)$.

Here is an example to illustrate the "volatility" of the sign of the second derivative under a coordinate change near a regular point. Let

$$y = f(x) = e^x, \quad x = h(u) = \tfrac{1}{2}\ln(u+1); \quad \text{then } y = g(u) = \sqrt{u+1}.$$

The two graphs make the point immediately: the exponential function has a graph that is everywhere concave up but the square root function has a graph that is everywhere concave down. Let us go through the analytic details. First note that the origin is not a critical point, because $f'(0) = 1$ and $g'(0) = \tfrac{1}{2}$. (The values of the derivatives need not agree, but one cannot be zero unless the other is.) Second, we have $h'(0) = \tfrac{1}{2}$ and $h''(0) = -\tfrac{1}{2}$. Finally, for the second derivatives we have $f''(0) = 1$ and

$$g''(0) = f''(0) \cdot \left(h'(0)\right)^2 + f'(0) \cdot h''(0) = 1 \cdot \tfrac{1}{4} + 1 \cdot -\tfrac{1}{2} = -\tfrac{1}{4}.$$

One of the main objects of this book is to bring to the fore the geometric character of functions and maps. The geometric attributes of a map are the ones left unchanged when the coordinates change. The eigenvalues of a linear map are geometric in this way, and so are its rank and nullity. For a nonlinear function, we tend to concentrate on local behavior, for then we can hope to bring calculus to bear. Thus, critical points are genuine geometric features of a function: if the first derivative equals zero in one coordinate system, it will be zero in every other. The concavity of a function at a critical point is likewise geometric: if the critical point is a minimum in one coordinate system, it will be a minimum in every other.

Geometry and local behavior

But the concavity of a function at a noncritical point is not geometric. This does not mean we cannot calculate concavity. We can; it is given by the sign of the second derivative. Concavity is nongeometric because the graphs that represent the same function in two different coordinate systems can have opposite concavities at the same (noncritical) point. An individual representative will have a particular concavity at a point, but other—equally valid—representatives will have the opposite concavity.

A Taylor expansion gives us a good way to think about the role and the significance of the various derivatives of a function. Expanding $y = f(x)$ in window coordinates near $x = a$ gives

Significance of the various derivatives

$$\Delta y = f'(a)\Delta x + \tfrac{1}{2}f''(a)(\Delta x)^2 + \tfrac{1}{6}f'''(a)(\Delta x)^3 + \cdots.$$

The first nonzero term dominates. Thus, if $f'(a) \neq 0$, then the local behavior of f near a is entirely determined by $f'(a)$: the inverse function theorem implies there is a coordinate change $\Delta x = h(\Delta u)$ for which

$$\Delta y = f'(a)\Delta u.$$

The linear term dominates; all the other terms vanish, implying all the higher derivatives, including the second derivative, have become zero in the new coordinate system.

If, by contrast, the linear term is missing, $f'(a) = 0$, then dominance is transferred to the quadratic term. If $f''(a) \neq 0$, that is exactly what happens. Morse's lemma implies there is a coordinate change $\Delta x = k(\Delta v)$ for which

$$\Delta y = \tfrac{1}{2} f''(a) (\Delta v)^2.$$

In summary: $f'(a)$ determines the local behavior of f when $f'(a) \neq 0$; $f''(a)$ is geometrically irrelevant. But if $f'(a) = 0$, then $f''(a)$ determines local behavior, at least if $f''(a) \neq 0$. If $f''(a) = 0$, then the cubic term should dominate, and so forth.

7.2 Functions of two variables

The local behavior of a function of one variable near a critical point is determined by the single quadratic term in the Taylor expansion, if that term is present. However, critical points of a function of two or more variables are more complicated: the local behavior of a function may not be determined by the quadratic terms, even when they are all present. Let us see how this can happen.

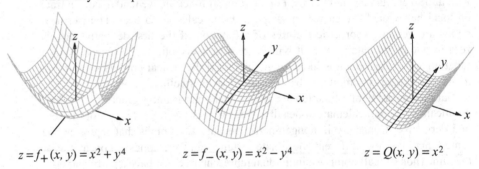

$$z = f_+(x, y) = x^2 + y^4 \qquad z = f_-(x, y) = x^2 - y^4 \qquad z = Q(x, y) = x^2$$

Example 1: bowls,
saddles, and gutters

Consider first the pair of functions

$$f_+(x,y) = x^2 + y^4 \text{ and } f_-(x,y) = x^2 - y^4.$$

Each has a critical point at the origin, and each function serves as its own Taylor expansion there. In both cases, the quadratic part of the expansion is $Q_f(x,y) = x^2$; the y-variable is absent. If local behavior at a critical point were always determined by the quadratic terms, we would have to conclude that f_+ and f_- have the same local behavior at the origin. But they obviously do not: the graph of f_+ is a bowl, and f_+ has a minimum at the origin. The graph of f_- is a saddle, and f_- has a "minimax" there.

The crucial distinction between f_+ and f_- lies in the way the y-variable appears in their formulas, but Q_f has no y-terms so it cannot "see" that distinction. The missing terms mean that, although $z = Q_f(x,y)$ does have a minimum at the origin, the minimum is nonisolated: all points along the y-axis are minima. (By contrast, the minimum of f_+ is an *isolated* critical point.) As a result, the graph of Q_f is neither a bowl nor a saddle; it has a new shape that we call a "*gutter*." If the bottom of the gutter were to be bent up (e.g., by the addition of $+y^4$), it becomes a bowl; bent down (e.g., by adding $-y^4$), it becomes a saddle.

It appears we can attribute the degeneracy of the critical point of f_+ or f_- to this defect in Q_f. In fact, this is true, but it is not the whole story: we now show that, even if all three quadratic terms are nonzero, those terms may still not determine local behavior. Transform f_+ and f_- by rotating coordinates $45°$ (dilating by $\sqrt{2}$, to keep the formulas simple). That is, let

Example 2: rotate example 1

$$L : \begin{cases} x = u - v, \\ y = u + v, \end{cases}$$

and let

$$g_+(u,v) = f_+(L(u,v)) = u^2 - 2uv + v^2 + u^4 + 4u^3v + 6u^2v^2 + 4uv^3 + v^4,$$
$$g_-(u,v) = f_-(L(u,v)) = u^2 - 2uv + v^2 - u^4 - 4u^3v - 6u^2v^2 - 4uv^3 - v^4.$$

Each of the new functions still has a critical point at the origin, and each formula still serves as its own Taylor expansion there. There is no qualitative change, either: g_+, like f_+, has a minimum at the origin, and g_-, like f_-, has a saddle. Because the quadratic parts of the new functions are identical,

$$Q_g(u,v) = u^2 - 2uv + v^2,$$

the new Q_g does no better at determining local behavior than the original Q_f did, even though all three quadratic terms are present in Q_g.

The formula for Q_g is different from the formula for Q_f; however, its graph is not, because the rotation–dilation that transforms $f_\pm$ into $g_\pm$ also transforms Q_f into Q_g. The graph of Q_g is just the graph of Q_f rotated $45°$, a gutter whose bottom lies along the line $v = u$. Thus, without referring directly to the connection between g_+ and f_+, we can still attribute the degeneracy of the critical point of g_+ at the origin to the fact that the graph of Q_g is a gutter. In geometric terms, Q_g has the same defect as Q_f.

Q_g has the same defect as Q_f

In analytic terms, the defect arises because there is a coordinate change that transforms Q_g into a single square, $z = \pm x^2$, so that the other variable is completely missing. To determine when a critical point is degenerate, we must therefore decide when a general function of the form

$$Q(x,y) = Ax^2 + 2Bxy + Cy^2$$

can be transformed into a single square. To do this it helps to use vector and matrix notation.

Quadratic forms

Definition 7.1 *A **quadratic form** in two variables is a function of the form*

$$Q(x,y) = Ax^2 + 2Bxy + Cy^2 = (x\ y) \begin{pmatrix} A & B \\ B & C \end{pmatrix} \begin{pmatrix} x \\ y \end{pmatrix} = \mathbf{x}^\dagger M \mathbf{x} = Q(\mathbf{x}).$$

Quadratic forms and matrices

The symmetric matrix M is called the **matrix of the quadratic form**. There is a 1–1 correspondence: $Q \leftrightarrow M$. That is, every symmetric matrix determines a unique quadratic form, and every quadratic form determines a unique symmetric matrix. The symmetry is essential for uniqueness, because, for example,

$$2xy = (x\ y) \begin{pmatrix} 0 & 2 \\ 0 & 0 \end{pmatrix} \begin{pmatrix} x \\ y \end{pmatrix} = (x\ y) \begin{pmatrix} 0 & 1 \\ 1 & 0 \end{pmatrix} \begin{pmatrix} x \\ y \end{pmatrix}.$$

This points up the fact that if we start with any 2×2 matrix A, the formula $Q(\mathbf{x}) = \mathbf{x}^\dagger A \mathbf{x}$ defines a unique quadratic form. However, if we start instead with the form Q, there is only one symmetric matrix M for which $\mathbf{x}^\dagger M \mathbf{x} = Q(\mathbf{x})$. (Thus we write the xy coefficient of Q as $2B$ to simplify splitting it into two equal parts on the "off-diagonal" of M, to make M symmetric.)

Transforming a quadratic form

Suppose L is an invertible 2×2 matrix so $\mathbf{x} = L\mathbf{u}$ is a linear coordinate change. Then, in terms of the new coordinates $\mathbf{u} = (u, v)$, the quadratic form $Q(\mathbf{x}) = \mathbf{x}^\dagger M \mathbf{x}$ is transformed into

$$\widehat{Q}(\mathbf{u}) = Q(L\mathbf{u}) = (L\mathbf{u})^\dagger M (L\mathbf{u}) = \mathbf{u}^\dagger (L^\dagger M L) \mathbf{u}.$$

Thus $\widehat{Q}$ is also a quadratic form. Furthermore, $L^\dagger M L$ is symmetric (here $L^\dagger$ is the transpose of L) because $(L^\dagger M L)^\dagger = L^\dagger M^\dagger L^{\dagger\dagger} = L^\dagger M L$; therefore $\widehat{M} = L^\dagger M L$ is the matrix of $\widehat{Q}$. For example, if

$$Q \leftrightarrow \begin{pmatrix} 5 & 3 \\ 3 & -1 \end{pmatrix} \quad \text{and} \quad L = \begin{pmatrix} 1 & 2 \\ 1 & -1 \end{pmatrix}, \quad \text{then } \widehat{Q} \leftrightarrow \begin{pmatrix} 10 & 14 \\ 14 & 7 \end{pmatrix};$$

that is, L transforms $Q = 5x^2 + 6xy - y^2$ to $\widehat{Q} = 10u^2 + 28uv + 7v^2$; see the exercises. Note that, because L is invertible by definition, $L^\dagger M L$ is invertible if and only if M is. The following theorem identifies the quadratic forms that have the defect we have come to associate with degenerate critical points. Although the theorem is a special case of Theorem 7.10 (see below, p. 244), we give it its own proof.

Theorem 7.3. *Let $Q(\mathbf{x}) = \mathbf{x}^\dagger M \mathbf{x}$ be a quadratic form. Suppose a linear coordinate change $\mathbf{x} = L\mathbf{u}$ can be chosen so that the variable u does not appear in the formula*

$$\widehat{Q}(u, v) = \widehat{Q}(\mathbf{u}) = \mathbf{u}^\dagger L^\dagger M L \mathbf{u}$$

for the transformed quadratic form. Then the matrix M of Q is noninvertible and conversely.

Proof. Let us first suppose M is noninvertible. Then there is a nonzero vector $\mathbf{r}$ in its kernel: $M\mathbf{r} = \mathbf{0}$. Choose a second vector $\mathbf{s}$ so that $\{\mathbf{r}, \mathbf{s}\}$ form a basis for $\mathbb{R}^2$, and let L be the invertible matrix whose columns are the vectors $\mathbf{r}$ and $\mathbf{s}$. If we write Q as transformed by L in the form

$$\widehat{Q}(u,v) = \mathbf{u}^\dagger L^\dagger M L \mathbf{u} = (u \ v) \begin{pmatrix} \alpha & \beta \\ \beta & \gamma \end{pmatrix} \begin{pmatrix} u \\ v \end{pmatrix},$$

then the variable u will be missing from this expression if $\alpha = \beta = 0$, that is, if the entries in the first row and the first column of $L^\dagger M L$ equal 0.

To show that $L^\dagger M L$ has this property, first write L and $L^\dagger$ in the form

$$L = (\mathbf{r} \ \mathbf{s}), \quad L^\dagger = \begin{pmatrix} \mathbf{r}^\dagger \\ \mathbf{s}^\dagger \end{pmatrix}.$$

(Note that $\mathbf{r}^\dagger$ and $\mathbf{s}^\dagger$ are *row* vectors.) Then matrix multiplication allows us to write ML in a similar way, as

$$ML = (M\mathbf{r} \ M\mathbf{s}) = (\mathbf{0} \ M\mathbf{s}).$$

It follows that

$$L^\dagger M L = \begin{pmatrix} \mathbf{r}^\dagger \\ \mathbf{s}^\dagger \end{pmatrix} (\mathbf{0} \ M\mathbf{s}) = \begin{pmatrix} \mathbf{r}^\dagger \mathbf{0} & \mathbf{r}^\dagger M\mathbf{s} \\ \mathbf{s}^\dagger \mathbf{0} & \mathbf{s}^\dagger M\mathbf{s} \end{pmatrix} = \begin{pmatrix} 0 & \mathbf{r}^\dagger M\mathbf{s} \\ 0 & \mathbf{s}^\dagger M\mathbf{s} \end{pmatrix}.$$

The entries in the first column of $L^\dagger M L$ are therefore zero, and because the matrix is symmetric, the entries in its first row must be zero as well.

To prove the converse, we suppose that one of the variables in $\mathbf{u} = (u,v)$ is missing from the expression

$$\widehat{Q}(\mathbf{u}) = \mathbf{u}^\dagger L^\dagger M L \mathbf{u}.$$

Then $\widehat{M} = L^\dagger M L$ has a row (and a column) of zeros, so $\det \widehat{M} = 0$, implying $\widehat{M}$ is noninvertible. Consequently, $M = (L^\dagger)^{-1} \widehat{M} L^{-1}$ is noninvertible, as well. □

As a result of Theorem 7.3, we find that the natural way to distinguish between quadratic forms is provided by the following definition.

Definition 7.2 *A quadratic form $Q(\mathbf{x}) = \mathbf{x}^\dagger M \mathbf{x}$ is **nondegenerate** if its matrix M is invertible, and is **degenerate** otherwise.*

Corollary 7.4 *The quadratic form $Q(x,y) = Ax^2 + 2Bxy + Cy^2$ is nondegenerate if and only if $AC \neq B^2$.*

Proof. The determinant of the matrix of Q is $AC - B^2$; Q is nondegenerate if and only if this determinant is nonzero. □

To connect these general results about quadratic forms back to the local behavior of a function at a critical point, we introduce the *Hessian*.

The Hessian

Definition 7.3 *Suppose the function $z = f(x,y)$ has continuous second derivatives on a neighborhood of a critical point $(x,y) = (a,b)$. The **Hessian of f at (a,b)** is the symmetric matrix of second derivatives*

$$H_{(a,b)} = \begin{pmatrix} f_{xx}(a,b) & f_{xy}(a,b) \\ f_{yx}(a,b) & f_{yy}(a,b) \end{pmatrix}.$$

*The **Hessian form of f at (a,b)** is the quadratic form associated with the Hessian.*

Continuity of the second derivatives guarantees that $H_{(a,b)}$ is symmetric. Because there is usually no chance for confusion, we use the symbol $H_{(a,b)}$ for the Hessian form as well; thus

$$H_{(a,b)}(x,y) = f_{xx}(a,b)x^2 + 2f_{xy}(a,b)xy + f_{yy}(a,b)y^2.$$

Local behavior and the Hessian

Now assume that f has continuous third derivatives near (a,b) so we can write the second-order Taylor expansion of f at (a,b). In terms of window coordinates $\Delta x = x - a$, $\Delta y = y - b$ and $\Delta z = f(a + \Delta x, b + \Delta y) - f(a,b)$ and the Hessian form, the expansion is simply

$$\Delta z = \tfrac{1}{2}H_{(a,b)}(\Delta x, \Delta y) + O(3).$$

This tells us the local behavior of f near (a,b), so we ask: when, and how, does the Hessian determine that local behavior? In other words, when does the quadratic form $H_{(a,b)}(\Delta x, \Delta y)$ dominate the higher-order terms represented by $O(3)$? The answer is provided by Morse's lemma.

Definition 7.4 *Suppose the function $z = f(x,y)$ has continuous second derivatives near the critical point (a,b). Then (a,b) is **nondegenerate** if the Hessian $H_{(a,b)}$ of f at (a,b) is nondegenerate, and is **degenerate** otherwise.*

Theorem 7.5 (Morse's lemma). *Suppose $z = f(x,y)$ has continuous third derivatives in a neighborhood of a nondegenerate critical point (a,b). Then, in a sufficiently small window centered at (a,b), there is a coordinate change $(\Delta u, \Delta v) = \mathbf{h}(\Delta x, \Delta y)$ (nonlinear, in general) for which*

$$\Delta z = \pm(\Delta u)^2 \pm (\Delta v)^2.$$

The signs of $(\Delta u)^2$ and $(\Delta v)^2$ are the signs of the eigenvalues of the Hessian $H_{(a,b)}$ of f at (a,b).

Proof. See the proof of the n-variable version, Theorem 7.16 (p. 248), in the next section.   □

Analogies

The two eigenvalues of the Hessian are analogous to the single second derivative in the one-variable version (Theorem 7.1, p. 221). The Hessian is symmetric; therefore its eigenvalues are real (Exercise 2.14.a, p. 60). The critical point is nondegenerate; therefore the eigenvalues are nonzero; the sign of each is either positive or negative.

Corollary 7.6 (Second derivative test) *Suppose $z = f(x,y)$ has continuous third derivatives in a neighborhood of a critical point $(x,y) = (a,b)$. Then the nature of the critical point depends on the values of the second partial derivatives of f, (all evaluated at (a,b)), as follows.*

- *A saddle point if $f_{xx}f_{yy} - f_{xy}^2 < 0$*

- *A local minimum if $f_{xx}f_{yy} - f_{xy}^2 > 0$ and $f_{xx} + f_{yy} > 0$*

- *A local maximum if $f_{xx}f_{yy} - f_{xy}^2 > 0$ and $f_{xx} + f_{yy} < 0$*

If $f_{xx}f_{yy} - f_{xy}^2 = 0$, the test is inconclusive.

Proof. According to Morse's lemma, the nature of the critical point is determined by the signs of the eigenvalues, as follows. If the eigenvalues have opposite signs, then $\Delta z = \pm\big((\Delta u)^2 - (\Delta v)^2\big)$, a saddle; if both are positive, then $\Delta z = (\Delta u)^2 + (\Delta v)^2$, a local minimum; if both are negative, then $\Delta z = -(\Delta u)^2 - (\Delta v)^2$, a local maximum; if either is zero, Morse's lemma does not apply.

If λ_1 and λ_2 are the eigenvalues of $H_{(a,b)}$, then

$$\lambda_1\lambda_2 = \det H_{\mathbf{a}} = f_{xx}f_{yy} - f_{xy}^2, \quad \lambda_1 + \lambda_2 = \operatorname{tr} H_{\mathbf{a}} = f_{xx} + f_{yy}.$$

All the assertions of the test now follow, including the final one about an inconclusive result.   □

We now work through the details of a rather rich and varied example to see how Morse's lemma applies. The example begins with the function

Example: the wine bottle

$$z = f(x,y) = (x^2 + y^2 - 1)^2.$$

First of all, because x and y appear only in the form $x^2 + y^2$, the graph must be rotationally symmetric around the z-axis. Furthermore, $z \geq 0$ (because z equals a positive square), and z attains its minimum value, $z = 0$, everywhere on the circle $x^2 + y^2 = 1$. (These minima are thus nonisolated critical points; cf. page 225.) If (x,y) is near the origin, but $(x,y) \neq (0,0)$, then $z < 1$. But $z = 1$ when $(x,y) = (0,0)$, so z has a local maximum at the origin. The graph of f therefore resembles the base of a wine bottle. (The sediment that precipitates out of an old wine will settle into the small space along the ring of minima.)

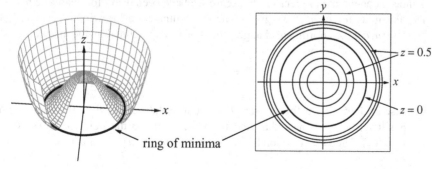

The level curves reflect the circular symmetry; they are all concentric with the origin. Each level $0 < z < 1$ consists of a pair of circles on either side of the level $z = 0$ (the unit circle). Each level above $z = 1$ is a single circle that lies outside the unit circle.

Analyzing the critical points of f

Let us carry out a standard analysis of the critical points of f. We have

$$\frac{\partial f}{\partial x} = 4x(x^2 + y^2 - 1), \quad \frac{\partial f}{\partial y} = 4y(x^2 + y^2 - 1),$$

so $(x, y) = (0, 0)$ is a critical point in addition to each of the points where $x^2 + y^2 = 1$. To apply the second derivative test, we need the Hessian, which equals

$$H_{(a,b)} = \begin{pmatrix} 4(a^2 + b^2 - 1) + 8a^2 & 8ab \\ 8ab & 4(a^2 + b^2 - 1) + 8b^2 \end{pmatrix}$$

at an arbitrary point (a, b). At the origin,

$$H_{(0,0)} = \begin{pmatrix} -4 & 0 \\ 0 & -4 \end{pmatrix},$$

so the test succeeds and tells us that the origin is a (nondegenerate) local maximum. At any point on $a^2 + b^2 = 1$, however, the Hessian reduces to

$$H_{(a,b)} = \begin{pmatrix} 8a^2 & 8ab \\ 8ab & 8b^2 \end{pmatrix} \text{ but } \det H_{(a,b)} = 64a^2b^2 - 64a^2b^2 = 0,$$

so the test fails. All points on the ring $a^2 + b^2 = 1$ of minima are degenerate critical points of f. Consider now what this means for the Hessian form:

$$H_{(a,b)}(\Delta x, \Delta y) = 8a^2(\Delta x)^2 + 16ab\Delta x\Delta y + 8b^2(\Delta y)^2 = 8(a\Delta x + b\Delta y)^2.$$

The Hessian form is degenerate

The Hessian form involves only the square of a single quantity, $a\Delta x + b\Delta y$. Thus, if we introduce the new variables

$$\Delta u = a\Delta x + b\Delta y, \quad \Delta v = -b\Delta x + a\Delta y,$$

then the Hessian form is just $8(\Delta u)^2$. The variable Δv is missing here, so the Hessian is indeed degenerate in precisely the sense we have been using for quadratic forms. The formulas for Δu and Δv give us new coordinates in the window centered at $(x, y) = (a, b)$; the coordinate change is the linear map defined by the matrix

$$P = \begin{pmatrix} a & b \\ -b & a \end{pmatrix}.$$

The missing variable in the Hessian

Because $a^2 + b^2 = 1$, it follows that P is a pure rotation. Let θ be the angle from the positive x axis to the radial line from the origin to the point (a, b) (so $\theta = \arctan(b/a)$). Then P is rotation by the angle $\arctan(-b/a) = -\arctan(b/a) = -\theta$.

This means that the positive Δx-axis lies at the angle $-\theta$ from the positive Δu-axis; see Exercise 5.16, page 180. Consequently, the Δu-axis points in the same direction as the vector (a,b)—the radial direction—so the Δv-axis is tangent to the ring of minima. Compare this to our previous example: when the Hessian form had no y-component, it had a line of critical points in the direction of the y-axis.

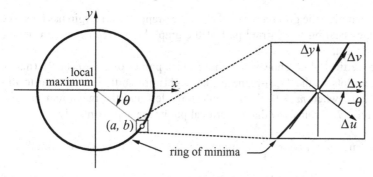

The purpose of our extended example is to see how Morse's lemma illuminates the structure of a function near a nondegenerate critical point. But the critical points of the ring of minima are degenerate, so Morse's lemma does not apply to them. (Morse's lemma does apply to the isolated maximum at the origin, but the character of that critical point is already evident.) We have more success by first altering the function so its ring of minima "breaks up" into just two isolated critical points. We can do this by tipping the graph slightly, as in the figure in the margin, below. The base, which had been sitting on the entire ring of minima, now shifts to rest on a single point. This point is the absolute minimum of the new function. As we show presently, the opposite point on the ring will shift into a saddle point. There are no other critical points (besides the local maximum that persists near the origin). All this happens no matter how slightly the graph is tipped.

Although it is easier to think of the tipping as a rotation—for example, a rotation of the (x,z)-plane about the y-axis—the formula for the altered function will be simpler if the tipping is done by a *shear*—again, of the (x,z)-plane; see the example in the margin. A vertical shear with slope m is given by

Modify the function by tipping its graph

$$S_m : \begin{cases} \text{new}\, x = x, \\ \text{new}\, y = y, \\ \text{new}\, z = mx + z. \end{cases}$$

$z = f(x, 0)$

$\xrightarrow{\;S_m\;}$

$z = f_m(x, 0)$

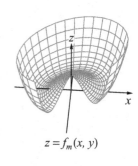

$z = f_m(x, y)$

The shear in the figure uses $m = 0.4$. We see it has the right sort of action on the grid of squares and, at the same time, we see what it does to (a vertical slice of) the graph of f. The formula for the sheared function f_m is

$$z = f_m(x,y) = mx + f(x,y) = (x^2 + y^2 - 1)^2 + mx.$$

Notice that even the grid on the surface of the graph in the margin has been sheared. Also, the shearing has carried part of the graph below the negative x-axis, as we would expect.

The vertical slice shows that the critical points of $z = f_m(x,0)$ (marked by the open dots in the (x,z)-plane on the right) are shifted in relation to those of $z = f(x,0)$: when $m > 0$, the minima move left and the maximum moves right. In Exercise 7.3, you show that the critical points are approximately

$$\text{maximum} : x \approx \frac{m}{4}, \quad \text{minima} : x \approx -\frac{m}{8} \pm 1, \quad \text{when } m \text{ is small.}$$

Critical points of f_m

Let us now analyze the critical points of $z = f_m(x,y)$, where m is small but nonzero. We must have

$$\frac{\partial f_m}{\partial x} = 4x(x^2 + y^2 - 1) + m = 0, \quad \frac{\partial f_m}{\partial y} = 4y(x^2 + y^2 - 1) = 0.$$

For $\partial f_m/\partial y = 0$ to hold, either $y = 0$ or $x^2 + y^2 - 1 = 0$. If we assume the second of these equations, then $\partial f_m/\partial x = 0$ reduces to $m = 0$; but this contradicts our assumption that $m \neq 0$. Hence, no point on the ring of minima of the original f is a critical point of the new function f_m. So let us assume instead that $y = 0$. Then $\partial f_m/\partial x = 0$ reduces to

$$4x(x^2 - 1) + m = 4x^3 - 4x + m = 0.$$

When m is small, this cubic has three real roots, p_1, p_2, p_3; see Exercise 7.3.

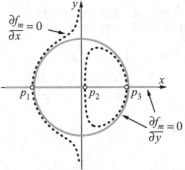

The figure above shows an alternate geometric approach to locating the critical points. They appear as the points of intersection of the critical curves on which $\partial f_m/\partial x = 0$ (shown dotted in the figure; $m = 0.3$) and $\partial f_m/\partial y = 0$ (the circle-plus-line shown in gray). The curves intersect in the three points p_1, p_2, p_3 on the x-axis.

To determine the type of each critical point, we calculate the Hessian, restricting ourselves to points of the form $(p,0)$:

$$H_{(p,0)} = \begin{pmatrix} 12p^2 - 4 & 0 \\ 0 & 4p^2 - 4 \end{pmatrix}.$$

When m is sufficiently small and positive, the three critical points satisfy

$$p_1 < -1, \quad 0 < p_2 < 1/\sqrt{3}, \quad 1/\sqrt{3} < p_3 < 1.$$

This allows us to make the following inferences about their Hessians:

$$H_{(p_1,0)} = \begin{pmatrix} + & 0 \\ 0 & + \end{pmatrix}, \quad H_{(p_2,0)} = \begin{pmatrix} - & 0 \\ 0 & - \end{pmatrix}, \quad H_{(p_3,0)} = \begin{pmatrix} + & 0 \\ 0 & - \end{pmatrix}.$$

It follows that p_1 is a (local) minimum, p_2 a (local) maximum, and p_3 a saddle. (What happens if $m < 0$?)

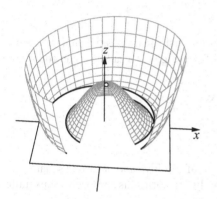
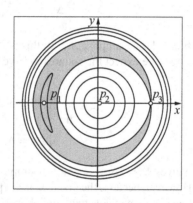

Think of the figure on the left, above, as showing the graph of f_m filled with liquid up to the level of the saddle point p_3. The level curve of f_m at that level is a thin crescent that has the characteristic "X" shape (albeit elongated and bent) where it passes through the saddle point itself. In the contour plot on the right, the liquid surface is shown in gray. Outside the crescent, the spacing between successive level curves is still $\Delta z = 0.25$, as it was for the original function in the contour plot on page 229. However, at that spacing, no further level curves will be found inside the crescent; the minimum point p_1 lies only about 0.2 units below the saddle. The single curve that is shown inside (in the shaded crescent) is about 0.18 units below the level of the saddle. As $m \to 0$, the shaded crecent shape gets thinner, converging to the ring of minima when $m = 0$, and this contour plot becomes the one on page 229.

Now let us see what Morse's lemma tells us about the saddle point $(p_3,0)$. Fundamentally, it provides new curvilinear coordinates $(\Delta u, \Delta v)$ that will reduce the window equation to $(\Delta u)^2 - (\Delta v)^2$. To understand this, we begin by constructing the window equation at any point $(p,0)$ on the x-axis:

$$\Delta z = f_m(p + \Delta x, \Delta y) - f_m(p, 0)$$
$$= (4p^3 - 4p + m)\,\Delta x + (6p^2 - 2)(\Delta x)^2 + (2p^2 - 2)(\Delta y)^2$$
$$+ 4p\,(\Delta x)^3 + 4p\,\Delta x\,(\Delta y)^2 + (\Delta x)^4 + 2(\Delta x)^2(\Delta y)^2 + (\Delta y)^4.$$

Completing the square   At a critical point, $4p^3 - 4p + m = 0$, so Δz loses its linear term (as we expect). If the window equation were purely quadratic, of the form

$$\Delta z = A\,(\Delta x)^2 + 2B\,\Delta x\,\Delta y + C\,(\Delta y)^2,$$

with A, B, C constants, then we could make Δz a sum of two squares by the familiar process of completing the square (assuming $A \neq 0$):

$$\Delta z = A\left((\Delta x)^2 + 2\frac{B}{A}\Delta x\,\Delta y + \frac{B^2}{A^2}(\Delta y)^2\right) - \frac{B^2}{A}(\Delta y)^2 + C(\Delta y)^2$$
$$= A\left(\Delta x + \frac{B}{A}\Delta y\right)^2 - \left(\frac{B^2}{A} - C\right)(\Delta y)^2.$$

To finish, let us suppose $A > 0$; this makes the first square positive and the second negative (we expect the squares to have different signs at a saddle). The coordinate change

$$\mathbf{h}: \begin{cases} \Delta u = \sqrt{A}\,\Delta x + \dfrac{B}{\sqrt{A}}\Delta y, \\[2ex] \Delta v = \Delta y\,\sqrt{\dfrac{B^2}{A} - C}, \end{cases}$$

then gives $\Delta z = (\Delta u)^2 - (\Delta v)^2$, a simple sum of (positive and negative) squares.

Morse's observations   However, because the given Δz is not purely quadratic, this approach seems futile. But now Morse makes two crucial observations:

- The validity of the change of coordinates $\mathbf{h}$ does not depend on the coefficients A, B, and C being constants; a quadratic form with variable coefficients can work, too.

- The window equation at any critical point can be "disassembled" properly into a quadratic form with variable coefficients.

He then provides (remarkably simple) instructions for disassembling the window equation into the proper components. We define equivalent instructions below (Lemma 7.3 p. 249), and they give us the following (see p. 251).

$$A = 6p^2 - 2 + 4p\,\Delta x + (\Delta x)^2 + \tfrac{1}{3}(\Delta y)^2,$$
$$B = \tfrac{4}{3}p\,\Delta y + \tfrac{2}{3}\Delta x\,\Delta y,$$
$$C = 2p^2 - 2 + \tfrac{4}{3}p\,\Delta x + \tfrac{1}{3}(\Delta x)^2 + (\Delta y)^2.$$

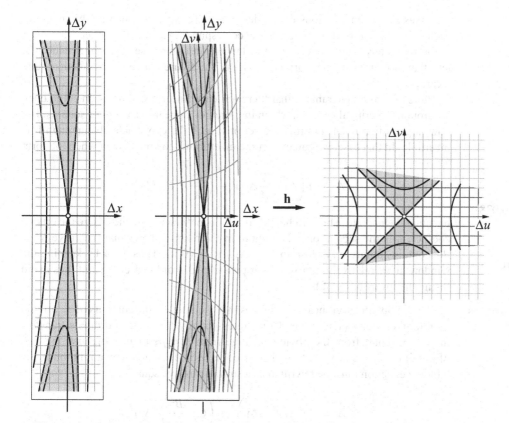

We see above the action of the map $\mathbf{h} : (\Delta x, \Delta y) \mapsto (\Delta u, \Delta v)$, using the expressions for A, B, and C just given, with $p = p_3 = 0.9872574766623532$. On the left is the window with its "native" $(\Delta x, \Delta y)$-coordinates. In the middle is the same $(\Delta x, \Delta y)$-window but now overlaid with the curvilinear coordinates $(\Delta u, \Delta v)$ pulled back by $\mathbf{h}$. On the right is the (curved) image of the window as pushed forward by $\mathbf{h}$ to the $(\Delta u, \Delta v)$-plane. The windows are very small; the spacing in both coordinate grids is 0.005. It is clear that $\mathbf{h}$ "squares up" the contours: the zero-level $\Delta z = 0$ becomes the pair of perpendicular straight lines $\Delta v = \pm \Delta u$. Notice that the zero-level intersects the $(\Delta u, \Delta v)$-grid lines in exactly the same places in both windows. The other two contours are not equally spaced with $\Delta z = 0$ but are instead chosen at levels (namely, $\Delta z = -0.0002$ and $\Delta z = 0.0006$) that show up well in the original (thin) window.

"Squaring up" contours near the saddle point

It remains to verify that $\mathbf{h}$ is indeed a valid coordinate change—that is, an invertible map—on some neighborhood of $(\Delta x, \Delta y) = (0,0)$. The functions that appear in $\mathbf{h}$ are smooth where they are defined; thus the inverse function theorem says it is sufficient to show that the derivative $\mathbf{dh}_{(0,0)}$ is invertible. This follows (cf. Exercise 7.4) from

$\mathbf{h}$ is invertible

$$\mathbf{dh}_{(0,0)} = \begin{pmatrix} \sqrt{6p^2 - 2} & 0 \\ 0 & \sqrt{2 - 2p^2} \end{pmatrix} \approx \begin{pmatrix} 1.96165 & 0 \\ 0 & 0.225045 \end{pmatrix}.$$

The axes are eigendirections of the derivative $d\mathbf{h}_{(0,0)}$; consequently, the image of each axis under $\mathbf{h}$ itself is tangent to the corresponding axis in the target. Moreover, $\mathbf{h}$ approximately doubles horizontal distances but compresses vertical distances to less than a quarter of their original length. The figure above shows all this quite clearly.

Morse's lemma guarantees that there are curvilinear coordinates on some open set around the critical point on which the function appears as a sum of squares. But how large is that open set? It is the set on which the coordinate change map $\mathbf{h}$ is invertible. In this case, we can expect the invertibility to break down when the form

$$\Delta z = A \left(\Delta x + \frac{B}{A} \Delta y \right)^2 - \left(\frac{B^2}{A} - C \right) (\Delta y)^2$$

becomes degenerate. This will happen if either coefficient vanishes. Here the crucial coefficient is the second one. The figure in the margin shows that the curve $B^2 = AC$ contains points very close to $p_3 : (\Delta x, \Delta y) = (0,0)$. Thus, only by keeping the window at p_3 relatively narrow was it possible to avoid that curve and thus avoid losing the invertibility of $\mathbf{h}$.

We can obtain curvilinear coordinates that "square up" the contours of f_m around its minimum point $(x,y) = (p_1,0)$ using essentially the same coordinate transformation $\mathbf{h}$. Apart from the obvious change from $p = p_3$ to $p = p_1$, just one pair of modifications is needed. First, because the critical point is now a minimum, the window equation must be rewritten as a sum of positive squares,

$$\Delta z = A \left(\Delta x + \frac{B}{A} \Delta y \right)^2 + \left(C - \frac{B^2}{A} \right) (\Delta y)^2.$$

Note the change in the form of the coefficient of $(\Delta y)^2$; this forces a corresponding alteration in the formula for Δv:

$$\Delta v = \Delta y \sqrt{C - \frac{B^2}{A}}.$$

The result is shown below. The "native" window, on the left, has the same proportions as the one we used for the saddle point, but it is half again as large. The source and the target of $\mathbf{h}$ are drawn to the same scale (and a grid square in the $(\Delta u, \Delta v)$-plane is 0.01 units on a side), making it evident that $\mathbf{h}$ stretches the horizontal direction but compresses the vertical. What the grids actually show us are the effects of the pullback by $\mathbf{h}^{-1}$: horizontal compression and vertical elongation. The contours of f are equally spaced, at the levels $\Delta z = 0.0005, 0.0010, 0.0015, 0.0020$. Notice that each contour meets points of the $(\Delta u, \Delta v)$-grid in exactly the same places in both windows.

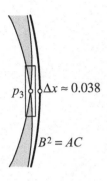

The domain of invertibility

p_3 $\Delta x \approx 0.038$

$B^2 = AC$

Curvilinear coordinates near the minimum

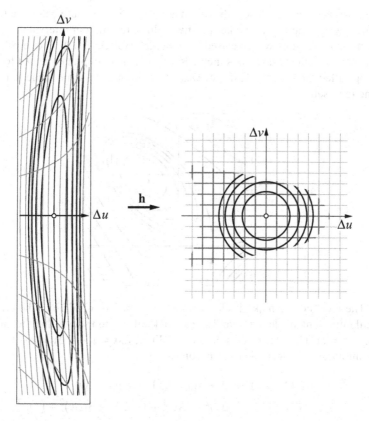

Details for the derivative $\mathbf{dh}_{(0,0)}$ are very similar to those for the saddle point; note that the slight change in the definition of Δv has caused $\sqrt{2-2p^2}$ to be replaced by $\sqrt{2p^2-2}$.

Invertiblility of **h**

$$\mathbf{dh}_{(0,0)} = \begin{pmatrix} \sqrt{6p^2-2} & 0 \\ 0 & \sqrt{2p^2-2} \end{pmatrix} \approx \begin{pmatrix} 2.0367 & 0 \\ 0 & 0.2222 \end{pmatrix}.$$

Because $\mathbf{dh}_{(0,0)}$ is once again a diagonal matrix, the image of each axis under **h** is tangent to its corresponding axis in the target. Horizontal lengths are approximately doubled and vertical ones are compressed by the factor 2/9. We conclude that **h** is locally invertible, giving valid curvilinear coordinates $(\Delta u, \Delta v)$ in some suitably restricted window centered at the minimum point $(x,y) = (p_1,0)$.

We now consider briefly a second function that is simpler than the wine bottle but nevertheless illustrates new aspects of Morse's lemma. The function is one introduced by Descartes:

Example: the folium of Descartes

$$z = f(x,y) = x^3 + y^3 - 3xy.$$

Some of its level curves near the origin are shown in the figure below. In the shaded region, where the function takes positive values, the contour interval is $\Delta z = 1.5$; in the unshaded region, we have used a smaller interval: $\Delta z = 0.2$. The zero-level curve that separates the two regions includes a leaf-shaped loop that has led to the curve being called the *folium* ("leaf") of Descartes. We use the same name to refer to the function itself.

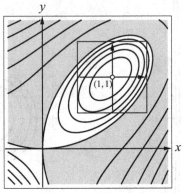

The level curves make it clear that $z = f(x,y)$ has a saddle at the origin and a local minimum inside the "leaf," and a quick calculation shows that the minimum is at $(x,y) = (1,1)$. In terms of window coordinates $\Delta x = x - 1$, $\Delta y = y - 1$ centered at the minimum, the formula for f becomes

$$z = (1 + \Delta x)^3 + (1 + \Delta y)^3 - 3(1 + \Delta x)(1 + \Delta y)$$
$$= 1 + 3\Delta x + 3(\Delta x)^2 + (\Delta x)^3 + 1 + 3\Delta y + 3(\Delta y)^2 + (\Delta y)^3$$
$$- 3 - 3\Delta x - 3\Delta y - 3\Delta x \Delta y$$
$$= -1 + 3(\Delta x)^2 - 3\Delta x \Delta y + 3(\Delta y)^2 + (\Delta x)^3 + (\Delta y)^3.$$

This reduces to

$$\Delta z = A(\Delta x)^2 + 2B\Delta x \Delta y + C(\Delta y)^2,$$

with $\Delta z = z + 1$ and

$$A = 3 + \Delta x, \quad B = -3/2, \quad C = 3 + \Delta y,$$

Action of the coordinate change h

The standard coordinate change

$$\mathbf{h} : \begin{cases} \Delta u = \sqrt{A}\left(\Delta x + \dfrac{B}{A}\Delta y\right), \\[2mm] \Delta v = \Delta y \sqrt{C - \dfrac{B^2}{A}}, \end{cases}$$

in the window then transforms Δz into

$$\Delta z = (\Delta u)^2 + (\Delta v)^2.$$

The contours in the original $(\Delta x, \Delta y)$-window are roughly elliptical. The map $\mathbf{h}$ carries them to concentric circles in the target $(\Delta u, \Delta v)$-plane. The figure below helps us to follow the details. The curvilinear $(\Delta u, \Delta v)$ coordinates that are overlaid on the source on the left are the ones pulled back from the target by $\mathbf{h}$. Therefore, the intersections between the original contours and the curvilinear grid in the source match exactly the intersections between the image circles and the square grid in the target.

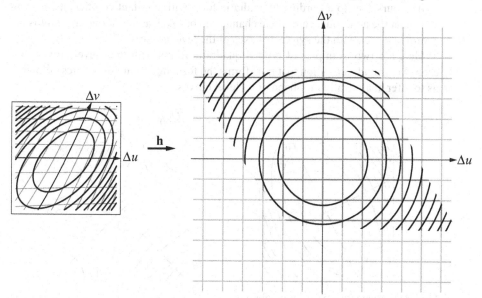

At this scale (the source window is a unit square), the contours are close to ellipses, and $\mathbf{h}$ looks almost linear. Its linear approximation at the origin is

$$\mathbf{dh}_{(0,0)} = \begin{pmatrix} \sqrt{3} & -\sqrt{3}/2 \\ 0 & 3/2 \end{pmatrix}.$$

Action of $\mathbf{h}$

The map resembles a horizontal shear that pushes points that lie above the horizontal axis to the left and points below to the right. Horizontal distances are increased by a factor of about $\sqrt{3} \approx 1.7$, and vertical ones by a factor of about 1.5. The effect of the dilation is to make the ellipses both larger and somewhat wider; the effect of the shear is then to turn them into circles.

Notice that the Δu- and Δv-axes we see overlaid on the source do not line up with the major and minor axes of the nested ellipses. This points to the main difference between the folium and the wine bottle examples. At the minimum of the tipped wine bottle, the curvilinear coordinate axes were aligned with the principal axes of the (approximate) ellipses, so the coordinate change $\mathbf{h}$ had a simpler action there: to turn the ellipses into circles, it just stretched the ellipses by two different factors along their principal axes. As a consequence, the derivative $\mathbf{dh}_{(0,0)}$ was a diagonal matrix, representing a pure strain in the coordinate directions.

Comparing the folium and the wine bottle

At the minimum point of the folium, however, the derivative is not a diagonal matrix. Appearances to the contrary notwithstanding, it is a pure strain, though, (rather than the shear it appears to be) because its eigenvalues, $\sqrt{3}$ and $3/2$, are real and unequal (cf. Theorem 2.6, p. 40). Hence we can convert $\mathbf{dh}_{(0,0)}$ into a diagonal matrix by using a further coordinate change that will align the new coordinate axes with the strain directions, that is, with the principal axes of the ellipses. In fact, if we restrict ourselves to an ordinary quadratic form with constant coefficients, we can show that the additional coordinate change can be taken as a rotation (that aligns the coordinate axes with the symmetry axes of the level curves).

Quadratic forms under rotations

Here are two examples of typical quadratic forms with their level curves. For each, we provide a rotation that transforms the form into a sum of squares, allowing us to infer analytically the shape of its level curves.

$$Q_{\text{ell}} = 6x^2 - 4xy + 3y^2, \qquad\qquad Q_{\text{hyp}} = x^2 + 6xy + y^2.$$

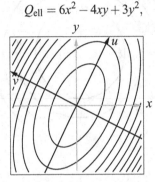
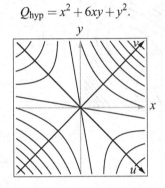

Under the respective coordinate changes

$$\mathbf{h}_{\text{ell}} : \begin{cases} x = \dfrac{u - 2v}{\sqrt{5}}, \\[2mm] y = \dfrac{2u + v}{\sqrt{5}}, \end{cases} \qquad \mathbf{h}_{\text{hyp}} : \begin{cases} x = \dfrac{u + v}{\sqrt{2}}, \\[2mm] y = \dfrac{-u + v}{\sqrt{2}}, \end{cases}$$

the two quadratic forms pull back to

$$Q_{\text{ell}}^* = 2u^2 + 7v^2, \qquad Q_{\text{hyp}}^* = -2u^2 + 4v^2.$$

The map $\mathbf{h}_{\text{ell}}$ is rotation by $\theta = \arctan 2$, and $\mathbf{h}_{\text{hyp}}$ is rotation by $\theta = -45°$. The rotations cause the (u, v)-coordinates to line up with what appear to be the symmetry axes of the level curves. We say that the quadratic forms have been *transformed to principal axes*.

The ellipses

The equation $Q_{\text{ell}}^* = r$ $(r > 0)$ describes an ellipse whose principal axes are the coordinate axes. All the different ellipses (i.e., for different $r > 0$) have the same proportions; that is, they are similar figures in the sense of Euclidean geometry. Because rotation preserves lengths and angles, we conclude that the level curves of Q_{ell} are nested similar ellipses that share their principal axes.

The hyperbolas

Likewise, $Q_{\text{hyp}}^* = r$ describes a hyperbola whose principal axes are the coordinate axes. All the different hyperbolas have the same asymptotes; these are the straight lines ("degenerate hyperbolas") defined by $Q^*_{\text{hyp}} = Q_{\text{hyp}} = 0$. In the (u,v)-coordinates, the asymptotes have the equations $v = \pm u/\sqrt{2}$; in the original (x,y)-coordinates, we can get the equations either by substitution using $\mathbf{h}_{\text{hyp}}$ or by completing the square:

$$-8x^2 + (3x+y)^2 = 0 \text{ or } y = (-3 \pm \sqrt{8})x.$$

Because rotation preserves lengths and angles, we conclude that the level curves of Q_{hyp} are hyperbolas that share asymptotes and principal axes.

Not only do the signs of the coefficients in the formulas for Q^*_{ell} and Q^*_{hyp} have geometric meaning, their ratio does, too: it determines the "aspect ratio" of the level curves. For example, in the first figure, seven ellipses cross the v-axis in the same distance that just two cross the u-axis. In the second figure, six hyperbolas cross the v-axis in the distance that three cross the u-axis. Call the ratio of the numbers in each pair the *aspect ratio* of the curves. This ratio is the same as (the absolute value of) the ratio of the eigenvalues of the symmetric matrices that define the forms:

The coefficients are eigenvalues

$$M_{\text{ell}} = \begin{pmatrix} 6 & -2 \\ -2 & 3 \end{pmatrix}, \qquad M_{\text{hyp}} = \begin{pmatrix} 1 & 3 \\ 3 & 1 \end{pmatrix},$$

$$\begin{aligned} p_{\text{ell}}(\lambda) &= \lambda^2 - 9\lambda + 14 & p_{\text{hyp}}(\lambda) &= \lambda^2 - 2\lambda - 8 \\ &= (\lambda - 2)(\lambda - 7), & &= (\lambda + 2)(\lambda - 4). \end{aligned}$$

Thus, the ratio of the eigenvalues determines the geometry of the level curves: the sign of the ratio indicates the kind of curves ($+$ for ellipses, $-$ for hyperbolas) and its magnitude indicates their aspect ratio.

As we have seen (cf. p. 226), when a linear map $\mathbf{x} = L\mathbf{u}$ is used to change coordinates in the quadratic form $Q(\mathbf{x}) = \mathbf{x}^\dagger M \mathbf{x}$ defined by a symmetric 2×2 matrix M, the matrix of the transformed quadratic form is $M^* = L^\dagger M L$:

$$Q^*(\mathbf{u}) = Q(L\mathbf{u}) = (L\mathbf{u})^\dagger M (L\mathbf{u}) = \mathbf{u}^\dagger L^\dagger M L \mathbf{u} = \mathbf{u}^\dagger M^* \mathbf{u}.$$

But the rotation matrices we are now using for coordinate changes have a special property: the transpose of a rotation is its inverse:

$$R_\theta^{-1} = R_{-\theta} = R_\theta^\dagger.$$

Therefore, when a rotation R is used to transform a quadratic form, we can write the relation between the two matrices defining the forms in a new way: $M^* = R^{-1}MR$. In particular, if the transformed Q^* is a sum of squares, then its matrix M^* is a diagonal matrix, and we have the following result.

Theorem 7.7. *If the rotation* $\mathbf{x} = R\mathbf{u}$ *transforms the quadratic form* $Q(\mathbf{x}) = \mathbf{x}^\dagger M \mathbf{x}$ *into* $Q^*(\mathbf{u}) = \mathbf{u}^\dagger D \mathbf{u}$, *where D is a diagonal matrix, then the diagonal elements of D are the eigenvalues of M and the columns of R are corresponding eigenvectors.*

Proof. Let the diagonal elements of D be α_1 and α_2, and let $\mathbf{e}_1 = (1,0)^\dagger$ and $\mathbf{e}_2 = (0,1)^\dagger$ be the standard basis vectors in $\mathbb{R}^2$. Then

$$D\mathbf{e}_1 = \alpha_1\mathbf{e}_1, \quad D\mathbf{e}_2 = \alpha_2\mathbf{e}_2,$$

so α_i is an eigenvalue of D with eigenvector $\mathbf{e}_i$, $i = 1, 2$. By assumption, $D = R^\dagger MR = R^{-1}MR$, or $RD = MR$. Let $\mathbf{v}_i = R\mathbf{e}_i$; this is the i-th column of R. We find

$$\alpha_i\mathbf{v}_i = R(\alpha_i\mathbf{e}_i) = RD\mathbf{e}_i = MR\mathbf{e}_i = M\mathbf{v}_i, \quad i = 1, 2,$$

implying that α_i is an eigenvalue of M with eigenvector $\mathbf{v}_i$. $\square$

Transforming to principal axes

Using Theorem 7.7 as a guide, we now have a way to transform a quadratic form to principal axes, that is, a way to construct a rotation $\mathbf{x} = R\mathbf{u}$ that will align the $\mathbf{u}$-coordinate axes with the principal axes of the curves $Q(\mathbf{x}) = $ constant and reduce the form to a sum of squares.

Theorem 7.8 (Principal axes theorem). *For any quadratic form $Q(\mathbf{x}) = \mathbf{x}^\dagger M\mathbf{x}$, there is a rotation $\mathbf{x} = R\mathbf{u}$ that transforms Q into a sum of squares $Q^*(\mathbf{u}) = \lambda_1 u^2 + \lambda_2 v^2$, where λ_1 and λ_2 are the eigenvalues of M.*

Proof. We use a proof that extends naturally to quadratic forms in n variables. We know M has a real eigenvalue λ_1 with an eigenvector $\mathbf{v}$ that we can assume to be a unit vector. Let $\mathbf{w}$ be a unit vector orthogonal to $\mathbf{v}$, chosen so the square $\mathbf{v} \wedge \mathbf{w}$ has positive orientation (cf. p. 41). Let R be the matrix whose columns are $\mathbf{v}$ and $\mathbf{w}$, in that order:

$$R = \begin{pmatrix} \mathbf{v} & \mathbf{w} \end{pmatrix}, \quad R^\dagger = \begin{pmatrix} \mathbf{v}^\dagger \\ \mathbf{w}^\dagger \end{pmatrix}, \quad R^\dagger R = \begin{pmatrix} \mathbf{v}^\dagger\mathbf{v} & \mathbf{v}^\dagger\mathbf{w} \\ \mathbf{w}^\dagger\mathbf{v} & \mathbf{w}^\dagger\mathbf{w} \end{pmatrix} = \begin{pmatrix} 1 & 0 \\ 0 & 1 \end{pmatrix},$$

so $R^\dagger = R^{-1}$. Then $MR = \begin{pmatrix} M\mathbf{v} & M\mathbf{w} \end{pmatrix} = \begin{pmatrix} \lambda_1\mathbf{v} & M\mathbf{w} \end{pmatrix}$, and

$$R^{-1}MR = R^\dagger MR = \begin{pmatrix} \lambda_1\mathbf{v}^\dagger\mathbf{v} & \mathbf{v}^\dagger M\mathbf{w} \\ \lambda_1\mathbf{w}^\dagger\mathbf{v} & \mathbf{w}^\dagger M\mathbf{w} \end{pmatrix} = \begin{pmatrix} \lambda_1 & 0 \\ 0 & \beta \end{pmatrix} = D,$$

where $\beta = \mathbf{w}^\dagger M\mathbf{w}$. The lower-left term of D is zero because $\mathbf{v}$ and $\mathbf{w}$ are orthogonal; the upper-right term is zero because $D = R^\dagger MR$ is symmetric. The proof of Theorem 7.7 shows that $\beta = \lambda_2$, the second eigenvalue of M, and that $\mathbf{w}$ is a corresponding eigenvector.

Finally, because $\mathbf{v}$ lies on the unit circle and $\mathbf{w}$ lies $90°$ counterclockwise from it,

$$\mathbf{v} = \begin{pmatrix} \cos\theta \\ \sin\theta \end{pmatrix}, \quad \mathbf{w} = \begin{pmatrix} -\sin\theta \\ \cos\theta \end{pmatrix},$$

for some $0 \le \theta < 2\pi$. Thus $R = R_\theta$. $\square$

Corollary 7.9 *Level curves of the quadratic form $Q(\mathbf{x}) = \mathbf{x}^\dagger M\mathbf{x}$ are ellipses if $\det M > 0$ and are hyperbolas if $\det M < 0$.*

Proof. Transforming Q to principal axes gives $Q^*(\mathbf{u}) = \lambda_1 u^2 + \lambda_2 v^2$, where $\det M = \lambda_1 \lambda_2$. The level curves are ellipses when this product is positive and hyperbolas when it is negative. $\qquad\square$

7.3 Morse's lemma

In this section we show that the local behavior of a function $z = f(x_1,\ldots,x_n)$ at a nondegenerate critical point is determined by its Hessian matrix of second derivatives at the critical point. The key step is Morse's lemma, which provides a coordinate change that reduces the function to a pure sum of squares near the critical point. We begin by transferring to n dimensions all the terms and concepts introduced in the previous section.

Definition 7.5 *A **quadratic form** in n variables is a function of the form*   Quadratic forms

$$Q(\mathbf{x}) = \mathbf{x}^\dagger M \mathbf{x},$$

where $\mathbf{x} = (x_1,\ldots,x_n)$ *(treated as a column vector) and M is an $n \times n$ matrix.*

Note that the matrix M need not be symmetric, nor is it uniquely defined by Q;   Antisymmetric matrices
see the exercises. In fact, adding a antisymmetric matrix R to M does not alter Q, because R by itself defines the quadratic form that is identically zero. The next lemma is the converse; it says that only the antisymmetric matrices define the zero form. (An antisymmetric matrix is also said to be *skew-symmetric*.)

Lemma 7.1. *Suppose the quadratic form $Q_0(\mathbf{x}) = \mathbf{x}^\dagger R \mathbf{x}$ is identically zero; then the matrix $R = (r_{ij})$ is* antisymmetric; *that is, $r_{ji} = -r_{ij}$ for every $i, j = 1,\ldots,n$.*

Proof. We evaluate $Q_0(\mathbf{x})$ for particular vectors $\mathbf{x}$. First take $\mathbf{x} = \mathbf{e}_i$, the ith standard basis vector in $\mathbb{R}^n$. Then $0 = Q_0(\mathbf{e}_i) = r_{ii}$. Next, take $\mathbf{x} = \mathbf{e}_i + \mathbf{e}_j$, $i \neq j$; then $0 = Q_0(\mathbf{e}_i + \mathbf{e}_j) = r_{ij} + r_{ji}$. $\qquad\square$

According to the next lemma, with each quadratic form Q we can associate a   Symmetric matrices
unique symmetric matrix M that defines the form: $Q(\mathbf{x}) = \mathbf{x}^\dagger M \mathbf{x}$. We write $Q \leftrightarrow M$ to indicate this association.

Lemma 7.2. *Suppose $Q(\mathbf{x}) = \mathbf{x}^\dagger M \mathbf{x}$ is a quadratic form, where M is an arbitrary $n \times n$ matrix. Then $\tilde{M} = (M + M^\dagger)/2$ is symmetric and defines the same quadratic form. Moreover, if S is symmetric and $Q(\mathbf{x}) = \mathbf{x}^\dagger S \mathbf{x}$, then $S = \tilde{M}$.*

Proof. Let $Q^\dagger(\mathbf{x}) = \mathbf{x}^\dagger M^\dagger \mathbf{x}$ be the quadratic form defined by the transpose matrix $M^\dagger$. Because $Q^\dagger(\mathbf{x})$ is just a scalar (a 1×1 matrix), it is equal to its own transpose; thus

$$Q^\dagger(\mathbf{x}) = (\mathbf{x}^\dagger M^\dagger \mathbf{x})^\dagger = \mathbf{x}^\dagger M \mathbf{x} = Q(\mathbf{x}).$$

In other words, even when M and $M^\dagger$ are different, they define the same quadratic form. Now let $\tilde{Q}$ be the quadratic form defined by $\tilde{M}$. Then

$$\widetilde{Q}(\mathbf{x}) = \mathbf{x}^{\dagger} \tfrac{1}{2}(M + M^{\dagger})\mathbf{x} = \tfrac{1}{2}(\mathbf{x}^{\dagger}M\mathbf{x} + \mathbf{x}^{\dagger}M^{\dagger}\mathbf{x}) = \tfrac{1}{2}(Q(\mathbf{x}) + Q^{\dagger}(\mathbf{x})) = Q(\mathbf{x}).$$

If $Q(\mathbf{x}) = \mathbf{x}^{\dagger}S\mathbf{x}$; then the quadratic form Q_0 defined by the symmetric matrix $R = S - \widetilde{M}$ must be identically zero: $Q_0(\mathbf{x}) = \mathbf{x}^{\dagger}R\mathbf{x} \equiv 0$. By the previous lemma, R is also antisymmetric, so R must be the zero matrix, implying that $S = \widetilde{M}$.   □

We now single out the degenerate quadratic forms as the ones that are either missing a variable or can be so transformed by a suitable linear coordinate change. We show, as we did in the two-variable case, that a form is degenerate in this sense precisely when its associated matrix is noninvertible.

Theorem 7.10. *Let $Q(\mathbf{x}) = \mathbf{x}^{\dagger}M\mathbf{x}$ be a quadratic form in n variables, where M is the symmetric matrix associated with Q. Suppose a linear coordinate change $\mathbf{x} = L\mathbf{u}$ can be chosen so that the variable u_1 does not appear in the formula*

$$\widehat{Q}(u_1, \dots, u_n) = \widehat{Q}(\mathbf{u}) = Q(L\mathbf{u}) = \mathbf{u}^{\dagger}L^{\dagger}ML\mathbf{u}$$

for the transformed quadratic form. Then M is noninvertible, and conversely.

Proof. Let us first suppose M is noninvertible. Then there is a nonzero vector $\mathbf{r}$ in its kernel: $M\mathbf{r} = \mathbf{0}$. Choose additional vectors $\mathbf{s}_2, \dots, \mathbf{s}_n$ so that the n vectors $\{\mathbf{r}, \mathbf{s}_2, \dots, \mathbf{s}_n\}$ form a basis for $\mathbb{R}^n$, and let L be the invertible matrix whose columns are the vectors $\mathbf{r}, \mathbf{s}_2, \dots, \mathbf{s}_n$, in that order. The variable u_1 will be missing from

$$\widehat{Q}(\mathbf{u}) = \mathbf{u}^{\dagger}L^{\dagger}ML\mathbf{u}$$

if all the entries in the first row and the first column of the matrix $L^{\dagger}ML$ are zero.

To show that $L^{\dagger}ML$ has this property, first write L and $L^{\dagger}$ in the form

$$L = \begin{pmatrix} \mathbf{r} & \mathbf{s}_2 & \cdots & \mathbf{s}_n \end{pmatrix} \quad \text{and} \quad L^{\dagger} = \begin{pmatrix} \mathbf{r}^{\dagger} \\ \mathbf{s}_2^{\dagger} \\ \vdots \\ \mathbf{s}_n^{\dagger} \end{pmatrix}.$$

Then matrix multiplication allows us to write the $n \times n$ matrix ML is a similar way, as

$$ML = \begin{pmatrix} M\mathbf{r} & M\mathbf{s}_2 & \cdots & M\mathbf{s}_n \end{pmatrix} = \begin{pmatrix} \mathbf{0} & M\mathbf{s}_2 & \cdots & M\mathbf{s}_n \end{pmatrix}.$$

In that case,

$$L^{\dagger}ML = \begin{pmatrix} \mathbf{r}^{\dagger} \\ \mathbf{s}_2^{\dagger} \\ \vdots \\ \mathbf{s}_n^{\dagger} \end{pmatrix} \begin{pmatrix} \mathbf{0} & M\mathbf{s}_2 & \cdots & M\mathbf{s}_n \end{pmatrix} = \begin{pmatrix} \mathbf{r}^{\dagger}\mathbf{0} & \mathbf{r}^{\dagger}M\mathbf{s}_2 & \cdots & \mathbf{r}^{\dagger}M\mathbf{s}_n \\ \mathbf{s}_2^{\dagger}\mathbf{0} & \mathbf{s}_2^{\dagger}M\mathbf{s}_2 & \cdots & \mathbf{s}_2^{\dagger}M\mathbf{s}_n \\ \vdots & \vdots & \ddots & \vdots \\ \mathbf{s}_n^{\dagger}\mathbf{0} & \mathbf{s}_n^{\dagger}M\mathbf{s}_2 & \cdots & \mathbf{s}_n^{\dagger}M\mathbf{s}_n \end{pmatrix}.$$

Every entry in the first column is 0; but $L^{\dagger}ML$ is symmetric, so every entry in the first row is 0, as well. Thus the first variable, u_1, is everywhere missing from $\widehat{Q}(\mathbf{u})$.

To prove the converse, suppose that the jth variable is missing from the expression of a quadratic form. Then the jth row and jth column of the matrix associated with the form contain only zeros, implying the determinant of the matrix is zero and the matrix is noninvertible.   □

Definition 7.6 *A quadratic form Q is **nondegenerate** if its associated symmetric matrix is invertible, and is **degenerate** otherwise.*

Nondegeneracy

Corollary 7.11 *A quadratic form is nondegenerate if and only if the eigenvalues of its associated symmetric matrix are all nonzero.*

Proof. The determinant of a matrix equals the product of its eigenvalues, so the matrix is invertible if and only if all its eigenvalues are nonzero.   □

There is more we must say about the eigenvalues of the symmetric matrix M of a quadratic form. Because we obtain eigenvalues as the roots of a polynomial, in general those eigenvalues are complex numbers, even when the entries of M are all real numbers. However, the eigenvalues associated with a quadratic form via its symmetric matrix are all real.

Eigenvalues of a
symmetric matrix

Theorem 7.12. *If M is a symmetric $n \times n$ matrix with real entries, then all the eigenvalues of M are real numbers.*

Proof. Let $p(\lambda)$ be the characteristic polynomial of M (Definition 2.1, p. 35); by the fundamental theorem of algebra, there are n (not necessarily distinct) complex numbers $\lambda_1, \ldots, \lambda_n$ that are the roots of $p(\lambda) = 0$. For each distinct root $\lambda = \alpha + i\beta$ (with α and β real), there is a complex eigenvector $\mathbf{z} = \mathbf{x} + i\mathbf{y}$ such that $M\mathbf{z} = \lambda\mathbf{z}$ and $\mathbf{z} \neq \mathbf{0}$. Each of these has a **complex conjugate**:

$$\overline{\lambda} = \alpha - i\beta, \quad \overline{\mathbf{z}} = \mathbf{x} - i\mathbf{y}.$$

If $\overline{\lambda} = \lambda$, then $\beta = 0$ so $\lambda = \alpha$, a real number.

Thus, to prove the theorem, we show $\overline{\lambda} = \lambda$; to do this, we calculate the matrix product $\overline{\mathbf{z}}^\dagger M\mathbf{z}$ two ways. First,

$$\overline{\mathbf{z}}^\dagger M\mathbf{z} = \overline{\mathbf{z}}^\dagger (\lambda\mathbf{z}) = \lambda (\overline{\mathbf{z}}^\dagger \mathbf{z}).$$

In the second calculation, we use the fact that $M^\dagger = M = \overline{M}$, because M is symmetric and real, and we equate $\overline{\mathbf{z}}^\dagger M\mathbf{z}$ and $\mathbf{z}^\dagger \overline{\mathbf{z}}$ with their transposes because they are scalars:

$$\overline{\mathbf{z}}^\dagger M\mathbf{z} = (\overline{\mathbf{z}}^\dagger M\mathbf{z})^\dagger = \mathbf{z}^\dagger M^\dagger \overline{\mathbf{z}} = \mathbf{z}^\dagger \overline{M}\overline{\mathbf{z}} = \mathbf{z}^\dagger \overline{\lambda}\overline{\mathbf{z}} = \overline{\lambda}(\mathbf{z}^\dagger \overline{\mathbf{z}}) = \overline{\lambda}(\mathbf{z}^\dagger \overline{\mathbf{z}})^\dagger = \overline{\lambda}(\overline{\mathbf{z}}^\dagger \mathbf{z}).$$

Thus $\overline{\lambda}(\overline{\mathbf{z}}^\dagger \mathbf{z}) = \lambda (\overline{\mathbf{z}}^\dagger \mathbf{z})$, and because $\overline{\mathbf{z}}^\dagger \mathbf{z} = \mathbf{x}^\dagger \mathbf{x} + \mathbf{y}^\dagger \mathbf{y} = \|\mathbf{x}\|^2 + \|\mathbf{y}\|^2 > 0$, we can divide by $\overline{\mathbf{z}}^\dagger \mathbf{z}$ and conclude $\overline{\lambda} = \lambda$.   □

Although the eigenvalues of a real symmetric matrix must be real, the eigenvectors need not be. For example, every nonzero complex vector is an eigenvector of the identity matrix (with real eigenvalue 1). However, we can show that, in a sense, the complex eigenvectors are superfluous: there is always a real eigenvector associated with each real eigenvalue of a real matrix, symmetric or otherwise.

Eigenvectors with
real eigenvalues

Theorem 7.13. *Suppose* $\mathbf{z}$ *is a complex eigenvector of the real matrix* M, *associated with the real eigenvalue* λ. *Then the real and imaginary parts of* $\mathbf{z}$ *are (real) eigenvectors of* M *associated with* λ.

Proof. Write $\mathbf{z} = \mathbf{x} + i\mathbf{y}$; then $\lambda(\mathbf{x} + i\mathbf{y}) = \lambda\mathbf{z} = M\mathbf{z} = M(\mathbf{x} + i\mathbf{y})$. The real and imaginary parts of this equation hold separately; because λ and M are real, the real and imaginary are

$$\lambda\mathbf{x} = M\mathbf{x}, \quad \lambda\mathbf{y} = M\mathbf{y}. \qquad \square$$

The Hessian

Thus all the eigenvalues of a symmetric matric are real, and each distinct eigenvalue has a corresponding real eigenvector. We are now ready to introduce the Hessian of a function of n variables and begin the local analysis of that function near a critical point.

Definition 7.7 *Suppose the function* $z = f(\mathbf{x})$ *has continuous second derivatives on a neighborhood of a critical point* $\mathbf{x} = \mathbf{a}$. *The* **Hessian of** f **at** $\mathbf{a}$ *is the symmetric matrix of second derivatives*

$$H_{\mathbf{a}} = \begin{pmatrix} f_{11}(\mathbf{a}) & \cdots & f_{1n}(\mathbf{a}) \\ \vdots & \ddots & \vdots \\ f_{n1}(\mathbf{a}) & \cdots & f_{nn}(\mathbf{a}) \end{pmatrix}.$$

The **Hessian form of** f **at** $\mathbf{a}$ *is the quadratic form associated with the Hessian.*

As we noted already in the two-variable case, continuity of the second derivatives guarantees that $H_{\mathbf{a}}$ is symmetric. Moreover, we continue to use the symbol $H_{\mathbf{a}}$ for the Hessian form as well; thus

$$H_{\mathbf{a}}(x_1, \ldots, x_n) = f_{11}(\mathbf{a})x_1^2 + 2f_{12}(\mathbf{a})x_1 x_2 + \cdots + f_{nn}(\mathbf{a})x_n^2.$$

Definition 7.8 *Suppose the function* $z = f(\mathbf{x})$ *has continuous second derivatives near the critical point* $\mathbf{a}$. *Then* $\mathbf{a}$ *is* **nondegenerate** *if the Hessian* $H_{\mathbf{a}}$ *of* f *at* $\mathbf{a}$ *is nondegenerate, and is* **degenerate** *otherwise.*

The effect of coordinate changes

Our goal is to show that coordinate changes can put a function into a particularly simple form near a nondegenerate critical point. But we must ask: can a coordinate change eliminate a critical point, or can it convert a nondegenerate critical point into a degenerate one? We now show that criticality and nondegeneracy are geometric properties of functions, unaltered by coordinate changes.

Theorem 7.14. *Suppose the coordinate change* $\mathbf{x} = \mathbf{h}(\mathbf{u})$ *transforms* $f(\mathbf{x})$ *into* $g(\mathbf{u})$: $f(\mathbf{x}) = f(\mathbf{h}(\mathbf{u})) = g(\mathbf{u})$. *Then* $z = f(\mathbf{x})$ *has a critical point at* $\mathbf{x} = \mathbf{a} = \mathbf{h}(\mathbf{b})$ *if and only if* $z = g(\mathbf{u})$ *has a critical point at* $\mathbf{u} = \mathbf{b}$.

Proof. By the chain rule, $d g_{\mathbf{b}} = d f_{\mathbf{a}} \circ d\mathbf{h}_{\mathbf{b}}$. Because $d\mathbf{h}_{\mathbf{b}}$ is invertible because $\mathbf{h}$ is a coordinate change,

$$d g_{\mathbf{b}} = \mathbf{0} \iff d f_{\mathbf{a}} = \mathbf{0}. \qquad \square$$

Theorem 7.15. *Suppose the coordinate change* $\mathbf{x} = \mathbf{h}(\mathbf{u})$ *transforms* $f(\mathbf{x})$ *into* $g(\mathbf{u})$, *where* f *and* g *have continuous third derivatives. Then* $\mathbf{u} = \mathbf{b}$ *is a nondegenerate critical point of* $z = g(\mathbf{u})$ *if and only if* $\mathbf{x} = \mathbf{a} = \mathbf{h}(\mathbf{b})$ *is a nondegenerate critical point of* $z = f(\mathbf{x})$.

Proof. We have $f(\mathbf{x}) = f(\mathbf{h}(\mathbf{u})) = g(\mathbf{u})$. Let $H_\mathbf{a}$ be the Hessian matrix of f at $\mathbf{a}$, and let $H_\mathbf{b}^*$ be the Hessian matrix of g at $\mathbf{b}$; we must establish a connection between $H_\mathbf{a}$ and $H_\mathbf{b}^*$ that implies one is invertible precisely when the other is.

The Hessians appear in the respective Taylor expansions of f and g:

$$f(\mathbf{a} + \Delta\mathbf{x}) - f(\mathbf{a}) = \tfrac{1}{2}\Delta\mathbf{x}^\dagger H_\mathbf{a}\,\Delta\mathbf{x} + O((\Delta\mathbf{x})^3),$$

$$g(\mathbf{b} + \Delta\mathbf{u}) - g(\mathbf{b}) = \tfrac{1}{2}\Delta\mathbf{u}^\dagger H_\mathbf{b}^*\,\Delta\mathbf{u} + O((\Delta\mathbf{u})^3).$$

However,

$$f(\mathbf{a} + \Delta\mathbf{x}) - f(\mathbf{a}) = \Delta z = g(\mathbf{b} + \Delta\mathbf{u}) - g(\mathbf{b}),$$

so we can begin to connect the two Hessians by writing

$$2\Delta z = \Delta\mathbf{x}^\dagger H_\mathbf{a}\,\Delta\mathbf{x} + O((\Delta\mathbf{x})^3) = \Delta\mathbf{u}^\dagger H_\mathbf{b}^*\,\Delta\mathbf{u} + O((\Delta\mathbf{u})^3).$$

Now express $\Delta\mathbf{x}$ in terms of $\Delta\mathbf{u}$ by using the differentiability of $\mathbf{h}$ at $\mathbf{b}$:

$$\Delta\mathbf{x} = \mathbf{x} - \mathbf{a} = \mathbf{h}(\mathbf{b} + \Delta\mathbf{u}) - \mathbf{h}(\mathbf{b}) = d\mathbf{h}_\mathbf{b}(\Delta\mathbf{u}) + o(\Delta\mathbf{u}) = L\Delta\mathbf{u} + o(\Delta\mathbf{u}).$$

For visual clarity we have set $d\mathbf{h}_\mathbf{b} = L$ here; the remainder is "little oh" of $\Delta\mathbf{u}$. By Exercise 3.28 (p. 104), $L\Delta\mathbf{u} = O(\Delta\mathbf{u})$, so $\Delta\mathbf{x} = O(\Delta\mathbf{u})$.

For every $\Delta\mathbf{u} \neq \mathbf{0}$, write $\Delta\mathbf{u} = s\Delta\mathbf{y}$ with $\Delta\mathbf{y}$ a unit vector and a suitable $s > 0$. Then $O((\Delta\mathbf{u})^3) = O(s^3)$,

$$\Delta\mathbf{x} = sL\Delta\mathbf{y} + o(s) = s\big(L\Delta\mathbf{y} + o(s)/s\big), \quad O((\Delta\mathbf{x})^3) = O(s^3),$$

and we can write the two expressions for $2\Delta z$ as

$$s^2\big(L\Delta\mathbf{y} + o(s)/s\big)^\dagger H_\mathbf{a}\big(L\Delta\mathbf{y} + o(s)/s\big) + O(s^3) = s^2\,\Delta\mathbf{y}^\dagger H_\mathbf{b}^*\Delta\mathbf{y} + O(s^3).$$

Now divide the equation by s^2 and take the limit as $s \to 0$, using $o(s)/s \to 0$ and $O(s^3)/s^2 \to 0$. The result is

$$(L\Delta\mathbf{y})^\dagger H_\mathbf{a}(L\Delta\mathbf{y}) = \Delta\mathbf{y}^\dagger\big(L^\dagger H_\mathbf{a} L\big)\Delta\mathbf{y} = \Delta\mathbf{y}^\dagger H_\mathbf{b}^*\Delta\mathbf{y}$$

for every $\Delta\mathbf{y} \neq \mathbf{0}$. This implies $L^\dagger H_\mathbf{a} L = H_\mathbf{b}^*$ and hence

$$\det H_\mathbf{b}^* = \det H_\mathbf{a}\,(\det L)^2.$$

Because $\det L = \det d\mathbf{h}_\mathbf{b} \neq 0$ because $\mathbf{h}$ is a coordinate change, $\det H_\mathbf{b}^* \neq 0$ if and only if $\det H_\mathbf{a} \neq 0$. $\qquad\square$

The equations $dg_\mathbf{b} = df_\mathbf{a} \circ d\mathbf{h}_\mathbf{a}$ and $\det H_\mathbf{b}^* = \det H_\mathbf{a}\,(\det d\mathbf{h}_\mathbf{b})^2$ in the last two proofs are the multivariable analogues of the earlier equations $g'(0) = f'(0)\,h'(0)$

Analogous equations

and $g''(0) = f''(0)(h'(0))^2$ that showed criticality and nondegeneracy were geometric properties of single-variable functions (p. 222).

We are now ready to state and prove the main theorem. It first appears in an important paper on the topological properties of multivariable functions that Marston Morse published in 1925 [13]. Because the theorem was just one of several technical facts he needed to establish the paper's main results (now called *Morse theory*), it was natural for him to label this fact as a lemma. For us, however, the fact is central, though it is still always called *Morse's lemma*: at a nondegenerate critical point, a function can always be converted into a sum of squares.

Theorem 7.16 (Morse's lemma). *Suppose $z = f(\mathbf{x})$ has continuous third derivatives on an open set X^n, the point $\mathbf{x} = \mathbf{a}$ in X^n is a nondegenerate critical point of f, and the Hessian matrix $H_\mathbf{a}$ has r negative eigenvalues. Then, in a sufficiently small window $W_\mathbf{a}$ centered at $\mathbf{a}$, there is a coordinate change $\Delta\mathbf{u} = \mathbf{h}(\Delta\mathbf{x})$ for which*

$$\Delta z = f(\mathbf{a} + \Delta\mathbf{x}) - f(\mathbf{a})$$
$$= -(\Delta u_1)^2 - \cdots - (\Delta u_r)^2 + (\Delta u_{r+1})^2 + \cdots + (\Delta u_n)^2.$$

Because the Hessian $H_\mathbf{a}$ is symmetric and the critical point is nondegenerate, the eigenvalues of $H_\mathbf{a}$ are all real and nonzero. If all are positive (i.e., $r = 0$ in the statement of the theorem), then there are no negative squares in the sum. If all eigenvalues are negative (i.e., $r = n$), then there are no positive squares in the sum.

The proof of Morse's lemma breaks up naturally into three parts. In the first part (Theorem 7.18), a coordinate change reduces the window equation for a function at an nondegenerate critical point into a simple quadratic form with variable coefficients. In the second part (Theorem 7.19), a further coordinate change "diagonalizes" the quadratic form. This means that the form becomes a sum of positive and negative squares (and the symmetric matrix associated with it becomes a diagonal matrix). But significantly, it also means that the coefficients of the quadratic form become constants. In other words, any function "looks like" a sum of squares near a nondegenerate critical point. The third part of the proof of Morse's lemma (Theorem 7.25) shows that the number of negative squares in the sum does not depend on the way the coordinate changes were chosen, but is always equal to the number of negative eigenvalues in the Hessian of the given function at its critical point.

Morse begins the proof of the Morse lemma by expanding a function into linear and quadratic terms in a way that is uncannily similar to Taylor's expansion. Taylor's formula splits the function into three simple pieces—a constant, a linear form, and a quadratic form—plus a fourth piece that contains the remaining "complexity" of the function. Morse recasts the formula so there is no separate remainder; the coefficients of the quadratic form become variable, and contain all the complexity that Taylor's formula puts into the remainder. We have already seen Morse's expansion put to use: on pages 219–220 we used it to determine the local behavior of a function of one variable near a critical point. Here, then, for the sake of comparison are the theorems that provide the expansions of Taylor and Morse.

Theorem 7.17 (Taylor). *Suppose $z = f(\mathbf{x})$ has continuous third derivatives on an open set that contains the line segment from $\mathbf{a}$ to $\mathbf{a} + \Delta\mathbf{x}$. Then*

$$f(\mathbf{a} + \Delta\mathbf{x}) = f(\mathbf{a}) + \sum_{i=1}^{n} \frac{\partial f}{\partial x_i}(\mathbf{a})\Delta x_i + \frac{1}{2}\sum_{i,j=1}^{n} \frac{\partial^2 f}{\partial x_i \partial x_j}(\mathbf{a})\Delta x_i \Delta x_j + O((\Delta\mathbf{x})^3). \qquad \square$$

Theorem 7.18 (Morse). *Suppose $z = f(\mathbf{x})$ has continuous third derivatives on an open set that contains the line segment from $\mathbf{a}$ to $\mathbf{a} + \Delta\mathbf{x}$. Then there are continuously differentiable functions $h_{ij}(\Delta\mathbf{x}) = h_{ji}(\Delta\mathbf{x})$ for which*

$$f(\mathbf{a} + \Delta\mathbf{x}) = f(\mathbf{a}) + \sum_{i=1}^{n} \frac{\partial f}{\partial x_i}(\mathbf{a})\Delta x_i + \sum_{i,j=1}^{n} h_{ij}(\Delta\mathbf{x})\Delta x_i \Delta x_j,$$

and $h_{ij}(\mathbf{0}) = \dfrac{1}{2}\dfrac{\partial^2 f}{\partial x_i \partial x_j}(\mathbf{a})$.

Proof. For clarity, we separate the proof into a number of steps. One of our aims is to provide explicit instructions for constructing the coefficients $h_{ij}(\Delta\mathbf{x})$ of the quadratic form. Note, in what follows, similarities with the proof of Taylor's theorem.

With the following lemma, we are able to build all the terms in Morse's formula, including the crucial coefficients h_{ij}.

Step 1

Lemma 7.3. *Suppose $z = F(\mathbf{x})$ has continuous derivatives of order $k + 1$ on an open set that contains the line segment from $\mathbf{a}$ to $\mathbf{a} + \Delta\mathbf{x}$. Then there are functions $p_i(\Delta\mathbf{x})$ with continuous derivatives of order k for which*

$$F(\mathbf{a} + \Delta\mathbf{x}) = F(\mathbf{a}) + \sum_{i=1}^{n} p_i(\Delta\mathbf{x})\Delta x_i,$$

and $p_i(\mathbf{0}) = \dfrac{\partial F}{\partial x_i}(\mathbf{a})$, $i = 1,\ldots,n$.

Proof. We express the difference $\Delta z = F(\mathbf{a} + \Delta\mathbf{x}) - F(\mathbf{a})$ as an integral, as in the beginning of the proof of Taylor's theorem for a single-variable function (cf. p. 79):

$$\int_0^1 \frac{d}{dt}F(\mathbf{a} + t\Delta\mathbf{x})\,dt = F(\mathbf{a} + t\Delta\mathbf{x})\bigg|_0^1 = F(\mathbf{a} + \Delta\mathbf{x}) - F(\mathbf{a}) = \Delta z.$$

In this multivariable setting, the chain rule gives us

$$\frac{d}{dt}F(\mathbf{a} + t\Delta\mathbf{x}) = \sum_{i=1}^{n} \frac{\partial F}{\partial x_i}(\mathbf{a} + t\Delta\mathbf{x})\Delta x_i,$$

so

$$\Delta z = \sum_{i=1}^{n} \left(\int_0^1 \frac{\partial F}{\partial x_i}(\mathbf{a} + t\Delta\mathbf{x})\,dt\right)\Delta x_i.$$

Therefore we take

$$p_i(\Delta\mathbf{x}) = \int_0^1 \frac{\partial F}{\partial x_i}(\mathbf{a} + t\Delta\mathbf{x})\,dt.$$

Because $\partial F/\partial x_i$ has continuous derivatives of order k, so does p_i. Moreover,

$$p_i(\mathbf{0}) = \int_0^1 \frac{\partial F}{\partial x_i}(\mathbf{a} + t\mathbf{0})\,dt = \frac{\partial F}{\partial x_i}(\mathbf{a})\int_0^1 dt = \frac{\partial F}{\partial x_i}(\mathbf{a}). \qquad\square$$

Step 2   Now apply Lemma 7.3 to the function f itself to obtain functions $g_i(\Delta\mathbf{x})$ for which

$$f(\mathbf{a} + \Delta\mathbf{x}) = f(\mathbf{a}) + \sum_{i=1}^n g_i(\Delta\mathbf{x})\,\Delta x_i.$$

According to the same lemma, each function g_i has continuous second derivatives, and

$$g_i(\mathbf{0}) = \frac{\partial f}{\partial x_i}(\mathbf{a}),$$

which gives us a start on Morse's expansion.

Step 3   Apply Lemma 7.3 again to each $g_i(\Delta\mathbf{x})$, $i = 1,\ldots,n$, this time taking $\mathbf{a} = \mathbf{0}$. We get functions $\widetilde{h}_{ij}(\Delta\mathbf{x})$, $j = 1,\ldots,n$, with continuous first derivatives for which

$$g_i(\Delta\mathbf{x}) = g_i(\mathbf{0}) + \sum_{j=1}^n \widetilde{h}_{ij}(\Delta\mathbf{x})\,\Delta x_j = \frac{\partial f}{\partial x_i}(\mathbf{a}) + \sum_{j=1}^n \widetilde{h}_{ij}(\Delta\mathbf{x})\,\Delta x_j,$$

$$\text{and } \widetilde{h}_{ij}(\mathbf{0}) = \frac{\partial g_i}{\partial x_j}(\mathbf{0}).$$

Comment: Nominally, each g_i is a function of the window variables Δx_j, but because Δx_j and x_j differ merely by a constant ($\Delta x_j = x_j - a_j$), the differential operators

$$\frac{\partial}{\partial(\Delta x_j)} \quad\text{and}\quad \frac{\partial}{\partial x_j}$$

have the same action. For simplicity we therefore write

$$\frac{\partial g_i}{\partial x_j} \quad\text{instead of}\quad \frac{\partial g_i}{\partial(\Delta x_j)}$$

here and in all the following work.

Step 4   Now substitute the expression for $g_i(\Delta\mathbf{x})$ into the formula for $f(\mathbf{a} + \Delta\mathbf{x})$ in Step 2:

$$f(\mathbf{a} + \Delta\mathbf{x}) = f(\mathbf{a}) + \sum_{i=1}^n \frac{\partial f}{\partial x_i}(\mathbf{a})\,\Delta x_i + \sum_{i=1}^n \sum_{j=1}^n \widetilde{h}_{ij}(\Delta\mathbf{x})\,\Delta x_i \Delta x_j.$$

This looks like Morse's expansion; in particular, the last term is a quadratic form with variable coefficients $\widetilde{h}_{ij}(\Delta\mathbf{x})$. But nothing in Lemma 7.3 ensures that $\widetilde{h}_{ji}(\Delta\mathbf{x}) =$

$\widetilde{h}_{ij}(\Delta\mathbf{x})$ for every $i,j = 1,\ldots,n$, as required by the theorem. (In other words, the matrix $\widetilde{H}(\Delta\mathbf{x}) = \left(\widetilde{h}_{ij}(\Delta\mathbf{x})\right)$ that defines the quadratic form need not be symmetric.)

But we can use Lemma 7.2 to replace the matrix $\widetilde{H}$ by the symmetric matrix $H = (\widetilde{H} + \widetilde{H}^{\dagger})/2$ without altering the quadratic form. That is, if we let

Step 5

$$h_{ij}(\Delta\mathbf{x}) = \frac{\widetilde{h}_{ij}(\Delta\mathbf{x}) + \widetilde{h}_{ji}(\Delta\mathbf{x})}{2},$$

then

$$f(\mathbf{a} + \Delta\mathbf{x}) = f(\mathbf{a}) + \sum_{i=1}^{n} \frac{\partial f}{\partial x_i}(\mathbf{a})\Delta x_i + \sum_{i=1}^{n}\sum_{j=1}^{n} h_{ij}(\Delta\mathbf{x})\Delta x_i \Delta x_j$$

and $h_{ji}(\Delta\mathbf{x}) = h_{ij}(\Delta\mathbf{x})$ for all $i,j = 1,\ldots,n$.

In remains only to verify that $h_{ij}(\mathbf{0}) = \frac{1}{2}\frac{\partial^2 f}{\partial x_i \partial x_j}(\mathbf{a})$. We claim that, in fact,

Step 6

$$\widetilde{h}_{ij}(\mathbf{0}) = \frac{1}{2}\frac{\partial^2 f}{\partial x_j \partial x_i}(\mathbf{a}) = \frac{1}{2}\frac{\partial^2 f}{\partial x_i \partial x_j}(\mathbf{a}) = \widetilde{h}_{ji}(\mathbf{0}).$$

To prove the claim, note (Step 3) that $\widetilde{h}_{ij}(\mathbf{0}) = \frac{\partial g_i}{\partial x_j}(\mathbf{0})$. Therefore, because

$$g_i(\Delta\mathbf{x}) = \int_0^1 \frac{\partial f}{\partial x_i}(\mathbf{a} + t\Delta\mathbf{x})\,dt,$$

we can link $\widetilde{h}_{ij}$ to f by calculating the appropriate partial derivative of g_i. This involves differentiation under the integral sign, a delicate matter but one that is allowed here because the integrand is continuously differentiable; see an introductory text on real analysis. We have (by the chain rule)

$$\frac{\partial g_i}{\partial x_j}(\Delta\mathbf{u}) = \int_0^1 \frac{\partial}{\partial x_j}\left(\frac{\partial f}{\partial x_i}(\mathbf{a} + t\Delta\mathbf{x})\right)dt = \int_0^1 \frac{\partial^2 f}{\partial x_j \partial x_i}(\mathbf{a} + t\Delta\mathbf{x})\,t\,dt,$$

from which it follows that

$$\widetilde{h}_{ij}(\mathbf{0}) = \frac{\partial g_i}{\partial x_j}(\mathbf{0}) = \int_0^1 \frac{\partial^2 f}{\partial x_j \partial x_i}(\mathbf{a})\,t\,dt$$

$$= \frac{\partial^2 f}{\partial x_j \partial x_i}(\mathbf{a})\int_0^1 t\,dt = \frac{1}{2}\frac{\partial^2 f}{\partial x_j \partial x_i}(\mathbf{a}).$$

This completes the proof of Theorem 7.18, and incidentally shows that, even though the matrix $\widetilde{H}(\Delta\mathbf{x})$ may not be symmetric in general, at least it is when $\Delta\mathbf{x} = \mathbf{0}$. $\square$

Let us see how the instructions provided in this proof (in Steps 2 and 3) give us formulas for the functions h_{ij} at a critical point of the "tipped wine bottle" function

Example: constructing the h_{ij}

$f_m(x,y) = (x^2 + y^2 - 1)^2 + mx$ (Chapter 7.2, pages 231–237). These are the formulas for A, B, and C that appear on page 234.

To find the h_{ij}, we must first construct the g_i, and these involve partial derivatives of the expression $f_m(\mathbf{x})$ (evaluated at $\mathbf{x} = \mathbf{a} + t\Delta\mathbf{x}$). But the window function

$$\Delta z = f_m(\mathbf{x}) - f_m(\mathbf{a})$$

has the same partial derivatives; the two functions merely differ by a constant. Furthermore, when we restrict $\mathbf{a}$ to the form $(p, 0)$ (because we are interested only in critical points), we can use the expression for Δz we have already computed on page 234:

$$\Delta z = (6p^2 - 2)(\Delta x)^2 + (2p^2 - 2)(\Delta y)^2$$
$$+ 4p(\Delta x)^3 + 4p\Delta x(\Delta y)^2 + (\Delta x)^4 + 2(\Delta x)^2(\Delta y)^2 + (\Delta y)^4.$$

Finally, keeping in mind the comment in Step 3 of the last proof, that derivatives with respect to Δx_i and x_i are interchangeable, we can now compute

$$\frac{\partial(\Delta z)}{\partial(\Delta x)} = 2(6p^2 - 2)\Delta x + 12p(\Delta x)^2 + 4p(\Delta y)^2 + 4(\Delta x)^3 + 4\Delta x(\Delta y)^2,$$

$$\frac{\partial(\Delta z)}{\partial(\Delta y)} = 2(2p^2 - 2)\Delta y + 8p\Delta x\Delta y + 4(\Delta x)^2\Delta y + 4(\Delta y)^3.$$

Thus,

$$g_1(\Delta x, \Delta y) = \int_0^1 \frac{\partial(\Delta z)}{\partial(\Delta x)}(t\Delta x, t\Delta y)\,dt$$

$$= \int_0^1 \left\{ 2t(6p^2 - 2)\Delta x + 4t^2\left[3p(\Delta x)^2 + p(\Delta y)^2\right] + 4t^3\left[(\Delta x)^3 + \Delta x(\Delta y)^2\right] \right\} dt$$

$$= (6p^2 - 2)\Delta x + 4p(\Delta x)^2 + \tfrac{4}{3}p(\Delta y)^2 + (\Delta x)^3 + \Delta x(\Delta y)^2.$$

In a similar way,

$$g_2(\Delta x, \Delta y) = \int_0^1 \frac{\partial(\Delta z)}{\partial(\Delta y)}(t\Delta x, t\Delta y)\,dt$$

$$= (2p^2 - 2)\Delta y + \tfrac{8}{3}p\Delta x\Delta y + (\Delta x)^2\Delta y + (\Delta y)^3.$$

We are now ready to compute the four functions h_{ij}. By definition, $h_{11} = \widetilde{h}_{11}$, so

$$h_{11}(\Delta x, \Delta y) = \int_0^1 \frac{\partial g_1}{\partial(\Delta x)}(t\Delta x, t\Delta y)\,dt$$

$$= \int_0^1 \left\{ [6p^2 - 2] + t[8p\Delta x] + t^2[3(\Delta x)^2 + (\Delta y)^2] \right\} dt$$

$$= (6p^2 - 2) + 4p\Delta x + (\Delta x)^2 + \tfrac{1}{3}(\Delta y)^2.$$

(This is the function $A(\Delta x, \Delta y)$ given on page 234). Next, notice that

$$\frac{\partial g_1}{\partial(\Delta y)} = \tfrac{8}{3}p\,\Delta y + 2\,\Delta x\,\Delta y = \frac{\partial g_2}{\partial(\Delta x)},$$

implying $\widetilde{h}_{12} = \widetilde{h}_{21} = h_{12}$. We have

$$\begin{aligned}
h_{12}(\Delta x, \Delta y) &= \int_0^1 \frac{\partial g_1}{\partial(\Delta y)}(t\Delta x, t\Delta y)\,dt \\
&= \int_0^1 \left\{ t\left[\tfrac{8}{3}p\,\Delta y\right] + t^2\left[2\,\Delta x\,\Delta y\right] \right\} dt = \tfrac{4}{3}p\,\Delta y + \tfrac{2}{3}\Delta x\,\Delta y \\
&= B(\Delta x, \Delta y).
\end{aligned}$$

Finally, because $h_{22} = \widetilde{h}_{22}$, we have

$$\begin{aligned}
h_{22}(\Delta x, \Delta y) &= \int_0^1 \frac{\partial(\Delta g_2)}{\partial(\Delta y)}(t\Delta x, t\Delta y)\,dt \\
&= \int_0^1 \left\{ [2p^2 - 2] + t\left[\tfrac{8}{3}p\,\Delta x\right] + t^2\left[(\Delta x)^2 + 3(\Delta y)^2\right] \right\} dt \\
&= (2p^2 - 2) + \tfrac{4}{3}p\,\Delta x + \tfrac{1}{3}(\Delta x)^2 + (\Delta y)^2.
\end{aligned}$$

Because this is the function $C(\Delta x, \Delta y)$ given earlier, we have completed the example.

We now move on to the next part of the proof of Morse's lemma. We can assume, by Theorem 7.18, that our function is already written in window coordinates as a quadratic form with variable coefficients:

$$\Delta z = f(\mathbf{a} + \Delta\mathbf{x}) - f(\mathbf{a}) = \sum_{i,j=1}^{n} h_{ij}(\Delta\mathbf{x})\,\Delta x_i\,\Delta x_j,$$

where $h_{ji}(\Delta\mathbf{x}) = h_{ij}(\Delta\mathbf{x})$ and

$$h_{ij}(\mathbf{0}) = \frac{1}{2}\frac{\partial^2 f}{\partial u_i\,\partial u_j}(\mathbf{a}).$$

Our goal is to "diagonalize" this quadratic form. If the coefficients were constants instead of functions, then linear algebra would provide a standard diagonalization method that involves changing coordinates, one variable at a time, by "completing the square." We actually use this method because, as Morse pointed out, it works just as well with variable coefficients.

The first step in completing the square is to divide by the leading coefficient (this is h_{11} in the quadratic form we are dealing with, and was A in the example we worked through on pages 234–237); therefore that coefficient must be nonzero. Of course, we have no reason a priori to expect $h_{11} \neq 0$. Even in the simple example

$$Q(\Delta x_1, \Delta x_2) = 2\,\Delta x_1\,\Delta x_2,$$

Morse's lemma, part 2: diagonalizing the quadratic form

The leading coefficient

the leading coefficient is zero ($h_{11} = h_{22} = 0$, $h_{12} = h_{21} = 1$). In this case, though, we can fix the problem with an obvious coordinate change:

$$\Delta x_1 = \Delta y_1 - \Delta y_2, \quad \Delta x_2 = \Delta y_1 + \Delta y_2.$$

Then

$$Q(\Delta x_1, \Delta x_2) = 2\,\Delta x_1\,\Delta x_2 = (\Delta y_1)^2 - (\Delta y_2)^2 = Q^*(\Delta y_1, \Delta y_2),$$

Making the leading coefficient nonzero

so the coefficients of the form Q^* that results from the coordinate change are $h_{11}^* = 1$, $h_{12}^* = h_{21}^* = 0$, $h_{22}^* = -1$. In fact, the following lemma says we can always make the leading coefficient nonzero, at least if the form is nondegenerate. The lemma concerns a quadratic form with variable coefficients

$$Q(\Delta \mathbf{x}) = \sum_{i,j=1}^{n} h_{ij}(\Delta \mathbf{x})\,\Delta x_i\,\Delta x_j, \quad h_{ji}(\Delta \mathbf{x}) = h_{ij}(\Delta \mathbf{x}).$$

Lemma 7.4. *Suppose the matrix $h_{ij}(\mathbf{0})$ is invertible. Then there is a linear coordinate change $\Delta \mathbf{x} = \mathbf{L}(\Delta \mathbf{y})$ for which*

$$Q(\mathbf{L}(\Delta \mathbf{y})) = Q^*(\Delta \mathbf{y}) = \sum_{i,j=1}^{n} h_{ij}^*(\Delta \mathbf{y})\,\Delta y_i\,\Delta y_j,$$

with $h_{11}^*(\mathbf{0}) \neq 0$.

Proof. Let $H(\Delta \mathbf{x})$ be the symmetric matrix with entry $h_{ij}(\Delta \mathbf{x})$ in the ith row and jth column. Suppose first that one of the diagonal elements of $H(\mathbf{0})$ is nonzero, say $h_{JJ}(\mathbf{0}) \neq 0$. Define

$$\mathbf{L}: \Delta x_1 = \Delta y_J, \quad \Delta x_J = \Delta y_1, \quad \Delta x_k = \Delta y_k, \quad k \neq 1, J.$$

This is a transposition permutation and is its own inverse (and if $J = 1$, it is the identity). In terms of the new variables, $h_{11}^*(\Delta \mathbf{y}) = h_{JJ}(\Delta \mathbf{x})$, so $h_{11}^*(\mathbf{0}) \neq 0$, and we are done.

The alternative is that all diagonal elements of $H(\mathbf{0})$ are zero. In that case, some other element $h_{1j}(\mathbf{0})$, $j = 2, \ldots, n$ in the first row of $H(\mathbf{0})$ must be nonzero. Otherwise, $\det H(\mathbf{0}) = 0$, which is contrary to hypothesis. So suppose $h_{1J}(\mathbf{0}) = h_{J1}(\mathbf{0}) \neq 0$, where $J \neq 1$. Define

$$\mathbf{L}: \Delta x_1 = \Delta y_1 - \Delta y_J, \quad \Delta x_J = \Delta y_1 + \Delta y_J, \quad \Delta x_k = \Delta y_k, \quad k \neq 1, J.$$

This $\mathbf{L}$ is also invertible; in fact, it is a rotation–dilation of the $(\Delta x_1, \Delta x_J)$-plane. To determine $h_{11}^*(\Delta \mathbf{y})$, we need to determine all places where $(\Delta y_1)^2$ appears as a quadratic factor in the form $Q^*(\Delta \mathbf{y}) = Q(\Delta \mathbf{x})$. There are three such places: in $(\Delta x_1)^2$, in $(\Delta x_J)^2$, and in $\Delta x_1 \Delta x_J$. We find

$$h_{11}^*(\Delta \mathbf{y}) = h_{11}(\Delta \mathbf{x}) + h_{JJ}(\Delta \mathbf{x}) + 2h_{1J}(\Delta \mathbf{x}),$$

and thus

$$h_{11}^*(\mathbf{0}) = h_{11}(\mathbf{0}) + h_{JJ}(\mathbf{0}) + 2h_{1J}(\mathbf{0}).$$

The first two terms are diagonal elements of $H(\mathbf{0})$ and hence, by assumption, are zero; the remaining term gives $h_{11}^*(\mathbf{0}) = 2h_{1J}(\mathbf{0}) \neq 0$. $\qquad\square$

The following theorem carries out the diagonalization process that gives Δz as a sum of positive and negative squares. It is the heart of Morse's lemma but does not complete the proof because it does not determine how many of the squares are positive and how many are negative.

Reducing the function
to a sum of squares

Theorem 7.19. *Suppose* $z = f(\mathbf{x})$ *has continuous third derivatives on an open set* X^n, *and the point* $\mathbf{a}$ *in* X^n *is a nondegenerate critical point of* f. *Then, in a sufficiently small window* $W_{\mathbf{a}}$ *centered at* $\mathbf{a}$, *there is a coordinate change* $\Delta\mathbf{u} = \mathbf{h}(\Delta\mathbf{x})$ *so that*

$$\Delta z = f(\mathbf{a} + \Delta\mathbf{x}) - f(\mathbf{a}) = \pm(\Delta u_1)^2 \pm \cdots \pm (\Delta u_n)^2.$$

Proof. We can assume by Theorem 7.18 (p. 249) that we have already written Δz as a quadratic form with variable coefficients:

$$\Delta z = \sum_{i,j=1}^{n} h_{ij}(\Delta\mathbf{x})\,\Delta x_i \Delta x_j,$$

where $h_{ji}(\Delta\mathbf{x}) = h_{ij}(\Delta\mathbf{x})$ and

$$h_{ij}(\mathbf{0}) = \frac{1}{2}\frac{\partial^2 f}{\partial x_i \partial x_j}(\mathbf{a}).$$

Moreover, because f has continuous third derivatives, the same theorem tells us that the coefficients $h_{ij}(\Delta\mathbf{x})$ have continuous first derivatives.

This proof also goes in stages; at each stage, a coordinate change "splits off" one more variable as a perfect square. In other words, we claim that after k stages, the window equation will look like

Proof by induction

$$\Delta z = \pm(\Delta v_1)^2 \pm \cdots \pm (\Delta v_k)^2 + \sum_{i,j=k+1}^{n} h_{ij}^*(\Delta\mathbf{v})\,\Delta v_i \Delta v_j,$$

and that the new coefficients h_{ij}^* are rational functions of the coefficients from the previous stage. Because each stage is like every other, the proof is a mathematical induction. Thus, we assume that we have already reached the stage where $k = M - 1$ squares have been "split off" from the quadratic form, and deduce that the next stage, $k = M$, also holds. (The initial step in the induction is just the one where $M = 1$, so we do not need to prove it separately.)

Thus we focus on the residual quadratic form

$$Q_M(\Delta\mathbf{v}) = \sum_{i,j=M}^{n} h_{ij}^*(\Delta\mathbf{v})\,\Delta v_i \Delta v_j.$$

Note that this is a quadratic form in just the variables $\Delta v_M, \ldots, \Delta v_n$, although the coefficients h_{ij}^* remain functions of all the variables $\Delta\mathbf{v} = (\Delta v_1, \ldots, \Delta v_n)$. The leading coefficient is h_{MM}^*, and we can assume (by Lemma 7.4) that $h_{MM}^*(\mathbf{0}) \neq 0$. If we separate out all appearances of Δv_M as a quadratic factor, we get

$$Q_M(\Delta\mathbf{v}) = h_{MM}^*(\Delta\mathbf{v}) \left((\Delta v_M)^2 + 2\Delta v_M \sum_{j=M+1}^{n} \frac{h_{Mj}^*(\Delta\mathbf{v})}{h_{MM}^*(\Delta\mathbf{v})} \Delta v_j \right)$$
$$+ \sum_{i,j=M+1}^{n} h_{ij}^*(\Delta\mathbf{v}) \Delta v_i \Delta v_j.$$

Completing the square   Completing the square then gives us

$$Q_M(\Delta\mathbf{v}) = h_{MM}^*(\Delta\mathbf{v}) \left(\Delta v_M + \sum_{j=M+1}^{n} \frac{h_{Mj}^*(\Delta\mathbf{v})}{h_{MM}^*(\Delta\mathbf{v})} \Delta v_j \right)^2$$
$$- h_{MM}^*(\Delta\mathbf{v}) \left(\sum_{j=M+1}^{n} \frac{h_{Mj}^*(\Delta\mathbf{v})}{h_{MM}^*(\Delta\mathbf{v})} \Delta v_j \right)^2 + \sum_{i,j=M+1}^{n} h_{ij}^*(\Delta\mathbf{v}) \Delta v_i \Delta v_j.$$

Together, the terms on the second line constitute a new quadratic form in the variables $\Delta v_{M+1}, \ldots, \Delta v_n$ alone, but with variable coefficients that still depend on all the Δv_i, in general. We write that new form as

$$Q_{M+1}(\Delta\mathbf{v}) = \sum_{i,j=M+1}^{n} \widehat{h}_{ij}(\Delta\mathbf{v}) \Delta v_i \Delta v_j.$$

The formulas show that the new coefficients $\widehat{h}_{ij}$ are rational functions of the h_{ij}^* (in which only h_{MM}^* appears in the denominator). Therefore, the new $\widehat{h}_{ij}$ have continuous first derivatives wherever the denominator $h_{MM}^*(\Delta\mathbf{v})$ does not vanish. This confirms the assertion about the coefficients that is part of the induction.

The new residual form   Completion of the square leads us to the coordinate change

$$\mathbf{h}_M : \begin{cases} \Delta w_M = \sqrt{|h_{MM}^*(\Delta\mathbf{v})|} \left(\Delta v_M + \displaystyle\sum_{j=M+1}^{n} \frac{h_{Mj}^*(\Delta\mathbf{v})}{h_{MM}^*(\Delta\mathbf{v})} \Delta v_j \right), \\ \Delta w_i = \Delta v_i, \qquad i \neq M, \end{cases}$$

that transforms Q_M into

$$Q_M(\Delta\mathbf{v}) = \pm(\Delta w_M)^2 + Q_{M+1}(\Delta\mathbf{w}).$$

The new coordinates split off one more variable as a perfect square, and leave a new residue Q_{M+1} that is again a quadratic form, but with one less quadratic variable. The coefficients of the new residual form are continuously differentiable functions of the new coordinates.

The coordinate change   The induction is therefore completed as soon as we prove that the map $\mathbf{h}_M$ is a valid coordinate change. We use the inverse function theorem. First, note that the components of $\mathbf{h}_M$ have continuous first derivatives near the origin, because they are rational functions of $\sqrt{|h^*_{MM}(\Delta\mathbf{v})|}$ and $h^*_{Mj}(\Delta\mathbf{v})$, $j = M+1,\ldots,n$. Next,

$$d(\mathbf{h}_M)_0 = \begin{pmatrix} 1 & \cdots & 0 & 0 & 0 & \cdots & 0 \\ \vdots & \ddots & \vdots & \vdots & \vdots & \ddots & \vdots \\ 0 & \cdots & 1 & 0 & 0 & \cdots & 0 \\ 0 & \cdots & 0 & \sqrt{|h^*_{MM}(0)|} & \dfrac{h^*_{M,M+1}(0)}{\sqrt{|h^*_{MM}(0)|}} & \cdots & \dfrac{h^*_{M,n}(0)}{\sqrt{|h^*_{MM}(0)|}} \\ 0 & \cdots & 0 & 0 & 1 & \cdots & 0 \\ \vdots & \ddots & \vdots & \vdots & \vdots & \ddots & \vdots \\ 0 & \cdots & 0 & 0 & 0 & \cdots & 1 \end{pmatrix},$$

so $\det d(\mathbf{h}_M)_0 = \sqrt{|h^*_{MM}(0)|} \neq 0$ and the linear map $d(\mathbf{h}_M)_0$ is invertible. The inverse function theorem then implies that $\mathbf{h}_M$ itself is invertible on some window W_M centered at $\mathbf{a}$, and the inverse is continuously differentiable on the image $U_M = \mathbf{h}_M(W_M)$.

The coordinate change $\Delta\mathbf{u} = \mathbf{h}(\Delta\mathbf{x})$ that will carry out the entire diagonalization is the composite $\mathbf{h} = \mathbf{h}_n \circ \cdots \circ \mathbf{h}_1$ that carries out the individual changes, one after another. The proof makes it reasonably clear that the composite is well defined, that is, that we can always carry out the next coordinate change in the sequence. Alternatively, note that each successive pair $\mathbf{h}_{i+1} \circ \mathbf{h}_i$ of changes is defined on the open set $U_i \cap W_{i+1}$, which is certainly nonempty because it contains $\mathbf{0} = \mathbf{h}_i(\mathbf{0})$. Finally, by the chain rule, the composite $\mathbf{h}$ is continuously differentiable. □

We now come to the final part of the proof of Morse's lemma, where we show that the number of negative squares in the new formula

$$\Delta z = \pm(\Delta u_1)^2 \pm \cdots \pm (\Delta u_n)^2$$

for f is equal to the number of negative eigenvalues in the Hessian of f at $\mathbf{x} = \mathbf{a}$. In particular, it follows that this number does not depend on the choices we made in constructing the coordinate change $\Delta\mathbf{u} = \mathbf{h}(\Delta\mathbf{x})$.

In terms of the new coordinates, Δz is a particularly simple quadratic form. For clarity, let us assume there are s negative squares and the coordinates have been rearranged so all the negative squares come first in the new formula; then

$$\Delta z = Q_K(\Delta\mathbf{u}) = \Delta\mathbf{u}^\dagger K \Delta\mathbf{u},$$

where the symmetric matrix K representing the form is

$$K = \begin{pmatrix} -I_s & \\ & I_{n-s} \end{pmatrix},$$

and the off-diagonal entries are all zero. We can also write Δz as the Taylor expansion

$$\Delta z = \tfrac{1}{2} H_{\mathbf{a}}(\Delta\mathbf{x}) + O((\Delta\mathbf{x})^3) = \tfrac{1}{2}\Delta\mathbf{x}^\dagger H_{\mathbf{a}}\Delta\mathbf{x} + O((\Delta\mathbf{x})^3),$$

Morse's lemma, part 3: role of the Hessian

in which $H_\mathbf{a}$ is the Hessian of f at the critical point in terms of the original $\mathbf{x}$ coordinates. The next theorem provides the first link between the matrices K and $H_\mathbf{a}$.

Theorem 7.20. *Let $\Delta\mathbf{u} = \mathbf{h}(\Delta\mathbf{x})$ be the coordinate change of Theorem 7.19, and let $L = \mathbf{dh}_0$ be the derivative of $\mathbf{h}$ at $\Delta\mathbf{x} = \mathbf{0}$; then*

$$L^\dagger K L = \tfrac{1}{2} H_\mathbf{a}.$$

Proof. The proof is left as an exercise; it is similar to the earlier proof (Theorem 7.15, p. 247) that connects the Hessians of equivalent functions at corresponding critical points.   □

Corollary 7.21 *The matrices $\tfrac{1}{2}H_\mathbf{a}$ and K represent the same quadratic form in the coordinates $\Delta\mathbf{x}$ and $\Delta\mathbf{u}$, respectively.*

Proof. Let $L = \mathbf{dh}_0$, as in Theorem 7.20; then the linear coordinate change $\Delta\mathbf{u} = L\Delta\mathbf{x}$ converts $Q_K(\Delta\mathbf{u}) = \Delta\mathbf{u}^\dagger K \Delta\mathbf{u}$ into

$$\widehat{Q}_K(\Delta\mathbf{x}) = Q_K(L\Delta\mathbf{x}) = (L\Delta\mathbf{x})^\dagger K(L\Delta\mathbf{x}) = \mathbf{x}^\dagger L^\dagger K L \Delta\mathbf{x} = \tfrac{1}{2}\Delta\mathbf{x}^\dagger H_\mathbf{a} \Delta\mathbf{x}.$$ □

Different coordinate representations of a quadratic form

The corollary leads us to regard a quadratic form as a fixed geometric object that has different representations in different coordinate systems. In other words, if we give a "geometric" vector $\mathbf{v}$ coordinates in two different ways,

$$\Delta\mathbf{x} \longleftrightarrow \mathbf{v} \longleftrightarrow \Delta\mathbf{u},$$

then there is a function $\mathbf{Q}$ (the "geometric" quadratic form) defined on such vectors for which

$$\widehat{Q}_K(\Delta\mathbf{x}) = \mathbf{Q}(\mathbf{v}) = Q_K(\Delta\mathbf{u}).$$

What properties do Q_K and $\widehat{Q}_K$ have in common? These are the geometric properties of the underlying function $\mathbf{Q}$.

Definition 7.9 *We say the quadratic form Q is **negative definite** (respectively, **positive definite**) on a set S in $\mathbb{R}^n$ if $Q(\mathbf{v}) < 0$ (respectively, $Q(\mathbf{v}) > 0$) for every $\mathbf{v} \neq \mathbf{0}$ in S.*

Definition 7.10 *The **index** of the quadratic form Q is the maximum dimension of a subspace N of $\mathbb{R}^n$ on which Q is negative definite.*

The index is a geometric property

The index of a quadratic form is defined without reference to its representation in a particular coordinate system, so it is a geometric property of the form.

Theorem 7.22. *The index of the quadratic form*

$$Q_K(\Delta\mathbf{u}) = -(\Delta u_1)^2 - \cdots - (\Delta u_s)^2 + (\Delta u_{s+1})^2 + \cdots (\Delta u_n)^2$$

is equal to s.

Proof. Let N^s be the s-dimensional linear subspace of vectors of the form

$$\Delta \mathbf{u} = (\Delta u_1, \dots, \Delta u_s, 0, \dots, 0).$$

Then

$$Q_K(\Delta \mathbf{u}) = -(\Delta u_1)^2 - \cdots - (\Delta u_s)^2 < 0$$

for every nonzero $\Delta \mathbf{u}$ in N^s, implying that the index of Q_K is at least equal to s.

We are done if we show that the index of Q_K is at most equal to s. First note that, by a similar argument, Q_K is positive definite on the $(n-s)$-dimensional linear subspace P^{n-s} of vectors of the form

$$\Delta \mathbf{u} = (0, \dots, 0, \Delta u_{s+1}, \dots, \Delta u_n).$$

Now suppose the index of Q_K were greater than s. Then there would be a linear subspace N^{s+1} of dimension $s+1$ on which Q_K were negative definite. But then the intersection $P^{n-s} \cap N^{s+1}$ would be a linear subspace of dimension at least 1, and would thus contain nonzero vectors. The value of Q_K on such a vector would be both positive and negative, an absurdity we attribute to the assumption that the index could be greater than s. We reject the assumption and conclude that the index is exactly s. $\qquad \square$

By definition, the index of a quadratic form is independent of the coordinate representation. Therefore, because $\widehat{Q}_K$ and Q_K represent the same form in different coordinates, $\widehat{Q}_K$ and Q_K must have the same index. Consequently, the Hessian form

$$\widehat{Q}_K(\Delta \mathbf{x}) = \tfrac{1}{2} \Delta \mathbf{x}^\dagger H_{\mathbf{a}} \Delta \mathbf{x}$$

Q_K and $\widehat{Q}_K$ both have index r

must have index s. It remains to show that s equals the number of negative eigenvalues of $H_{\mathbf{a}}$. This involves "transforming $H_{\mathbf{a}}$ to principal axes" (cf. p. 242): using an n-dimensional rotation to reduce the Hessian form $\widehat{Q}_K$ to a sum of squares in which the eigenvalues of $H_{\mathbf{a}}$ appear as coefficients.

Definition 7.11 *An $n \times n$ invertible matrix P is **orthogonal** if its transpose equals its inverse: $P^\dagger = P^{-1}$.*

An orthogonal matrix gets its name from the fact that its columns are mutually orthogonal unit vectors. That is, if we write

$$P^\dagger = \begin{pmatrix} \mathbf{w}_1^\dagger \\ \vdots \\ \mathbf{w}_n^\dagger \end{pmatrix}, \quad P = \begin{pmatrix} \mathbf{w}_1 & \cdots & \mathbf{w}_n \end{pmatrix},$$

then the condition $P^\dagger P = I$ implies $\|\mathbf{w}_i\|^2 = \mathbf{w}_i^\dagger \mathbf{w}_i = 1$ for every $i = 1, \dots, n$, and $\mathbf{w}_i^\dagger \mathbf{w}_j = 0$ for every $i \neq j$.

Let $\mathbf{e}_1 \wedge \cdots \wedge \mathbf{e}_n$ be the unit n-cube whose edges are the standard basis vectors in $\mathbb{R}^n$. Then

$$P(\mathbf{e}_1 \wedge \cdots \wedge \mathbf{e}_n) = P(\mathbf{e}_1) \wedge \cdots \wedge P(\mathbf{e}_n) = \mathbf{w}_1 \wedge \cdots \wedge \mathbf{w}_n,$$

so (Definition 2.5, p. 46) $\mathrm{vol}\, \mathbf{w}_1 \wedge \cdots \wedge \mathbf{w}_n = \det P = \pm 1$ for every orthogonal matrix P.

Definition 7.12 *An orthogonal matrix P is a **rotation** if $\det P = +1$.*

Rotations and
orthogonal matrices

Thus, a rotation is an orthogonal matrix that preserves orientation. If P is orthogonal but $\det P = -1$, P can be converted into a rotation by changing the signs of the entries in any one of its columns.

Theorem 7.23 (Principal axes theorem). *If $Q(\mathbf{x}) = \mathbf{x}^\dagger M \mathbf{x}$ is a quadratic form in n variables, then there is a rotation $\mathbf{x} = R\mathbf{u}$ of $\mathbb{R}^n$ that transforms Q into a sum of squares*

$$Q(R\mathbf{u}) = Q^*(\mathbf{u}) = \lambda_1 u_1^2 + \cdots + \lambda_n u_n^2,$$

where $\lambda_1, \ldots, \lambda_n$ are the eigenvalues of M.

The theorem stated
in terms of matrices

Proof. The theorem asserts that, for any $n \times n$ symmetric matrix M, there is a rotation R for which

$$R^{-1}MR = R^\dagger MR = D$$

is a diagonal matrix whose diagonal elements are the eigenvalues of M. We prove the theorem in this form by using mathematical induction on n.

If $n = 1$ there is nothing to do; we can take R to be the 1×1 identity matrix. Now assume that any $(n-1) \times (n-1)$ symmetric matrix can be diagonalized by a suitable rotation on $\mathbb{R}^{n-1}$, and consider an $n \times n$ symmetric matrix M.

Let $\mathbf{u}_1$ be an eigenvector of M, $M\mathbf{u}_1 = \lambda_1 \mathbf{u}_1$, and take $\mathbf{u}_1$ to be a unit vector. Extend $\mathbf{u}_1$ to an orthonormal basis $\{\mathbf{u}_1, \mathbf{w}_1, \ldots, \mathbf{w}_{n-1}\}$ of $\mathbb{R}^n$. We may assume (by changing the sign of $\mathbf{w}_{n-1}$, if necessary) that the n-cube $\mathbf{u}_1 \wedge \mathbf{w}_1 \wedge \cdots \wedge \mathbf{w}_{n-1}$ has positive orientation, and thus that the matrix

$$R_1 = \begin{pmatrix} \mathbf{u}_1 & \mathbf{w}_1 & \cdots & \mathbf{w}_{n-1} \end{pmatrix}$$

is a rotation. Then

$$MR_1 = \begin{pmatrix} M\mathbf{u}_1 & M\mathbf{w}_1 & \cdots & M\mathbf{w}_{n-1} \end{pmatrix} = \begin{pmatrix} \lambda_1 \mathbf{u}_1 & M\mathbf{w}_1 & \cdots & M\mathbf{w}_{n-1} \end{pmatrix},$$

$$R_1^\dagger MR_1 = \begin{pmatrix} \lambda_1 \mathbf{u}_1^\dagger \mathbf{u}_1 & \mathbf{u}_1^\dagger M\mathbf{w}_1 & \cdots & \mathbf{u}_1^\dagger M\mathbf{w}_{n-1} \\ \lambda_1 \mathbf{w}_1^\dagger \mathbf{u}_1 & \mathbf{w}_1^\dagger M\mathbf{w}_1 & \cdots & \mathbf{w}_1^\dagger M\mathbf{w}_{n-1} \\ \vdots & \vdots & \ddots & \vdots \\ \lambda_1 \mathbf{w}_{n-1}^\dagger \mathbf{u}_1 & \mathbf{w}_{n-1}^\dagger M\mathbf{w}_1 & \cdots & \mathbf{w}_{n-1}^\dagger M\mathbf{w}_{n-1} \end{pmatrix}$$

$$= \begin{pmatrix} \lambda_1 & 0 & \cdots & 0 \\ 0 & m_{11}^* & \cdots & m_{1,n-1}^* \\ \vdots & \vdots & \ddots & \vdots \\ 0 & m_{n-1,1}^* & \cdots & m_{n-1,n-1}^* \end{pmatrix}, \quad m_{ij}^* = \mathbf{w}_i^\dagger M\mathbf{w}_j.$$

The zeros appear in the first column because every $\mathbf{w}_i \perp \mathbf{u}_1$, and in the first row because $R_1^\dagger M R_1$ is symmetric. Also,

$$m_{ij}^* = \mathbf{w}_i^\dagger M \mathbf{w}_j = (\mathbf{w}_i^\dagger M \mathbf{w}_j)^\dagger = \mathbf{w}_j^\dagger M \mathbf{w}_i = m_{ji}^*,$$

so $M^* = (m_{ij}^*)$ is an $(n-1) \times (n-1)$ symmetric matrix.

By the induction hypothesis, there is a rotation R^* that diagonalizes M^*; that is, $(R^*)^\dagger M^* R^* = D^*$. Let

$$R_2 = \begin{pmatrix} 1 & \\ & R^* \end{pmatrix};$$

this is the $n \times n$ matrix with 1 and R^* on the diagonal, and with all off-diagonal elements not shown equal to zero. Then

$$R_2^\dagger = \begin{pmatrix} 1 & \\ & (R^*)^\dagger \end{pmatrix} \quad \text{and} \quad R_2^\dagger R_2 = \begin{pmatrix} 1 & \\ & (R^*)^\dagger R^* \end{pmatrix} = \begin{pmatrix} 1 & \\ & I_{n-1} \end{pmatrix} = I_n,$$

the $n \times n$ identity matrix, thus R_2 is orthogonal. Moreover, $\det R_2 = 1 \times \det R^* = 1$, so R_2 is a rotation.

Lemma 7.5. *The matrix $R = R_2 R_1$ is a rotation and diagonalizes M.*

Proof. We know R is a rotation because it is a product of rotations. Moreover,

$$R^\dagger M R = R_2^\dagger R_1^\dagger M R_1 R_2 = R_2^\dagger \begin{pmatrix} \lambda_1 & \\ & M^* \end{pmatrix} R_2 = \begin{pmatrix} 1 & \\ & (R^*)^\dagger \end{pmatrix} \begin{pmatrix} \lambda_1 & \\ & M^* \end{pmatrix} \begin{pmatrix} 1 & \\ & R^* \end{pmatrix}$$

$$= \begin{pmatrix} \lambda_1 & \\ & (R^*)^\dagger M^* R^* \end{pmatrix} = \begin{pmatrix} \lambda_1 & \\ & D^* \end{pmatrix};$$

this is an $n \times n$ diagonal matrix. $\qquad\square$

Lemma 7.6. *If M is symmetric, P is orthogonal, and $P^\dagger M P$ is a diagonal matrix with diagonal elements α_i, $i = 1, \ldots, n$, then α_i is an eigenvalue of M and the i-th column of P is a corresponding eigenvector.*

Proof. This is the n-dimensional version of Theorem 7.7, page 241, and has a similar proof. The key is that $P^\dagger M P = D$ implies $MP = PD$ because $P^\dagger = P^{-1}$; see the exercises. $\qquad\square$

Thus the diagonal elements in the diagonal matrix of Lemma 7.5 are the eigenvalues of M, and the proof of the principal axes theorem is complete. $\qquad\square$

Our proof of the principal axes theorem indicates that the eigenvectors of a symmetric matrix have properties not shared by matrices in general. Before returning to the analysis of critical points, we pause to establish some of those properties.

Interlude

Definition 7.13 *The eigenvalue α of the matrix M has **multiplicity k** if the the factor $\lambda - \alpha$ appears k times in a factorization of the characteristic polynomial of M.*

Although each distinct real eigenvalue of an $n \times n$ real matrix M has a real eigen-vector associated with it (Theorem 7.13, p. 246), in general a repeated eigenvalue may not possess additional linearly independent eigenvectors. Consequently, the eigenvectors of such an $n \times n$ matrix may not span $\mathbb{R}^n$. The first such examples we saw were the shears M_7 and M_8 of Chapter 2 (pp. 38ff.). However, a symmetric matrix has a "complete" set of eigenvectors.

Corollary 7.24 *Suppose M is a symmetric matrix with an eigenvalue α of multiplicity k. Then the eigenvectors associated with α form a k-dimensional subspace E_α of $\mathbb{R}^n$.*

Proof. Sums and scalar multiples of eigenvectors associated with α are again eigen-vectors associated with α, so they form a subspace E_α of $\mathbb{R}^n$. Because α has multiplicity k, precisely k columns of the orthogonal matrix P of Lemma 7.6 are eigenvectors associated with α. Those columns are linearly independent, so the dimension of E_α is at least k.

We must now show the dimension of E_α is not greater than k. Let $\mathbf{v}_1, \ldots, \mathbf{v}_n$ be the columns of P; assume, by rearranging them if necessary, that the first k columns, $\mathbf{v}_1, \ldots, \mathbf{v}_k$, are the eigenvectors associated with α. Suppose $\mathbf{w}$ is in E_α; because $\{\mathbf{v}_1, \ldots, \mathbf{v}_n\}$ is an orthonormal basis, we can write

$$\mathbf{w} = v_1 \mathbf{v}_1 + \cdots + v_n \mathbf{v}_n,$$

where $v_j = \mathbf{w}^\dagger \mathbf{v}_j = \mathbf{v}_j^\dagger \mathbf{w}$ by orthonormality of the basis. Let α_j be the eigenvector associated with $\mathbf{v}_j$; then $\alpha_j = \alpha$ if and only if $j = 1, \ldots, k$. We have

$$\alpha_j v_j = \alpha_j \mathbf{w}^\dagger \mathbf{v}_j = \mathbf{w}^\dagger M \mathbf{v}_j = (\mathbf{w}^\dagger M \mathbf{v}_j)^\dagger = \mathbf{v}_j^\dagger M \mathbf{w} = \alpha \mathbf{v}_j^\dagger \mathbf{w} = \alpha v_j;$$

$M\mathbf{w} = \alpha \mathbf{w}$ because $\mathbf{w}$ is in E_α. Thus $(\alpha_j - \alpha)v_j = 0$. This forces $v_j = 0$ for $j > k$, implying that $\{\mathbf{v}_1, \ldots, \mathbf{v}_k\}$ spans E_α. Hence $\dim E_\alpha = k$.   □

The subspace E_α is sometimes called the **eigenspace** associated with α. Thus, for a *symmetric* matrix, the dimension of the eigenspace associated with an eigenvalue equals the multiplicity of that eigenvalue.

To complete the third part of the proof of Morse's lemma, we must show that, whenever a coordinate change reduces Δz to a sum of squares, the number of negative squares in that sum is always equal to the number of negative eigenvalues of the Hessian matrix $H_\mathbf{a}$. For a quadratic form with constant coefficients under a linear coordinate change, the invariance of the number of negative squares and positive squares was shown by J. J. Sylvester in 1852 [18]. He characterized the result as "... a law to which my view of the physical meaning of quantity of matter inclines me, upon the ground of analogy, to give the name of the Law of Inertia for Quadratic Forms, as expressing the fact of the existence of an invariable number inseparably attached to such forms."

Sometimes, to underscore the invariant nature of the index of a quadratic form, we add Sylvester's term and call it the **index of inertia**.

Theorem 7.25. *Suppose $z = f(\mathbf{x})$ has continuous third derivatives on an open set X^n, $\mathbf{x} = \mathbf{a}$ is a nondegenerate critical point of f in X^n, and the Hessian matrix $H_\mathbf{a}$ of f at $\mathbf{a}$ has r negative eigenvalues. If $\Delta\mathbf{u} = \mathbf{h}(\Delta\mathbf{x})$ is a coordinate change in a window centered at $\mathbf{a}$ that reduces Δz to a sum of squares,*

$$\Delta z = -(\Delta u_1)^2 - \cdots - (\Delta u_s)^2 + (\Delta u_{s+1})^2 + \cdots + (\Delta u_n)^2,$$

then $s = r$.

Proof. By Theorem 7.20 (p. 258), the linear map $\Delta\mathbf{u} = \mathbf{dh_0}(\Delta\mathbf{x})$ converts the quadratic form

$$\Delta z = Q(\Delta\mathbf{u}) = -(\Delta u_1)^2 - \cdots - (\Delta u_s)^2 + (\Delta u_{s+1})^2 + \cdots + (\Delta u_n)^2$$

into

$$\Delta z = \widehat{Q}(\Delta\mathbf{x}) = \tfrac{1}{2}\Delta\mathbf{x}^\dagger H_\mathbf{a}\,\Delta\mathbf{x}.$$

Therefore, Q and $\widehat{Q}$ are just different coordinate representations of the same (geometric) quadratic form $\mathbf{Q}$, and must therefore have the same index. By Theorem 7.22, the index of Q is s. By transforming $\widehat{Q}$ to principal axes (Theorem 7.23), we see the index of $\widehat{Q}$ is r. Thus $s = r$. $\square$

This completes the third part of the proof of Morse's lemma, and thus completes the entire proof. $\square$

Proof is complete

One of the consequences of Morse's lemma is that the second derivatives of a function at a nondegenerate critical point determine the type of that point. In fact, the type is completely characterized by a single number: the *index of inertia* of its Hessian form. This leads to the following definition and theorem.

Definition 7.14 *The **index**, or **index of inertia**, of a nondegenerate critical point of a function is the index of its Hessian, that is, the number of negative eigenvalues of the Hessian matrix at that point.*

Index of a critical point

Theorem 7.26 (Second derivative test). *Suppose $\mathbf{x} = \mathbf{a}$ is a nondegenerate critical point of a function $z = f(\mathbf{x})$ that possesses continuous third derivatives. If r is the index of $\mathbf{a}$, then*

- $\mathbf{a}$ *is a **local minimum** if $r = 0$.*

- $\mathbf{a}$ *is a **local maximum** if $r = n$.*

- $\mathbf{a}$ *is a **saddle** if $0 < r < n$.* $\square$

Morse's lemma and the second derivative test classify nondegenerate critical points: there are $n + 1$ classes, one for each possible index. Two critical points are in the same class if a coordinate change will transform one into the other. For degenerate critical points, the situation is very different. There are infinitely many classes, and no complete classification exists, although there are partial results. The analysis of (degenerate) critical points is part of the larger study of *singularities of mappings*, an active area of current research.

Classifying critical points

One useful observation we can make is that a nonisolated critical point—for example, a point on the ring of minima of the wine bottle—is necessarily degenerate. The proof (by Morse) is a nice application of the inverse function theorem.

Theorem 7.27. *Suppose* $\mathbf{x} = \mathbf{a}$ *is a nondegenerate critical point of a function* $z = f(\mathbf{x})$ *that has continuous second derivatives in some neighborhood* X^n *of* $\mathbf{a}$. *Then* $\mathbf{a}$ *is isolated, in the sense that there is some nonempty open ball* B_ε *centered at* $\mathbf{a}$ *that contains no other critical point of* f.

Proof. The gradient of f defines a map $\nabla f : X^n \to \mathbb{R}^n$,

$$\nabla f : \begin{cases} u_1 = \dfrac{\partial f}{\partial x_1}(\mathbf{x}), \\ \quad\vdots \\ u_n = \dfrac{\partial f}{\partial x_n}(\mathbf{x}), \end{cases}$$

that is continuously differentiable, because f is twice continuously differentiable.

By construction, a point $\mathbf{b}$ is a critical point of f if and only if $\nabla f(\mathbf{b}) = \mathbf{0}$; in particular, $\nabla f(\mathbf{a}) = \mathbf{0}$. Furthermore, the matrix of the derivative $d(\nabla f)_\mathbf{a}$ coincides with the Hessian matrix $H_\mathbf{a}$, so the nondegeneracy of $\mathbf{a}$ implies that $d(\nabla f)_\mathbf{a}$ is invertible. The inverse function theorem then implies that the map ∇f itself is invertible on some open ball B_ε centered at $\mathbf{a}$. In particular, ∇f is 1–1 there, so no point $\mathbf{b} \neq \mathbf{a}$ is mapped to $\mathbf{0}$. That is, no point $\mathbf{b} \neq \mathbf{a}$ in B_ε is a critical point of f.   □

Earlier (see pp. 222–224), we observed that the value of the second derivative at a regular point of a single-variable function could be transformed into any new value whatsoever by a suitable coordinate change. At a critical point, this degree of volatility does not occur: the sign of the second derivative cannot be changed. In effect, the convexity of a function graph is a geometric invariant at a critical point but not at a regular point. There is a similar distinction between the regular and the critical points of a function of several variables. For suppose $\mathbf{x} = \mathbf{a}$ is a regular point of $z = f(\mathbf{x})$. By the implicit function theorem (in particular, Corollary 6.8, p. 198), local coordinates $(\Delta u_1, \ldots, \Delta u_n)$ can be chosen near $\mathbf{a}$ so that $\Delta z = \Delta u_n$. Thus, in terms of the new variables, the function is linear, and all of its second derivatives are identically zero. Whatever information we thought might be conveyed by the original derivatives $\partial^2 f / \partial x_i \partial x_j$ has vanished with the coordinate change.

By contrast, suppose $\mathbf{x} = \mathbf{a}$ is a critical point of $z = f(\mathbf{x})$. When $\mathbf{a}$ is nondegenerate, Morse's lemma tells us the index of inertia of the Hessian of f at $\mathbf{a}$ is a geometric invariant. That is, if $\mathbf{x} = \mathbf{h}(\mathbf{u})$ is a local coordinate change near $\mathbf{a} = \mathbf{h}(\mathbf{b})$ that transforms f into $g(\mathbf{u}) = f(\mathbf{h}(\mathbf{u})) = f(\mathbf{x})$, then

- $\mathbf{u} = \mathbf{b}$ is a nondegenerate critical point of $z = g(\mathbf{u})$.

- The index of inertia of the Hessian of g at $\mathbf{b}$ equals the index of inertia of f at $\mathbf{a}$.

If $\mathbf{a}$ is degenerate, then the rank of the Hessian is not maximal. In this case, though, an extension of our methods can show that the rank and the index of inertia are both geometric invariants.

Exercises

7.1. Suppose $y = f(x)$ has a continuous second derivative and $f'(0) \neq 0$.

    a. For any value of B, the function $x = h_B(u) = u + Bu^2$ is a coordinate change near the origin; explain why.

    b. Let $g(u) = f(h_B(u))$ represent f under the coordinate change. Show that B can be chosen so that $g''(0) = A$, where A is an arbitrary number. Write a formula that expresses how B depends on A.

7.2. Construct the symmetric matric that corresponds to each of the following quadratic forms.

    a. $Q(x,y) = 5x^2 + 18xy - 2y^2$.

    b. $Q(x,y) = xy - x^2 - y^2$

    c. $Q(x,y) = (2x - y)(2y - x)$;

    d. $Q(x,y) = \begin{pmatrix} x & y \end{pmatrix} \begin{pmatrix} 1 & 5 \\ -1 & -5 \end{pmatrix} \begin{pmatrix} x \\ y \end{pmatrix}$.

7.3. Show that, when m is small, the roots of $4x^3 - 4x + m = 0$ (cf. p. 232) are approximately $m/4$ and $\pm 1 - m/8$. Specifically, you can show that

$$\frac{\partial f_m}{\partial x}\left(\pm 1 - \frac{m}{8}, 0\right) = O(m^2), \qquad \frac{\partial f_m}{\partial x}\left(\frac{m}{4}, 0\right) = O(m^3).$$

7.4. Verify that the derivative $d\mathbf{h}_{(0,0)}$ of the coordinate change map given on page 234 has the form shown on page 235.

7.5. Construct the symmetric matric that corresponds to each of the following quadratic forms.

    a. $Q(x,y,z) = 10xy - 2yz + zx$.

    b. $Q(x,y,z) = (x - y + z)(x + y - z)$.

    c. $Q(x_1, \ldots, x_n) = \sum_{i=1}^{n} (i - 5)x_i^2$.

    d. $Q(x_1, \ldots, x_n) = \sum_{i=1}^{n} \sum_{j=1}^{n} (i + j)x_i x_j$.

    e. $Q(x_1, \ldots, x_n) = \sum_{i=1}^{n} \sum_{j=1}^{n} (i - j)x_i x_j$.

7.6. Let $f(x,y) = 3x^2 - x^3 - y^2$.

    a. Verify that $(0,0)$ and $(2,0)$ are critical points of f.

    b. Find the second-order Taylor polynomial for f at $(2,0)$; call it $P(x,y)$. Graph together $f(x,y)$ and $P(x,y)$ on a small neighborhood of $(2,0)$; specifically, use $1.9 \leq x \leq 2.1$, $-0.1 \leq y \leq 0.1$.

    c. Does P have a critical point at $(2,0)$? What kind? Do the graphs show that P and f have the same type of critical point at $(2,0)$? What kind of critical point does f have at $(2,0)$?

    d. Find the second-order Taylor polynomial for f at $(0,0)$; call it $Q(x,y)$. Graph together $f(x,y)$ and $Q(x,y)$ on a small neighborhood of $(0,0)$; specifically, use $-0.1 \leq x \leq 0.1$, $-0.1 \leq y \leq 0.1$.

e. What kind of critical point does Q have at $(0,0)$. Do the graphs show that Q and f have the same type of critical point at $(0,0)$? What kind of critical point does f have at $(0,0)$?

7.7. a. Find all critical points of $f(x,y) = x^3 + y^3 - 3x - 12y$.

   b. At each critical point P of f, construct the second-order Taylor polynomial T_P of f. Does $T_P(x,y)$ also have a critical point at P? What kind?

   c. In a small neighborhood of each of the critical points P, sketch the graph of f together with the Taylor polynomial T_P. Does f resemble T_P near P? Is P the same type of critical point for f that it is for T_P?

   d. Conclusion: List the critical points of f, and indicate the type of each.

7.8. a. Find all critical points of $f(x,y) = x^3 - 3xy^2 - x^2 + 3y^2$.

   b. At each critical point P of f, construct the second-order Taylor polynomial T_P of f. Does $T_P(x,y)$ also have a critical point at P? What kind?

   c. In a small neighborhood of each of the critical points P, sketch the graph of f together with the Taylor polynomial T_P. Does f resemble T_P near P? Is P the same type of critical point for f that it is for T_P?

   d. Conclusion: List the critical points of f, and indicate the type of each.

7.9. Locate the critical point of $Q(x,y) = ax^2 + 2bxy + cy^2 + dx + ey + k$ and determine its type. On which of the six parameters does the location depend, and on which does the type depend?

7.10. Locate all the critical points of $\Phi(\theta, v) = 1 - \cos\theta + \frac{1}{2}v^2$, and determine the type of each.

7.11. Let $z = f(x,y) = p^2x^2 + q^2y^2$, $0 < p^2 < q^2$, and let $D_a(x,y)$ be the square of the distance from the point $(0,0,a)$ on the z-axis to the point $(x,y,f(x,y))$ on the graph of f.

   a. Make a sketch.

   b. Show that D_a has a critical point at the origin, for every a.

   c. For two values of a, that critical point is degenerate; determine those values.

   d. At all other points a, determine how the type of the critical point depends on a.

7.12. Show that the formula originally used for the quadratic terms in Taylor's expansion (see Theorem 3.18, p. 94, and the discussion leading up to it) gives the same value as the formula using the Hessian. That is, show

$$(\Delta\mathbf{u} \cdot \nabla)^2 f(\mathbf{a}) = (\Delta\mathbf{u})^\dagger H_\mathbf{a} \Delta\mathbf{u},$$

where ∇ is the gradient differential operator.

7.13. Let M be an $n \times n$ real symmetric matrix, and let λ be an eigenvalue of M. The purpose of this exercise is to prove that λ is real. So suppose the contrary; let $\lambda = a + bi$ (with $b \neq 0$), and let $Z = X + iY$ (where X and Y are real $n \times 1$ vectors) be a complex eigenvector for λ: $MZ = \lambda Z$ and $Z \neq 0$.

   a. Let $\overline{\lambda} = a - bi$ be the complex conjugate of λ, and let $\overline{Z} = X - iY$ be the complex conjugate of Z. Show that $\overline{\lambda}$ is also an eigenvalue of M (Hint: What is $\overline{M}$?) with eigenvector $\overline{Z}$.

   b. Show that the 1×1 matrix $\overline{Z}^{\dagger} MZ$ equals $\lambda \|Z\|^2$.

   c. The **conjugate transpose** of the $k \times l$ matrix A is the $l \times k$ matrix $\overline{A}^{\dagger}$. Show that the conjugate transpose of $\overline{Z}^{\dagger} MZ$ equals $\overline{\lambda} \|Z\|^2$.

   d. Compare $\overline{Z}^{\dagger} MZ$ and its conjugate transpose to conclude that $\overline{\lambda} = \lambda$, showing λ is real.

7.14. Let M be an $n \times n$ real symmetric matrix. The pupose of this exercise is to show that eigenvectors of different eigenvalues must be orthogonal. So suppose X_1 is an eigenvector of M with eigenvalue λ_1 and X_2 is an eigenvector with eigenvalue $\lambda_2 \neq \lambda_1$.

   a. Show that $X_2^{\dagger} MX_1 = \lambda_1 (X_2 \cdot X_1)$, where $X_2 \cdot X_1$ is the ordinary "dot product" of vectors.

   b. Use $(MX_2)^{\dagger} = X_2^{\dagger} M$ (why is this true?) to show that $X_2^{\dagger} MX_1 = \lambda_2 (X_2 \cdot X_1)$. Conclude that $X_2 \cdot X_1 = 0$.

7.15. a. Find the functions $p_i(\Delta x, \Delta y)$, $i = 1, 2$ provided by Lemma 7.3 for the function $F(x,y) = e^x \sin y$ when $(a,b) = (0,0)$.

   b. Verify that $e^x \sin y = p_1(x,y)x + p_2(x,y)y$.

7.16. The folium of Descartes $f(x,y)$ (p. 237) evidently has a saddle point at the origin. This exercise provides new local coordinates (u,v) in which the folium takes the form $-u^2 + v^2$.

   a. Determine $\varphi(\xi, \eta) = f(\xi - \eta, \xi + \eta)$; this is the form the folium takes under a (global) $45°$ rotation and dilation $\mathbf{c}(\xi, \eta)$.

   b. Show that φ can be writen in the form $\alpha(\xi)\xi^2 + \beta(\xi)\eta^2$ and determine $\alpha(\xi)$ and $\beta(\xi)$.

   c. Introduce a local coordinate change $(u,v) = \mathbf{k}(\xi, \eta)$ near $(\xi, \eta) = (0,0)$ that reduces φ to $-u^2 + v^2$. Prove that $\mathbf{k}$ is a coordinate change near the origin.

   d. Let $\mathbf{h} = \mathbf{k} \circ \mathbf{c}^{-1}$. Use a suitable graphing uility to sketch the pullback of a coordinate grid in the (u,v)-plane by $\mathbf{h}$ to show that the pullback carries level curves of $-u^2 + v^2$ to level curves of $f(x,y)$. Compare your result with the figure on page 239.

7.17. a. Sketch the zero-level of the function $f(x,y) = (xy^2 - 1)(x^2y - 1)$ in the first quadrant and infer that f has a saddle point at $(x,y) = (1,1)$.

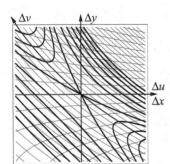

b. Express f in terms of window coordinates at $(1,1)$; that is, determine $\Delta z = f(1+\Delta x, 1+\Delta y) - f(1,1)$ as a (sixth-degree) polynomial in Δx and Δy.

c. Show that the functions of Morse's decomposition (Theorem 7.18) at the saddle point are

$$h_{11} = 2 + \Delta x + \tfrac{8}{3}\Delta y + \tfrac{3}{2}\Delta x\,\Delta y + \tfrac{3}{2}\Delta y^2 + \tfrac{9}{10}\Delta x\,\Delta y^2 + \tfrac{3}{10}\Delta y^3 + \tfrac{1}{5}\Delta x\,\Delta y^3,$$

$$h_{12} = \tfrac{5}{2} + \tfrac{8}{3}\Delta x + \tfrac{8}{3}\Delta y + \tfrac{3}{4}\Delta x^2 + 3\Delta x\,\Delta y + \tfrac{3}{4}\Delta y^2 + \tfrac{9}{10}\Delta x^2\,\Delta y + \tfrac{9}{10}\Delta x\,\Delta y^2$$
$$+ \tfrac{3}{10}\Delta x^2\,\Delta y^2,$$

$$h_{22} = 2 + \tfrac{8}{3}\Delta x + \Delta y + \tfrac{3}{2}\Delta x^2 + \tfrac{3}{2}\Delta x\,\Delta y + \tfrac{3}{10}\Delta x^3 + \tfrac{9}{10}\Delta x^2\,\Delta y + \tfrac{1}{5}\Delta x^3\,\Delta y.$$

d. Verify, by direct computation, that $\Delta z = h_{11}\,\Delta x^2 + 2h_{12}\,\Delta x\,\Delta y + h_{22}\,\Delta y^2$.

e. Complete the square to obtain the coordinate change $(\Delta u, \Delta v) = \mathbf{h}(\Delta x, \Delta y)$ that reduces Δz to the simple diagonal form $\Delta z = \Delta u^2 - \Delta v^2$. Prove that $\mathbf{h}$ is a coordinate change near the origin in the $(\Delta x, \Delta y)$ window.

f. Use a suitable graphing utility to sketch the pullback of a coordinate grid in the $(\Delta u, \Delta v)$ window to show that level curves of $\Delta u^2 - \Delta v^2$ pull back to level curves of Δz in the $(\Delta x, \Delta y)$ window. The figure in the margin shows the $(\Delta u, \Delta v)$ coordinate grid in the $(\Delta x, \Delta y)$ window, together with level curves of f. (Levels in the Δu direction are twice as far apart as those in the Δv direction.)

7.18. a. The function $g(x,y) = (x^2 - y^2)^3 - 2x(x^2 - y^2) + 1$ has a saddle point at $(x,y) = (1,0)$. (A $45°$ rotation–dilation converts the function f of the preceding exercise into g.) Carry out all the steps of the preceding exercise to obtain the local coordinate change $(\Delta u, \Delta v) = \mathbf{h}(\Delta x, \Delta y)$ given by Morse's decomposition that reduces g to $\Delta u^2 - \Delta v^2$ in a window centered at $(1,0)$.

b. Prove that the local coordinate change provided by Morse's decomposition in this exercise is not a rotation–dilation of the local coordinate change of the preceding exercise. Suggestion: Consider the derivative of each coordinate change at the origin.

7.19. Find the functions $h_{ij}(x_1, x_2)$ in Morse's decomposition (Theorem 7.18) for the function $f(x_1, x_2) = \cos x_1 \cos x_2$ at the point $(a_1, a_2) = (0,0)$. Verify that

$$h_{ij}(0,0) = \frac{1}{2}\frac{\partial^2 f}{\partial x_i \partial x_j}(0,0)$$

for every i, j.

Chapter 8
Double Integrals

Abstract Double integrals arise in a variety of scientific contexts, essentially as a way to calculate the product of quantities that vary. They are introduced in the first multivariable calculus course, together with the iterated (repeated) integrals that are often used to evaluate them. This chapter concentrates on definitions and properties, and begins with a problem in gravitational attraction that leads to double integrals. It then introduces a precise notion of area called *Jordan content*, and uses that to define the integral. The next chapter concentrates on evaluation, using iterated integrals, curvilinear coordinates, and the change of variables formula.

8.1 Example: gravitational attraction

Newton's law of **universal gravitation** says that between any two masses there is an attractive force that is proportional to each of the masses and to the inverse square of the distance between them.

The gravitational field

Force is a vector quantity. To describe the force that one mass m exerts on another μ, we begin with the vector $\mathbf{r}$ that gives the position of m with respect to μ. Then the force acts in the direction of the unit vector $\mathbf{u} = \mathbf{r}/\|\mathbf{r}\|$, and its magnitude is proportional to μ and to $m/\|\mathbf{r}\|^2$. If we let G denote the proportionality constant, as customary, then we can write

$$\text{force} = \mu\mathbf{f}, \quad \text{where } \mathbf{f} = \frac{Gm}{\|\mathbf{r}\|^3}\mathbf{r}.$$

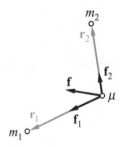

According to Newton's second law of motion, the vector $\mathbf{f}$ is the **acceleration** of the "test mass" μ; $\mathbf{f}$ depends only on the mass m and on the *position* of μ in relation to m. Such a vector function of position is called a *vector field*. This particular vector field is the **gravitational field** due to the mass m. To determine the gravitational force that m exerts on a test mass μ located anywhere in space, just multiply the acceleration field vector $\mathbf{f}$ at that point by the size μ of the test mass. The gravitational field

J.J. Callahan, *Advanced Calculus: A Geometric View*, Undergraduate Texts in Mathematics, DOI 10.1007/978-1-4419-7332-0_8, © Springer Science+Business Media, LLC 2010

defined by a collection of masses m_i, $i = 1, \ldots, N$ is just the vector sum of their individual fields:

$$\mathbf{f} = \sum_{i=1}^{N} \mathbf{f}_i = \sum_{i=1}^{N} \frac{Gm_i}{\|\mathbf{r}_i\|^3} \mathbf{r}_i.$$

The gravitational field of a large square plate

Our formula appears to define the gravitational field of only a finite number of masses that are concentrated at discrete points. However, by using limit processes (in which sums become integrals), we can extend the formula to continuous distributions of matter. To see how this happens, let us analyze the gravitational field of a large homogeneous plate of uniform thickness. For such a plate, the mass of any piece is simply proportional to its area A:

$$\text{mass} = \rho A.$$

The constant of proportionality ρ gives the mass per unit area, or the *mass density*, of the plate. We determine the gravitational field of the plate only for points directly above the center of the plate. (Eventually, we assume that the plate is so large that it is effectively infinite in extent. In that case, every location on the plate is like every other; we are able to think of any point on the plate as its "center.")

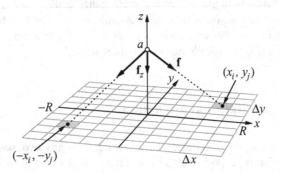

For now, let the plate be the square $-R \le x, y \le R$ in the (x,y)-plane. We want to determine the gravitational field at the point $(0,0,a)$, $a > 0$, on the z-axis. We can use the additivity of the field: divide the plate into a number of small cells, approximate the field due to each cell, and then add the results to get an estimate of the field due to the whole plate.

Approximating the field

We choose cells that form a grid of small congruent squares of dimensions $\Delta x = \Delta y = R/k$ and mass $\rho \, \Delta x \, \Delta y$. If R/k is small enough, we can approximate the field due to a single square by imagining all its mass is concentrated at its center. If the center is at $(x_i, y_j, 0)$, then

$$\mathbf{r} = (x_i, y_j, 0) - (0, 0, a) = (x_i, y_j, -a),$$

so the gravitational field due to this one square is approximately

$$\mathbf{f}_{ij} = \frac{G\rho \, \Delta x \, \Delta y}{(x_i^2 + y_j^2 + a^2)^{3/2}} (x_i, y_j, -a).$$

As the figure above indicates, the horizontal component of $\mathbf{f}_{ij}$ is canceled by the horizontal component of the field due to the square centered at symmetrically opposite point $(-x_i, -y_j)$ on the plate. Thus, for the whole field $\mathbf{f}$, only its vertical component is nonzero. In fact, the four points (x_i, y_j), $(x_i, -y_j)$, $(-x_i, y_j)$, and $(-x_i, -y_j)$ contribute the same vertical component, so we can restrict ourselves to squares in the first quadrant, writing the four contributions together as the scalar

$$4 \frac{G\rho \, \Delta x \, \Delta y}{(x_i^2 + y_j^2 + a^2)^{3/2}} \cdot (-a) = \frac{-4G\rho a \, \Delta x \, \Delta y}{(x_i^2 + y_j^2 + a^2)^{3/2}}.$$

Therefore, the vertical component of the field at $z = a$ that is due to the whole plate is approximately the double sum

$$\text{field} \approx \sum_{i=1}^{k} \sum_{j=1}^{k} \frac{-4G\rho a \, \Delta x \, \Delta y}{(x_i^2 + y_j^2 + a^2)^{3/2}}.$$

The coordinates (x_i, y_j) of the centers of the squares in the first quadrant start with $(x_1, y_1) = (\Delta x/2, \Delta y/2)$ and then increase by steps of Δx and Δy:

$$x_1 = \Delta x/2, \qquad\qquad y_1 = \Delta y/2,$$
$$x_i = x_{i-1} + \Delta x, \; i = 2, \ldots, k, \qquad y_j = y_{j-1} + \Delta y, \; j = 2, \ldots, k.$$

Using a simple program such as the the one below, we can compute the double sum for any given value of a. To simplify the computation, however, we first choose units that make $4G\rho = 1$. Then, keeping in mind that the dimensions of the plate should be large in comparison to the distance to a point in the field (i.e., $R \gg a$), we choose $R = 32$ and then do a sequence of calculations for $a = 0.2, 0.1$, and 0.05.

Computing the double sum

```
           PROGRAM: The gravitational field of a large plate
a = .2
R = 32
k = 64
dx = R / k
dy = dx
sum = 0
x = dx / 2
FOR i = 1 TO k
  y = dy / 2
  FOR j = 1 TO k
    sum = sum - a * dx * dy / (x ^ 2 + y ^ 2 + a ^ 2) ^ (3 / 2)
    y = y + dy
  NEXT j
  x = x + dx
NEXT i
PRINT k, sum
```

The results appear in the table below. Each column indicates how the estimate of field strength changes as the number of squares increases when a is fixed. (Note that there are k^2 squares, and thus more than a million when $k = 1024$.) It appears

How the sum varies with k

that the sums are converging to some fixed value as k increases. For example, when $a = 0.2$, the first seven or eight digits of the sum have "stabilized" when $k = 1024$, suggesting that a limit is emerging:

$$\text{field at } 0.2 = \lim_{k \to \infty} \text{sum} = -1.561\,957\ldots.$$

But for the same value of k, only the first four digits appear to have stabilized when $a = 0.1$, and only the first two when $a = 0.05$. However, when k is doubled, even the sum for 0.05 appears to have stabilized out to four digits. This suggests

$$\text{field at } 0.1 = -1.566\ldots, \qquad \text{field at } 0.05 = -1.568\ldots.$$

$a = 0.2$		$a = 0.1$		$a = 0.05$	
k	sum	k	sum	k	sum
64	-1.233	64	-0.757	128	-0.759
128	-1.526	128	-1.238	256	-1.240
256	$-1.561\,691$	256	$-1.530\,399$	512	-1.533
512	$-1.561\,957\,628$	512	$-1.566\,110$	1024	$-1.568\,320$
1024	$-1.561\,957\,637$	1024	$-1.566\,377$	2048	$-1.568\,587$

The Riemann integral as a limit of sums

We recognize that the double sums are, in fact, *Riemann* sums for the function

$$\varphi_a(x,y) = \frac{-4G\rho a}{(x^2 + y^2 + a^2)^{3/2}};$$

thus, the limiting value that we are seeking for each a is the Riemann integral of that function for the given a:

$$\text{field at } a = \lim_{k \to \infty} \sum_{i=1}^{k} \sum_{j=1}^{k} \varphi_a(x_i, y_j)\,\Delta x \Delta y = \iint_{\substack{0 \le x \le R, \\ 0 \le y \le R}} \varphi_a(x,y)\,dx\,dy.$$

In this expression, "$dx\,dy$" is sometimes called the **element of area**; it represents the limit of the area $\Delta x \Delta y$ of a rectangle in the Cartesian grid. When other coordinates are used, the element of area may have a different form. However, our expression for a double integral always contains an element of area.

"Integral as product"

The integral arises here in a typical way: it is a number that is essentially the product of two quantities. (For example, field strength is the product of a mass and the reciprocal of a distance squared.) But the quantities involved are variable, so their product cannot be found directly as a single number. The remedy is to restrict the quantities being multiplied to small cells on which they become nearly constant. (For example, restrict to small squares on the plate). Now calculate a representative product on each cell, and then add the results over all cells. The sum gives an approximation to the numerical value we seek. To get a better approximation, make the cells even smaller. If the sums tend to a limit as the cells get smaller, that limit is defined to be the integral. In Chapter 8.3, we use these ideas to define the integral and catalogue some of its properties.

Let us return to the estimates of the gravitational field provided by our calculations. Notice that field strength does not vary with the inverse square of the distance a to the plate. On the contrary, the calculations suggest that field strength is essentially constant for all $a \ll R$. If $R = \infty$, then $a \ll R$ for all finite a, so it seems reasonable to speculate that the gravitational field of an infinite plate is exactly constant. But we cannot check this directly: if $R = \infty$, the Riemann sums we use to estimate the field are not defined. (The BASIC program breaks down on its second line.) Indeed, we define the Riemann integral only for a bounded domain.

How the field varies with position

But there is a standard way out of the difficulty: determine the value of the integral as a function of R, and then see if the values tend to a well-defined limit as $R \to \infty$. When the limit exists, it is called the *improper integral*. Improper double integrals arise from unbounded functions as well as from unbounded domains; we define both kinds in Chapter 9. However, we can even now confirm that an infinite homogeneous plate produces a constant gravitational field; we just need to start with a different shape.

Improper integrals

Because we are interested ultimately in the field of the infinite plate, let us allow ourselves to change the shape of its finite approximation. Specifically, we change the plate from a square to a circle, and then exploit the circular symmetry by changing from Cartesian to polar coordinates. This change leads to a one-variable improper integral that determines the field of an infinite plate.

The gravitational field of a circular plate

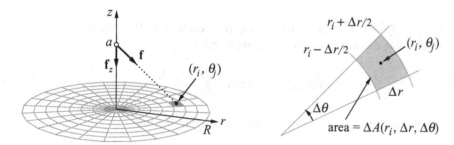

Let the plate have radius R and be centered at the origin; then it is given by the inequalities $0 \le r \le R$, $0 \le \theta < 2\pi$. To divide the plate into small cells, it is natural to use equally spaced concentric circles $r = $ constant and radial lines $\theta = $ constant, with spacings

$$\Delta r = \frac{R}{k}, \quad \Delta \theta = \frac{2\pi}{l}, \quad k, l \text{ positive integers.}$$

The cells are not uniform in size; their areas grow with r. Choose the representative point (r_i, θ_j) at the center of the cell. Each cell is the portion of a circular ring of thickness Δr that is cut out by a central angle $\Delta \theta$. The area of the whole ring between $r_i - \Delta r/2$ and $r_i + \Delta r/2$ is

Area of a cell

$$\pi \left(r_i + \frac{\Delta r}{2} \right)^2 - \pi \left(r_i - \frac{\Delta r}{2} \right)^2 = 2\pi r_i \Delta r.$$

A single cell occupies the fraction $\Delta\theta/2\pi$ of this ring, so its area is

$$\left(\frac{\Delta\theta}{2\pi}\right) 2\pi r_i \Delta r = r_i \Delta r \Delta\theta = \Delta A(r_i, \Delta r, \Delta\theta) = \Delta A,$$

and we can write its mass as $\rho\,\Delta A$. Assuming the mass is concentrated at (r_i, θ_j), we estimate the cell's contribution to the gravitational field at $z = a$ is approximately

$$\frac{-G\rho a\,\Delta A}{(r_i^2 + a^2)^{3/2}}.$$

Therefore, the field at $z = a$ that is due to the whole plate is now approximated by the double sum

$$\text{field at } a \approx \sum_{i=1}^{k}\sum_{j=1}^{l} \frac{-G\rho a\,\Delta A}{(r_i^2 + a^2)^{3/2}}.$$

Simplifications due to symmetry
    Notice that the values θ_j are absent: all cells in a given ring (i.e., with fixed r_i) make the same contribution. This symmetry with respect to θ means we can write the inner sum (where i, and hence r_i, is fixed) as

$$\sum_{j=1}^{l} \frac{-G\rho a\,\Delta A}{(r_i^2 + a^2)^{3/2}} = \frac{-G\rho a}{(r_i^2 + a^2)^{3/2}} \sum_{j=1}^{l} \Delta A = \frac{-G\rho a}{(r_i^2 + a^2)^{3/2}}\, 2\pi r_i \Delta r,$$

because $\sum \Delta A$ is just the area of an entire circular ring. Thus the field due to the whole plate reduces to a sum over a single index:

$$\text{field at } a \approx -2\pi G\rho a \sum_{i=1}^{k} \frac{r_i \Delta r}{(r_i^2 + a^2)^{3/2}}.$$

But this is just a Riemann sum for the one-variable function

$$h(r) = -2\pi G\rho a \frac{r}{(r^2 + a^2)^{3/2}}.$$

Therefore, as $k \to \infty$ and $\Delta r \to 0$, the sum becomes the ordinary (i.e., single) integral

$$\int_0^R h(r)\,dr = -2\pi G\rho a \int_0^R \frac{r\,dr}{(r^2 + a^2)^{3/2}}$$

$$= 2\pi G\rho a \left.\frac{1}{(r^2 + a^2)^{1/2}}\right|_0^R = 2\pi G\rho a \left(\frac{1}{(R^2 + a^2)^{1/2}} - \frac{1}{a}\right).$$

Because the sum becomes a better and better approximation to the field as $\Delta r \to 0$ (when $k \to \infty$), we write

$$\text{field at } a = -2\pi G\rho a \int_0^R \frac{r\,dr}{(r^2 + a^2)^{3/2}} = 2\pi G\rho\left(\frac{a}{\sqrt{R^2 + a^2}} - 1\right).$$

Let us compare this result with what we computed for the square. Thus we set $4G\rho = 1$ and $R = 32$, and then find

field at $0.2 = -1.56098$, ... at $0.1 = -1.56589$, ... at $0.05 = -1.56834$.

These values are virtually identical with the corresponding ones for the square. Now that we have an exact formula for the field, it is easy to see what happens as the plate becomes infinite in extent, that is, as we let $R \to \infty$. Because

$$\lim_{R \to \infty} \frac{a}{\sqrt{R^2 + a^2}} = 0,$$

we find that

$$\text{field} = 2\pi G\rho \cdot (-1) = -2\pi G\rho = \text{constant};$$

the field is indeed independent of the distance a from the plate. Furthermore, after the normalization $4G\rho = 1$, the field strength takes on the constant value $-\pi/2 = -1.570793\ldots$. In fact, we express the field of the infinite plate as an improper integral, that is, as the limit of a sequence of "proper" integrals:

$$\int_0^\infty \frac{r\,dr}{(r^2 + a^2)^{3/2}} = \lim_{R \to \infty} \int_0^R \frac{r\,dr}{(r^2 + a^2)^{3/2}} = \lim_{R \to \infty} \left(\frac{1}{(R^2 + a^2)^{1/2}} - \frac{1}{a} \right) = -\frac{1}{a}.$$

We evaluate improper double integrals in a similar way (cf. Chapter 9.2).

Let us return to the finite circular plane and the double sum formula

$$\sum_{i=1}^k \sum_{j=1}^l \frac{-G\rho a \Delta A}{(r_i^2 + a^2)^{3/2}}$$

that expresses the approximate field strength in polar coordinates. As we did with Cartesian coordinates, we can recognize that these are Riemann sums for the function

$$\psi_a(r, \theta) = \frac{-G\rho a}{(r^2 + a^2)^{3/2}}.$$

The exact value of the field strength is thus the Riemann integral that represents the limit of these sums:

$$\text{field at } a = \lim_{\substack{k \to \infty \\ l \to \infty}} \sum_{i=1}^k \sum_{j=1}^l \psi_a(r_i, \theta_j) \Delta A = \iint_{\substack{0 \le r \le R, \\ 0 \le \theta < 2\pi}} \psi_a(r, \theta)\,dA.$$

In this expression, dA is the element of area (cf. p. 272); it represents the limit of the area ΔA of a cell in the polar coordinate grid.

For a Cartesian grid, $dA = dx\,dy$ is just the product of the "elements of length" dx and dy for the individual coordinates. However, $dA \ne dr\,d\theta$ for a polar grid. On the contrary, we have already noted that $\Delta A = (\Delta r)(r\Delta\theta)$. Although $\Delta\theta$ is dimensionless, $r\Delta\theta$ does have the dimensions of a length: $r\Delta\theta$ is the length of the arc subtended by the angle $\Delta\theta$ on the circle of radius r. Informally, then, $r\,d\theta$ and dr are the "elements of length" whose product is the element of area $dA = r\,dr\,d\theta$. Thus we write

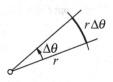

$$\text{field at } a = \iint\limits_{\substack{0\le r\le R,\\ 0\le\theta<2\pi}} \psi_a(r,\theta)\,r\,dr\,d\theta = -G\rho a \iint\limits_{\substack{0\le r\le R,\\ 0\le\theta<2\pi}} \frac{r\,dr\,d\theta}{(r^2+a^2)^{3/2}}.$$

8.2 Area and Jordan content

Riemann integration: a sketch

The definition of the Riemann integral of a function over a region is simple in outline. First, partition the region into many small pieces; then multiply the "size" of each piece by a value that the function takes on somewhere in that piece, and sum those products; finally, repeat the process with ever smaller pieces and take the limit of the computed sums. To convert this sketch into something useful and precise, one of the questions we must decide is what kind of pieces can be used to make up a partition.

Partitioning 2-dimensional regions

When the function depends on just a single real variable x, the answer is immediate: each small piece is an interval $a \le x \le b$ whose "size" is its length, $b - a$. But if the function depends on two real variables, x and y, the answer is not so clear. Certainly there is a 2-dimensional analogue for an interval; it is a rectangle $a \le x \le b$, $c \le y \le d$ in the (x,y)-plane with sides parallel to the axes, whose "size" is given by its area $A = (b-a)(d-c)$. But now consider what happens under a change of variables. On the line, a small interval is generally transformed into another small interval. On the plane, however, a small rectangle is transformed into a quadrilateral with curved sides (see Chapter 4), so these more general shapes appear in partitions as naturally as rectangles do. We therefore get a better answer to the question by focusing not on the shape of a small piece but on whether we can assign it a "size."

Jordan content

The *size* of a region will be its *area*, of course; we have to worry about admissible shapes because not every region in the plane has a well-defined area. For example, there is no consistent way to assign a nonzero area to the set of points in the unit square that have rational coordinates (the **rational points**); see below, page 279. If this is not immediately evident, however, it may be because our notion of *area* is more intuitive than precise. Thus, we construct a precise notion of *size* (called *Jordan content*) that captures our intuitive ideas about area, extends immediately to higher dimensions, and fits well with the process of integration.

Plane topology

Before we discuss Jordan content, though, we must establish some basic topology concerning the interior and the boundary of a set in the plane.

Definition 8.1 *The **open** (respectively, **closed**) disk of radius $r > 0$ centered at the point $\mathbf{p}$ in $\mathbb{R}^2$ is the set of all points $\mathbf{x}$ in $\mathbb{R}^2$ for which $\|\mathbf{x}-\mathbf{p}\| < r$ (respectively, $\|\mathbf{x}-\mathbf{p}\| \le r$).*

Definition 8.2 *A point* **p** *is an* ***interior point of a set S*** *in* $\mathbb{R}^2$ *if some open disk centered at* **p** *is contained entirely in S.*

Definition 8.3 *A point* **q** *is an* ***exterior point of S*** *if it is an interior point of* S^c*, the complement of S in* $\mathbb{R}^2$*. A point* **b** *is a* ***boundary point of S*** *if it is neither an interior nor an exterior point of S.*

Every interior point of S lies inside S, of course; what makes it an *interior* point, however, is the fact that it is surrounded by points that also lie inside S. Likewise, every exterior point lies outside S and is surrounded by points outside S. An individual boundary point may lie either inside or outside S, but every open disk centered at a boundary point contains at least one point in S and one point outside S (see Exercise 8.5). For example, the open and closed disks with a given radius and center have the same boundary points, namely the points on the circle with that radius and center. They represent two extremes: the closed disk contains all its boundary points, but the open disk contains none. These become the models for *closed* and *open* sets in general.

Open and closed sets

closed disk

boundary point

open disk

Definition 8.4 *A set is* ***closed*** *if it contains all its boundary points; it is* ***open*** *if it contains none of them.*

Thus every point of an open set is an interior point. It is more common to define a closed set, however, as one whose complement is open. The following theorem shows that our definition is equivalent to the usual one.

Theorem 8.1. *The set S is closed $\Leftrightarrow$ The complement S^c is open.*

Proof. S is closed $\Leftrightarrow S^c$ contains no boundary points of S
         $\Leftrightarrow S^c$ contains only exterior points of S
         $\Leftrightarrow$ all points of S^c are interior points of S^c
         $\Leftrightarrow S^c$ is open. $\square$

Definition 8.5 *The* ***interior of S***, *denoted* $^\circ S$, *is the set of interior points of S; the* ***boundary of S***, *denoted* ∂S, *is the set of boundary points of S; the* ***closure of S***, *denoted* $\overline{S}$, *is* $S \cup \partial S$.

Thus, S is open if $S = {^\circ S}$ and is closed if $S = \overline{S}$. We "open" S (i.e., create its interior) by removing from S all its boundary points; we "close" S (create its closure) by adding to S all its boundary points. The symbol we have introduced for *boundary* may have no good rationale at the moment, but its aptness should become clear when we reconsider Green's theorem in Chapter 10; see especially page 427.

When S is a familiar shape—such as a polygon—its interior, exterior, and boundary are what we expect. But when S is less familar, intuition may be a poor guide. For example, even though a polygon's boundary separates the interior from the exterior, this is not true for all sets. For one thing, the boundary may not be a finite collection of curves or line segments. Take S to be the plane minus the origin; then ∂S is just the origin and there are no exterior points at all. According to Exercise 8.5,

The boundary can be nonintuitive

every disk centered at a boundary point must contain a point outside S; in this case, the only such point is the center of the disk itself.

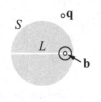

For an even more instructive example, take S to be the open unit disk centered at the origin, minus the portion L of the x-axis that lies within 1 unit of the origin. The exterior points $\mathbf{q}$ of S are those for which $\|\mathbf{q}\| > 1$. The boundary of S is a pair of curves: the unit circle and the line segment L. Any neighborhood of a point on the unit circle does indeed contain both interior and exterior points of S, but that is not true of L. At any point $\mathbf{b}$ of L, a sufficiently small open disk centered at $\mathbf{b}$ will contain no exterior points of S whatsoever. Therefore, we cannot say that L "separates" the interior of S from its exterior.

For a set that cannot be sketched easily, its interior and boundary may be even more nonintuitive. For example, take S to be the set of all points in the closed disk of radius 1 together with all *rational* points in the disk of radius 2, both centered at the origin. The interior of S is the open disk of radius 1, and the exterior of S consists of all points $\mathbf{q}$ with $\|\mathbf{q}\| > 2$. The boundary of S is everything else; it is the annulus of points $\mathbf{b}$ with $1 \leq \|\mathbf{b}\| \leq 2$. Here is a set whose boundary is "thick;" although ∂S does separate the interior and the exterior of S, it is not a simple 1-dimensional curve. Another example in the same vein is the set of all rational points in the plane. The interior and the exterior are both empty, so the boundary is the entire plane. As we show, sets such as these that have "thick" boundaries will always fail to have Jordan content.

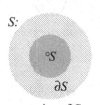

exterior of S

To define Jordan content, we use a method that we introduce now in an intuitive and ad hoc way to find the areas of two particular sets. In one case, the method succeeds; in the other, it does not, illustrating how a set can fail to have Jordan content.

Areas from squares

The fundamental shape whose area we know is a square. Consider how we might use only squares to approximate the area of the closed unit disk S. Cover the plane with a grid of squares of width w. If L of them lie entirely inside S, then the area of S must be at least Lw^2. If U of them meet S, then the area of S can be no more than Uw^2. For example, if $w = 1/5$ and the origin is one of the grid intersection points (below, left), then we can use 3–4–5 right triangles to show that $L = 60$ gray squares (15 in each quadrant) lie inside S, and $U = 104$ squares altogether meet S, implying

$$2.4 = \frac{60}{25} = Lw^2 < \text{area} \, S < Uw^2 = \frac{104}{25} = 4.16.$$

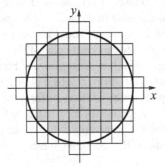
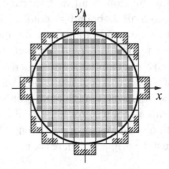

These are lower and upper bounds for the area, but neither is a good estimate for the correct value, $\pi \approx 3.14$. The smaller fails to count area that it should and the larger counts area that it should not. The difference between the bounds—which indicates the coarseness of the estimates—equals the total area $44/25 = 1.76$ of the white squares in the figure on the left.

To get better estimates, use smaller squares; then more of the area inside S will be counted, and less outside. Take $w = 1/10$, giving us four small squares in each large square (above, right). We find nine additional small squares (darker gray) inside S in each quadrant, so $L = 4 \times 60 + 36 = 276$. Furthermore, 14 new small squares (hatched) in each quadrant now lie completely outside S, so $U = 4 \times 104 - 56 = 360$, and the new bounds are

Refine the grid to get a better estimate

$$2.76 = Lw^2 < \text{area}\,S < Uw^2 = 3.6.$$

The difference between the new bounds (which is the area of the 84 small white squares in the figure on the right) is less than half what it was in the previous stage; in this sense, the new estimates are twice as good as the old. (Their average, 3.18, is within $1\frac{1}{4}\%$ of the true area.)

Calculations made with further refinements of the grid (see Exercise 8.17) suggest that the difference between the bounds—as measured by the area of the white squares—shrinks to zero, forcing our estimates of the area of S toward a single value. Notice that those white squares also cover the boundary circle, implying that the area of the boundary is zero. The circle is, of course, a pair of graphs; as we show (Theorem 8.10, p. 284), the graph of a continuous function on a bounded interval will always have zero area. In summary, S has an area, and its boundary is "thin" enough to have zero area.

Area of the boundary

For our second example, take S to be the set of rational points in the unit square that lies in the first quadrant with a corner at the origin. Let us use the same method of counting squares to find the area of S, assuming it has an area. As before, we cover the plane with a grid of squares of width w. But now, because S has no interior points, no grid square lies entirely inside S. In other words, there are no solid gray squares; L is always zero. On the other hand, many grid squares have a point in common with S, so $U > 0$. For example, if $w = 1/5$, then we can count 25 such grid squares inside the unit square itself, plus 5 more meeting the unit square along each of its four sides, plus 1 more at each each corner; thus $U = 25 + 4 \times 5 + 4 = 49$. If S does have an area, then

A set with no area

$$0 = Lw^2 < \text{area}\,S < Uw^2 = \frac{49}{25} = 1.96.$$

When we set $w = 1/10$ to refine the grid, then $U = 100 + 4 \times 10 + 4 = 144$ and the bounds become

$$0 = Lw^2 < \text{area}\,S < Uw^2 = \frac{144}{100} = 1.44.$$

More generally, if $w = 1/2^n$, then $U = 2^{2n} + 4 \times 2^n + 4$ and the bounds are

$$0 = Lw^2 < \text{area}\, S < Uw^2 = 1 + \frac{1}{2^{n-2}} + \frac{1}{2^{2n-2}}.$$

No matter how small we make w, the bounds never get any better than

$$0 < \text{area}\, S \le 1.$$

Our method of counting squares thus fails to assign a meaningful value to the area of S.

A boundary with positive area

How is this failure connected to the size, or "thickness," of ∂S? For the closed disk, the difference $Uw^2 - Lw^2$ was the area of the white squares that covered the boundary circle; it served as an upper bound on the area of that circle. For the rational points in the square, the difference is

$$Uw^2 - Lw^2 = Uw^2 > 1$$

for all $w > 0$; in particular, the difference does not converge to zero. Indeed, the area of ∂S is 1. To see why, recall from our earlier examples (p. 278) that $\partial S = \overline{S}$ is the closed unit square, so area $\partial S = \text{area}\, \overline{S} = 1$. Thus, in contrast with the disk, the set of rational points in the square does not have an area, but its boundary is "thick" and does have positive area.

Defining Jordan content

We can now formalize the method we use to define the *Jordan content* of a set. There are three steps. First, we count grid squares to get monotonic sequences of "inner" and "outer" areas. Second, we compute the limiting areas as the grids are refined. Third, we see whether the two limits are equal; if they are, the set has Jordan content equal to the common value. Although it might seem reasonable to choose the grids based on how well they are adapted to a given set, it is not a priori evident that the value obtained from one sequence of grids would then equal the value obtained from another. Thus, to eliminate ambiguity, we always use just one collection of grids, $\mathcal{J}_0, \mathcal{J}_1, \mathcal{J}_2, \dots$. Later (pp. 287ff.), we in fact introduce other grids and prove that they yield the same value given by the grids $\mathcal{J}_k$.

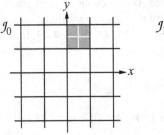 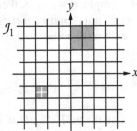 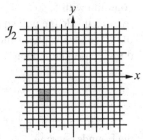

The grids $\mathcal{J}_k$

The grid $\mathcal{J}_0$ consists of the closed unit squares in the (x, y)-plane that are bounded by the vertical lines $x = $ integer and $y = $ integer. To get the squares of the next grid $\mathcal{J}_1$, divide each unit square into four congruent subsquares. Because every square of $\mathcal{J}_1$ lies entirely in a single square of $\mathcal{J}_0$, we say $\mathcal{J}_1$ is a *refinement* of $\mathcal{J}_0$.

Use the same procedure to get $\mathcal{J}_2$ as a refinement of $\mathcal{J}_1$, and so on. The squares in the grid $\mathcal{J}_k$ at stage k have width $w = 1/2^k$.

Definition 8.6 *Let $\underline{J}_k(S)$ denote the total area of the squares in $\mathcal{J}_k$ that are entirely contained in the bounded set S, and let $\overline{J}_k(S)$ denote the total area of the squares in $\mathcal{J}_k$ that intersect S.*

$\underline{J}_k(S)$ and $\overline{J}_k(S)$

The "inner" and "outer" area estimates for S ($\underline{J}_k(S)$ and $\overline{J}_k(S)$, respectively) at the various stages are nested together in the following chain:

$$0 \le \underline{J}_0(S) \le \cdots \le \underline{J}_k(S) \le \underline{J}_{k+1}(S) \le \cdots \le \overline{J}_{l+1}(S) \le \overline{J}_l(S) \le \cdots \le \overline{J}_0(S) < \infty.$$

To see this, note first that $0 \le \underline{J}_0(S)$ because our squares all have positive area. Also, $\overline{J}_0(S)$ is finite because the bounded set S can meet only a finite number of unit squares. The remaining inequalities in the chain follow from the next two lemmas.

Lemma 8.1. *For any bounded set S and integer $k \ge 0$, $\underline{J}_k(S) \le \underline{J}_{k+1}(S)$ and $\overline{J}_{k+1}(S) \le \overline{J}_k(S)$.*

Proof. If a square is counted in $\underline{J}_k(S)$, then its four subsquares, with the same total area, are counted in $\underline{J}_{k+1}(S)$; hence $\underline{J}_k(S) \le \underline{J}_{k+1}(S)$.

Similarly, if a square is counted in $\overline{J}_{k+1}(S)$, then the square in $\mathcal{J}_k$ that contains it is counted in $\overline{J}_k(S)$; that implies $\overline{J}_{k+1}(S) \le \overline{J}_k(S)$. □

Lemma 8.2. *For any bounded set S and integers $k, l \ge 0$, $\underline{J}_k(S) \le \overline{J}_l(S)$.*

Proof. First note that $\underline{J}_j(S) \le \overline{J}_j(S)$ for every $j \ge 0$, because any square counted in $\underline{J}_j(S)$ is also counted in $\overline{J}_j(S)$. Then (using Lemma 8.1),

$$k \ge l \quad \Rightarrow \quad \underline{J}_k(S) \le \overline{J}_k(S) \le \overline{J}_l(S);$$
$$k < l \quad \Rightarrow \quad \underline{J}_k(S) \le \underline{J}_l(S) \le \overline{J}_l(S). \qquad □$$

The nested inequalities imply that the sequence $\underline{J}_k(S)$ of "inner areas" is monotonic increasing and bounded above; hence it has a finite limit. The sequence $\overline{J}_k(S)$ is monotonic decreasing and bounded below, so it has a finite limit, too.

Definition 8.7 *The **inner** and **outer Jordan content** of the bounded set S are the respective limits*

Inner and outer Jordan content

$$\underline{J}(S) = \lim_{k \to \infty} \underline{J}_k(S) \quad \text{and} \quad \overline{J}(S) = \lim_{k \to \infty} \overline{J}_k(S).$$

Lemma 8.2 implies that $\underline{J}(S) \le \overline{J}(S)$, but $\underline{J}(S)$ and $\overline{J}(S)$ need not be equal. For example, when S is the set of rational points in the unit square, our earlier work shows that $\underline{J}(S) = 0$ and $\overline{J}(S) = 1$.

Definition 8.8 *If $\underline{J}(S) = \overline{J}(S)$, then we say S **is Jordan measurable**, or is a **J-set**, and its **Jordan content** is $J(S) = \underline{J}(S) = \overline{J}(S)$.*

The definition of Jordan content is now clear; however, it is not yet evident that the Jordan content of a set equals its usual area, not even for one of the original grid squares! Most of the rest of this section is devoted to establishing the properties of Jordan content. From that emerges its connection with area. (Exercise 8.7, for example, establishes that the Jordan content of a square in $\mathcal{J}_k$ is indeed its area, $1/2^{2k}$.) The first property is the fundamental one we saw in the two illustrative examples: a set has Jordan content precisely when its boundary is "thin" enough to have Jordan content equal to zero.

Theorem 8.2. *The set S is Jordan measurable $\Leftrightarrow J(\partial S) = 0$.*

Proof. We make use of the fact that

$$S \text{ is Jordan measurable} \Leftrightarrow \lim_{k \to \infty} \left(\bar{J}_k(S) - \underline{J}_k(S) \right) = 0.$$

The number $\bar{J}_k(S) - \underline{J}_k(S)$ is the total area of the squares that meet S but are not entirely contained in S. Each such square thus contains a point $\mathbf{p}$ in S and a point $\mathbf{q}$ not in S. Also, it is a convex set that contains the entire line segment connecting $\mathbf{p}$ and $\mathbf{q}$. Therefore, by Exercise 8.6, this square contains a point of ∂S, so it is counted in $\bar{J}_k(\partial S)$; hence

$$\bar{J}_k(S) - \underline{J}_k(S) \leq \bar{J}_k(\partial S).$$

Conversely, suppose a square Q_1 in $\mathcal{J}_k$ contains a point $\mathbf{b}$ in ∂S. We claim $\mathbf{b}$ must also lie in one of the squares Q_2 that is counted in $\bar{J}_k(S) - \underline{J}_k(S)$. If $\mathbf{b}$ is an interior point of the square Q_1, then every sufficiently small open disk centered at $\mathbf{b}$ lies in Q_1. But by Exercise 8.5, every such disk contains at least one point in S and at least one point not in S. Thus we can take $Q_2 = Q_1$.

If, on the contrary, $\mathbf{b}$ lies on either a side or a corner of Q_1, then it also meets either one or three squares adjacent to Q_1 in $\mathcal{J}_k$. Because $\mathbf{b}$ is in ∂S, at least one of these (two or four) squares contains a point in S and at least one contains a point not in S. For suppose each square contained exclusively one kind of point or the other. Because $\mathbf{b}$ is in each of these squares, it must be both in S and not in S. This is impossible, so at least one square Q_2 contains both kinds of points; Q_2 is counted in $\bar{J}_k(S) - \underline{J}_k(S)$.

Thus, each square Q_1 counted in $\bar{J}_k(\partial S)$ is either equal to a square Q_2 that is counted in $\bar{J}_k(S) - \underline{J}_k(S)$, or it is one of the eight neighbors of Q_2 in the grid $\mathcal{J}_k$. The total area of squares Q_1 is therefore not larger than nine times the total area of squares Q_2:

$$\bar{J}_k(\partial S) \leq 9 \left(\bar{J}_k(S) - \underline{J}_k(S) \right).$$

The two displayed inequalities now allow us to write

$$S \text{ is Jordan measurable} \Leftrightarrow \lim_{k \to \infty} \left(\bar{J}_k(S) - \underline{J}_k(S) \right) = 0$$
$$\Leftrightarrow \lim_{k \to \infty} \bar{J}_k(\partial S) = 0$$
$$\Leftrightarrow J(\partial S) = 0. \qquad \square$$

The next several results concern primarily outer Jordan content and sets whose Jordan content is zero. They are useful on their own and they culminate in the theorem (Theorem 8.10) that the graph of continuous function on a bounded interval has Jordan content zero.

Outer content and Jordan content zero

Theorem 8.3. *Let S and T be bounded sets with $S \subseteq T$; then $\bar{J}(S) \leq \bar{J}(T)$.*

Proof. Every square in $\mathcal{J}_k$ that meets S also meets T, so $\bar{J}_k(S) \leq \bar{J}_k(T)$. The inequality is preserved in the limit as $k \to \infty$, so $\bar{J}(S) \leq \bar{J}(T)$.   □

Corollary 8.4 *If T has Jordan content zero, then so does every subset of T.*

Proof. If $S \subseteq T$ and $\bar{J}(T) = 0$, then $\bar{J}(S) = 0$.   □

Theorem 8.5. *If S and T are bounded sets, then $\bar{J}(S \cup T) \leq \bar{J}(S) + \bar{J}(T)$.*

Proof. Every square in $\mathcal{J}_k$ that meets $S \cup T$ meets either S or T (or both); thus

$$\bar{J}_k(S \cup T) \leq \bar{J}_k(S) + \bar{J}_k(T).$$

The inequality is preserved in the limit as $k \to \infty$.   □

Corollary 8.6 *If $S_1, \ldots, S_p$ are bounded sets, then*

$$\bar{J}(S_1 \cup \cdots \cup S_p) \leq \bar{J}(S_1) + \cdots + \bar{J}(S_p).$$ □

Corollary 8.7 *The union of a finite number of sets that have Jordan content zero also has Jordan content zero.*   □

In particular, every finite set of points has Jordan content zero.

Corollary 8.8 *Suppose that, for any $\varepsilon > 0$, the set S is contained in a finite number of sets whose total Jordan content is less than ε. Then S has Jordan content zero.*

Proof. Suppose $S \subseteq T_1 \cup \cdots \cup T_p$ and

$$\bar{J}(T_1) + \cdots + \bar{J}(T_p) = J(T_1) + \cdots + J(T_p) < \varepsilon.$$

Then $\bar{J}(S) < \varepsilon$ for every $\varepsilon > 0$, so $\bar{J}(S) = 0$.   □

Theorem 8.9. *The Jordan content of a square in $\mathcal{J}_k$ is its ordinary area, $1/2^{2k}$, and the Jordan content of the rectangle $[a,b] \times [c,d]$ is its ordinary area $(b-a)(d-c)$.*

Proof. See the exercises.   □

We now introduce the notions of the *floor* and *ceiling* of a real number as convenient tools for our work. They are used immediately in the next proof.

Floor and ceiling

Definition 8.9 *The **floor of the real number x**, denoted $\lfloor x \rfloor$, is the largest integer m for which $m \leq x$. The **ceiling of x**, denoted $\lceil x \rceil$, is the smallest integer M for which $x \leq M$.*

Theorem 8.10. *The graph of a continuous function on a bounded interval has Jordan content zero.*

Proof. Let $y = f(x)$ be continuous on the interval $A \leq x \leq B$. We show that the graph of f is contained in a finite number of rectangles whose total area is less than any preassigned $\varepsilon > 0$. Corollary 8.8 then implies that the graph has Jordan content zero.

Because the interval is closed and bounded, f is *uniformly* continuous (see a text on analysis), which means that for any $\varepsilon > 0$, there is a $\delta > 0$ for which

$$|u - v| < \delta \quad \Rightarrow \quad |f(u) - f(v)| < \frac{\varepsilon}{4(B - A)},$$

for any $A \leq u, v \leq B$. A bound like $\varepsilon/4(B - A)$ is chosen with hindsight, of course; the reason emerges below. Let

$$\alpha = \left\lfloor \frac{A}{\delta} \right\rfloor, \quad \beta = \left\lceil \frac{B}{\delta} \right\rceil,$$

so that $\alpha\delta \leq A < (\alpha + 1)\delta$ and $(\beta - 1)\delta < B \leq \beta\delta$. Now partition the x-axis into nonoverlapping closed intervals of length δ, beginning at $\alpha\delta$ and ending at $\beta\delta$. Then A lies in the first interval and B in the last. For clarity, we want these two to be different intervals, so we require $(\alpha + 1)\delta < (\beta - 1)\delta$; it is sufficient to take $2\delta < B - A$ (Exercise 8.9).

Suppose u and v lie in the same interval. If $\bar{x}$ is the midpoint of that interval, then $|u - \bar{x}| \leq \delta/2 < \delta$, $|v - \bar{x}| \leq \delta/2 < \delta$, so

$$|f(u) - f(v)| \leq |f(u) - f(\bar{x})| + |f(\bar{x}) - f(v)|$$

$$< \frac{\varepsilon}{4(B - A)} + \frac{\varepsilon}{4(B - A)} = \frac{\varepsilon}{2(B - A)}.$$

The inequalities imply that the graph of $y = f(x)$ is entirely contained inside $\beta - \alpha$ closed rectangles, each of which is δ units wide and $\varepsilon/2(B - A)$ units tall. (As the figure below demonstrates, it may be necessary to take $\bar{x}$ to be A or B in the first or last interval, so there may be some overlapping at the ends.) The total area of the rectangles is

$$(\beta - \alpha) \times \delta \times \frac{\varepsilon}{2(B - A)}.$$

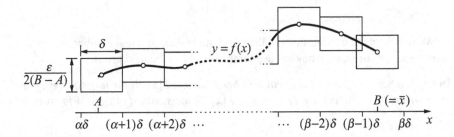

But $(\beta - \alpha)\delta < B - A + 2\delta$; because we have also taken $2\delta < B - A$, it follows that $(\beta - \alpha)\delta < 2(B - A)$ and thus

$$\text{total area} = \frac{(\beta - \alpha)\delta}{2(B - A)}\varepsilon < \varepsilon.$$

The graph of f is therefore contained in a finite number of sets whose total Jordan content can be made smaller than any preassigned positive number ε. By Corollary 8.8, the graph of f has Jordan content zero. $\qquad\square$

Corollary 8.11 *Suppose S is a bounded set in the (x,y)-plane whose boundary consists of a finite number of curves, each of which is the graph of a continuous function $y = f(x)$ or $x = \varphi(y)$. Then S is Jordan measurable.*

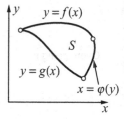

Proof. Each graph has Jordan content zero. There are only finitely many in ∂S, so ∂S likewise has Jordan content zero. By Theorem 8.2 (p. 282), S itself is Jordan measurable. $\qquad\square$

Theorem 8.12. *If S and T are Jordan measurable sets, then so are $S \cup T$ and $S \cap T$, and*

$$J(S \cup T) \leq J(S) + J(T),$$
$$J(S \cap T) \leq J(S), \quad J(S \cap T) \leq J(T).$$

Proof. Each boundary point of either $S \cup T$ or $S \cap T$ is a boundary point of S or of T:

$$\partial(S \cup T) \subseteq \partial S \cup \partial T, \quad \partial(S \cap T) \subseteq \partial S \cup \partial T.$$

By hypothesis, ∂S and ∂T have Jordan content zero; hence, so do their subsets $\partial(S \cup T)$ and $\partial(S \cap T)$ (Corollary 8.4). Consequently, $S \cup T$ and $S \cap T$ are both Jordan measurable. Theorem 8.5 then implies

$$J(S \cup T) = \bar{J}(S \cup T) \leq \bar{J}(S) + \bar{J}(T) = J(S) + J(T).$$

Because $S \cap T \subseteq S$, Theorem 8.3 implies $J(S \cap T) = \bar{J}(S \cap T) \leq \bar{J}(S) = J(S)$. Finally, $S \cap T \subseteq T$, so the same argument gives $J(S \cap T) \leq J(T)$. $\qquad\square$

Definition 8.10 *Two sets **overlap** if their interiors have a nonempty intersection. They are **nonoverlapping** if their interiors are disjoint.*

Theorem 8.13. *If S and T are bounded Jordan measurable sets that do not overlap, then*
$$J(S \cup T) = J(S) + J(T).$$

Proof. Because S and T have disjoint interiors, a grid square Q_1 that counts in the area $\underline{J}_k(S)$ cannot be entirely contained in T, so it does not count in $\underline{J}_k(T)$. Similarly, a grid square that counts in $\underline{J}_k(T)$ does not count in $\underline{J}_k(S)$. Of course, every grid square that counts in one or the other counts in $\underline{J}_k(S \cup T)$, so

$$\underline{J}_k(S) + \underline{J}_k(T) \le \underline{J}_k(S \cup T).$$

In the limit, $\underline{J}(S) + \underline{J}(T) \le \underline{J}(S \cup T)$, and then $J(S) + J(T) \le J(S \cup T)$ because S, T, and $S \cup T$ are Jordan measurable. Finally, with Theorem 8.12 we have

$$J(S) + J(T) = J(S \cup T). \qquad \square$$

Finite additivity of Jordan content

This leads immediately to the **finite additivity** of Jordan content.

Corollary 8.14 *If $S_1, \ldots, S_p$ are Jordan measurable sets, and no two overlap, then $S_1 \cup \cdots \cup S_p$ is Jordan measurable and*

$$J(S_1 \cup \cdots \cup S_p) = J(S_1) + \cdots + J(S_p). \qquad \square$$

Overlapping sets

When sets overlap, there is still a definite relation between the area of their union and their individual areas. We make use of the **set difference** $T \setminus S$; this is the set of points in T that are not in S (i.e., $T \setminus S = T \cap S^c$). The definition does not assume $S \subseteq T$.

Lemma 8.3. *Suppose T and $S \subseteq T$ are Jordan measurable; then $T \setminus S$ is Jordan measurable and $J(T \setminus S) = J(T) - J(S)$.*

Proof. Becuase $\partial(T \setminus S) \subseteq \partial T \cup \partial S$ (see Exercise 8.14.b), $\partial(T \setminus S)$ has area zero and $T \setminus S$ is Jordan measurable. Because $T = (T \setminus S) \cup S$ and $T \setminus S$ and S do not overlap (they are disjoint), we have

$$J(T) = J(T \setminus S) + J(S). \qquad \square$$

Theorem 8.15. *Suppose S and T are Jordan measurable; then*

$$J(S \cup T) + J(S \cap T) = J(S) + J(T).$$

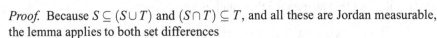

Proof. Because $S \subseteq (S \cup T)$ and $(S \cap T) \subseteq T$, and all these are Jordan measurable, the lemma applies to both set differences

$$(S \cup T) \setminus S = T \setminus (S \cap T).$$

Because the set differences are equal, the lemma implies

$$J(S \cup T) - J(S) = J(T) - J(S \cap T);$$

to get the theorem, just rearrange the terms. $\qquad \square$

Another way to state the theorem that perhaps indicates more explicitly how the overlap $S \cap T$ affects the area of the union is

$$J(S \cup T) = J(S) + J(T) - J(S \cap T).$$

Area under a graph

We can equate the integral of a function with the Jordan content of the region under its graph, making a useful connection between area and Jordan content.

Theorem 8.16. *Let* $y = f(x)$ *be continuous and nonnegative on* $a \leq x \leq b$, *and let* S *be the region in the* (x,y)-*plane bounded by the vertical lines* $x = a$ *and* $x = b$, *the* x-*axis, and the graph of* f. *Then* S *is Jordan measurable and*

$$J(S) = \int_a^b f(x)\,dx.$$

Proof. The boundary of S has Jordan content zero (Theorem 8.10), so S is Jordan measurable.

To get estimates for the integral of f, subdivide the interval $a \leq x \leq b$ into K equal pieces $I_1, \ldots, I_K$, each of length $\Delta x = (b - a)/K$. Let m_j and M_j the the minimum and maximum values of $y = f(x)$ on I_j; then

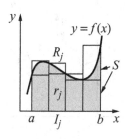

$$m_1 \Delta x + \cdots + m_K \Delta x \leq \int_a^b f(x)\,dx \leq M_1 \Delta x + \cdots + M_K \Delta x,$$

and these bounds converge to the value of the integral as $K \to \infty$.

Now let r_j be the rectangle with base I_j and height m_j, and let R_j be the rectangle with the same base but height M_j. Then $r_1, \ldots, r_K$ are nonoverlapping rectangles whose union is contained in S, and $R_1, \ldots, R_K$ are nonoverlapping rectangles whose union contains S:

$$r_1 \cup \cdots \cup r_K \subseteq S \subseteq R_1 \cup \cdots \cup R_K.$$

By the finite additivity of Jordan content (Corollary 8.14),

$$J(r_1 \cup \cdots \cup r_K) = J(r_1) + \cdots + J(r_k),$$
$$J(R_1 \cup \cdots \cup R_K) = J(R_1) + \cdots + J(R_K),$$

and the set inclusions imply

$$J(r_1) + \cdots + J(r_k) \leq J(S) \leq J(R_1) + \cdots + J(R_K).$$

But $J(r_j) = m_j \Delta x$ and $J(R_j) = M_j \Delta x$, so

$$m_1 \Delta x + \cdots + m_K \Delta x \leq J(S) \leq M_1 \Delta x + \cdots + M_K \Delta x.$$

Thus, $J(S)$ has the same bounds as the integral; these bounds therefore converge to $J(S)$ as well as to the integral. $\square$

In elementary geometry, congruent figures have the same area; we now prove they have the same Jordan content, too. We need to show that Jordan content is preserved by the translations, rotations, and reflections that link congruent figures. Such invariance is not immediately obvious, because we have restricted ourselves to a single collection of grids $\mathcal{J}_k$. For example, a translate of S does not, in general, have the same relation to $\mathcal{J}_k$ that S itself does. However, if we translate the grid as well and can show that the translated grid yields the same Jordan content as the original grid, then it follows that Jordan content is invariant under translations. This leads us to the task of showing that the Jordan content of a set can be determined

Jordan content
via other grids

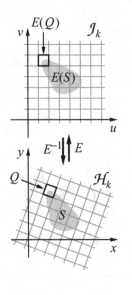

just as well by many other grids. We concentrate on grids obtained from $\mathcal{I}_k$ by a Euclidean motion or, more generally, by a coordinate change.

For simplicity, we begin with a Euclidean motion $E : \mathbb{R}^2 \to \mathbb{R}^2$. The action of E is given by an orthogonal matrix R (either a rotation or a reflection) followed by a translation. We can write formulas for E and its inverse as

$$\mathbf{u} = E(\mathbf{x}) = R\mathbf{x} + \mathbf{a}, \quad \mathbf{x} = E^{-1}(\mathbf{u}) = R^{-1}(\mathbf{u} - \mathbf{a}) = R^\dagger \mathbf{u} + \mathbf{b},$$

where $\mathbf{b} = -R^\dagger \mathbf{a}$ and $R^\dagger = R^{-1}$ because R is orthogonal. If S is a bounded set in the plane, then its image $E(S)$ under a Euclidean motion stands in the same relation to the grid $\mathcal{I}_k$ that S itself does to the new grid $\mathcal{H}_k = E^{-1}(\mathcal{I}_k)$. Each time a square Q in $\mathcal{H}_k$ counts in estimating the area of S, its image $E(Q)$ in $\mathcal{I}_k$ counts in estimating the area of $E(S)$.

But there is still a problem, because we do not know that Q has the same Jordan content as $E(Q)$. (This is precisely the question we are trying to settle: the invariance of Jordan content under Euclidean motions!) Nevertheless, Q is certainly Jordan measurable, because ∂Q is just a finite collection of line segments with Jordan content zero. Making no assumption about the value of $J(Q)$, we now construct the analogues of $\underline{J}_k$, $\overline{J}_k$, $\underline{J}$, and $\overline{J}$ for the grids $\mathcal{H}_k$ (cf. Definition 8.6, page 281).

$\underline{H}_k$ and $\overline{H}_k$ for $\mathcal{H}_k$

Definition 8.11 Let $\underline{H}_k(S)$ denote the total Jordan content of the squares in $\mathcal{H}_k$ that are entirely contained in the bounded set S, and let $\overline{H}_k(S)$ denote the total Jordan content of the squares in $\mathcal{H}_k$ that intersect S.

We can express these very compactly using set and summation notation:

$$\underline{H}_k(S) = \sum_{\substack{Q \in \mathcal{H}_k \\ Q \subseteq S}} J(Q), \quad \overline{H}_k(S) = \sum_{\substack{Q \in \mathcal{H}_k \\ Q \cap S \neq \phi}} J(Q)$$

The values of $\underline{H}_k(S)$ and $\overline{H}_l(S)$ are nested in the same way as the values of $\underline{J}_k(S)$ and $\overline{J}_l(S)$; this gives us the limits

$$\underline{H}(S) = \lim_{k \to \infty} \underline{H}_k(S), \quad \overline{H}(S) = \lim_{k \to \infty} \overline{H}_k(S),$$

and the inequality $\underline{H}(S) \leq \overline{H}(S)$.

H content

Definition 8.12 If $\underline{H}(S) = \overline{H}(S)$, we say that S is H measurable and we define the H content of S to be $H(S) = \underline{H}(S) = \overline{H}(S)$.

Lemma 8.4. Suppose S is Jordan measurable, then $\underline{H}_k(S) \leq J(S) \leq \overline{H}_k(S)$ for every $k = 0, 1, 2, \ldots$.

Proof. Suppose $Q_1, \ldots, Q_p$ are the squares of $\mathcal{H}_k$ that are counted in $\underline{H}_k(S)$. Because they are nonoverlapping and $Q_1 \cup \cdots \cup Q_p \subseteq S$, the finite additivity of J implies

$$\underline{H}_k(S) = J(Q_1) + \cdots + J(Q_p) = J(Q_1 \cup \cdots \cup Q_p) \leq J(S).$$

The inequality $J(S) \leq \overline{H}_k(S)$ is proven in a similar way. $\square$

Corollary 8.17 *Suppose a Jordan measurable set S is also H measurable, then* $H(S) = J(S)$. $\qquad\qquad\qquad\qquad\qquad\qquad\qquad\qquad\qquad\qquad\qquad\qquad\qquad\square$

But must a J-measurable set also be H measurable? The corollary directs our attention to the difference $\overline{H}_k(S) - \underline{H}_k(S)$, because

$$S \text{ is } H \text{ measurable} \Leftrightarrow \lim_{k\to\infty}\left(\overline{H}_k(S) - \underline{H}_k(S)\right) = 0$$

(cf. the comment at the beginning of the proof of Theorem 8.2, p. 282). Therefore, if we can show that $\overline{H}_k(S) - \underline{H}_k(S)$ is smaller than any preassigned $\varepsilon > 0$ when k is sufficiently large, we shall have shown that $H(S) = J(S)$. To complete our argument, it is useful to introduce the notion of a *tubular neighborhood*.

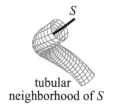

tubular
neighborhood of S

Definition 8.13 *If S is a bounded set, the **tubular neighborhood of S of width** $w > 0$ is the set of points T that are within distance w of some point of S.*

The tubular neighborhood of S of width w is the union of the open disks of radius w centered at all the points of S. If S is a smooth curve in space, and w is small enough, then the tubular neighborhood looks like a tube with S at its core; together, they resemble a coaxial cable.

Lemma 8.5. *Suppose S has Jordan content zero and* $\varepsilon > 0$ *is given. Then S has a tubular neighborhood T of some width* $\delta > 0$ *for which* $\overline{J}(T) < \varepsilon$.

Proof. Because $J(S) = 0$, we know $\overline{J}_k(S) \to 0$ as $k \to \infty$. Choose K so large that $\overline{J}_K(S) < \varepsilon/9$. The squares Q in $\mathcal{J}_K$ that are counted in $\overline{J}_K(S)$ cover S and have total Jordan content less than $\varepsilon/9$.

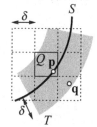

Define T to be the tubular neighborhood of S of width $\delta = 1/2^K$, and let $\mathbf{q}$ be a point in T. By definition, $\mathbf{q}$ is within distance $\delta = 1/2^K$ of some point $\mathbf{p}$ in S. Let Q be a square counted in $\overline{J}_K(S)$ that contains $\mathbf{p}$. Because the squares in $\mathcal{J}_K$ are closed and have width $\delta = 1/2^K$, the point $\mathbf{q}$ is either in Q or one of the eight neighbors of Q. Now $\mathbf{q}$ is an arbitrary point of T, so T is covered by the squares Q and their immediate neighbors, whose total Jordan content is less than $9 \times \varepsilon/9 = \varepsilon$; hence $\overline{J}(T) \le \overline{J}_K(T) < \varepsilon$. $\qquad\qquad\qquad\qquad\qquad\qquad\qquad\square$

Theorem 8.18. *Suppose S is Jordan measurable; then it is H measurable and* $J(S) = H(S)$.

Proof. By Corollary 8.17 and the discussion following it, we need only show that, given any $\varepsilon > 0$, there is a K such that

$$\overline{H}_K(S) - \underline{H}_K(S) < \varepsilon.$$

Note: the sequence $\overline{H}_k(S) - \underline{H}_k(S)$ decreases monotonically as k increases, so we also have $\overline{H}_k(S) - \underline{H}_k(S) < \varepsilon$ for all $k \ge K$.

By hypothesis, S is Jordan measurable, so ∂S has Jordan content zero. Therefore, using the given $\varepsilon > 0$ and Lemma 8.5, we know ∂S has a tubular neighborhood T of some width $\delta > 0$ for which $\overline{J}(T) < \varepsilon$. Now choose K so that the *diameter* (see below, Definition 8.14) of any grid square Q in $\mathcal{H}_K$ is less than δ. The diameter of Q is its diagonal length $\sqrt{2}/2^K < 1/2^{K-1}$, so it is sufficient to choose

$$K > 1 + \log_2(1/\delta).$$

Consider the difference $\overline{H}_K(S) - \underline{H}_K(S)$. Adapting the arguments of the proof of Theorem 8.2 from $\mathcal{J}_k$ to $\mathcal{H}_K$, we draw two conclusions:

- $\overline{H}_K(S) - \underline{H}_K(S)$ is the total Jordan content of the squares Q in $\mathcal{H}_K$ that meet S but are not entirely contained in S.

- Each such square Q contains a point of ∂S.

The diameter of Q is less than δ (by construction); thus the entire square Q lies within the tubular neighborhood T. Hence, the total Jordan content of the squares Q counted in $\overline{H}_K(S) - \underline{H}_K(S)$ is less than the outer Jordan content of T; that is,

$$\overline{H}_K(S) - \underline{H}_K(S) < \varepsilon. \qquad \square$$

Jordan content of congruent figures

One of our main objectives, to show that congruent figures have the same Jordan content, now follows as an immediate corollary.

Corollary 8.19 *If S is Jordan measurable and $E : \mathbb{R}^2 \to \mathbb{R}^2$ is a Euclidean motion, then $E(S)$ is Jordan measurable and $J(E(S)) = J(S)$.*

Proof. $J(E(S)) = H(S) = J(S)$. $\qquad \square$

Jordan content and ordinary area

Thus, if a rectangle R with sides of length l and w lies anywhere in the (x,y)-plane, a Euclidean motion E will transform it into the rectangle $E(R) : 0 \le x \le l, 0 \le y \le w$. By Exercise 8.8, $J(E(R)) = lw$; therefore, $J(R) = lw$. If P is a parallelogram with base b and height h, then P can be decomposed into nonoverlapping sets A and B so that A and a translate $E(B)$ are nonoverlapping and form a rectangle R with the same base and height. Therefore,

$$J(P) = J(A) + J(B) = J(A) + J(E(B)) = J(R) = bh.$$

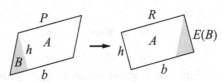

If T is a triangle with base b and height h, then similar geometric arguments show that $J(T) = \frac{1}{2}bh$. Because a polygon can be written as a union of nonoverlapping triangles, it follows that the Jordan content of any polygon equals its ordinary area.

General grids $\mathcal{G}_k$ and G content

Under what conditions will a more general collection $\mathcal{G}_k$ $(k = 0, 1, 2, \ldots)$ of grids on the plane determine Jordan content? We assume that each grid is a refinement of its predecessor, and also that the individual cells P of a grid are closed, connected, nonoverlapping sets with positive Jordan content that together cover $\mathbb{R}^2$. The cells need not be congruent or even straight-sided. We define:

- $\underline{G}_k(S)$ is the total Jordan content of the cells P of $\mathcal{G}_k$ that are entirely contained in S;

- $\overline{G}_k(S)$ is the total Jordan content of the cells P of $\mathcal{G}_k$ that meet S.

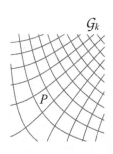

$\mathcal{G}_k$

$$\underline{G}_k(S) = \sum_{\substack{P \in \mathcal{G}_k \\ P \subseteq S}} J(P), \quad \overline{G}_k(S) = \sum_{\substack{P \in \mathcal{G}_k \\ P \cap S \neq \phi}} J(P).$$

Because $\mathcal{G}_k$ refines its predecessor, $\underline{G}_k(S)$ increases monotonically with k to its limiting value $\underline{G}(S)$, the inner G content of S. Likewise, $\overline{G}_k(S)$ decreases monotonically to the outer G content, $\overline{G}(S)$, and

$$\underline{G}_k(S) \leq \underline{G}(S) \leq \overline{G}(S) \leq \overline{G}_k(S).$$

If the inner and outer G content are equal, we say S is **G measurable** and has **G content** $G(S) = \underline{G}(S) = \overline{G}(S)$. Under what conditions will $G(S) = J(S)$? For H content, the answer was provided by Theorem 8.18, whose proof involved the *linear* dimensions (i.e., diameters) of the grid elements.

Definition 8.14 *The **diameter** of the set S, denoted $\delta(S)$, is the maximum distance between any two points in its closure $\overline{S}$. The **mesh size** $\|\mathcal{G}\|$ of a grid $\mathcal{G}$ is the smallest upper bound on the size of the diameters of the elements P of $\mathcal{G}$.*

Diameter $\delta(S)$ and
mesh size $\|\mathcal{G}\|$

Theorem 8.20. *If $\|\mathcal{G}_k\| \to 0$ as $k \to \infty$, then every Jordan-measurable set S is G measurable, and $J(S) = G(S)$.*

Proof. This argument imitates the proof of Theorem 8.18 and the lemmas preceding it. First of all, because S is Jordan measurable,

$$\underline{G}_k(S) \leq J(S) \leq \overline{G}_k(S)$$

(cf. Lemma 8.4). Hence, to prove the theorem it suffices to show

$$\lim_{k \to \infty} \left(\overline{G}_k(S) - \underline{G}_k(S) \right) = 0.$$

Suppose $\varepsilon > 0$ is given; we wish to show that there is an integer $K = K(\varepsilon)$ for which

$$\overline{G}_k(S) - \underline{G}_k(S) < \varepsilon,$$

for all $k > K$. Because S is Jordan measurable, $J(\partial S) = 0$ and ∂S has a tubular neighborhood T of some width δ for which $\overline{J}(T) < \varepsilon$ (Lemma 8.5). Because $\|\mathcal{G}_k\| \to 0$ as $k \to \infty$, we can choose K so that $\|\mathcal{G}_k\| < \delta$ for all $k > K$. Each cell P in the grid $\mathcal{G}_k$ contains a point $\mathbf{p}$ in S and a point $\mathbf{q}$ not in S. Because P is connected, there is a continuous path in P from $\mathbf{p}$ to $\mathbf{q}$, and that path must contain a point of ∂P (Exercise 8.6). If $k > K$, then the diameter of P is less than δ, so P is entirely contained within the tubular neighborhood T. Hence the total Jordan content of the cells P counted in $\overline{G}_k(S) - \underline{G}_k(S)$ is less than $\overline{J}(T) < \varepsilon$; that is,

$$\overline{G}_k(S) - \underline{G}_k(S) < \varepsilon. \qquad \qquad \square$$

We are now able to extend to Jordan content our earlier observation about the area magnification factor of a linear map (Theorem 2.8, p. 42). We make frequent use of the fact that the Jordan content of a polygon P is its ordinary (absolute) area; thus, if $L(P)$ is its polygonal image under a linear map, then

$$J(L(P)) = |\det L|\, J(P).$$

Lemma 8.6. *If* $L : \mathbb{R}^2 \to \mathbb{R}^2$ *is linear and* $J(S) = 0$, *then* $J(L(S)) = 0$.

Proof. If L is not invertible, the proof is immediate, because the whole image $L(\mathbb{R}^2)$ is a line, so $L(S)$ is a finite line segment and automatically has Jordan content zero.

If L is invertible (i.e., $\det L \neq 0$), then we show that $L(S)$ is contained in a union of sets whose total Jordan content is less than any positive number ε. The lemma then follows from Corollary 8.8 (p. 283).

By hypothesis, $J(S) = 0$ so $\overline{J}_K(S) < \varepsilon/|\det L|$ for some integer K. That is, S is covered by squares Q whose total Jordan content is less than $\varepsilon/|\det L|$. Therefore $L(S)$ is covered by the images $L(Q)$ of those squares. Because $J(L(Q)) = |\det L|\, J(Q)$, the total Jordan content of the sets covering $L(S)$ is less than ε. $\qquad \square$

Corollary 8.21 *The image of a Jordan-measurable set under a linear map is Jordan measurable.* $\qquad \square$

We can now show that the Jordan content multiplier of a linear map is the absolute value of its determinant.

Theorem 8.22. *Suppose* S *is Jordan measurable and* $L : \mathbb{R}^2 \to \mathbb{R}^2$ *is linear; then* $J(L(S)) = |\det L|\, J(S)$.

Proof. If L is not invertible, then $\det L = 0$, and $L(S)$ is a bounded subset of the line $L(\mathbb{R}^2)$. Thus $J(L(S)) = 0 = |\det L|\, J(S)$.

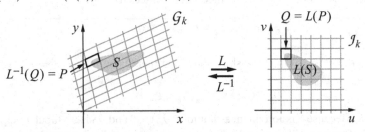

For an invertible map L, we adapt the argument we used to prove that a Euclidean motion preserves Jordan content (pp. 287–290). The key to the argument is to note that $L(S)$ has the same relation to the grid $\mathcal{J}_k$ that S itself does to the new grid $\mathcal{G}_k = L^{-1}(\mathcal{J}_k)$ (see the figure below). Of course, to use the G-content functions associated with $\mathcal{G}_k$ (as defined above, p. 291), we want Theorem 8.20 to hold. We must therefore check that the maximum diameter $\|\mathcal{G}_k\|$ of a cell P of $\mathcal{G}_k$ tends to zero as $k \to \infty$. Exercise 8.16 establishes that the diameters of Q and $P = L^{-1}(Q)$ are linked by a constant σ that depends only on L^{-1} and not on k: $\delta(P) = \sigma\delta(Q)$. Thus all cells P of G_k have the same diameter: $\|G_k\| = \sigma\delta(Q)$. Because $\delta(Q) = \sqrt{2}/2^k$, it follows that $\|G_k\| \to 0$ as $k \to \infty$. By Theorem 8.20, the given Jordan-measurable set S is also G measurable, and $J(S) = G(S)$.

As P is contained in S precisely when $Q = L(P)$ is contained in $L(S)$, we have (using $P = L^{-1}(Q)$ as well)

$$\underline{G}_k(S) = \sum_{\substack{P \in G_k \\ P \subseteq S}} J(P) = \sum_{\substack{Q \in \mathcal{I}_k \\ Q \subseteq L(S)}} J(L^{-1}(Q)).$$

Because Q is a polygon, $J(L^{-1}(Q)) = |\det L^{-1}| J(Q)$, so the last sum becomes

$$|\det L^{-1}| \sum_{\substack{Q \in \mathcal{I}_k \\ Q \subseteq L(S)}} J(Q) = |\det L^{-1}| \underline{J}_k(L(S)),$$

or just $\underline{G}_k(S) = |\det L^{-1}| \underline{J}_k(L(S))$. Because $|\det L^{-1}| = |\det L|^{-1}$, we have

$$\underline{J}_k(L(S)) = |\det L| \underline{G}_k(S).$$

In the limit as $k \to \infty$, $J(L(S)) = |\det L| G(S) = |\det L| J(S)$. □

When we take up integration in the next section, we need an even larger class of grids than sequences $G_0, G_1, G_2, \ldots$ of successive refinements of the sort we have considered so far. The nonnegative integers that we use to index these grids have a natural order that is imparted to the grids themselves: there is a "first" grid, then a "second," and so on. When we say that each grid refines its predecessor, we make implicit use of that order.

So when we enlarge the class of grids, such a larger collection $\{G\}$ typically has no natural ordering. Thus, even though one grid in the collection may be a refinement of another, the notion of "predecessor" is now missing, and we are no longer able to say that a grid refines its predecessor. Nevertheless, we still assume the cells P in any grid are closed, connected, nonoverlapping sets with positive Jordan content, and together they cover $\mathbb{R}^2$. Then we define (exactly as we did for the more restricted class of grids G_k):

- $\underline{G}_G(S)$ is the total Jordan content of the cells P of G that are entirely contained in S.

- $\overline{G}_G(S)$ is the total Jordan content of the cells P of G that meet S.

$$\underline{G}_G(S) = \sum_{\substack{P \in G \\ P \subseteq S}} J(P), \quad \overline{G}_G(S) = \sum_{\substack{P \in G \\ P \cap S \neq \phi}} J(P)$$

We should next get inner and outer G content (i.e., $\underline{G}(S)$ and $\overline{G}(S)$). When grids were indexed by integers $k = 0, 1, 2, \ldots$, we just took limits of $\underline{G}_k(S)$ and $\overline{G}_k(S)$ as $k \to \infty$ (and monotonicity guaranteed that the limits existed). But for an arbitrary

Grids for integration

$\underline{G}_G(S)$ and $\overline{G}_G(S)$ vary with $\|G\|$

collection of grids $\{\mathcal{G}\}$, there is no index that supplies an ordering. Nevertheless, if we consider how $\underline{G}_{\mathcal{G}}(S)$ and $\overline{G}_{\mathcal{G}}(S)$ vary with mesh size $\|\mathcal{G}\|$, we see that they do have well-defined limits, at least if S has Jordan content. We then have a way, once again, to define G content.

To see how this happens, suppose S is Jordan measurable. Then, for any grid $\mathcal{G}$,

$$\underline{G}_{\mathcal{G}}(S) \le J(S) \le \overline{G}_{\mathcal{G}}(S)$$

(cf. Lemma 8.4). Therefore, if we can show that

$$\lim_{\|\mathcal{G}\|\to 0} \left(\overline{G}_{\mathcal{G}}(S) - \underline{G}_{\mathcal{G}}(S) \right) = 0,$$

then we know the following limits exist:

$$\underline{G}(S) = \lim_{\|\mathcal{G}\|\to 0} \underline{G}_{\mathcal{G}}(S) = J(S), \quad \overline{G}(S) = \lim_{\|\mathcal{G}\|\to 0} \overline{G}_{\mathcal{G}}(S) = J(S).$$

We can then say

- S is G measurable.

- $G(S) = \underline{G}(S) = \overline{G}(S)$.

- $J(S) = G(S)$.

Hence, for any given $\varepsilon > 0$, we must show there is a $\delta > 0$ for which

$$\|\mathcal{G}\| < \delta \quad \Rightarrow \quad \overline{G}_{\mathcal{G}}(S) - \underline{G}_{\mathcal{G}}(S) < \varepsilon.$$

By Lemma 8.5 we know that ∂S has a tubular neighborhood T of some positive width δ for which $\overline{J}(T) < \varepsilon$. Now suppose $\|\mathcal{G}\| < \delta$. Then, by essentially the same argument as in the proof of Theorem 8.20, the total Jordan content of the cells P that are counted in $\overline{G}_{\mathcal{G}}(S) - \underline{G}_{\mathcal{G}}(S)$ is less than $\overline{J}(T) < \varepsilon$. In other words,

$$\|\mathcal{G}\| < \delta \quad \Rightarrow \quad \overline{G}_{\mathcal{G}}(S) - \underline{G}_{\mathcal{G}}(S) < \varepsilon,$$

as required. We state the conclusion as a theorem.

Theorem 8.23. *If $\{\mathcal{G}\}$ is an infinite collection of integration grids whose mesh sizes $\|\mathcal{G}\|$ come arbitrarily close to zero, then G content is defined for all sets S that have Jordan content, and $G(S) = J(S)$.* □

Use *area* to denote *Jordan content*

As we have seen, the Jordan content of any plane figure of elementary Euclidean geometry is its ordinary area. For that reason, we now go back to the simpler and more familiar term *area*. Thus, we say that S **has area** if it is Jordan measurable; in that case, the **area** of S is denoted $A(S) = J(S)$. For a more general set S, its **inner area** $\underline{A}(S)$ is its inner Jordan content $\underline{J}(S)$, and its **outer area** $\overline{A}(S)$ is its outer Jordan content $\overline{J}(S)$.

Volume in $\mathbb{R}^3$

With grids of cubes instead of squares, it requires virtually no alterations to transfer the theory of Jordan content from $\mathbb{R}^2$ to $\mathbb{R}^3$. Let us assume that has been done. Then, following what we just did in $\mathbb{R}^2$, we call the Jordan content of a set S in $\mathbb{R}^3$ its **volume**, and write $V(S) = J(S)$. In fact, we can use the same method to get the analogue of area or volume in any dimension.

8.3 Riemann and Darbou integrals

We now introduce double integrals and establish their properties; in the next chapter we develop methods for evaluating them. We define the Riemann integral of a function $z = f(x,y)$ on a closed bounded set S that has area (i.e., is Jordan measurable). We assume f is bounded on S and is extended to all of $\mathbb{R}^2$ by setting $f(x,y) = 0$ when (x,y) is not in S.

Let G be a grid of the sort we considered near the end of the previous section. Thus, the cells Q of G are closed, bounded, nonoverlapping sets that have area. We let $A(Q)$ denote the area of Q and $\delta(Q)$ its diameter (cf. p. 291); the diameters have a finite bound $\|G\|$, the *mesh size* of G. The cells of G must cover S, but they need not cover all of $\mathbb{R}^2$. Furthermore, those cells need not be congruent, nor need they have straight sides. We call G an **integration grid**.

Integration grids

Let $Q_1, \ldots, Q_N$ be all the cells of G that meet S; we write the area $A(Q_i)$ as ΔA_i. A **Riemann sum for f over S** is an expression of the form

Riemann sums

$$\sum_{i=1}^{N(G)} f(x_i, y_i)\, \Delta A_i,$$

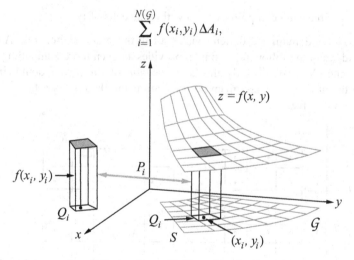

where (x_i, y_i) is a point of Q_i, $i = 1, \ldots, N$. Note that the sum depends upon the grid G and the points (x_i, y_i), as well as on f and S. By writing $N = N(G)$ as well, we call attention to the fact that the number of cells that meet S depends on G.

To interpret such a sum it helps to let f be positive and continuous, as in the figure above. Then $f(x_i, y_i)$ is approximately the height of the prism P_i that has base Q_i, vertical sides, and an irregular top formed by the graph of f; $f(x_i, y_i)\, \Delta A_i$ is approx-

Approximating volumes

imately its volume. The Riemann sum therefore approximates the total volume of
the solid that lies above S and under the graph. To get a better aproximation, make
the individual cells smaller; more exactly, use a new grid G with a smaller mesh
size $\|G\|$. In fact, we expect all Riemann sums will be as close as we wish to the
actual volume, as long as the mesh size is sufficiently small, *independently of the
way the points are chosen in a grid*. This leads us to the definition of the *Riemann
integral*.

The Riemann integral

Definition 8.15 *If the Riemann sums for f over S have a limit that is independent
of the points (x_i, y_i) as $\|G\| \to 0$, then we say **f is integrable over S** and the **integral**
is that limit:*

$$\iint_S f(x,y)\, dA = \lim_{\|G\| \to 0} \sum_{i=1}^{N(G)} f(x_i, y_i)\, \Delta A_i.$$

More exactly, this a *Riemann* integral, which is different from the Darboux integral
that we introduce presently, as well as from other kinds of integrals that we do not
consider. To make it clear that convergence to the limit is uniform with respect to the
points chosen in the cells of a grid, we put the definition more formally, as follows.
Given an $\varepsilon > 0$, there is a $\delta > 0$ such that, for every grid G with $\|G\| < \delta$,

$$\left| \sum_{i=1}^{N(G)} f(x_i, y_i)\, \Delta A_i - \iint_S f(x,y)\, dA \right| < \varepsilon,$$

regardless of the choice of points (x_i, y_i) within the cells of G.

Absolute (unoriented)
double integrals

Because the domain is 2-dimensional, we call this a *double* integral. (A rectangular grid, as in the following example, provides an even more compelling reason
for the name.) More particularly, this is an *absolute*, or *unoriented*, double integral,
because the grid cells Q_i are given no orientation and their areas $\Delta A_i = A(Q_i)$ are
always nonnegative.

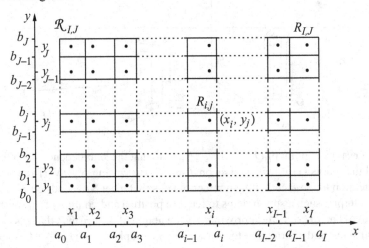

Here is a special class of grids that are frequently used to construct Riemann sums. First partition the x- and y-axes into nonoverlapping intervals:

$$[a_{i-1}, a_i] : a_{i-1} \leq x \leq a_i, \quad i = 1, \ldots I,$$
$$[b_{j-1}, b_j] : b_{j-1} \leq y \leq b_j, \quad j = 1, \ldots J.$$

Let $\Delta x_i = a_i - a_{i-1}$ and $\Delta y_j = b_j - b_{j-1}$ denote the lengths of these intervals. Now form the rectangles $R_{i,j} : [a_{i-1}, a_i] \times [b_{j-1}, b_j]$; the area of $R_{i,j}$ is the product $A(R_{i,j}) = \Delta x_i \Delta y_j$. These rectangles are cells of a grid $\mathcal{R}_{I,J}$ that is natural to index with a *pair* of integers I, J (in contrast, e.g., to the grid $\mathcal{I}_k$).

Let x_i be a point in the x-interval $[a_{i-1}, a_i]$, and let y_j be a point in the y-interval $[b_{j-1}, b_j]$. Then (x_i, y_j) is a point in $R_{i,j}$, and a Riemann sum for f over S naturally takes the form of a *double* sum:

$$\sum_{i=1}^{I} \sum_{j=1}^{J} f(x_i, y_j) \Delta x_i \Delta y_j.$$

If f is integrable, then these sums approach the integral of f as the grid mesh size tends to zero. In that case it is natural to replace the "element of area" dA by $dx\,dy$ and to write the limit itself as

$$\iint_S f(x,y)\,dx\,dy = \lim_{\|\mathcal{R}\| \to 0} \sum_{i=1}^{I(\mathcal{R})} \sum_{j=1}^{J(\mathcal{R})} f(x_i, y_j) \Delta x_i \Delta y_j.$$

The two summation signs on the right now explain why we use a pair of integral signs to denote the integral, and they also suggest why we call it a *double* integral.

As we have already noted, the terms in a Riemann sum are products of lengths $f(x_i, y_i)$ and areas ΔA_i, so we usually think of a Riemann sum and the resulting integral as *volumes*. We pursue this below, but first we make a connection between double integrals and areas.

Theorem 8.24. *A constant function $f(x,y) = c$ is integrable over every set S that has area, and*

$$\iint_S c\,dA = cA(S) = c \times \text{area}\,S.$$

Proof. Let $\{\mathcal{G}\}$ be an infinite collection of integration grids whose mesh sizes $\|\mathcal{G}\|$ get arbitrarily close to zero. Let $\overline{G}_G$ be the outer content function associated with $\mathcal{G}$ (cf. p. 293). Because $f(x,y) = 0$ outside S, by construction, the grid cells Q_i of $\mathcal{G}$ that make a nonzero contribution to a Riemann sum for f are precisely the ones that meet S; thus

$$\sum_{i=1}^{N} c\Delta A_i = c \sum_{Q_i \cap S \neq \phi} A(Q_i) = c\,\overline{G}_G(S).$$

The collection $\{\mathcal{G}\}$ satisfies the hypotheses of Theorem 8.23; therefore, because S has area, it is G measurable and $G(S) = A(S)$. Hence, $\overline{G}_G(S) \to A(S)$ as $\|\mathcal{G}\| \to 0$.

This limit does not depend on the collection $\{G\}$; therefore we conclude all Riemann sums have the same limit, namely $cA(S)$, and f is thus integrable. $\square$

Theorem 8.25. *Any bounded function f on a set S of zero area is integrable, and*

$$\iint_S f(x,y)\, dA = 0.$$

Proof. Suppose B is a bound on f: $|f(x,y)| \le B$ for all (x,y) in $\mathbb{R}^2$. Let $\varepsilon > 0$ be given. By Lemma 8.5, we can construct a tubular neighborhood of S of width $\delta > 0$ for which $\bar{J}(T) < \varepsilon/B$. Let G be any grid for which $\|G\| < \delta$. In a Riemann sum, the cells Q_i of G that meet S are contained entirely in T; thus, for any choice of $\mathbf{p}_i$ in Q_i, we have

$$\left| \sum_{i=1}^N f(\mathbf{p}_i)\Delta A_i - 0 \right| \le \sum_{i=1}^N |f(\mathbf{p}_i)|\Delta A_i \le B \sum_{i=1}^N \Delta A_i \le B\,\bar{J}(T) < \varepsilon.$$

This shows that f is integrable and the value of the integral is 0. $\square$

Properties of double integrals

Arguments similar to those in the last proof can be used to establish the following general properties of double integrals; see the exercises. These properties are formally the same as those of integrals of a single-variable function.

Theorem 8.26. *Suppose f and g are integrable over S; then so are cf and $f \pm g$, and*

$$\iint_S cf\, dA = c\iint_S f\, dA, \quad \iint_S (f \pm g)\, dA = \iint_S f\, dA \pm \iint_S g\, dA. \qquad \square$$

Theorem 8.27. *Suppose f is integrable over the nonoverlapping sets R and S; then f is integrable over $R \cup S$ and*

$$\iint_{R\cup S} f\, dA = \iint_R f\, dA + \iint_S f\, dA. \qquad \square$$

Additivity and linearity of the integral

The second theorem says that the integral is additive over sets in the same sense that area (Jordan content) is; see Corollary 8.14, page 286. The first theorem says that the integral acts as a linear operator on functions.

Theorem 8.28. *Suppose f is integrable over S and $f(x,y) \ge 0$ on S; then*

$$\iint_S f(x,y)\, dA \ge 0. \qquad \square$$

Corollary 8.29 *Suppose f and g are integrable over S, and $f(x,y) \le g(x,y)$ for every point (x,y) in S; then*

$$\iint_S f(x,y)\, dA \le \iint_S g(x,y)\, dA. \qquad \square$$

Corollary 8.30 *Suppose f is integrable over S and $m \leq f(x,y) \leq M$ for every point (x,y) in S; then*

$$m A(S) \leq \iint_S f(x,y)\,dA \leq MA(S). \qquad \square$$

Proof. By Theorem 8.26, $g(x,y) = f(x,y) - m \geq 0$ is integrable, and

$$0 \leq \iint_S g(x,y)\,dA = \iint_S f(x,y)\,dA - \iint_S m\,dA = \iint_S f(x,y)\,dA - mA(S),$$

which leads to the first of the stated inequalities. The second is obtained in a similar way. $\qquad \square$

We have a standing assumption that integration applies only to bounded functions. That assumption is essential in the last corollary. For example, let S be the unit interval on the x-axis, so $A(S) = 0$ as a region in the plane. Let

Integrate only bounded functions

$$f(x,y) = \begin{cases} 1/\sqrt{x} & 0 < x \leq 1,\ y = 0, \\ 0 & \text{otherwise.} \end{cases}$$

Now consider Riemann sums for f on S constructed with the grids $\mathcal{I}_n$ used to define Jordan content. Let Q_1 be the square $[0, 1/2^n] \times [0, 1/2^n]$ and let $\mathbf{p}_1 = (1/2^{6n}, 0)$. Then

$$f(\mathbf{p}_1)\Delta A_1 = \frac{1}{\sqrt{1/2^{6n}}} \cdot \frac{1}{2^{2n}} = 2^{3n} \cdot \frac{1}{2^{2n}} = 2^n \to \infty \quad \text{as } n \to \infty,$$

so no Riemann sum that contains this term can converge to a finite value. In other words, f is not integrable, even though its domain has area zero.

We now turn to the Darboux integral. It is constructed from sums involving upper and lower bounds of a function on each cell of a grid. Although the Darboux and Riemann integrals have similar definitions (and we eventually show they have the same value), the Darboux integral is defined more in the style of Jordan content: there are analogues of inner and outer areas on a grid and inner and outer content as limits.

Bounds and the Darboux integral

Suppose f is bounded on S, and $\mathcal{G}$ is an integration grid whose cells $Q_1, \ldots, Q_N$ cover S. We do not assume f is Riemann integrable over S. Let M_i be the smallest of the upper bounds (the "least upper bound") for f on Q_i, and let m_i be the largest of the lower bounds (the "greatest lower bound"). In other words,

Least upper bound; greatest lower bound

$$m_i \leq f(\mathbf{p}_i) \leq M_i$$

for all points $\mathbf{p}_i$ in Q_i, and these bounds are the best possible. To see what "best possible" means here, consider the following example:

$$f(x) = \begin{cases} 1 & x = 0, \\ x & 0 < x \leq 1, \end{cases}$$

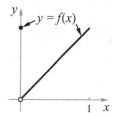

where $Q : 0 \leq x \leq 1$. Then 0 is a lower bound for f on Q, but no larger number $\varepsilon > 0$ is a lower bound, because we can always choose $\bar{x}$ in Q so that $f(\bar{x}) < \varepsilon$ (e.g., take $\bar{x} = \varepsilon/2$). Thus $m = 0$ is the *greatest* lower bound, which is also called an *infimum*. Because there is no "smallest" positive real number, the set of values $f(x)$ has no minimum, but it does have an *infimum*.

inf and sup

Definition 8.16 *Any nonempty set of numbers Z that is bounded below has a **greatest lower bound**, glb Z, or **infimum**, inf Z; if Z is bounded above, it has a **least upper bound**, lub Z, or **supremum**, sup Z.*

Lower and upper
Darboux sums

Definition 8.17 *The **lower** and **upper Darboux sums** for f over S and the grid $\mathcal{G}$ are, respectively,*

$$\underline{D}_{\mathcal{G}}(f,S) = \sum_{i=1}^{N} m_i \Delta A_i, \quad \overline{D}_{\mathcal{G}}(f,S) = \sum_{i=1}^{N} M_i \Delta A_i.$$

Lower and upper Darboux sums give us lower and upper bounds on all possible Riemann sums that can be constructed with the grid $\mathcal{G}$; that is,

$$\underline{D}_{\mathcal{G}}(f,S) \leq \sum_{i=1}^{N} f(\mathbf{p}_i) \Delta A_i \leq \overline{D}_{\mathcal{G}}(f,S),$$

no matter how $\mathbf{p}_i$ is chosen in Q_i. The following lemma, which says no lower sum is larger than any upper sum, plays the same role here that Lemma 8.2, page 281, does for the inner and outer area estimates of Jordan content.

Lemma 8.7. *For every pair of integration grids $\mathcal{H}$ and $\mathcal{G}$,*

$$\underline{D}_{\mathcal{H}}(f,S) \leq \overline{D}_{\mathcal{G}}(f,S).$$

Proof. We construct the **common refinement**, $\mathcal{K}$, of $\mathcal{H}$ and $\mathcal{G}$. The cells of $\mathcal{K}$ consist of the intersections $P \cap Q$, where P is a cell in $\mathcal{H}$ and Q is a cell in $\mathcal{G}$. Then $\mathcal{K}$ does indeed refine both $\mathcal{H}$ and $\mathcal{G}$, so the usual arguments about refinements imply

$$\underline{D}_{\mathcal{H}}(f,S) \leq \underline{D}_{\mathcal{K}}(f,S) \leq \overline{D}_{\mathcal{K}}(f,S) \leq \overline{D}_{\mathcal{G}}(f,S). \qquad \square$$

Thus each upper sum is an upper bound for all lower sums, and each lower sum is a lower bound for all upper sums. Consequently, the following least upper bound and the greatest lower bound are well-defined.

Lower and upper
Darboux integrals

Definition 8.18 *The **lower Darboux integral** $\underline{D}(f,S)$ **of** f **over** S is the least upper bound of the numbers $\underline{D}_{\mathcal{G}}(f,S)$, over all grids $\mathcal{G}$. Similarly, the **upper Darboux integral** $\overline{D}(f,S)$ is the greatest lower bound of the numbers $\overline{D}_{\mathcal{G}}(f,S)$.*

Theorem 8.31. $\underline{D}(f,S) \leq \overline{D}(f,S)$.

Proof. Choose $\mathcal{G}$ arbitrarily; by Lemma 8.7, $\overline{D}_{\mathcal{G}}(f,S)$ is an upper bound for all possible lower sums, so it is at least as large as their least upper bound:

$$\underline{D}(f,S) \le \overline{D}_G(f,S).$$

By this inequality, the lower integral $\underline{D}(f,S)$ is a lower bound for all possible upper sums (because G is arbitrary), so it is at least as small as their greatest lower bound:

$$\underline{D}(f,S) \le \overline{D}(f,S). \qquad \square$$

Definition 8.19 *If $\underline{D}(f,S) = \overline{D}(f,S)$, then f is **Darboux integrable** over S, and its Darboux integral is $D(f,S) = \underline{D}(f,S) = \overline{D}(f,S)$.*

The Darboux integral

The next two theorems establish that the two notions of integral are equivalent.

Theorem 8.32. *If f is Riemann integrable on S, then it is also Darboux integrable, and the two integrals are equal.*

Proof. Because f is bounded, its upper and lower Darboux sums are defined for all grids. To prove the theorem, it is enough to show that, for any given $\varepsilon > 0$, there is a grid G for which

$$\overline{D}_G(f,S) - \underline{D}_G(f,S) < \varepsilon.$$

Because $\varepsilon > 0$ is arbitrary, it then follows that $\overline{D}(f,S) - \underline{D}(f,S) = 0$ and that f is Darboux integrable. Moreover, because every Riemann sum is trapped between $\underline{D}_G(f,S)$ and $\overline{D}_G(f,S)$, so is the Riemann integral. The Darboux integral is trapped the same way, so the two integrals must be equal.

Using the given $\varepsilon > 0$ and the hypothesis that f is Riemann integrable, choose $\delta > 0$ so that, for every integration grid G with $\|G\| < \delta$,

$$\left| \sum_{i=1}^{N} f(\mathbf{p_i})\Delta A_i - \iint_S f(x,y)\, dA \right| < \frac{\varepsilon}{4},$$

regardless of how $\mathbf{p}_i$ is chosen in the cell Q_i of the grid G (cf. Definition 8.15). What we take from this is the fact that the difference between any two Riemann sums for f with the grid G is less than $\varepsilon/2$.

Fix a grid G for which $\|G\| < \delta$, and let $Q_1,\ldots,Q_N$ be the cells of G that meet S. Construct the the lower and upper Darboux sums

$$\underline{D}_G(f,S) = \sum_{i=1}^{N} m_i \Delta A_i, \quad \overline{D}_G(f,S) = \sum_{i=i}^{N} M_i \Delta A_i,$$

in the usual way, and set

$$A = \sum_{i=i}^{N} \Delta A_i = \overline{G}_G(S),$$

the outer G content of S with respect to the grid G.

Because m_i is the greatest lower bound of $f(\mathbf{x})$ on Q_i, $m_i + \varepsilon/4A$ is not a lower bound. In other words, there is a point $\mathbf{p}_i$ in each Q_i for which

$$f(\mathbf{p}_i) < m_i + \frac{\varepsilon}{4A}.$$

Therefore,

$$\sum_{i=1}^{N} f(\mathbf{p}_i) \Delta A_i < \sum_{i=1}^{N} m_i \Delta A_i + \frac{\varepsilon}{4A} \sum_{i=i}^{N} \Delta A_i, = \underline{D}_G(f,S) + \frac{\varepsilon}{4}.$$

In a similar way, there is a point $\mathbf{q}_i$ in each Q_i for which

$$M_i - \frac{\varepsilon}{4A} < f(\mathbf{q}_i),$$

and a similar argument shows that

$$\overline{D}_G(f,S) - \frac{\varepsilon}{4} < \sum_{i=1}^{N} f(\mathbf{q}_i) \Delta A_i.$$

Subtracting the first inequality from the second, we find

$$\overline{D}_G(f,S) - \underline{D}_G(f,S) - \frac{\varepsilon}{2} < \sum_{i=1}^{N} f(\mathbf{q}_i) \Delta A_i - \sum_{i=1}^{N} f(\mathbf{p}_i) \Delta A_i < \frac{\varepsilon}{2}.$$

The last inequality in this sequence is just the fact that any two Riemann sums differ by less than $\varepsilon/2$. Hence $\overline{D}_G(f,S) - \underline{D}_G(f,S) < \varepsilon$; by what was said above, this completes the proof.   □

Theorem 8.33. *If f is Darboux integrable on S, then it is also Riemann integrable, and the two integrals are equal.*

Proof. Let D be the value of the Darboux integral of f on S. We must show that, for any given $\varepsilon > 0$, there is a $\delta > 0$ so that, for any integration grid G with $\| G \| < \delta$,

$$\left| \sum_{i=1}^{N} f(\mathbf{p}_i) \Delta A_i - D \right| < \varepsilon,$$

regardless of how the point $\mathbf{p}_i$ is chosen in the cell Q_i of G.

Every Darboux integrable function is bounded, by definition; choose B so that $|f(x,y)| \le B$ on S. The definition also implies that upper and lower Darboux sums for f get arbitrarily close to D. Thus, for the ε given above, we can select a particular grid $\mathcal{H}$ for which

$$D - \frac{\varepsilon}{2} < \underline{D}_\mathcal{H}(f,S) \text{ and } \overline{D}_\mathcal{H}(f,S) < D + \frac{\varepsilon}{2}.$$

Suppose the cells of $\mathcal{H}$ that cover S are $P_1, \ldots, P_J$. Let ∂P denote the set of boundary points of all these cells. Because each P_j has area, $A(\partial P_j) = 0$ and thus $A(\partial P) = 0$. By Lemma 8.5, page 289, ∂P has a tubular neighborhood T of some width $\delta > 0$ for which

$$\overline{J}(T) < \frac{\varepsilon}{4B}.$$

Now let $\mathcal{G}$ be any integration grid for which $\|\mathcal{G}\| < \delta$. We claim that

$$D - \varepsilon < \underline{D}_{\mathcal{G}}(f, S) \quad \text{and} \quad \overline{D}_{\mathcal{G}}(f, S) < D + \varepsilon.$$

Every Riemann sum constructed with the grid $\mathcal{G}$ lies between $\underline{D}_{\mathcal{G}}(f, S)$ and $\overline{D}_{\mathcal{G}}(f, S)$; thus it follows from the claim that

$$D - \varepsilon < \sum_{i=1}^{N} f(\mathbf{p}_i) \Delta A_i < D + \varepsilon,$$

which is equivalent to what we need to prove.

We now prove the first of the inequalities in the claim; the second can be proven by essentially the same argument. We divide the cells of $\mathcal{G}$ into two classes, as follows.

Proving $D - \varepsilon < \underline{D}_{\mathcal{G}}(f, S)$

- $R_1, \ldots, R_K$ lie entirely within the tubular neighborhood T.

- $Q_1, \ldots, Q_N$ contain points outside T.

Now let $\mathcal{K}$ be the common refinement of $\mathcal{H}$ and $\mathcal{G}$; by definition, the cells of $\mathcal{K}$ are $P_j \cap Q_i$ and $P_j \cap R_l$. But because the diameter of each Q_i is less than δ, Q_i does not meet ∂P, and thus lies entirely in a single cell P_j of $\mathcal{H}$. In other words, $P_j \cap Q_i$ is either empty or it is just Q_i. The cells of $\mathcal{K}$ are therefore

$$Q_1, \ldots, Q_N, \quad \text{and} \quad \begin{array}{ccc} P_1 \cap R_1, & \ldots, & P_1 \cap R_K, \\ P_2 \cap R_1, & \ldots, & P_2 \cap R_K, \\ \vdots & \ddots & \vdots \\ P_J \cap R_1, & \ldots, & P_J \cap R_K. \end{array}$$

We have

$$\sum_{j=1}^{J} A(P_j \cap R_k) = A(R_k), \quad \sum_{k=1}^{K} A(R_k) < \overline{J}(T) < \frac{\varepsilon}{4B}.$$

We now construct the Darboux lower sums associated with $\mathcal{G}$ and $\mathcal{K}$. For this we need the greatest lower bound of f on each cell of each of these grids:

$$m_i = \inf_{\mathbf{p} \in Q_i} f(\mathbf{p}), \quad \widehat{m}_k = \inf_{\mathbf{p} \in R_k} f(\mathbf{p}), \quad m_{jk} = \inf_{\mathbf{p} \in P_j \cap R_k} f(\mathbf{p}).$$

Then

$$\underline{D}_{\mathcal{K}}(f, S) = \sum_{i=1}^{N} m_i A(Q_i) + \sum_{k=1}^{K} \sum_{j=1}^{J} m_{jk} A(P_j \cap R_k),$$

$$\underline{D}_{\mathcal{G}}(f, S) = \sum_{i=1}^{N} m_i A(Q_i) + \sum_{k=1}^{K} \widehat{m}_k A(R_k) = \sum_{i=1}^{N} m_i A(Q_i) + \sum_{k=1}^{K} \sum_{j=1}^{J} \widehat{m}_k A(P_j \cap R_k).$$

Subtracting, we find

$$\underline{D}_{\mathcal{K}}(f,S) - \underline{D}_{G}(f,S) = \sum_{k=1}^{K}\sum_{j=1}^{J}(m_{jk} - \widehat{m}_{k})A(p_{j} \cap R_{k})$$

$$\leq 2B\sum_{k=1}^{K}\sum_{j=1}^{J}A(P_{j} \cap R_{k}) \leq 2B\sum_{k=1}^{K}A(R_{k}) < 2B \cdot \frac{\varepsilon}{4B} = \frac{\varepsilon}{2},$$

or $\underline{D}_{\mathcal{K}}(f,S) < \underline{D}_{G}(f,S) + \varepsilon/2$. Because $\mathcal{K}$ is a refinement of $\mathcal{H}$,

$$\underline{D}_{\mathcal{H}}(f,S) \leq \underline{D}_{\mathcal{K}}(f,S).$$

We thus have a sequence of inequalities,

$$D - \frac{\varepsilon}{2} < \underline{D}_{\mathcal{H}}(f,S) \leq \underline{D}_{\mathcal{K}}(f,S) < \underline{D}_{G}(f,S) + \frac{\varepsilon}{2},$$

that together establish the first claim, $D - \varepsilon < \underline{D}_{G}(f,S)$. $\quad\square$

The Riemann-Darboux integral

The common value produced by the two definitions is sometimes called the **Riemann–Darboux integral**. However, we usually just call it the *integral*, and employ the two definitions interchangeably, depending on which is more useful in a particular situation. For example, the proof of the next theorem uses the Darboux characterization of the integral.

Theorem 8.34. *Suppose f is integrable on S; then so is $|f|$ and*

$$\left| \iint_{S} f(x,y)\, dA \right| \leq \iint_{S} |f(x,y)|\, dA.$$

Proof. First we prove $|f|$ is integrable; it is enough to show that, given any $\varepsilon > 0$, there is a grid $\mathcal{H}$ for which

$$\overline{D}_{\mathcal{H}}(|f|,S) - \underline{D}_{\mathcal{H}}(|f|,S) < \varepsilon.$$

To analyze the upper and lower Darboux sums for both $|f|$ and f, let $Q_{1}, \dots, Q_{N}$ be the cells in an arbitrary grid G, and let

$$m_{i}^{*} = \inf_{\mathbf{p}\in Q_{i}} |f(\mathbf{p})|, \qquad m_{i} = \inf_{\mathbf{p}\in Q_{i}} f(\mathbf{p}),$$

$$M_{i}^{*} = \sup_{\mathbf{p}\in Q_{i}} |f(\mathbf{p})|, \qquad M_{i} = \sup_{\mathbf{p}\in Q_{i}} f(\mathbf{p}).$$

Now it is always true that $M_{i}^{*} - m_{i}^{*} \leq M_{i} - m_{i}$ (see Exercise 8.20); thus, for any grid G,

$$\overline{D}_{G}(|f|,S) - \underline{D}_{G}(|f|,S) = \sum_{i=1}^{N}(M_{i}^{*} - m_{i}^{*})\Delta A_{i},$$

$$\leq \sum_{i=1}^{N}(M_{i} - m_{i})\Delta A_{i} = \overline{D}_{G}(f,S) - \underline{D}_{G}(f,S).$$

Because f is integrable, by hypothesis, there is always a grid $\mathcal{H}$ for which

$$\overline{D}_{\mathcal{H}}(f,S) - \underline{D}_{\mathcal{H}}(f,S) < \varepsilon;$$

then $\overline{D}_{\mathcal{H}}(|f|,S) - \underline{D}_{\mathcal{H}}(|f|,S) < \varepsilon$ as well, so $|f|$ is integrable.

Finally, because $-|f(x,y)| \le f(x,y) \le |f(x,y)|$ and the integral is monotone (cf. Corollary 8.29),

$$-\iint_S |f(x,y)|\, dA \le \iint_S f(x,y)\, dA \le \iint_S |f(x,y)|\, dA. \qquad \square$$

One of the fundamental results of calculus is that a continuous function is integrable. However, because we extend every function to the whole plane by setting it equal to zero outside its given domain, even a continuous function becomes, in general, discontinuous across the boundary of that domain. This causes difficulties in proving integrability, because some cells of a grid straddle the boundary; in those cells, the function can take widely different values, even if the cell is small. In proving that a continuous function is integrable, we, however, take all this into account, and even allow certain other discontinuities in the function.

The integral of a continuous function

Theorem 8.35. *Let Z be a subset of S with $A(Z) = 0$. If f is bounded and continuous on $S \setminus Z$, then f is integrable on S.*

Proof. We show that, given any $\varepsilon > 0$, there is a grid $\mathcal{J}_k$ (one of the original grids of congruent squares) for which

$$\overline{D}_{\mathcal{J}_k}(f,S) - \underline{D}_{\mathcal{J}_k}(f,S) < \varepsilon.$$

Let B be a global bound for f; that is, $|f(x,y)| \le B$ for all points (x,y) in $\mathbb{R}^2$. Let T be a tubular neighborhood of $Z \cup \partial S$ of positive width, chosen so that $\overline{J}(T) < \varepsilon/4B$.

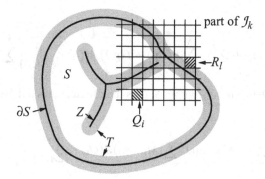

Now T contains all points of discontinuity of f, so f is continuous everywhere on $S \setminus T$. Furthermore, T is open so $S \setminus T$ is closed and bounded, implying f is uniformly continuous there. Thus, for the given ε, we can choose a $\delta > 0$ so that if $\mathbf{p}$ and $\mathbf{q}$ are in $S \setminus T$, then

$$\|\mathbf{p} - \mathbf{q}\| < \delta \;\Rightarrow\; |f(\mathbf{p}) - f(\mathbf{q})| < \frac{\varepsilon}{2A(S)}.$$

Here $A(S)$ is the area of S; by Theorem 8.25, we may assume that $A(S) > 0$.

Now consider any grid $\mathcal{J}_k$ for which $\|\mathcal{J}_k\| < \delta$ ($k > 1 + \log_2(1/\delta)$ suffices), and divide the squares of $\mathcal{J}_k$ into two classes:

- $Q_1, \ldots, Q_N$ lie entirely within $S \setminus T$.

- $R_1, \ldots, R_L$ meet the tubular neighborhood T.

Let

$$m_i = \inf_{\mathbf{p} \in Q_i} f(\mathbf{p}), \qquad \widehat{m}_l = \inf_{\mathbf{p} \in R_l} f(\mathbf{p}),$$

$$M_i = \sup_{\mathbf{p} \in Q_i} f(\mathbf{p}), \qquad \widehat{M}_l = \sup_{\mathbf{p} \in R_l} f(\mathbf{p}).$$

Then the difference between the upper and lower Darboux sums over $\mathcal{J}_k$ is

$$\overline{D}_{\mathcal{J}_k}(f,S) - \underline{D}_{\mathcal{J}_k}(f,S) = \sum_{i=1}^{N}(M_i - m_i)A(Q_i) + \sum_{l=1}^{L}(\widehat{M}_l - \widehat{m}_l)A(R_l).$$

Consider the first sum on the right. Because f is continuous on each closed bounded set Q_i (because Q_i lies entirely in $S \setminus T$), Q_i contains points $\mathbf{q}_i$ and $\mathbf{p}_i$ at which f attains its supremum and its infimum, respectively:

$$M_i = f(\mathbf{q}_i), \quad m_i = f(\mathbf{p}_i).$$

But because Q_i has diameter less than δ, we have $\|\mathbf{q}_i - \mathbf{p}_i\| < \delta$, so

$$M_i - m_i = f(\mathbf{q}_i) - f(\mathbf{p}_i) < \frac{\varepsilon}{2A(S)}.$$

Therefore,

$$\sum_{i=1}^{N}(M_i - m_i)A(Q_i) = \sum_{i=1}^{N}\big(f(\mathbf{q}_i) - f(\mathbf{p}_i)\big)A(Q_i) < \frac{\varepsilon}{2A(S)}\sum_{i=1}^{N}A(Q_i).$$

All the squares Q_i lie entirely within S; thus their total area is not greater than $A(S)$, implying

$$\sum_{i=1}^{N}A(Q_i) \le A(S) \quad \text{and} \quad \sum_{i=1}^{N}(M_i - m_i)A(Q_i) < \frac{\varepsilon}{2A(S)}A(S) = \frac{\varepsilon}{2}.$$

Now consider the second sum, the one involving the cells R_l. Because $\widehat{M}_l$ and $-\widehat{m}_l$ are both bounded by B,

$$\sum_{l=1}^{L} (\widehat{M}_l - \widehat{m}_l) A(R_l) \leq 2B \sum_{l=1}^{L} A(R_l).$$

Because the squares R_l cover T and they all meet T, they are precisely the squares involved in computing the outer area of T with the grid $\mathcal{J}_k$:

$$\sum_{l=1}^{L} A(R_l) = \bar{J}_k(T).$$

By the definition of outer Jordan content, we know $\bar{J}_k(T)$ decreases monotonically to $\bar{J}(T)$ as $k \to \infty$. Because $\bar{J}(T) < \varepsilon/4B$ by construction, we must have $\bar{J}_K(T) < \varepsilon/4B$ as well when K is sufficiently large, implying

$$\sum_{l=1}^{L} (\widehat{M}_l - \widehat{m}_l) A(R_l) < 2B \cdot \frac{\varepsilon}{4B} = \frac{\varepsilon}{2}.$$

Therefore, if $k > K$ and $k > 1 + \log_2(1/\delta)$, then

$$\overline{D}_{\mathcal{J}_k}(f,S) - \underline{D}_{\mathcal{J}_k}(f,S) < \frac{\varepsilon}{2} + \frac{\varepsilon}{2} = \varepsilon. \qquad \square$$

For functions of a single variable, there is an analogue of the preceding theorem that is particularly useful because it provides us with a large class of integrable functions. For example, it implies that a function $y = f(x)$ with only a finite number of finite jump discontinuities is integrable. To prove it, just adapt—and simplify—the preceding proof; see Exercise 8.18.

Theorem 8.36. *Suppose $f(x)$ is bounded and continuous on a closed interval $[a,b]$ minus a finite set of points; then f is integrable on $[a,b]$.* $\qquad \square$

The proof of Theorem 8.35 also implies that **restricted** Riemann sums, using only the cells contained in the interior of the domain of integration, have the same limit as unrestricted sums. The following corollary provides the details.

Corollary 8.37 *Suppose f is bounded and integrable on S; then*

$$\iint_S f(x,y)\, dA = \lim_{\|G\| \to 0} \sum_{Q_j \subset {}^{\circ}S} f(\mathbf{q}_j) A(Q_j),$$

where the sum is taken over only those cells Q_j of G that lie within ${}^{\circ}S$, the interior of S.

Proof. Let $\varepsilon > 0$ be given. Then, because f is integrable over S, there is a $\delta > 0$ such that any grid G with $\|G\| < \delta$ has

$$\left| \iint_S f(x,y)\, dA - \sum_{P_i \cap S \neq \phi} f(\mathbf{p}_i) A(P_i) \right| < \frac{\varepsilon}{2}.$$

Integrating single-variable functions

Restricted Riemann sums

(The sum on the right is an ordinary, unrestricted Riemann sum over all cells P_i of G that meet S.) Suppose $|f(x,y)| \leq B$ for all (x,y) in S. Let T be a tubular neighborhood of ∂S for which $\bar{J}(T) < \varepsilon/2B$, and suppose that T has width $w > 0$ (cf. Definition 8.13, p. 289). Further restrict G, if necessary, so that $\|G\| < w$; then all the cells P_i of G that appear in the unrestricted Riemann sum above are contained in $S \cup T$. Divide these cells into two classes:

- $Q_1, \ldots, Q_J$ lie entirely within $S \setminus \partial S = {}^\circ S$.

- $R_1, \ldots, R_K$ meet ∂S (and hence lie entirely within T).

Then, for any $\mathbf{r}_k$ in R_k, $k = 1, \ldots, K$,

$$\left| \sum_{R_k} f(\mathbf{r}_k) A(R_k) \right| \leq \sum_{R_k} |f(\mathbf{r}_k)| A(R_k) \leq B \sum_{R_k} A(R_k),$$

but because $R_1 \cup \ldots R_K \subseteq T$, we have

$$\sum_{R_k} A(R_k) \leq \bar{J}(T) < \frac{\varepsilon}{2B} \quad \text{and} \quad \left| \sum_{R_k} f(\mathbf{r}_k) A(R_k) \right| < \frac{\varepsilon}{2}.$$

Consequently, for any $\mathbf{q}_j$ in Q_j $(j = 1, \ldots, J)$,

$$\left| \iint_S f(x,y)\, dA - \sum_{Q_j} f(\mathbf{q}_j) A(Q_j) \right|$$

$$\leq \left| \iint_S f(x,y)\, dA - \sum_{P_i} f(\mathbf{p}_i) A(P_i) \right| + \left| \sum_{R_k} f(\mathbf{r}_k) A(R_k) \right| < \varepsilon,$$

where $\mathbf{p}_i = \mathbf{q}_j$ when $P_i = Q_j$ and $\mathbf{p}_i = \mathbf{r}_k$ when $P_i = R_k$. This proves that the integral is the limit of restricted Riemann sums. $\square$

Integrals as volumes

Theorem 8.35 also provides a way to connect double integrals to volumes. In the following theorem, the volume of a solid W is its 3-dimensional Jordan content, denoted $V(W)$. We make use of the analogue of Theorem 8.9, that the volume of the rectangular parallelepiped $[a, a+l] \times [b, b+w] \times [c, c+h]$ is the product lwh.

Theorem 8.38. *Suppose $f \geq 0$ is bounded and integrable on a closed bounded set S that has area, and W is the solid region in $\mathbb{R}^3$ that lies between S and the graph of $z = f(x,y)$. Then W has volume, and*

$$V(W) = \iint_S f(x,y)\, dA.$$

Proof. We show that, given any $\varepsilon > 0$, there is a grid $\mathcal{J}_k$ of squares in the plane for which

$$\underline{D}_{\mathcal{J}_k}(f, S) - \varepsilon < \underline{V}(W) \leq \overline{V}(W) \leq \overline{D}_{\mathcal{J}_k}(f, S).$$

(Here $\underline{V}(W)$ is the **inner volume** of W—that is, the 3-dimensional inner Jordan content $\underline{J}(W)$; similarly, $\overline{V}(W) = \overline{J}(W)$ is the **outer volume**.) Because f is integrable over S and ε can be any positive number, the inequalities show that $V(W)$ exists and equals the integral of f over S.

We begin by choosing a global bound B for f: $|f(x,y)| \leq B$ on $\mathbb{R}^2$. Then, because $A(\partial S) = 0$, we can choose k so large that the squares $R_1, \ldots, R_L$ of $\mathcal{J}_k$ that meet ∂S have total area less than ε/B. Let $Q_1, \ldots, Q_N$ be the remaining squares of $\mathcal{J}_k$ that meet S; they lie entirely within $S \setminus \partial S$. Let

$$m_i = \inf_{\mathbf{p} \in Q_i} f(\mathbf{p}), \qquad \widehat{m}_l = \inf_{\mathbf{p} \in R_l} f(\mathbf{p}),$$

$$M_i = \sup_{\mathbf{p} \in Q_i} f(\mathbf{p}), \qquad \widehat{M}_l = \sup_{\mathbf{p} \in R_l} f(\mathbf{p}).$$

Taking into account the fact that every $\widehat{m}_l \leq B$ and that the total area of the squares R_l is less than ε/B, we find that the lower Darboux sum for f over S is

$$\underline{D}_{\mathcal{J}_k}(f, S) = \sum_{i=1}^{N} m_i A(Q_i) + \sum_{l=1}^{L} \widehat{m}_l A(R_l) \leq \sum_{i=1}^{N} m_i A(Q_i) + B \sum_{l=1}^{L} A(R_l)$$

$$< \sum_{i=1}^{N} m_i A(Q_i) + B \cdot \frac{\varepsilon}{B} = \sum_{i=1}^{N} m_i A(Q_i) + \varepsilon,$$

or

$$\underline{D}_{\mathcal{J}_k}(f, S) - \varepsilon < \sum_{i=1}^{N} m_i A(Q_i) = \sum_{i=1}^{N} V(P_i).$$

In the last sum, P_i is the parallelepiped with base Q_i and height m_i; its volume is $m_i A(Q_i)$. These parallelepipeds are nonoverlapping sets that are entirely contained in W, so their total volume is not larger than the inner volume of W:

$$\sum_{i=1}^{N} V(P_i) \leq \underline{V}(W).$$

This gives $\underline{D}_{\mathcal{J}_k}(f, S) - \varepsilon < \underline{V}(W)$, the first of the two inequalities we must establish.

The second inequality is more straightforward. In the formula for the upper Darboux sum,

$$\overline{D}_{\mathcal{J}_k}(f, S) = \sum_{i=1}^{N} M_i A(Q_i) + \sum_{l=1}^{L} \widehat{M}_l A(R_l),$$

each term is the volume of a parallelepiped based on one of the squares Q_i or R_l. These parallelepipeds are nonoverlapping and their union entirely contains W. Consequently, their total volume $\overline{D}_{\mathcal{J}_k}(f, S)$ is at least as large as the outer volume of W:

$$\overline{V}(W) \leq \overline{D}_{\mathcal{J}_k}(f, S). \qquad \square$$

With cubes replacing squares, we can define and calculate the Jordan content of a region D in $\mathbb{R}^3$ (cf. p. 295). Then, modifying the exposition at the beginning of this section by using a grid $\mathcal{G}$ whose cells are cubes Q_i instead of squares we can define the Riemann triple integral of a function $f(x,y,z)$ over a 3-dimensional region D as

$$\iiint\limits_D f(x,y,z)\,dV = \lim_{\|\mathcal{G}\|\to 0} \sum_{i=1}^{N(\mathcal{G})} f(x_i,y_i,z_i)\,\Delta V_i,$$

where $\Delta V_i = J(Q_i)$, the Jordan content, or volume, of Q_i. Compare this to Definition 8.15 for double integrals. All the theorems and corollaries of this section have natural extensions to triple integrals. In particular (cf. Theorem 8.35), a function that is bounded and continuous on a region $D \setminus Z$, where D has volume and Z has volume zero, is integrable. Having made these observations, we now assume that triple integrals are available for our future work.

Jordan content is an example of a **set function**: it assigns a real number to each of the sets in a certain collection. There are numerous other examples, including integrals themselves. In many cases, a set function even has a derivative. We end this section by showing that the derivative of a suitable set function is a point function whose integral equals the original set function. This is, in fact, a version of the fundamental theorem of calculus. To fix ideas, we first explore some examples.

Imagine a thin flat plate that lies over a portion of the (x,y)-plane, and suppose it has a continuous but nonuniform mass distribution. Let S be a subset of the plane with positive area, and let $M(S)$ be the total mass of the portion of the plate that lies over S. If $A(S)$ is the area of S, then

$$\frac{M(S)}{A(S)} = \text{average mass density over } S.$$

Intuitively, the mass density $\rho(x,y)$ of the plate at the point (x,y) should be the limit of $M(S)/A(S)$ as the set S "shrinks down" to (x,y), in the sense that $\delta(S) \to 0$ for sets S that contain (x,y); $\delta(S)$ is the diameter of S (Definition 8.14, p. 291). Thus, mass *distribution* is a set function, mass *density* is a point function, and the second is the derivative of the first. That is,

$$\text{mass density at } (x,y) = \rho(x,y) = M'(x,y) = \lim_{\delta(S)\to 0} \frac{M(S)}{A(S)},$$

for (x,y) in S, if the limit exists. A related example is a 3-dimensional solid with a continuous but nonuniform mass distribution. Let D be any region of positive volume $V(D)$ in $\mathbb{R}^3$ that contains the point (x,y,z), and let $M(D)$ be the mass of the portion of the solid that lies in D; then

$$\rho(x,y,z) = M'(x,y,z) = \lim_{\delta(D)\to 0} \frac{M(D)}{V(D)}$$

is the mass density of the solid at (x,y,z), if the limit exists.

Another physical example is the hydrostatic force a liquid applies to the walls of its container. The force $F(S)$ on any portion S of the surface of the container is a set function. If S has area, then $F(S)/A(S)$ is the average pressure (force per unit area) on S; as S shrinks down to a point, this ratio approaches the pressure at that point. Electric charge on a plate, and the related charge density, show that a set function can take negative, as well as positive, values.

For a different kind of example (using subsets of $\mathbb{R}$ instead of $\mathbb{R}^2$, for simplicity), suppose X is a random variable (cf. p. 20) that takes real values. For any subset S of $\mathbb{R}$ that has a length $L(S)$, we define

Example: probability and probability density

$$P(S) = \text{probability that } X \text{ lies in } S.$$

Probability is a set function. The corresponding probability density function $p(x)$ should be the limit of $P(S)/L(S)$ as S "shrinks down" to x. A common example is the normal density function

$$P'(x) = p(x) = \frac{e^{-x^2/2}}{\sqrt{2\pi}},$$

which determines the normal probability function

$$P(S) = \frac{1}{\sqrt{2\pi}} \int_S e^{-x^2/2}\, dx.$$

Here we find a set function that is the integral of its derivative.

Integrals provide a very general class of set functions. Define

Example: set functions of integral type

$$F(S) = \iint_S f(x,y)\, dA,$$

where $f(x,y)$ is a fixed function that is bounded and continuous on some fixed open set Ω in $\mathbb{R}^2$. Then F is a set function; it assigns a real number to each subset S of Ω that has area.

Let (x,y) be a point in Ω, and let S_n be a collection of closed subsets of Ω with positive area that all contain (x,y). Let m_n and M_n be the minimum and maximum values, respectively, of f on S_n. Then, by Corollary 8.29, page 298,

$$m_n \le \frac{F(S_n)}{A(S_n)} \le M_n.$$

Suppose $\delta(S_n) \to 0$ as $n \to \infty$. The continuity of f implies that $m_n \to f(x,y)$ and $M_n \to f(x,y)$ as $n \to \infty$. In other words,

$$\lim_{n \to \infty} \frac{F(S_n)}{A(S_n)} = f(x,y).$$

Because this limit is independent of the choice of the sets S_n used to compute it, we define it to be the **derivative** of F at (x,y), and write

$$F'(x,y) = \lim_{n \to \infty} \frac{F(S_n)}{A(S_n)}.$$

Although F is a set function, its derivative $F' = f$ is a point function. We call F a *set function of integral type*. The following theorem summarizes our observations.

Theorem 8.39. *A set function of integral type has a deriviative, and the set function is equal to the integral of its derivative.*   □

A set function may not be of integral type

Note that continuous mass distributions and normal probability are both set functions of integral type:

$$M(S) = \iint_S \rho(x,y)\,dA, \quad M'(x,y) = \rho(x,y);$$

$$P(S) = \int_S \frac{e^{-x^2/2}}{\sqrt{2\pi}}\,dx, \qquad P'(x) = \frac{e^{-x^2/2}}{\sqrt{2\pi}}.$$

But not all set functions are. Here is a simple example to the contrary. Let

$$\widetilde{M}(S) = \begin{cases} 1 & \text{if } S \text{ contains the origin,} \\ 0 & \text{otherwise.} \end{cases}$$

You can think of $\widetilde{M}$ as the set function associated with a unit point mass concentrated at the origin on $\mathbb{R}$. To show $\widetilde{M}$ is not of integral type, suppose the contrary. That is, suppose

$$\widetilde{M}(S) = \int_S g(x)\,dx$$

for some integrable function $g(x)$ (that need not even be continuous). Now suppose $Q_1 = [-1,0]$, $Q_2 = [0,1]$, and $S = [-1,1]$; by definition of $\widetilde{M}$,

$$\widetilde{M}(Q_1) = \widetilde{M}(Q_2) = \widetilde{M}(S) = 1.$$

However, by assumption we have

$$\widetilde{M}(S) = \int_{-1}^{1} g(x)\,dx = \int_{-1}^{0} g(x)\,dx + \int_{0}^{1} g(x)\,dx = \widetilde{M}(Q_1) + \widetilde{M}(Q_2) = 2,$$

a contradiction. The contradiction arises because the integral is additive on nonoverlapping sets (Theorem 8.27, p. 298), but $\widetilde{M}$ is not:

$$\widetilde{M}(S_1 \cup S_2) \neq \widetilde{M}(S_1) + \widetilde{M}(S_2).$$

A set function cannot be of integral type unless it possesses, at the outset, all the relevant properties of a Riemann integral.

Exercises

8.1. a. Adapt the program that estimates the gravitational field of a large plate (p. 271) to a version of BASIC or a similar language and use it to reproduce the table of values of the field that are found in the text.

   b. In the original computation of the gravitational field, we assumed the plate density was constant and took $4G\rho = 1$. Recompute all the tabular values assuming that $4G\rho = 1/(1+x^2+y^2)$. This provides estimates for the double integral

$$\iint\limits_{\substack{0 \le x \le R, \\ 0 \le y \le R}} \frac{-a\,dx\,dy}{(1+x^2+y^2)(x^2+y^2+a^2)^{3/2}}.$$

   c. In which case is the plate less massive, and in which case is the gravitational field weaker? Does the less massive plate have the weaker field?

8.2. In Chapter 9.3, pages 342–343, the area of the curved region $1 \le x^2 - y^2 \le 2$, $1 \le 2xy \le 3$, in the (x,y)-plane is given by the double integral

$$\iint\limits_{\substack{1 \le u \le 2, \\ 1 \le v \le 3}} \frac{du\,dv}{4\sqrt{u^2+v^2}}.$$

   a. Approximate the integral by a Riemann sum using a 2×4 grid of squares with the integrand evaluated at the center of each square. Use a modification of the BASIC program in the previous question to show the value of the Riemann sum is 0.204 806.

   b. Obtain additional approximations using a 20×40 grid and a 200×400 grid. How close are these to the estimate 0.205 213 found anlytically on page 342?

8.3. Adapt the previous BASIC program to estimate the value of the integral

$$\iint_S 4(x^2+y^2)\,dx\,dy$$

on the square $S: 0.2 \le x \le 1, 0.2 \le y \le 1$ (cf. Exercise 9.38.c, p. 384). Evaluate the function at the center of each grid square. Show that, with a 4×4 grid, the value is 2.0992 and with a 20×20 grid the value is 2.1156. How large must the grid be to make the value 2.11626?

8.4. Is the interior of the complement of S equal to the complement of the interior of S? If not, does either of these sets always contain the other?

8.5. Suppose **b** is a boundary point of S. Show that every open disk centered at **b** contains at least one point **p** in S and also at least one point **q** that is not in S.

8.6. Suppose **p** is a point in S and **q** is a point not in S. Show that at least one point on any continuous curve from **p** to **q** is in ∂S.

8.7. Let Q be a square in the grid $\mathcal{I}_k$, and let $m > k$. Show that

$$\underline{J}_m(Q) = \frac{1}{2^{2k}} \quad \text{and} \quad \bar{J}_m(Q) = \frac{1}{2^{2k}} + 4\frac{2^{m-k}}{2^{2m}} + 4\frac{1}{2^{2m}}.$$

Conclude that Q is Jordan measurable and $J(Q) = 1/2^{2k}$.

8.8. Show that the rectangle $a \le x \le b, c \le y \le d$ has Jordan content $(b-a)(d-c)$.

8.9. Suppose $\delta > 0$, $\alpha\delta \le a < (\alpha+1)\delta$, and $(\beta-1)\delta < b \le \beta\delta$. Show that $2\delta < b - a$ implies $(\alpha+1)\delta < (\beta-1)\delta$.

8.10. Show that $\underline{J}(S) = 0 \Leftrightarrow {}^\circ S = \phi$ (i.e., the interior of S is empty).

8.11. Suppose S is Jordan measurable and ${}^\circ S \subseteq T \subseteq \bar{S}$. Show that T is Jordan measurable and $J(T) = J({}^\circ S) = J(S) = J(\bar{S})$.

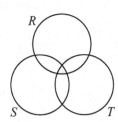

8.12. Suppose R, S, and T are Jordan measurable; show that

$$J(R \cup S \cup T) = J(R) + J(S) + J(T)$$
$$- J(R \cap S) - J(S \cap T) - J(T \cap R) + J(R \cap S \cap T).$$

This includes showing that all the sets on the right-hand side of the equation are Jordan measurable.

8.13. Generalize the result in the previous exercise to four sets, and then to p sets $S_1, \dots, S_p$.

8.14. Let S and T denote Jordan-measurable bounded subsets of the plane.

a. Give an example in which $\partial(T \setminus S)$ has a point in ∂S that is not in ∂T.
b. Prove that $\partial(T \setminus S) \subseteq \partial T \cup \partial S$.

8.15. Modify the proof of Lemma 8.5 so that it works with cubes and sets in $\mathbb{R}^3$.

8.16. Let the linear map $L : \mathbb{R}^2 \to \mathbb{R}^2$ be given by the matrix

$$L = \begin{pmatrix} a & b \\ c & d \end{pmatrix},$$

and suppose Q is the square $0 \le x \le w$, $0 \le y \le w$. Show that the ratio $\sigma = \delta(L(Q))/\delta(Q)$ of the diameters of Q and its image is the larger of the two numbers

$$\frac{a^2 + b^2 + c^2 + d^2 \pm 2(ab + cd)}{\sqrt{2}},$$

and thus depends only on L, not on Q. Make a sketch to illustrate how these numbers are connected to $L(Q)$.

8.17. Confirm (e.g., by writing suitable programs) that the inner and outer areas, $\underline{J}_k$ and $\bar{J}_k$, of the unit disk have the values indicated in the following table.

k	Inner Squares	Outer Squares	$\underline{J}_k$	$\bar{J}_k$
0	0	12	0	12
1	4	24	1	6
2	32	68	2	4.25
3	164	232	2.5625	3.625
4	732	864	2.859375	3.375
5	3080	3340	3.007813	3.261719
6	12596	13112	3.075195	3.201172
7	50920	51948	3.107910	3.170654
8	204836	206888	3.125549	3.156860
9	821424	825524	3.133484	3.149124

8.18. Prove Theorem 8.36.

8.19. Let S be the unit square in $\mathbb{R}^2$, and let

$$f(x,y) = \begin{cases} 1 & \text{if } x \text{ and } y \text{ are irrational,} \\ 0 & \text{otherwise.} \end{cases}$$

For an arbitrary grid G determine the upper and lower Darboux sums $\bar{D}_G(f,S)$ and $\underline{D}_G(f,S)$. What are the values of the upper and lower Darboux integrals of f on S? Is f Darboux integrable on S?

8.20. Suppose f is integrable on a closed cell Q, and

$$m^* = \inf_{\mathbf{p}\in Q} |f(\mathbf{p})|, \qquad m = \inf_{\mathbf{p}\in Q} f(\mathbf{p}),$$
$$M^* = \sup_{\mathbf{p}\in Q} |f(\mathbf{p})|, \qquad M = \sup_{\mathbf{p}\in Q} f(\mathbf{p}).$$

Show that $M^* - m^* \le M - m$.

8.21. Suppose a thin flat plate is a disk of radius R centered at the origin of $\mathbb{R}^2$. Suppose its mass distribution is circularly symmetric and that the mass of the disk of radius α centered at the origin is $\alpha/(1+\alpha)$, for every $0 \le \alpha \le R$.

a. What is the mass of an annulus whose radii are $a - \Delta r/2$ and $a + \Delta r/2$?

b. What is the mass $M(S)$ of the piece S of this annulus cut off by radial lines $\theta = b - \Delta\theta/2$ and $\theta = b + \Delta\theta/2$. What is the area $A(S)$.

c. Determine the mass density at the point (a,b) on the plate as

$$\rho(a,b) = \lim_{\substack{\Delta r \to 0 \\ \Delta\theta \to 0}} \frac{M(S)}{A(S)}.$$

d. Using ρ, verify that the mass of the disk of radius α has the value it should; that is, verify

$$\iint\limits_{x^2+y^2\le\alpha^2} \rho(x,y)\, dA = \frac{\alpha}{1+\alpha}.$$

e. Repeat all the previous analysis assuming that the mass of the disk of radius α centered at the origin is just α.

8.22. Let S be a closed bounded set with area in the (x,y)-plane. The **moment** of S about the y-axis is the integral

$$\iint_S x\, dA.$$

Estimate the moment of the square $S : 0.2 \le x \le 1, 0.2 \le y \le 1$ about the y-axis by adapting the BASIC program of Exercise 8.3, above. Use a 4×4 grid and a 20×20 grid.

Chapter 9
Evaluating Double Integrals

Abstract Although the definition of the integral reflects its origins in scientific problems, its evaluation relies on a considerable range of mathematical concepts and tools. Most fundamental is the change of variables formula; the single-variable version ("u-substitution") is perhaps the core technique of integration in the introductory calculus course. By contrast, the method of iterated integrals has no single-variable analogue; it evaluates a double integral by "partial integration" of one variable at a time. This chapter connects double and iterated integrals, establishes the change of variables formula, and discusses Green's theorem as a tool for evaluating double integrals and as a reason for orienting them.

9.1 Iterated integrals

We define iterated integrals in their own terms, independently of double integrals. First, suppose that S is the region in the (x,y)-plane that lies between the graphs $y = \gamma(x)$ and $y = \delta(x)$ when $a \leq x \leq b$. We assume γ and δ are continuous and $\gamma(x) \leq \delta(x)$ everywhere on this interval; we can write

Partial integration

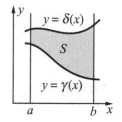

$$S: \quad \begin{matrix} a \leq x \leq b, \\ \gamma(x) \leq y \leq \delta(x). \end{matrix}$$

Now let $f(x,y)$ be a continuous function on S; for each x in $[a,b]$, compute the *"partial integral" of $f(x,y)$ with respect to y* from $\gamma(x)$ to $\delta(x)$:

$$F_2(x) = \int_{\gamma(x)}^{\delta(x)} f(x,y)\,dy, \quad a \leq x \leq b.$$

This is a continuous function of x. As an example, let $f(x,y) = x^2 y^3$ and let S be the region between $y = \gamma(x) = 1/2$ and $y = \delta(x) = \sqrt{x}$ when $1/4 \leq x \leq 1$. Then the partial integral is

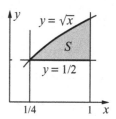

J.J. Callahan, *Advanced Calculus: A Geometric View*, Undergraduate Texts in Mathematics, DOI 10.1007/978-1-4419-7332-0_9, © Springer Science+Business Media, LLC 2010

$$F_2(x) = \int_{1/2}^{\sqrt{x}} x^2 y^3 \, dy = \frac{x^2 y^4}{4} \Big|_{1/2}^{\sqrt{x}} = \frac{x^2(\sqrt{x})^4}{4} - \frac{x^2/16}{4} = \frac{x^4}{4} - \frac{x^2}{64}.$$

We can reverse the roles of the two variables if we start with a region T described in the following way,

$$T: \begin{array}{l} c \le y \le d, \\ \alpha(y) \le x \le \beta(y). \end{array}$$

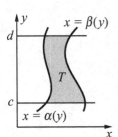

Then, for each y in the interval $[\alpha, \beta]$, the partial integral of $f(x,y)$ with respect to x from $\alpha(y)$ to $\beta(y)$ is

$$F_1(y) = \int_{\alpha(y)}^{\beta(y)} f(x,y) \, dx, \quad c \le y \le d.$$

This is continuous if α and β are. It can happen that a particular region can be described both ways. This is true for our example above:

$$S: \begin{array}{l} 1/4 \le x \le 1, \\ 1/2 \le y \le \sqrt{x}; \end{array} \quad \text{and also} \quad S: \begin{array}{l} 1/2 \le y \le 1, \\ y^2 \le x \le 1. \end{array}$$

Therefore

$$F_1(y) = \int_{y^2}^{1} x^2 y^3 \, dx = \frac{x^3 y^3}{3} \Big|_{y^2}^{1} = \frac{y^3 - y^9}{3},$$

so the two partial integrals of $x^2 y^3$ are certainly different; they are even functions of different variables.

Iterated integrals

Now integrate each of the partial integrals $F_2(x)$ or $F_1(y)$ over its own domain:

$$\int_a^b F_2(x) \, dx = \int_a^b \left(\int_{\gamma(x)}^{\delta(x)} f(x,y) \, dy \right) dx,$$

$$\int_c^d F_1(y) \, dy = \int_c^d \left(\int_{\alpha(y)}^{\beta(y)} f(x,y) \, dx \right) dy.$$

Note that in each we have performed a repeated, or iterated, integration of the original function $f(x,y)$, first with respect to one variable and then the other. These are the **iterated integrals** of $f(x,y)$.

The iterated integrals have the same value

To illustrate, let us return to our example $f(x,y) = x^2 y^3$ over the region S. We have

$$\int_{1/4}^{1} \left(\int_{1/2}^{\sqrt{x}} x^2 y^3 \, dy \right) dx = \int_{1/4}^{1} \left(\frac{x^4}{4} - \frac{x^2}{64} \right) dx = \frac{x^5}{20} - \frac{x^3}{192} \Big|_{1/4}^{1} = \frac{459}{10240}$$

when the iteration is performed in one order, and

$$\int_{1/2}^{1} \left(\int_{y^2}^{1} x^2 y^3 \, dx \right) dy = \int_{1/2}^{1} \left(\frac{y^3 - y^9}{3} \right) dy = \frac{y^4}{12} - \frac{y^{10}}{30} \Big|_{1/2}^{1} = \frac{459}{10240}$$

when it is performed in the other. These calculations suggest a general result: the two iterated integrals are always equal (Corollary 9.4, below). As we show, this happens because the iterated integrals, taken in either order, equal the double integral. Here is the statement and proof when the domain S is a rectangle.

Theorem 9.1. *Suppose $f(x,y)$ is continuous on the rectangle R defined by $a \leq x \leq b$, $c \leq y \leq d$; then*

$$\iint_R f(x,y)\,dA = \int_a^b \left(\int_c^d f(x,y)\,dy \right) dx = \int_c^d \left(\int_a^b f(x,y)\,dx \right) dy.$$

Proof. We prove the double integral equals the first of the two iterated integrals; to show it also equals the second, interchange x and y. Let

$$F_2(x) = \int_c^d f(x,y)\,dy;$$

then we show

$$\iint_R f(x,y)\,dA = \int_a^b F_2(x)\,dx$$

by proving that, for any $\varepsilon > 0$,

$$\left| \iint_R f(x,y)\,dA - \int_a^b F_2(x)\,dx \right| < \varepsilon.$$

We begin by subdividing R with a grid of congruent rectangles. For positive integers I and J, let

$$\Delta x = \frac{b-a}{I}, \quad \Delta y = \frac{d-c}{J},$$

and then define

$$x_1 = a + \Delta x, \qquad\qquad y_1 = c + \Delta y,$$
$$x_i = x_{i-1} + \Delta x, \quad i = 2, \ldots, I, \quad y_j = y_{j-1} + \Delta y, \quad j = 2, \ldots, J.$$

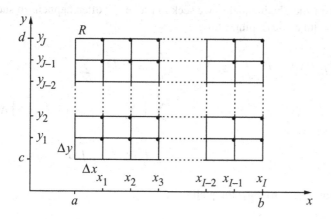

Each point (x_i, y_j) is the upper right corner of its cell. The mesh size of this grid, $\delta = \sqrt{(\Delta x)^2 + (\Delta y)^2}$, can be made as small as we wish by choosing both I and J sufficiently large.

Now suppose $\varepsilon > 0$ is given. Because f is continuous on R, it is integrable there (Theorem 8.35, p. 305). Consequently, all Riemann sums constructed using a grid with a sufficiently small mesh will be arbitrarily close to the value of the double integral of f over R. Therefore, we can choose I and J so large that

$$\left| \iint_R f(x,y)\, dA - \sum_{i=1}^{I} \sum_{j=1}^{J} f(x_i, y_j)\, \Delta x \Delta y \right| < \frac{\varepsilon}{3}.$$

The continuous function F_2 is likewise integrable over $[a,b]$, so Riemann sums for its integral are also arbitrarily close to the value of the integral if the partition of the x-interval $[a,b]$ is fine enough. Therefore, by increasing the size of I, if necessary, we can make the inequality

$$\left| \int_a^b F_2(x)\, dx - \sum_{i=1}^{I} F_2(x_i)\, \Delta x \right| < \frac{\varepsilon}{3}$$

hold as well. Finally, each $F_2(x_i)$ is itself an integral,

$$F_2(x_i) = \int_c^d f(x_i, y)\, dy, \quad i = 1, \ldots, I,$$

and thus has Riemann sum approximations. Therefore, after a sufficiently large I has been fixed, we can then increase the size of J, if necesary, so that all Riemann sums for each of the integrals $F_2(x_1), \ldots, F_2(x_I)$ will be arbitrarily close to the value of that integral:

$$\left| F_2(x_i) - \sum_{j=1}^{J} f(x_i, y_j)\, \Delta y \right| < \frac{\varepsilon}{3(b-a)}, \quad i = 1, \ldots, I.$$

Now consider the inequality we seek to prove. As often happens in such a proof, we begin with a telescoping sum,

$$\iint_R f(x,y)\, dA - \int_a^b F_2(x)\, dx = \iint_R f(x,y)\, dA - \sum_{i=1}^{I} \sum_{j=1}^{J} f(x_i, y_j)\, \Delta x \Delta y$$

$$+ \sum_{i=1}^{I} \left(\sum_{j=1}^{J} f(x_i, y_j)\, \Delta y - F_2(x_i) \right) \Delta x$$

$$+ \sum_{i=1}^{I} F_2(x_i)\, \Delta x - \int_a^b F_2(x)\, dx,$$

and then apply the triangle inequality:

$$\left| \iint_R f(x,y)\, dA - \int_a^b F_2(x)\, dx \right| \leq \left| \iint_R f(x,y)\, dA - \sum_{i=1}^{I} \sum_{j=1}^{J} f(x_i, y_j)\, \Delta x\, \Delta y \right|$$

$$+ \left| \sum_{i=1}^{I} \left(\sum_{j=1}^{J} f(x_i, y_j)\, \Delta y - F_2(x_i) \right) \Delta x \right|$$

$$+ \left| \sum_{i=1}^{I} F_2(x_i)\, \Delta x - \int_a^b F_2(x)\, dx \right|.$$

For I and J large enough, the first term on the right is bounded by $\varepsilon/3$, and so is the third term; we claim the same is true for the second. We have

$$\left| \sum_{i=1}^{I} \left(\sum_{j=1}^{J} f(x_i, y_j)\, \Delta y - F_2(x_i) \right) \Delta x \right| \leq \sum_{i=1}^{I} \left| \sum_{j=1}^{J} f(x_i, y_j)\, \Delta y - F_2(x_i) \right| \Delta x$$

$$< \sum_{i=1}^{I} \frac{\varepsilon}{3(b-a)}\, \Delta x = \frac{\varepsilon}{3(b-a)} \cdot I \Delta x = \frac{\varepsilon}{3},$$

as claimed. By what has been said above, this proves the theorem.   □

Corollary 9.2 *In the iterated integration of a continuous function with constant limits of integration, the order of integration can be reversed.*   □

Order of integration

Theorem 9.3. *Let S be the region defined by $a \leq x \leq b$, $\gamma(x) \leq y \leq \delta(x)$, where $\gamma(x)$ and $\delta(x)$ are continuous functions of x on $[a,b]$. Let $f(x,y)$ be continuous on S; then*

$$\iint_S f(x,y)\, dA = \int_a^b \left(\int_{\gamma(x)}^{\delta(x)} f(x,y)\, dy \right) dx.$$

Proof. This theorem is similar to Theorem 9.1, and can be proven in essentially the same way. We begin by constructing a rectangle

$$R: \begin{array}{l} a \leq x \leq b, \\ c \leq y \leq d, \end{array}$$

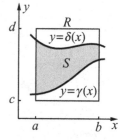

where $c \leq \gamma(x)$ and $\delta(x) \leq d$, for all x in $[a,b]$. Now R contains S, and if we extend $f(x,y)$ in the usual way by having $f(x,y) = 0$ outside S, then f is continuous everywhere in R, except (in general) on the graphs $y = \gamma(x)$ and $y = \delta(x)$.

By Theorem 8.10 (p. 284), these graphs form a set Z of area zero. Because f is continuous on $R \setminus Z$, it is integrable on R by Theorem 8.35 (p. 305). Moreover, because S and $R \setminus S$ are nonoverlapping sets (on each of which f is integrable), we have

$$\iint_R f(x,y)\, dA = \iint_S f(x,y)\, dA + \iint_{R \setminus S} f(x,y)\, dA = \iint_S f(x,y)\, dA,$$

because the integral is additive (Theorem 8.27) and $f = 0$ on $R \setminus S$.

Now fix x; then $f(x,y)$ is a bounded function of y on the interval $[c,d]$, and is continuous except at the two points where $y = \gamma(x)$ and $y = \delta(x)$. Therefore, the partial integral of f with respect to y over $[c,d]$ exists and equals the integral over the smaller interval $[\gamma(x), \delta(x)]$ (because $f = 0$ outside that smaller interval). Let

$$F_2(x) = \int_c^d f(x,y)\,dy = \int_{\gamma(x)}^{\delta(x)} f(x,y)\,dy$$

denote the common value; then $F_2(x)$ is a continuous function of x on $[a,b]$. To prove the theorem it is sufficient to show that

$$\iint_R f(x,y)\,dA = \int_a^b \left(\int_c^d f(x,y)\,dy \right) dx = \int_a^b F_2(x)\,dx.$$

Although this equation does not follow directly from the statement of Theorem 9.1 (because f is not continuous everywhere on R), we can show that it does follow from the proof.

Cover R with a grid of rectangles whose width is $\Delta x = (b-a)/I$ and whose height is $\Delta y = (d-c)/J$, and define (x_i,y_j) for $i = 1,\ldots,I$, $j = 1,\ldots,J$, as in that proof (cf. p. 319). With these choices we can now construct the various Riemann sums that appear in the following inequality, taken from the same proof:

$$\left| \iint_R f(x,y)\,dA - \int_a^b F_2(x)\,dx \right| \le \left| \iint_R f(x,y)\,dA - \sum_{i=1}^I \sum_{j=1}^J f(x_i,y_j)\Delta x \Delta y \right|$$

$$+ \left| \sum_{i=1}^I \left(\sum_{j=1}^J f(x_i,y_j)\Delta y - F_2(x_i) \right) \Delta x \right|$$

$$+ \left| \sum_{i=1}^I F_2(x_i)\Delta x - \int_a^b F_2(x)\,dx \right|.$$

Now choose $\varepsilon > 0$. Then, because f is integrable on R, because $f(x_i,y)$ is integrable with respect to y for each $i = 1,\ldots,I$, and because F_2 is integrable on $[a,b]$, we can choose I and J large enough that each of the terms on the right is less than $\varepsilon/3$. Because $\varepsilon > 0$ is arbitrary, the left-hand side of the inequality must equal zero. By what has been said above, this completes the proof. $\square$

Interchanging the variables

The theorem holds with the roles of x and y reversed. That is, if $f(x,y)$ is continuous over the region $T : c \le y \le d$, $\alpha(y) \le x \le \beta(y)$, then

$$\iint_T f(x,y)\,dA = \int_c^d \left(\int_{\alpha(y)}^{\beta(y)} f(x,y)\,dx \right) dy.$$

This implies the following corollary.

Corollary 9.4 *Suppose $f(x,y)$ is continuous over a region S that has the alternate descriptions*

$$S: \begin{array}{l} a \leq x \leq b, \\ \gamma(x) \leq y \leq \delta(x), \end{array} \qquad S: \begin{array}{l} c \leq y \leq d, \\ \alpha(y) \leq x \leq \beta(y); \end{array}$$

then

$$\int_a^b \left(\int_{\gamma(x)}^{\delta(x)} f(x,y)\, dy \right) dx = \int_c^d \left(\int_{\alpha(y)}^{\beta(y)} f(x,y)\, dx \right) dy.$$

Proof. Both iterated integrals equal the double integral $\iint_S f(x,y)\, dA$. $\qquad\qquad\square$

Here are two common ways to write an iterated integral that dispense with the large parentheses:

Notation

$$\int_a^b \left(\int_{\gamma(x)}^{\delta(x)} f(x,y)\, dy \right) dx = \int_a^b \int_{\gamma(x)}^{\delta(x)} f(x,y)\, dy\, dx = \int_a^b dx \int_{\gamma(x)}^{\delta(x)} f(x,y)\, dy;$$

$$\int_c^d \left(\int_{\alpha(y)}^{\beta(y)} f(x,y)\, dx \right) dy = \int_c^d \int_{\alpha(y)}^{\beta(y)} f(x,y)\, dx\, dy = \int_c^d dy \int_{\alpha(y)}^{\beta(y)} f(x,y)\, dx.$$

Most often, we use the first; the order of dx and dy indicates the order in which the partial integrations are to be carried out.

A good example of the way we can evaluate a double integral with iterated single integrals is provided by the gravitational field of a square plate (pp. 270–272) at a point above the center of the plate:

The gravitational field by iterated integrals

$$\text{field at } a = \iint_S \frac{-a\,dA}{(x^2+y^2+a^2)^{3/2}}, \qquad S: \begin{array}{l} 0 \leq x \leq R, \\ 0 \leq y \leq R. \end{array}$$

(In this expression we have used $4G\rho = 1$ and we have written the element of area as dA.) As an iterated integral, the field is

$$\text{field at } a = \int_0^R \int_0^R \frac{-a\,dy}{(x^2+y^2+a^2)^{3/2}}\, dx.$$

The first integration, with respect to y, can be done with the pullback substitution $y = \sqrt{x^2+a^2}\tan\theta$ (see Exercise 9.1); the result is

$$\int_0^R \frac{-a\,dy}{(x^2+y^2+a^2)^{3/2}} = \frac{-ay}{(x^2+a^2)(x^2+y^2+a^2)^{1/2}}\Bigg|_0^R$$
$$= \frac{-aR}{(x^2+a^2)(x^2+R^2+a^2)^{1/2}}.$$

The antidifferentiation needed for the second integration is more readily done with a table or a computer algebra system:

A closed-form formula for field strength

$$\text{field at } a = \int_0^R \frac{-aR\,dx}{(x^2+a^2)(x^2+R^2+a^2)^{1/2}}$$

$$= -\arctan\left(\frac{Rx}{a\sqrt{x^2+R^2+a^2}}\right)\bigg|_0^R = -\arctan\left(\frac{R^2}{a\sqrt{2R^2+a^2}}\right).$$

This gives us a closed-form expression for the field that can shed light on the numerical results we found in Chapter 8. First note (see Exercise 9.2) that

$$\frac{R^2}{a\sqrt{2R^2+a^2}} = \frac{R^2}{aR\sqrt{2}\sqrt{1+(a/2R)^2}} = \frac{R}{a\sqrt{2}} + O(a/R) \text{ as } a/R \to 0;$$

from this we obtain the approximation

$$\text{field at } a \approx -\arctan\left(\frac{R}{a\sqrt{2}}\right).$$

Comparing estimates of field strength

The following table gives the field strength (with $R = 32$) as determined by a Riemann sum (the numerical estimate from Chapter 8.1), by the approximation immediately above, and by the complete formula derived from the iterated integrals.

a	Numerical Estimate	$-\arctan\left(\dfrac{R}{a\sqrt{2}}\right)$	$-\arctan\left(\dfrac{R^2}{a\sqrt{2R^2+a^2}}\right)$
0.2	$-1.561\,957\ldots$	$-1.561\,957\,722$	$-1.561\,957\,636$
0.1	$-1.566\ldots$	$-1.566\,376\,938$	$-1.566\,376\,927$
0.05	$-1.568\ldots$	$-1.568\,586\,622$	$-1.568\,586\,620$

We also noted in Chapter 8 that the computations suggest the field becomes constant as the size of the plate increases, that is, as $R \to \infty$. The formula confirms this: because

$$\lim_{R\to\infty} \frac{R^2}{a\sqrt{2R^2+a^2}} = \lim_{R\to\infty} \frac{R}{a\sqrt{2}} = \infty,$$

we can say

$$\text{field of infinite plate at height } a = -\arctan(\infty) = -\frac{\pi}{2} = -1.570\,796\,327,$$

a value that is independent of the height a.

The circular plate in Cartesian coordinates

To continue our illustration, let us evaluate the field of a circular plate using Cartesian coordinates. Of course polar coordinates lead to a simpler evaluation; however, we have already done that (cf. pp. 273–275), and got the result

$$\text{field at } a = 2\pi G\rho\left(\frac{a}{\sqrt{R^2+a^2}} - 1\right).$$

Cartesian coordinates give us the chance to use iterated integrals to compare calculations.

We can define the region occupied by the circular plate as

$$S: \quad \begin{matrix} -R \leq x \leq R, \\ -\sqrt{R^2 - x^2} \leq y \leq \sqrt{R^2 - x^2}; \end{matrix}$$

therefore,

$$\text{field at } a = \iint\limits_{S} \frac{-G\rho a\, dA}{(x^2 + y^2 + a^2)^{3/2}} = -G\rho a \int_{-R}^{R} \int_{-\sqrt{R^2 - x^2}}^{\sqrt{R^2 - x^2}} \frac{dy}{(x^2 + y^2 + a^2)^{3/2}} \, dx.$$

We have already (p. 323) computed the inner antiderivative, and found

$$\int_{-\sqrt{R^2 - x^2}}^{\sqrt{R^2 - x^2}} \frac{dy}{(x^2 + y^2 + a^2)^{3/2}} = \left. \frac{y}{(x^2 + a^2)(x^2 + y^2 + a^2)^{1/2}} \right|_{-\sqrt{R^2 - x^2}}^{\sqrt{R^2 - x^2}}$$

$$= \frac{2\sqrt{R^2 - x^2}}{(x^2 + a^2)(R^2 + a^2)^{1/2}}.$$

Consequently (again resorting to tables or a computer algebra system to find the antiderivative),

$$\text{field at } a = \frac{-2G\rho a}{\sqrt{R^2 + a^2}} \int_{-R}^{R} \frac{\sqrt{R^2 - x^2}}{x^2 + a^2} \, dx$$

$$= \frac{-2G\rho a}{\sqrt{R^2 + a^2}} \left(-\arctan\left(\frac{x}{\sqrt{R^2 - x^2}} \right) + \frac{\sqrt{R^2 + a^2}}{a} \arctan\left(\frac{x\sqrt{R^2 + a^2}}{a\sqrt{R^2 - x^2}} \right) \right) \Bigg|_{-R}^{R}$$

Because

$$\arctan\left(\frac{x}{\sqrt{R^2 - x^2}} \right) \Bigg|_{x = \pm R} = \arctan(\pm\infty) = \pm\frac{\pi}{2},$$

and likewise

$$\arctan\left(\frac{x\sqrt{R^2 + a^2}}{a\sqrt{R^2 - x^2}} \right) \Bigg|_{x = \pm R} = \arctan(\pm\infty) = \pm\frac{\pi}{2},$$

we have

$$\text{field at } a = \frac{-2G\rho a}{\sqrt{R^2 + a^2}} \left(-\frac{\pi}{2} + \frac{\sqrt{R^2 + a^2}}{a} \cdot \frac{\pi}{2} - \frac{\pi}{2} + \frac{\sqrt{R^2 + a^2}}{a} \cdot \frac{\pi}{2} \right)$$

$$= \frac{2\pi G\rho a}{\sqrt{R^2 + a^2}} \left(1 - \frac{\sqrt{R^2 + a^2}}{a} \right) = 2\pi G\rho \left(\frac{a}{\sqrt{R^2 + a^2}} - 1 \right),$$

in agreement with the computation using polar coordinates.

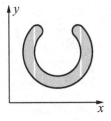

Theorem 9.3 allows us to reduce a double integral to iterated single integrals; however, it holds only for a restricted class of regions. For example, the horseshoe-shaped region shown in the margin does not meet the restriction (at least for Cartesian coordinates). Nevertheless, it is the union of a finite number of nonoverlapping sets (e.g., the five whose boundaries are shown by the white lines) that, separately, do meet the restriction. The following theorem asserts that this is enough.

Theorem 9.5. *Suppose $f(x,y)$ is continuous on a bounded region R that is the union of nonoverlapping sets $S_1, \ldots, S_p, T_1, \ldots, T_q$ of the form*

$$S_i: \begin{array}{c} a_i \leq x \leq b_i, \\ \gamma_i(x) \leq y \leq \delta_i(x), \end{array} \qquad T_j: \begin{array}{c} c_j \leq x \leq d_j, \\ \alpha_j(y) \leq x \leq \beta_j(y); \end{array}$$

then

$$\iint_R f(x,y)\,dA = \int_{a_1}^{b_1} \int_{\gamma_1(x)}^{\delta_1(x)} f(x,y)\,dy\,dx + \cdots + \int_{a_p}^{b_p} \int_{\gamma_p(x)}^{\delta_p(x)} f(x,y)\,dy\,dx$$

$$+ \int_{c_1}^{d_1} \int_{\alpha_1(y)}^{\beta_1(y)} f(x,y)\,dx\,dy + \cdots + \int_{c_q}^{d_q} \int_{\alpha_q(y)}^{\beta_q(y)} f(x,y)\,dx\,dy.$$

Proof. Because $R = S_1 \cup \cdots \cup S_p \cup T_1 \cup \cdots \cup T_q$ is a decomposition into nonoverlapping sets, we have

$$\iint_R f(x,y)\,dA = \iint_{S_1} f(x,y)\,dA + \cdots + \iint_{S_p} f(x,y)\,dA$$

$$+ \iint_{T_1} f(x,y)\,dA + \cdots + \iint_{T_q} f(x,y)\,dA$$

by the additivity of the integral (Theorem 8.27, p. 298). The result now follows by reducing each double integral on the right to the appropriate iterated integral.   □

9.2 Improper integrals

The integral of a function is defined using values of the function and the sizes of small regions, so it is natural to deal only with bounded functions over closed bounded regions. However, scientific and mathematical questions just as naturally involve unbounded functions and unbounded regions, so it is important to extend the process of integration to these more general settings. Such extensions are called *improper integrals*; they are evaluated as limits of "proper" integrals.

A function without a *proper* integral

By a **proper** integral we mean one whose value is determined, in principle, as a limit of Riemann sums. For example,

$$\int_0^1 \frac{dx}{\sqrt{x}}$$

is not a proper integral in this sense. Of course, in any Riemann sum for $f(x) = 1/\sqrt{x}$ we must avoid the point $x = 0$ because $f(0)$ is undefined. However, this is not the heart of the problem. To see what is, subdivide the interval $(0,1]$ into equal subintervals of length $\Delta x = 1/k$, for any positive integer k. Then form a Riemann sum Σ whose first term is

$$f(1/k^3) \cdot \Delta x = \frac{1}{\sqrt{1/k^3}} \cdot \frac{1}{k} = \sqrt{k}.$$

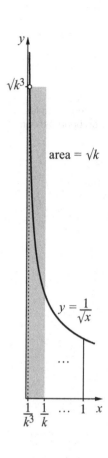

area = $\sqrt{k}$

$y = \dfrac{1}{\sqrt{x}}$

For the remaining terms in Σ, make any valid choices. Because those terms are all positive, we have

$$\Sigma \geq \sqrt{k} \to \infty \text{ as } k \to \infty.$$

Because these Riemann sums do not converge, there is no integral. The problem is not that $f(0)$ is undefined but that $f(x)$ is unbounded on that first interval $(0, \Delta x]$. (We have already used this example in a slightly different form for a similar purpose on page 299.) Theorem 8.36 (p. 307) confirms this; it says that if $f(x)$ were bounded on $(0,1]$, it would be integrable on $[0,1]$.

On any smaller interval $[a,1] \subset (0,1]$, $f(x)$ is bounded and continuous and therefore integrable. Because this integral gives the area under the part of the graph that lies above the interval $[a,1]$ on the x-axis, and because that area increases monotonically as $[a,1] \to (0,1]$, it seems reasonable to define

$$\int_0^1 \frac{dx}{\sqrt{x}} = \lim_{a \to 0} \int_a^1 \frac{dx}{\sqrt{x}}$$

if the values on the right converge to a finite limit. In fact,

$$\int_a^1 \frac{dx}{\sqrt{x}} = 2\sqrt{x} \Big|_a^1 = 2\big(1 - \sqrt{a}\big) \to 2.$$

Thus we can say the improper integral "converges" and has the value 2:

$$\int_0^1 \frac{dx}{\sqrt{x}} = 2.$$

Is this argument unnecesarily elaborate and painstaking? It would appear that we could just write

$$\int_0^1 \frac{dx}{\sqrt{x}} = 2\sqrt{x} \Big|_0^1 = 2$$

and get the correct value. However, this computation uses the fundamental theorem of calculus, which says that

$$\int_a^b f(x)\,dx = F(x) \Big|_a^b$$

The need for improper integrals

when $f(x)$ is continuous on $[a,b]$ and $F'(x) = f(x)$ there. But in our case the fundamental theorem fails to apply, because $f(x) = 1\sqrt{x}$ is not continuous on $[0,1]$. To integrate $1/\sqrt{x}$ over $(0,1]$, we must extend the definition of *integral* in some fashion; the ordinary, or "proper," integral does not exist.

Unbounded functions on bounded domains

More generally, to define the improper integral of a function $f(x)$ that is continuous but unbounded on the open interval $a < x < b$, first take $a < \alpha < \beta < b$ and compute the ordinary integral

$$I(\alpha,\beta) = \int_{\alpha}^{\beta} f(x)\,dx$$

as a function of its endpoints α and β. (The integral exists because f is bounded and continuous on $[\alpha,\beta]$.) If the values $I(\alpha,\beta)$ have a finite limit as $\alpha \to a$ and $\beta \to b$, then the **improper integral** converges and its value is that limit:

$$\int_{a}^{b} f(x)\,dx = \lim_{\substack{\alpha \to a \\ \beta \to b}} \int_{\alpha}^{\beta} f(x)\,dx.$$

More generally, if $f(x)$ is continuous on the closed interval $[a,b]$ except for the points $c_1 < c_2 < \cdots < c_k$ at which it becomes unbounded (and $a \leq c_1, c_k \leq b$), then we define the **improper integral**

$$\int_{a}^{b} f(x)\,dx = \int_{a}^{c_1} f(x)\,dx + \int_{c_1}^{c_2} f(x)\,dx + \cdots + \int_{c_k}^{b} f(x)\,dx$$

if all the intermediate improper integrals on the right converge.

Note that the intermediate improper integrals must converge separately and independently of each other. For example, the integral of $1/x$ over $[-1,1]$ is improper because $1/x$ is unbounded as $x \to 0$, so we must write

$$\int_{-1}^{1} \frac{dx}{x} = \int_{-1}^{0} \frac{dx}{x} + \int_{0}^{1} \frac{dx}{x}.$$

But this fails to converge, because neither improper integral on the right converges:

$$\int_{-1}^{0} \frac{dx}{x} = \lim_{\beta \to 0} \int_{-1}^{\beta} \frac{dx}{x} = \lim_{\beta \to 0} \ln|\beta| = -\infty,$$

$$\int_{0}^{1} \frac{dx}{x} = \lim_{\alpha \to 0} \int_{\alpha}^{1} \frac{dx}{x} = \lim_{\alpha \to 0} -\ln\alpha = +\infty.$$

If we were to link the two intermediate integrals in the following way,

$$\int_{-1}^{1} \frac{dx}{x} = \lim_{\alpha \to 0} \left(\int_{-1}^{-\alpha} \frac{dx}{x} + \int_{\alpha}^{1} \frac{dx}{x} \right),$$

we would reach the false conclusion

$$\int_{-1}^{1} \frac{dx}{x} = \lim_{\alpha \to 0} \left(\ln \alpha - \ln \alpha \right) = 0.$$

There are improper integrals for unbounded domains as well as for unbounded functions. If we try to calculate a Riemann sum for a function over an unbounded domain, one of the cells must have infinite size. However, there is a natural way to define an improper integral. Assuming that f is bounded and integrable on every finite subinterval of $a \leq x < \infty$ or $-\infty < x \leq b$, respectively, we set

$$\int_{a}^{\infty} f(x)\, dx = \lim_{B \to \infty} \int_{a}^{B} f(x)\, dx, \qquad \int_{-\infty}^{b} f(x)\, dx = \lim_{A \to -\infty} \int_{A}^{b} f(x)\, dx.$$

Thus, for example,

$$\int_{0}^{\infty} \frac{dx}{1+x^2} = \lim_{B \to \infty} \int_{0}^{B} \frac{dx}{1+x^2} = \lim_{B \to \infty} \arctan x \Big|_{0}^{B} = \lim_{B \to \infty} \arctan B = \frac{\pi}{2}.$$

We sometimes find a sequence like this abbreviated to

$$\int_{0}^{\infty} \frac{dx}{1+x^2} = \arctan x \Big|_{0}^{\infty} = \arctan \infty = \frac{\pi}{2},$$

but we must always understand that the briefer calculation depends on the validity of the longer one.

We can now turn from single to double integrals. Suppose R is a closed bounded region with area in $\mathbb{R}^2$ and Z a set of area zero. If $f(x,y)$ is bounded and continuous on $R \setminus Z$, then we know f is "properly" integrable over R (Theorem 8.35, p. 305). But suppose we allow f to become unbounded on $R \setminus Z$ while remaining continuous there; can we define the *improper* integral of f over R?

Single integrals suggest that we consider a monotonically increasing sequence $S_1 \subseteq S_2 \subseteq \cdots$ of closed subsets of $R \setminus Z$ for which $A(S_k) \to A(R)$ as $k \to \infty$. On each S_k, f is continuous and bounded, so it has a "proper" integral

$$I_k = \iint_{S_k} f(x,y)\, dA.$$

If the sequence $I_1, I_2, \ldots$ has a finite limit, I, we would like to say that the improper integral of f over R converges and has the value I.

However, in any definition that involves choices (as this does with the sequence $S_1, S_2, \ldots$), we must make certain that the outcome does not depend on the choices made. Thus, if $T_1 \subseteq T_2 \subseteq \cdots$ is another sequence of closed subsets of $R \setminus Z$ for which $A(T_m) \to A(R)$, and

$$J_m = \iint_{T_m} f(x,y)\, dA,$$

then we must verify that the sequence $J_1, J_2, \ldots$ also has a finite limit, J, and then that $J = I$.

Here is an example that illustrates how much variability there can be in the outcome. Consider $f(x,y) = 1/x$ on the unit square $R : -1 \leq x,y \leq 1$. Of course, f is undefined on the y-axis $Z : x = 0$, and is continuous but unbounded on $R \setminus Z$.

Let V_k be the infinite strip $-1/k < x < 1/k$ that is centered symmetrically on the y-axis, and let $S_k = R \setminus V_k$. The sets S_k are nested monotonically and $A(S_k) = 4 - 4/k \to 4 = A(R)$; furthermore,

$$I_k = \iint_{S_k} \frac{dA}{x} = \int_{-1}^{1} \int_{-1}^{-1/m} \frac{dx}{x}\, dy + \int_{-1}^{1} \int_{1/m}^{1} \frac{dx}{x}\, dy = 0$$

by symmetry. In fact,

$$\int_{-1}^{1} \int_{-1}^{-1/m} \frac{dx}{x}\, dy = -\int_{-1}^{1} \int_{1/m}^{1} \frac{dx}{x}\, dy;$$

that is, the contributions to I_k from the left-half plane and the right-half plane exactly cancel.

By contrast, let W_m be the *asymmetric* strip $-1/m < x < 1/m^2$, and let $T_m = R \setminus W_m$. Now $A(T_m) = 4 - 2/m - 2/m^2$, so we still have $A(T_m) \to A(R)$ as $m \to \infty$. But this time the cancellation is incomplete; the integral over T_m reduces to

$$J_m = \iint_{T_m} \frac{dA}{x} = \int_{-1}^{1} \int_{1/m^2}^{1/m} \frac{dx}{x}\, dy = 2\left(\ln 1/m - \ln 1/m^2\right) = 2\ln m.$$

(See the exercises.) Because $J_m \to \infty$, the two sets of integrals do not converge the same way, so the improper integral fails to exist.

We can attribute the variability of the outcomes to the way f changes sign on R. If we replace f by $|f|$ then that variability disappears. In fact, the integrals

$$\widehat{I}_k = \iint_{S_k} |f(x,y)|\, dA \ \text{ and } \ \widehat{J}_m = \iint_{T_m} |f(x,y)|\, dA$$

are now both unbounded monotonic increasing sequences of numbers.

A comparison with infinite series

Compare what is happening with the integrals to what can happen with certain infinite series. For example,

$$1 - \tfrac{1}{2} + \tfrac{1}{3} - \tfrac{1}{4} + \tfrac{1}{5} - \tfrac{1}{6} + \tfrac{1}{7} - \tfrac{1}{8} + \tfrac{1}{9} + \cdots = \ln 2;$$

that is, the sequence of partial sums 1, $1 - \tfrac{1}{2}$, $1 - \tfrac{1}{2} + \tfrac{1}{3}$, ... has the limiting value $\ln 2$. But a rearrangement of the terms can change the sum:

$$1 + \tfrac{1}{3} - \tfrac{1}{2} + \tfrac{1}{5} + \tfrac{1}{7} - \tfrac{1}{4} + \tfrac{1}{9} + \tfrac{1}{11} - \tfrac{1}{6} + \cdots = \tfrac{3}{2}\ln 2.$$

(Instead of strictly alternating positive and negative terms, the new series includes two positive terms for every negative one; see the exercises.) Choosing the order of terms here is analogous to choosing how the subsets S_k and T_m expand to fill out $R \setminus Z$. In both cases, different choices lead to different outcomes. Finally, replacing

f by $|f|$ is analogous to making all terms in the series be positive. In this case we get the harmonic series:

$$1 + \tfrac{1}{2} + \tfrac{1}{3} + \tfrac{1}{4} + \tfrac{1}{5} + \tfrac{1}{6} + \tfrac{1}{7} + \tfrac{1}{8} + \tfrac{1}{9} + \cdots = \infty.$$

The harmonic series diverges; its sequence of partial sums is monotonically increasing and unbounded. Because the alternating series for $\ln 2$ converges but the related series of absolute values (the harmonic series) does not, we say the series for $\ln 2$ is **conditionally convergent**. By contrast, both the alternating (geometric) series

Conditional convergence

$$1 - \tfrac{1}{2} + \tfrac{1}{4} - \tfrac{1}{8} + \tfrac{1}{16} - \tfrac{1}{32} + \tfrac{1}{64} - \tfrac{1}{128} + \cdots = \tfrac{2}{3}$$

and the corresponding series of absolute values

$$1 + \tfrac{1}{2} + \tfrac{1}{4} + \tfrac{1}{8} + \tfrac{1}{16} + \tfrac{1}{32} + \tfrac{1}{64} + \tfrac{1}{128} + \cdots = 2$$

converge, so we say the alternating series for $\tfrac{2}{3}$ is **absolutely convergent**. An absolutely convergent series is more robust: rearranging its terms does not change its value.

Absolute convergence

The improper integral we define is the analogue of an absolutely convergent series; its value will not change when we change the way the region $R \setminus Z$ is "filled up" by closed subsets S_k or T_m.

Theorem 9.6. *Let R be a closed bounded set with area, let Z be a set with area zero, and let $S_1, S_2, \ldots$ be a monotonic increasing sequence of closed subsets of $R \setminus Z$ for which $A(S_k) \to A(R)$ as $k \to \infty$. Suppose $f(x,y)$ is continuous but unbounded on $R \setminus Z$, but*

$$\widehat{I}_k = \iint_{S_k} |f(x,y)| \, dA \leq B$$

for some bound B and for all k. Then the numbers

$$I_k = \iint_{S_k} f(x,y) \, dA$$

have a finite limit I as $k \to \infty$. Furthermore, the value of I is independent of the way the closed subsets S_k are chosen.

Proof. To show that various limits exist we use the *Cauchy convergence criterion*: the sequence $a_1, a_2, \ldots$ of real numbers has a finite limit if and only if, for every $\varepsilon > 0$, there is an $N = N(\varepsilon)$ such that

$$i, j > N \implies |a_i - a_j| < \varepsilon.$$

The criterion says that the limit exists if all the a_i are arbitrarily close to one another when i is sufficiently large; for a proof see a text on analysis.

We first show that the integrals I_k converge for a particular collection of sets S_k. Suppose $i > j$; then $S_i \supseteq S_j$ so we have

$$\iint\limits_{S_i} f(x,y)\,dA = \iint\limits_{S_j} f(x,y)\,dA + \iint\limits_{S_i\setminus S_j} f(x,y)\,dA,$$

and similarly for $|f(x,y)|$. Because $|f(x,y)| \geq 0$, the sequence $\widehat{I}_1, \widehat{I}_2, \ldots$ is monotonic increasing; by hypothesis, it is bounded above so it has a finite limit. Therefore, by the Cauchy convergence criterion, we know that for any $\varepsilon > 0$, we can find an N for which

$$\widehat{I}_i - \widehat{I}_j = |\widehat{I}_i - \widehat{I}_j| < \varepsilon$$

whenever $i > j > N$. But

$$\widehat{I}_i - \widehat{I}_j = \iint\limits_{S_i} |f(x,y)|\,dA - \iint\limits_{S_j} |f(x,y)|\,dA = \iint\limits_{S_i\setminus S_j} |f(x,y)|\,dA,$$

so

$$\iint\limits_{S_i\setminus S_j} |f(x,y)|\,dA < \varepsilon.$$

For any closed set Q in $R \setminus Z$, we have

$$\left| \iint_Q f(x,y)\,dA \right| \leq \iint_Q |f(x,y)|\,dA.$$

Therefore, when $i > j > N$ we have

$$|I_i - I_j| = \left| \iint_{S_i} f(x,y)\,dA - \iint_{S_j} f(x,y)\,dA \right| = \left| \iint\limits_{S_i\setminus S_j} f(x,y)\,dA \right| \leq \iint\limits_{S_i\setminus S_j} |f(x,y)|\,dA < \varepsilon.$$

By the Cauchy convergence criterion, the sequence $I_1, I_2, \ldots$ converges to a finite limit.

Now let $T_1 \subseteq T_2 \subseteq \cdots$ be another sequence of closed sets with $A(T_m) \to A(R)$. We claim

$$\widehat{J}_m = \iint_{T_m} |f(x,y)|\,dA \leq B$$

for the same bound B. The foregoing proof would then imply that the sequence

$$J_m = \iint_{T_m} f(x,y)\,dA$$

also has a limit, J.

To prove the claim, let T be any one of the sets T_m. We know $f(x,y)$ is bounded on T: $|f(x,y)| \leq M$ for some M (that depends on T). Because $T \setminus (T \cap S_k) = T \setminus S_k$, we have

$$\iint\limits_T g(x,y)\,dA - \iint\limits_{T\cap S_k} g(x,y)\,dA = \iint\limits_{T\setminus S_k} g(x,y)\,dA,$$

where $g(x,y)$ stands for either $f(x,y)$ or $|f(x,y)|$. Therefore,

$$\left| \iint\limits_{T} g(x,y)\,dA - \iint\limits_{T \cap S_k} g(x,y)\,dA \right| \le \iint\limits_{T \setminus S_k} |g(x,y)|\,dA$$

$$= \iint\limits_{T \setminus S_k} |f(x,y)|\,dA \le M \cdot A(T \setminus S_k)$$

by Corollary 8.30. Now $A(T \setminus S_k) \to 0$ as $k \to \infty$, because $T \setminus S_k \subseteq R \setminus S_k$ and we have

$$A(T \setminus S_k) \le A(R \setminus S_k) = A(R) - A(S_k)$$

by Lemma 8.3, and $A(S_k) \to A(R)$ by hypothesis. It follows that

$$\iint\limits_{T} f(x,y)\,dA = \lim_{k \to \infty} \iint\limits_{T \cap S_k} f(x,y)\,dA \quad \text{and} \quad \iint\limits_{T} |f(x,y)|\,dA = \lim_{k \to \infty} \iint\limits_{T \cap S_k} |f(x,y)|\,dA.$$

Using the second equation and $T \cap S_k \subseteq S_k$, we find

$$\iint\limits_{T} |f(x,y)|\,dA = \lim_{k \to \infty} \iint\limits_{T \cap S_k} |f(x,y)|\,dA \le \lim_{k \to \infty} \iint\limits_{S_k} |f(x,y)|\,dA \le B,$$

proving the claim and showing that the limit J exists.

To prove that $I = J$, we begin by noting that

$$\left| \iint\limits_{T} f(x,y)\,dA - \iint\limits_{T \cap S_j} f(x,y)\,dA \right| = \lim_{i \to \infty} \left| \iint\limits_{T \cap S_i} f(x,y)\,dA - \iint\limits_{T \cap S_j} f(x,y)\,dA \right|$$

$$= \lim_{i \to \infty} \left| \iint\limits_{T \cap (S_i \setminus S_j)} f(x,y)\,dA \right|$$

$$\le \lim_{i \to \infty} \iint\limits_{S_i \setminus S_j} |f(x,y)|\,dA \le \varepsilon,$$

a result that holds for all $j > N$, where N was the number provided by the Cauchy convergence criterion for the sequence $\widehat{I}_1, \widehat{I}_2, \ldots$. (In particular, this N is independent of the choice of T.) The initial equality uses

$$\iint\limits_{T} f(x,y)\,dA = \lim_{i \to \infty} \iint\limits_{T \cap S_i} f(x,y)\,dA.$$

Furthermore, because the sequence $\widehat{J}_1, \widehat{J}_2, \ldots$ also converges, there is a number L for which

$$|\widehat{J}_l - \widehat{J}_m| < \varepsilon$$

for all $l > m > L$. Reversing the roles of S_k and T_m we can therefore conclude that

$$\left| \iint_S f(x,y)\,dA - \iint_{S \cap T_m} f(x,y)\,dA \right| \leq \varepsilon$$

for every $m > L$ and any closed subset S of $R \setminus Z$.

Finally, the telescoping sum

$$I - J = I - \iint_{S_k} f(x,y)\,dA + \iint_{S_k} f(x,y)\,dA - \iint_{S_k \cap T_m} f(x,y)\,dA$$

$$+ \iint_{S_k \cap T_m} f(x,y)\,dA - \iint_{T_m} f(x,y)\,dA + \iint_{T_m} f(x,y)\,dA - J$$

leads to the triangle inequality

$$|I - J| \leq \left| I - \iint_{S_k} f(x,y)\,dA \right| + \left| \iint_{S_k} f(x,y)\,dA - \iint_{S_k \cap T_m} f(x,y)\,dA \right|$$

$$+ \left| \iint_{S_k \cap T_m} f(x,y)\,dA - \iint_{T_m} f(x,y)\,dA \right| + \left| \iint_{T_m} f(x,y)\,dA - J \right|.$$

If we choose k and m sufficiently large, each of the four terms on the right will be bounded by ε, so $|I - J| \leq 4\varepsilon$. Because $\varepsilon > 0$ is arbitrary, $I = J$. □

Improper integral of an unbounded function

Definition 9.1 *Suppose R is a closed bounded set with area, Z a set with area zero, and S_k is a monotonic increasing collection of closed subsets of $R \setminus Z$ for which $A(S_k) \to A(R)$ as $k \to \infty$. Suppose $f(x,y)$ is continuous but unbounded on $R \setminus Z$, and the integrals*

$$\iint_{S_k} |f(x,y)|\,dA$$

*are uniformly bounded in k. Then the **improper integral of f over R** is*

$$\iint_R f(x,y)\,dA = \lim_{k \to \infty} \iint_{S_k} f(x,y)\,dA.$$

When the improper integral exists, we often say that it **converges**.

How "unbounded" can a function be and still have a convergent improper integral? For example,

$$f(x,y) = \frac{1}{r^p}, \quad r = \sqrt{x^2 + y^2}$$

is unbounded near the origin when $p > 0$; for which values of p does

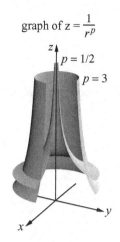

graph of $z = \dfrac{1}{r^p}$

$$\iint\limits_{x^2+y^2 \le 1} \frac{dA}{r^p}$$

converge? The figure suggests $1/r^p$ becomes unbounded more rapidly as p becomes larger; thus, the integral—thought of as the volume under the graph—is more likely to converge for smaller values of p. To answer the question, let S_k be the set of points (x,y) in the ring $1/k^2 \le x^2 + y^2 \le 1$. Then, changing to polar coordinates, we find

$$I_k = \iint_{S_k} \frac{dA}{r^p} = \int_0^{2\pi} \int_{1/k}^1 \frac{r\,dr}{r^p}\,d\theta = 2\pi \frac{r^{2-p}}{2-p}\bigg|_{1/k}^1 = \frac{2\pi}{2-p}(1+k^{p-2}).$$

This formula does not allow $p = 2$, so we deal with that case separately.

First, the sequence $I_1, I_2, \ldots$ has a finite limit as $k \to \infty$ only if $k^{p-2} \to 0$, that is, only if $p - 2 < 0$, or $p < 2$. If $p = 2$, then

$$I_k = \iint_{S_k} \frac{dA}{r^2} = \int_0^{2\pi} \int_{1/k}^1 \frac{dr}{r}\,d\theta = 2\pi \ln r\bigg|_{1/k}^1 = 2\pi \ln k \to \infty.$$

Thus the improper integral converges if and only if $0 < p < 2$, in which case it has the value

$$\iint\limits_{x^2+y^2 \le 1} \frac{dA}{r^p} = \frac{2\pi}{2-p}.$$

(Of course, this formula also gives the value of the "proper" integral that exists for all $p \le 0$.) Our example has the following useful generalization.

Because $|f(x,y)| = 1/r^p = f(x,y)$, the improper integral of f converges "absolutely" or not at all; we do not need to test separately whether the integrals of $|f(x,y)|$ are uniformly bounded, as stipulated in Theorem 9.6.

Testing for absolute convergence

For another example, take $g(x,y) = \ln r$. It is also unbounded near the origin, but because $\ln r < 0$ when $0 < r < 1$, we should first consider $|g(x,y)| = -\ln r$. Thus, on the disks $S_k : 1/k^2 \le x^2 + y^2 \le 1$ that we just used,

$$\widehat{I}_k = \iint\limits_{1/k^2 \le x^2+y^2 \le 1} |g(x,y)|\,dA = \int_0^{2\pi} \int_{1/k}^1 -r\ln r\,dr\,d\theta.$$

The function $z = -r\ln r$ is continuous and bounded by $1/e$ on $0 < r \le 1$, so we have

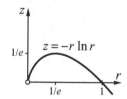

$$\int_{1/k}^1 -r\ln r\,dr \le \frac{1}{e}\left(1-\frac{1}{k}\right) \le \frac{1}{e},$$

implying the uniform bound $\widehat{I}_k \le 2\pi/e$ for all k. Therefore, by Theorem 9.6, the improper integral

$$\iint\limits_{x^2+y^2\le 1} \ln\sqrt{x^2+y^2}\,dA$$

converges. (It converges to $-\pi/2$; see Exercise 9.22).

Unbounded regions   An integral will also be improper when its domain R is unbounded. As in the earlier case of an unbounded function, choose a monotonic increasing sequence of closed bounded subsets $S_1 \subseteq S_2 \subseteq \cdots$ of R. Because it no longer makes sense to require $A(S_k) \to A(R)$ (because the area of R may be infinite), we achieve what we really want—that the sets S_k eventually cover R—by stipulating instead that each closed bounded subset of R be contained in some S_k. As in the earlier case, we also require absolute convergence.

Theorem 9.7. *Let R be an unbounded set in $\mathbb{R}^2$, Z a set with area zero, and $S_1, S_2, \ldots$ a monotonic increasing sequence of closed bounded subsets of R such that, given any closed bounded subset W of R, $S_k \supseteq W$ for some k. Suppose $f(x,y)$ is bounded and continuous on $R \setminus Z$, and*

$$\widehat{I_k} = \iint_{S_k} |f(x,y)|\,dA \le B$$

for some bound B and for all k. Then the numbers

$$I_k = \iint_{S_k} f(x,y)\,dA$$

have a finite limit I as $k \to \infty$. Furthermore, the value of I is independent of the way the closed subsets S_k are chosen.

Proof. This proof has many parallels with the previous one; we focus on the points where the two differ. To begin, because f and $|f|$ are bounded and continuous on $R \setminus Z$, the same is true on each $S_k \setminus Z$, so f and $|f|$ are integrable on each S_k. This is enough to establish, as in the earlier proof, that the sequence $I_1, I_2, \ldots$ has a finite limit.

The next step is to consider a second monotonic increasing sequence of closed bounded subsets $T_1, T_2, \ldots$ that exhaust R the same way the sequence $S_1, S_2, \ldots$ does. Each T_m is a closed bounded subset of R; thus it is entirely contained in some S_k, by hypothesis. Hence,

$$\widehat{J_m} = \iint_{T_m} |f(x,y)|\,dA \le \iint_{S_k} |f(x,y)|\,dA \le B.$$

As noted in the earlier proof, this implies that the integrals

$$J_m = \iint_{T_m} f(x,y)\,dA$$

converge to a finite limit, J.

To prove that $I = J$, the earlier proof first established that

$$\left| \iint_{T_m} f(x,y)\, dA - \iint_{T_m \cap S_k} f(x,y)\, dA \right| \le \varepsilon$$

for all $k > N$, and uniformly for all T_m. There, the key step was that

$$\iint_{T_m} f(x,y)\, dA = \lim_{i \to \infty} \iint_{T_m \cap S_i} f(x,y)\, dA;$$

here, this holds for the simple reason that $T_m \cap S_i = T_m$ for all i sufficiently large. Reversing the roles of S_k and T_m we likewise conclude that

$$\left| \iint_{S_k} f(x,y)\, dA - \iint_{S_k \cap T_m} f(x,y)\, dA \right| \le \varepsilon$$

for every $m > L$ and for all S_k. Then $|I - J| \le 4\varepsilon$ as before. $\qquad\square$

Definition 9.2 *When the conditions of the previous theorem are met, then the **improper integral of f over R converges** to*

Improper integral over an unbounded domain

$$\iint_R f(x,y)\, dA = \lim_{k \to \infty} \iint_{S_k} f(x,y)\, dA.$$

In Chapter 1, we met one of the standard examples of an improper integral over an unbounded interval:

The normal density function of statistics

$$\iint_{\mathbb{R}^2} e^{(-x^2 - y^2)/2}\, dA = 2\pi.$$

We used this to evaluate another improper integral,

$$\int_{-\infty}^{\infty} e^{-(x-\mu)^2/2\sigma^2}\, dx = \sigma\sqrt{2\pi},$$

that relates to the density function of the normal probability distribution of statistics.

9.3 The change of variables formula

This book begins with the change of variables formula for single integrals. It says that when there is an invertible pullback function $x = \varphi(s)$, then

The formula for single integrals

$$\int_a^b f(x)\, dx = \int_{\varphi^{-1}(a)}^{\varphi^{-1}(b)} f(\varphi(s))\, \varphi'(s)\, ds.$$

Our goal here is to state and prove the analogous formula for double integrals. As we show in a moment, the single integral takes orientation into account; however, we have not yet defined double integrals with orientation. At this stage, therefore, we must suppress the information about orientation in the single-integral formula to carry through an analogy. Here is an example that illustrates both the problem and the solution.

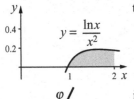

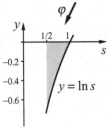

Let $f(x) = \ln x/x^2$; we can see by eye that the value of the integral

$$\int_1^2 \frac{\ln x}{x^2}\,dx$$

is positive but less than 0.2. (The vertical scales of the graphs shown in the margin have been doubled for clarity.) To find the value using the change of variables formula, consider the pullback $x = \varphi(s) = 1/s$. Then $\varphi'(s) = -1/s^2$, and

$$\int_1^2 \frac{\ln x}{x^2}\,dx = \int_1^{1/2} s^2 \ln \frac{1}{s} \left(\frac{-ds}{s^2} \right) = \int_1^{1/2} -\ln \frac{1}{s}\,ds = \int_1^{1/2} \ln s \, ds.$$

Note that the new integrand, $\ln s$, is *negative* on the new interval $[1/2, 1]$ More significantly, the new integration is carried out in the *negative* sense, from 1 backwards to 1/2. These two "negative" aspects of the new integral combine to produce a positive value,

$$\int_1^{1/2} \ln s \, ds = s \ln s - s \bigg|_1^{1/2} = \tfrac{1}{2} \ln \tfrac{1}{2} - \tfrac{1}{2} - (\ln 1 - 1) = \tfrac{1}{2}(1 - \ln 2) \approx 0.15.$$

The effect of reversing orientation

What we see in this example always happens when φ is *orientation-reversing*: $\varphi'(s) < 0$ changes the sign of the integrand, and the oriented interval $a \to b$ is transformed into the oriented interval $\varphi^{-1}(a) \leftarrow \varphi^{-1}(b)$; that is, $a < b$ implies $\varphi^{-1}(a) > \varphi^{-1}(b)$.

An orientation-free change of variables formula

We therefore need to reformulate the way we write a single integral so as to suppress this information about orientation. The unoriented version of the change of variables formula has the following form,

$$\int_I f(x)\,dx = \int_{\varphi^{-1}(I)} f(\varphi(s))\,|\varphi'(s)|\,ds.$$

In this formula, I stands for the *unoriented* set of real numbers x that lie between a and b, inclusive, and $\varphi^{-1}(I)$ is the unoriented set of real numbers s for which $\varphi(s)$ lies in I. The integral over the *unoriented* domain I is defined in terms of the usual integral, as follows.

$$\int_I f(x)\,dx = \begin{cases} \displaystyle\int_a^b f(x)\,dx & \text{if } I = [a,b], \\[2ex] \displaystyle\int_b^a f(x)\,dx & \text{if } I = [b,a]. \end{cases}$$

Before we prove that the new, unoriented, change of variables formula is the correct modification of the original one, let us verify that it works on the example $f(x) = \ln x/x^2$ with $x = \varphi s = 1/s$. Because $\varphi([1,2]) = [1/2,1]$ and $|\varphi'(s)| = +1/s^2$,

$$\int_{[1,2]} \frac{\ln x}{x^2}\, dx = \int_{[1/2,1]} s^2 \ln\frac{1}{s}\left(\frac{ds}{s^2}\right) = \int_{1/2}^1 -\ln s\, ds = s - s\ln s\, \Big|_{1/2}^1 = \tfrac{1}{2}(1 - \ln 2).$$

To prove that the oriented change of variables formula leads to the unoriented one, let us assume $a < b$; we can use a similar argument if $b < a$. If φ is orientation-preserving, then $|\varphi'(s)| = \varphi'(s)$ and $\varphi^{-1}(a) < \varphi^{-1}(b)$, so

$$\int_{\varphi^{-1}(I)} f(\varphi(s))\,|\varphi'(s)|\,ds = \int_{\varphi^{-1}(a)}^{\varphi^{-1}(b)} f(\varphi(s))\,\varphi'(s)\,ds = \int_a^b f(x)\,dx = \int_I f(x)\,dx.$$

If φ is orientation-reversing, then $|\varphi'(s)| = -\varphi'(s)$ and $\varphi^{-1}(b) < \varphi^{-1}(a)$, so

$$\int_{\varphi^{-1}(I)} f(\varphi(s))\,|\varphi'(s)|\,ds = \int_{\varphi^{-1}(b)}^{\varphi^{-1}(a)} f(\varphi(s))\,(-\varphi'(s)\,ds)$$

$$= \int_{\varphi^{-1}(a)}^{\varphi^{-1}(b)} f(\varphi(s))\,\varphi'(s)\,ds = \int_a^b f(x)\,dx = \int_I f(x)\,dx.$$

The new formula holds in both cases.

We can now formulate an analogous change of variables formula for double integrals. Let $f(x,y)$ be a continuous function on a domain D in $\mathbb{R}^2$, and assume that D is a closed bounded set that has area. Then the integral

$$\iint_D f(x,y)\,dx\,dy$$

Change of variables for double integrals

exists (Theorem 8.35, p. 305). We use "$dx\,dy$" here in place of "dA" as a way to keep track of the variables that appear in different integrals. If the change of variables is given by the pullback substitution

$$\boldsymbol{\varphi} : \begin{cases} x = x(s,t), \\ y = y(s,t), \end{cases}$$

then we show that the integral is transformed by

$$\iint_D f(x,y)\,dx\,dy = \iint_{\boldsymbol{\varphi}^{-1}(D)} f(\boldsymbol{\varphi}(s,t))\,\big|\det d\boldsymbol{\varphi}_{(s,t)}\big|\,ds\,dt.$$

In particular, $\big|\det d\boldsymbol{\varphi}_{(s,t)}\big|$ corresponds to $|\varphi'(s)|$; this is the absolute value of the *Jacobian*,

$$J_{\boldsymbol{\varphi}}(s,t) = \det d\boldsymbol{\varphi}_{(s,t)} = \begin{vmatrix} \dfrac{\partial x}{\partial s}(s,t) & \dfrac{\partial x}{\partial t}(s,t) \\[2mm] \dfrac{\partial y}{\partial s}(s,t) & \dfrac{\partial y}{\partial t}(s,t) \end{vmatrix} = \dfrac{\partial(x,y)}{\partial(s,t)}.$$

Consequently,

$$\iint\limits_{\boldsymbol{\varphi}^{-1}(D)} f(\boldsymbol{\varphi}(s,t)) \left| \det d\boldsymbol{\varphi}_{(s,t)} \right| ds\,dt = \iint\limits_{\boldsymbol{\varphi}^{-1}(D)} f(x(s,t),y(s,t)) \left| \dfrac{\partial(x,y)}{\partial(s,t)} \right| ds\,dt$$

gives us an alternate expression for the transformed integral that is useful in the following work. Moreover, the notation suggests that, when the variables themselves change, then

$$dx\,dy \text{ changes to } \left| \dfrac{\partial(x,y)}{\partial(s,t)} \right| ds\,dt.$$

For the moment, this is just a mnemonic. Before proving the theorem that estab-lishs the change of variables formula for double integrals, let us first explore some examples.

Most familiar is the change to polar coordinates that we have already used several times:

$$\boldsymbol{\varphi}: \begin{cases} x = r\cos\theta, \\ y = r\sin\theta; \end{cases} \quad \dfrac{\partial(x,y)}{\partial(r,\theta)} = \begin{vmatrix} \cos\theta & -r\sin\theta \\ \sin\theta & r\cos\theta \end{vmatrix} = r \geq 0;$$

$$\iint\limits_{D} f(x,y)\,dx\,dy = \iint\limits_{\boldsymbol{\varphi}^{-1}(D)} f(r\cos\theta, r\sin\theta)\,r\,dr\,d\theta.$$

Changing variables to simplify domains   With single integrals, we change variables to simplify the integrand. With dou-ble integrals, there is a second reason: to simplify the domain. One example is an integral with circular symmetry; it is often recast into polar coordinates. A second example is the integral

$$\iint\limits_{D} x^2 + y^2 \ dx\,dy,$$

where the domain D is the curvilinear quadrilateral in the first quadrant whose points (x,y) satisfy the inequalities

$$D: \begin{array}{l} 1 \leq x^2 - y^2 \leq 2, \\ 1 \leq 2xy \leq 3. \end{array}$$

The sides of D are hyperbolic arcs. The quadratic map

$$\mathbf{g}: \begin{cases} u = x^2 - y^2, \\ v = 2xy, \end{cases}$$

that we analyzed on pages 116–121 straightens these arcs. For example, the hyperbola $x^2 - y^2 = 1$ in the first quadrant of the (x,y)-plane becomes the line $u = 1$ in the first quadrant of the (u,v)-plane. The quadrilateral D becomes the rectangle

$$R = \mathbf{g}(D): \quad \begin{array}{l} 1 \le u \le 2, \\ 1 \le v \le 3. \end{array}$$

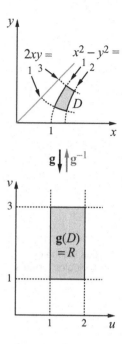

Unfortunately, $\mathbf{g}$ is a *push-forward* map, not a *pullback*, so the change of variables formula does not apply directly. But $\mathbf{g}$ is invertible on the first quadrant, and $\mathbf{g}^{-1}$ does indeed pull back (x,y) to (u,v), so we let $\mathbf{g}^{-1}$ play the role of $\boldsymbol{\varphi}$ and write

$$\iint\limits_{D} f(x,y)\,dx\,dy = \iint\limits_{\mathbf{g}(D)} f(\mathbf{g}^{-1}(u,v))\left|\det d\mathbf{g}^{-1}_{(u,v)}\right|\,du\,dv.$$

To evaluate the right-hand side we can use formulas for the components of $\mathbf{g}^{-1}(u,v)$ that appear in Exercise 4.13 (p. 144). However, we can actually determine everything we need without recourse to those formulas. To begin,

$$u^2 = x^4 - 2x^2y^2 + y^4 \quad \text{and} \quad v^2 = 4x^2y^2,$$

so $u^2 + v^2 = x^4 + 2x^2y^2 + y^4 = (x^2+y^2)^2$ and

$$f(x,y) = x^2 + y^2 = \sqrt{u^2+v^2} = f(\mathbf{g}^{-1}(u,v)).$$

Next,

$$\det d\mathbf{g}^{-1} = \frac{1}{\det d\mathbf{g}} = \frac{1}{4(x^2+y^2)} = \frac{1}{4\sqrt{u^2+v^2}} > 0,$$

so the integrand of the transformed integral is just

$$f(\mathbf{g}^{-1}(u,v))\left|\det d\mathbf{g}^{-1}_{(u,v)}\right| = \sqrt{u^2+v^2}\,\frac{1}{4\sqrt{u^2+v^2}} = \frac{1}{4}.$$

Therefore, the change of variables formula gives

$$\iint\limits_{D} x^2 + y^2\,dx\,dy = \iint\limits_{R} \frac{1}{4}\,du\,dv = \frac{1}{4}\,\text{area}\,R = \frac{1}{2}.$$

(Of course we could have tried to evaluate the original double integral by reducing it to iterated integrals, but they lack the simplicity and elegance of the transformed double integral.)

We can now formulate the change of variables formula for double integrals under a general push-forward substitution $\mathbf{g}$ *that has an inverse*:

Change of variables
with a push-forward

$$\mathbf{g}: \begin{cases} u = u(x,y), \\ v = v(x,y); \end{cases} \qquad \mathbf{g}^{-1}: \begin{cases} x = x(u,v), \\ y = y(u,v). \end{cases}$$

If we write the Jacobian as

$$J_{\mathbf{g}^{-1}}(u,v) = \det d\mathbf{g}^{-1} = \frac{\partial(x,y)}{\partial(u,v)},$$

then the change of variables formula takes the form

$$\iint_D f(x,y)\,dx\,dy = \iint_{\mathbf{g}(D)} f(x(u,v),y(u,v)) \left| \frac{\partial(x,y)}{\partial(u,v)} \right| du\,dv.$$

Note that we have expressed the transformed integral in terms of the component functions $x(u,v)$ and $y(u,v)$ of $\mathbf{g}^{-1}$. But usually only the components of $\mathbf{g}$ itself are given; it may be difficult or impossible to find closed-form expressions for $x(u,v)$ and $y(u,v)$. This can make it impractical to transform an integral by a push-forward substitution.

Areas via the change of variables formula

The change of variables formula for double integrals also gives us a way to determine areas. To continue the last example, we have

$$\text{area}\,D = \iint_D 1\,dx\,dy = \iint_R \frac{1}{4\sqrt{u^2+v^2}}\,du\,dv.$$

One way to continue is to convert the double integral into an iterated integral:

$$\iint_R \frac{1}{4\sqrt{u^2+v^2}}\,du\,dv = \frac{1}{4}\int_1^3 \int_1^2 \frac{du}{\sqrt{u^2+v^2}}\,dv.$$

This can be evaluated using a computer algebra system (or a table of integrals). We can also use the pullback substitution $u = v \cdot \sinh(s)$ (see Exercise 1.16, p. 23) to get

$$\int_1^2 \frac{du}{\sqrt{u^2+v^2}} = \operatorname{arcsinh} \frac{u}{v} \Big|_1^2 = \operatorname{arcsinh} \frac{2}{v} - \operatorname{arcsinh} \frac{1}{v}.$$

Integration by parts (when $v > 0$) gives

$$\int 1 \cdot \operatorname{arcsinh} \frac{a}{v}\,dv = v \cdot \operatorname{arcsinh} \frac{a}{v} - \int \frac{-a}{\sqrt{v^2+a^2}}\,dv$$

$$= v \cdot \operatorname{arcsinh} \frac{a}{v} + a \cdot \operatorname{arcsinh} \frac{v}{a},$$

so we get

$$\int_1^3 \left(\operatorname{arcsinh} \frac{2}{v} - \operatorname{arcsinh} \frac{1}{v} \right) dv$$

$$= v \cdot \operatorname{arcsinh} \frac{2}{v} + 2 \cdot \operatorname{arcsinh} \frac{v}{2} - v \cdot \operatorname{arcsinh} \frac{1}{v} - \operatorname{arcsinh} v \Big|_1^3 \approx 0.820\,853.$$

Finally, incorporating the factor $1/4$, we find $\text{area}\,D \approx 0.205\,213$.

We can also get an approximation to area D by approximating the value of the double integral by a Riemann sum. The integrand $J(u,v) = 1/4\sqrt{u^2+v^2}$ is, of course, the local area magnification factor for the map $\mathbf{g}^{-1} : R \to D$. Therefore, we can estimate the area of D as follows. Divide R into small rectangular cells Q of area $\Delta u \, \Delta v$; multiply that area by the value of J at the center of C to approximate the area of the image $\mathbf{g}^{-1}(Q)$ in D; add the results. The table below does this with R partitioned into eight squares of area $1/4$. The accumulated sum of $J \Delta u \Delta v$ is tallied in the right column; it yields the estimate $0.204\,806$ for the area of D (cf. Exercise 8.2, p. 313).

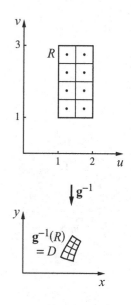

u	v	$J(u,v)$	Sum
1.25	1.25	0.141421	0.035355
1.25	1.75	0.116248	0.064417
1.25	2.25	0.097129	0.086994
1.25	2.75	0.082761	0.109390
1.75	1.25	0.116248	0.138451
1.75	1.75	0.101102	0.163705
1.75	2.25	0.087706	0.185632
1.75	2.75	0.076696	0.204806

For a *linear* map L, $|\det L|$ is the magnification factor for areas (by Theorem 8.22, p. 292):

Area magnification for nonlinear maps

$$A(L(S)) = |\det L| A(S),$$

when S is any subset of the plane that has area. (Here *area* is nonnegative; in the following section we consider oriented regions that are assigned negative area.) For a *nonlinear* map $\boldsymbol{\varphi}(s,t)$, the connection between $A(\boldsymbol{\varphi}(S))$ and $A(S)$ is not so simple, but the change of variables formula still allows us to write

$$A(\boldsymbol{\varphi}(S)) = \iint_S |\det d\boldsymbol{\varphi}_{(s,t)}| \, ds \, dt = \iint_S |J_{\boldsymbol{\varphi}}(s,t)| \, ds \, dt.$$

Using the language of set functions (see below, p. 352), we show how this equation makes $|J_{\boldsymbol{\varphi}}(s,t)|$ the local area magnification factor for $\boldsymbol{\varphi}$. Here, it is the crucial "base case" of the change of variables formula for double integrals. We state it now as a theorem.

Theorem 9.8. *Let Ω be a bounded open set in $\mathbb{R}^2$, and let $\boldsymbol{\varphi} : \Omega \to \mathbb{R}^2$ be a continuously differentiable map that has a continuously differentiable inverse $\boldsymbol{\varphi}^{-1} : \boldsymbol{\varphi}(\Omega) \to \Omega$. Suppose the set S has area and its closure $\overline{S} = S \cup \partial S$ lies within Ω; then $\boldsymbol{\varphi}(S)$ has area and*

Jacobian as local area magnification factor

$$A(\boldsymbol{\varphi}(S)) = \iint_S |J_{\boldsymbol{\varphi}}(s,t)| \, ds \, dt,$$

where $J_{\boldsymbol{\varphi}}(s,t)$ is the Jacobian of $\boldsymbol{\varphi}$ at (s,t).

Proof. The proof is simple in principle; it follows an argument given by J. Schwartz [16]. First, partition S into small pieces Q_i. On Q_i, choose a representative value for the area magnification factor $|J_{\boldsymbol{\varphi}}| = |\det d\boldsymbol{\varphi}|$. Then the area of the image $\boldsymbol{\varphi}(Q_i)$ is approximately $|J_{\boldsymbol{\varphi}}|A(Q_i)$, and the sum of such terms approximates the area of $\boldsymbol{\varphi}(S)$.

However, the details of the proof are numerous and lengthy; they involve several steps that we write as separate lemmas. We first need an open set U containing $\overline{S}$ on whose closure the functions $\boldsymbol{\varphi}$, $d\boldsymbol{\varphi}$, and $J_{\boldsymbol{\varphi}}$ are uniformly continuous.

Lemma 9.1. *There is an open set U for which $\overline{S} \subset U \subset \overline{U} \subset \Omega$.*

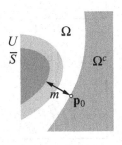

Proof. Let $\mathbf{p}$ be a point in $\mathbb{R}^2$; define the function

$$d(\mathbf{p}) = \min_{\mathbf{s} \in \overline{S}} \|\mathbf{p} - \mathbf{s}\|$$

that gives the distance from $\mathbf{p}$ to the closed set $\overline{S}$. Then $d(\mathbf{p}) = 0$ if and only if $\mathbf{p}$ is in $\overline{S}$; in particular, $d(\Omega^c) > 0$ because $\overline{S} \subset \Omega$. But Ω^c is closed and d is continuous, so $(\mathbf{p})$ attains its minimum $m > 0$ at some point $\mathbf{p}_0$ in Ω^c: $d(\Omega^c) \geq d(\mathbf{p}_0) = m > 0$. To complete the proof of the lemma, we can take

$$U = \{\mathbf{p} : d(\mathbf{p}) < m/2\}, \quad \overline{U} = \{\mathbf{p} : d(\mathbf{p}) \leq m/2\}. \qquad \square$$

The next lemma makes the first step in connecting $A(\boldsymbol{\varphi}(Q))$ to the integral of $|J_{\boldsymbol{\varphi}}|$ over Q, where Q is a square in one of the original grids $\mathcal{J}_k$ used to define Jordan content. It says that the outer area of the image $\boldsymbol{\varphi}(Q)$ is bounded by the maximum value of $|J_{\boldsymbol{\varphi}}|$ on Q.

Lemma 9.2. *For any given $\varepsilon > 0$, there is a positive integer K such that if $Q \subseteq \overline{U}$ is a square of $\mathcal{J}_k$ and $k \geq K$, then*

$$\overline{A}(\boldsymbol{\varphi}(Q)) \leq (M + O(\varepsilon))A(Q),$$

where M is the maximum value of $|J_{\boldsymbol{\varphi}}|$ on Q.

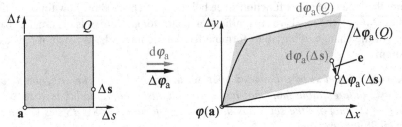

Proof. The idea of the proof is to compare the action of $\boldsymbol{\varphi}$ to the action of its linear approximation $d\boldsymbol{\varphi}_\mathbf{a}$ taken at the lower-left hand corner $\mathbf{a}$ of Q. In terms of local (or "window") coordinates

$$\Delta\mathbf{s} = \mathbf{s} - \mathbf{a} \quad \text{and} \quad \Delta\mathbf{x} = \mathbf{x} - \boldsymbol{\varphi}(\mathbf{a})$$

centered at this corner and its image, the two maps are

$$\Delta \mathbf{x} = \Delta \boldsymbol{\varphi}_\mathbf{a}(\Delta \mathbf{s}) = \boldsymbol{\varphi}(\mathbf{a} + \Delta \mathbf{s}) - \boldsymbol{\varphi}(\mathbf{a}) \text{ and } \Delta \mathbf{x} = d\boldsymbol{\varphi}_\mathbf{a}(\Delta \mathbf{s}).$$

Because $d\boldsymbol{\varphi}_\mathbf{a}$ is linear, the image $d\boldsymbol{\varphi}_\mathbf{a}(Q)$ is a parallelogram, and

$$A(d\boldsymbol{\varphi}_\mathbf{a}(Q)) = |\det d\boldsymbol{\varphi}_\mathbf{a}| A(Q) = |J_{\boldsymbol{\varphi}}(\mathbf{a})| A(Q) \leq M \times A(Q).$$

We now use the continuity of $\boldsymbol{\varphi}$ and $d\boldsymbol{\varphi}_\mathbf{a}$ to show that, when Q is sufficiently small, the two images $d\boldsymbol{\varphi}_\mathbf{a}(Q)$ and $\Delta \boldsymbol{\varphi}_\mathbf{a}(Q)$ are close enough so that our bound on the area of the first leads to a (slightly larger) bound on the outer area of the second.

For any point $\Delta \mathbf{s}$, we want to determine the difference

$$\mathbf{e} = \Delta \boldsymbol{\varphi}_\mathbf{a}(\Delta \mathbf{s}) - d\boldsymbol{\varphi}_\mathbf{a}(\Delta \mathbf{s}).$$

It is convenient to work with the component functions of $\Delta \boldsymbol{\varphi}_\mathbf{a}$ and $d\boldsymbol{\varphi}_\mathbf{a}$:

$$\Delta \boldsymbol{\varphi}_\mathbf{a} : \begin{cases} \Delta x = x(a + \Delta s, b + \Delta t) - x(a,b), \\ \Delta y = y(a + \Delta s, b + \Delta t) - y(a,b), \end{cases} \quad d\boldsymbol{\varphi}_\mathbf{a} : \begin{cases} \Delta x = x_s(a,b)\Delta s + x_t(a,b)\Delta t, \\ \Delta y = y_s(a,b)\Delta s + y_t(a,b)\Delta t. \end{cases}$$

Applying the law of the mean (Theorem 3.5, p. 75) to each component of $\Delta \boldsymbol{\varphi}_\mathbf{a}$, we get

$$\Delta \boldsymbol{\varphi}_\mathbf{a} : \begin{cases} \Delta x = x_s(a_1,b_1)\Delta s + x_t(a_1,b_1)\Delta t, \\ \Delta y = y_s(a_2,b_2)\Delta s + y_t(a_2,b_2)\Delta t, \end{cases}$$

where (a_1,b_1) and (a_2,b_2) are two properly chosen points on the line connecting (a,b) and $(a + \Delta s, b + \Delta t)$. This allows us to write $\mathbf{e} = \Delta \boldsymbol{\varphi}_\mathbf{a}(\Delta \mathbf{s}) - d\boldsymbol{\varphi}_\mathbf{a}(\Delta \mathbf{s})$ as

$$\mathbf{e} : \begin{cases} \Delta x = \big(x_s(a_1,b_1) - x_s(a,b)\big)\Delta s + \big(x_t(a_1,b_1) - x_t(a,b)\big)\Delta t, \\ \Delta y = \big(y_s(a_2,b_2) - y_s(a,b)\big)\Delta s + \big(y_t(a_2,b_2) - y_t(a,b)\big)\Delta t. \end{cases}$$

To get a bound on $\mathbf{e}$, we use the continuity of $d\boldsymbol{\varphi}_\mathbf{s}$ as a function of the point $\mathbf{s}$. On the closed bounded set $\overline{U}$ (Lemma 9.1), the four components x_s, x_t, y_s, y_t of $d\boldsymbol{\varphi}_\mathbf{s}$ are uniformly continuous. Thus, for $\varepsilon > 0$ as given in the statement of the lemma, there is a $\delta > 0$ such that

$$\|(s_1 - s_2, t_1 - t_2)\| < \delta \quad \Longrightarrow \quad \|x_s(s_1,t_1) - x_s(s_2,t_2)\| < \varepsilon,$$

and likewise for x_t, y_s, and y_t. Now choose K so that the mesh size $\|\mathcal{I}_K\| = \sqrt{2}/2^K$ is less than δ (Definition 8.14, p. 291). Then, for any $k \geq K$, we have

$$\|\mathcal{I}_k\| = \frac{\sqrt{2}}{2^k} \leq \frac{\sqrt{2}}{2^K} < \delta.$$

Now let $Q \subseteq \overline{U}$ be a square of $\mathcal{I}_k$, $k \geq K$; then

$$\|(a_1 - a, b_1 - b)\| < \delta \text{ and } \|(a_2 - a, b_2 - b)\| < \delta,$$

and $0 \le \Delta s, \Delta t \le 1/2^k$. Therefore,

$$\|\mathbf{e}\| : \begin{cases} |\Delta x| \le |x_s(a_1,b_1) - x_s(a,b)| \Delta s + |x_t(a_1,b_1) - x_t(a,b)| \Delta t < \dfrac{2\varepsilon}{2^k}, \\[4mm] |\Delta y| \le |y_s(a_2,b_2) - y_s(a,b)| \Delta s + |y_t(a_2,b_2) - y_t(a,b)| \Delta t < \dfrac{2\varepsilon}{2^k}, \end{cases}$$

so $\|\mathbf{e}\| < 2\varepsilon\sqrt{2}/2^k = W$.

Thus, every point of $\Delta\boldsymbol{\varphi}_{\mathbf{a}}(Q) = \boldsymbol{\varphi}(Q)$ lies either in the parallelogram $d\boldsymbol{\varphi}_{\mathbf{a}}(Q)$ or within the distance W from one of the four lines that make up its boundary. So $\boldsymbol{\varphi}(Q)$ is contained in the union of the parallelogram and four rectangles of length $L + 2W$ and width $2W$, where L is the length of the longer side of the parallelogram. A bound on the length L will thus lead to a bound on the outer area of $\boldsymbol{\varphi}(Q)$.

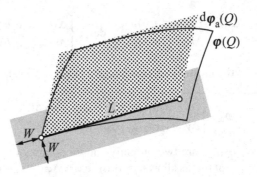

To get a bound on L it is enough to consider the two sides of $d\boldsymbol{\varphi}_{\mathbf{a}}(Q)$ that meet at the corner $\boldsymbol{\varphi}(a,b)$. These are the vectors

$$d\boldsymbol{\varphi}_{\mathbf{a}} \begin{pmatrix} 1/2^k \\ 0 \end{pmatrix} = \frac{1}{2^k} \begin{pmatrix} x_s(a,b) \\ x_t(a,b) \end{pmatrix} \quad \text{and} \quad d\boldsymbol{\varphi}_{\mathbf{a}} \begin{pmatrix} 0 \\ 1/2^k \end{pmatrix} = \frac{1}{2^k} \begin{pmatrix} y_s(a,b) \\ y_t(a,b) \end{pmatrix}.$$

Because $d\boldsymbol{\varphi}_{\mathbf{s}}$ is a continuous function of $\mathbf{s}$ and Q lies in the closed bounded set $\overline{U}$, the four component functions of $d\boldsymbol{\varphi}_{\mathbf{s}}$ are uniformly bounded on $\overline{U}$: for some $B > 0$,

$$|x_s(s,t)| \le B, \quad |x_t(s,t)| \le B, \quad |y_s(s,t)| \le B, \quad |y_t(s,t)| \le B,$$

for all (s,t) in $\overline{U}$. This implies $L \le B\sqrt{2}/2^k$; thus, for each of the four rectangles,

$$\text{area} = 2W(L + 2W) \le \frac{4\varepsilon\sqrt{2}}{2^k} \left(\frac{B\sqrt{2}}{2^k} + \frac{4\varepsilon\sqrt{2}}{2^k} \right) = \frac{8\varepsilon(B + 4\varepsilon)}{2^{2k}}.$$

Because $\varepsilon(B + 4\varepsilon) = O(\varepsilon)$ as $\varepsilon \to 0$, we have

$$\overline{A}(\boldsymbol{\varphi}(Q)) \le A(d\boldsymbol{\varphi}_{\mathbf{a}}(Q)) + 4\frac{8\varepsilon(B + 4\varepsilon)}{2^{2k}} \le M\,A(Q) + O(\varepsilon)A(Q). \qquad \square$$

Lemma 9.3. *If S has area and $\overline{S} \subset \Omega$, then*

$$\overline{A}(\boldsymbol{\varphi}(S)) \leq \iint_S |J_{\boldsymbol{\varphi}}(s,t)| \, ds \, dt.$$

Proof. To prove this lemma, we construct upper Darboux sums for $|J_{\boldsymbol{\varphi}}(s,t)|$ on S (cf. Definition 8.17, p. 300).

By the proof of Lemma 9.1, $\overline{S}$ is contained in the interior of a bounded closed set $\overline{U} \subset \Omega$ whose points lie within distance $m/2$ of $\overline{S}$. Fix $\varepsilon > 0$ and choose K to satisfy both the previous lemma and the condition that the mesh size $\|\mathcal{J}_K\|$ is less than $m/2$. Then, if $k \geq K$, any square Q of $\mathcal{J}_k$ that meets S will lie within $\overline{U}$. Let $Q_1, \ldots, Q_I$ be the squares of $\mathcal{J}_k$ that meet S. Set

$$M_i = \max_{(s,t) \in Q_i} |J_{\boldsymbol{\varphi}}(s,t)|, \quad i = 1, \ldots, I;$$

then, by the previous lemma,

$$\overline{A}(\boldsymbol{\varphi}(Q_i)) \leq (M_i + O(\varepsilon)) A(Q_i), \quad i = 1, \ldots, I.$$

Because $S \subseteq Q_1 \cup \cdots \cup Q_I$ and thus $\boldsymbol{\varphi}(S) \subseteq \boldsymbol{\varphi}(Q_1) \cup \cdots \cup \boldsymbol{\varphi}(Q_I)$, we have

$$\overline{A}(\boldsymbol{\varphi}(S)) \leq \sum_{i=1}^{I} \overline{A}(\boldsymbol{\varphi}(Q_i)) \leq \sum_{i=1}^{I} (M_i + O(\varepsilon)) A(Q_i)$$

$$= \sum_{i=1}^{I} M_i A(Q_i) + \sum_{i=1}^{I} O(\varepsilon) A(Q_i).$$

The first term is the upper Darboux sum for $|J_{\boldsymbol{\varphi}}(s,t)|$ over S and the grid $\mathcal{J}_k$; the second is $O(\varepsilon)$ times the outer area of S over the same grid; that is,

$$\overline{A}(\boldsymbol{\varphi}(S)) \leq \overline{D}_{\mathcal{J}_k}(|J_{\boldsymbol{\varphi}}(s,t)|, S) + O(\varepsilon) \overline{A}_k(S).$$

Because $|J_{\boldsymbol{\varphi}}(s,t)|$ is integrable over S, the upper Darboux sums converge to the integral (and the outer areas to the area) as $k \to \infty$; thus

$$\overline{A}(\boldsymbol{\varphi}(S)) \leq \iint_S |J_{\boldsymbol{\varphi}}(s,t)| \, ds \, dt + O(\varepsilon) A(S).$$

This inequality holds for any $\varepsilon > 0$; therefore it continues to hold as $\varepsilon \to 0$ (and hence as $O(\varepsilon) \to 0$), so

$$\overline{A}(\boldsymbol{\varphi}(S)) \leq \iint_S |J_{\boldsymbol{\varphi}}(s,t)| \, ds \, dt. \qquad \square$$

Part of the assertion of the main theorem is that $\boldsymbol{\varphi}(S)$ has area (implying we are able to replace $\overline{A}(\boldsymbol{\varphi}(S))$ by $A(\boldsymbol{\varphi}(S))$ in the lemma just proven). The next two lemmas establish that $\boldsymbol{\varphi}(S)$ does indeed have area by showing that its boundary $\partial(\boldsymbol{\varphi}(S))$ has outer area equal to zero.

Lemma 9.4. $\partial(\boldsymbol{\varphi}(S)) \subseteq \boldsymbol{\varphi}(\partial S)$.

Proof. Let $\mathbf{x} = \boldsymbol{\varphi}(\mathbf{s})$ be a boundary point of $\boldsymbol{\varphi}(S)$; we must show that $\mathbf{s}$ is a boundary point of S. We use the criterion established in Exercise 8.5 (p. 313): every open disk centered at a boundary point of a set T contains at least one point in T and one point not in T.

Let $D_1 \subset \Omega$ be an open disk centered at $\mathbf{s}$. By a corollary to the inverse function theorem (Corollary 5.3, p. 174), $\mathbf{x}$ is an interior point of $\boldsymbol{\varphi}(D_1)$, so there is an open disk D_2 centered at $\mathbf{x}$ for which $D_2 \subseteq \boldsymbol{\varphi}(D_1)$. But $\mathbf{x}$ is boundary point of $\boldsymbol{\varphi}(S)$, so D_2 contains a point $\mathbf{p}_2$ in $\boldsymbol{\varphi}(S)$ and another point $\mathbf{q}_2$ that is not in $\boldsymbol{\varphi}(S)$. But then the point $\mathbf{p}_1 = \boldsymbol{\varphi}^{-1}(\mathbf{p}_2)$ in D_1 is in S and the point $\mathbf{q}_1 = \boldsymbol{\varphi}^{-1}(\mathbf{q}_2)$ in D_1 is not in S. (Draw a picture.) $\qquad\square$

Lemma 9.5. $\overline{A}(\partial(\boldsymbol{\varphi}(S))) = 0$.

Proof. Apply Lemma 9.3 to the zero-area set ∂S. If $|J_{\boldsymbol{\varphi}}(s,t)| \leq B$ on ∂S; then

$$\overline{A}(\partial(\boldsymbol{\varphi}(S))) \leq \overline{A}(\boldsymbol{\varphi}(\partial S)) \leq \iint_{\partial S} |J_{\boldsymbol{\varphi}}(s,t)| \, ds \, dt \leq B\, A(\partial S) = 0. \qquad\square$$

Corollary 9.9 $A(\boldsymbol{\varphi}(S)) \leq \iint_S |J_{\boldsymbol{\varphi}}(s,t)| \, ds \, dt.$ $\qquad\square$

Lemma 9.6. $A(\boldsymbol{\varphi}(S)) \geq \iint_S |J_{\boldsymbol{\varphi}}(s,t)| \, ds \, dt.$

Proof. The idea of the proof is to apply the previous arguments to the inverse map $\boldsymbol{\varphi}^{-1}$ and a set $T = \boldsymbol{\varphi}(R)$, where R has area and $R \cup \partial R \subset \Omega$. By Corollary 9.9, T has area. Furthermore,

$$\overline{T} = T \cup \partial T \subseteq \boldsymbol{\varphi}(R \cup \partial R) \subset \boldsymbol{\varphi}(\Omega),$$

so we can write

$$A(R) = A(\boldsymbol{\varphi}^{-1}(T)) \leq \iint_{\boldsymbol{\varphi}(R)} |J_{\boldsymbol{\varphi}^{-1}}(x,y)| \, dx \, dy.$$

Now let R be a square Q of the grid $\mathcal{I}_k$, and let μ be the minimum value of $|J_{\boldsymbol{\varphi}}(s,t)|$ on Q. Note that $\mu > 0$, because $d\boldsymbol{\varphi}_{\mathbf{s}}$ is invertible and a uniformly continuous function of $\mathbf{s}$ on Q. Using Corollary 4.13, page 138, we find

$$|J_{\boldsymbol{\varphi}^{-1}}(x,y)| = \frac{1}{|J_{\boldsymbol{\varphi}}(s(x,y),t(x,y))|} \leq \frac{1}{\mu}$$

for all (x,y) in $\boldsymbol{\varphi}(Q)$. Therefore,

$$A(Q) \leq \iint_{\boldsymbol{\varphi}(Q)} |J_{\boldsymbol{\varphi}^{-1}}(x,y)| \, dx \, dy \leq \frac{A(\boldsymbol{\varphi}(Q))}{\mu}.$$

or $A(\boldsymbol{\varphi}(Q)) \geq \mu A(Q)$.

Let $P_1, \ldots, P_J$ be the squares of the grid $\mathcal{I}_k$ that are entirely contained in S, and let

$$\mu_j = \min_{(s,t) \in P_j} |J_{\boldsymbol{\varphi}}(s,t)|, \quad j = 1, \ldots, J.$$

Because $S \supseteq P_1 \cup \cdots \cup P_J$ and $\boldsymbol{\varphi}(S) \supseteq \boldsymbol{\varphi}(P_1) \cup \cdots \cup \boldsymbol{\varphi}(P_J)$, we have

$$A(\boldsymbol{\varphi}(S)) \geq \sum_{j=1}^{J} A(\boldsymbol{\varphi}(P_j)) \geq \sum_{j=1}^{J} \mu_j A(P_j) = \underline{D}_{\mathcal{I}_k}(|J_{\boldsymbol{\varphi}}|, S),$$

the lower Darboux sum for $|J_{\boldsymbol{\varphi}}|$ over S and the grid $\mathcal{I}_k$. In the limit as $k \to \infty$, the Darboux sum becomes the integral:

$$A(\boldsymbol{\varphi}(S)) \geq \iint_S |J_{\boldsymbol{\varphi}}(s,t)| \, ds \, dt. \qquad \square$$

This completes the proof of Theorem 9.8. $\qquad \square$

We have already constructed new integration grids from old ones using invertible linear maps (cf. pp. 287–293): if the grid G has cells Q_i and $L : \mathbb{R}^2 \to \mathbb{R}^2$ is invertible, then the images $L(Q_i)$ form the cells of a new grid $\mathcal{H} = L(G)$. Furthermore, L determines a constant σ that relates the mesh sizes of G and $\mathcal{H}$: $\|\mathcal{H}\| = \sigma \|G\|$. Theorem 9.8 creates the possibility of creating new grids using invertible nonlinear maps.

New grids from nonlinear maps

Let $\boldsymbol{\varphi} : \Omega \to \boldsymbol{\varphi}(\Omega)$ be continuously differentiable with a continuously differentiable inverse on $\boldsymbol{\varphi}(\Omega)$. Let G be a grid whose cells Q_i lie within Ω. By definition, the Q_i are closed nonoverlapping sets with area. By Theorem 9.8, the sets $P_i = \boldsymbol{\varphi}(Q_i)$ are likewise closed nonoverlapping sets with area. We define them to be the cells of the grid $\mathcal{H} = \boldsymbol{\varphi}(G)$.

For a nonlinear map $\boldsymbol{\varphi}$, there is no general analogue to the scale factor σ. However, suppose S has area and is a closed subset of Ω. Then the sets

Restricting a grid to a closed set

$$\widehat{Q}_i = Q_i \cap S \quad \text{and} \quad \widehat{P}_i = P_i \cap \boldsymbol{\varphi}(S) = \boldsymbol{\varphi}(\widehat{Q}_i)$$

are closed nonoverlapping sets with area, so they constitute the cells of grids that we denote G_S and $\mathcal{H}_{\boldsymbol{\varphi}(S)}$, respectively. For these special grids, there is a natural link between their mesh sizes. To find it, note that the continuous map $\boldsymbol{\varphi}$ is uniformly continuous on S. Therefore, given any $\varepsilon > 0$ there is a $\delta > 0$ such that

$$\|\mathbf{s}_1 - \mathbf{s}_2\| < \delta \quad \Longrightarrow \quad \|\boldsymbol{\varphi}(\mathbf{s}_1) - \boldsymbol{\varphi}(\mathbf{s}_2)\| < \varepsilon.$$

This implies

$$\|G_S\| < \delta \quad \Longrightarrow \quad \|\mathcal{H}_{\boldsymbol{\varphi}(S)}\| < \varepsilon;$$

in other words, we can make $\|\mathcal{H}_{\boldsymbol{\varphi}(S)}\|$ as small as we wish by making $\|G_S\|$ sufficiently small.

When calculating Riemann sums for a continuous function f defined only on a set S, we first define f to be zero outside S in order to allow for the possibility that f is evaluated on the part of a cell that lies outside S. Because f on this larger domain is usually not continuous, a delicate argument is needed to show that the Riemann sums converge. The grids $\mathcal{G}_S$ eliminate this problem, because their cells lie entirely in S. The following theorem shows that Riemann sums constructed with these restricted grids still converge to the integral.

Theorem 9.10. *Suppose f is continuous on a closed bounded set S that has area. Then Riemann sums constructed with grids $\mathcal{G}_S$ converge to the integral of f as $\|\mathcal{G}_S\| \to 0$.*

Proof. Let $\varepsilon > 0$ be given; we must find a $\delta > 0$ so that

$$\left| \iint_S f(x,y)\, dA - \sum_{i=1}^{I} f(x_i,y_i) A(\widehat{Q}_i) \right| < \varepsilon$$

for any grid $\mathcal{G}_S$ with $\|\mathcal{G}_S\| < \delta$, and for any point (x_i,y_i) in the cell $\widehat{Q}_i$ of $\mathcal{G}_S$, for each $i = 1,\ldots,I$.

Because f is uniformly continuous on S, we can choose $\delta > 0$ so that

$$\|(x,y) - (x',y')\| < \delta \quad \Longrightarrow \quad |f(x,y) - f(x',y')| < \frac{\varepsilon}{A(S)}.$$

Now let $\mathcal{G}_S$ be any grid for which $\|\mathcal{G}_S\| < \delta$. Then, because

$$\iint_S f(x,y)\, dA = \sum_{i=1}^{I} \iint_{\widehat{Q}_i} f(x,y)\, dA \quad \text{and} \quad f(x_i,y_i) A(\widehat{Q}_i) = \iint_{\widehat{Q}_i} f(x_i,y_i)\, dA,$$

we have

$$\left| \iint_S f(x,y)\, dA - \sum_{i=1}^{I} f(x_i,y_i) A(\widehat{Q}_i) \right| \leq \left| \sum_{i=1}^{I} \iint_{\widehat{Q}_i} f(x,y)\, dA - \sum_{i=1}^{I} \iint_{\widehat{Q}_i} f(x_i,y_i)\, dA \right|$$

$$\leq \sum_{i=1}^{I} \iint_{\widehat{Q}_i} |f(x,y) - f(x_i,y_i)|\, dA < \frac{\varepsilon}{A(S)} \sum_{i=1}^{I} \iint_{\widehat{Q}_i} dA = \varepsilon. \qquad \square$$

Change of variables in double integrals

We can use this result immediately, to prove the main formula on the *change of variables* in double integrals (by continuing to follow the argument of J. Schwartz in [16]).

Theorem 9.11 (Change of variables). *Let Ω be a bounded open set in $\mathbb{R}^2$, and let $\boldsymbol{\varphi} : \Omega \to \mathbb{R}^2 : (s,t) \to (x,y)$ be a continuously differentiable map that has a continuously differentiable inverse $\boldsymbol{\varphi}^{-1} : \boldsymbol{\varphi}(\Omega) \to \Omega$. Suppose the function $f(x,y)$ is continuous on a closed set $D \subset \boldsymbol{\varphi}(\Omega)$ that has area; then*

$$\iint_D f(x,y)\, dx\, dy = \iint_{\boldsymbol{\varphi}^{-1}(D)} f(x(s,t),y(s,t)) \left| \frac{\partial(x,y)}{\partial(s,t)} \right| ds\, dt.$$

Proof. By Theorem 8.35, page 305, $f(x,y)$ is integrable on D. By Theorem 9.8, $S = \boldsymbol{\varphi}^{-1}(D)$ has area and the function

$$f(x(s,t),y(s,t)) \left| \frac{\partial(x,y)}{\partial(s,t)} \right|$$

is bounded and continuous on $S = \boldsymbol{\varphi}^{-1}(D)$, so it is integrable there. To prove that the two integrals in the statement of the theorem are equal, we show they differ by less than any preassigned $\varepsilon > 0$.

Let G_S be an arbitrary integration grid whose cells Q_i, $i = 1,\ldots,I$, partition S, and let $\mathcal{H}_D = \boldsymbol{\varphi}(G_S)$ be the image grid; its cells $P_i = \boldsymbol{\varphi}(Q_i)$ partition $D = \boldsymbol{\varphi}(S)$. Let (s_i,t_i) be a point in Q_i, and let $(x_i,y_i) = \boldsymbol{\varphi}(s_i,t_i)$ be the corresponding point in P_i. Consider the following, obtained by applying the triangle inequality to a rather lengthy telescoping sum:

$$\left| \iint_D f(x,y)\,dx\,dy - \iint_S f(\boldsymbol{\varphi}(s,t)) \, |J_{\boldsymbol{\varphi}}(s,t)|\,ds\,dt \right|$$

$$\leq \left| \iint_D f(x,y)\,dx\,dy - \sum_{i=1}^I f(x_i,y_i)A(P_i) \right|$$

$$+ \left| \sum_{i=1}^I f(x_i,y_i)A(P_i) - \sum_{i=1}^I \iint_{Q_i} f(\boldsymbol{\varphi}(s_i,t_i)) \, |J_{\boldsymbol{\varphi}}(s,t)|\,ds\,dt \right|$$

$$+ \left| \sum_{i=1}^I \iint_{Q_i} (f(\boldsymbol{\varphi}(s_i,t_i)) - f(\boldsymbol{\varphi}(s,t))) \, |J_{\boldsymbol{\varphi}}(s,t)|\,ds\,dt \right|$$

$$+ \left| \sum_{i=1}^I \iint_{Q_i} f(\boldsymbol{\varphi}(s,t)) \, |J_{\boldsymbol{\varphi}}(s,t)|\,ds\,dt - \iint_S f(\boldsymbol{\varphi}(s,t)) \, |J_{\boldsymbol{\varphi}}(s,t)|\,ds\,dt \right|.$$

Now consider each of the four terms on the right-hand side of the inequality.

The first term contains a Riemann sum for f on D and the grid $\mathcal{H}_D$. Because f is integrable, there is a $\delta_1 > 0$ that makes that term less than $\varepsilon/2$ whenever $\|\mathcal{H}_D\| < \delta_1$ and (x_i,y_i) is an arbitrary point in P_i. As we saw above (p. 349), we have $\|\mathcal{H}_D\| < \delta_1$ when $\|G_S\| < \delta_2$ for some properly chosen $\delta_2 > 0$.

The second term is zero by Theorem 9.8:

$$A(P_i) = A(\boldsymbol{\varphi}(Q_i)) = \iint_{Q_i} |J_{\boldsymbol{\varphi}}(s,t)|\,ds\,dt, \quad i = 1,\ldots,I.$$

For the third term, first note that

$$\left| \sum_{i=1}^I \iint_{Q_i} (f(\boldsymbol{\varphi}(s_i,t_i)) - f(\boldsymbol{\varphi}(s,t))) \, |J_{\boldsymbol{\varphi}}(s,t)|\,ds\,dt \right|$$

$$\leq \sum_{i=1}^I \iint_{Q_i} |f(\boldsymbol{\varphi}(s_i,t_i)) - f(\boldsymbol{\varphi}(s,t))| \, |J_{\boldsymbol{\varphi}}(s,t)|\,ds\,dt.$$

Because $f(\boldsymbol{\varphi}(s,t))$ is uniformly continuous on S, there is a $\delta_3 > 0$ for which

$$\|(s,t) - (s',t')\| < \delta_3 \quad \Longrightarrow \quad |f(\boldsymbol{\varphi}(s,t)) - f(\boldsymbol{\varphi}(s',t'))| < \frac{\varepsilon}{2A(D)}.$$

Therefore, if G_S is any partition of S for which $\|G_S\| < \delta_3$, then

$$\sum_{i=1}^{I} \iint_{Q_i} |f(\boldsymbol{\varphi}(s_i,t_i)) - f(\boldsymbol{\varphi}(s,t))| \, |J_{\boldsymbol{\varphi}}(s,t)| \, ds \, dt$$

$$< \frac{\varepsilon}{2A(D)} \sum_{i=1}^{I} \iint_{Q_i} |J_{\boldsymbol{\varphi}}(s,t)| \, ds \, dt = \frac{\varepsilon}{2A(D)} \iint_{S} |J_{\boldsymbol{\varphi}}(s,t)| \, ds \, dt = \frac{\varepsilon}{2}.$$

The last equality in this chain is provided by Theorem 9.8:

$$\iint_{S} |J_{\boldsymbol{\varphi}}(s,t)| \, ds \, dt = A(\boldsymbol{\varphi}(S)) = A(D).$$

The fourth term, like the second, is zero. Therefore, the two integrals differ by less than ε whenever the partition G_S satisfies $\|G_S\| < \min \delta_2, \delta_3$. Because $\varepsilon > 0$ is arbitrary, the two integrals must be equal. $\square$

Local area magnification

Because $|\det L|$ is the area magnification factor for a linear map L of the plane, we have

$$\frac{A(L(S))}{A(S)} = |\det L|$$

for any subset of the plane that has area. For the nonlinear map $\boldsymbol{\varphi}$, we introduce the set function (cf. pp. 310–312)

$$M(S) = A(\boldsymbol{\varphi}(S)) = \iint_{S} |J_{\boldsymbol{\varphi}}(s,t)| \, ds \, dt$$

By Theorem 8.39, page 312, the derivative of M is

$$M'(s,t) = |J_{\boldsymbol{\varphi}}(s,t)|$$

In other words, if S contains the point (s,t), then

$$\frac{A(\boldsymbol{\varphi}(S))}{A(S)} \approx |J_{\boldsymbol{\varphi}}(s,t)|$$

as closely as we wish by making the diameter of S (p. 291) sufficiently small. It is in this sense that we consider

$$|J_{\boldsymbol{\varphi}}(s,t)| = |\det d\boldsymbol{\varphi}_{(s,t)}|$$

to be the **local area magnification factor** for $\boldsymbol{\varphi}$ when *area* is understood to be nonnegative Jordan content.

9.4 Orientation

In the next section, we introduce Green's theorem as an additional tool to evaluate double integrals. However, the integrals in Green's theorem are oriented. In this section, therefore, we say what it means for a 2-dimensional region to be oriented, and then define *oriented* double integrals. Finally, we extend the change of variables formula to oriented integrals.

Orientation in the plane involves, either explicitly or implicitly, comparison with the coordinate axes (or with the standard basis vectors $\mathbf{e}_1$, $\mathbf{e}_2$ that determine them). Consider first an ordered pair of linearly independent vectors $\{\mathbf{v}_1, \mathbf{v}_2\}$ in $\mathbb{R}^2$. To compare $\{\mathbf{v}_1, \mathbf{v}_2\}$ with $\{\mathbf{e}_1, \mathbf{e}_2\}$, let $L : \mathbb{R}^2 \to \mathbb{R}^2$ be the unique linear map for which $L(\mathbf{e}_1) = \mathbf{v}_1$, $L(\mathbf{e}_2) = \mathbf{v}_2$. Then we say the pair $\{\mathbf{v}_1, \mathbf{v}_2\}$ has the **same orientation** as $\{\mathbf{e}_1, \mathbf{e}_2\}$ if $\det L > 0$; otherwise, we say it has the **opposite orientation**. In particular, if we reverse the order of the vectors, orientation is reversed, as well: $\{\mathbf{v}_1, \mathbf{v}_2\}$ and $\{\mathbf{v}_2, \mathbf{v}_1\}$ always have opposite orientations. We write $\{\mathbf{v}_2, \mathbf{v}_1\} = -\{\mathbf{v}_1, \mathbf{v}_2\}$. The components of $\mathbf{v}_1$ and $\mathbf{v}_2$ with respect to the standard basis determine the orientation of $\{\mathbf{v}_1, \mathbf{v}_2\}$. That is, if

$$\mathbf{v}_1 = \begin{pmatrix} v_{11} \\ v_{21} \end{pmatrix}, \quad \mathbf{v}_2 = \begin{pmatrix} v_{12} \\ v_{22} \end{pmatrix},$$

then the matrix of L in terms of the standard basis is

$$L = \begin{pmatrix} v_{11} & v_{12} \\ v_{21} & v_{22} \end{pmatrix}, \quad \text{and} \quad \det L = v_{11}v_{22} - v_{12}v_{21}.$$

We also say that an ordered pair that has the same orientation as the standard basis is **positively oriented**; otherwise, we say it is **negatively oriented**. The figure in the margin helps make the point that orientation is always determined by reference to the coordinate axes: it is relative, not absolute. The pair $\{\mathbf{v}_1, \mathbf{v}_2\}$ illustrated is positively oriented.

The figure also shows that the ordering of a pair of linearly independent vectors implicitly defines a sense of rotation, namely, rotation from the first to the second through the smaller of the two angles determined by the vectors. "Sense of rotation" therefore gives us a second way to represent orientation. For example, we can confirm that the pair $\{\mathbf{v}_1, \mathbf{v}_2\}$ in the margin figure is positively oriented because it defines the same clockwise sense of rotation as the basis vectors.

To orient a region S with area in $\mathbb{R}^2$, orient each point $\mathbf{p}$ of S by assigning to $\mathbf{p}$ an ordered pair of linearly independent vectors $\{\mathbf{v}_1(\mathbf{p}), \mathbf{v}_2(\mathbf{p})\}$ in such a way that both $\mathbf{v}_1(\mathbf{p})$ and $\mathbf{v}_2(\mathbf{p})$ vary continuously with $\mathbf{p}$ over S. To indicate that S has acquired an orientation, we write it as $\vec{S}$. (On page 7, we introduced a similar notation for curves: $\vec{C}$ denotes a curve C together with an orientation.) There are now two different definitions of orientation when $\vec{S}$ is a parallelogram, but they agree if we assign to each point of $\mathbf{v} \wedge \mathbf{w}$ the ordered pair $\{\mathbf{v}, \mathbf{w}\}$. For each point $\mathbf{p}$ in $\vec{S}$, let $L_{\mathbf{p}} : \mathbb{R}^2 \to \mathbb{R}^2$ be the unique linear map for which $L_{\mathbf{p}}(\mathbf{e}_i) = \mathbf{v}_i(\mathbf{p})$, $i = 1, 2$. Then, by what we said above, the function

Orientation of an ordered pair of vectors

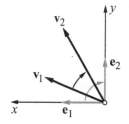

Positive and negative orientations

Orientation and sense of rotation

Orienting a region in the plane

$$\det L_{\mathbf{p}} = v_{11}(\mathbf{p})v_{22}(\mathbf{p}) - v_{12}(\mathbf{p})v_{21}(\mathbf{p})$$

varies continuously with $\mathbf{p}$ and is never zero. We write $-\vec{S}$ to denote $\vec{S}$ with its orientation reversed at every point. That is, if $\{\mathbf{v}_1(\mathbf{p}), \mathbf{v}_2(\mathbf{p})\}$ defines $\vec{S}$, then $-\{\mathbf{v}_1(\mathbf{p}), \mathbf{v}_2(\mathbf{p})\} = \{\mathbf{v}_2(\mathbf{p}), \mathbf{v}_1(\mathbf{p})\}$ defines $-\vec{S}$.

Theorem 9.12. *On any pathwise-connected component of $\vec{S}$ (i.e., a largest subset in which any two points can be joined by a continuous path in $\vec{S}$), all points have the same orientation.*

Proof. Let $\mathbf{p}$ and $\mathbf{q}$ be joined by the continuous path $\mathbf{s}(t)$, $a \le t \le b$, with $\mathbf{s}(a) = \mathbf{p}$ and $\mathbf{s}(b) = \mathbf{q}$. Then $\det L_{\mathbf{s}(t)}$ is a continuous and nonzero function of t on the interval $[a,b]$, so it cannot change sign. Therefore $\{\mathbf{v}_1(\mathbf{p}), \mathbf{v}_2(\mathbf{p})\}$ and $\{\mathbf{v}_1(\mathbf{q}), \mathbf{v}_2(\mathbf{q})\}$ have the same orientation.   □

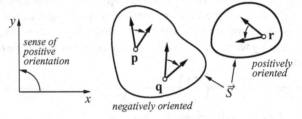

negatively oriented

Sense of rotation
at every point of $\vec{S}$

   In the figure, $\vec{S}$ has two components, one with positive orientation and the other with negative. As the figure suggests, the orientation of $\vec{S}$ always defines a "sense of rotation" at each point of $\vec{S}$, and all points in any connected component of $\vec{S}$ have the same sense of rotation.

Definition 9.3 *Two assignments $\mathbf{p} \mapsto \{\mathbf{w}_1(\mathbf{p}), \mathbf{w}_2(\mathbf{p})\}$, $\mathbf{p} \mapsto \{\mathbf{v}_1(\mathbf{p}), \mathbf{v}_2(\mathbf{p})\}$ define the **same orientation of S** if the unique linear map $M_{\mathbf{p}}(\mathbf{v}_i(\mathbf{p})) = \mathbf{w}_i(\mathbf{p})$, $i = 1,2$, has $\det M_{\mathbf{p}} > 0$ for all $\mathbf{p}$ in S.*

Positive and negative
orientations

   Two different assignments of ordered pairs of vectors define the same orientation precisely when they give the same sense of rotation. Thus, each component of a region S can be given exactly two orientations, either agreeing or disagreeing with the sense of rotation of the coordinate axes. If S has k components, it has 2^k possible orientations. We say $\vec{S}$ is **positively oriented** if all its components are positively oriented, and is **negatively oriented** if all its components are negatively oriented. Usually, S has just one component.

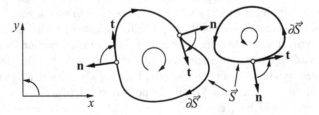

If ∂S is a piecewise-smooth curve, then an orientation of S *induces* an orientation of ∂S. To see how, let $\vec{S}$ be oriented. As the previous definition indicates, there is considerable freedom in choosing the orientation vectors at a point. Thus, orient each point of $\partial \vec{S}$ where it is smooth by the pair of vectors $\{\mathbf{n}, \mathbf{t}\}$, where $\mathbf{n}$ is the outward-pointing unit normal and $\mathbf{t}$ is one of the two unit tangents to $\partial \vec{S}$, choosing $\mathbf{t}$ so $\{\mathbf{n}, \mathbf{t}\}$ gives the local sense of rotation on that component of $\vec{S}$. In the figure above, the pair $\{\mathbf{n}, \mathbf{t}\}$ is oriented in the clockwise sense on one component and in the counterclockwise sense on the other. The choice of a tangent vector orients a piecewise-smooth curve. We use the tangent vector $\mathbf{t}$ to define the **orientation of ∂S induced by $\vec{S}$**.

Orientation induced
on a boundary

A map can also induce an orientation of its image. Let Ω be a bounded open set in $\mathbb{R}^2$, and let $\boldsymbol{\varphi} : \Omega \to \mathbb{R}^2$ be a continuously differentiable map that has a continuously differentiable inverse $\boldsymbol{\varphi}^{-1} : \boldsymbol{\varphi}(\Omega) \to \Omega$. Suppose $\vec{S}$ is an oriented set that has area and its closure $\overline{S} = S \cup \partial S$ lies within Ω. Suppose the ordered pair $\{\mathbf{v}_1(\mathbf{p}), \mathbf{v}_2(\mathbf{p})\}$ defines the orientation of $\vec{S}$ at the point $\mathbf{p}$. Then we orient the point $\boldsymbol{\varphi}(\mathbf{p})$ in $\boldsymbol{\varphi}(\vec{S})$ with the ordered pair of vectors

Orientation induced
on an image

$$\{\mathrm{d}\boldsymbol{\varphi}_\mathbf{p}(\mathbf{v}_1(\mathbf{p})), \mathrm{d}\boldsymbol{\varphi}_\mathbf{p}(\mathbf{v}_2(\mathbf{p}))\}.$$

Because $\boldsymbol{\varphi}$ is invertible, each image point $\boldsymbol{\varphi}(\mathbf{p})$ is assigned only one such pair. Because $\mathrm{d}\boldsymbol{\varphi}_\mathbf{p}$ is invertible, the image vectors are linearly independent. Finally, because $\boldsymbol{\varphi}$ is continuously differentiable, the assignment varies continuously with $\boldsymbol{\varphi}(\mathbf{p})$.

In Chapter 4 we first observed informally that the sign of the Jacobian of a map determines whether it preserves or reverses orientation. Now that we have defined the orientation of a region, we can state this observation as a theorem and prove it.

Orientation and
the Jacobian

Theorem 9.13. *The regions $\vec{S}$ and $\boldsymbol{\varphi}(\vec{S})$ have the same orientation if and only if the Jacobian $\det \mathrm{d}\boldsymbol{\varphi}_\mathbf{p}$ is everywhere positive.*

Proof. Suppose the ordered pair $\{\mathbf{v}_1(\mathbf{p}), \mathbf{v}_2(\mathbf{p})\}$ defines the orientation of $\vec{S}$ at $\mathbf{p}$; then $\{\mathrm{d}\boldsymbol{\varphi}_\mathbf{p}(\mathbf{v}_1(\mathbf{p})), \mathrm{d}\boldsymbol{\varphi}_\mathbf{p}(\mathbf{v}_2(\mathbf{p}))\}$ defines the induced orientation of $\boldsymbol{\varphi}(\vec{S})$ at $\boldsymbol{\varphi}(\mathbf{p})$. Let

$$L_\mathbf{p}(\mathbf{e}_i) = \mathbf{v}_i(\mathbf{p}) \text{ and } M_\mathbf{p}(\mathbf{e}_i) = \mathrm{d}\boldsymbol{\varphi}_\mathbf{p}(\mathbf{v}_i(\mathbf{p})), \quad i = 1, 2,$$

define the linear maps that determine the orientations. Then $\boldsymbol{\varphi}(\vec{S})$ has the same orientation at $\boldsymbol{\varphi}(\mathbf{p})$ that $\vec{S}$ does at $\mathbf{p}$ if and only if the determinants $\det M_\mathbf{p}$ and $\det L_\mathbf{p}$ have the same sign. But

$$M_\mathbf{p} = \mathrm{d}\boldsymbol{\varphi}_\mathbf{p} \circ L_\mathbf{p}, \quad \det M_\mathbf{p} = \det \mathrm{d}\boldsymbol{\varphi}_\mathbf{p} \det L_\mathbf{p},$$

so $\det M_\mathbf{p}$ and $\det L_\mathbf{p}$ have the same sign if and only if $\det \mathrm{d}\boldsymbol{\varphi}_\mathbf{p} > 0$. $\square$

We can now introduce *oriented* integrals, that is, double integrals defined over oriented regions. We begin with a closed, bounded, and *unoriented* subset S of the (x, y)-plane. Assume S has area and $f(x, y)$ is a function that is integrable over S; then we have the ordinary Riemann integral

Oriented integrals

$$\iint_S f(x,y)\,dA$$

as defined in Chapter 8.3. This integral is monotonic in the sense that $f(x,y) \ge 0$ on S implies $I \ge 0$ (Theorem 8.28, p. 298).

Definition 9.4 *If $\vec{S}$ has either positive or negative orientation, then the* **oriented integral of f over $\vec{S}$** *is*

$$\iint_{\vec{S}} f(x,y)\,dx\,dy = \operatorname{sgn}\vec{S} \iint_S f(x,y)\,dA,$$

where $\operatorname{sgn}\vec{S} = +1$ *when the orientation of $\vec{S}$ is positive and* $\operatorname{sgn}\vec{S} = -1$ *when it is negative.*

The oriented integral uses $dx\,dy$ rather than as dA as the "element of area" in order to help convey orientation, in a way we explain below.

Properties of oriented integrals

The definition has several immediate consequences. First, because $-\vec{S}$ is negatively oriented when $\vec{S}$ is positively oriented, and vice versa, their oriented integrals have opposite signs:

$$\iint_{-\vec{S}} f(x,y)\,dx\,dy = -\iint_{\vec{S}} f(x,y)\,dx\,dy.$$

Second, an oriented integral over a positively oriented region $\vec{S}$ is monotonic:

$$\iint_{\vec{S}} f(x,y)\,dx\,dy \ge 0 \ \text{ if } f \ge 0 \text{ on } \vec{S}.$$

Third, we can define the *signed* area of a positively or negatively oriented region $\vec{S}$ as

$$\operatorname{area}\vec{S} = \iint_{\vec{S}} dx\,dy = \operatorname{sgn}\vec{S} \times A(S),$$

where

$$A(S) = \iint_S dA$$

is the ordinary area (i.e., the Jordan measure) of the unoriented region S. Oriented area reverses sign with the orientation of the region:

$$\operatorname{area}(-\vec{S}) = -\operatorname{area}(\vec{S}).$$

$dy\,dx = -dx\,dy$

Writing the element of area in an oriented integral as $dx\,dy$ rather than dA gives us the opportunity to convey differences in orientation. If we take $dx\,dy$ as representing the coordinate axes in their usual order (i.e., positive orientation), then $-dx\,dy$ and $dy\,dx$ should both represent the opposite order (negative orientation). Let us, therefore, adopt the symbolic convention

$$dy\,dx = -dx\,dy,$$

so that

$$\iint_{\vec{S}} f(x,y)\,dy\,dx = -\iint_{\vec{S}} f(x,y)\,dx\,dy = \iint_{-\vec{S}} f(x,y)\,dx\,dy$$

for any positively or negatively oriented region $\vec{S}$. In particular,

$$\iint_{\vec{S}} dy\,dx = -\iint_{\vec{S}} dx\,dy = -\text{area}\,\vec{S}.$$

It is important to note that the sign change when switching from $dy\,dx$ to $dx\,dy$ in an oriented integral does not happen when we reverse the order of integration in iterated integrals (Corollary 9.2, p. 321). For example, if we integrate $f(x,y)$ over the rectangle $R : a \le x \le b, c \le y \le d$ and assume $\vec{R}$ is negatively oriented, then

Order in iterated integrals

$$\iint_{\vec{R}} f(x,y)\,dx\,dy = -\iint_R f(x,y)\,dA$$
$$= -\int_c^d \int_a^b f(x,y)\,dx\,dy = -\int_a^b \int_c^d f(x,y)\,dy\,dx.$$

As an example, let us compute the oriented integral of $f(x,y) = x^2$ over the triangle $\vec{S}$ with vertices $(0,0)$, $(0,1)$, and $(1,1)$, taken in that order. The boundary path indicates that $\vec{S}$ is negatively oriented. Because $f(x,y) \ge 0$ on $\vec{S}$, we therefore expect the value of the integral to be negative. We first express the oriented integral as an ordinary (unoriented) double integral, and then convert that to iterated integrals. For the last step, we can describe the unoriented set S by either of the following sets of inequalities:

Example

$$S:\ \begin{array}{l} 0 \le x \le 1, \\ x \le y \le 1; \end{array} \qquad S:\ \begin{array}{l} 0 \le y \le 1, \\ 0 \le x \le y. \end{array}$$

Using the first set, we have

$$\iint_{\vec{S}} x^2\,dx\,dy = -\iint_S x^2\,dA = -\int_0^1 \left(\int_x^1 x^2\,dy \right) dx = -\int_0^1 x^2 y \Big|_x^1 dx$$
$$= -\int_0^1 (x^2 - x^3)\,dx = -\left(\frac{1}{3} - \frac{1}{4} \right) = -\frac{1}{12}.$$

The second set leads to the same result.

In the *oriented* form of the change of variables formula for single integrals,

Change of variables with orientation

$$\int_a^b f(x)\,dx = \int_{\varphi^{-1}(a)}^{\varphi^{-1}(b)} f(\varphi(s))\,\varphi'(s)\,ds,$$

the sign of the Jacobian $\varphi'(s)$ indicates whether the interval from a to b and the interval from $\varphi^{-1}(a)$ to $\varphi^{-1}(b)$ have the same or opposite orientations. Here is the analogous formula for oriented double integrals.

Theorem 9.14 (Oriented change of variables). *Let Ω be a bounded open set in $\mathbb{R}^2$, and let $\boldsymbol{\varphi} : \Omega \to \mathbb{R}^2 : (s,t) \to (x,y)$ be a continuously differentiable map that has*

a continuously differentiable inverse $\boldsymbol{\varphi}^{-1} : \boldsymbol{\varphi}(\Omega) \to \Omega$. *Suppose the function* $f(x,y)$
is continuous on $\vec{D} \subset \boldsymbol{\varphi}(\Omega)$, *a closed, oriented, and pathwise-connected region that*
has area; then

$$\iint\limits_{\vec{D}} f(x,y)\, dx\, dy = \iint\limits_{\boldsymbol{\varphi}^{-1}(\vec{D})} f(x(s,t),y(s,t))\, \frac{\partial(x,y)}{\partial(s,t)}\, ds\, dt.$$

Proof. Because $\vec{D}$ has only one pathwise-connected component, all of its points
have the same orientation (Theorem 9.12), so $\mathrm{sgn}(\vec{D})$ is defined. Furthermore, the
Jacobian $J_{\boldsymbol{\varphi}}$ cannot change sign on $\vec{D}$, so all points of $\boldsymbol{\varphi}^{-1}(\vec{D})$ have the same orien-
tation, and $\mathrm{sgn}\,\boldsymbol{\varphi}^{-1}(\vec{D}) = \mathrm{sgn}\,J_{\boldsymbol{\varphi}^{-1}}\,\mathrm{sgn}\,\vec{D} = \mathrm{sgn}\,J_{\boldsymbol{\varphi}}\,\mathrm{sgn}\,\vec{D}$. Thus (using $dA_{x,y}$ and $dA_{s,t}$
to denote the unoriented elements of area),

$$\iint\limits_{\vec{D}} f(x,y)\, dx\, dy = \mathrm{sgn}\,\vec{D} \iint\limits_{D} f(x,y)\, dA_{x,y}$$

$$= \mathrm{sgn}\,\vec{D} \iint\limits_{\boldsymbol{\varphi}^{-1}(D)} f(x(s,t),y(s,t)) \left| \frac{\partial(x,y)}{\partial(s,t)} \right| dA_{s,t} \quad \text{(Theorem 9.11)}$$

$$= \mathrm{sgn}\,\vec{D}\, \mathrm{sgn}\,J_{\boldsymbol{\varphi}} \iint\limits_{\boldsymbol{\varphi}^{-1}(D)} f(x(s,t),y(s,t))\, \frac{\partial(x,y)}{\partial(s,t)}\, dA_{s,t}$$

$$= \mathrm{sgn}\,\boldsymbol{\varphi}^{-1}(D) \iint\limits_{\boldsymbol{\varphi}^{-1}(D)} f(x(s,t),y(s,t))\, \frac{\partial(x,y)}{\partial(s,t)}\, dA_{s,t}$$

$$= \iint\limits_{\boldsymbol{\varphi}^{-1}(\vec{D})} f(x(s,t),y(s,t))\, \frac{\partial(x,y)}{\partial(s,t)}\, ds\, dt. \qquad \qquad \square$$

Summary

The following summary points out parallels between the ways that elements of oriented single and double integrals transform under a change of variables ($\vec{I}$ is an
oriented interval on the x-axis):

$$x \to x(s) \qquad\qquad dx \to \frac{dx}{ds}\, ds, \qquad\qquad \vec{I} \to \boldsymbol{\varphi}^{-1}(\vec{I})$$

$$\begin{cases} x \to x(s,t), \\ y \to y(s,t), \end{cases} \qquad dx\, dy \to \frac{\partial(x,y)}{\partial(s,t)}\, ds\, dt, \qquad \vec{D} \to \boldsymbol{\varphi}^{-1}(\vec{D}).$$

**Example 1: areas in
curvilinear coordinates**

To illustrate the use of the oriented change of variables formula, we first compute
the signed area of a curvilinear quadrilateral specified by curvilinear coordinates
$(s,t) \mapsto (x(s,t),y(s,t))$ in the (x,y)-plane. Let Ω be the infinite strip in the (s,t)-
plane given by $-\pi/2 < s < \pi/2$, and let $\boldsymbol{\varphi} : \Omega \to \mathbb{R}^2$ be

$$\boldsymbol{\varphi}: \begin{cases} x = \sin s \cosh t, \\ y = \cos s \sinh t; \end{cases} \qquad d\boldsymbol{\varphi}_{(s,t)} = \begin{pmatrix} \cos s \cosh t & \sin s \sinh t \\ -\sin s \sinh t & \cos s \cosh t \end{pmatrix}.$$

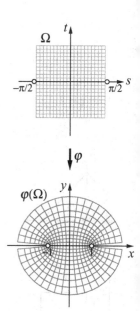

Everywhere on Ω, $\boldsymbol{\varphi}$ is orientation-preserving:

$$\frac{\partial(x,y)}{\partial(s,t)} = \cos^2 s \cosh^2 t + \sin^2 s \sinh^2 t = \cos^2 s + \sinh^2 t > 0.$$

Note: $\cosh^2 t = 1 + \sinh^2 t$). In fact, $\boldsymbol{\varphi}$ is a *conformal* map (Definition 4.2, p. 118) that "flares out" Ω in such a way that the image of the vertical line $s = \text{constant}$ lies on the hyperbola

$$\frac{x^2}{\sin^2 s} - \frac{y^2}{\cos^2 s} = 1.$$

The focal points of this hyperbola are $(\pm f, 0)$, where $f^2 = \sin^2 s + \cos^2 s = 1$ (see Exercise 9.24). The hyperbolas for various s are thus *confocal*. The image of the horizontal line $t = \text{constant}$ lies on the ellipse

$$\frac{x^2}{\cosh^2 s} + \frac{y^2}{\sinh^2 s} = 1.$$

Its focal points are $(\pm f, 0)$, where $f^2 = \cosh^2 t - \sinh^2 t = 1$. Thus the ellipses and hyperbolas are all simultaneously confocal; the image of Ω is the entire plane minus the two rays $|x| \geq 1$ on the x-axis.

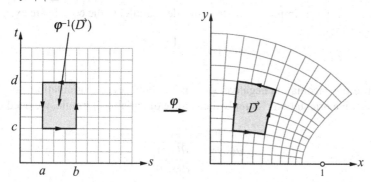

Let $\vec{D}$ be the positively oriented curvilinear quadrilateral in the (x,y)-plane bounded by

$$\frac{x^2}{\sin^2 a} - \frac{y^2}{\cos^2 a} = 1, \qquad \frac{x^2}{\cosh^2 c} + \frac{y^2}{\sinh^2 c} = 1,$$

$$\frac{x^2}{\sin^2 b} - \frac{y^2}{\cos^2 b} = 1, \qquad \frac{x^2}{\cosh^2 d} + \frac{y^2}{\sinh^2 d} = 1,$$

where $0 < a < b < \pi/2$ and $0 < c < d$. The rectangle $\boldsymbol{\varphi}^{-1}(\vec{D})$ is also positively oriented, and

$$\text{area}\,\vec{D} = \iint\limits_{\vec{D}} dx\,dy = \iint\limits_{\boldsymbol{\varphi}^{-1}(\vec{D})} (\cos^2 s + \sinh^2 t)\,ds\,dt = \int_a^b \int_c^d (\cos^2 s + \sinh^2 t)\,ds\,dt.$$

After some standard calculations (see Exercise 9.25), we find

$$\text{area}\,\vec{D} = (d-c)\frac{\sin 2b - \sin 2a}{4} + (b-a)\frac{\sinh 2d - \sinh 2c}{4}.$$

Example 2

For a second example, let us find

$$\iint_{\vec{D}} \frac{(x-y)^2}{1+x+y}\,dx\,dy,$$

where $\vec{D}$ is the rectangle with vertices $(1,-1)$, $(2,0)$, $(0,2)$, and $(-1,1)$, taken in that order. Thus $\vec{D}$ is positively oriented; as an unoriented set, it is given by the inequalities

$$D:\quad \begin{array}{c} 0 \le x+y \le 2, \\ -2 \le x-y \le 2. \end{array}$$

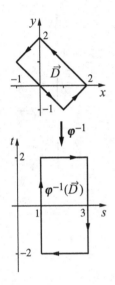

The form of the integrand and the expressions in these inequalities suggest that we set

$$\boldsymbol{\varphi}^{-1}: \begin{cases} s = 1+x+y, \\ t = x-y; \end{cases} \quad \frac{\partial(s,t)}{\partial(x,y)} = \begin{vmatrix} 1 & 1 \\ 1 & -1 \end{vmatrix} = -2.$$

Then $\boldsymbol{\varphi}^{-1}(\vec{D})$ is a simple rectangle with sides parallel to the coordinate axes:

$$\boldsymbol{\varphi}^{-1}(D): \quad \begin{array}{c} 1 \le s \le 3, \\ -2 \le t \le 2. \end{array}$$

Because the Jacobian is negative, $\boldsymbol{\varphi}^{-1}$ and $\boldsymbol{\varphi}$ both reverse orientation, so $\boldsymbol{\varphi}^{-1}(\vec{D})$ has negative orientation. To apply the change of variables formula, we need only

$$\frac{\partial(x,y)}{\partial(s,t)} = \frac{1}{\dfrac{\partial(s,t)}{\partial(x,y)}} = -\frac{1}{2}$$

(Corollary 4.13, p. 138). Therefore,

$$\iint\limits_{\vec{D}} \frac{(x-y)^2}{1+x+y}\,dx\,dy = \iint\limits_{\boldsymbol{\varphi}^{-1}(\vec{D})} \frac{t^2}{s}\left(-\frac{1}{2}\right)ds\,dt = +\frac{1}{2}\iint\limits_{\boldsymbol{\varphi}^{-1}(D)} \frac{t^2}{s}\,dA.$$

In the last equality, we convert the oriented integral over $\boldsymbol{\varphi}^{-1}(\vec{D})$ into an ordinary double integral over the unoriented set $\boldsymbol{\varphi}^{-1}(D)$; the sign change occurs because $\boldsymbol{\varphi}^{-1}(\vec{D})$ is negatively oriented (Definition 9.4). To complete the computation, we write

$$\frac{1}{2}\iint\limits_{\varphi^{-1}(D)}\frac{t^2}{s}\,dA = \frac{1}{2}\int_{-2}^{2}t^2\,dt\int_{1}^{3}\frac{ds}{s} = \frac{1}{2}\left.\frac{t^3}{3}\right|_{-2}^{2}\left.\ln s\right|_{1}^{3} = \frac{8\ln 3}{3}.$$

Our third example involves a similar integral,   Example 3

$$\iint_{\vec{D}}\frac{(x-y)^2(1+2y)}{1+x+y^2}\,dx\,dy;$$

$\vec{D}$ is the negatively oriented region that satisfies the equalities

$$0 \le x+y^2 \le 4, \quad 0 \le y \le x+2.$$

The integrand is positive on $\vec{D}$, but $\vec{D}$ is negatively oriented; therefore we expect the value of the integral to be negative. Guided once again by the form of the integrand and the shape of $\vec{D}$, we change coordinates with

$$\varphi^{-1}: \begin{cases} s = 1+x+y^2, \\ t = x-y; \end{cases} \qquad \frac{\partial(s,t)}{\partial(x,y)} = \begin{vmatrix} 1 & 2y \\ 1 & -1 \end{vmatrix} = -(1+2y).$$

Two of the factors in the integrand transform readily, but $1+2y$ has no simple expression in terms of s and t (but see Exercise 9.27). However, because $\partial(x,y)/\partial(s,t) = -1/(1+2y)$, we find

$$(1+2y)\,dx\,dy \to (1+2y)\frac{\partial(x,y)}{\partial(s,t)}\,ds\,dt = -ds\,dt$$

Therefore, because φ^{-1} is invertible on $1+2y > 0$ (see Exercise 9.27), the change of variables formula is valid and we have

$$\iint_{\vec{D}}\frac{(x-y)^2(1+2y)}{1+x+y^2}\,dx\,dy = -\iint_{\varphi^{-1}(\vec{D})}\frac{t^2}{s}\,ds\,dt,$$

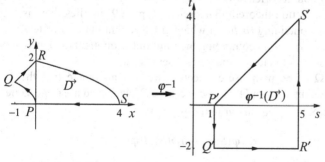

where the image $\varphi^{-1}(\vec{D})$ is the positively oriented trapezoid defined by $1 \le s \le 5$, $-2 \le t \le s-1$. Using iterated integrals and simple antiderivatives, we find

$$- \iint_{\boldsymbol{\varphi}^{-1}(\vec{D})} \frac{t^2}{s} \, ds \, dt = -\int_1^5 \int_{-2}^{s-1} \frac{t^2}{s} \, dt \, ds = -\int_1^5 \frac{t^3}{3s} \bigg|_{-2}^{s-1} \, ds = -\frac{52}{9} - \frac{7}{3} \ln 5.$$

Essential factors
in examples   Notice that the factor $1+2y$ was crucial; without it, the integral would not have been so easy to transform. In the integrals

$$\int (9+y+y^2)^{47} \, dy \quad \text{and} \quad \int (9+y+y^2)^{47}(1+2y) \, dy,$$

the same factor $1+2y$ plays the same role; the typographically simpler integral on the left is mathematically more ponderous. When examples are contrived for instructional purposes, they include such essential factors.

Local area multiplier
with orientation   In Chapter 4, we defined the *local area multiplier* (or *magnification factor*) of a nonlinear map $\boldsymbol{\varphi} : \Omega \to \mathbb{R}^2$ at a point (a,b) to be the area multiplier of the linear approximation $d\boldsymbol{\varphi}_{(a,b)}$ at that point (Definition 4.4, p. 115). In the previous section of this chapter, we justified that definition, at least up to sign. However, using the notions of orientation and signed area, we can now remove the sign restriction.

Theorem 9.15. *Suppose $\boldsymbol{\varphi}(s,t)$ is continuously differentiable on an open set U, and the Jacobian $J_{\boldsymbol{\varphi}}(a,b) \neq 0$ at some point (a,b) in U. Then*

$$\frac{\text{area}\,\boldsymbol{\varphi}(\vec{S})}{\text{area}\,\vec{S}} \to J_{\boldsymbol{\varphi}}(a,b)$$

as the diameter $\delta(\vec{S}) \to 0$, where the limit is taken over closed oriented sets $\vec{S}$ that have signed area and contain the point (a,b).

Proof. We show that the limit of the quotient of signed areas is the derivative of the ordinary set function (cf. pp. 310–312)

$$M(S) = \iint_S J_{\boldsymbol{\varphi}}(s,t) \, dA,$$

where S is $\vec{S}$ without its orientation.

The inverse function theorem (Theorem 5.2, p. 169) implies there is a smaller open set $\Omega \subseteq U$ containing (a,b) on which $\boldsymbol{\varphi}$ has a continuously differentiable inverse. Because $\delta(\vec{S}) \to 0$ in computing the limit of the quotient of signed areas, it is sufficient to restrict $\vec{S}$ to closed oriented subsets of Ω.

By taking Ω to be pathwise-connected, we can guarantee that the Jacobian $J_{\boldsymbol{\varphi}}(s,t)$ has constant sign on Ω. Hence, because $\vec{S}$ has signed area, the same is true of the image $\boldsymbol{\varphi}(\vec{S})$, and we can write (cf. p. 356)

$$\text{area}\,\boldsymbol{\varphi}(\vec{S}) = \text{sgn}\,\boldsymbol{\varphi}(\vec{S}) \, A(\boldsymbol{\varphi}(S)),$$

where $\text{area}\,\vec{D}$ denotes the signed area of the positively or negatively oriented region $\vec{D}$ (p. 356). From Theorem 9.8 and the fact that $\text{sgn}\,J_{\boldsymbol{\varphi}}$ is well defined on S, we get

$$A(\boldsymbol{\varphi}(S)) = \iint_S |J_{\boldsymbol{\varphi}}(s,t)|\, dA = \mathrm{sgn}\, J_{\boldsymbol{\varphi}} \iint_S J_{\boldsymbol{\varphi}}(s,t)\, dA = \mathrm{sgn}\, J_{\boldsymbol{\varphi}}\, M(S).$$

Thus,

$$\mathrm{area}\,\boldsymbol{\varphi}(\vec{S}) = \mathrm{sgn}\,\boldsymbol{\varphi}(\vec{S}) \times A(\boldsymbol{\varphi}(S)) = \mathrm{sgn}\,\boldsymbol{\varphi}(\vec{S}) \times \mathrm{sgn}\, J_{\boldsymbol{\varphi}} \times M(S) = \mathrm{sgn}\,\vec{S} \times M(S);$$

$\mathrm{sgn}\,\boldsymbol{\varphi}(\vec{S})\, \mathrm{sgn}\, J_{\boldsymbol{\varphi}}\, M(S) = \mathrm{sgn}\,\vec{S}$ follows from the proof of Theorem 9.14. We also have $\mathrm{area}\,\vec{S} = \mathrm{sgn}\,\vec{S} \times A(S)$; thus it follows that

$$\frac{\mathrm{area}\,\boldsymbol{\varphi}(\vec{S})}{\mathrm{area}\,\vec{S}} = \frac{M(S)}{A(S)} \to M'(a,b) = J_{\boldsymbol{\varphi}}(a,b)$$

for positively or negatively oriented sets $\vec{S}$ that contain (a,b) and for which $\delta(S) \to 0$ (Theorem 8.39, p. 312). □

The theorem implies that $\mathrm{area}\,\boldsymbol{\varphi}(\vec{S}) \approx J_{\boldsymbol{\varphi}}(a,b)\,\mathrm{area}\,\vec{S}$ for any sufficiently small positively or negatively oriented region containing the point (a,b). For this reason we say that the Jacobian $J_{\boldsymbol{\varphi}}(a,b)$ is the *local signed area magnification factor* for the map $\boldsymbol{\varphi}$ at (a,b).

$\mathrm{area}\,\boldsymbol{\varphi}(\vec{S}) \approx J_{\boldsymbol{\varphi}}\,\mathrm{area}\,\vec{S}$

The change of variables formulas we have established for double integrals and domains in $\mathbb{R}^2$ (Theorem 9.11 and Theorem 9.14) extend naturally to triple integrals and domains in $\mathbb{R}^3$. We state the extensions here with the understanding that they can be proved by adapting the proofs of the 2-dimensional versions. To help underscore the distinction between the oriented and unoriented cases, we use dV as the unoriented element of volume.

Changing variables in triple integrals

Theorem 9.16 (Change of variables in $\mathbb{R}^3$). *Let Ω be a bounded open set in $\mathbb{R}^3$, and let $\boldsymbol{\varphi} : \Omega \to \mathbb{R}^3 : (r,s,t) \to (x,y,z)$ be a continuously differentiable map that has a continuously differentiable inverse $\boldsymbol{\varphi}^{-1} : \boldsymbol{\varphi}(\Omega) \to \Omega$. Suppose the function $f(x,y,z)$ is continuous on $D \subset \boldsymbol{\varphi}(\Omega)$, a closed region that has volume; then*

$$\iiint_D f(x,y,z)\, dV_{x,y,z} = \iiint_{\boldsymbol{\varphi}^{-1}(D)} f(\boldsymbol{\varphi}(r,s,t)) \left| \frac{\partial(x,y,z)}{\partial(r,s,t)} \right| dV_{r,s,t}. \qquad \square$$

Theorem 9.17 (Oriented change of variables in $\mathbb{R}^3$).
Let Ω be a bounded open set in $\mathbb{R}^3$, and let $\boldsymbol{\varphi} : \Omega \to \mathbb{R}^3 : (r,s,t) \to (x,y,z)$ be a continuously differentiable map that has a continuously differentiable inverse $\boldsymbol{\varphi}^{-1} : \boldsymbol{\varphi}(\Omega) \to \Omega$. Suppose the function $f(x,y,z)$ is continuous on $\vec{D} \subset \boldsymbol{\varphi}(\Omega)$, a closed, oriented, and pathwise-connected region that has volume; then

$$\iiint_{\vec{D}} f(x,y,z)\, dx\, dy\, dz = \iiint_{\boldsymbol{\varphi}^{-1}(\vec{D})} f(\boldsymbol{\varphi}(r,s,t)) \frac{\partial(x,y,z)}{\partial(r,s,t)}\, dr\, ds\, dt. \qquad \square$$

9.5 Green's theorem

Green's theorem equates the double integral of a certain function over an oriented region in the plane to the path integral of a related expression over that region's boundary (with its induced orientation). Each integral can be used to evaluate the other. We state and prove several increasingly more general versions of Green's theorem, and then use the final version to extend the change of variables formula for oriented double integrals to settings in which the change of variables may not be invertible.

A special case of
Green's theorem

The first version of Green's theorem is a special case involving a positively oriented region $\vec{S}$ that can be described in both of the following ways:

$$\vec{S}: \quad \begin{matrix} a \le x \le b, \\ \gamma(x) \le y \le \delta(x); \end{matrix} \qquad \vec{S}: \quad \begin{matrix} c \le y \le d, \\ \alpha(y) \le x \le \beta(y); \end{matrix}$$

we assume $\gamma(x)$ and $\delta(x)$ are continuous functions of x on $[a,b]$, and $\alpha(y)$ and $\beta(y)$ are continuous functions of y on $[c,d]$. The orientation on $\vec{S}$ induces (p. 355) an orientation on $\partial \vec{S}$.

Theorem 9.18 (Green's theorem). *Suppose $P(x,y)$ and $Q(x,y)$ are continuously differentiable functions defined on the closure of the region $\vec{S}$, and $\partial \vec{S}$ has the orientation induced by $\vec{S}$; then*

$$\oint_{\partial \vec{S}} P\,dx + Q\,dy = \iint_{\vec{S}} \left(\frac{\partial Q}{\partial x} - \frac{\partial P}{\partial y} \right) dx\,dy.$$

Proof. We assume $\vec{S}$ has positive orientation, and prove half of the equality,

$$\oint_{\partial \vec{S}} P\,dx = \iint_{\vec{S}} -\frac{\partial P}{\partial y}\,dx\,dy,$$

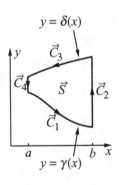

$y = \delta(x)$

using the first description of $\vec{S}$. The other half, involving Q and $\partial Q / \partial x$, is done in a similar way using the second description of $\vec{S}$.

First consider the path integral. As the figure indicates, we can partition the oriented path $\partial \vec{S}$ into four segments (or fewer: either vertical segment $\vec{C}_2$ or $\vec{C}_4$ may reduce to a point). Neither vertical segment contributes to the path integral, because x is constant and $dx = 0$ there. Consequently,

$$\oint_{\partial \vec{S}} P\,dx = \int_{\vec{C}_1} P\,dx + \int_{\vec{C}_3} P\,dx.$$

On $\vec{C}_1$ and $\vec{C}_3$ we can use x itself as the parameter (but make x "run backwards" from b to a for $\vec{C}_3$); then

$$\int_{\vec{C}_1} P\, dx = \int_a^b P(x, \gamma(x))\, dx,$$

$$\int_{\vec{C}_3} P\, dx = \int_b^a P(x, \delta(x))\, dx = -\int_a^b P(x, \delta(x))\, dx,$$

and hence

$$\oint_{\partial \vec{S}} P\, dx = \int_a^b \big(P(x, \gamma(x)) - P(x, \delta(x)) \big)\, dx.$$

That is, the path integral reduces to an ordinary single integral. We now show that the double integral reduces to the same ordinary single integral:

$$\iint_{\vec{S}} -\frac{\partial P}{\partial y}\, dx\, dy = \int_a^b \int_{\gamma(x)}^{\delta(x)} -\frac{\partial P}{\partial y}(x, y)\, dy\, dx$$

$$= \int_a^b -P(x, y)\Big|_{\gamma(x)}^{\delta(x)} dx = \int_a^b \big(P(x, \gamma(x)) - P(x, \delta(x)) \big)\, dx.$$

This completes half the proof; use a similar argument with Q and with the second description of $\vec{S}$ to prove the other half.   □

Below we consider how Green's theorem can be used as a tool for evaluating double integrals. More commonly, though, it is a tool for evaluating path integrals, and we consider this use first.

Evaluating the path integral

Corollary 9.19 *Suppose $P = p(x)$ is a function of x alone, and $Q = q(y)$ a function of y alone; then*

$$\oint_{\partial \vec{S}} p(x)\, dx + q(y)\, dy = 0.$$

Proof. $Q_x - P_y = 0$, so $\iint_{\vec{S}} (Q_x - P_y)\, dx\, dy = 0.$ □

Recall that $\Phi(x, y)$ is called a *potential* (cf. p. 25) for the vector field $\mathbf{F}(x, y) = (P(x, y), Q(x, y))$ if $\mathbf{F} = \operatorname{grad} \Phi$; that is, if $P = \partial \Phi / \partial x$, $Q = \partial \Phi / \partial y$.

Potential functions

Corollary 9.20 *Suppose the vector field $(P(x, y), Q(x, y))$ has a potential $\Phi(x, y)$ that has continuous second derivatives on $\vec{S}$; then*

$$\oint_{\partial \vec{S}} P\, dx + Q\, dy = 0.$$

Proof. $Q_x - P_y = \Phi_{yx} - \Phi_{xy} = 0$ on $\vec{S}$ when Φ has continuous second derivatives on $\vec{S}$.   □

The second corollary is a generalization of the first when $p(x)$ and $q(y)$ are continuously differentiable, because then we can take

$$\Phi(x, y) = \int p(x)\, dx + \int q(y)\, dy.$$

Green's theorem can be used to evaluate a double integral by reducing it to a path integral. Specifically, given

$$\iint_{\vec{S}} f(x,y)\, dx\, dy, \quad \text{set } F(x,y) = \int f(x,y)\, dx.$$

That is, F is a "partial integral" of $f(x,y)$ with respect to x, which means only that $\partial F/\partial x = f$. If we now take $P(x,y) = 0$ and $Q(x,y) = F(x,y)$, then Green's theorem gives

$$\iint_{\vec{S}} f(x,y)\, dx\, dy = \oint_{\partial \vec{S}} F(x,y)\, dy.$$

We can even write this as

$$\iint_{\vec{S}} f(x,y)\, dx\, dy = \oint_{\partial \vec{S}} \left(\int f(x,y)\, dx \right) dy,$$

a kind of iterated integral in which one of the iterates is a path integral.

To illustrate, let us compute (cf. the example on p. 357)

$$\iint_{\vec{S}} x^2\, dx\, dy$$

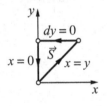

where $\vec{S}$ is the positively oriented triangle with vertices $(0,0)$, $(1,1)$, and $(0,1)$. We have

$$\iint_{\vec{S}} x^2\, dx\, dy = \oint_{\partial \vec{S}} \left(\int x^2\, dx \right) dy = \oint_{\partial \vec{S}} \frac{x^3}{3}\, dy.$$

The path $\partial \vec{S}$ has three segments, but the path integral vanishes along two of them: on the top, $dy = 0$; on the vertical side, $x = 0$. On the diagonal side, we can use $x = y$ as the parameter, so

$$\oint_{\partial \vec{S}} \frac{x^3}{3}\, dy = \int_0^1 \frac{y^3}{3}\, dy = \frac{1}{12}.$$

Incidentally, when we convert the double integral of $f(x,y)$ over $\vec{S}$ into the path integral of

$$F(x,y) = \int f(x,y)\, dx$$

over $\partial \vec{S}$, the partial integral $F(x,y)$ is determined only up to an additive "constant" with respect to the integration variable x. Such a "constant" is, in fact, an arbitrary function of y. However, this indeterminacy has no effect on the outcome. Adding an arbitrary function of y to the partial integral $x^3/3$ in the last example, we find

$$\iint_{\vec{S}} x^2\, dx\, dy = \oint_{\partial \vec{S}} \left(\int x^2\, dx \right) dy = \oint_{\partial \vec{S}} \left(\frac{x^3}{3} + q(y) \right) dy = \oint_{\partial \vec{S}} \frac{x^3}{3}\, dy,$$

because Corollary 9.19 implies

$$\oint_{\partial \vec{S}} q(y)\,dy = 0.$$

We turn now to the task of extending Green's theorem to more general oriented domains $\vec{S}$. Our first step in this direction is to assume that $\vec{S}$ no longer admits both descriptions that we use in computing iterated integrals, but only one of them. For example, suppose we know only that

Green's theorem for more general regions

$$\vec{S}: \quad \begin{array}{l} a \leq x \leq b, \\ \gamma(x) \leq y \leq \delta(x). \end{array}$$

Then our earlier proof of the equality

$$\oint_{\partial \vec{S}} P\,dx = \iint_{\vec{S}} -\frac{\partial P}{\partial y}\,dx\,dy$$

still holds, but we do need a new proof of the second half of the theorem,

$$\oint_{\partial \vec{S}} Q\,dy = \iint_{\vec{S}} \frac{\partial Q}{\partial x}\,dx\,dy,$$

because that depended on the now-absent second description of $\vec{S}$.

To construct a new proof, let F be a "partial integral" of Q with respect to y:

$$F(x,y) = \int Q(x,y)\,dy \text{ or } F_y(x,y) = Q(x,y).$$

Because we assume that Q has continuous first derivatives, F has continuous second derivatives, and $Q_x = F_{yx} = F_{xy}$. We can express the double integral of Q_x in terms of F:

$$\iint_{\vec{S}} Q_x(x,y)\,dx\,dy = \int_a^b \int_{\gamma(x)}^{\delta(x)} F_{xy}(x,y)\,dy\,dx = \int_a^b F_x(x,y)\Big|_{\gamma(x)}^{\delta(x)}\,dx$$

$$= \int_a^b \left(F_x(x,\delta(x)) - F_x(x,\gamma(x))\right)\,dx.$$

We now show that the integral of Q over the path $\partial \vec{S}$ reduces to the same expression, making separate calculations on each of the four segments $\vec{C}_1, \vec{C}_2, \vec{C}_3, \vec{C}_4$. On $\vec{C}_1$ we can use x as the parameter with $y = \gamma(x)$ and $a \leq x \leq b$:

$$\int_{\vec{C}_1} Q\,dy = \int_a^b Q(x,\gamma(x))\,\gamma'(x)\,dx$$

Now consider $F(x, \gamma(x))$; because

$$\frac{d}{dx}F(x,\gamma(x)) = F_x(x,\gamma(x)) + F_y(x,\gamma(x))\,\gamma'(x) = F_x(x,\gamma(x)) + Q(x,\gamma(x))\,\gamma'(x),$$

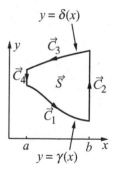

the chain rule and the defining condition $F_y = Q$ give us

$$\int_a^b Q(x, \gamma(x))\, \gamma'(x)\, dx = \int_a^b \frac{d}{dx} F(x, \gamma(x))\, dx - \int_a^b F_x(x, \gamma(x))\, dx.$$

The first integral on the right equals $F(b, \gamma(b)) - F(a, \gamma(a))$, so

$$\int_{\vec{C}_1} Q\, dy = F(b, \gamma(b)) - F(a, \gamma(a)) - \int_a^b F_x(x, \gamma(x))\, dx.$$

On $\vec{C}_2$, $x = b$ and we can use y as the parameter with $\gamma(b) \le y \le \delta(b)$:

$$\int_{\vec{C}_2} Q\, dy = \int_{\gamma(b)}^{\delta(b)} Q(b, y)\, dy = \int_{\gamma(b)}^{\delta(b)} F_y(b, y)\, dy = F(b, \delta(b)) - F(b, \gamma(b)).$$

On $\vec{C}_3$, we can again take x as the parameter, but now $y = \delta(x)$ and we must integrate with respect to x from b to a:

$$\int_{\vec{C}_3} Q\, dy = \int_b^a Q(x, \delta(x))\, \delta'(x)\, dx = - \int_a^b Q(x, \delta(x))\, \delta'(x)\, dx.$$

Using $F(x, \delta(x))$ and an argument similar to the one for $\vec{C}_1$, we find

$$\int_{\vec{C}_3} Q\, dy = -F(b, \delta(b)) + F(a, \delta(a)) + \int_a^b F_x(x, \delta(x))\, dx.$$

On $\vec{C}_4$, $x = a$ and we can again use y as the parameter, but must now integrate from $\delta(a)$ to $\gamma(a)$:

$$\int_{\vec{C}_4} Q\, dy = \int_{\delta(a)}^{\gamma(a)} Q(a, y)\, dy = \int_{\delta(a)}^{\gamma(a)} F_y(a, y)\, dy = F(a, \gamma(a)) - F(a, \delta(a)).$$

Therefore,

$$\oint_{\partial \vec{S}} Q\, dy = \int_{\vec{C}_1} Q\, dy + \int_{\vec{C}_2} Q\, dy + \int_{\vec{C}_3} Q\, dy + \int_{\vec{C}_4} Q\, dy$$

$$= \int_a^b F_x(x, \delta(x))\, dx - \int_a^b F_x(x, \gamma(x))\, dx = \iint_{\vec{S}} Q_x\, dx\, dy$$

when all the cancellations are taken into account.   □

The same arguments, *mutatis mutandis*, allow us to prove Green's theorem for a region $\vec{S}$ when we have only the single description

$$\vec{S}: \quad \begin{array}{c} c \le y \le d, \\ \alpha(y) \le x \le \beta(y). \end{array}$$

Our final version of Green's theorem uses oriented regions $\vec{S}$ that are finite unions of the two kinds we have already considered. In particular, we allow the boundary $\partial\vec{S}$ to have more than one component, although each component will still have the orientation induced by $\vec{S}$. Thus (cf. p. 355), if $\mathbf{n}$ is the outward-pointing unit normal at any smooth point of $\partial\vec{S}$, then the orienting unit tangent $\mathbf{t}$ for $\partial\vec{S}$ at that point is chosen so that the pair $\{\mathbf{n},\mathbf{t}\}$ agrees with the orientation of $\vec{S}$ itself. We assume, as in the figure below, that $\vec{S}$ is closed, bounded, and oriented, and that it can be subdivided into a finite number of nonoverlapping closed cells $\vec{S}_1,\dots,\vec{S}_N$ with the same orientation as $\vec{S}$. As the figure suggests, this can often be accomplished with properly placed vertical or horizontal lines.

Green's theorem on more general domains

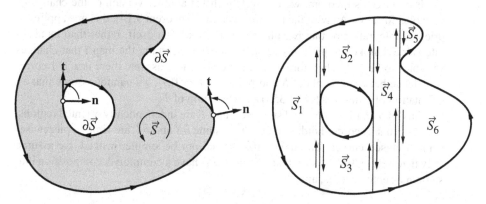

To prove that Green's theorem holds on $\vec{S}$, consider separately the double integral and the path integral. For the double integral we have

Combining the double integrals

$$\iint_{\vec{S}} (Q_x - P_y)\,dx\,dy = \iint_{\vec{S}_1} (Q_x - P_y)\,dx\,dy + \cdots + \iint_{\vec{S}_N} (Q_x - P_y)\,dx\,dy$$

immediately, by Theorem 8.27, page 298.

The path integrals combine in a more interesting way. If two cells $\vec{S}_i$ and $\vec{S}_j$ have a boundary segment $\vec{C}$ in common, then their outward normals point in opposite directions on $\vec{C}$, because $\vec{S}_i$ and $\vec{S}_j$ are on opposite sides of $\vec{C}$. Therefore, the orientation of $\vec{C}$ as part of $\partial\vec{S}_i$ is opposite its orientation as part of $\partial\vec{S}_j$, so the contributions that $\vec{C}$ makes to

Combining the path integrals

$$\oint_{\partial\vec{S}_i} P\,dx + Q\,dy \quad \text{and} \quad \oint_{\partial\vec{S}_j} P\,dx + Q\,dy$$

exactly cancel. The only contributions that do not cancel are from those boundary segments $\vec{C}$ that $\vec{S}_i$ shares with $\vec{S}$ itself. By construction, the orientation of $\vec{C}$ as part of $\partial\vec{S}_i$ is the same as its orientation as part of $\partial\vec{S}$. Therefore, after all the cancellations are taken into account,

$$\oint_{\partial\vec{S}} P\,dx + Q\,dy = \oint_{\partial\vec{S}_1} P\,dx + Q\,dy + \cdots + \oint_{\partial\vec{S}_N} P\,dx + Q\,dy.$$

Thus, because Green's theorem holds for each $\vec{S}_i$, it holds for $\vec{S}$:

$$\iint_{\vec{S}} (Q_x - P_y)\, dx\, dy = \oint_{\partial\vec{S}} P\, dx + Q\, dy. \qquad \Box$$

Of course, if the orientation of $\partial\vec{S}$ is *opposite* the orientation induced by the orientation of $\vec{S}$, then

$$\oint_{\partial\vec{S}} P\, dx + Q\, dy = -\iint_{\vec{S}} (Q_x - P_y)\, dx\, dy.$$

Change of variables via Green's Theorem

Using Green's theorem, we now establish yet another version of the change of variables formula. As with the previous version (Theorem 9.14), this one applies to oriented integrals. For that reason, it uses the Jacobian itself, rather than its absolute value. Unlike the previous version, it does not require the map $\mathbf{f}$ that changes variables to be 1–1. Therefore, because we no longer assume there is a 1–1 correspondence between points of S and points of $T = \mathbf{f}(S)$, we cannot assume that an orientation of S induces (cf. p. 355) an orientation of T.

Thus let S and $T = \mathbf{f}(S)$, and suppose $\vec{S}$ and $\vec{T}$ are independently oriented regions, with Green's Theorem holding on each. Assume $\partial\vec{S}$ and $\partial\vec{T}$ are simple, piecewise-smooth closed curves. The image $\mathbf{f}(\partial\vec{S})$ need not be simple; instead, we assume only that $\mathbf{f}(\partial\vec{S}) \subseteq \partial\vec{T}$ as sets, that $\partial\vec{S}$ and $\mathbf{f}(\partial\vec{S})$ have a common decomposition into smooth oriented curves, and

$$\mathbf{f}(\partial\vec{S}) = k\, \partial\vec{T}$$

as oriented paths. The integer k counts the number of times $\mathbf{f}$ wraps $\partial\vec{S}$ around $\partial\vec{T}$ in the positive direction, minus the number of times it wraps in the negative direction. In the figure below (where the images have been separated for clarity), $k = +1$.

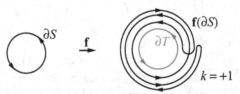

To compensate for the possibility that $\mathbf{f}$ is not 1–1, we require that it now have continuous second derivatives.

Theorem 9.21 (Change of variables with Green's theorem).
Suppose $\mathbf{f}(s,t) = (x(s,t), y(s,t))$ has continuous second derivatives on a bounded open set Ω in $\mathbb{R}^2$. Let $\vec{S} \subset \Omega$ and $\vec{T} = \mathbf{f}(\vec{S})$ be closed, independently oriented sets whose boundaries $\partial\vec{S}$ and $\partial\vec{T}$ are simple closed curves. Assume that Green's theorem holds for both $\vec{S}$ and $\vec{T}$, that $\partial\vec{S}$ and $\mathbf{f}(\partial\vec{S})$ have common decompositions into smooth oriented curves, and that $\mathbf{f}(\partial\vec{S}) = k\, \partial\vec{T}$ as oriented paths. Then, for any continuous function $g(x,y)$ on $\vec{T}$,

$$k \iint_{\vec{T}} g(x,y)\, dx\, dy = \iint_{\vec{S}} g(x(s,t), y(s,t)) \frac{\partial(x,y)}{\partial(s,t)}\, ds\, dt.$$

Proof. Because Green's theorem holds for the region $\vec{T}$, we can write

$$\iint_{\vec{T}} g(x,y)\,dx\,dy = \oint_{\partial\vec{T}} G(x,y)\,dy,$$

where $G(x,y)$ is some function for which $G_x(x,y) = g(x,y)$ (i.e., a "partial integral"). Because $k\,\partial\vec{T} = \mathbf{f}(\partial\vec{S})$, we have

$$k\iint_{\vec{T}} g(x,y)\,dx\,dy = k\underbrace{\oint G(x,y)\,dy}_{\partial(\vec{T})} = \underbrace{\oint G(x,y)\,dy}_{k\,\partial\vec{T}} = \underbrace{\oint G(x,y)\,dy}_{\mathbf{f}(\partial\vec{S})}.$$

Now apply Exercise 4.37 (p. 149) to $\mathbf{f}$ to transform the last path integral:

$$\underbrace{\oint G(x,y)\,dy}_{\mathbf{f}(\partial\vec{S})} = \underbrace{\oint G(x(s,t),y(s,t))(y_s\,ds + y_t\,dt)}_{\partial\vec{S}} = \underbrace{\oint G^*y_s\,ds + G^*y_t\,dt}_{\partial\vec{S}}.$$

(Here $G^*(s,t) = G(x(s,t),y(s,t))$.) Applying Green's Theorem a second time, we transform this new path integral over $\partial\vec{S}$ back into a double integral, but now one over $\vec{S}$:

$$\oint_{\partial\vec{S}} G^*y_s\,ds + G^*y_t\,dt = \iint_{\vec{S}}\left((G^*y_t)_s - (G^*y_s)_t\right)ds\,dt.$$

The terms in the new double integral are

$$(G^*y_t)_s = \frac{\partial}{\partial s}\left(G(x(s,t),y(s,t))\cdot y_t(s,t)\right) = (G_x^*x_s + G_y^*y_s)y_t + G^*y_{ts}$$

and

$$(G^*y_s)_t = \frac{\partial}{\partial t}\left(G(x(s,t),y(s,t))\cdot y_s(s,t)\right) = (G_x^*x_t + G_y^*y_t)y_s + G^*y_{st}.$$

Therefore,

$$(G^*y_t)_s - (G^*y_s)_t = G_x^*(x_sy_t - x_ty_s) + G^*(y_{ts} - y_{st}).$$

The second term vanishes because $y_{ts} - y_{st} = 0$; this is where we need the hypothesis that the map $\mathbf{f}$ has continuous second derivatives. Finally, because $G_x^* = g(x(s,t),y(s,t))$ and $x_sy_t - x_ty_s$ is the Jacobian of $\mathbf{f}$, we have

$$\iint_{\vec{S}}\left((G^*y_t)_s - (G^*y_s)_t\right)ds\,dt = \iint_{\vec{S}} g(x(s,t),y(s,t))\frac{\partial(x,y)}{\partial(s,t)}\,ds\,dt. \qquad \square$$

For our first example of a noninvertible change of variables, we use the quadratic map $\mathbf{f}: \mathbb{R}^2 \to \mathbb{R}^2$,

$$\mathbf{f}: \begin{cases} x = s^2 - t^2, \\ y = 2st, \end{cases} \qquad J(s,t) = \frac{\partial(x,y)}{\partial(s,t)} = \begin{vmatrix} 2s & -2t \\ 2t & 2s \end{vmatrix} = 4(s^2 + t^2),$$

Example 1: the quadratic map

that we analyzed in Chapter 4 (cf. pp. 116–121). We saw there that **f** squares distances from the origin and doubles polar angles. Away from the origin, $J > 0$ and **f** is locally 1–1, by the inverse function theorem.

Let S be the unit disk in the (s,t)-plane; its image $T = \mathbf{f}(S)$ is the unit disk in the (x,y)-plane. Note that S_1, the upper half of S, already covers all of T, and so does the lower half, S_2. The images of the boundaries of S_1 and S_2 are the same; however, in the figure below they have been separated slightly, for clarity.

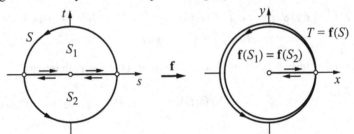

For the sake of illustration, let $\vec{S}$ have positive orientation and $\vec{T}$ negative. Then the image of $\partial\vec{S}$ wraps twice around $\partial\vec{T}$, but with the opposite orientation:

$$\mathbf{f}(\partial\vec{S}) = -2\,\partial\vec{T}.$$

For any continuous function $g(x,y)$ on $\vec{T}$, Theorem 9.21 asserts that

$$-2\iint_{\vec{T}} g(x,y)\,dx\,dy = \iint_{\vec{S}} g(s^2 - t^2, 2st)\, 4(s^2 + t^2)\,ds\,dt.$$

For instance, if $g(x,y) \equiv 1$, then the assertion reduces to

$$-2\,\text{area}\,\vec{T} = -2\iint_{\vec{T}} dx\,dy = \iint_{\vec{S}} 4(s^2 + t^2)\,ds\,dt.$$

To verify this, note that $\vec{T}$ is a negatively oriented unit disk, so area $\vec{T} = -\pi$ and the left-hand side is $+2\pi$. The right-hand side has the same value, as we can see by making a change to polar coordinates:

$$\iint_{\vec{S}} 4(s^2 + t^2)\,ds\,dt = \int_0^{2\pi} d\theta \int_0^1 4r^2 r\,dr = 2\pi r^4 \Big|_0^1 = 2\pi.$$

For a second instance, take $g(x,y) = x^2$; then we must verify that

$$-2\iint_{\vec{T}} x^2\,dx\,dy = \iint_{\vec{S}} (s^2 - t^2)^2\, 4(s^2 + t^2)\,ds\,dt.$$

Because $\vec{T}$ is negatively oriented, we have

$$-2 \iint_{\vec{T}} x^2 \, dx \, dy = +2 \iint_T x^2 \, dA = 2 \int_0^{2\pi} \int_0^1 r^2 \cos^2 \theta \, r \, dr \, d\theta$$

$$= 2 \int_0^{2\pi} \cos^2 \theta \, d\theta \int_0^1 r^3 \, dr = 2 \cdot \pi \cdot \frac{1}{4} = \frac{\pi}{2}.$$

The integral over $\vec{S}$ can also be evaluated by a change to polar coordinates in which $s = r\cos\theta$, $t = r\sin\theta$. Because

$$(s^2 - t^2)^2 = (r^2 \cos^2 \theta - r^2 \sin^2 \theta)^2 = r^4 \cos^2 2\theta,$$

we find

$$\iint_{\vec{S}} (s^2 - t^2)^2 \, 4(s^2 + t^2) \, ds \, dt = \int_0^{2\pi} \cos^2 2\theta \int_0^1 4r^7 \, dr = \pi \times \frac{1}{2} = \frac{\pi}{2}.$$

For our second example of a change of variables that transforms integrals using Green's theorem, we take

Example 2: a fold

$$\mathbf{f}: \begin{cases} x = s, \\ y = t^2; \end{cases} \qquad J(s,t) = \frac{\partial(x,y)}{\partial(s,t)} = \begin{vmatrix} 1 & 0 \\ 0 & 2t \end{vmatrix} = 2t.$$

The Jacobian $J(s,t)$ is positive in the upper half-plane and negative in the lower; it changes sign on the s-axis. The map $\mathbf{f}$ is a *fold* (cf. Exercise 4.21, p. 146). It folds the (s,t)-plane along the s-axis; points that are symmetrically placed across the s-axis have the same image.

Let $\vec{S}$ be the rectangle $0 \leq s \leq 1$, $-1 \leq t \leq 1$ with positive orientation. Where J changes sign, split $\vec{S}$ into two positively oriented nonoverlapping cells $\vec{S}_1$ $(t \geq 0)$ and $\vec{S}_2$ $(t \leq 0)$, so that we can write $\vec{S} = \vec{S}_1 + \vec{S}_2$. The image $T = \mathbf{f}(S)$ is the unit square $0 \leq x \leq 1$, $0 \leq y \leq 1$; If we make $\vec{T}$ positively oriented, then

$$\vec{T} = \mathbf{f}(\vec{S}_1) = -\mathbf{f}(\vec{S}_2).$$

Notice that the image of the boundary, $\mathbf{f}(\partial \vec{S})$, is a proper subset of the boundary of $\vec{T}$; it does not "wrap around" $\partial \vec{T}$. In fact, we can write $\mathbf{f}(\partial \vec{S})$ as $\vec{C} - \vec{C}$, where $\vec{C}$ goes around three sides of $\vec{T}$. This implies

$$\mathbf{f}(\partial \vec{S}) = 0 \times \partial \vec{T}$$

as an oriented path. Therefore, by Theorem 9.21,

$$\iint_{\vec{S}} g(s,t^2) \, 2t \, ds \, dt = 0 \times \iint_{\vec{T}} g(x,y) \, dx \, dy = 0$$

for any continuous function $g(x,y)$ defined on the unit square $\vec{T}$.

To see, from another perspective, why the integral of any function of the form $G(s,t) = 2t \, g(s,t^2)$ over $\vec{S}$ must automatically equal zero, we note two things. First, $G(s,t)$ is an odd function of t (i.e., $G(s,-t) = -G(s,t)$). Second, the region $\vec{S}$ is

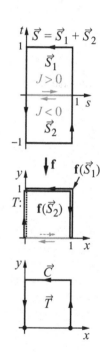

symmetric across the s-axis: the point (s,t) is $\vec{S}$ if and only if $(s,-t)$ is. Therefore, if we write

$$\iint_{\vec{S}} g(s,t^2)\,2t\,ds\,dt = \int_0^1 ds \int_{-1}^1 G(s,t)\,dt,$$

then we see that we must integrate an odd function of t over a t-interval that is symmetric about the origin. The value of such an integral is always zero (Exercise 9.26).

Example 3: folding a different region

Our third example again uses the fold map, but applies it to a region on which the boundary condition of Theorem 9.21 fails. The region is the parallelogram S shown below. Because parts of $\mathbf{f}(\partial S)$ lie in the interior of T, $\mathbf{f}(\partial S) \not\subseteq \partial T$. Therefore, no matter how $\vec{S}$ and $\vec{T}$ are oriented, no integer k can be found for which

$$\mathbf{f}(\partial \vec{S}) = k\,\partial \vec{T}.$$

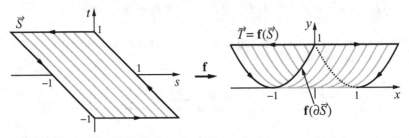

Example 4: a pleat

Our fourth example is the map

$$\mathbf{f}: \begin{cases} x = s, \\ y = t^3 - \tfrac{1}{3}st; \end{cases} \qquad J(s,t) = \frac{\partial(x,y)}{\partial(s,t)} = \begin{vmatrix} 1 & 0 \\ -\tfrac{1}{3}t & 3t^2 - \tfrac{1}{3}s \end{vmatrix} = 3t^2 - \tfrac{1}{3}s.$$

For reasons that soon become clear, this is called a **pleat**. Let $\vec{S}$ be the positively oriented rectangle $-1 \leq s \leq 2$, $-1 \leq t \leq 1$. Note that $\mathbf{f}$ preserves vertical lines, because $x = c$ when $s = c$. Horizontal lines are not preserved, but each is mapped to some straight line. Specifically, the horizontal line $t = k$ is mapped to the line $y = k^3 - kx/3$ (using $s = x$) with slope $-k/3$ and y-intercept k^3. Therefore, the image of $\vec{S}$ is the trapezoid

$$\vec{T}: \begin{matrix} -1 \leq x \leq 2, \\ \tfrac{1}{3}x - 1 \leq y \leq -\tfrac{1}{3}x + 1 \end{matrix}$$

that we define to be positively oriented. In that case, $\mathbf{f}(\partial \vec{S}) = +1 \times \partial \vec{T}$.

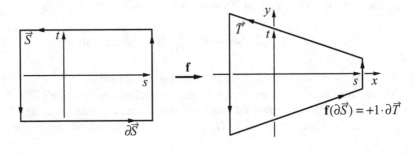

Let us compute the area of the trapezoid $\vec{T}$ first using a double integral provided by the change of variables formula, and then by elementary geometry. The integral gives

Area given by
a double integral

$$\text{area } \vec{T} = +1 \times \iint_{\vec{T}} dx\, dy = \iint_{\vec{S}} \left(3t^2 - \tfrac{1}{3}s\right) ds\, dt$$

$$= \int_{-1}^{2} \int_{-1}^{1} \left(3t^2 - \tfrac{1}{3}s\right) dt\, ds = \int_{-1}^{2} t^3 - \tfrac{1}{3}st \,\Big|_{-1}^{1} ds$$

$$= \int_{-1}^{2} \left(2 - \tfrac{2}{3}s\right) ds = 2s - \tfrac{1}{3}s^2 \,\Big|_{-1}^{2} = 4 - \tfrac{4}{3} - \left(-2 - \tfrac{1}{3}\right) = 5.$$

As a figure in elementary geometry, $\vec{T}$ has "height" $H = 3$ and "bases" $B_1 = 2\tfrac{2}{3}$, $B_2 = \tfrac{2}{3}$; therefore, we find once again that

$$\text{area } \vec{T} = H\left(\frac{B_1 + B_2}{2}\right) = 5.$$

The map $\mathbf{f}$ has subtleties that, among other things, make the validity of the transformation of integrals more surprising than we might at first imagine. The family of horizontal lines in the figure below shows clearly that $\mathbf{f}$ makes a "pleat" in the target. The pleat is made up of two folds that come together at the origin of the target. Inside the pleat, each target point is the image of three points in the source; outside, only one.

f makes a pleat

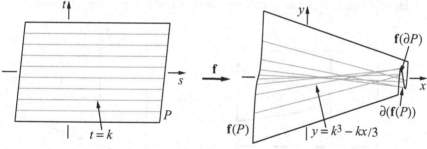

To make the pleating more apparent, we have sheared the rectangle into a parallelogram P. This makes the image of the right edge of P follow a cubic curve that shows how the pleat is folded. As a result, $\mathbf{f}(\partial P) \not\subseteq \partial(\mathbf{f}(P))$: part of that cubic curve lies in the interior of $\mathbf{f}(P)$, and the image of part of the interior of P lies on $\partial(\mathbf{f}(P))$. Although the boundary condition of Theorem 9.21 holds on S, it does not hold on P, even though the shear that changes S into P can be as slight as we wish.

The source is folded to make the pleat. Given any point in the source near that fold, there is another point (on the other side of the fold) that has the same image. In other words, $\mathbf{f}$ is never locally 1–1 at a fold point. But, by the inverse function theorem, $\mathbf{f}$ is locally 1–1 near any point (s,t) where $J(s,t) \neq 0$. Therefore, the fold points of $\mathbf{f}$ must occur where $J = 0$, on the parabola $s = 9t^2$. The pleat is the image

The fold locus
and its image

of this parabola; that is, the pleat is the set of points $\mathbf{f}(9t^2, t)$. It therefore has the parametric equations

$$x = 9t^2,$$
$$y = t^3 - \tfrac{1}{3}9t^3 = -2t^3.$$

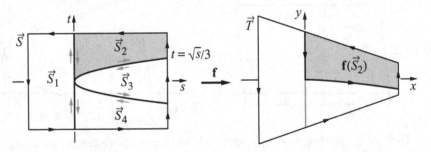

The pleat makes a cusp at the origin. We can see this afresh by setting $t = \sqrt[3]{y/-2}$, so that

$$x = 9\left(\sqrt[3]{y/-2}\right)^2 = \tfrac{9}{\sqrt[3]{4}}y^{2/3}.$$

This is a function whose graph is a cusp.

As we have seen, $\mathbf{f}$ is a threefold cover of the inside of the cusp. It would appear that calculating the area of the trapezoid $\mathbf{f}(S)$ by integrating over S would count the area inside the cusp three times, instead of just once. But note that $\mathbf{f}$ reverses the orientation of one of the three "sheets" that cover the inside of the cusp (namely, the inside of the parabola). Let us now see how this leads to the correct outcome.

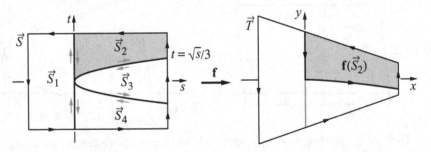

We partition S into the four nonoverlapping, positively oriented regions shown above, so that $\vec{S} = \vec{S}_1 + \vec{S}_2 + \vec{S}_3 + \vec{S}_4$. On each region, $\mathbf{f}$ is 1–1. The change of variables formula implies that the area of $\vec{T}$ is

$$\iint_{\vec{S}} J \, ds \, dt = \iint_{\vec{S}_1} J \, ds \, dt + \iint_{\vec{S}_2} J \, ds \, dt + \iint_{\vec{S}_3} J \, ds \, dt + \iint_{\vec{S}_4} J \, ds \, dt,$$

where $J = 3t^2 - \tfrac{1}{3}s$ in every case. The value of the first integral on the right is $2\tfrac{1}{3}$, the area of the trapezoid $\mathbf{f}(\vec{S}_1)$. For the second integral, we have

$$\iint_{\vec{S}_2} \left(3t^2 - \tfrac{1}{3}s\right) ds\, dt = \int_0^2 \int_{\sqrt{s}/3}^1 \left(3t^2 - \tfrac{1}{3}s\right) dt\, ds$$

$$= \int_0^2 t^3 - \tfrac{1}{3}st \Big|_{\sqrt{s}/3}^1 ds = \int_0^2 \left(1 - \tfrac{1}{3}s + \tfrac{2}{27}s^{3/2}\right) ds$$

$$= s - \tfrac{1}{6}s^2 + \tfrac{2\cdot 2}{27\cdot 5}s^{5/2} \Big|_0^2 = 2 - \tfrac{4}{6} + \tfrac{16\sqrt{2}}{135} = \tfrac{4}{3} + \tfrac{16\sqrt{2}}{135}.$$

The part of $\mathbf{f}(\vec{S}_2)$ that lies in the first quadrant is a trapezoid whose area is $4/3$; the remaining part, the curvilinear triangle in the fourth quadrant, must therefore account for the remaining area, namely $16\sqrt{2}/135 \approx 0.17$. By symmetry, the fourth integral has the same value, and breaks down in the same way:

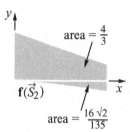

$$\iint_{\vec{S}_4} \left(3t^2 - \tfrac{1}{3}s\right) ds\, dt = \tfrac{4}{3} + \tfrac{16\sqrt{2}}{135}.$$

We see that $\mathbf{f}(\vec{S}_1)$ and the trapezoidal parts of $\mathbf{f}(\vec{S}_2)$ and $\mathbf{f}(\vec{S}_4)$ already completely cover $\vec{T}$, and their total area is $2\tfrac{1}{3} + 2\cdot\tfrac{4}{3} = 5$, the area of $\vec{T}$. The integrals over $\vec{S}_2$ and $\vec{S}_4$ therefore produce an excess of $+32\sqrt{2}/135$. With this in mind, consider the third integral:

$$\iint_{\vec{S}_3} \left(3t^2 - \tfrac{1}{3}s\right) ds\, dt = \int_0^2 \int_{-\sqrt{s}/3}^{\sqrt{s}/3} \left(3t^2 - \tfrac{1}{3}s\right) dt\, ds$$

$$= \int_0^2 t^3 - \tfrac{1}{3}st \Big|_{-\sqrt{s}/3}^{\sqrt{s}/3} ds = \int_0^2 -\tfrac{4}{27}s^{3/2} ds = -\tfrac{8}{135}2^{5/2}$$

$$= -\tfrac{32\sqrt{2}}{135}.$$

This is negative and exactly cancels the excess contributions made by $\vec{S}_2$ and $\vec{S}_4$. The sum of the four oriented integrals is 5.

Notice that although $\mathbf{f}(\partial\vec{S}) = \partial\vec{T}$ as sets and as oriented paths in Example 4, $\mathbf{f}$ is not a 1–1 map of $\partial\vec{S}$ to $\partial\vec{T}$: the image of ∂S "doubles back" on itself briefly inside the cusp. However, all we need (for the proof of the change of variables theorem to hold) is that $\mathbf{f}(\partial\vec{S})$ make a single traversal of $\vec{T}$ in the sense induced by $\vec{T}$, because then

$$\oint_{\mathbf{f}(\partial\vec{S})} P(x,y)\, dx + Q(x,y)\, dy = \oint_{\partial\vec{T}} P(x,y)\, dx + Q(x,y)\, dy$$

for any continuous integrands $P(x,y)$ and $Q(x,y)$.

f need not be 1–1

Exercises

9.1. Use the pullback substitution $y = \sqrt{x^2 + a^2}\tan\theta$ (here y is a function of θ; x is fixed) to show that

$$\int_0^R \frac{dy}{(x^2 + y^2 + a^2)^{3/2}} = \frac{R}{(x^2 + a^2)(x^2 + R^2 + a^2)^{1/2}}.$$

9.2. Show that

$$\frac{1}{\sqrt{1 + (a/2R)^2}} = 1 + O\left((a/R)^2\right) \quad \text{as } a/R \to 0$$

and then provide the details to show that

$$\frac{R^2}{a\sqrt{2R^2 + a^2}} = \frac{R}{a\sqrt{2}} + O(a/R) \quad \text{as } a/R \to 0.$$

9.3. Determine the value of each of the following iterated integrals. (Note the different orders "$dy\,dx$" and "$dx\,dy$". Which integrals are equal and which are not? Is that what you expect?

a. $\displaystyle\int_0^3 \int_1^5 (x^2 + xy^3)\,dy\,dx = \int_0^3 \left(\int_1^5 (x^2 + xy^3)\,dy \right) dx$

b. $\displaystyle\int_1^5 \int_0^3 (x^2 + xy^3)\,dx\,dy$

c. $\displaystyle\int_0^3 \int_1^5 (x^2 + xy^3)\,dx\,dy$

d. $\displaystyle\int_1^5 \int_0^3 (x^2 + xy^3)\,dy\,dx$

9.4. Evaluate:

a. $\displaystyle\int_{-1}^1 \int_{y-5}^{2y+1} xy\,dx\,dy$

c. $\displaystyle\int_0^1 \int_{x^2}^{\sqrt{x}} 1\,dy\,dx$

b. $\displaystyle\int_0^4 \int_{x/2}^{\sqrt{x}} (y^3 + x^2 y)\,dy\,dx$

d. $\displaystyle\int_0^1 \int_{y^2}^{\sqrt{y}} dx\,dy$

9.5. Sketch, in the (x,y)-plane, the domain of integration of each of integrals in Exercises 9.3 and 9.4.

9.6. Sketch each of the following regions in the (x,y)-plane and then describe it in the form

$$a \le x \le b, \quad \gamma(x) \le y \le \delta(x).$$

a. The unit circle (the circle of radius 1 centered at the origin).
b. The circle of radius 3 centered at $(5, -1)$.
c. The circle of radius R centered at (p, q).
d. The bounded region that lies between the graphs of $y = x^2$ and $y = 4$.

e. The bounded region in the first quadrant that lies between the graphs $y = x^2$ and $x = y^2$.

f. The triangle with vertices at $(0,0)$, $(5,5)$, and $(0,5)$.

g. The diamond-shaped (or lozenge-shaped) region whose vertices are at the points $(1,0)$, $(0,1)$, $(-1,0)$, and $(0,-1)$.

h. The region shown shaded in the margin. Write P as (p,q); you need not determine the values of p or q.

9.7. Describe each of the following regions in the form

$$c \leq y \leq d, \quad \alpha(y) \leq x \leq \beta(y).$$

a. The circle of radius R centered at $(x,y) = (p,q)$.

b. The bounded region that lies between the graphs of $y = x^2$ and $x = y^2$.

c. The triangle with vertices at $(0,0)$, $(5,5)$, and $(0,5)$.

d. The triangle with vertices at $(0,0)$, $(5,5)$, and $(5,0)$.

e. The region where $0 \leq x \leq 2$, $0 \leq y \leq x$. Sketch this!

f. The region where $0 \leq x \leq 2$, $x^3 \leq y \leq 10 - x$.

9.8. Reverse the order of integration in the following integral:

$$\int_0^2 \int_{x^3}^{10-x} f(x,y)\,dy\,dx.$$

That is, rewrite it an iterated integral with the integration done first with respect to x, and then with respect to y (i.e., "$dx\,dy$" instead of "$dy\,dx$").

9.9. By reversing the order of integration, show that

$$\int_0^1 \int_y^1 e^{x^2}\,dx\,dy = \frac{e-1}{2}.$$

What happens when you try to determine the integral directly, without reversing the order of integration?

9.10. In each of the following, reverse the order of integration and then calculate the new integral.

a. $\int_0^2 \int_{-\sqrt{4-x^2}}^{\sqrt{4-x^2}} dy\,dx$

b. $\int_0^1 \int_{x^2}^{\sqrt{x}} dy\,dx$

c. $\int_0^1 \int_x^{\sqrt[3]{x}} y^2\,dy\,dx$

d. $\int_0^4 \int_{x/2}^{\sqrt{x}} (y^3 + x^2 y)\,dy\,dx$

9.11. Evaluate the given double integral using iterated integrals.

a. $\iint_R (x^2 + xy^3)\,dA,$ R: rectangle with vertices $(0,1)$, $(3,1)$, $(3,5)$, $(0,5)$

b. $\iint_D xy\, dA,$ D: bounded region between the graphs $y = x^2$ and $x = y^2$

c. $\iint\limits_{\substack{-1 \le x \le 1 \\ 0 \le y \le 3}} y^2\, dA$ d. $\iint\limits_{\substack{0 \le y \le 2 \\ y \le x \le 4-y}} x\, dA$

e. $\iint_K x\, dA,$ K: disk of radius 2 centered at $(0,0)$

f. $\iint_K x\, dA,$ K: disk of radius r centered at (p,q)

g. $\iint_T xy\, dA,$ T: triangle with vertices $(0,0)$, $(4,4)$, $(0,4)$

9.12. Express, as a double integral, the volume of the solid bounded by the planes $x = 0, y = 0, z = 0$, and $x+y+z = 4$. Then determine the volume by evaluating the integral.

9.13. The double integral

$$\iint\limits_{(x-p)^2+(y-q)^2 \le R^2} H\, dA, \qquad H \text{ constant,}$$

is the volume of a familiar solid shape. Describe the shape quite precisely and use that knowledge (rather than an iterated integral) to determine the value of the integral.

9.14. The double integral

$$\iint\limits_{x^2+y^2 \le R^2} \sqrt{R^2-x^2-y^2}\, dA$$

is the volume of a familiar solid shape. What is the shape? Use that knowledge (rather that an iterated integral) to determine the value of the integral.

9.15. Show that

$$\iint_D \frac{\partial^2 f}{\partial x\, \partial y}\, dA = f(P) - f(Q) + f(R) - f(S),$$

where $P = (a,c)$, $Q = (b,c)$, $R = (b,d)$, and $S = (a,d)$ are the vertices of the rectangle D.

9.16. In Exercise 9.15, take $f(x,y) = xy$ and $a = c = 0$. Then show, using separate observations or arguments, that the expression $f(P) - f(Q) + f(R) - f(S)$ and the double integral both equal the area of D.

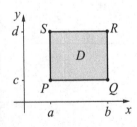

9.17. The volume of the "ravioli" $z = \sin x \sin y$ shown in the margin is

$$\int_0^\pi \int_0^\pi \sin x \sin y\, dy\, dx.$$

Evaluate this integral using the result in Exercise 9.15. What is $f(x,y)$ in this case?

$z = \sin(x)\,\sin(y)$

9.18. a. What is the mean, or average, value $\bar{z}$ (cf. Definition 3.1, p. 75) of the function $ax + by + c$ on the circle $x^2 + y^2 \leq R^2$. Does the value depend on the coefficients a or b?

b. The double integral

$$\iint\limits_{x^2+y^2 \leq R^2} (ax + by + c)\, dA$$

is the volume of a certain solid shape. Describe the shape and make a general sketch (take $c > R > 0$). Determine the volume directly from your knowledge of $\bar{z}$, without calculating the integral again.

c. Without calculating the following integral, explain why

$$\iint\limits_{x^2+y^2 \leq R^2} (ax + by)\, dA = 0,$$

for $R > 0$ and for any a and b.

9.19. a. (This concerns a function of a single variable, and it appears here to provide a comparison with the result of the previous exercise.) What is the average value $\bar{y}$ of $y = mx + b$ on the interval $-R \leq x \leq R$?

b. Does $\bar{y}$ depend on m? Draw a graph of $y = mx + b$ that explains your answer. (To make a concrete graph, try $m = 1/2$ and $b = 3$.)

c. Your work on this and the previous exercise should now allow you to make a general statement about the average value of a linear function (of either one or two variables) on a domain that is symmetric with respect to the origin. What is the statement?

9.20. a. Show that the average value $\bar{z}$ of $z = \sqrt{R^2 - x^2 - y^2}$ on the disk $x^2 + y^2 \leq R^2$ is $2R/3$.

b. Sketch the graph of $z = \sqrt{R^2 - x^2 - y^2}$ and describe it in words.

c. On the same axes, sketch the horizontal plane $z = \bar{z}$. This defines a cylinder over the disk $x^2 + y^2 \leq R^2$ and the volume of this cylinder should equal the volume of the solid $z \leq \sqrt{R^2 - x^2 - y^2}$ over the same disk. Why? Do the volumes appear equal, or nearly so, in your sketch? (This fact was discovered by the Greek mathematician Archimedes (c. 287–212 B.C.E.); it implies that the volume of a ball of radius R is $4\pi R^3/3$.)

9.21. (Here is another single-variable problem that is provides a comparison with an analogous result for a function of two variables.)

a. Show that the average value $\bar{y}$ of $y = \sqrt{R^2 - x^2}$ on the line $-R \leq x \leq R$ is $\pi R/4$.

b. Sketch the graph of $y = \sqrt{R^2 - x^2}$ and describe it in words. On the same plane sketch the graph $y = \bar{y}$.

c. According to the definition of $\bar{y}$, the area of the rectangle under the horizontal line $y = \bar{y}$ should equal the area under the graph of $y = \sqrt{R^2 - x^2}$; why? Does your graph show this?

9.22. a. Show that $\displaystyle\iint\limits_{\varepsilon^2 \le x^2 + y^2 \le 1} \ln\sqrt{x^2 + y^2}\, dA = -\pi(1 - \varepsilon^2 + 2\varepsilon^2 \ln \varepsilon)/2.$

b. Show the improper integral $\displaystyle\iint\limits_{x^2 + y^2 \le 1} \ln\sqrt{x^2 + y^2}\, dA$ converges to $-\pi/2$.

9.23. Use the pullback $x = \cos s$ to show $\displaystyle\int_{1/2}^{1} \frac{dx}{\sqrt{x^2 - x^4}} = \ln(2 + \sqrt{3}).$

9.24. The aim here is analyze the focal points of the hyperbola $x^2/a^2 - y^2/b^2 = 1$ and the ellipse $x^2/a^2 + y^2/b^2 = 1$.

a. By definition, a hyperbola is the locus of points for which the difference of the distances to two fixed points (its *focal points*) is a constant. Show that the focal points of the hyperbola are $\mathbf{p}_\pm = (\pm\sqrt{a^2 + b^2}, 0)$ in the following way: Parametrize the part of the hyperbola for which $x > 0$ as $\mathbf{x} = (x, y) = (a\cosh t, b\sinh t)$, and set $D_\pm = \|\mathbf{x} - \mathbf{p}_\pm\|$. Show by direct computation that $D_\pm = \sqrt{a^2 + b^2}\cosh t \mp a$ and thus that $D_- - D_+ = 2a$.

b. Conclude that the hyperbolas $x^2/\sin^2 s - y^2/\cos^2 s = 1$ (with s arbitrary) are confocal with focal points $(\pm 1, 0)$.

c. By definition, an ellipse is the locus of points for which the sum of the distances to the two focal points is a constant. Show that, when $a > b > 0$, the focal points of the ellipse are $\mathbf{p}_\pm = (\pm\sqrt{a^2 - b^2}, 0)$. Adapt the aproach you took for the hyperbola.

d. Conclude that the ellipses $x^2/\cosh^2 s + y^2/\sinh^2 s = 1$ (with s arbitrary) are confocal with focal points $(\pm 1, 0)$.

9.25. Use $\cos^2 s + \sinh^2 t = \frac{1}{2}(\cos 2s + \cosh 2t)$ (for example) to compute

$$\int_a^b \int_c^d (\cos^2 s + \sinh^2 t)\, ds\, dt.$$

9.26. Suppose $f(t)$ is an odd function ($f(-t) = -f(t)$) that is integrable on the interval $-a \le t \le a$. Use the change of variable $t = -s$ to show

$$\int_{-a}^0 f(t)\, dt = \int_a^0 f(s)\, ds \text{ and hence } \int_{-a}^a f(t)\, dt = 0.$$

9.27. The aim is to show that the map φ^{-1} is invertible on the half plane $y > -1/2$, where

$$\varphi^{-1} : \begin{cases} s = 1 + x + y^2, \\ t = x - y. \end{cases}$$

a. Show that the image of the line $y = a$ under $\boldsymbol{\varphi}^{-1}$ is the line $t = s + b$, where $b = -1 - a - a^2$. Show that, for any two values of $a < -1/2$, the image lines are different.

b. Show that $\boldsymbol{\varphi}^{-1}$ is 1–1 on each line $y = a$. Conclude $\boldsymbol{\varphi}^{-1}$ is 1–1 on the entire half plane $y > -1/2$.

9.28. Show that $$\iiint_{s(D)} f(x,y,z)\,dx\,dy\,dz =$$

$$\iiint_D f(\rho \cos\theta \cos\varphi, \rho \sin\theta \cos\varphi, \rho \sin\varphi)\rho^2 \cos\varphi \, d\rho\, d\theta\, d\varphi$$

is the change of variables formula for triple integrals under the spherical coordinate map $\mathbf{s}$ (Exercise 5.10, p. 178).

9.29. Determine the change of variables formula for fourfold integrals under the map σ (Exercise 5.25, p. 183) that is the analogue in $\mathbb{R}^4$ of the spherical coordinate change in $\mathbb{R}^3$.

9.30. Compute both the path integral and the double integral of Green's theorem for $P = xy$, $Q = y$, and R the unit square in the (x,y)-plane.

9.31. a. Use the result of Exercise 9.18.c to show that

$$\oint_{x^2+y^2=R^2} f(x)\,dx + (ax^2 + bxy + cy^2)\,dy = 0$$

for any function $f(x)$ and any values of a, b, and c.

b. Use the same idea to explain why

$$\oint_{x^2+y^2=R^2} f(x)\,dx + (ax^2 + bxy + cy^2 + \alpha x + \beta y + \gamma)\,dy = \alpha\pi R^2,$$

when $x^2 + y^2 = R^2$ has positive orientation and $f(x)$, a, b, c, α, β, and γ are arbitrary.

9.32. Use Green's theorem to evaluate each of the following path integrals.

a. $\oint_C 5y\,dx + 2x\,dy$, $\quad C$: triangle with vertices $(1,5)$, $(9,2)$, $(8,8)$.

b. $\oint_{\vec{C}} (x^2 - x^3)\,dx + (x^3 + y^2)\,dy$, $\quad \vec{C}$: counterclockwise unit circle.

c. $\oint_{\vec{C}} ye^x\,dx + xe^y\,dy$, $\quad \vec{C}$: rectangle with vertices $(-1,1),(7,1),(7,5),(-1,5)$.

9.33. Show that path integral $\oint_{\partial\vec{R}} x\,dy$ equals the area of the oriented region $\vec{R}$.

9.34. Show that the path integrals $\oint_{\partial \vec{R}} -y\,dx$ and $\frac{1}{2}\oint_{\partial \vec{R}} x\,dy - y\,dx$ both *also* equal the area of $\vec{R}$.

9.35. Let $\vec{D}$ be the elliptical "disk" $x^2/a^2 + y^2/b^2 \leq 1$ with positive orientation.

   a. Sketch $\vec{D}$ when $a = 5$, $b = 3$.
   b. Find parametric equations $x = x(t)$, $y = y(t)$ for the boundary ellipse $\partial \vec{D}$.
   c. Use $\oint_{\partial \vec{D}} x\,dy$ and the parametrizations of $\partial \vec{D}$ to show that area $D = \pi ab$.

   Note that if $b = a$ then $\vec{D}$ is a *circular* disk with radius $a = b$ and area πa^2.

9.36. Suppose that $H(x,y)$ is a **harmonic function**; that is, H satisfies the Laplace equation:

$$\frac{\partial^2 H}{\partial x^2} + \frac{\partial^2 H}{\partial y^2} = 0.$$

   Show that $\oint_{\vec{C}} \frac{\partial H}{\partial y}\,dx - \frac{\partial H}{\partial x}\,dy = 0$ for any closed curve $\vec{C}$.

9.37. Show that, under the maps

$$\mathbf{q}: \begin{cases} x = s^3 - 3st^2, \\ y = 3s^2t - t^3, \end{cases} \qquad \mathbf{s}: \begin{cases} x = s^4 - 6s^2t^2 + t^4, \\ y = 4s^3t - 4st^3, \end{cases}$$

   the positively oriented unit disk $\vec{S}$ covers the positively oriented unit disk $\vec{D}$ three and four times, respectively, and the analogous integrals

$$\iint_{\vec{S}} J_{\mathbf{q}}(s,t)\,ds\,dt \quad \text{and} \quad \iint_{\vec{S}} J_{\mathbf{s}}(s,t)\,ds\,dt$$

   (where $J_{\mathbf{q}}$ and $J_{\mathbf{s}}$ are the Jacobians of $\mathbf{q}$ and $\mathbf{s}$) have the values 3π and 4π.

9.38. Let $\mathbf{g}: \mathbb{R}^2 \to \mathbb{R}^2 : (x,y) \to (u,v)$ be the quadratic map (p. 340), and let the **arrowhead** $A = \mathbf{g}(S)$ be the image of the square $S: 0.2 \leq x \leq 1, 0.2 \leq y \leq 1$.

   a. Sketch A in the (u,v)-plane.
   b. Show that $\partial(x,y)/\partial(u,v) = 1/4\sqrt{u^2 + v^2}$.
   c. Determine area $A = \iint_A du\,dv = \iint_S 4(x^2 + y^2)\,dx\,dy = 3968/1875$; cf. Exercise 8.3, p. 313.
   d. Show that $\iint_A \frac{du\,dv}{\sqrt{u^2 + v^2}} = \iint_S 4\,dx\,dy = 4\,\text{area}\,S = 2.56$.
   e. Determine the moments (cf. Exercise 8.22, p. 316) of the *arrowhead* around the v- and u-axes:

$$\iint_A u\,du\,dv, \quad \iint_A v\,du\,dv.$$

f. Determine $\displaystyle\iint_A \frac{du\,dv}{v}$ and $\displaystyle\iint_A \frac{du\,dv}{u^2+v^2}$.

9.39. Let D be the quarter-disk $0 \leq x^2 + y^2 \leq 1$, $0 \leq x$, $0 \leq y$ in the first quadrant, and let $\mathbf{g}$ be the quadratic map from the previous exercise. Determine

$$\text{area}\,\mathbf{g}(D) = \iint_{\mathbf{g}(D)} du\,dv \text{ and also } \iint_{\mathbf{g}(D)} \sqrt{u^2+v^2}\,du\,dv.$$

The next four exercises are intended to explore the question: How 'infinite' is $1/r$ at the origin in various dimensions? That is, although $1/r$ is infinite, its integral may or may not be, depending on the dimension of the space in which the calculation is done. The following exercise asks you to explore the same question for $1/r^2$ and then to compare your two sets of results.

9.40. Let B^1 be the interval $[-1,1]$ on the x-axis. Let $r = \sqrt{x^2} = |x|$. Show that

$$\int_{B^1} \frac{1}{r}\,dx = \int_{-1}^{1} \frac{1}{|x|}\,dx = 2\int_0^1 \frac{1}{x}\,dx = \infty.$$

(Think of B^1 as the "unit ball" in one dimension.)

9.41. Let B^2 be the unit disk in the (x,y)-plane: $r^2 = x^2 + y^2 \leq 1$. (Think of B^2 as the "unit ball" in two dimensions.) Is

$$\iint_{B^2} \frac{1}{r}\,dx\,dy$$

finite or infinite? Did you make a coordinate change to calculate the integral?

9.42. Let B^3 be the unit ball in (x,y,z)space: $r^2 = x^2 + y^2 + z^2 \leq 1$. Use an appropriate change of variables to determine whether

$$\iiint_{B^3} \frac{1}{r}\,dx\,dy\,dz$$

is finite or infinite. Make a conjecture about the integral of $1/r$ over the unit ball in $\mathbb{R}^n$.

9.43. Integrate $1/r^2$ over the unit ball in $\mathbb{R}^n$, $n = 1, 2, 3$.

a. How does the finiteness of the integral of $1/r^2$ depend on n?

b. In each dimension, how does the integral of $1/r$ compare to the integral of $1/r^2$? Is there some sense in which $1/r$ is either "more infinite" or "less infinite" than $1/r^2$ at the origin?

9.44. Determine

$$\iiint_{x^2+y^2+z^2\leq 1} \frac{1}{1+x^2+y^2+z^2}\,dx\,dy\,dz.$$

Suggestion: Show that $\dfrac{A}{1+A} = 1 - \dfrac{1}{1+A}$, and then use this fact.

9.45. a. Determine

$$\iiint\limits_{R^2 \le x^2+y^2+z^2 \le (R+\Delta R)^2} dx\,dy\,dz,$$

where $R > 0$ is fixed and $\Delta R \ll R$ is small. (The domain here is called a "thin shell".)

b. Show that the integral, which is the volume of the thin shell, equals $4\pi R^2 \Delta R + O(\Delta R^2)$.

9.46. Determine

$$\iiint\limits_{R^2 \le x^2+y^2+z^2 \le (R+\Delta R)^2} \frac{1}{x^2+y^2+z^2}\,dx\,dy\,dz, \quad \Delta R \ll R.$$

Chapter 10
Surface Integrals

Abstract We turn now to integrals over curved surfaces in space. They are analogous, in several ways, to integrals over curved paths. Both arise in scientific problems as ways to express the product of quantities that vary. The first surface integral we consider measures *flux*, the amount of fluid flowing through a surface. The integrand of a surface integral, like a path integral, can be either a scalar or a vector function: flux is the integral of a vector function, whereas area—another surface integral—is the integral of a scalar. Also, orientation matters, at least when the integrand is a vector function.

10.1 Measuring flux

How much fluid will pass through a plane region S in space? If fluid moves with constant velocity $\mathbf{v}$, then during a time interval Δt it will fill out an oblique cylinder with base S and generator $\mathbf{v}\Delta t$. The volume of that cylinder is the product of the area of its base with the height h perpendicular to that base. Now h equals the length of the projection of the generator on $\mathbf{n}$, the unit normal to S in the direction of flow: $h = \mathbf{v}\Delta t \cdot \mathbf{n}$. Therefore, if we denote the area of S by ΔA, then the volume of fluid is

$$\text{volume} = \mathbf{v} \cdot \mathbf{n}\,\Delta A\,\Delta t.$$

To determine the amount of fluid—that is, its mass—we just need to factor in its mass density ρ:

$$\text{mass} = \rho\mathbf{v} \cdot \mathbf{n}\,\Delta A\,\Delta t.$$

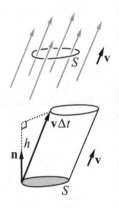

Flux density
and total flux

The vector quantity $\mathbb{V} = \rho\mathbf{v}$ is called the **flux density** (or *flow density*) of the fluid. Flux density is a *rate*; when ρ is measured in kilograms per cubic meter and velocity in meters per second, flux density is measured in kilograms per square meter per second. Its magnitude is the mass of fluid, in kilograms, that flows perpendicularly through a unit area in unit time. The mass of fluid that crosses the region S in unit time is called the **total flux** (or *total flow*) **through S**; it is the product

J.J. Callahan, *Advanced Calculus: A Geometric View*, Undergraduate Texts in Mathematics,   387
DOI 10.1007/978-1-4419-7332-0_10, © Springer Science+Business Media, LLC 2010

$$\text{total flux} = \mathbb{V} \cdot \mathbf{n} \Delta A \text{ kilograms per second.}$$

In general, we allow flux density to vary continuously from point to point, but require it to be constant in time at any given point: $\mathbb{V} = \mathbb{V}(x,y,z)$. Physically, $\mathbb{V}$ is called a *steady flow*; mathematically, it is a continuous *vector field* on (a portion of) $\mathbb{R}^3$. We usually call $\mathbb{V}$ a **flow field**.

From which side does the fluid cross S?
Our expression for total flux does not yet tell us from which side the fluid crosses S. However, if we fix one of the two unit normals $\mathbf{n}$ in advance—that is, before we consider any given fluid flow—

$$\mathbb{V} \cdot \mathbf{n} > 0 \qquad\qquad \mathbb{V} \cdot \mathbf{n} < 0$$

then total flux $\mathbb{V} \cdot \mathbf{n} \Delta A$ becomes a signed quantity whose value is negative precisely when the fluid crosses S in the direction opposite $\mathbf{n}$.

Normals and orientation
Assigning a unit normal to a plane region S in space is equivalent to orienting it. To make the connection, we must first explain what it means to orient S in space. Essentially, it is the same as orienting it in the plane (p. 353): assign to each point $\mathbf{p}$ of S an ordered pair $\{\mathbf{v}_1(\mathbf{p}), \mathbf{v}_2(\mathbf{p})\}$ of linearly independent vectors that vary continuously with $\mathbf{p}$. The vectors are now in $\mathbb{R}^3$, of course, but we constrain them to be tangent to S at the point $\mathbf{p}$. Following earlier practice, we let $\vec{S}$ denote S with an orientation.

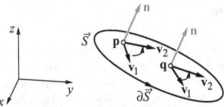

Orientation determines the normal
Next, we must make the connection between orienting S and choosing a unit normal for it. Suppose the ordered pair $\{\mathbf{v}_1(\mathbf{p}), \mathbf{v}_2(\mathbf{p})\}$ orients $\vec{S}$ at $\mathbf{p}$. Then, as in the figure above, we choose

$$\mathbf{n}(\mathbf{p}) = \frac{\mathbf{v}_1(\mathbf{p}) \times \mathbf{v}_2(\mathbf{p})}{\|\mathbf{v}_1(\mathbf{p}) \times \mathbf{v}_2(\mathbf{p})\|}$$

to be the unit normal to $\vec{S}$ at $\mathbf{p}$. On any pathwise-connected component of $\vec{S}$, both $\mathbf{n}(\mathbf{p})$ and the orientation of $\vec{S}$ are constant (Theorem 9.12, p. 354).

If we think of orientation as defining a "sense of rotation" on $\vec{S}$ (cf. p. 353), then, from the side of $\vec{S}$ on which $\mathbf{n}$ lies, that rotation is counterclockwise. This assumes that the coordinate frame in $\mathbb{R}^3$ is right-handed, for then the sense of rotation in the (x,y)-plane, as viewed from the positive z-axis, is counterclockwise.

The normal determines the orientation
It is equally straightforward to connect the choice of a unit normal to the choice of an orientation. Once the unit normal $\mathbf{n}$ for S is given, choose any two linearly

independent vectors $\mathbf{v}_1$ and $\mathbf{v}_2$ that are perpendicular to $\mathbf{n}$ and such that

$$\{\mathbf{v}_1, \mathbf{v}_2, \mathbf{n}\} \text{ or, equivalently } \{\mathbf{n}, \mathbf{v}_1, \mathbf{v}_2\},$$

is a positively oriented triple of vectors in $\mathbb{R}^3$. Then $\mathbf{v}_1$ and $\mathbf{v}_2$ are everywhere tangent to S, so we can orient $\vec{S}$ by assigning $\mathbf{v}_i(\mathbf{p}) = \mathbf{v}_i$, $i = 1, 2$, at every point $\mathbf{p}$ in $\vec{S}$ where $\mathbf{n}$ is the orienting normal.

The figure above also indicates that the orientation of $\vec{S}$ induces an orientation of $\partial\vec{S}$, just as in $\mathbb{R}^2$. When we view $\vec{S}$ from the side toward which the orienting normal $\mathbf{n}$ points, then $\vec{S}$ lies on the left as $\partial\vec{S}$ is traversed in the positive direction.

Induced orientation
on the boundary

Definition 10.1 *Let a fluid have constant flux density* $\mathbb{V}$, *and let* $\vec{S}$ *be a plane region in space that has finite area* ΔA *and orientation given by the unit normal* $\mathbf{n}$. *The total flux of the fluid through* $\vec{S}$ *in unit time is*

$$\Phi = \mathbb{V} \cdot \mathbf{n} \Delta A.$$

This formula has important special cases. Let $\mathbb{V} = (X, Y, Z)$, and suppose $\vec{S} = \vec{S}_x$ lies in the plane $x = \alpha$, has area $\Delta A = \Delta A_x$, and is oriented by the positive x-axis: $\mathbf{n} = (1, 0, 0)$. Then total flux through $\vec{S}_x$ is

Regions parallel to the
coordinate planes

$$\Phi = \Phi_x = (X, Y, Z) \cdot (1, 0, 0) \Delta A_x = X \Delta A_x.$$

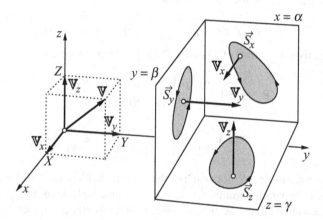

Here total flux depends only on the component X of the flow field $\mathbb{V}$ that is perpendicular to the plane $x = \alpha$; the other two components, Y and Z, in directions parallel to that plane, have no effect. In other words, $\mathbb{V}$ and $\mathbb{V}_x = (X, 0, 0)$ have the same total flux through $\vec{S}_x$. For a region $\vec{S}_y$ in the plane $y = \beta$ or $\vec{S}_z$ in $z = \gamma$, we find, respectively,

$$\Phi_y = Y \Delta A_y, \quad \Phi_z = Z \Delta A_z.$$

The figure above also shows how a region parallel to each coordinate plane is oriented when the remaining positive axis is used as the defining normal (and the three axes together have their usual right-hand orientation):

Plane	Normal	Order of Axes
(y,z)	x-axis	$y \to z$
(x,z)	y-axis	$z \to x$
(y,z)	z-axis	$y \to z$

In two cases, the plane is oriented by its axes in alphabetical order, but in the third, by the opposite order. As the figure in the margin shows, the correct order is the cyclic one $\cdots \to x \to y \to z \to x \to \cdots$ that the three coordinate axes have when they are viewed from the positive orthant (i.e., from the region where $x > 0$, $y > 0$, and $z > 0$).

Total flux through a parallelogram

If $\vec{S}$ is the oriented parallelogram $\mathbf{p} \wedge \mathbf{q}$, then its orienting unit normal is

$$\mathbf{n} = \frac{\mathbf{p} \times \mathbf{q}}{\|\mathbf{p} \times \mathbf{q}\|}$$

(if $\mathbf{p} \times \mathbf{q} \neq \mathbf{0}$) and its area is $\Delta A = \|\mathbf{p} \times \mathbf{q}\|$. Therefore, $\mathbf{n}\Delta A = \mathbf{p} \times \mathbf{q}$, and total flux through $\mathbf{p} \wedge \mathbf{q}$ takes the simple form

$$\Phi = \mathbb{V} \cdot \mathbf{p} \times \mathbf{q} = \mathbf{p} \times \mathbf{q} \cdot \mathbb{V},$$

the scalar triple product of $\mathbf{p}$, $\mathbf{q}$, and $\mathbb{V}$ (cf. p. 43). To compute $\mathbf{n}$, we need $\mathbf{p} \times \mathbf{q} \neq \mathbf{0}$; however, if $\mathbf{p} \times \mathbf{q} = \mathbf{0}$, then $\Delta A = 0$ and $\Phi = 0$, so the formula $\Phi = \mathbb{V} \cdot \mathbf{p} \times \mathbf{q}$ is still valid. If

$$\mathbb{V} = (X, Y, Z), \quad \mathbf{p} = (p_1, p_2, p_3), \quad \mathbf{q} = (q_1, q_2, q_3),$$

then (e.g., from the proof of Theorem 2.11, p. 43), we have

$$\mathbf{p} \times \mathbf{q} = \left(\begin{vmatrix} p_2 & p_3 \\ q_2 & q_3 \end{vmatrix}, \begin{vmatrix} p_3 & p_1 \\ q_3 & q_1 \end{vmatrix}, \begin{vmatrix} p_1 & p_2 \\ q_1 & q_2 \end{vmatrix} \right),$$

$$\Phi = X \begin{vmatrix} p_2 & p_3 \\ q_2 & q_3 \end{vmatrix} + Y \begin{vmatrix} p_3 & p_1 \\ q_3 & q_1 \end{vmatrix} + Z \begin{vmatrix} p_1 & p_2 \\ q_1 & q_2 \end{vmatrix}.$$

Components of total flux

Suppose we project $\vec{S}$ onto each of the coordinate planes $x = 0$, $y = 0$, and $z = 0$; the images are parallelograms $\vec{S}_x$, $\vec{S}_y$, and $\vec{S}_z$, respectively, whose areas are the 2×2 determinants that appear as the components of the vector $\mathbf{p} \times \mathbf{q}$ (see p. 44):

$$\Delta A_x = \begin{vmatrix} p_2 & p_3 \\ q_2 & q_3 \end{vmatrix}, \quad \Delta A_y = \begin{vmatrix} p_3 & p_1 \\ q_3 & q_1 \end{vmatrix}, \quad \Delta A_z = \begin{vmatrix} p_1 & p_2 \\ q_1 & q_2 \end{vmatrix}.$$

Each of these is the *signed* area of an oriented parallelogram whose orientation is determined by the coordinate plane in which it lies. From the discussion above, we know that the value of the total flux through each of $\vec{S}_x$, $\vec{S}_y$, and $\vec{S}_z$ can be written as

$$\Phi_x = X\Delta A_x, \quad \Phi_y = Y\Delta A_y, \quad \Phi_z = Z\Delta A_z.$$

These are the **components of total flux through $\vec{S}$**: $\Phi = \Phi_x + \Phi_y + \Phi_z$.

An example helps clarify these ideas. To simplify the picture as much as possible, we work with triangles instead of parallelograms. Let $\vec{T}$ be the triangle spanned by a pair of vectors **p** and **q** and oriented by **p** × **q**. In the figure below,

$$\mathbf{p} = (-6, 0, -2), \quad \mathbf{q} = (-6, 4, 0), \quad \text{and} \quad \mathbb{V} = (-1, 1, 3);$$

p and **q** are placed so that each edge of $\vec{T}$ lies in one of the coordinate planes. Consequently, $\vec{T}$ and its projections $\vec{T}_x$, $\vec{T}_y$, and $\vec{T}_z$ form a tetrahedron. We have

$$\mathbf{p} \times \mathbf{q} = \left(\begin{vmatrix} 0 & -2 \\ 4 & 0 \end{vmatrix}, \begin{vmatrix} -2 & -6 \\ 0 & -6 \end{vmatrix}, \begin{vmatrix} -6 & 0 \\ -6 & 4 \end{vmatrix} \right) = (8, 12, -24) = 4(2, 3, -6),$$

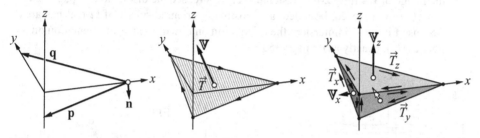

The triangle $\vec{T}$ has half the area of the parallelogram $\mathbf{p} \wedge \mathbf{q}$; therefore total flux through $\vec{T}$ is

$$\Phi = \tfrac{1}{2} \mathbb{V} \cdot \mathbf{p} \times \mathbf{q} = \tfrac{1}{2}(-8 + 12 - 72) = -34.$$

Notice that, in the figure, the boundary of $\vec{T}$ has clockwise, or negative, orientation as we view it from the side on which $\mathbb{V}$ lies. This confirms that Φ must be negative. Furthermore, $\|\mathbf{p} \times \mathbf{q}\| = 28$, so

$$\mathbf{n} = \tfrac{1}{7}(2, 3, -6) \quad \text{and} \quad \Delta A = \text{area } T = \|\mathbf{p} \times \mathbf{q}\|/2 = 14.$$

We can read off the signed areas of $\vec{T}_x$, $\vec{T}_y$, and $\vec{T}_z$ as one-half of the corresponding component of $\mathbf{p} \times \mathbf{q}$:

$$\Delta A_x = 4, \quad \Delta A_y = 6, \quad \Delta A_z = -12.$$

The signs here confirm our direct observations: the boundaries of $\vec{T}_x$ and $\vec{T}_y$ have positive (counterclockwise) orientation with respect to the positive x- and y-axes, but $\vec{T}_z$ has negative (clockwise) orientation with respect to the positive z-axis. The total flux through each of these faces is

$$\Phi_x = -1 \times \Delta A_x = -4, \quad \Phi_y = +1 \times \Delta A_y = +6, \quad \Phi_z = +3 \times \Delta A_z = -36.$$

Of the three faces, total flux is positive only through $\vec{T}_y$, because only on that face does the component of $\mathbb{V}$ (shown in outline in the figure, lying inside the tetrahedron) point in the same direction as the orienting normal. Finally, because flux density $\mathbb{V}$ is constant, no fluid accumulates in the tetrahedron: the fluid that flows through the one face $\vec{T}$ must equal the total that flows through the other three that have the same boundary as $\vec{T}$:

$$\Phi = \Phi_x + \Phi_y + \Phi_z.$$

Flux through a
***curved* surface**

Now suppose that the oriented surface $\vec{S}$ is curved rather than flat. To be definite, let $\vec{S}$ be a parametrized surface patch. Thus we begin with a continuously differentiable 1–1 immersion $\mathbf{f} : \Omega \to \mathbb{R}^3$, where Ω is an open set in $\mathbb{R}^2$. The condition that $\mathbf{f}$ be an immersion means (Definition 6.8, p. 212) that the derivative $d\mathbf{f}_{(a,b)}$ is itself 1–1 (or, in this case, has maximal rank 2) for every (a,b) in Ω. This guarantees that the image of $\mathbf{f}$ is fully 2-dimensional everywhere; see the discussion on page 128.

Let $\vec{U}$ be a closed, bounded, and positively oriented subset of Ω that has area. Because $\mathbf{f}$ is a 1–1 immersion, the orientation on $\vec{U}$ will induce an orientation on its image $\mathbf{f}(\vec{U})$, exactly as on page 355.

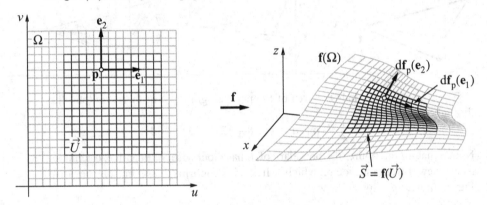

Induced orientation

Theorem 10.1. *If the vectors* $\{\mathbf{v}_1(\mathbf{p}), \mathbf{v}_2(\mathbf{p})\}$ *determine the orientation of* $\vec{U}$, *then their images*

$$\{d\mathbf{f}_{\mathbf{p}}(\mathbf{v}_1(\mathbf{p})), d\mathbf{f}_{\mathbf{p}}(\mathbf{v}_2(\mathbf{p}))\}$$

determine an orientation of $\mathbf{f}(\vec{U})$ *that is called the **induced orientation**.*

Proof. First of all, the image vectors are tangent to $\mathbf{f}(\vec{U})$ at $\mathbf{f}(\mathbf{p})$. Second, they vary continuously with the point $\mathbf{f}(\mathbf{p})$ because $\mathbf{v}_i(\mathbf{p})$ vary continuously with $\mathbf{p}$, and $\mathbf{f}$ is continuously differentiable. Third, they are linearly independent because $\mathbf{f}$ is an immersion at $\mathbf{p}$. □

Oriented surface patch

Definition 10.2 *We say* $\vec{S}$ *is an **oriented surface patch** if* $\vec{S} = \mathbf{f}(\vec{U})$, *where the map* $\mathbf{f} : \Omega \to \mathbb{R}^3$ *is a continuously differentiable 1–1 immersion on an open set* Ω *in* $\mathbb{R}^2$, $\vec{U} \subset \Omega$ *is a closed, bounded, positively oriented set with area, and* $\vec{S}$ *has the induced orientation.*

By the discussion on pages 388–389, we can always replace an ordered pair $\{d\mathbf{f}_{(a,b)}(\mathbf{v}_1), d\mathbf{f}_{(a,b)}(\mathbf{v}_2)\}$ of tangent vectors by the normal

$$d\mathbf{f}_{(a,b)}(\mathbf{v}_1) \times d\mathbf{f}_{(a,b)}(\mathbf{v}_2),$$

to orient $\vec{S}$ at the point (a,b). Let us orient $\vec{U}$ in $\mathbb{R}^2$ using just the standard basis vectors, by assigning the pair $\{\mathbf{e}_1, \mathbf{e}_2\}$ to every point $\mathbf{p}$ (as in the figure above). Then the vectors $d\mathbf{f}_{(a,b)}(\mathbf{e}_1)$ and $d\mathbf{f}_{(a,b)}(\mathbf{e}_2)$ that orient $\vec{S}$ are the columns of the matrix $d\mathbf{f}_{(a,b)}$, in that order. Hence, if

$$\mathbf{f}: \begin{cases} x = x(u,v), \\ y = y(u,v), \\ z = z(u,v), \end{cases} \qquad d\mathbf{f}_{(a,b)} = \begin{pmatrix} x_u(a,b) & x_v(a,b) \\ y_u(a,b) & y_v(a,b) \\ z_u(a,b) & z_v(a,b) \end{pmatrix},$$

parametrizes the oriented surface patch $\vec{S}$, then the cross-product of the column vectors of $d\mathbf{f}_{(a,b)}$ determines the orienting normal for $\vec{S}$ at $\mathbf{f}(a,b)$:

Orienting normal
of the parametrization

$$N_{\mathbf{f}}(a,b) = \left(\begin{vmatrix} y_u(a,b) & y_v(a,b) \\ z_u(a,b) & z_v(a,b) \end{vmatrix}, \begin{vmatrix} z_u(a,b) & z_v(a,b) \\ x_u(a,b) & x_v(a,b) \end{vmatrix}, \begin{vmatrix} x_u(a,b) & x_v(a,b) \\ y_u(a,b) & y_v(a,b) \end{vmatrix} \right)$$

$$= \left(\frac{\partial(y,z)}{\partial(u,v)}, \frac{\partial(z,x)}{\partial(u,v)}, \frac{\partial(x,y)}{\partial(u,v)} \right)_{(a,b)}.$$

Because $\mathbf{f}$ is an immersion everywhere on Ω, the columns of $d\mathbf{f}_{(a,b)}$ are linearly independent and, therefore, $N_{\mathbf{f}}(a,b) \neq \mathbf{0}$.

From a parametrization $\mathbf{f}$ of $\vec{S}$ we can always construct a parametrization $\mathbf{f}^*$ of the oppositely oriented patch $-\vec{S}$ by reversing the order of the parameters. Specifically, let $L : \mathbb{R}^2 \to \mathbb{R}^2 : (s,t) \to (u,v)$ be the reflection

Parametrizing $-\vec{S}$

$$L: \begin{cases} u = t, \\ v = s, \end{cases} \qquad L = \begin{pmatrix} 0 & 1 \\ 1 & 0 \end{pmatrix},$$

and let $\Omega^* = L^{-1}(\Omega)$ and $U^* = L^{-1}(U)$ as sets. Let $\vec{U}$ and $\vec{U}^*$ both be positively oriented; then, because L reverses orientation, $L(\vec{U}^*) = -\vec{U}$. The final step is to let $\mathbf{f}^* = \mathbf{f} \circ L$; then $\mathbf{f}^*$ is defined on Ω^* and

$$\mathbf{f}^*(\vec{U}^*) = \mathbf{f} \circ L(\vec{U}^*) = \mathbf{f}(-\vec{U}) = -\vec{S}.$$

Now let $\vec{S}$ be an oriented surface patch in (x,y,z)-space, and suppose a fluid with continuously varying flux density $\mathbb{V}(x,y,z)$ flows through $\vec{S}$. Our goal is to determine the total flux of $\mathbb{V}$ through $\vec{S}$. If we first approximate $\vec{S}$ by a collection of oriented parallelograms, then total flux through those parallelograms gives us an estimate of the total flux through $\vec{S}$. To get the parallelograms, partition the parameter domain $\vec{U}$ with one of the grids $\mathcal{J}_k$ that are used to define Jordan content in the plane (cf. p. 281), and let $\vec{Q}$ be the square cell of $\mathcal{J}_k$ whose lower-left corner is at the point (a,b), positively oriented as a part of $\vec{U}$. The image of $\vec{Q}$ under the linear map $d\mathbf{f}_{(a,b)}$ is an oriented parallelogram $\vec{P}$ in $\mathbb{R}^3$ that is tangent to $\vec{S}$ at $\mathbf{f}(a,b)$ and has one corner there. (If k is large enough, every $\vec{Q}$ that meets $\vec{U}$ will lie entirely within Ω, so $d\mathbf{f}_{(a,b)}$ will be defined.) See the next figure.

Estimating total flux

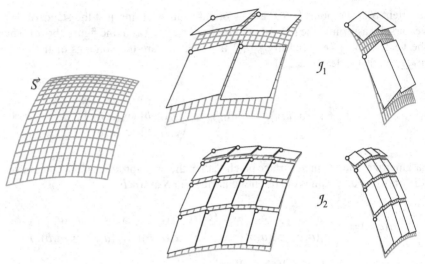

Approximating $\vec{S}$ by parallelograms

As $\vec{Q}$ ranges over all the cells of $\mathcal{J}_k$ that meet $\vec{U}$, the image parallelograms make up a collection of plates attached to $\vec{S}$ at their points of tangency, as in the figure above. The plates resemble the scales that cover the skin of a reptile or armadillo. The figure shows the surface patch $\vec{S}$ first by itself, then with the parallelograms from $\mathcal{J}_1$ attached, and finally with the parallelograms from $\mathcal{J}_2$ attached. We see each set of parallelograms from two different viewpoints. The figures suggest that the plates give us a rough approximation to the surface, an approximation that improves as the plates become smaller and more numerous, that is, as k increases.

Total flux through a single parallelogram

To estimate the total flux through a single parallelogram $\vec{P} = \mathbf{df}_{(a,b)}(\vec{Q})$, note that the edges of $\vec{Q}$ are multiples of the basis vectors $\mathbf{e}_1$ and $\mathbf{e}_2$ in $\mathbb{R}^2$. If we write those edges as

$$\Delta u\, \mathbf{e}_1 \quad \text{and} \quad \Delta v\, \mathbf{e}_2,$$

where $\Delta u = \Delta v = 1/2^k > 0$ when $\vec{Q}$ is a cell in $\mathcal{J}_k$, then we can then write the edges of $\vec{P}$ as

$$\mathbf{p} = \Delta u\, \mathbf{df}_{(a,b)}(\mathbf{e}_1) = \Delta u \begin{pmatrix} x_u(a,b) \\ y_u(a,b) \\ z_u(a,b) \end{pmatrix}, \quad \mathbf{q} = \Delta v\, \mathbf{df}_{(a,b)}(\mathbf{e}_2) = \Delta v \begin{pmatrix} x_v(a,b) \\ y_v(a,b) \\ z_v(a,b) \end{pmatrix}.$$

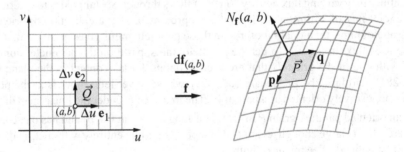

Therefore,

$$\mathbf{p} \times \mathbf{q} = N_{\mathbf{f}}(a,b)\,\Delta u\,\Delta v = \mathbf{n}\,\Delta A,$$

where

$$N_{\mathbf{f}}(a,b) = \left(\frac{\partial(y,z)}{\partial(u,v)}, \frac{\partial(z,x)}{\partial(u,v)}, \frac{\partial(x,y)}{\partial(u,v)} \right)_{(a,b)}$$

is the orienting normal for $\vec{S}$ at $\mathbf{f}(a,b)$ (p. 393). By the geometric definition of the cross product $\mathbf{p} \times \mathbf{q}$, $\mathbf{n}$ is the unit normal in the same direction as $N_{\mathbf{f}}(a,b)$ and ΔA is the ordinary area of the unoriented parallelogram P.

It remains for us to apply the formula $\Phi = \mathbb{V} \cdot \mathbf{p} \times \mathbf{q}$ on $\vec{P}$, but this requires the flow field $\mathbb{V}$ to be constant. We can get a constant by replacing

$$\mathbb{V}(x,y,z) = (X(x,y,z), Y(x,y,z), Z(x,y,z))$$

everywhere on $\vec{P}$ by the single value $\mathbb{V}(\mathbf{f}(a,b))$ that $\mathbb{V}$ takes on at the corner where $\vec{P}$ is attached to $\vec{S}$. By hypothesis, $\mathbb{V}(x,y,z)$ is continuous, so the error caused by this replacement can be made as small as we wish by taking $\vec{Q}$ sufficiently small. We now find

$$\Phi \approx \left(X \frac{\partial(y,z)}{\partial(u,v)} + Y \frac{\partial(z,x)}{\partial(u,v)} + Z \frac{\partial(x,y)}{\partial(u,v)} \right)_{(a,b)} \Delta u\,\Delta v.$$

The right-hand side is a constant determined by (a,b): the three Jacobians are evaluated at $(u,v) = (a,b)$, and X, Y, and Z are evaluated at the point $(x,y,z) = (x(a,b), y(a,b), z(a,b))$.

We can now estimate total flux through the oriented surface patch $\vec{S}$ by adding up the contributions from all the cells $\vec{Q}_1, \ldots, \vec{Q}_I$ that meet the domain $\vec{U}$. Let the lower-left corner of $\vec{Q}_i$ be at the point (u_i, v_i); then

Total flux for a given parametrization of $\vec{S}$

$$\Phi \approx \sum_{i=1}^{I} \left(X \frac{\partial(y,z)}{\partial(u,v)} + Y \frac{\partial(z,x)}{\partial(u,v)} + Z \frac{\partial(x,y)}{\partial(u,v)} \right)_{(u_i,v_i)} \Delta u\,\Delta v.$$

This is a Riemann sum for the oriented integral

$$\Phi_{\mathbf{f}} = \iint_{\vec{U}} \left(X \frac{\partial(y,z)}{\partial(u,v)} + Y \frac{\partial(z,x)}{\partial(u,v)} + Z \frac{\partial(x,y)}{\partial(u,v)} \right)_{(u,v)} du\,dv.$$

Because the integrand is continuous, the integral exists and the Riemann sums converge to it as $k \to \infty$ (Theorem 8.35, p. 305).

Definition 10.3 *Suppose $(x,y,z) = \mathbf{f}(u,v)$ parametrizes the oriented surface patch $\vec{S} = \mathbf{f}(\vec{U})$, and $\mathbb{V} = (X,Y,Z)$ is a continuous flow field defined on an open set containing $\vec{S}$, then the **total flux of $\mathbb{V}$ through $\vec{S}$ for the given parametrization** is the oriented integral*

$$\Phi_{\mathbf{f}} = \iint_{\vec{U}} \left(X \frac{\partial(y,z)}{\partial(u,v)} + Y \frac{\partial(z,x)}{\partial(u,v)} + Z \frac{\partial(x,y)}{\partial(u,v)} \right)_{(u,v)} du\,dv.$$

The notation suggests that the value of $\Phi_{\mathbf{f}}$ depends on the parametrization $\mathbf{f}$. However, if total flux through a surface is to be physically meaningful, its value should be independent of the parametrization of that surface. Before we show that it is, we calculate Φ for two examples.

Example 1: radial flow out of a sphere

Let us determine the total flux of $\mathbb{V} = (X, Y, Z) = (Cx, Cy, Cz)$ (where C is a constant) through the unit sphere $\vec{S}$, parametrized as

$$\mathbf{f}: \begin{cases} x = \cos\theta\cos\varphi, \\ y = \sin\theta\cos\varphi, \\ z = \sin\varphi; \end{cases} \qquad \vec{U}: \begin{array}{l} -\pi \le \theta \le \pi, \\ -\pi/2 \le \varphi \le \pi/2. \end{array}$$

(Strictly, speaking, a surface patch can cover only a portion of the sphere; $\mathbf{f}$ is not 1–1. However, no essential error is introduced by using this parametrization; see pages 417–419. It is simpler to compute flux through the whole sphere.) The flow field $\mathbb{V}$ is radial; each vector points away from the origin with a magnitude proportional to its distance from the origin. Thus, although $\mathbb{V}$ varies, it is everywhere normal to the sphere and has constant magnitude $\|\mathbb{V}\| = C$ there. It follows directly—without calculating the integral—that

$$\Phi = \|\mathbb{V}\| \times \text{area}\, S = 4\pi C.$$

Let us compare this with the value provided by the integral. Because

$$\frac{\partial(y,z)}{\partial(\theta,\varphi)} = \begin{vmatrix} \cos\theta\cos\varphi & -\sin\theta\sin\varphi \\ 0 & \cos\varphi \end{vmatrix} = \cos\theta\cos^2\varphi,$$

$$\frac{\partial(z,x)}{\partial(\theta,\varphi)} = \begin{vmatrix} 0 & \cos\varphi \\ -\sin\theta\cos\varphi & -\cos\theta\sin\varphi \end{vmatrix} = \sin\theta\cos^2\varphi,$$

$$\frac{\partial(x,y)}{\partial(\theta,\varphi)} = \begin{vmatrix} -\sin\theta\cos\varphi & -\cos\theta\sin\varphi \\ \cos\theta\cos\varphi & -\sin\theta\sin\varphi \end{vmatrix} = \sin\varphi\cos\varphi,$$

the integrand is

$$C\cos\theta\cos\varphi \cdot \cos\theta\cos^2\varphi + C\sin\theta\cos\varphi \cdot \sin\theta\cos^2\varphi + C\sin\varphi \cdot \sin\varphi\cos\varphi$$
$$= C(\cos^2\theta\cos^3\varphi + \sin^2\theta\cos^3\varphi + \sin^2\varphi\cos\varphi)$$
$$= C(\cos^2\varphi + \sin^2\varphi)\cos\varphi = C\cos\varphi,$$

and therefore the integral equals

$$\iint_{\vec{U}} C\cos\varphi\, d\theta\, d\varphi = \int_{-\pi}^{\pi} C\left(\int_{-\pi/2}^{\pi/2} \cos\varphi\, d\varphi \right) d\theta = \int_{-\pi}^{\pi} 2C\, d\theta = 4\pi C.$$

Note that the orientation normal given by $\mathbf{f}$ at the point (x, y, z) is

$$N_f = \left(\frac{\partial(y,z)}{\partial(\theta,\varphi)}, \frac{\partial(z,x)}{\partial(\theta,\varphi)}, \frac{\partial(x,y)}{\partial(\theta,\varphi)} \right) = \cos\varphi \cdot (x,y,z).$$

This is, of course, a multiple of the radius vector (x,y,z). Moreover, it is a positive multiple (at least when $-\pi/2 < \varphi < \pi/2$), so N_f is an *outward* normal on the sphere.

For a second example, consider a constant flow $\mathbb{V} = (A,B,C)$ through the same sphere with the same parametrization. Because the flow is constant, all the fluid that enters on one side of the sphere exits on the other. That is, *inflow* equals *outflow*, so we expect the net flux through the whole sphere to be zero: $\Phi = 0$. The integral is

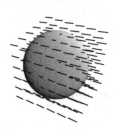

Example 2: constant flow through a sphere

$$\int_{-\pi}^{\pi} \int_{-\pi/2}^{\pi/2} (A\cos\theta\cos^2\varphi + B\sin\theta\cos^2\varphi + C\sin\varphi\cos\varphi)\, d\varphi\, d\theta,$$

and can be dealt with one term at a time. The first is

$$A \int_{-\pi}^{\pi} \cos\theta\, d\theta \int_{-\pi/2}^{\pi/2} \cos^2\varphi\, d\varphi = A \times 0 \times \pi/2 = 0.$$

For similar reasons, the second and third terms also equal zero, so $\Phi = 0$.

Our calculation of Φ for a given surface is tied to a parametrization of that surface. If we change the parametrization, will Φ change as well? Consider what happens when we revisit Example 1 with a different parametrization for the sphere. Let $\mathbf{g} : \mathbb{R}^2 \to \mathbb{R}^3$ be given by

Does Φ depend on the parametrization of S?

$$\mathbf{g} : \begin{cases} x = \dfrac{2u}{1+u^2+v^2}, \\[2mm] y = \dfrac{2v}{1+u^2+v^2}, \\[2mm] z = \dfrac{1-u^2-v^2}{1+u^2+v^2}. \end{cases}$$

Because $\|\mathbf{g}(u,v)\|^2 = 1$ for every (u,v) in $\mathbb{R}^2$, the image of $\mathbf{g}$ is some part of the unit sphere. In fact, $\mathbf{g}(\mathbb{R}^2)$ covers the entire sphere except for the south pole $(x,y,z) = (0,0,-1)$ (see Exercise 10.6). After some calculations (and setting $D = 1 + u^2 + v^2$ to simplify the expressions), we find

$$\frac{\partial(y,z)}{\partial(u,v)} = \frac{8u}{D^3}, \quad \frac{\partial(z,x)}{\partial(u,v)} = \frac{8v}{D^3}, \quad \frac{\partial(x,y)}{\partial(u,v)} = \frac{4(1-u^2-v^2)}{D^3}.$$

This means that the orienting normal for $\mathbf{g}$ at the point (x,y,z) is

$$N_g = \left(\frac{\partial(y,z)}{\partial(u,v)}, \frac{\partial(z,x)}{\partial(u,v)}, \frac{\partial(x,y)}{\partial(u,v)} \right) = \frac{4}{D^2} \cdot (x,y,z).$$

Because $4/D^2 > 0$, N_g is a positive multiple of the radius vector (x,y,z) and is thus, like N_f, an outward normal. That is, $\mathbf{g}$ and $\mathbf{f}$ induce the same orientation of the sphere.

Let us now calculate Φ using **g** instead of **f**. The integrand is

$$\frac{16Cu^2 + 16Cv^2 + 4C(1 - u^2 - v^2)^2}{D^4} = \frac{4C(1 + u^2 + v^2)^2}{D^4} = \frac{4C}{D^2},$$

so the integral itself is

$$\iint_{\mathbb{R}^2} \frac{4C\,du\,dv}{\left(1 + (u^2 + v^2)^2\right)^2} = \int_0^{2\pi} d\theta \int_0^\infty \frac{4C\rho\,d\rho}{(1 + \rho^2)^2} = 2\pi \left.\frac{-2C}{1 + \rho^2}\right|_0^\infty = 4\pi C,$$

under a change to polar coordinates. We find that the two parametrizations of $\vec{S}$ give the same value for Φ.

Preview:
invariance of Φ

    What we have just seen is true in general: total flux is independent of the parametrization, at least for parametrizations that induce the same orientation. To prove this, we first show that an orientation-preserving coordinate change in the source gives a new parametrization that induces the same orientation on $\vec{S}$ and yields the same value for total flux. Then we show that any two parametrizations that induce the same orientation on $\vec{S}$ are related by an orientation-preserving coordinate change in their sources (and thus give the same total flux).

Comparing
parametrizations

    In the following theorems, $\mathbf{f} : \Omega \to \mathbb{R}^3$ and $\mathbf{g} : \Omega^* \to \mathbb{R}^3$ both parametrize the oriented surface patch $\vec{S}$; they have coordinate functions

$$\mathbf{f} : \begin{cases} x = x(u,v), \\ y = y(u,v), \\ z = z(u,v), \end{cases} \qquad \mathbf{g} : \begin{cases} x = \xi(s,t), \\ y = \eta(s,t), \\ z = \zeta(s,t), \end{cases}$$

and orienting normals

$$N_{\mathbf{f}}(u,v) = \left(\frac{\partial(y,z)}{\partial(u,v)}, \frac{\partial(z,x)}{\partial(u,v)}, \frac{\partial(x,y)}{\partial(u,v)}\right), \quad N_{\mathbf{g}}(s,t) = \left(\frac{\partial(\eta,\zeta)}{\partial(s,t)}, \frac{\partial(\zeta,\xi)}{\partial(s,t)}, \frac{\partial(\xi,\eta)}{\partial(s,t)}\right).$$

Thus **f** and **g** are 1–1 immersions on their domains, $\mathbf{f}(\Omega) = \mathbf{g}(\Omega^*)$, and $\vec{S} = \mathbf{f}(\vec{U}) = \mathbf{g}(\vec{U}^*)$, where $\vec{U}$ and $\vec{U}^*$ both have area and are closed, bounded, and positively oriented subsets of Ω and Ω^*, respectively. For the flow field $\mathbb{V} = (X(x,y,z), Y(x,y,z), Z(x,y,z))$, we define

$$\Phi_{\mathbf{f}} = \iint_{\vec{U}} \left(X(\mathbf{f}(u,v))\frac{\partial(y,z)}{\partial(u,v)} + Y(\mathbf{f}(u,v))\frac{\partial(z,x)}{\partial(u,v)} + Z(\mathbf{f}(u,v))\frac{\partial(x,y)}{\partial(u,v)}\right) du\,dv,$$

$$\Phi_{\mathbf{g}} = \iint_{\vec{U}^*} \left(X(\mathbf{g}(s,t))\frac{\partial(\eta,\zeta)}{\partial(s,t)} + Y(\mathbf{g}(s,t))\frac{\partial(\zeta,\xi)}{\partial(s,t)} + Z(\mathbf{g}(s,t))\frac{\partial(\xi,\eta)}{\partial(s,t)}\right) ds\,dt.$$

In the first theorem, **g** is constructed from **f** by a coordinate change, that is, by a map $\boldsymbol{\varphi} : \Omega^* \to \Omega : (s,t) \to (u,v)$ that is continuously differentiable on an open set Ω^* in $\mathbb{R}^2$ and has a continuously differentiable inverse on Ω.

Theorem 10.2. *Suppose* $\mathbf{f} : \Omega \to \mathbb{R}^3 : (u,v) \to (x,y,z)$ *parametrizes the surface patch* $\vec{S} = \mathbf{f}(\vec{U})$, *and let* $\boldsymbol{\varphi} : \Omega^* \to \Omega : (s,t) \to (u,v)$ *be a coordinate change. Set* $\mathbf{g}(s,t) = \mathbf{f}(\boldsymbol{\varphi}(s,t))$ *on* Ω^*. *If the Jacobian of* $\boldsymbol{\varphi}$, $\partial(u,v)/\partial(s,t)$, *is everywhere positive on* Ω^*, *then* $\vec{U}^* = \boldsymbol{\varphi}^{-1}(\vec{U})$ *is positively oriented,* $\mathbf{g}$ *parametrizes* $\vec{S}$ *as an oriented surface patch, and* $\Phi_{\mathbf{g}} = \Phi_{\mathbf{f}}$.

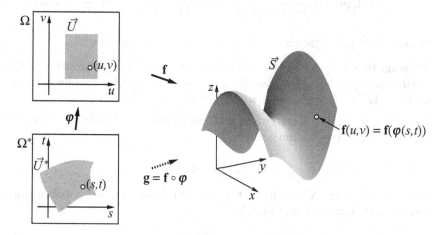

Proof. Because $\boldsymbol{\varphi}$ is a coordinate change, $\mathbf{g}$ is a 1–1 immersion on Ω^* and $\vec{U}^* = \boldsymbol{\varphi}^{-1}(\vec{U})$ has area. Furthermore, $\boldsymbol{\varphi}$ preserves orientation because its Jacobian is positive; hence $\vec{U}^*$ is positively oriented (Theorem 9.13, p. 355). Because $\mathbf{g}(\vec{U}^*) = \mathbf{f}(\vec{U}) = \vec{S}$, $\mathbf{g}$ parametrizes $\vec{S}$ as an oriented surface patch.

It remains for us to prove that $\Phi_{\mathbf{g}} = \Phi_{\mathbf{f}}$. For this, it is helpful to write $(x,y,z) = \mathbf{g}(s,t) = \mathbf{f}(\boldsymbol{\varphi}(s,t))$ in terms of coordinates:

$$\begin{cases} x = \xi(s,t) = x(u(s,t),v(s,t)), \\ y = \eta(s,t) = y(u(s,t),v(s,t)), \\ z = \zeta(s,t) = z(u(s,t),v(s,t)). \end{cases}$$

The chain rule then implies the following about the various Jacobians:

$$\frac{\partial(\eta,\zeta)}{\partial(s,t)} = \frac{\partial(y,z)}{\partial(u,v)} \frac{\partial(u,v)}{\partial(s,t)}, \qquad \frac{\partial(\zeta,\xi)}{\partial(s,t)} = \frac{\partial(z,x)}{\partial(u,v)} \frac{\partial(u,v)}{\partial(s,t)},$$

$$\frac{\partial(\xi,\eta)}{\partial(s,t)} = \frac{\partial(x,y)}{\partial(u,v)} \frac{\partial(u,v)}{\partial(s,t)}.$$

We now show that the first terms of $\Phi_{\mathbf{g}}$ and $\Phi_{\mathbf{f}}$ are equal:

$$\iint_{\vec{U}^*} X(\mathbf{g}(s,t)) \frac{\partial(\eta,\zeta)}{\partial(s,t)} \, ds \, dt = \iint_{\vec{U}} X(\mathbf{f}(u,v)) \frac{\partial(y,z)}{\partial(u,v)} \, du \, dv.$$

Equality of the other two pairs of terms can be established the same way. We begin with the substitutions

$$\iint_{\bar{U}^*} X(\mathbf{g}(s,t)) \frac{\partial(\eta,\zeta)}{\partial(s,t)}\,ds\,dt = \iint_{\bar{U}^*} X(\mathbf{f}(\boldsymbol{\varphi}(s,t))) \frac{\partial(y,z)}{\partial(u,v)}\frac{\partial(u,v)}{\partial(s,t)}\,ds\,dt.$$

The oriented change of variables formula (Theorem 9.14, p. 357) then implies

$$\iint_{\bar{U}^*} X(\mathbf{f}(\boldsymbol{\varphi}(s,t))) \frac{\partial(y,z)}{\partial(u,v)}\frac{\partial(u,v)}{\partial(s,t)}\,ds\,dt = \iint_{\boldsymbol{\varphi}(\bar{U}^*)} X(\mathbf{f}(u,v)) \frac{\partial(y,z)}{\partial(u,v)}\,du\,dv.$$

Because $\boldsymbol{\varphi}(\vec{U}^*) = \vec{U}$, the proof is complete, by what we said above. $\square$

Coordinate changes from parametrizations

Theorem 10.3. *Suppose* $\mathbf{f} : \Omega \to \mathbb{R}^3$ *and* $\mathbf{g} : \Omega^* \to \mathbb{R}^3$ *both parametrize the oriented surface patch* $\vec{S}$. *Then there is an orientation-preserving coordinate change* $\boldsymbol{\varphi} : \Omega^* \to \Omega$ *for which* $\mathbf{g}(s,t) = \mathbf{f}(\boldsymbol{\varphi}(s,t))$ *for all* (s,t) *in* Ω^*.

Proof. The map $\mathbf{f}$ is 1–1 everywhere on Ω, so its inverse $\mathbf{f}^{-1}$ is defined on $\mathbf{f}(\Omega) = \mathbf{g}(\Omega^*)$. Consequently, we can define

$$\boldsymbol{\varphi}(s,t) = \mathbf{f}^{-1}(\mathbf{g}(s,t))$$

for every (s,t) in Ω^*. Although $\boldsymbol{\varphi}$ is 1–1 because $\mathbf{f}^{-1}$ and $\mathbf{g}$ are, it is not obvious that it is also differentiable. The chain rule,

$$d\boldsymbol{\varphi}_{(s,t)} = d\mathbf{f}^{-1}_{\mathbf{g}(s,t)} \circ d\mathbf{g}_{(s,t)}$$

fails here, because the needed derivative of $\mathbf{f}^{-1}$ is not available. To see why, recall that derivatives are linearizations. Because $\mathbf{g}$ maps an open subset of $\mathbb{R}^2$ to $\mathbb{R}^3$, its linearization $d\mathbf{g}_{(s,t)}$ at any point (s,t) is a map from $\mathbb{R}^2$ to $\mathbb{R}^3$. For the chain rule to work, the linearization $d\mathbf{f}^{-1}_{\mathbf{g}(s,t)}$ would have to map $\mathbb{R}^3$ back to $\mathbb{R}^2$, and that would require $\mathbf{f}^{-1}$ itself to be defined on an open subset of $\mathbb{R}^3$. Unfortunately, $\mathbf{f}^{-1}$ is undefined off $\mathbf{f}(\Omega)$: $d\mathbf{f}^{-1}_{\mathbf{g}(s,t)}$ does not exist.

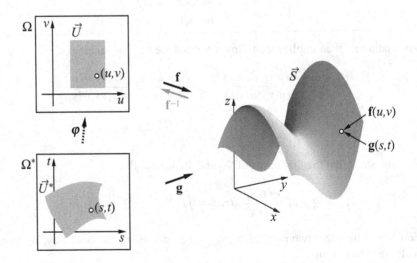

But differentiability is a local condition; let us give $\boldsymbol{\varphi}$ a new local formulation that makes its differentiability evident. Fix a point (s_0, t_0) in Ω^*, and let $(u_0, v_0) = \boldsymbol{\varphi}(s_0, t_0)$ and $(x_0, y_0, z_0) = \mathbf{g}(s_0, t_0) = \mathbf{f}(u_0, v_0)$. Because $\mathbf{f}$ is an immersion at (u_0, v_0), Theorem 6.20 (p. 212) provides a coordinate change $\mathbf{h} : N^3 \to \mathbb{R}^3$ defined on a neighborhood of (x_0, y_0, z_0) that makes $\mathbf{h} \circ \mathbf{f}$ an *injection*. That is, for every (u, v) near (u_0, v_0),

$$\mathbf{h} \circ \mathbf{f}(u, v) = (u, v, 0).$$

If $\Pi : \mathbb{R}^3 \to \mathbb{R}^2$ is the linear projection map that discards the third coordinate (i.e., $\Pi(x, y, z) = (x, y)$), then

$$\Pi \circ \mathbf{h} \circ \mathbf{f}(u, v) = (u, v).$$

In other words, $\Pi \circ \mathbf{h} : N^3 \to \mathbb{R}^2$ plays the role of the inverse of $\mathbf{f}$ on $\vec{S}$ near (x_0, y_0, z_0), but with the advantage that it is defined on a full 3-dimensional neighborhood of (x_0, y_0, z_0). Therefore, we set

$$\boldsymbol{\varphi}(s, t) = \Pi \circ \mathbf{h} \circ \mathbf{g}(s, t)$$

for all (s, t) near (s_0, t_0), and then have

$$d\boldsymbol{\varphi}_{(s_0, t_0)} = \Pi \circ d\mathbf{h}_{(x_0, y_0, z_0)} \circ d\mathbf{g}_{(s_0, t_0)}.$$

Because $\mathbf{h}$ and $\mathbf{g}$ are continuously differentiable and (s_0, t_0) was an arbitrary point in Ω^*, $\boldsymbol{\varphi}$ is continuously differentiable on Ω^*. A similar argument shows that $\boldsymbol{\varphi}^{-1}$ is continuously differentiable. Because $\vec{U}^*$ and $\vec{U} = \boldsymbol{\varphi}(\vec{U}^*)$ are both positively oriented (by definition of $\vec{S}$), $\boldsymbol{\varphi}$ preserves orientation (Theorem 9.13, p. 355). $\quad\square$

Corollary 10.4 *Suppose* $\mathbf{f} : \Omega \to \mathbb{R}^3$ *and* $\mathbf{g} : \Omega^* \to \mathbb{R}^3$ *both parametrize the oriented surface patch* $\vec{S}$. *Then* $\Phi_{\mathbf{f}} = \Phi_{\mathbf{g}}$; *total flux of a fluid through* $\vec{S}$ *is independent of the parametrization.* $\quad\square$

Invariance of Φ

Although the formula for Φ gives the same value no matter which parametrization is used to compute it, that formula is nevertheless bound to a parametrization. The invariance of Φ would be reflected better by an intrinsic formula, one not bound to a parametrization. The existing expression,

$$\iint_{\vec{U}} \left(X(\mathbf{f}(u, v)) \frac{\partial(y, z)}{\partial(u, v)} + Y(\mathbf{f}(u, v)) \frac{\partial(z, x)}{\partial(u, v)} + Z(\mathbf{f}(u, v)) \frac{\partial(x, y)}{\partial(u, v)} \right) du\, dv,$$

is a double integral over a portion of the (u, v) parameter plane; an intrinsic formula would eliminate those parameters. The three Jacobians that appear here are the sort that would be "transformed away" when we change variables in an oriented double integral (e.g., using Theorem 9.14, p. 357). For example,

$$\frac{\partial(y, z)}{\partial(u, v)} du\, dv \quad \text{would be replaced by} \quad dy\, dz.$$

If we make that replacement here, and similarly replace

$$\frac{\partial(z,x)}{\partial(u,v)}\,du\,dv \quad \text{by} \quad dz\,dx, \quad \frac{\partial(x,y)}{\partial(u,v)}\,du\,dv \quad \text{by} \quad dx\,dy,$$

$\mathbf{f}(u,v)$ by (x,y,z), and $\vec{U}$ by $\vec{S}$, then no trace of the original parameters remains, and we are left with

$$\Phi = \iint_{\vec{S}} X(x,y,z)\,dy\,dz + Y(x,y,z)\,dz\,dx + Z(x,y,z)\,dx\,dy.$$

Surface integrals

This is a new kind of object called a *surface integral*. It provides the intrinsic formula we seek, expressing Φ solely in terms of the oriented surface patch $\vec{S}$ and the flow field $\mathbb{V}$ (by its component functions X, Y, and Z).

Definition 10.4 *Suppose the vector field* $\mathbb{V}(x,y,z) = (X,Y,Z)$ *is defined and continuous on the oriented surface patch* $\vec{S}$; *the **surface integral of** $\mathbb{V}$ **over** $\vec{S}$ *is the expression*

$$\iint_{\vec{S}} X(x,y,z)\,dy\,dz + Y(x,y,z)\,dz\,dx + Z(x,y,z)\,dx\,dy$$

whose value is given by the double integral

$$\iint_{\vec{U}} \left(X(\mathbf{f}(u,v))\,\frac{\partial(y,z)}{\partial(u,v)} + Y(\mathbf{f}(u,v))\,\frac{\partial(z,x)}{\partial(u,v)} + Z(\mathbf{f}(u,v))\,\frac{\partial(x,y)}{\partial(u,v)} \right) du\,dv,$$

where $\mathbf{f}: \Omega \to \mathbb{R}^3$ *is any parametrization of* $\vec{S} = \mathbf{f}(\vec{U})$.

In effect, the parametrization pulls back the surface integral from $\vec{S}$ in $\mathbb{R}^3$ to a double integral on $\vec{U}$ in $\mathbb{R}^2$. Corollary 10.4 implies that the value of the surface integral is independent of the paramatrization of $\vec{S}$.

Theorem 10.5. *When the orientation of* $\vec{S}$ *is reversed, the surface integral changes sign:*

$$\iint_{-\vec{S}} X\,dy\,dz + Y\,dz\,dx + Z\,dx\,dy = -\iint_{\vec{S}} X\,dy\,dz + Y\,dz\,dx + Z\,dx\,dy.$$

Proof. Suppose $\mathbf{f}: \Omega \to \mathbb{R}^3$ parametrizes the oriented surface patch $-\vec{S}$; then, by definition, $-\vec{S} = \mathbf{f}(\vec{U})$ for some positively oriented region $\vec{U} \subset \Omega$. Let $L: \mathbb{R}^2 \to \mathbb{R}^2$ be the orientation-reversing linear map (reflection)

$$(u,v) = L(s,t) = (t,s),$$

and let $\Omega^* = L^{-1}(\Omega)$. Because L^{-1} reverses orientation, the induced orientation (p. 355) on the image $\vec{U}^* = L^{-1}(-\vec{U})$ is positive. Define $\mathbf{g} = \mathbf{f} \circ L$; then $\mathbf{g}: \Omega^* \to \mathbb{R}^3$ parametrizes

$$\mathbf{g}(\vec{U}^*) = \mathbf{f}(L(\vec{U}^*)) = \mathbf{f}(-\vec{U}) = -\mathbf{f}(\vec{U}) = \vec{S}$$

itself, because $\vec{U}^*$ is positively oriented.

The expressions involved in proving the two surface integrals equal are long and complicated. To simplify our work, we deal only with the first terms of the integrals; the second and third terms can be dealt with the same way. First use the parametrization **f** to get

$$\iint_{-\vec{S}} X\,dy\,dz = \iint_{\vec{U}} X(\mathbf{f}(u,v)) \frac{\partial(y,z)}{\partial(u,v)}\,du\,dv.$$

Now use the change of variables $(u,v) = L(s,t)$, $\mathbf{f}(u,v) = \mathbf{f}(L(s,t)) = \mathbf{g}(s,t)$ and the oriented change of variables formula (Theorem 9.14, p. 357) to write

$$\iint_{\vec{U}} X(\mathbf{f}(u,v)) \frac{\partial(y,z)}{\partial(u,v)}\,du\,dv = \iint_{L^{-1}(\vec{U})} X(\mathbf{f}(L(s,t))) \frac{\partial(y,z)}{\partial(u,v)} \frac{\partial(u,v)}{\partial(s,t)}\,ds\,dt$$

$$= \iint_{-\vec{U}^*} X(\mathbf{g}(s,t)) \frac{\partial(y,z)}{\partial(s,t)}\,ds\,dt = -\iint_{\vec{U}^*} X(\mathbf{g}(s,t)) \frac{\partial(y,z)}{\partial(s,t)}\,ds\,dt.$$

The last integral is a parametric representation of the surface integral

$$-\iint_{\vec{S}} X\,dy\,dz;$$

by what was said above, this proves the theorem.   □

Note the similarity in form between a surface integral and a path integral (for a path that lies in space):

The form of path and surface integrals

$$\iint_{\vec{S}} X\,dy\,dz + Y\,dz\,dx + Z\,dx\,dy \quad \text{versus} \quad \int_{\vec{C}} P\,dx + Q\,dy + R\,dz.$$

In these integrals, the physical vectors, *flux density* $\mathbb{V} = (X,Y,Z)$ and *force* $\mathbf{F} = (P,Q,R)$, are represented by their components. However, for the path integral there is an alternate form in which **F** itself appears:

$$\text{work} = \int_{\vec{C}} P\,dx + Q\,dy + R\,dz = \int_C \mathbf{F} \cdot \mathbf{t}\,ds.$$

Here **t** is the unit tangent that orients the path $\vec{C}$ (see p. 19), and ds is the "element of arc length" for the unoriented path C. This alternate form is an *unoriented* path integral; information about the orientation of $\vec{C}$ has been transferred to the integrand, to the factor **t**.

The surface integral also has an alternate form that is analogous to the second path integral. We can derive that new form by reconstructing our estimates for total flux through $\vec{S}$. In the original construction, we began with a collection of oriented parallelograms $\vec{P}_1, \ldots, \vec{P}_I$ that approximated $\vec{S}$; $\Phi(\vec{P}_i)$ was the total flux through $\vec{P}_i$, and the sum

An alternate form for a surface integral

$$\sum_{i=1}^{I} \Phi(\vec{P}_i)$$

estimated total flux through $\vec{S}$ itself. If $\vec{P}_i = \mathbf{p}_i \wedge \mathbf{q}_i$ and $\mathbb{V}_i$ was the value of $\mathbb{V}$ at the corner (u_i, v_i) of $\vec{P}_i$, then (p. 395)

$$\Phi(\vec{P}_i) \approx \mathbb{V}_i \cdot \mathbf{p}_i \times \mathbf{q}_i = \left(X \frac{\partial(y,z)}{\partial(u,v)} + Y \frac{\partial(z,x)}{\partial(u,v)} + Z \frac{\partial(x,y)}{\partial(u,v)} \right)_{(u_i,v_i)} \Delta u \, \Delta v.$$

The expressions on the right are the terms in a convergent Riemann sum; in the limit they give the surface integral

$$\Phi = \iint_{\vec{S}} X \, dy \, dz + Y \, dz \, dx + Z \, dx \, dy.$$

Rewriting the cross-product   To construct the second, alternate, form of the surface integral, note that our estimate for $\Phi(\vec{P}_i)$ used the component form of the cross-product: $\mathbf{p}_i \times \mathbf{q}_i$. We now switch to the geometric form,

$$\mathbf{p}_i \times \mathbf{q}_i = \mathbf{n}_i \, \Delta A_i,$$

in which $\Delta A_i \geq 0$ is the absolute area of $\vec{P}_i$ and $\mathbf{n}_i$ is its orienting unit normal. In terms of these geometric variables,

$$\sum_{i=1}^{I} \Phi(\vec{P}_i) = \sum_{i=1}^{I} \mathbb{V}_i \cdot \mathbf{n}_i \, \Delta A_i.$$

This is the Riemann sum; if we follow the usual pattern in expressing its limit as an integral, we get

$$\Phi = \lim_{\substack{I \to \infty \\ \Delta A_i \to 0}} \sum_{i=1}^{I} \mathbb{V}_i \cdot \mathbf{n}_i \, \Delta A_i = \iint_{S} \mathbb{V} \cdot \mathbf{n} \, dA.$$

This is the alternate form for a surface integral; that is,

$$\iint_{\vec{S}} X \, dy \, dz + Y \, dz \, dx + Z \, dx \, dy = \iint_{S} \mathbb{V} \cdot \mathbf{n} \, dA.$$

The integrand $\mathbb{V} \cdot \mathbf{n}$ is the normal component of flux density on $\vec{S}$; the domain of integration is the *unoriented* surface patch S. We call dA the **element of surface area** for S. Information about the orientation of $\vec{S}$ has been transferred from the domain of integration to the integrand. Compare the new integral to the original expession $\mathbb{V} \cdot \mathbf{n} \, \Delta A$ for total flux through a plane region (cf. pp. 387–389 and Definition 10.1).

Area and scalar integrals   In the next section, we discuss integrals of the general form

$$\iint_{S} f(\mathbf{x}) \, dA,$$

where f is a scalar function defined on a region in space that contains the surface patch S. In particular, $f(\mathbf{x}) \equiv 1$ leads to a notion of area for S and indicates why we think of dA as the element of surface area for S.

10.2 Surface area and scalar integrals

In this section, we define the area of a curved surface in space as an integral. Using that as a basis, we then define the integral of a scalar function over a curved surface. Surface area is analogous to arc length, and scalar integrals over surfaces are similarly analogous to scalar integrals over curved paths.

To define the area of an unoriented surface patch S, we begin with a parametrization $\mathbf{f} : \Omega \to \mathbb{R}^3$ of S. Thus $\mathbf{f}$ is a continuously differentiable 1–1 immersion on an open set Ω, $U \subset \Omega$ is a closed, bounded, unoriented set with area, and $S = \mathbf{f}(U)$. Using $\mathbf{f}$ and its derivative, we can approximate S by a collection of parallelograms whose total area will give us an estimate for the area of S. This is essentially the procedure described beginning on page 393. Pick one of the grids $\mathcal{J}_k$ used to define Jordan content in the plane, and select the square cells $Q_1, \ldots, Q_I$ of $\mathcal{J}_k$ that meet U. Let (u_i, v_i) be the lower-left corner of Q_i, and let

Approximating a surface patch

$$P_i = d\mathbf{f}_{(u_i, v_i)}(Q_i).$$

This is a parallelogram tangent to S at $\mathbf{f}(u_i, v_i)$; the parallelograms $P_1, \ldots, P_I$ together give us an approximation of S that improves as $k \to \infty$. The edges of Q_i are multiples of the standard basis vectors that we can write as

$$\Delta u\, \mathbf{e}_1 \quad \text{and} \quad \Delta v\, \mathbf{e}_2,$$

where $\Delta u = \Delta v = 1/2^k$. The corresponding edges of P_i are

$$\mathbf{p}_i = \Delta u\, d\mathbf{f}_{(u_i, v_i)}(\mathbf{e}_1), \quad \mathbf{q}_i = \Delta v\, d\mathbf{f}_{(u_i, v_i)}(\mathbf{e}_2);$$

they are multiples of the columns of $d\mathbf{f}_{(u_i, v_i)}$. If

$$\mathbf{f} : \begin{cases} x = x(u,v), \\ y = y(u,v), \\ z = z(u,v), \end{cases} \qquad d\mathbf{f}_{(a,b)} = \begin{pmatrix} x_u(a,b) & x_v(a,b) \\ y_u(a,b) & y_v(a,b) \\ z_u(a,b) & z_v(a,b) \end{pmatrix}$$

then (p. 393)

$$\mathbf{p}_i \times \mathbf{q}_i = \left(\frac{\partial(y,z)}{\partial(u,v)}, \frac{\partial(z,x)}{\partial(u,v)}, \frac{\partial(x,y)}{\partial(u,v)} \right)_{(u_i, v_i)} \Delta u\, \Delta v,$$

and the area of P_i is

Area of P_i

$$\|\mathbf{p}_i \times \mathbf{q}_i\| = \sqrt{ \left[\frac{\partial(y,z)}{\partial(u,v)} \right]^2 + \left[\frac{\partial(z,x)}{\partial(u,v)} \right]^2 + \left[\frac{\partial(x,y)}{\partial(u,v)} \right]^2 } \Bigg|_{(u_i, v_i)} \Delta u\, \Delta v.$$

This is the ordinary nonnegative area of an unoriented region. The total area of the parallelograms that approximate S is therefore

$$\sum_{i=1}^{I} \sqrt{\left[\frac{\partial(y,z)}{\partial(u,v)}\right]^2 + \left[\frac{\partial(z,x)}{\partial(u,v)}\right]^2 + \left[\frac{\partial(x,y)}{\partial(u,v)}\right]^2} \Bigg|_{(u_i,v_i)} \Delta u \, \Delta v.$$

Area for a given parametrization

This is a Riemann sum for the double integral

$$\mathrm{area}_\mathbf{f}(S) = \iint_U \sqrt{\left[\frac{\partial(y,z)}{\partial(u,v)}\right]^2 + \left[\frac{\partial(z,x)}{\partial(u,v)}\right]^2 + \left[\frac{\partial(x,y)}{\partial(u,v)}\right]^2} \, du \, dv.$$

Because **f** has continuous first derivatives, the integrand is continuous and the Riemann sums converge to the integral as $k \to \infty$. If we consider

$$M(U) = \mathrm{area}_\mathbf{f}(S) = \iint_U \sqrt{\left[\frac{\partial(y,z)}{\partial(u,v)}\right]^2 + \left[\frac{\partial(z,x)}{\partial(u,v)}\right]^2 + \left[\frac{\partial(x,y)}{\partial(u,v)}\right]^2} \, du \, dv$$

to be a set function defined on closed bounded subsets $U \subset \Omega$ that have area (cf. pp. 310–312), then its derivative is

$$M'(u,v) = \sqrt{\left[\frac{\partial(y,z)}{\partial(u,v)}\right]^2 + \left[\frac{\partial(z,x)}{\partial(u,v)}\right]^2 + \left[\frac{\partial(x,y)}{\partial(u,v)}\right]^2}$$

Local area multiplier

(Theorem 8.39, p. 312). This implies that

$$\frac{\mathrm{area}_\mathbf{f}(S)}{A(U)} \approx \sqrt{\left[\frac{\partial(y,z)}{\partial(u,v)}\right]^2 + \left[\frac{\partial(z,x)}{\partial(u,v)}\right]^2 + \left[\frac{\partial(x,y)}{\partial(u,v)}\right]^2}$$

as closely as we wish by making the diameter of U sufficiently small. It is for this reason that we defined (Definition 4.9, p. 139)

$$\sqrt{\left[\frac{\partial(y,z)}{\partial(u,v)}\right]^2 + \left[\frac{\partial(z,x)}{\partial(u,v)}\right]^2 + \left[\frac{\partial(x,y)}{\partial(u,v)}\right]^2}$$

to be the local area multiplier for **f**.

Does the value of the integral for surface area depend on the parametrization? Let $\mathbf{g} : \Omega^* \to \mathbb{R}^3$ be another parametrization of S, where

$$(x,y,z) = \mathbf{g}(s,t) = (\xi(s,t), \eta(s,t), \zeta(s,t)).$$

There is a closed bounded set $U^* \subset \Omega^*$ with area, for which $\mathbf{g}(U^*) = S$ and

$$\mathrm{area}_\mathbf{g}(S) = \iint_{U^*} \sqrt{\left[\frac{\partial(\eta,\zeta)}{\partial(s,t)}\right]^2 + \left[\frac{\partial(\zeta,\xi)}{\partial(s,t)}\right]^2 + \left[\frac{\partial(\xi,\eta)}{\partial(s,t)}\right]^2} \, ds \, dt.$$

Invariance of surface area

Theorem 10.6. *Surface area is independent of the parametrization used to compute it:* $\mathrm{area}_\mathbf{g}(S) = \mathrm{area}_\mathbf{f}(S)$.

Proof. If we set orientation aside, the proof of Theorem 10.3 provides a coordinate change $\boldsymbol{\varphi} : \Omega^* \to \Omega$, $(u,v) = \boldsymbol{\varphi}(s,t)$, for which $\mathbf{g}(U^*) = U$ and $\mathbf{g}(s,t) = \mathbf{f}(\boldsymbol{\varphi}(s,t))$ for all (s,t) in Ω^*. The chain rule implies

$$\frac{\partial(\eta,\zeta)}{\partial(s,t)} = \frac{\partial(y,z)}{\partial(u,v)}\frac{\partial(u,v)}{\partial(s,t)}, \quad \frac{\partial(\zeta,\xi)}{\partial(s,t)} = \frac{\partial(z,x)}{\partial(u,v)}\frac{\partial(u,v)}{\partial(s,t)}, \quad \frac{\partial(\xi,\eta)}{\partial(s,t)} = \frac{\partial(x,y)}{\partial(u,v)}\frac{\partial(u,v)}{\partial(s,t)}.$$

Therefore, on Ω^* we have

$$\sqrt{\left[\frac{\partial(\eta,\zeta)}{\partial(s,t)}\right]^2 + \left[\frac{\partial(\zeta,\xi)}{\partial(s,t)}\right]^2 + \left[\frac{\partial(\xi,\eta)}{\partial(s,t)}\right]^2}$$
$$= \sqrt{\left[\frac{\partial(y,z)}{\partial(u,v)}\right]^2 + \left[\frac{\partial(z,x)}{\partial(u,v)}\right]^2 + \left[\frac{\partial(x,y)}{\partial(u,v)}\right]^2} \left|\frac{\partial(u,v)}{\partial(s,t)}\right|$$

Now make this substitution in the formula for $\text{area}_{\mathbf{g}}(S)$, and then use the basic change of variables formula (Theorem 9.11, p. 350) to get

$$\text{area}_{\mathbf{g}}(S) = \iint_{U^*} \sqrt{\left[\frac{\partial(\eta,\zeta)}{\partial(s,t)}\right]^2 + \left[\frac{\partial(\zeta,\xi)}{\partial(s,t)}\right]^2 + \left[\frac{\partial(\xi,\eta)}{\partial(s,t)}\right]^2} \, ds\,dt$$
$$= \iint_{U^*} \sqrt{\left[\frac{\partial(y,z)}{\partial(u,v)}\right]^2 + \left[\frac{\partial(z,x)}{\partial(u,v)}\right]^2 + \left[\frac{\partial(x,y)}{\partial(u,v)}\right]^2} \left|\frac{\partial(u,v)}{\partial(s,t)}\right| \, ds\,dt$$
$$= \iint_{U} \sqrt{\left[\frac{\partial(y,z)}{\partial(u,v)}\right]^2 + \left[\frac{\partial(z,x)}{\partial(u,v)}\right]^2 + \left[\frac{\partial(x,y)}{\partial(u,v)}\right]^2} \, du\,dv = \text{area}_{\mathbf{f}}(S). \qquad \square$$

Definition 10.5 *Let* $\mathbf{f} : \Omega \to \mathbb{R}^3 : (u,v) \to (x,y,z)$ *be any parametrization of the surface patch* $S = \mathbf{f}(U)$; *then the* **surface area of** S *is*   Surface area of S

$$A(S) = \iint_{U} \sqrt{\left[\frac{\partial(y,z)}{\partial(u,v)}\right]^2 + \left[\frac{\partial(z,x)}{\partial(u,v)}\right]^2 + \left[\frac{\partial(x,y)}{\partial(u,v)}\right]^2} \, du\,dv.$$

Although the value of $A(S)$ is independent of the parametrization used to compute it, our expression for $A(S)$ is still bound to a parametrization. As with total flux, the invariance of $A(S)$ would be reflected better by an intrinsic formula, one not bound to a parametrization. We can get that formula by looking at the areas of small cells on S.

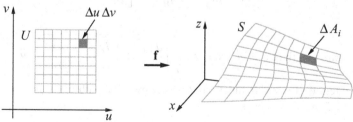

Let $\mathbf{f} : \Omega \to \mathbb{R}^3$ be a parametrization of S with $\mathbf{f}(U) = S$, and let Q_i be a cell of the grid $\mathcal{I}_k$ that meets U. Suppose the image $\mathbf{f}(Q_i)$ has area ΔA_i as given by Definition 10.5. Then, using the local area muliplier for $\mathbf{f}$ (p. 406), we have

$$\Delta A_i \approx \sqrt{\left[\frac{\partial(y,z)}{\partial(u,v)}\right]^2 + \left[\frac{\partial(z,x)}{\partial(u,v)}\right]^2 + \left[\frac{\partial(x,y)}{\partial(u,v)}\right]^2}\Bigg|_{(u_i,v_i)} \Delta u \, \Delta v,$$

$$\text{so} \quad \sum_{i=1}^{I} \Delta A_i \approx \sum_{i=1}^{I} \sqrt{\left[\frac{\partial(y,z)}{\partial(u,v)}\right]^2 + \left[\frac{\partial(z,x)}{\partial(u,v)}\right]^2 + \left[\frac{\partial(x,y)}{\partial(u,v)}\right]^2}\Bigg|_{(u_i,v_i)} \Delta u \, \Delta v.$$

An intrinsic formula for surface area

These sums both converge to $A(S)$. We write the limit on the left as an integral, following the usual pattern:

$$\lim_{\substack{I \to \infty \\ \Delta A_i \to 0}} \sum_{i=1}^{I} \Delta A_i = \iint_S dA.$$

This gives us the simple intrinsic expression

$$A(S) = \iint_S dA$$

for the surface area of S. Comparing the intrinsic with the parametric expression for $A(S)$, we can see why

$$dA = \sqrt{\left[\frac{\partial(y,z)}{\partial(u,v)}\right]^2 + \left[\frac{\partial(z,x)}{\partial(u,v)}\right]^2 + \left[\frac{\partial(x,y)}{\partial(u,v)}\right]^2} \, du \, dv$$

is described as the *element of surface area* on S.

Comparison with arc length

There are striking similarities between surface area and arc length. If the path C in $\mathbb{R}^3$ is parametrized by

$$\mathbf{f}(u) = (x(u), y(u), z(u)), \quad a \le u \le b,$$

then the *element of arc length* on C is

$$ds = \sqrt{\left[\frac{dx}{du}\right]^2 + \left[\frac{dy}{du}\right]^2 + \left[\frac{dz}{du}\right]^2} \, du,$$

and

$$\text{arc length of } C = \int_C ds = \int_a^b \sqrt{\left[\frac{dx}{du}\right]^2 + \left[\frac{dy}{du}\right]^2 + \left[\frac{dz}{du}\right]^2} \, du,$$

For the integral of a scalar function $H(x,y,z)$ over the path C (Definition 1.6 and Theorem 1.5, p. 18), we have

$$\int_C H(x,y,z)\,ds = \int_a^b H(\mathbf{f}(u)) \sqrt{\left[\frac{dx}{du}\right]^2 + \left[\frac{dy}{du}\right]^2 + \left[\frac{dz}{du}\right]^2}\,du.$$

As noted after the proof of Theorem 1.5, the value of the integral on the left does not depend on either the orientation of C or the parametrization of C used in the integral on the right.

Definition 10.6 *Let S be a surface patch in $\mathbb{R}^3$, and let $H(x,y,z)$ be a continuous function defined on S. We set*

Surface integral of
a scalar function

$$\iint_S H(x,y,z)\,dA = \iint_U H(\mathbf{f}(u,v)) \sqrt{\left[\frac{\partial(y,z)}{\partial(u,v)}\right]^2 + \left[\frac{\partial(z,x)}{\partial(u,v)}\right]^2 + \left[\frac{\partial(x,y)}{\partial(u,v)}\right]^2}\,du\,dv,$$

where $\mathbf{f}:\Omega \to \mathbb{R}^3$ is a parametrization of $S = \mathbf{f}(U)$.

For the surface integral on the left to be well defined, its value must be independent of the parametrization used for S on the right.

Theorem 10.7. *Let $\mathbf{f}:\Omega \to \mathbb{R}^3$ and $\mathbf{g}:\Omega^* \to \mathbb{R}^3$ be two parametrizations of $S = \mathbf{f}(U) = \mathbf{g}(U^*)$, with $U \subset \Omega$ and $U^* \subset \Omega^*$. Then*

Invariance of the
scalar integral

$$\iint_U H(\mathbf{f}(u,v)) \sqrt{\left[\frac{\partial(y,z)}{\partial(u,v)}\right]^2 + \left[\frac{\partial(z,x)}{\partial(u,v)}\right]^2 + \left[\frac{\partial(x,y)}{\partial(u,v)}\right]^2}\,du\,dv$$

$$= \iint_{U^*} H(\mathbf{g}(s,t)) \sqrt{\left[\frac{\partial(y,z)}{\partial(s,t)}\right]^2 + \left[\frac{\partial(z,x)}{\partial(s,t)}\right]^2 + \left[\frac{\partial(x,y)}{\partial(s,t)}\right]^2}\,ds\,dt,$$

for any continuous function $H(x,y,z)$ defined on S.

Proof. See Exercise 10.7. □

For example, if $H = \rho$ is mass density at a point of S, then the integral of H is the total mass of S. If H is density of electric charge (a signed quantity that may be negative), then the integral of H is the total electric charge on S. If $H = \mathbf{V} \cdot \mathbf{n}$, where $\mathbf{V}$ is a flux density and $\mathbf{n}$ is an orienting unit normal for the oriented surface patch $\vec{S}$, then the integral of H over S is total flux of $\mathbf{V}$ through $\vec{S}$.

The gravitational field of a hollow sphere is yet another example of a surface integral, one that we now construct by adapting the work we did in Chapter 8.1 on the field of a flat plate. Let the sphere be a unit sphere S centered at the origin of (x,y,z)-space; suppose it has negligible thickness and has uniform density ρ (mass per unit area). By symmetry, it is enough to determine the vertical component of the gravitational field at a test point $\mathbf{a} = (0,0,a)$ on the positive z-axis. Symmetry guarantees that the x- and y-components of the field at $\mathbf{a}$ are zero; as we find, it matters whether the test point is inside or outside the sphere.

The gravitational field
of a hollow sphere

Let S be divided into small regions $S_1,\ldots,S_I$; suppose S_i has area ΔA_i and contains the point $\mathbf{p}_i = (x_i,y_i,z_i)$. Let

$$\mathbf{r}_i = \mathbf{p}_i - \mathbf{a} = (x_i, y_i, z_i - a)$$

be the vector from the test point to $\mathbf{p}_i$. Then the gravitational field on the test point that is due to the region S_i is approximately

$$\mathbf{f}_i \approx \frac{G\rho \, \Delta A_i}{\|\mathbf{r}_i\|^3} \, \mathbf{r}_i = \frac{G\rho \, \Delta A_i}{(x_i^2 + y_i^2 + (z_i - a)^2)^{3/2}} (x_i, y_i, z_i - a),$$

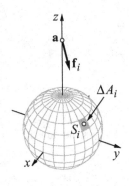

where G is the usual gravitational constant. We can therefore approximate the z-component of the gravitational field for the whole sphere S by the (scalar) sum

$$\text{field} \approx G\rho \sum_{i=1}^{I} \frac{z_i - a}{(x_i^2 + y_i^2 + (z_i - a)^2)^{3/2}} \Delta A_i.$$

Now let $I \to \infty$ and let the maximum diameter of S_i tend to zero; in the limit, the sum becomes the surface integral

$$\text{field} = G\rho \iint_S \frac{z - a}{(x^2 + y^2 + (z - a)^2)^{3/2}} \, dA = G\rho \iint_S \frac{z - a}{(1 + a^2 - 2az)^{3/2}} \, dA$$

(because $x^2 + y^2 + z^2 = 1$ on S).

To evaluate the surface integral, we use the parametrization

$$\begin{cases} x = \cos\theta \cos\varphi, \\ y = \sin\theta \cos\varphi, \\ z = \sin\varphi; \end{cases} \qquad U : \begin{array}{l} -\pi \le \theta \le \pi, \\ -\pi/2 \le \varphi \le \pi/2, \end{array}$$

of the unit sphere, keeping in mind the caveat we made when using the parametrization calculate total flux through the sphere (p. 396). From that earlier work, we find

$$dA = \sqrt{\left[\frac{\partial(y, z)}{\partial(\theta, \varphi)}\right]^2 + \left[\frac{\partial(z, x)}{\partial(\theta, \varphi)}\right]^2 + \left[\frac{\partial(x, y)}{\partial(\theta, \varphi)}\right]^2} \, d\theta \, d\varphi = \cos\varphi \, d\theta \, d\varphi.$$

We can now compute the field (e.g., using a table of integrals or a computer algebra system):

$$\text{field} = G\rho \int_{-\pi}^{\pi} d\theta \int_{-\pi/2}^{\pi/2} \frac{(\sin\varphi - a)\cos\varphi}{(1 + a^2 - 2a\sin\varphi)^{3/2}} \, d\varphi$$

$$= 2\pi G\rho \cdot \frac{1 - a\sin\varphi}{a^2 \sqrt{1 + a^2 - 2a\sin\varphi}} \Bigg|_{-\pi/2}^{\pi/2}$$

$$= \frac{2\pi G\rho}{a^2} \left(\frac{1 - a}{\sqrt{(1 - a)^2}} - \frac{1 + a}{\sqrt{(1 + a)^2}} \right) = \frac{2\pi G\rho}{a^2} \left(\frac{1 - a}{|1 - a|} - \frac{1 + a}{|1 + a|} \right).$$

By assumption, $a \geq 0$; therefore $|1+a| = 1+a$ and $(1+a)/|1+a| = 1$. The first term is more interesting:

$$\frac{1-a}{|1-a|} = \begin{cases} 1 & \text{if } a < 1, \\ -1 & \text{if } a > 1. \end{cases}$$

Therefore,

$$\text{field} = \begin{cases} 0 & \text{if } a < 1, \\ -\dfrac{4\pi G\rho}{a^2} & \text{if } a > 1. \end{cases}$$

Inside, the sphere induces no gravitational field whatsoever. Outside, the sphere acts as if all its mass $4\pi\rho$ were concentrated at the origin. On the sphere itself, the field is discontinuous. When $a = 1$, the value of the field is the average of its inside and outside values; see Exercise 10.8.

The field vanishes inside the sphere

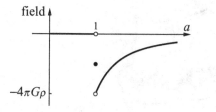
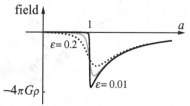

The discontinuity occurs where the z-axis passes through the sphere. If we put small holes at the north and south poles, then there is no matter on the z-axis, so the field should become continuous. The graphs on the right, above, show what happens. To determine the new field, we can use the same surface integral, parametrized the same way, but with φ restricted to

A sphere with a small hole at each pole

$$-\pi/2 + \varepsilon \leq \varphi \leq \pi/2 - \varepsilon,$$

where $\varepsilon > 0$ is some small number. Thus,

$$\text{field} = 2\pi G\rho \left. \frac{1 - a\sin\varphi}{a^2\sqrt{1+a^2 - 2a\sin\varphi}} \right|_{-\pi/2+\varepsilon}^{\pi/2-\varepsilon}$$

$$= \frac{2\pi G\rho}{a^2} \left(\frac{1 - a\cos\varepsilon}{\sqrt{1+a^2 - 2a\cos\varepsilon}} - \frac{1 + a\cos\varepsilon}{\sqrt{1+a^2 + 2a\cos\varepsilon}} \right).$$

(Note that $\sin(\pm(\pi/2 - \varepsilon)) = \pm\sin(\pi/2 - \varepsilon) = \pm\cos\varepsilon$.) This is indeed a continuous function of a; the graphs show $\varepsilon = 0.2$, 0.075, and 0.01. It is evident that the field strength fades away inside the sphere as the holes close up (i.e., as $\varepsilon \to 0$), and the field develops a discontinuity where the z-axis meets the sphere.

There is also an ingenious geometric argument (see, for e.g., the Feynman Lectures [6]) that explains why the field vanishes inside a hollow sphere. However, that argument relies on the symmetry of the sphere and cannot be easily modified when

we break the symmetry with holes at the poles. By contrast, the surface integral for
the field still works.

Limitations of
surface patches

Not every surface can be represented as a single surface patch. For one thing, the
parametrization that defines a patch is 1–1, and the domain of the parametrization
has a boundary; therefore the patch itself must have a boundary. A surface without
boundary (e.g., a sphere or a torus) cannot be a surface patch. Furthermore, because
a surface patch has a well-defined tangent plane at each point, no surface with edges
or corners (e.g., a cube) can be a surface patch, either. However, because each of
these examples can be assembled from a finite collection of surface patches, we are
led to define a surface S as a union of surface patches $S_1, \ldots, S_k$ satisfying certain
conditions.

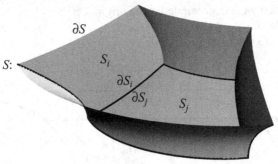

A union of
surface patches

To identify those conditions, we can be guided by the surface S illustrated above.
For a start, we must have
$$S = S_1 \cup S_2 \cup \cdots \cup S_k,$$
where each S_i is a surface patch defined by a continuously differentiable parametrization
$$\mathbf{f}_i : \Omega_i \to \mathbb{R}^3, \quad U_i \subset \Omega_i, \quad S_i = \mathbf{f}_i(U_i), \quad \partial S_i = \mathbf{f}_i(\partial U_i).$$
Recall (Definition 10.2, p. 392; see also p. 405) that the domain of a parametrization
is always an open set Ω_i that extends beyond the closed bounded set U_i that is
mapped to the patch S_i.

How the patches
fit together

To ensure that the patches fit together properly, we require that each boundary
∂U_i, and hence each ∂S_i, is a piecewise-smooth closed curve (cf. p. 9), made up of
a finite number of smooth arcs. Then we require that any two patches S_i and S_j meet
only along their boundaries, and that any three patches meet in, at most, a finite
number of isolated points. As the figure shows, some arcs are part of the boundary
of two patches; the remaining arcs, each of which lies in only one patch, together
form the boundary, ∂S, of S. In the following definition, no orientation is assumed,
and "smooth" is used in the sense of *continuously differentiable*.

Definition 10.7 *A subset S of $\mathbb{R}^3$ is a **piecewise-smooth surface** if it can be decomposed into a finite number of surface patches that fit together as above.*

For example, we can decompose the unit sphere into four surface patches:
S_1, eastern belt; S_2, western belt; S_3, northern cap; S_4, southern cap. The figure

in the margin shows the first three patches, separated slightly for clarity. Latitude and longitude parametrize the belts S_1 and S_2; only the longitude ranges differ:

$$\mathbf{f}_{1,2}: \begin{cases} x = \cos\theta\cos\varphi, \\ y = \sin\theta\cos\varphi, \\ z = \sin\varphi; \end{cases} \quad U_1: \begin{array}{l} 0 \le \theta \le \pi, \\ -\alpha \le \varphi \le \alpha; \end{array} \quad U_2: \begin{array}{l} -\pi \le \theta \le 0, \\ -\alpha \le \varphi \le \alpha. \end{array}$$

The angle $0 < \alpha < \pi/2$ gives the latitude of the northern boundary of the belts. The polar caps are just the graphs of

$$z = \pm\sqrt{1 - x^2 - y^2}, \quad U_{3,4}: x^2 + y^2 \le \cos^2\alpha.$$

A surface can be decomposed into a finite number of surface patches in many ways. Another way to decompose the unit sphere (cf. p. 397) uses just its northern and southern hemispheres ($S_\pm$):

$$\mathbf{g}_\pm: \begin{cases} x = \dfrac{2u}{1 + u^2 + v^2}, \\[2mm] y = \dfrac{2v}{1 + u^2 + v^2}, \\[2mm] z = \pm\dfrac{1 - u^2 - v^2}{1 + u^2 + v^2}; \end{cases} \quad U_\pm: u^2 + v^2 \le 1.$$

Because the definition of the integral of a scalar function on a surface patch (Definition 10.6) does not depend on any orientation of the patch, we can use it to define the integral of a scalar function on an unoriented piecewise-smooth surface.

Integrating a scalar function on a piecewise smooth surface

Definition 10.8 *Suppose $H(x,y,z)$ is a continuous function defined on a piecewise-smooth surface S; if $S = S_1 \cup \cdots \cup S_k$ is a decomposition of S into surface patches, then*

$$\iint_S H(x,y,z)\,dA = \sum_{i=1}^{k} \iint_{S_i} H(x,y,z)\,dA.$$

In particular, if $H(x,y,z) \equiv 1$, then the integrals define surface area:

$$\iint_S dA = \text{area}\,S = \sum_{i=1}^{k} \text{area}\,S_i.$$

A piecewise-smooth surface S has many different decompositions into surface patches; therefore an integral over S will be well defined only if its value is independent of that decomposition.

The integral is independent of the decomposition

Theorem 10.8. *Suppose $S = S_1 \cup \cdots \cup S_k = T_1 \cup \cdots \cup T_m$ gives two decompositions of the piecewise-smooth surface S into surface patches. Then*

$$\sum_{i=1}^{k} \iint_{S_i} H(x,y,z)\,dA = \sum_{j=1}^{m} \iint_{T_j} H(x,y,z)\,dA.$$

Proof. Suppose S_i is parametrized by $(x,y,z) = \mathbf{f}_i(u,v)$, with $\mathbf{f}_i(U_i) = S_i$ and

$$J_i(u,v) = \sqrt{\left[\frac{\partial(y,z)}{\partial(u,v)}\right]^2 + \left[\frac{\partial(z,x)}{\partial(u,v)}\right]^2 + \left[\frac{\partial(x,y)}{\partial(u,v)}\right]^2},$$

the local area magnification factor for $\mathbf{f}_i$. Then, by definition,

$$\iint_{S_i} H\,dA = \iint_{U_i} H(\mathbf{f}_i(u,v))J_i(u,v)\,du\,dv.$$

Let $R_{ij} = S_i \cap T_j$; this is a "common refinement" (cf. p. 300) of the decompositions given by S_i and by T_j. (Of course, some sets R_{ij} may be empty.) Because $\mathbf{f}_i$ is continuous and 1–1 on U_i, the sets

$$(U_i)_j = \mathbf{f}_i^{-1}(R_{ij}) \subseteq U_i$$

are closed, bounded, and have area. Also, because $S_i = \cup_{j=1}^m R_{ij}$ and the R_{ij} are nonoverlapping,

$$U_i = \cup_{j=1}^m (U_i)_j$$

is a decomposition into nonoverlapping sets. Therefore, when i is fixed, each R_{ij} with $j = 1,\ldots,m$ is a (possibly empty) surface patch parametrized by $\mathbf{f}_i$, with $R_{ij} = \mathbf{f}_i((U_i)_j)$. Hence,

$$\iint_{U_i} H(\mathbf{f}_i(u,v))J_i(u,v)\,du\,dv = \sum_{j=1}^m \iint_{(U_i)_j} H(\mathbf{f}_i(u,v))J_i(u,v)\,du\,dv$$

$$= \sum_{j=1}^m \iint_{R_{ij}} H\,dA,$$

and thus

$$\sum_{i=1}^k \iint_{S_i} H\,dA = \sum_{i=1}^k \sum_{j=1}^m \iint_{R_{ij}} H\,dA.$$

A similar argument, beginning with T_j, shows that

$$\sum_{j=1}^m \iint_{T_j} H\,dA = \sum_{j=1}^m \sum_{i=1}^k \iint_{R_{ij}} H\,dA. \qquad \Box$$

From piecewise-smooth to smooth

By definition, a piecewise-smooth surface can have edges and corners; at such points, the surface fails to be smooth, that is, to have a well-defined tangent plane. But edges and corners can occur only where two surface patches of a decomposition meet. Thus, in our first decomposition above, the unit sphere could fail to be smooth only at a point on the boundary of one of the belts (S_1, S_2) or one of the caps (S_3, S_4). However, these are all interior points of the hemispheres $S_\pm$ of our second decomposition (except for the two points on the x-axis), so they must be smooth points of the sphere, after all. (The two points on the x-axis are interior points of a third decomposition that uses the polar caps and the equatorial belts rotated 90° around the z-axis.) Because every point on the sphere is interior to some surface patch in a decomposition, the sphere has no edges or corners: it is smooth.

Definition 10.9 *A piecewise-smooth surface S is **smooth** if every point of S that is not in ∂S is an interior point of a surface patch that appears in some decomposition of S.*

Smooth surfaces

We turn now to the question of *orientation*. On pages 388–389, we described two equivalent ways to orient a surface S in space. First, to each point $\mathbf{p}$ in S we assigned an ordered pair of linearly independent tangent vectors $\{\mathbf{v}_1(\mathbf{p}), \mathbf{v}_2(\mathbf{p})\}$ to S at $\mathbf{p}$ in such a way that each $\mathbf{v}_i(\mathbf{p})$ varied continuously with $\mathbf{p}$. Second, we assigned to $\mathbf{p}$ one of the unit normals $\mathbf{n}(\mathbf{p})$ to S at $\mathbf{p}$ in such a way that $\mathbf{n}(\mathbf{p})$ varied continuously with $\mathbf{p}$.

Ways to orient
a surface

A particular surface may admit no orientation whatsoever. One impediment is the presence of an edge or a corner, where there is no well-defined tangent plane or normal vector. It would appear, then, that a surface that is piecewise smooth but not smooth cannot be oriented. In fact, we see below that this impediment can sometimes be overcome. A different sort of impediment affects even some smooth surfaces; the overall shape of the surface may preclude orientation. Perhaps the simplest example is the *Möbius strip*.

Impediments to
orientation

The following **Möbius strip** is the smooth surface M formed by the union of two surface patches M_1 and M_2 parametrized by the same functions and with adjoining domains U_1 and U_2:

The Möbius strip

$$\mathbf{f} : \begin{cases} x = (5 - v\cos u)\cos 2u, \\ y = (5 - v\cos u)\sin 2u, \\ z = -v\sin u; \end{cases} \quad U_1 : \begin{array}{l} 0 \le u \le \pi/2, \\ -1 \le v \le 1; \end{array} \quad U_2 : \begin{array}{l} \pi/2 \le u \le \pi, \\ -1 \le v \le 1. \end{array}$$

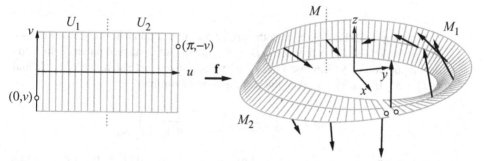

The boundary points $(0, v)$ and $(\pi, -v)$ have the same image (although the figure shows them separated slightly for clarity):

$$\mathbf{f}(0, v) = (5 - v, 0, 0) = \mathbf{f}(\pi, -v).$$

In effect, the rectangles U_1 and U_2 form a single ribbon that $\mathbf{f}$ bends into a loop, joining one end to the other after giving the ribbon a half-twist. Thus, if we follow a continuously varying normal vector along the center of the ribbon, we find that the normals at the two ends, on the seam where the ends of the ribbon join, point in opposite directions.

The Möbius strip
cannot be oriented

This direction reversal is shown by the parametrization normal $N_{\mathbf{f}}(u,v)$, whose components are the continuous functions

$$\frac{\partial(y,z)}{\partial(u,v)} = -v\sin 2u - 2(5 - v\cos u)\sin u\cos 2u,$$

$$\frac{\partial(z,x)}{\partial(u,v)} = v\cos 2u - 2(5 - v\cos u)\sin u\sin 2u,$$

$$\frac{\partial(x,y)}{\partial(u,v)} = 2(5 - v\cos u)\cos u.$$

At each point $\mathbf{f}(0,v) = \mathbf{f}(\pi,-v)$ on the seam, the parametrization normals are in conflict:

$$N_{\mathbf{f}}(\pi,-v) = (0,-v,-2(5-v)) = -N_{\mathbf{f}}(0,v).$$

It is therefore impossible to define a nonzero normal that varies continuously over all of M: the Möbius strip cannot be oriented.

Nevertheless, it is still possible to integrate a scalar function over M. In particular, M has a well-defined area (that we can compute using a numerical integrator, for example):

$$\text{area } M = \iint_{M} dA = \iint_{U_1+U_2} \sqrt{\left[\frac{\partial(y,z)}{\partial(u,v)}\right]^2 + \left[\frac{\partial(z,x)}{\partial(u,v)}\right]^2 + \left[\frac{\partial(x,y)}{\partial(u,v)}\right]^2}\, du\, dv$$

$$= \int_{-1}^{1}\int_{0}^{\pi} \sqrt{v^2 + 4(5 - v\cos u)^2}\, du\, dv$$

$$\approx 62.9377.$$

Unit normals on
a smooth surface

Now suppose that S is a smooth surface and $\mathbf{p}$ is a point in the interior of S— that is, in $S \setminus \partial S$. By definition, $\mathbf{p}$ is in the interior of a surface patch S^* in some decomposition of S. Suppose $\mathbf{f} : \Omega \to \mathbb{R}^3$ parametrizes that patch, with $U \subset \Omega$ and $\mathbf{f}(U) = S^*$. If $f(a,b) = \mathbf{p}$, then the parametrization normal

$$N_{\mathbf{f}}(a,b) = \left(\frac{\partial(y,z)}{\partial(u,v)}, \frac{\partial(z,x)}{\partial(u,v)}, \frac{\partial(x,y)}{\partial(u,v)}\right)_{(a,b)}$$

is nonzero and is normal to S at $\mathbf{p}$. From it we can construct the two unit normals $\pm\mathbf{n}(\mathbf{p})$ to S at $\mathbf{p}$.

Orientability of a
smooth surface

Definition 10.10 *Suppose S is a smooth surface that is pathwise connected; S is said to be **orientable** if a unit normal vector* $\mathbf{n}(\mathbf{p})$ *can be chosen that varies continuously over all points* $\mathbf{p}$ *in* $S \setminus \partial S$. *Such an assignment is called **an orientation of S**; we say that **S is oriented**, and write* $\vec{S}$.

A surface is pathwise connected if any two points can be joined by a continuous path that lies on the surface. Because an orientable surface S is pathwise connected, the orienting normal at any point determines the orienting normal at every other point, by continuity. Thus, S has two just orientations, which we can denote as $\vec{S}$ and $-\vec{S}$.

Every surface patch is orientable, by definition. Any surface patch S_i in a decomposition of a smooth oriented surface $\vec{S}$ has an orientation induced as a subset of that surface. We write

$$\vec{S} = \vec{S}_1 + \cdots + \vec{S}_k$$

to represent a decomposition of a smooth oriented surface into surface patches with their induced orientations. We can now define the total flux of of a vector field through a smooth oriented surface.

Surface integrals

Definition 10.11 *If* $\vec{S}$ *is a smooth oriented surface and* $\mathbb{V} = (X, Y, Z)$ *is a vector field defined on* $\vec{S}$, *then the **surface integral of** $\mathbb{V}$ **over** $\vec{S}$ is*

$$\iint_{\vec{S}} X\,dy\,dz + Y\,dz\,dx + Z\,dx\,dy = \sum_{i=1}^{k} \iint_{\vec{S}_i} X\,dy\,dz + Y\,dz\,dx + Z\,dx\,dy,$$

where $\vec{S} = \vec{S}_1 + \cdots + \vec{S}_k$ *is a decomposition into oriented surface patches.*

An individual surface integral on the right is computed using a parametrization of the given oriented surface patch. However, the surface integral over $\vec{S}$ will be well defined only if the sum on the right is independent of the decompositon of $\vec{S}$ into oriented surface patches. The following theorem ensures this; it is similar to Theorem 10.8 and can be proven that same way.

Theorem 10.9. *Suppose* $\vec{S} = \vec{S}_1 + \cdots + \vec{S}_k = \vec{T}_1 + \cdots + \vec{T}_m$ *gives two decompositions of the smooth oriented surface S into oriented surface patches. Then*

$$\sum_{i=1}^{k} \iint_{\vec{S}_i} X\,dy\,dz + Y\,dz\,dx + Z\,dx\,dy = \sum_{j=1}^{m} \iint_{\vec{T}_j} X\,dy\,dz + Y\,dz\,dx + Z\,dx\,dy. \qquad \square$$

Radial flow out of
a sphere, again

To illustrate, let us go back and recalculate the total flux of the field $\mathbb{V} = (Cx, Cy, Cz)$ through the unit sphere $\vec{S}$ oriented by its outward normal (Example 1, p. 396), decomposing $\vec{S}$ (as on p. 412) into an eastern belt $\vec{S}_1$, a western belt $\vec{S}_2$, a northern cap $\vec{S}_3$, and a southern cap $\vec{S}_4$. For the belts $\vec{S}_1$ and $\vec{S}_2$, we can just use the parametrization from Example 1, replacing the domain $-\pi/2 \le \varphi \le \pi/2$ by $-\alpha \le \varphi \le \alpha$:

$$\iint_{\vec{S}_1 + \vec{S}_2} Cx\,dy\,dz + Cy\,dz\,dx + Cz\,dx\,dy = \int_{-\pi}^{\pi} C\,d\theta \int_{-\alpha}^{\alpha} \cos\varphi\,d\varphi = (2\pi C)(2\sin\alpha).$$

We parametrize the northern cap $\vec{S}_3$ as

$$\mathbf{f}_3 : \begin{cases} x = u, \\ y = v, \\ z = \sqrt{1 - u^2 - v^2}, \end{cases} \qquad \vec{U}_3 : u^2 + v^2 \leq \cos^2 \alpha;$$

then

$$\frac{\partial(y,z)}{\partial(u,v)} = \begin{vmatrix} 0 & 1 \\ -u/z & -v/z \end{vmatrix} = \frac{u}{z}, \quad \frac{\partial(z,x)}{\partial(u,v)} = \begin{vmatrix} -u/z & -v/z \\ 1 & 0 \end{vmatrix} = \frac{v}{z}, \quad \frac{\partial(x,y)}{\partial(u,v)} = \begin{vmatrix} 1 & 0 \\ 0 & 1 \end{vmatrix} = 1,$$

and

$$Cx\,dy\,dz + Cy\,dz\,dx + Cz\,dx\,dy = \left(\frac{Cu^2}{z} + \frac{Cv^2}{z} + Cz \right) du\,dv = \frac{C}{z}\,du\,dv.$$

Therefore (introducing polar coordinates $u = \rho \cos \theta$, $v = \rho \sin \theta$),

$$\iint\limits_{\vec{S}_3} Cx\,dy\,dz + Cy\,dz\,dx + Cz\,dx\,dy = \iint\limits_{u^2+v^2 \leq \cos^2 \alpha} \frac{C}{\sqrt{1 - u^2 - v^2}}\,du\,dv$$

$$= C \int_0^{2\pi} d\theta \int_0^{\cos \alpha} \frac{\rho\,d\rho}{\sqrt{1 - \rho^2}} = 2\pi C \left(-\sqrt{1 - \rho^2} \,\Big|_0^{\cos \alpha} \right) = 2\pi C (1 - \sin \alpha).$$

If we were to parametrize the southern cap $\vec{S}_4$ by just changing the sign of z,

$$\mathbf{g}_4 : \begin{cases} x = u, \\ y = v, \\ z = -\sqrt{1 - u^2 - v^2}, \end{cases} \qquad \vec{U}_4 : u^2 + v^2 \leq \cos^2 \alpha,$$

then we would have

$$\frac{\partial(y,z)}{\partial(u,v)} = \begin{vmatrix} 0 & 1 \\ -u/z & -v/z \end{vmatrix} = \frac{u}{z}, \quad \frac{\partial(z,x)}{\partial(u,v)} = \begin{vmatrix} -u/z & -v/z \\ 1 & 0 \end{vmatrix} = \frac{v}{z}, \quad \frac{\partial(x,y)}{\partial(u,v)} = \begin{vmatrix} 1 & 0 \\ 0 & 1 \end{vmatrix} = 1.$$

These are the components of the orientation normal

$$N_{\mathbf{g}_4} = \left(\frac{u}{z}, \frac{v}{z}, 1 \right) = \frac{1}{z}(u, v, z) = \frac{1}{z}(x, y, z).$$

$\mathbf{g}_4$ parametrizes
$-\vec{S}_4$, not $+\vec{S}_4$

But because $1/z < 0$, this is a negative multiple of the radius vector (x, y, z) at a point on $\vec{S}_4$, and hence points inward. However, the orientation of $\vec{S}$ requires an outward normal here. The remedy is to reverse u and v:

$$\mathbf{f}_4 : \begin{cases} x = v, \\ y = u, \\ z = -\sqrt{1 - u^2 - v^2}, \end{cases} \qquad \vec{U}_4 : u^2 + v^2 \leq \cos^2 \alpha.$$

Now

$$\frac{\partial(y,z)}{\partial(u,v)} = \begin{vmatrix} 1 & 0 \\ -u/z & -v/z \end{vmatrix} = -\frac{v}{z}, \quad \frac{\partial(z,x)}{\partial(u,v)} = \begin{vmatrix} -u/z & -v/z \\ 0 & 1 \end{vmatrix} = -\frac{u}{z},$$

$$\frac{\partial(x,y)}{\partial(u,v)} = \begin{vmatrix} 0 & 1 \\ 1 & 0 \end{vmatrix} = -1,$$

and

$$N_{\mathbf{f}_4} = \left(-\frac{v}{z}, -\frac{u}{z}, -1 \right) = -\frac{1}{z}(v,u,z) = -\frac{1}{z}(x,y,z).$$

This a positive multiple of the radius vector (x,y,z), and hence an outward normal. Furthermore,

$$Cx\,dy\,dz + Cy\,dz\,dx + Cz\,dx\,dy = \left(-\frac{Cv^2}{z} - \frac{Cu^2}{z} - Cz \right) du\,dv$$

$$= -\frac{C}{z}du\,dv = +\frac{C}{\sqrt{1-u^2-v^2}}du\,dv,$$

as with $\vec{S}_3$, so

$$\iint_{\vec{S}_4} Cx\,dy\,dz + Cy\,dz\,dx + Cz\,dx\,dy = 2\pi C(1 - \sin\alpha).$$

Total flux through $\vec{S}_4$ equals total flux through $\vec{S}_3$, as is already evident by symmetry. Total flux out of the whole sphere is therefore

$$\sum_{i=1}^{4} \iint_{\vec{S}_i} Cx\,dy\,dz + Cy\,dz\,dx + Cz\,dx\,dy = 4\pi C,$$

precisely the value we found earlier, when we assumed the whole sphere could be treated as if it were a single surface patch. Although this analysis does not justify that assumption, it does show why the earlier computation worked. As $\alpha \to \pi/2$, $\sin\alpha \to 1$; therefore, total flux through the polar caps approaches 0 and total flux through the two belts approaches $4\pi C$. These conclusions are also clear on physical grounds.

Why the earlier computation worked

At an edge or a corner, a piecewise-smooth surface does not have a well-defined normal (or a pair of linearly independent tangent vectors), so it cannot be oriented the same way as a smooth surface. However, a surface patch is always orientable, and because the boundary of any surface patch used in a decomposition of a piecewise-smooth surface is a piecewise-smooth curve, the orientation of the

Orientation with edges and corners

patch induces an orientation of its boundary. As we see in the figure below, it may be possible to orient all the surface patches in a decomposition so that their common boundary arcs have opposite orientations and thus cancel each other.

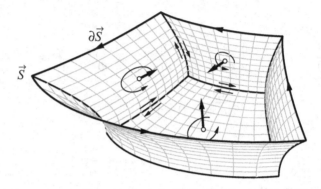

Orientability of a piecewise-smooth surface

Definition 10.12 *Suppose S is a piecewise-smooth surface that is pathwise connected, and $S = \vec{S}_1 \cup \cdots \cup \vec{S}_k$ is a decomposition into oriented surface elements. Suppose that whenever two surface elements $\vec{S}_i$ and $\vec{S}_j$ have a common boundary arc, $\partial \vec{S}_i$ and $\partial \vec{S}_j$ have opposite orientations there. Then we say $\boldsymbol{S}$ is orientable and is oriented by those surface elements. We write*

$$\vec{S} = \vec{S}_1 + \cdots + \vec{S}_k, \quad \partial \vec{S} = \partial \vec{S}_1 + \cdots + \partial \vec{S}_k.$$

The equation for $\partial \vec{S}$ reflects the cancellations that occur on the arcs that pairs of different $\partial \vec{S}_i$ have in common. The unpaired arcs that remain have a well-defined orientation and make up the oriented boundary of $\vec{S}$. The surface integral of a vector field over a piecewise-smooth oriented surface is defined exactly as for a smooth oriented surface; moreover, the definition is independent of the way the surface is decomposed into patches.

Surface integrals

Definition 10.13 *If $\vec{S}$ is a piecewise-smooth oriented surface and $\mathbb{V} = (X, Y, Z)$ is a vector field defined on $\vec{S}$, then the **surface integral of** $\mathbb{V}$ **over** $\vec{S}$ is*

$$\iint_{\vec{S}} X\,dy\,dz + Y\,dz\,dx + Z\,dx\,dy = \sum_{i=1}^{k} \iint_{\vec{S}_i} X\,dy\,dz + Y\,dz\,dx + Z\,dx\,dy,$$

where $\vec{S} = \vec{S}_1 + \cdots + \vec{S}_k$ is a decomposition into oriented surface patches.

Theorem 10.10. *Suppose $\vec{S} = \vec{S}_1 + \cdots + \vec{S}_k = \vec{T}_1 + \cdots + \vec{T}_m$ gives two decompositions of the piecewise-smooth oriented surface S into oriented surface patches. Then*

$$\sum_{i=1}^{k} \iint_{\vec{S}_i} X\,dy\,dz + Y\,dz\,dx + Z\,dx\,dy = \sum_{j=1}^{m} \iint_{\vec{T}_j} X\,dy\,dz + Y\,dz\,dx + Z\,dx\,dy. \qquad \square$$

To illustrate, let us determine the total flux of $\mathbb{V} = (x+y, y-x, 0)$ out of the unit cube in (x,y,z)-space. Thus, we orient each face of the cube with the outward normal, show this gives us an orientation of the surface $\vec{S}$ of the entire cube, integrate $\mathbb{V}$ over each face, and then add the results.

We can parametrize the six faces with affine functions, but it takes some care to get the orientations right. To parametrize $\vec{S}_{x=1}$, the face that lies in the plane $x = 1$, let $\vec{U}$ be the positively oriented unit square in the (u,v)-plane, and let $\mathbf{f}_{x=1} : \vec{U} \to \mathbb{R}^3$ be $(x,y,z) = (1,u,v)$. The orientation normal is

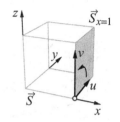

$$N_{x=1} = \left(\frac{\partial(y,z)}{\partial(u,v)}, \frac{\partial(z,x)}{\partial(u,v)}, \frac{\partial(x,y)}{\partial(u,v)} \right) = (1,0,0);$$

it does indeed point out of the cube on $\vec{S}_{x=1}$. Total flux through $\vec{S}_{x=1}$ is

$$\iint_{\vec{S}_{x=1}} (x+y)\,dy\,dz + (y-x)\,dz\,dx = \iint_{\vec{U}} (1+u) \times 1\,du\,dv = \int_0^1 dv \int_0^1 (1+u)\,du = \frac{3}{2}.$$

To parametrize the face $\vec{S}_{x=0}$ that lies in the plane $x = 0$, suppose we were to use $\mathbf{g} : (x,y,z) = (0,u,v)$ with (u,v) again in the oriented unit square $\vec{U}$. The orientation normal is the same as on $\vec{S}_{x=1}$,

$$N_{\mathbf{g}} = (1,0,0),$$

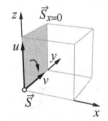

but $\vec{S}_{x=0}$ is on the other side of the cube, so $N_{\mathbf{g}}$ points into the cube. Thus $\mathbf{g}$ gives the wrong orientation. We get the correct orientation by reversing u and v: let $\mathbf{f}_{x=0} : (x,y,z) = (0,v,u)$. Then

$$\frac{\partial(y,z)}{\partial(u,v)} = -1,$$

leading to an orientation normal

$$N_{x=0} = (-1,0,0)$$

that points out of the cube. With $\mathbf{f}_{x=0}$ we find that total flux through $\vec{S}_{x=0}$ is

$$\iint_{\vec{S}_{x=0}} (x+y)\,dy\,dz + (y-x)\,dz\,dx = \iint_{\vec{U}} (0+v) \times -1\,du\,dv = \int_0^1 du \int_0^1 -v\,dv = -\frac{1}{2}.$$

To parametrize the face $\vec{S}_{y=1}$, a bit of experimentation suggests that we use $\mathbf{f}_{y=1} : (x,y,z) = (v,1,u)$. Then

$$\frac{\partial(z,x)}{\partial(u,v)} = \begin{vmatrix} 1 & 0 \\ 0 & 1 \end{vmatrix} = 1,$$

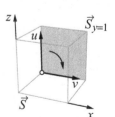

giving an orientation normal

$$N_{y=1} = (0,1,0)$$

that points out of the cube. Total flux through $\vec{S}_{y=1}$ is

$$\iint_{\vec{S}_{y=1}} (x+y)\,dy\,dz + (y-x)\,dz\,dx = \iint_{\vec{U}} (1-v)\times 1\,du\,dv = \int_0^1 du \int_0^1 (1-v)\,dv = \frac{1}{2}.$$

The parametrization $\mathbf{f}_{y=0} : (x,y,z) = (u,0,v)$ of the face $\vec{S}_{y=0}$ has the orientation normal

$$N_{y=0} = (0,-1,0)$$

that points out of the cube. Total flux through $\vec{S}_{y=0}$ is

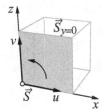

$$\iint_{\vec{S}_{y=0}} (x+y)\,dy\,dz + (y-x)\,dz\,dx = \iint_{\vec{U}} (0-u)\cdot -1\,du\,dv = \int_0^1 dv \int_0^1 u\,du = \frac{1}{2}.$$

We can parametrize $\vec{S}_{z=1}$ with $\mathbf{f}_{z=1} : (x,y,z) = (u,v,1)$ and $\vec{S}_{z=0}$ with $\mathbf{f}_{z=0} : (x,y,z) = (v,u,0)$. Then

$$N_{z=1} = (0,0,1), \quad N_{z=0} = (0,0,-1),$$

as required. However total flux is zero through both faces:

$$\iint_{\vec{S}_{z=1,0}} (x+y)\,dy\,dz + (y-x)\,dz\,dx = \iint_{\vec{U}} 0\,du\,dv = 0.$$

This is already clear because the flow is everywhere parallel to the (x,y)-plane, so $\mathbb{V}\cdot N_{z=1,0} = 0$. Addition now gives us the total flux out of the whole cubical surface $\vec{S}$:

$$\iint_{\vec{S}} (x+y)\,dy\,dz + (y-x)\,dz\,dx = 2.$$

Alternate form for a surface integral   The integral of the vector field $\mathbb{V} = (X,Y,Z)$ over an oriented piecewise-smooth surface $\vec{S}$ has the alternate form

$$\iint_S \mathbb{V}\cdot \mathbf{n}\,dA = \iint_{\vec{S}} X\,dy\,dz + Y\,dz\,dx + Z\,dx\,dy$$

that integrates the scalar function $\mathbb{V}\cdot \mathbf{n}$ over the unoriented surface S. If S is smooth, the unit normal $\mathbf{n}$ that appears here is the one that defines the orientation of $\vec{S}$. If S is only piecewise-smooth, then $\mathbf{n}$ is not defined everywhere. But if $\vec{S} = \vec{S}_1 + \cdots + \vec{S}_k$ is a decomposition into oriented surface patches, then we define

$$\iint_S \mathbb{V}\cdot \mathbf{n}\,dA = \sum_{i=1}^{k} \iint_{S_i} \mathbb{V}\cdot \mathbf{n}_i\,dA,$$

where $\mathbf{n}_i$ is the orienting unit normal on $\vec{S}_i$. (On the interior of $\vec{S}_i$, $\mathbf{n}_i$ is the orienting unit normal on $\vec{S}$, as well.)

10.3 Differential forms

The integrands of path and surface integrals, and of oriented single and double integrals, are *differential forms*. We generate new forms using algebraic operations and differentiation; in particular, these operations give us a simple connection between the forms that appear in the path and double integrals of Green's theorem. The books by H. Flanders [7] and H. Edwards [5] provide more extensive treatments of differential forms.

To fix ideas, we begin with differential forms in $\mathbb{R}^3$. In (x,y,z)-space, there are three "basic" differentials: dx, dy, and dz. A **differential k-form**, or an **exterior k-form**, or just a **k-form** $\alpha = \alpha(x,y,z)$, is a sum of "monomials" that contain exactly k of these differentials, as follows:

Forms in $\mathbb{R}^3$

k	general k-form
0	$g(x,y,z)$
1	$P(x,y,z)\,dx + Q(x,y,z)\,dy + R(x,y,z)\,dz$
2	$X(x,y,z)\,dy\,dz + Y(x,y,z)\,dz\,dx + Z(x,y,z)\,dx\,dy$
3	$H(x,y,z)\,dx\,dy\,dz$
> 3	0

A general k-form is thus a linear combination of certain **basic k-forms**

$$1,\ dx,\ dy,\ dz,\ dy\,dz,\ dz\,dx,\ dx\,dy,\ dx\,dy\,dz;$$

we require the coefficient functions to have continuous second derivatives. A 1-form is the integrand of a path integral, so it is integrated over an oriented 1-dimensional domain. A 2-form is integrated over an oriented 2-dimensional domain, and a 3-form is integrated over an oriented 3-dimensional domain. Even a 0-form fits this pattern; see below, pages 428–429.

The sum of two k-forms is another k-form in the usual way, and the product of a k-form by a function is another k-form. We do not define the sum of a k-form and an l-form when $k \neq l$; for one thing, such a sum could not be an integrand. However, we can define the product of a k-form α and an l-form β. It is a $(k+l)$-form $\alpha \wedge \beta$, called the **exterior**, or **wedge**, **product** of α and β. On the basic differentials, the exterior product is *anticommutative*:

Algebra; exterior product

$$dx \wedge dy = -dy \wedge dx = dx\,dy,$$
$$dy \wedge dz = -dz \wedge dy = dy\,dz,$$
$$dz \wedge dx = -dx \wedge dz = dz\,dx,$$
$$dx \wedge dx = dy \wedge dy = dz \wedge dz = 0.$$

Anticommutativity implies the last line; for example, interchanging the first dx with the second gives $dx \wedge dx = -dx \wedge dx$, so $2(dx \wedge dx) = 0$.

The definition says that each basic 2-form is just an exterior product; for example, $dx\,dy$ stands for $dx \wedge dy$. For the basic 3-form, we have

$$dx\,dy\,dz = dx \wedge dy \wedge dz = dy \wedge dz \wedge dx = dz \wedge dx \wedge dy$$
$$= -dy \wedge dx \wedge dz = -dz \wedge dy \wedge dx = -dx \wedge dz \wedge dy.$$

Because anticommutativity forces the exterior product of basic differentials to be zero unless they are distinct, there is no nonzero k-form in $\mathbb{R}^3$ when $k > 3$. Note that the exterior product is not anticommutative in all cases:

$$(dy\,dz) \wedge dx = dx\,dy\,dz = dx \wedge (dy\,dz).$$

For completeness, we define the exterior product with a 0-form—that is, an ordinary function $g = g(x,y,z)$—as

$$g \wedge \alpha = \alpha \wedge g = g(x,y,z)\,\alpha(x,y,z)$$

for any k-form α.

General products

We can now compute the exterior product of any two forms by using the distributive law. For the 1-forms

$$\alpha = \alpha(x,y,z) = P\,dx + Q\,dy + R\,dz, \quad \theta = \theta(x,y,z) = F\,dx + G\,dy + H\,dz,$$

we have

$$\alpha \wedge \theta = (P\,dx + Q\,dy + R\,dz) \wedge (F\,dx + G\,dy + H\,dz)$$
$$= PG\,dx \wedge dy + PH\,dx \wedge dz + QF\,dy \wedge dx$$
$$+ QH\,dy \wedge dz + RF\,dz \wedge dx + RG\,dz \wedge dy$$
$$= (QH - RG)\,dy\,dz + (RF - PH)\,dz\,dx + (PG - QF)\,dx\,dy.$$

A similar calculation of $\theta \wedge \alpha$ would then show that $\theta \wedge \alpha = -\alpha \wedge \theta$. For the 2-form

$$\beta = X\,dy\,dz + Y\,dz\,dx + Z\,dx\,dy,$$

the wedge product $\alpha \wedge \beta$ has only three nonzero terms:

$$\alpha \wedge \beta = (PX + QY + RZ)\,dx\,dy\,dz = \beta \wedge \alpha.$$

Integrating a differential

Suppose $\vec{C}$ is an oriented curve that we parametrize as

$$\mathbf{x}(t) = (x(t), y(t), z(t)), \quad a \le t \le b.$$

Consider the simple 1-form $\alpha = dx$ on $\vec{C}$; we have

$$\int_{\vec{C}} \alpha = \int_{\vec{C}} dx = \int_a^b x'(t)\,dt = x(t)\Big|_a^b = x(b) - x(a),$$

which is the change in x along $\vec{C}$:

$$\int_{\vec{C}} dx = \Delta x \text{ along } \vec{C}.$$

Now let $\alpha = g_x\,dx + g_y\,dy + g_z\,dz$, where g_x, g_y, and g_z are the partial derivatives of a continuously differentiable function $g(x,y,z)$. (We call g a *potential function* for α; cf. p. 25.) Then, using the chain rule to convert $g_x x' + g_y y' + g_z z'$ into $dg(\mathbf{x}(t))/dt$, we have

$$\int_{\vec{C}} g_x\,dx + g_y\,dy + g_z\,dz = \int_a^b (g_x x' + g_y y' + g_z z')\,dt$$

$$= \int_a^b \frac{d}{dt} g(\mathbf{x}(t))\,dt = g(\mathbf{x}(t))\Big|_a^b = g(\mathbf{x}(b)) - g(\mathbf{x}(a))$$

which is the change in g along $\vec{C}$. Analogy with the simple differential dx suggests we set

Differential of
a function

$$g_x\,dx + g_y\,dy + g_z\,dz = dg,$$

and call this the **differential of g**, because then we have

$$\int_{\vec{C}} dg = \Delta g \text{ along } \vec{C}.$$

Using the fact that dg is defined for any 0-form g, we now define the **differential**, or **exterior derivative**, $d\alpha$ of any k-form α. There are two rules. First, if $\alpha = g \wedge \beta$, where β is a basic k-form, then

Exterior derivative

$$d\alpha = dg \wedge \beta,$$

a $(k+1)$-form. Second, for any k-forms α and ω,

$$d(\alpha \pm \omega) = d\alpha \pm d\omega.$$

It follows that the exterior derivative of any k-form is a $(k+1)$-form.
For a general 1-form $\alpha = P\,dx + Q\,dy + R\,dz$, we have

$$\begin{aligned} d\alpha &= dP \wedge dx + dQ \wedge dy + dR \wedge dz \\ &= (P_x\,dx + P_y\,dy + P_z\,dz) \wedge dx + (Q_x\,dx + Q_y\,dy + Q_z\,dz) \wedge dy \\ &\quad + (R_x\,dx + R_y\,dy + R_z\,dz) \wedge dz \\ &= (R_y - Q_z)\,dy\,dz + (P_z - R_x)\,dz\,dx + (Q_x - P_y)\,dx\,dy. \end{aligned}$$

For a general 2-form $\omega = X\,dy\,dz + Y\,dz\,dx + Z\,dx\,dy$, the calculation is briefer:

$$d\omega = (X_x + Y_y + Z_z)\,dx\,dy\,dz.$$

For a 3-form $\gamma = H\,dx\,dy\,dz$, the exterior derivative $d\gamma$ is a 4-form, and hence is automatically 0.

$d^2 = 0$

Theorem 10.11. *For every k-form α in $\mathbb{R}^3$, $d^2\alpha = d(d\alpha) = 0$.*

Proof. Suppose α is a 0-form, $\alpha = g$; then

$$\begin{aligned}
d^2\alpha &= d(g_x\,dx + g_y\,dy + g_z\,dz)\\
&= (g_{xy}\,dy + g_{xz}\,dz)\,dx + (g_{yx}\,dx + g_{yz}\,dz)\,dy + (g_{zx}\,dx + g_{zy}\,dy)\,dz\\
&= (g_{zy} - g_{yz})\,dy\,dz + (g_{xz} - g_{zx})\,dz\,dx + (g_{yx} - g_{xy})\,dx\,dy\\
&= 0.
\end{aligned}$$

All the coefficients vanish by the "equality of mixed partials" for functions with continuous second derivatives.

Suppose α is a 1-form: $\alpha = P\,dx + Q\,dy + R\,dz$; then (see above)

$$d\alpha = (R_y - Q_z)\,dy\,dz + (P_z - R_x)\,dz\,dx + (Q_x - P_y)\,dx\,dy.$$

This is a 2-form whose exterior derivative is

$$\begin{aligned}
d^2\alpha &= \big((R_y - Q_z)_x + (P_z - R_x)_y + (Q_x - P_y)_z\big)\,dx\,dy\,dz\\
&= \big(R_{yx} - Q_{zx} + P_{zy} - R_{xy} + Q_{xz} - P_{yz}\big)\,dx\,dy\,dz\\
&= 0.
\end{aligned}$$

Again the "equality of mixed partials" implies that the coefficient vanishes. Finally, if α is a k-form with $k \geq 2$, then $d^2\alpha$ is a $(k+2)$-form and hence vanishes automatically. Thus $d^2 = 0$ on all differential forms. $\qquad\square$

Anticommutativity in $d^2 = 0$

Note that $d^2 = 0$ is a consequence of the anticommutativity of the exterior product on basic differentials. For example, in

$$d^2 g = d(g_x\,dx + g_y\,dy + g_z\,dz),$$

the first term contributes

$$g_{xy}\,dy \wedge dx$$

and the second contributes

$$g_{yx}\,dx \wedge dy = g_{xy}\,dx \wedge dy = -g_{xy}\,dy \wedge dx.$$

The anticommutativity, in turn, is a reflection of the fact that $dy \wedge dx$ represents an oriented element of area for a double integral, and $dx \wedge dy$ represents the element of area for the opposite orientation. All these relations are nicely illustrated by differential forms in the plane.

Differential forms in $\mathbb{R}^2$

In the (x,y)-plane, there are just four "basic differentials,"

$$1, \quad dx, \quad dy, \quad dx\,dy,$$

and the general 1-form and 2-form are, respectively,

$$\alpha(x,y) = P(x,y)\,dx + Q(x,y)\,dy \ \text{ and } \ \theta(x,y) = H(x,y)\,dx\,dy.$$

The exterior derivative is defined as for forms in $\mathbb{R}^3$; the exterior derivative of the 1-form $\alpha = P\,dx + Q\,dy$ is

$$\begin{aligned}
d\alpha &= dP \wedge dx + dQ \wedge dy \\
&= (P_x\,dx + P_y\,dy) \wedge dx + (Q_x\,dx + Q_y\,dy) \wedge dy \\
&= (Q_x - P_y)\,dx\,dy.
\end{aligned}$$

This means that Green's theorem,

$$\oint_{\partial\vec{R}} P\,dx + Q\,dy = \iint_{\vec{R}} (Q_x - P_y)\,dx\,dy,$$

becomes, in the language of differential forms,

$$\oint_{\partial\vec{R}} \alpha = \iint_{\vec{R}} d\alpha.$$

We can write this equation in an even more striking way by regarding an oriented integral as a function, or map, that assigns a number (the value of the integral) to each pair of objects of a particular sort: the first object is an oriented k-dimensional region $\vec{D}$; the second is a k-form ω. To emphasize how an integral is the "pairing" of a region and a form, let us write it in symbolic fashion as

$$\langle \vec{D}, \omega \rangle.$$

For example, $\vec{D}$ could be the interval $[a,b]$ and $\omega(x) = g(x)\,dx$; then

$$\langle \vec{D}, \omega \rangle = \langle [a,b], g(x)\,dx \rangle = \int_a^b g(x)\,dx.$$

But $\vec{D}$ could just as well be a piecewise-smooth oriented surface in space and $\omega(x,y,z) = X\,dy\,dz + Y\,dz\,dx + Z\,dx\,dy$; then (Definition 10.13)

$$\langle \vec{D}, \omega \rangle = \iint_{\vec{D}} X\,dy\,dz + Y\,dz\,dx + Z\,dx\,dy.$$

In terms of this symbolic pairing, Green's theorem has the form

$$\langle \partial\vec{R}, \alpha \rangle = \langle \vec{R}, d\alpha \rangle.$$

The operator d assigns the 2-form $d\alpha$ to the 1-form α; the operator ∂ (the symbol is a cursive "d" in the Cyrillic alphabet) assigns the 1-dimensional region $\partial\vec{R}$ to the 2-dimensional region $\vec{R}$. The symbolic content of Green's theorem is that each of these operators turns into the other when it "moves across" the pairing. In this context, we say that the operators d and ∂ are **adjoints**. The boundary operator is written as a d (albeit as a Russian ∂) because it is the adjoint of the exterior derivative, d.

The adjoint relation between ∂ and d supports the fact that $d^2 = 0$ (Theorem 10.11), because it is independently clear that $\partial^2 = 0$ (i.e., the boundary of a boundary is empty: $\partial(\partial D) = \emptyset$).

Green's theorem in dimension 1

Green's theorem, in its symbolic form $\langle \partial \vec{R}, \alpha \rangle = \langle \vec{R}, d\alpha \rangle$, has a remarkable 1-dimensional analogue. On the x-axis, there are just two basic differential forms, 1 and dx, generating the 0-forms $G(x)$ and 1-forms $g(x)\,dx$. In the fundamental theorem of calculus,

$$\int_a^b G'(x)\,dx = G(b) - G(a),$$

the left-hand side is the integral of the 1-form $d\alpha = G'\,dx$, where α itself is the 0-form $\alpha = G$. Can we make the right-hand side into a kind of "0-dimensional integral" of α?

0-dimensional regions

The basic 0-dimensional object is a single point. But the fundamental theorem involves a pair of points, the boundary points of $[a, b]$. To include this case, we take a 0-dimensional "region" to be any finite collection of points $D : \{a_1, a_2, \ldots, a_n\}$. Because a 0-form $G(x)$ takes a value on each of these points—and integrals represent sums—one possibility is to define the symbolic 0-dimensional integral to be

$$\langle D, G \rangle = G(a_1) + G(a_2) + \cdots + G(a_n).$$

However, this fails to produce the minus sign we see in $G(b) - G(a)$.

Orientation

To get the needed minus sign, we introduce orientation. Because $\vec{I} = [a, b]$ is itself oriented, we say it induces the orientation $\{-a, +b\}$ on $\partial \vec{I}$. The signs convey the orientation: the minus sign indicates where $\vec{I}$ begins; the plus sign where it ends.

The oppositely oriented $-\vec{I}$ induces the correct orientation of $-\partial \vec{I} : \{+a, -b\}$, regarded as $\partial \vec{I}$ with the opposite orientation. We convey the same information about $\partial \vec{I}$ if we write it not as a set but as a sum:

$$\partial \vec{I} = -a + b.$$

To see the advantage of this change, let

$$\vec{J} = [b, c], \quad \vec{K} = \vec{I} + \vec{J} = [a, b] + [b, c] = [a, c].$$

Then $\partial \vec{J} = -b + c$ and

$$\partial \vec{I} + \partial \vec{J} = -a + b - b + c = -a + c = \partial \vec{K}.$$

Also, $\partial(\vec{I} - \vec{I}) = -a + b + a - b = 0$, which is consistent with $\vec{I} - \vec{I} = 0$. We can therefore convert a general $D : \{a_1, \ldots, a_l\}$ into an oriented 0-dimensional region by attaching an integer m_i to each a_i, and writing the result as

$$\vec{D} = m_1 a_1 + \cdots + m_l a_l.$$

If we change either a point a_i or a "weight" m_i, then $\vec{D}$ is a different oriented 0-dimensional region. Finally, if G is a 0-form, we define the **0-dimensional oriented integral**

$$\langle \vec{D}, G \rangle = m_1 G(a_1) + \cdots + m_l G(a_l).$$

In these terms, the fundamental theorem of calculus takes the form

$$\langle \vec{I}, d\alpha \rangle = \langle \partial \vec{I}, \alpha \rangle,$$

Fundamental theorem:
$\langle \vec{I}, d\alpha \rangle = \langle \partial \vec{I}, \alpha \rangle$

where $\alpha = G$ is a 0-form and $\vec{I}$ is an oriented interval. The fundamental theorem and Green's theorem thus make the same assertion about a k-dimensional region and a k-form, only for different values of k. They are instances of a more general result, called Stokes' theorem, that we consider in the next chapter.

What happens to a differential form under a change of variables? For example, consider the change to polar coordinates with

Differential forms in
polar coordinates

$$\varphi : \begin{cases} x = r \cos \theta, \\ y = r \sin \theta. \end{cases}$$

Then, because we can treat x and y as functions of r and θ,

$$\begin{cases} dx = \cos \theta \, dr - r \sin \theta \, d\theta, \\ dy = \sin \theta \, dr + r \cos \theta \, d\theta. \end{cases}$$

The element of area, $dx\,dy$, is transformed into

$$\begin{aligned} dx\,dy &= (\cos \theta \, dr - r \sin \theta \, d\theta) \wedge (\sin \theta \, dr + r \cos \theta \, d\theta) \\ &= r \cos^2 \theta \, dr \wedge d\theta - r \sin^2 \theta \, d\theta \wedge dr \\ &= r \, dr \, d\theta, \end{aligned}$$

the element of area in polar coordinates. For a second example, consider the 1-form

$$\alpha(x,y) = \frac{-y}{x^2 + y^2} \, dx + \frac{x}{x^2 + y^2} \, dy.$$

If we use $\varphi^* \alpha(r, \theta)$ to denote the new form after the polar coordinate change φ is applied, then

$$\begin{aligned} \varphi^* \alpha(r, \theta) &= \frac{-r \sin \theta}{r^2} (\cos \theta \, dr - r \sin \theta \, d\theta) + \frac{r \cos \theta}{r^2} (\sin \theta \, dr + r \cos \theta \, d\theta) \\ &= \frac{-\sin \theta \cos \theta + \sin \theta \cos \theta}{r} \, dr + (\sin^2 \theta + \cos^2 \theta) \, d\theta \\ &= d\theta. \end{aligned}$$

In other words, $\alpha(x,y)$ is the differential of the function

$$\theta(x,y) = \arctan\frac{y}{x}, \quad (x,y) \neq (0,0),$$

as we can verify directly:

$$\frac{\partial \theta}{\partial x} = \frac{-y/x^2}{1+(y/x)^2} = \frac{-y}{x^2+y^2}, \quad \frac{\partial \theta}{\partial y} = \frac{1/x}{1+(y/x)^2} = \frac{x}{x^2+y^2}.$$

The graph of $\theta(x,y)$ is an infinite spiral ramp

The graph of the function $z = \theta(x,y)$ is a spiral ramp, as shown below. The z-axis is not part of the graph, because $(x,y) \neq (0,0)$. The ray $\theta = c$ in the (x,y)-plane is carried to the level $z = c$ but also to $z = c + 2n\pi$ for every integer n. The polar angle θ is therefore *multiple-valued* in a particular way; the graph reflects this by spiraling around the origin infinitely many times, with successive levels separated by $\Delta z = 2\pi$.

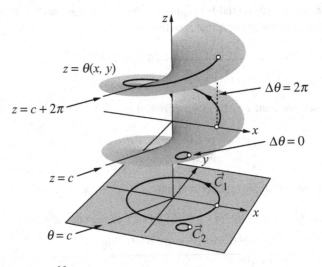

Because $\alpha = d\theta$,

$$\int_{\vec{C}} \alpha = \int_{\vec{C}} \frac{-y}{x^2+y^2}\,dx + \frac{x}{x^2+y^2}\,dy = \Delta\theta \text{ on } \vec{C}.$$

Ordinarily, the integral of the differential dg of a function g is zero on a closed path, because $\Delta g = 0$ there. The same is true of the integral of $d\theta$ on a path like $\vec{C}_2$ that does not enclose the origin. However, on a path like $\vec{C}_1$ that does enclose the origin, $z = \theta$ does not return to its starting value (it "climbs the ramp"), and $\Delta\theta \neq 0$.

Winding number

Definition 10.14 *Let $\vec{C}$ be an oriented closed path that does not meet the origin; the **winding number of $\vec{C}$** is*

$$W(\vec{C}) = \frac{1}{2\pi}\int_{\vec{C}} \frac{-y}{x^2+y^2}\,dx + \frac{x}{x^2+y^2}\,dy.$$

By what we've said, $W(\vec{C}) = \Delta\theta/2\pi$ is an integer; it counts the net number of times $\vec{C}$ "winds around" the origin in the positive sense minus the number of times in the negative sense. Furthermore, $W(-\vec{C}) = -W(\vec{C})$.

We now want to determine how a differential form is transformed when a map introduces new variables (possibly more general than an invertible change of variables). First, consider a map from (an open set in) $\mathbb{R}^2$ to $\mathbb{R}^2$:

(margin: How a mapping transforms a k-form)

$$\mathbf{f}: \begin{cases} x = x(u,v), \\ y = y(u,v), \end{cases} \qquad \begin{cases} dx = x_u\,du + x_v\,dv, \\ dy = y_u\,du + y_v\,dv. \end{cases}$$

We assume that $\mathbf{f}$ is continuously differentiable, but do not assume, for the moment, that it is invertible. In other words, $\mathbf{f}$ need not be a coordinate change. A general 0-form $\alpha(x,y) = g(x,y)$ is transformed into

$$\mathbf{f}^*\alpha(u,v) = g^*(u,v) = g(x(u,v), y(u,v)).$$

A general 1-form $\alpha(x,y) = P(x,y)\,dx + Q(x,y)\,dy$ is transformed into

$$\begin{aligned} \mathbf{f}^*\alpha(u,v) &= P(x(u,v),y(u,v))(x_u\,du + x_v\,dv) \\ &\quad + Q(x(u,v),y(u,v))(y_u\,du + y_v\,dv) \\ &= (P^*x_u + Q^*y_u)\,du + (P^*x_v + Q^*y_v)\,dv. \end{aligned}$$

For some purposes, the functions in differential forms should have continuous second derivatives. For $\mathbf{f}^*\alpha$ to meet that requirement, the components of $\mathbf{f}$ should have continuous third derivatives, because $\mathbf{f}^*\alpha$ contains the first derivatives of those components.

The basic 2-form $\alpha(x,y) = dx\,dy$ is transformed into

$$\begin{aligned} \mathbf{f}^*\alpha(u,v) &= (x_u\,du + x_v\,dv) \wedge (y_u\,du + y_v\,dv) \\ &= x_u y_v\,du \wedge dv + x_v y_u\,dv \wedge du = \frac{\partial(x,y)}{\partial(u,v)}\,du\,dv. \end{aligned}$$

Therefore, if $\alpha(x,y) = g(x,y)\,dx\,dy$, the general 2-form in $\mathbb{R}^2$, then

$$\mathbf{f}^*\alpha(u,v) = g^*(u,v)\frac{\partial(x,y)}{\partial(u,v)}\,du\,dv.$$

Notice that, although $\mathbf{f}$ maps the (u,v)-plane to the (x,y)-plane, the map $\mathbf{f}^*$ "goes the other way:" it maps differential forms on the (x,y)-plane to differential forms on the (u,v)-plane. Thus $\mathbf{f}^*$ **pulls back** forms from the (x,y)-plane to the (u,v)-plane; we call it the **pullback on forms** defined by $\mathbf{f}$.

(margin: The pullback on differential forms)

$$\text{forms in } (u,v) \xleftarrow{\;\;\mathbf{f}^*\;\;} \text{forms in } (x,y)$$
$$(u,v) \xrightarrow{\;\;\mathbf{f}\;\;} (x,y)$$

Pulling back a map
$\mathbf{f} : \mathbb{R}^2 \to \mathbb{R}^3$
The map $\mathbf{f}$ need not preserve dimension; indeed, $\mathbf{f} : \mathbb{R}^2 \to \mathbb{R}^3$ is an important case:

$$\mathbf{f} : \begin{cases} x = x(u,v), \\ y = y(u,v), \\ z = z(u,v). \end{cases} \qquad \begin{cases} dx = x_u\,du + x_v\,dv, \\ dy = y_u\,du + y_v\,dv, \\ dz = z_u\,du + z_v\,dv. \end{cases}$$

Pullbacks of 0- and 1-forms are similar to those for maps from $\mathbb{R}^2$ to $\mathbb{R}^2$. To be specific, if $\alpha(x,y,z) = g(x,y,z)$, then

$$\mathbf{f}^*\alpha(u,v) = g^*(u,v) = g(x(u,v),y(u,v),z(u,v));$$

and if $\alpha = P(x,y,z)\,dx + Q(x,y,z)\,dy + R(x,y,z)\,dz$, then

$$\mathbf{f}^*\alpha(u,v) = (P^*x_u + Q^*y_u + R^*z_u)\,du + (P^*x_v + Q^*y_v + R^*z_v)\,dv.$$

There are similarities for 2-forms, as well; we just need to take into account that there are now three basic 2-forms in $\mathbb{R}^3$: $dx\,dy$, $dy\,dz$, and $dz\,dx$. However, in $\mathbb{R}^2$ there is only one basic 2-form, so the previous case of maps from $\mathbb{R}^2$ to $\mathbb{R}^2$ need merely be applied three times:

$$\mathbf{f}^*dx\,dy = \frac{\partial(x,y)}{\partial(u,v)}\,du\,dv, \quad \mathbf{f}^*dy\,dz = \frac{\partial(y,z)}{\partial(u,v)}\,du\,dv, \quad \mathbf{f}^*dz\,dx = \frac{\partial(z,x)}{\partial(u,v)}\,du\,dv.$$

Thus, for the general 2-form $\alpha = X\,dy\,dz + Y\,dz\,dx + Z\,dx\,dy$, the pullback is

$$\begin{aligned} \mathbf{f}^*\alpha &= \left(X^*\frac{\partial(y,z)}{\partial(u,v)} + Y^*\frac{\partial(z,x)}{\partial(u,v)} + Z^*\frac{\partial(x,y)}{\partial(u,v)} \right) du\,dv \\ &= \left(X(\mathbf{f}(u,v))\frac{\partial(y,z)}{\partial(u,v)} + Y(\mathbf{f}(u,v))\frac{\partial(z,x)}{\partial(u,v)} + Z(\mathbf{f}(u,v))\frac{\partial(x,y)}{\partial(u,v)} \right) du\,dv. \end{aligned}$$

The final case to consider is the general 3-form $\alpha = H(x,y,z)\,dx\,dy\,dz$, but its pullback is zero because every 3-form in two variables must reduce to zero.

Surface integrals
reformulated
The language of differential forms and pullbacks gives us a vivid and succinct way to reformulate the definition of a surface integral (Definition 10.4, p. 402). Thus we are given an oriented surface patch $\vec{S}$ and a 2-form

$$\omega = X\,dy\,dz + Y\,dz\,dx + Z\,dx\,dy$$

that is defined and continuous on $\vec{S}$. If $\mathbf{f} : \Omega^2 \to \mathbb{R}^3 : \vec{U}^2 \to \vec{S}$ parametrizes $\vec{S}$ (Definition 10.2, p. 392), then

$$\iint_{\vec{S}} \omega = \iint_{\mathbf{f}(\vec{U})} \omega \quad \text{is by definition equal to} \quad \iint_{\vec{U}} \mathbf{f}^*\omega.$$

The pullback on $\mathbb{R}^3$
Let us now determine the pullback map for a continuously differentiable map from $\mathbb{R}^3$ to $\mathbb{R}^3$:

$$\mathbf{f}: \begin{cases} x = x(u,v,w), \\ y = y(u,v,w), \\ z = z(u,v,w), \end{cases} \qquad \begin{cases} dx = x_u\,du + x_v\,dv + x_w\,dw, \\ dy = y_u\,du + y_v\,dv + y_w\,dw, \\ dz = z_u\,du + z_v\,dv + z_w\,dw. \end{cases}$$

Again, pullbacks of 0- and 1-forms are similar to what we have already seen. For the general 0-form $\alpha(x,y,z) = g(x,y,z)$,

$$\mathbf{f}^*g = g^*(u,v,w) = g(x(u,v,w), y(u,v,w), z(u,v,w));$$

for the general 1-form $\alpha = P\,dx + Q\,dy + R\,dz$,

$$\mathbf{f}^*\alpha(u,v,w) = (P^*x_u + Q^*y_u + R^*z_u)\,du + (P^*x_v + Q^*y_v + R^*z_v)\,dv \\ + (P^*x_w + Q^*y_w + R^*z_w)\,dw.$$

With 2-forms, there are complications we have not seen before, because there are three basic 2-forms in the source. We begin with the basic 2-form $dx\,dy$ in the target, and write

$$\begin{aligned} \mathbf{f}^*dx\,dy &= (x_u\,du + x_v\,dv + x_w\,dw) \wedge (y_u\,du + y_v\,dv + y_w\,dw) \\ &= x_uy_v\,du \wedge dv + x_uy_w\,du \wedge dw + x_vy_u\,dv \wedge du \\ &\quad + x_vy_w\,dv \wedge dw + x_wy_u\,dw \wedge du + x_wy_v\,dw \wedge dv \\ &= \frac{\partial(x,y)}{\partial(u,v)}\,du\,dv + \frac{\partial(x,y)}{\partial(v,w)}\,dv\,dw + \frac{\partial(x,y)}{\partial(w,u)}\,dw\,du. \end{aligned}$$

Using similar results for the other basic 2-forms (i.e., $dy\,dz$ and $dz\,dx$; see the exercises), we find that the general 2-form

$$\alpha = X\,dy\,dz + Y\,dz\,dx + Z\,dx\,dy$$

is transformed into

$$\begin{aligned} \mathbf{f}^*\alpha(u,v,w) &= \left(X^*\frac{\partial(y,z)}{\partial(v,w)} + Y^*\frac{\partial(z,x)}{\partial(v,w)} + Z^*\frac{\partial(x,y)}{\partial(v,w)} \right)dv\,dw \\ &\quad + \left(X^*\frac{\partial(y,z)}{\partial(w,u)} + Y^*\frac{\partial(z,x)}{\partial(w,u)} + Z^*\frac{\partial(x,y)}{\partial(w,u)} \right)dw\,du \\ &\quad + \left(X^*\frac{\partial(y,z)}{\partial(u,v)} + Y^*\frac{\partial(z,x)}{\partial(u,v)} + Z^*\frac{\partial(x,y)}{\partial(u,v)} \right)du\,dv. \end{aligned}$$

The pullback of the general 3-form $\alpha = H\,dx\,dy\,dz$ is straightforward:

$$\mathbf{f}^*\alpha(u,v,w) = H^*\frac{\partial(x,y,z)}{\partial(u,v,w)}\,du\,dv\,dw.$$

Differential forms are involved in the calculation of physical quantities (e.g., work and total flux) whose values should be independent of the coordinate frames

Comparing integrals of α and $\boldsymbol{\varphi}^*\alpha$

in which they are computed. This prompts us to determine how the integrals of α and its pullback $\boldsymbol{\varphi}^*\alpha$ are related when $\boldsymbol{\varphi}$ is a coordinate change, that is, an invertible map with a continuously differentiable inverse.

Transforming the
integral of a 1-form

Theorem 10.12. *If* $\alpha(\mathbf{x}) = P\,dx + Q\,dy + R\,dz$ *is a 1-form,* $\vec{C}$ *is an oriented curve in* $\mathbf{u}$*-space, and* $\mathbf{x} = \boldsymbol{\varphi}(\mathbf{u})$ *is a coordinate change, then*

$$\int_{\boldsymbol{\varphi}(\vec{C})} \alpha = \int_{\vec{C}} \boldsymbol{\varphi}^*\alpha.$$

Proof. For simplicity, take $\alpha = P\,dx$; the other terms $Q\,dy$ and $R\,dz$ can be handled the same way. Then we know that

$$\boldsymbol{\varphi}^*\alpha(\mathbf{u}) = P^*x_u\,du + P^*x_v\,dv + P^*x_w\,dw,$$

where $P^*(\mathbf{u}) = P(\boldsymbol{\varphi}(\mathbf{u}))$. Let $\vec{C}$ be parametrized as $(u,v,w) = \mathbf{u}(t)$, with $a \le t \le b$. Then $\boldsymbol{\varphi}(\vec{C})$ is parametrized by $(x,y,z) = \boldsymbol{\varphi}(\mathbf{u}(t))$, $a \le t \le b$, and we have

$$\int_{\boldsymbol{\varphi}(\vec{C})} \alpha = \int_a^b P(\boldsymbol{\varphi}(\mathbf{u}(t))) \cdot \frac{d}{dt}x(\mathbf{u}(t))\,dt$$

$$= \int_a^b P^*(\mathbf{u}(t))(x_u \cdot u' + x_v \cdot v' + x_w \cdot w')\,dt = \int_{\vec{C}} \boldsymbol{\varphi}^*\alpha. \qquad \square$$

An abbreviated version of the same proof shows that the theorem is true for 1-forms $\alpha(x,y)$ on the plane.

Transforming the
integral of a 2-form

If $\alpha(x,y) = g\,dx\,dy$ is a 2-form, $\vec{D}$ an oriented region in the (u,v)-plane, and $\boldsymbol{\varphi}$ is an invertible coordinate change, then

$$\iint_{\boldsymbol{\varphi}(\vec{D})} \alpha = \iint_{\boldsymbol{\varphi}(\vec{D})} g\,dx\,dy = \iint_{\vec{D}} g^* \frac{\partial(x,y)}{\partial(u,v)}\,du\,dv = \iint_{\vec{D}} \boldsymbol{\varphi}^*\alpha.$$

The second equality is the change of variables formula for oriented double integrals (Theorem 9.14, p. 357). The next theorem deals with 2-forms in space.

Theorem 10.13. *. If* $\alpha(\mathbf{x}) = X\,dy\,dz + Y\,dz\,dx + Z\,dx\,dy$ *is a 2-form,* $\vec{S}$ *is an oriented surface patch in* $\mathbf{u}$*-space, and* $\mathbf{x} = \boldsymbol{\varphi}(\mathbf{u})$ *is a coordinate change, then* $\boldsymbol{\varphi}(\vec{S})$ *is an oriented surface patch in* $\mathbf{x}$*-space and*

$$\iint_{\boldsymbol{\varphi}(\vec{S})} \alpha = \iint_{\vec{S}} \boldsymbol{\varphi}^*\alpha.$$

Proof. Because $\vec{S}$ is an oriented surface patch in (u,v,w)-space, it has a parametrization $\mathbf{f} : \Omega \to \mathbb{R}^3 : (s,t) \to (u,v,w)$ with $\vec{S} = \mathbf{f}(\vec{U})$ for some closed, bounded, positively oriented set $\vec{U} \subset \Omega$ with area (Definition 10.2, p. 392). Because $\boldsymbol{\varphi}$ is a coordinate change in $\mathbb{R}^3$, the map

$$\boldsymbol{\varphi} \circ \mathbf{f} : \Omega \to \mathbb{R}^3 : (s,t) \to (x,y,z)$$

serves to parametrize $\boldsymbol{\varphi}(\vec{S}) = (\boldsymbol{\varphi} \circ \mathbf{f})(\vec{U})$, which is therefore an oriented surface patch in (x, y, z)-space. The two surface integrals that appear in the theorem can therefore be defined using the pullbacks of $\mathbf{f}$ and $\boldsymbol{\varphi} \circ \mathbf{f}$. The following lemma indicates how these pullbacks are related.

Lemma 10.1. *For any 2-form α, $(\boldsymbol{\varphi} \circ \mathbf{f})^* \alpha = (\mathbf{f}^* \circ \boldsymbol{\varphi}^*)\alpha = \mathbf{f}^*(\boldsymbol{\varphi}^* \alpha)$.*

Proof. For simplicity, take $\alpha = X \, dy \, dz$; the other terms $Y \, dz \, dx$ and $Z \, dx \, dy$ can be analyzed similarly. We have

$$(\boldsymbol{\varphi} \circ \mathbf{f})^* \alpha(\mathbf{s}) = X(\boldsymbol{\varphi}(\mathbf{f}(\mathbf{s}))) \frac{\partial(y, z)}{\partial(s, t)} \, ds \, dt.$$

We also have

$$\boldsymbol{\varphi}^* \alpha(\mathbf{u}) = X(\boldsymbol{\varphi}(\mathbf{u})) \left(\frac{\partial(y, z)}{\partial(v, w)} \, dv \, dw + \frac{\partial(y, z)}{\partial(w, u)} \, dw \, du + \frac{\partial(y, z)}{\partial(u, v)} \, du \, dv \right),$$

from which it follows that

$$\mathbf{f}^*(\boldsymbol{\varphi}^* \alpha)(\mathbf{s})$$
$$= X(\boldsymbol{\varphi}(\mathbf{f}(\mathbf{s}))) \left(\frac{\partial(y, z)}{\partial(v, w)} \overset{*}{\frac{\partial(v, w)}{\partial(s, t)}} + \frac{\partial(y, z)}{\partial(w, u)} \overset{*}{\frac{\partial(w, u)}{\partial(s, t)}} + \frac{\partial(y, z)}{\partial(u, v)} \overset{*}{\frac{\partial(u, v)}{\partial(s, t)}} \right) ds \, dt.$$

The three Jacobians marked with asterisks (which are usually not written explicitly) are understood to be functions of s and t via pullbacks. By Exercise 10.27,

$$\frac{\partial(y, z)}{\partial(v, w)} \frac{\partial(v, w)}{\partial(s, t)} + \frac{\partial(y, z)}{\partial(w, u)} \frac{\partial(w, u)}{\partial(s, t)} + \frac{\partial(y, z)}{\partial(u, v)} \frac{\partial(u, v)}{\partial(s, t)} = \frac{\partial(y, z)}{\partial(s, t)}. \qquad \square$$

To complete the proof of the theorem, we use the lemma and twice invoke the new formulation (p. 432) of the definition of a surface integral as the ordinary double integral of a pullback:

$$\iint_{\boldsymbol{\varphi}(\vec{S})} \alpha = \iint_{\boldsymbol{\varphi}(\mathbf{f}(\vec{U}))} \alpha = \iint_{\vec{U}} (\boldsymbol{\varphi} \circ \mathbf{f})^* \alpha = \iint_{\vec{U}} \mathbf{f}^* \circ \boldsymbol{\varphi}^* \alpha = \iint_{\mathbf{f}(\vec{U})} \boldsymbol{\varphi}^* \alpha = \iint_{\vec{S}} \boldsymbol{\varphi}^* \alpha. \qquad \square$$

Corollary 10.14 *Suppose that $\vec{S}$ is a piecewise-smooth oriented surface; then so is $\boldsymbol{\varphi}(\vec{S})$, and*

Allow $\vec{S}$ to be piecewise smooth

$$\iint_{\boldsymbol{\varphi}(\vec{S})} \alpha = \iint_{\vec{S}} \boldsymbol{\varphi}^* \alpha.$$

Proof. Let $\vec{S} = \vec{S}_1 + \cdots + \vec{S}_k$ be a decomposition into oriented surface patches, and suppose $\vec{S}_i$ is parametrized by $\mathbf{f}_i : \Omega_i \to \mathbb{R}^3$, with $\mathbf{f}_i(\vec{U}_i) = \vec{S}_i$. Then, by the proof of the theorem, $\boldsymbol{\varphi}(\vec{S}_i)$ is an oriented surface patch parametrized by $\boldsymbol{\varphi} \circ \mathbf{f}_i$. Therefore, because $\boldsymbol{\varphi}$ is 1–1,

$$\boldsymbol{\varphi}(\vec{S}) = \boldsymbol{\varphi}(\vec{S}_1) + \cdots + \boldsymbol{\varphi}(\vec{S}_k)$$

is a piecewise-smooth oriented surface. Finally, using the theorem on each pair of surface patches $\boldsymbol{\varphi}(\vec{S}_i)$ and $\vec{S}_i$, we obtain

$$\iint_{\boldsymbol{\varphi}(\vec{S})} \alpha = \sum_{i=1}^{k} \iint_{\boldsymbol{\varphi}(\vec{S}_i)} \alpha = \sum_{i=1}^{k} \iint_{\vec{S}_i} \boldsymbol{\varphi}^* \alpha = \iint_{\vec{S}} \boldsymbol{\varphi}^* \alpha. \qquad \square$$

Transforming the integral of a 3-form

The final possibility we must consider is how the integral of a 3-form $\alpha = H(\mathbf{x}) \, dx \, dy \, dz$ transforms under a coordinate change $\mathbf{x} = \boldsymbol{\varphi}(\mathbf{u})$. We know

$$\boldsymbol{\varphi}^* \alpha = H^*(\mathbf{u}) \frac{\partial(x,y,z)}{\partial(u,v,w)} \, du \, dv \, dw,$$

where $H^*(\mathbf{u}) = H(\boldsymbol{\varphi}(\mathbf{u}))$. The integrals here are triple integrals over an oriented region $\vec{D}$ in $\mathbf{u}$-space and its image $\boldsymbol{\varphi}(\vec{D})$ in $\mathbf{x}$-space. The change of variables formula for oriented triple integrals (Theorem 9.16, p. 363) yields

$$\iiint_{\boldsymbol{\varphi}(\vec{D})} \alpha = \iiint_{\boldsymbol{\varphi}(\vec{D})} H(\mathbf{x}) \, dx \, dy \, dz$$

$$= \iiint_{\vec{D}} H(\boldsymbol{\varphi}(\mathbf{x})) \frac{\partial(x,y,z)}{\partial(u,v,w)} \, du \, dv \, dw = \iiint_{\vec{D}} \boldsymbol{\varphi}^* \alpha.$$

Theorem 10.15. *The integral of a differential k-form in n variables (where $k \leq n$ and $n = 1,2,3$) is invariant under a coordinate change.* $\qquad \square$

$\boldsymbol{\varphi}$ and $\boldsymbol{\varphi}^*$ are adjoints

In terms of the symbolic integral pairing (p. 427), all the statements about the invariance of integrals under coordinate changes have the form

$$\langle \boldsymbol{\varphi}(\vec{D}), \alpha \rangle = \langle \vec{D}, \boldsymbol{\varphi}^* \alpha \rangle,$$

where α is a k-form in n variables, $n = 1,2,3$. In other words, the map $\boldsymbol{\varphi}$ and its pullback $\boldsymbol{\varphi}^*$ are adjoints (p. 428). *This is the essential content of the change of variables formulas* (wherein we take $k = n$). Incidentally, it is easy to check that the pairings are also equal when $n = 0$.

d and $\mathbf{f}^*$ commute

How does exterior differentiation interact with a pullback? Does it make a difference if we apply the exterior derivative before or after applying a map? In other words, do d and $\mathbf{f}^*$ commute? To explore this question, let us return to differential forms in just two variables. First consider the 0-form $\alpha = g(x,y)$; then

$$\mathbf{f}^* \alpha = g(x(u,v), y(u,v)),$$

so

$$d(\mathbf{f}^*\alpha) = \frac{\partial}{\partial u}g(x(u,v),y(u,v))\,du + \frac{\partial}{\partial v}g(x(u,v),y(u,v))\,dv$$

$$= \big(g_1(x(u,v),y(u,v))x_u + g_2(x(u,v),y(u,v))y_u\big)\,du$$

$$+ \big(g_1(x(u,v),y(u,v))x_v + g_2(x(u,v),y(u,v))y_v\big)\,dv$$

$$= (g_1^*x_u + g_2^*y_u)\,du + (g_1^*x_v + g_2^*y_v)\,dv.$$

Here g_i is the partial derivative of $g(x,y)$ with respect to its ith variable, and $g_i^* = (g_i)^* = \mathbf{f}^*(g_i)$ (so $g_i^* \neq (g^*)_i$, in general). On the other hand, $d\alpha = g_1\,dx + g_2\,dy$, and

$$\mathbf{f}^*(d\alpha) = g_1^* \cdot (x_u\,du + x_v\,dv) + g_2^* \cdot (y_u\,du + y_v\,dv)$$

$$= (g_1^*x_u + g_2^*y_u)\,du + (g_1^*x_v + g_2^*y_v)\,dv$$

$$= d(\mathbf{f}^*\alpha).$$

Next, consider the 1-form $\alpha = P\,dx$; then

$$\mathbf{f}^*\alpha = P^*x_u\,du + P^*x_u\,dv,$$

so

$$d(\mathbf{f}^*\alpha) = \frac{\partial}{\partial v}\big(P^*x_u\big)\,dv \wedge du + \frac{\partial}{\partial u}\big(P^*x_v\big)\,du \wedge dv$$

$$= \big[-(P_1^*x_v + P_2^*y_v)x_u - P^*x_{uv} + (P_1^*x_u + P_2^*y_u)x_v + P^*x_{vu}\big]\,du\,dv$$

$$= -P_2^*(y_v x_u - y_u x_v)\,du\,dv$$

$$= -P_2^*\frac{\partial(x,y)}{\partial(u,v)}\,du\,dv.$$

On the other hand, $d\alpha = -P_2\,dx\,dy$, and (from p. 431)

$$\mathbf{f}^*(d\alpha) = -P_2^*\frac{\partial(x,y)}{\partial(u,v)}\,du\,dv = d(\mathbf{f}^*\alpha).$$

Analysis of $\alpha = Q\,dy$ is similar. If $\alpha(x,y)$ is a k-form with $k \geq 2$, then $d\alpha = 0 = d(\mathbf{f}^*\alpha)$.

Theorem 10.16. *For any differentiable map $(x,y) = \mathbf{f}(u,v)$ and k-form $\alpha(x,y)$, $\mathbf{f}^*(d\alpha) = d(\mathbf{f}^*\alpha)$.* □

In fact, this theorem holds for differential forms $\alpha(x_1,\ldots,x_n)$ in any number of variables. In Exercise 10.28, you are asked to give a proof for $n = 3$.

The language of differential forms and symbolic pairings for integrals gives us a new way to look at the proofs of some earlier theorems. For example, consider the change of variables via Green's theorem (Theorem 9.21, p. 370):

The change of variables formula with integral pairings

Suppose $\mathbf{f}(s,t) = (x(s,t),y(s,t))$ has continuous second derivatives on a bounded open set Ω in $\mathbb{R}^2$. Let $\vec{S} \subset \Omega$ and $\vec{T} = \mathbf{f}(\vec{S})$ be closed oriented sets whose boundaries $\partial\vec{S}$ and $\partial\vec{T}$ are simple closed curves. Assume that Green's theorem holds for both

$\vec{S}$ and $\vec{T}$, that $\partial\vec{S}$ and $\mathbf{f}(\partial\vec{S})$ *have common decompositions into smooth oriented curves, and that* $\mathbf{f}(\partial\vec{S}) = k\cdot\partial\vec{T}$ *as oriented paths. Then, for any continuous function* $g(x,y)$ *on* $\vec{T}$,

$$k\iint_{\vec{T}} g(x,y)\,dx\,dy = \iint_{\vec{S}} g(x(s,t),y(s,t))\,\frac{\partial(x,y)}{\partial(s,t)}\,ds\,dt.$$

Proof. As in the original proof, choose $G(x,y)$ so that $G_x(x,y) = g(x,y)$, and let $\alpha = G\,dy$, $d\alpha = g\,dx\,dy$. A key step there was to transfer the path integral of α over $\mathbf{f}(\partial\vec{S})$ to the path integral of

$$\mathbf{f}^*\alpha(s,t) = G(x(s,t),y(s,t))(y_s\,ds + y_t\,dt) = G^*y_s\,ds + G^*y_t\,dt$$

over $\partial\vec{S}$. In the language of symbolic pairings, $\langle\mathbf{f}(\partial\vec{S}),\alpha\rangle = \langle\partial\vec{S},\mathbf{f}^*\alpha\rangle$, indicating that $\mathbf{f}$ and $\mathbf{f}^*$ are adjoints. The original proof invoked the results of an exercise. For future use, we restate these results in the language of differential forms and pullbacks.

f and f* are
adjoints on 1-forms

Lemma 10.2. *Let* $\mathbf{f}: U^2 \to \mathbb{R}^2$ *be continuously differentiable, and suppose* $\vec{C}$ *and* $\mathbf{f}(\vec{C})$ *are piecewise-smooth oriented curves with a common decomposition into smooth oriented curves:*

$$\vec{C} = \vec{C}_1 + \cdots + \vec{C}_m, \quad \mathbf{f}(\vec{C}) = \mathbf{f}(\vec{C}_1) + \cdots + \mathbf{f}(\vec{C}_m).$$

Then, for any 1-form α, $\langle\mathbf{f}(\vec{C}),\alpha\rangle = \langle\vec{C},\mathbf{f}^*\alpha\rangle$.

Proof. See Exercise 4.37, page 149. $\qquad\qquad\qquad\qquad\qquad\qquad\qquad$ □

The proof of the change of variables theorem using Green's theorem now follows from this sequence of equalities.

$$
\begin{aligned}
k\iint_{\vec{T}} g(x,y)\,dx\,dy &= k\langle\vec{T},d\alpha\rangle \\
&= k\langle\partial\vec{T},\alpha\rangle && \text{Green's theorem on } \vec{T} \\
&= \langle k\partial\vec{T},\alpha\rangle \\
&= \langle\mathbf{f}(\partial\vec{S}),\alpha\rangle && \text{hypothesis} \\
&= \langle\partial\vec{S},\mathbf{f}^*\alpha\rangle && \mathbf{f} \text{ and } \mathbf{f}^* \text{ are adjoints} \\
&= \langle\vec{S},d(\mathbf{f}^*\alpha)\rangle && \text{Green's theorem on } \vec{S} \\
&= \langle\vec{S},\mathbf{f}^*(d\alpha)\rangle && d \text{ and } \mathbf{f}^* \text{ commute} \\
&= \iint_{\vec{S}} g(x(s,t),y(s,t))\,\frac{\partial(x,y)}{\partial(s,t)}\,ds\,dt,
\end{aligned}
$$

because $\mathbf{f}^*(d\alpha) = g^*(s,t)\,\dfrac{\partial(x,y)}{\partial(s,t)}\,ds\,dt.$ $\qquad\qquad\qquad\qquad\qquad$ □

It is possible to construct differential k-forms in any number of variables. In $(x_1, x_2, \ldots, x_n)$-space, there are n basic differentials, dx_1, dx_2, $\ldots$, dx_n. For each multi-index $I = (i_1, \ldots, i_k)$ with $1 \leq i_1 < \cdots < i_k \leq n$, we define the basic k-form

$$dx_I = dx_{i_1} dx_{i_2} \cdots dx_{i_k} = dx_{i_1} \wedge dx_{i_2} \wedge \cdots \wedge dx_{i_k}.$$

There are as many basic k-forms as there are ways of choosing k distinct elements from a set of n elements; this number is

$$\binom{n}{k} = \frac{n!}{k!(n-k)!}$$

("n choose k"). A general k-form is a linear combination

$$\alpha = \sum_I P_I(x_1, \ldots, x_n) \, dx_I$$

of the $\binom{n}{k}$ basic k-forms in which the coefficient functions P_I all have continuous second derivatives. Products can then be calculated using the anticommutativity relations on the basic 1-forms:

$$dx_i \wedge dx_j = -dx_j \wedge dx_i, \quad i, j = 1, \ldots, n.$$

As we pointed out above, anticommutativity here implies

$$dx_i \wedge dx_i = 0, \quad i = 1, \ldots, n.$$

If σ is a k-form and τ is an l-form, then anticommutativity on the basic 1-forms implies

$$\sigma \wedge \tau = (-1)^{kl} \tau \wedge \sigma.$$

The exterior derivative of the 0-form $g(x_1, \ldots, x_n)$ is the 1-form

$$dg = \sum_{i=1}^{n} g_i \, dx_i,$$

where $g_i = \partial g / \partial x_i$. For a general k-form

$$\alpha = \sum_I P_I \, dx_I,$$

the exterior derivative is the $(k+1)$-form

$$d\alpha = \sum_I dP_I \wedge dx_I,$$

obtained using the exterior derivatives dP_I of the coefficient functions of α.

Theorem 10.17. *For every k-form α in $\mathbb{R}^n$, $d^2 \alpha = d(d\alpha) = 0$.*

Proof. As in the essentially identical Theorem 10.11, the proof reduces to the "equality of mixed partials" for functions with continuous second derivatives. If $\alpha = \sum_I P_I d\mathbf{x}_I$, then

$$d\alpha = \sum_I dP_I \wedge d\mathbf{x}_I = \sum_I \left(\sum_{i=1}^{n} \frac{\partial P_I}{\partial x_i} dx_i \right) \wedge d\mathbf{x}_I,$$

and

$$d^2\alpha = \sum_I \left(\sum_{i=1}^{n} d\left(\frac{\partial P_I}{\partial x_i} \right) dx_i \right) \wedge d\mathbf{x}_I = \sum_I \left(\sum_{i,j=1}^{n} \frac{\partial^2 P_I}{\partial x_j \partial x_i} dx_j \wedge dx_i \right) \wedge d\mathbf{x}_I.$$

The inner sum consists of n^2 terms. For each of the n terms with $j = i$,

$$\frac{\partial^2 P_I}{\partial x_i \partial x_i} dx_i \wedge dx_i = 0.$$

Now pair each remaining term with the term in which i and $j \neq i$ are interchanged:

$$\frac{\partial^2 P_I}{\partial x_j \partial x_i} dx_j \wedge dx_i + \frac{\partial^2 P_I}{\partial x_i \partial x_j} dx_i \wedge dx_j.$$

Because

$$\frac{\partial^2 P_I}{\partial x_j \partial x_i} = \frac{\partial^2 P_I}{\partial x_i \partial x_j} \quad \text{and} \quad dx_j \wedge dx_i = -dx_i \wedge dx_j,$$

each pair sums to zero, so the entire inner sum equals zero. $\square$

The product rule
for differentials

Theorem 10.18 (Product Rule). *Suppose α and θ are differential forms in $\mathbb{R}^n$, and α is a k-form. Then $d(\alpha \wedge \theta) = d\alpha \wedge \theta + (-1)^k \alpha \wedge d\theta$.*

Proof. It is sufficient to show this for "monomials"

$$\alpha = P d\mathbf{x}_I \quad \text{and} \quad \theta = Q d\mathbf{x}_J$$

that have disjoint multi-indices I and J. Then $\alpha \wedge \theta = PQ d\mathbf{x}_I \wedge d\mathbf{x}_J$, and we have

$$d(\alpha \wedge \theta) = \sum_m (P_m Q + P Q_m) \, dx_m \wedge d\mathbf{x}_I \wedge \mathbf{x}_J$$

$$= \sum_m P_m Q \, dx_m \wedge d\mathbf{x}_I \wedge \mathbf{x}_J + \sum_m P Q_m \, dx_m \wedge d\mathbf{x}_I \wedge \mathbf{x}_J,$$

where $P_m = \partial P / \partial x_m$ and the summation can be restricted to those indices m that do not occur in either I or J. The first sum is $d\alpha \wedge \theta$, because

$$d\alpha = \sum_m P_m \, dx_m \wedge d\mathbf{x}_I \quad \text{and} \quad d\alpha \wedge \theta = \sum_m P_m Q \, dx_m \wedge d\mathbf{x}_I \wedge d\mathbf{x}_J.$$

The second sum is $(-1)^k \alpha \wedge d\theta$, because $d\theta = \sum_m Q_m \, dx_m \wedge d\mathbf{x}_J$ and

$$\alpha \wedge d\theta = \sum_m PQ_m\, d\mathbf{x}_I \wedge dx_m \wedge d\mathbf{x}_J = (-1)^k \sum_m PQ_m\, dx_m \wedge d\mathbf{x}_I \wedge d\mathbf{x}_J.$$

The last equality is a consequence of the anticommutativity of basic 1-forms:

$$
\begin{aligned}
d\mathbf{x}_I \wedge \underline{dx_m} &= dx_{i_1} \wedge \cdots \wedge dx_{i_{k-1}} \wedge dx_{i_k} \wedge \underline{dx_m} \\
&= (-1)\, dx_{i_1} \wedge \ldots dx_{i_{k-1}} \wedge \underline{dx_m} \wedge dx_{i_k} \\
&= (-1)^2\, dx_{i_1} \wedge \cdots \wedge \underline{dx_m} \wedge dx_{i_{k-1}} \wedge dx_{i_k} \\
&\;\;\vdots \\
&= (-1)^k \underline{dx_m} \wedge dx_{i_1} \wedge \cdots \wedge dx_{i_{k-1}} \wedge dx_{i_k} \\
&= (-1)^k \underline{dx_m} \wedge d\mathbf{x}_I. \qquad \square
\end{aligned}
$$

Let us see how the coefficients of a $(k-1)$-form α in n variables determine the coefficients of its exterior derivative $\omega = d\alpha$, a k-form. We use the k-multi-index $I = (i_1, \ldots, i_k)$, $1 \le i_1 < \cdots < i_k \le n$ for the coefficients of ω,

Coefficients of $d\alpha$

$$\omega = \sum_I P_I\, d\mathbf{x}_I.$$

For the coefficients of α, we use the $(k-1)$-multi-index

$$\widehat{I}_s = (i_1, \ldots, \widehat{i_s}, \ldots, i_k), \quad s = 1, \ldots, k;$$

the circumflex over i_s means that i_s is deleted, so each $\widehat{I}_s$ contains only $k-1$ indices. Thus we can write

$$\alpha = \sum_{\widehat{I}_s} A_{\widehat{I}_s}\, d\mathbf{x}_{\widehat{I}_s}, \quad d\alpha = \sum_{\widehat{I}_s} d(A_{\widehat{I}_s})\, d\mathbf{x}_{\widehat{I}_s};$$

There are $\binom{n}{k}$ k-mult-indices I and $\binom{n}{k-1}$ $(k-1)$-multi-indices $\widehat{I}_s$.

Given that $d\alpha = \omega$, we must determine what contribution the various terms of $d(A_{\widehat{I}_s})\, d\mathbf{x}_{\widehat{I}_s}$ make to the term $P_I\, d\mathbf{x}_I$. We have

$$d(A_{\widehat{I}_s})\, d\mathbf{x}_{\widehat{I}_s} = \sum_j \frac{\partial A_{\widehat{I}_s}}{\partial x_j}\, dx_j\, dx_{i_1} \cdots \widehat{dx_{i_s}} \cdots dx_{i_k},$$

and the sum has only one nonzero term, the one in which the summation index j equals i_s. It then takes $s-1$ successive transpositions to move $dx_j = dx_{i_s}$ from its initial position in that term to its proper position in the basic differential; that is,

$$dx_{i_s}\, dx_{i_1} \cdots \widehat{dx_{i_s}} \cdots dx_{i_k} = (-1)^{s-1} dx_{i_1} \cdots dx_{i_s} \cdots dx_{i_k} = (-1)^{s-1} d\mathbf{x}_I.$$

This proves the following theorem.

Theorem 10.19. *If* $\alpha = \sum_{\widehat{I_s}} A_{\widehat{I_s}} \, d\mathbf{x}_{\widehat{I_s}}$, *then* $d\alpha = \sum_I \left(\sum_{s=1}^k (-1)^{s-1} \dfrac{\partial A_{\widehat{I_s}}}{\partial x_{i_s}} \right) d\mathbf{x}_I.$ $\square$

Action of the pullback   Now let us consider how a differential form is transformed by the pullback of a differentiable map $\mathbf{f} : U^n \to \mathbb{R}^p$ with component functions

$$\mathbf{f} : \begin{cases} x_1 = x_1(u_1, \ldots, u_n), \\ x_2 = x_2(u_1, \ldots, u_n), \\ \quad \vdots \\ x_p = x_p(u_1, \ldots, u_n). \end{cases}$$

Here U^n is an open subset of $\mathbb{R}^n$, and we allow $n \neq p$. For a 0-form, $\alpha(\mathbf{x}) = g(\mathbf{x})$, the pullback is

$$\mathbf{f}^* \alpha(\mathbf{u}) = g(\mathbf{f}(\mathbf{u})) = g^*(\mathbf{u}).$$

For a basic 1-form dx_i, the pullback is

$$\mathbf{f}^* dx_i = \sum_{j=1}^n \frac{\partial x_i}{\partial u_j} du_j, \quad i = 1, \ldots, p.$$

For the basic 2-form $dx_1 \, dx_2$, we have

$$\mathbf{f}^*(dx_1 \wedge dx_2) = \mathbf{f}^* dx_1 \wedge \mathbf{f}^* dx_2 = \sum_{j \neq m} \frac{\partial x_1}{\partial u_j} \frac{\partial x_2}{\partial u_m} du_j \wedge du_m$$

$$= \sum_{j < m} \frac{\partial x_1}{\partial u_j} \frac{\partial x_2}{\partial u_m} du_j \wedge du_m + \sum_{j > m} \frac{\partial x_1}{\partial u_j} \frac{\partial x_2}{\partial u_m} du_j \wedge du_m$$

Anticommutivity simplifies this. If we transpose the dummy summation indices (i.e., $j \leftrightarrow m$) in the second sum, then that sum becomes

$$\sum_{m > j} \frac{\partial x_1}{\partial u_m} \frac{\partial x_2}{\partial u_j} du_m \wedge du_j = \sum_{j < m} -\frac{\partial x_1}{\partial u_m} \frac{\partial x_2}{\partial u_j} du_j \wedge du_m.$$

Recombining this with the first sum, we get

$$\mathbf{f}^*(dx_1 \, dx_2) = \sum_{j < m} \left(\frac{\partial x_1}{\partial u_j} \frac{\partial x_2}{\partial u_m} - \frac{\partial x_1}{\partial u_m} \frac{\partial x_2}{\partial u_j} \right) du_j \wedge du_m$$

$$= \sum_{j < m} \frac{\partial(x_1, x_2)}{\partial(u_j, u_m)} du_j \, du_m.$$

This is essentially the same as the earlier calculation of $\mathbf{f}^*(dx \, dy)$ on page 433. More generally, if $I = (i_1, i_2)$ is any pair with $1 \leq i_1 < i_2 \leq p$, then

$$\mathbf{f}^*(dx_{i_1}\,dx_{i_2}) = \sum_{j<m} \frac{\partial(x_{i_1},x_{i_2})}{\partial(u_j,u_m)}\, du_j\,du_m.$$

We can write this in a way that is both more compact and more striking:

$$\mathbf{f}^*dx_I = \sum_J \frac{\partial \mathbf{x}_I}{\partial \mathbf{u}_J}\, d\mathbf{u}_J.$$

The summation multi-index J consists of all pairs $J = (j_1, j_2)$ with $1 \leq j_1 < j_2 \leq n$, and

$$\frac{\partial \mathbf{x}_I}{\partial \mathbf{u}_J} = \frac{\partial(x_{i_1},x_{i_2})}{\partial(u_{j_1},u_{j_2})}.$$

In fact, the same formula holds when dx_I is a basic k-form (where $I = (i_1,\ldots,i_k)$ and $1 \leq i_1 < \cdots < i_k \leq p$):

$$\mathbf{f}^*dx_I = \sum_J \frac{\partial \mathbf{x}_I}{\partial \mathbf{u}_J}\, d\mathbf{u}_J,$$

where now $J = (j_1,\ldots,j_k)$ with $1 \leq j_1 < \cdots < j_k \leq n$, and

$$\frac{\partial \mathbf{x}_I}{\partial \mathbf{u}_J} = \frac{\partial(x_{i_1},\ldots,x_{i_k})}{\partial(u_{j_1},\ldots,u_{j_k})}.$$

Theorem 10.20. *If* $\alpha = \sum_I P_I(\mathbf{x})\, dx_I$ *is a general k-form, then*

$$\mathbf{f}^*\alpha = \sum_I \sum_J P_I^*(\mathbf{u})\, \frac{\partial \mathbf{x}_I}{\partial \mathbf{u}_J}\, d\mathbf{u}_J. \qquad \square$$

Pullback of a general k-form

Exercises

10.1. Suppose there is a steady flow of matter given by the vector $\mathbb{V} = (2,-7,1)$ kilograms per second per square meter. (All space coordinates are given in meters.)

  a. In 1 second, how much matter passes through a unit square in the (x,y)-plane in the positive z-direction? Through a unit square in the (y,z)-plane in the positive x-direction? Through a unit square in the (z,x)-plane in the positive y-direction?

  b. How much matter passes through a triangle with area 12 meters^2 in the (x,y)-plane in the positive z-direction in 7 seconds?

  c. In 10 seconds, how much matter passes through the rectangle with vertices

$$(5,0,0), \quad (5,3,0), \quad (5,3,6), \quad (5,0,6)$$

in the direction in which x increases.

d. In unit time, how much matter passes through a unit square in the plane $x+y+z=1$ in the "upward" direction (i.e., the direction in which z increases)?

e. Calculate how much matter passes through each of the six faces of the unit cube Q in the first octant, *in the outward direction on each face* in unit time. The sum of these six numbers is zero; why?

10.2. Determine the total flux of the flow field $\mathbb{V} = (0,z,x)$ through:

a. The unit square in the (y,z)-plane, oriented in the positive x-direction.
b. The unit square in the (x,y)-plane, oriented in the negative z-direction.
c. The triangle with vertices $(2,2,0)$, $(0,2,2)$, $(2,0,2)$, using this ordering of the vertices to orient the boundary and thus the triangle itself.

10.3. Determine the flux of the flow field $\mathbb{V} = (x,y,z)$ through the surface S given by

$$\begin{aligned} x &= u+v \\ y &= u^2 - v^2 \\ z &= 2uv \end{aligned} \qquad \begin{aligned} 0 &\le u \le 3 \\ 0 &\le v \le 1 \end{aligned}$$

Assume that S inherits the positive orientation of the (u,v)-plane.

10.4. Calculate the flux of $\mathbb{V} = (x,y,z)$ out of the sphere S of radius R centered at the origin $(x,y,z) = (0,0,0)$ to show

$$\iint_S \mathbb{V} \cdot \mathbf{n} \, dA = 4\pi R^3.$$

10.5. Calculate the flux of $\mathbb{V} = (-y,x,0)$ out of the rectangular parallelepiped P in (x,y,z)-space given by $0 \le x \le 5$, $0 \le y \le 3$, $0 \le z \le 2$.

10.6. Let $\mathbf{g} : \mathbb{R}^2 \to \mathbb{R}^3 : (u,v) \to (x,y,z)$ be the map defined on page 397:

$$x = \frac{2u}{1+u^2+v^2}, \quad y = \frac{2u}{1+u^2+v^2}, \quad z = \frac{1-u^2-v^2}{1+u^2+v^2}.$$

a. Let $\widehat{S}$ be the unit sphere $x^2+y^2+z^2 = 1$ minus the "south pole" $(0,0,-1)$. Show that $\mathbf{g}(\mathbb{R}^2) \subseteq \widehat{S}$.

b. Show that $\mathbf{g}$ maps the (u,v)-plane onto $\widehat{S}$ by expressing (u,v) in terms of (x,y,z) when $\mathbf{g}(u,v) = (x,y,z)$. (That is, "invert" $\mathbf{g}$ on $\widehat{S}$.)

10.7. Prove Theorem 10.7 (e.g., by modifying the proof of Theorem 10.6).

10.8. When $a = 1$, the integral expression for the gravitational field of the hollow sphere (p. 410) involves the improper integral

$$\int_{-\pi/2}^{\pi/2} \frac{(\sin\varphi - 1)\cos\varphi}{(2 - 2\sin\varphi)^{3/2}} \, d\varphi.$$

Show that the improper integral converges and has the value -1, implying that the z-component of the field is $-2\pi G\rho$ when $a = 1$.

10.9. Determine the surface area of the torus

$$x = (R + a\cos v)\cos u, \quad y = (R + a\cos v)\sin u, \quad z = a\sin v,$$

where $R > a > 0$ and $0 \le u, v \le 2\pi$.

10.10. Calculate the differential dg when

a. $g(x,y) = x^3 - 3xy^2$;

b. $g(x,y) = \sin(xy)$;

c. $g(x,y) = x\cos y - y\sin x$;

d. $g(x,y,z) = \ln\sqrt{x^2 + y^2}$;

e. $g(x,y,z) = xy + yz + zx$;

f. $g(\rho,\varphi,\theta) = \rho\sin\varphi\cos\theta$;

g. $g(x,y) = \arctan(y/x)$;

h. $g(x,y,u,v) = xu - yv$;

i. $g(x,y,u,v) = xu/yv$;

j. $g(x_1, x_2, \ldots, x_n) = x_1 x_2 \cdots x_n$.

10.11. Calculate the differential of each of the following k-forms.

a. $\omega(x,y) = y\,dx - x\,dy$;

b. $\omega(x,y) = (x^2 - y^2)\,dx - 2xy\,dy$;

c. $\omega(x,y) = dx/y - dy/x$;

d. $\omega(x,y,z) = (y - z)\,dx + (z - x)\,dy + (x - y)\,dz$;

e. $\omega(x_1,\ldots,x_n) = \displaystyle\sum_{j=2}^{n-1}(x_{j-1} - x_{j+1})\,dx_j$;

f. $\omega(u,v,w) = u^2\,dv\,dw + v^2\,dw\,du + w^2\,du\,dv$;

g. $\omega(x,y,u,v) = (ue^x - ve^y)\,dx\,dy + (x^2 + y^2)\,du\,dv$

h. $\omega(x,y,u,v) = \sinh u\cosh v\,dx\,dy + \sin x\cos y\,du\,dv$;

i. $\omega(q_1,q_2,\ldots,q_n,p_1,p_2,\ldots,p_n)) = \displaystyle\sum_{j=1}^{n} p_j\,dq_j$.

j. $\omega(x_1,x_2,\ldots,x_n) = \displaystyle\sum_{j=1}^{n}(-1)^{j-1}x_j\,dx_1\cdots\widehat{dx_j}\cdots dx_n$;

k. $\omega(x_1,x_2,\ldots,x_n) = \displaystyle\sum_{j=1}^{n} x_j\,dx_1\cdots\widehat{dx_j}\cdots dx_n$;

10.12. Consider the 1-form dg that you obtained in each part of Exercise 10.10. Determine its differential, the 2-form $d^2g = d(dg)$, and confirm $d^2g = 0$ in each case.

10.13. Condsider the $(k+1)$-form $d\omega$ that you obtained in part of Exercise 10.11; confirm that $d^2\omega = 0$ in each case.

10.14. Calculate $du \wedge dv$ when

    a. $du = dx, \quad dv = 2y\,dy$;

    b. $du = \cos\theta\,dr - r\sin\theta\,d\theta, \quad dv = \sin\theta\,dr + r\cos\theta\,d\theta$;

    c. $du = 2x\,dx - 2y\,dy, \quad dv = 2y\,dx + 2x\,dy$;

    d. $du = 3(x^2 - y^2)\,dx - 6xy\,dy, \quad dv = 6xy\,dx + 3(x^2 - y^2)\,dy$.

10.15. For each of the following 1-forms ω, first show that $d\omega = 0$ and then find a function f for which $\omega = df$. That is, show ω is the differential of a 0-form (or function).

    a. $\omega(x,y) = x\,dx + \cos y\,dy$.

    b. $\omega(x,y) = f(x)\,dx + g(y)\,dy$.

    c. $\omega(u,v) = 2v\,du + 2u\,dv$.

    d. $\omega(x,y,z) = yz\,dx + zx\,dy + xy\,dz$.

    e. $\omega(x,y,z) = (y+z)\,dx + (z+x)\,dy + (x+y)\,dz$

    f. $\omega(x,y) = \dfrac{1}{y}\,dx - \dfrac{x}{y^2}\,dy$.

    g. $\omega(x,y,u,v) = \dfrac{u}{yv}\,dx - \dfrac{xu}{y^2v}\,dy + \dfrac{x}{yv}\,du - \dfrac{xu}{yv^2}\,dv$.

10.16. For each of the following 2-forms α, first show that $d\alpha = 0$ and then find a 1-form ω for which $d\omega = \alpha$.

    a. $\alpha(x,y) = (x-y)\,dx \wedge dy$.

    b. $\alpha(x,y) = \varphi(x,y)\,dx \wedge dy$.

    c. $\alpha(x,y,z) = dx \wedge dy + dy \wedge dz + dz \wedge dx$.

10.17. Let $\omega = (x^2 + y^2)\,dx\,dy\,dz$. Determine $\alpha = P(x,y,z)\,dx\,dy$ and $\beta = Q(x,y,z)\,dy\,dz$ so that $\omega = d\alpha = d\beta$.

10.18. Let $\omega = \frac{1}{2}(-y\,dx + x\,dy)$, $\alpha = d\omega = dx\,dy$, and let (cf. p. 359)

$$\varphi : \begin{cases} x = \sin s\,\cosh t, \\ y = \cos s\,\sinh t. \end{cases}$$

Determine the pullbacks $\varphi^*\omega$ and $\varphi^*\alpha$ and confirm that $d\varphi^*\omega = \varphi^*\alpha$.

10.19. (Spherical coordinates). Let

$$\sigma : \begin{cases} x = \rho\sin\varphi\cos\theta \ (= r\cos\theta) \\ y = \rho\sin\varphi\sin\theta \ (= r\sin\theta), \\ z = \rho\cos\varphi. \end{cases}$$

These equations are similar to the *spherical coordinates* of Exercise 5.10, page 178. The difference is that here φ is **co-latitude**, measuring the angle

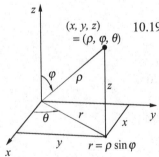

down from the positive z-axis. In the earlier exercise, φ was **latitude**, measuring the angle up from the (x,y)-plane. Here $(\rho,\varphi,\theta) \to (x,y,z)$ is seen to be orientation-preserving. (With φ representing *latitude*, however, the order needs to be (ρ,θ,φ): in Exercise 5.10, $\partial(x,y,z)/\partial(\rho,\theta,\varphi) \geq 0$.)

   a. Determine the Jacobian $\partial(x,y,z)/\partial(\rho,\varphi,\theta)$ as a function of ρ, φ, and θ.

   b. Determine the differentials dx, dy, and dz in terms of ρ, φ, θ, and their differentials.

   c. Determine the volume element $dx \wedge dy \wedge dz$ in terms of ρ, φ, θ, and the volume element $d\rho \wedge d\varphi \wedge d\theta$. Compare this to the Jacobian you obtained in part (a).

10.20. (Cylindrical coordinates: (r,θ,z)). These replace x and y by polar coordinates while leaving z unchanged: $x = r\cos\theta$, $y = r\sin\theta$, $z = z$.

   a. Determine the Jacobian $\partial(x,y,z)/\partial(r,\theta,z)$. Given the relation to polar coordinates in the plane, is this what you would expect?

   b. Determine the volume element $dx \wedge dy \wedge dz$ in terms of r, θ, z, and the volume element $dr \wedge d\theta \wedge dz$. Again, is this what you would expect?

10.21. Determine the pullback $\boldsymbol{\sigma}^*\alpha$ where $\boldsymbol{\sigma}$ is the spherical coordinates map of Exercise 10.19 and $\alpha = x\,dy\,dz + y\,dz\,dx + z\,dx\,dy$.

10.22. Let $\beta = x\,dy\,dz + y\,dz\,dx - 2z\,dx\,dy$, and let

$$\mathbf{f}: \begin{cases} x = \alpha\cos u\,\cosh v, \\ y = \alpha\sin u\,\cosh v, \\ x = v. \end{cases}$$

Determine the pullbacks $\mathbf{f}^*(dy\,dz)$, $\mathbf{f}^*(dz\,dx)$, $\mathbf{f}^*(dx\,dy)$, and $\mathbf{f}^*(\beta)$.

10.23. Let $\vec{S}$ be the surface defined parametrically by $x = u + v$, $y = u - v$, $z = v$, where $-1 \leq u \leq 3$, $0 \leq v \leq 2$ is positively oriented. Determine

$$\iint_{\vec{S}} xy\,dx\,dy + yz\,dy\,dz + zx\,dx\,dy.$$

10.24. Let $\vec{S}$ be parametrized by $x = a\cos u$, $y = a\sin u$, $z = v$, where $0 \leq u \leq 2\pi$, $0 \leq v \leq h$ is positively oriented.

   a. Show that $\vec{S}$ is a *cylinder* of radius a whose axis is the z-axis. Sketch $\vec{S}$, showing where the images of the u- and v-axes lie on the cylinder, and show how this indicates the orientation of $\vec{S}$.

   b. Determine the pullback of the 2-form $\alpha = (x^2 + y^2)\,dy \wedge dz$ to the (u,v)-plane and then determine

$$\iint_{\vec{S}} \alpha.$$

c. Show that $\iint_{\vec{S}} f(x,y,z)\, dx \wedge dy = 0$ for any function $f(x,y,z)$.

10.25. Let $\vec{S} = \mathbf{m}(\vec{U})$ is the oriented surface in $\mathbb{R}^4$ parametrized by

$$\mathbf{m}: \begin{cases} p = x^2 + y^2, \\ q = x - y, \\ r = xy, \\ s = x + y, \end{cases} \qquad \vec{U}: \begin{array}{l} 0 \le x \le 1, \\ 0 \le y \le 1. \end{array}$$

Let $\beta = pq\,dq \wedge dr + qr\,dp \wedge ds$.

a. Determine the pullback $\mathbf{m}^*(\beta)$.

b. Determine $\iint_{\vec{S}} \beta$.

10.26. Sketch the oriented curve $\vec{C}$,

$$(x(t), y(t)) = e^{\sin(t/2)}(\cos t, \sin t), \quad 0 \le t \le 4\pi,$$

and determine its winding number (Definition 10.14, p. 430).

10.27. Prove the claim

$$\frac{\partial(y,z)}{\partial(v,w)}\frac{\partial(v,w)}{\partial(s,t)} + \frac{\partial(y,z)}{\partial(w,u)}\frac{\partial(w,u)}{\partial(s,t)} + \frac{\partial(y,z)}{\partial(u,v)}\frac{\partial(u,v)}{\partial(s,t)} = \frac{\partial(y,z)}{\partial(s,t)}$$

made in the proof of Lemma 10.1.

10.28. Prove Theorem 10.16 for differential forms in three variables.

Chapter 11
Stokes' Theorem

Abstract Stokes' theorem equates the integral of one expression over a surface to the integral of a related expression over the curve that bounds the surface. A similar result, called Gauss's theorem, or the divergence theorem, equates the integral of a function over a 3-dimensional region to the integral of a related expression over the surface that bounds the region. The similarities are not accidental. Using the language of differential forms, we show these two theorems are instances (along with Green's theorem and the fundamental theorem of calculus) of a single theorem that connects one integral over a domain to a related one over its boundary. To explore the connections, we combine the "modern" approach, using differential forms to clarify statements and proofs, with the "classical" appoach, using vector fields to understand the individual theorems in the physical terms in which they arose.

11.1 Divergence

In this section, we analyze the flux of a continuously differentiable vector field $\mathbb{V} = (P, Q, R)$ through the boundary ∂D of a solid region D, in the direction of the normal on ∂D that points out of D. We saw, in an example worked out on pages 420–422, that the net flux could be nonzero. In other words, inward flow and outward flow need not always balance.

First take D to be a parallelpiped B whose edges are parallel to the coordinate axes (a box). Suppose B is centered at the point $(x, y, z) = (a, b, c)$ and has length Δx, width Δy, and height Δz. Its boundary ∂B consists of three pairs of plane parallel faces. The face S_{x+} of ∂B that lies in the plane $x = a + \Delta x/2$ has area $\Delta y \Delta z$ and outward normal $\mathbf{n}_x = (1, 0, 0)$. If the box is sufficiently small, we can approximate $\mathbb{V}$ everywhere on S_{x+} by its value at the center $(x, y, z) = (a + \Delta x/2, b, c)$ of S_{x+}. Under this assumption, total flux (Definition 10.1, p. 389) through S_{x+} is approximately

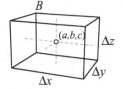

$$\Phi_{x+} \approx \mathbb{V}(a + \Delta x/2, b, c) \cdot \mathbf{n}_x \, \Delta y \Delta z = P(a + \Delta x/2, b, c) \, \Delta y \Delta z.$$

J.J. Callahan, *Advanced Calculus: A Geometric View*, Undergraduate Texts in Mathematics, DOI 10.1007/978-1-4419-7332-0_11, © Springer Science+Business Media, LLC 2010

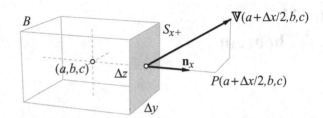

The parallel face S_{x-} that lies in the plane $x = a - \Delta x/2$ has the same area but the opposite outward normal $-\mathbf{n}_x$. Approximating $\mathbb{V}$ everywhere by its value at $(a - \Delta x/2, b, c)$, we can then write

$$\Phi_{x-} \approx \mathbb{V}(a - \Delta x/2, b, c) \cdot (-\mathbf{n}_x)\, \Delta y \Delta z = -P(a - \Delta x/2, b, c)\Delta y \Delta z.$$

Therefore,

$$\Phi_{x+} + \Phi_{x-} \approx \big(P(a + \Delta x/2, b, c) - P(a - \Delta x/2, b, c)\big)\Delta y \Delta z.$$

Approximate total flux through a pair of faces

By the microscope equation,

$$P(a + \Delta x/2, b, c) - P(a - \Delta x/2, b, c) \approx \frac{\partial P}{\partial x}(a, b, c)\Delta x$$

when $\Delta x \approx 0$, so

$$\Phi_{x+} + \Phi_{x-} \approx \frac{\partial P}{\partial x}(a, b, c)\Delta x \Delta y \Delta z = \frac{\partial P}{\partial x}(a, b, c)\Delta V,$$

where ΔV is the volume of the box B.

There are similar formulas for the other faces. For the pair $S_{y\pm}$ that lie in the planes $y = b \pm \Delta y/2$, the normals are $\pm \mathbf{n}_y = (0, \pm 1, 0)$, and we find

$$\Phi_{y+} + \Phi_{y-} \approx \big(Q(a, b + \Delta y/2, c) - Q(a, b - \Delta y/2, c)\big)\Delta z \Delta x$$

$$\approx \frac{\partial Q}{\partial y}(a, b, c)\Delta y \Delta z \Delta x = \frac{\partial Q}{\partial y}(a, b, c)\Delta V.$$

Similarly, for $S_{z\pm}$ we have

$$\Phi_{z+} + \Phi_{z-} \approx \big(R(a, b, c + \Delta z/2) - R(a, b, c - \Delta y/2)\big)\Delta z \Delta x \approx \frac{\partial R}{\partial z}(a, b, c)\Delta V.$$

Approximate total flux out of the box

Therefore, we estimate the total flux through ∂B in the outward direction to be

$$\Phi \approx \left(\frac{\partial P}{\partial x} + \frac{\partial Q}{\partial y} + \frac{\partial R}{\partial z}\right)_{(a,b,c)} \Delta V.$$

For boxes that are small enough for this formula to provide a good approximation, we find that total flux is proportional to the volume of the box. It is remarkable that any formula for Φ should involve 3-dimensional volume, because Φ measures flow through 2-dimensional surfaces. The proportionality factor that connects flux to volume is the scalar quantity

$$\frac{\partial P}{\partial x} + \frac{\partial Q}{\partial y} + \frac{\partial R}{\partial z},$$

evaluated at the center of the box. Thus, although total flux of $\mathbb{V}$ must certainly depend on $\mathbb{V}$, we find that when the surface is the complete boundary of a small box, total flux depends only on a certain scalar, called the *divergence*, derived from the components of $\mathbb{V}$.

Definition 11.1 *The **divergence** of the vector field $\mathbb{V} = (P, Q, R)$ is the scalar field (i.e., function)*

$$\operatorname{div} \mathbb{V} = \frac{\partial P}{\partial x} + \frac{\partial Q}{\partial y} + \frac{\partial R}{\partial z}.$$

The *divergence* of a vector field

To illustrate, consider our earlier example (pp. 420–422) of the total flux Φ of the vector field $\mathbb{V} = (x + y, y - x, 0)$ out of the unit cube. Here

$$\operatorname{div} \mathbb{V} = \frac{\partial}{\partial x}(x+y) + \frac{\partial}{\partial y}(y-x) + \frac{\partial}{\partial z}0 = 1 + 1 = 2,$$

a constant. The volume of the unit cube is $\Delta V = 1$; therefeore we obtain the estimate $\Phi \approx 2 \times 1 = 2$. In fact, we already found $\Phi = 2$ by direct calculation. Even though the unit cube is not "small," the estimate still works well because $\operatorname{div} \mathbb{V}$ is the same at all points.

The product $\operatorname{div} \mathbb{V} \Delta V$ in a small box leads us to a triple integral in a larger region. On page 309, we introduced the triple integral of a function $f(x, y, z)$ over a region D in (x, y, z)-space that has volume (3-dimensional Jordan content). In particular, if $f(x, y, z)$ is bounded and continuous on D, then

From $\operatorname{div} \mathbb{V} \Delta V$ to a triple integral

$$\iiint_D f(x, y, z) \, dV$$

exists (Theorem 8.35, p. 305, adapted from double to triple integrals). Furthermover, if $\delta(D)$, the diameter of D (Definition 8.14, p. 291) is sufficiently small, then it follows from Corollary 8.30, p. 299 that

$$f(a, b, c) \Delta V \approx \iiint_D f(x, y, z) \, dV,$$

where (a, b, c) is a point in D. Hence, when $\mathbb{V}$ is continuously differentiable, so that $\operatorname{div} \mathbb{V}$ is continuous, and B is a box with small diameter,

$$\operatorname{div} \mathbb{V}(a, b, c) \Delta V \approx \iiint_B \operatorname{div} \mathbb{V} \, dV.$$

But because we have just found that $\mathrm{div}\,\mathbb{V}(a,b,c)\,\Delta V$ approximates the total flux Φ through ∂B in the outward unit direction $\mathbf{n}$, we can also write

$$\mathrm{div}\,\mathbb{V}(a,b,c)\,\Delta V \approx \iint_{\partial B} \mathbb{V} \cdot \mathbf{n}\, dA, \ \text{ implying } \ \iiint_B \mathrm{div}\,\mathbb{V}\, dV \approx \iint_{\partial B} \mathbb{V} \cdot \mathbf{n}\, dA.$$

In fact, we show that these two integrals are actually equal: they both represent the total flux. More generally, we show that, for a large class of regions D, the triple integral of $\mathrm{div}\,\mathbb{V}$ over D equals the surface integral of $\mathbb{V} \cdot \mathbf{n}$ over ∂D. This equality is called the *divergence theorem*, or *Gauss's theorem*.

Theorem 11.1. *Let B be the unit cube, $0 \leq x,y,z \leq 1$, and $\mathbf{n}$ the outward unit normal on ∂B. Let $\mathbb{V}$ be a continuously differentiable vector field defined on an open set containing B; then*

$$\iiint_B \mathrm{div}\,\mathbb{V}\, dV = \iint_{\partial B} \mathbb{V} \cdot \mathbf{n}\, dA.$$

Proof. To make the proof clearer, we convert the integrands to differential forms. If $\mathbb{V} = (P,Q,R)$, then (cf. pp. 403–404)

$$\mathbb{V} \cdot \mathbf{n}\, dA = P\,dy\,dz + Q\,dz\,dx + R\,dx\,dy,$$
$$\mathrm{div}\,\mathbb{V}\, dV = (P_x + Q_y + R_z)\,dx\,dy\,dz.$$

Now that the integrands are differential forms, the domains of integration must be oriented. Let $\vec{B}$ have the positive orientation given by the standard basis vectors of $\mathbb{R}^3$ in their usual order. Let $\partial\vec{B}$ have the orientation induced by $\vec{B}$; this is the orientation given by $\mathbf{n}$, the outward unit normal on $\partial\vec{B}$.

Now we show that

$$\iiint_{\vec{B}} P_x\, dx\,dy\,dz = \iint_{\partial\vec{B}} P\, dy\,dz.$$

A similar approach can be used to show the other two pairs of components are equal, thus completing the proof.

Let us label the faces of $\partial\vec{B}$ using the notation from the example on pages 420–422. Thus, for example, $\vec{S}_{x=0}$ is the face (properly oriented) that lies in the plane $x = 0$, and

$$\partial\vec{B} = \vec{S}_{x=0} + \vec{S}_{x=1} + \vec{S}_{y=0} + \vec{S}_{y=1} + \vec{S}_{z=0} + \vec{S}_{z=1}.$$

Because $dy = 0$ on the faces $\vec{S}_{y=0}$ and $\vec{S}_{y=1}$, and because $dz = 0$ on the faces $\vec{S}_{z=0}$ and $\vec{S}_{z=1}$, we find

$$\iint_{\partial\vec{B}} P\, dy\,dz = \iint_{\vec{S}_{x=0}} P\, dy\,dz + \iint_{\vec{S}_{x=1}} P\, dy\,dz$$
$$= \int_0^1 \int_0^1 -P(0,y,z)\, dy\,dz + \int_0^1 \int_0^1 P(1,y,z)\, dy\,dz$$
$$= \int_0^1 \int_0^1 \big(P(1,y,z) - P(0,y,z)\big)\, dy\,dz.$$

As we saw with similar computations on page 421, to take the orientation of $S_{x=0}$ properly into account when we use y and z as parameters, we must include the minus sign in the integral of $P(0,y,z)$.

We can compute the triple integral as a simple (threefold) iterated integral:

$$\iiint_{\vec{B}} P_x \, dx\,dy\,dz = \int_0^1 \int_0^1 \left(\int_0^1 P_x \, dx \right) dy\,dz = \int_0^1 \int_0^1 \left(P(x,y,x) \Big|_{x=0}^{x=1} \right) dy\,dz$$

$$= \int_0^1 \int_0^1 \big(P(1,y,z) - P(0,y,z) \big) \, dy\,dz.$$

Thus the surface integral and the triple integral are equal; by the remark made above, this completes the proof. □

Theorem 11.2. *Let B be the unit ball, $x^2 + y^2 + z^2 \leq 1$, and $\mathbf{n}$ the outward unit normal on ∂B. Let $\mathbb{V}$ be a continuously differentiable vector field defined on an open set containing B; then*

Divergence theorem for a unit ball

$$\iiint_B \operatorname{div} \mathbb{V} \, dV = \iint_{\partial B} \mathbb{V} \cdot \mathbf{n} \, dA.$$

Proof. Again we convert the integrands to differential forms, orient the domains appropriately, and then show that

$$\iiint_{\vec{B}} P_x \, dx\,dy\,dz = \iint_{\partial \vec{B}} P \, dy\,dz;$$

similar arguments prove that the other two pairs of components are equal.

To determine the surface integral, let $\vec{S}_+$ and $\vec{S}_-$ be the graphs of the functions

$$\vec{S}_+ : x = +\sqrt{1 - y^2 - z^2} \quad \text{and} \quad \vec{S}_- : x = -\sqrt{1 - y^2 - z^2},$$

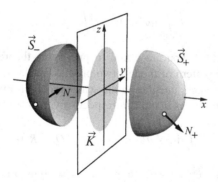

defined on the positively oriented disk $\vec{K} : y^2 + z^2 \leq 1$, and inheriting their orientations from $\vec{K}$. (In the figure, the surfaces are shown separated for clarity.) The orientation normals N_+ and N_- of both surfaces therefore point in the positive x-direction. Thus, N_+ points outward, but N_- points inward, so $\partial \vec{B} = \vec{S}_+ - \vec{S}_-$, and

$$\iint_{\partial \vec{B}} P\, dy\, dz = \iint_{\vec{S}_+} P\, dy\, dz - \iint_{\vec{S}_-} P\, dy\, dz$$

$$= \iint_{\vec{K}} P(\sqrt{1-y^2-z^2},y,z)\, dy\, dz - \iint_{\vec{K}} P(-\sqrt{1-y^2-z^2},y,z)\, dy\, dz$$

$$= \iint_{\vec{K}} \left(P(\sqrt{1-y^2-z^2},y,z) - P(-\sqrt{1-y^2-z^2},y,z) \right) dy\, dz$$

To determine the triple integral, let $\vec{B}$ be the positively oriented solid region given by the inequalties

$$\vec{B}: \quad \begin{array}{c} y^2+z^2 \le 1, \\ -\sqrt{1-y^2-z^2} \le x \le \sqrt{1-y^2-z^2}. \end{array}$$

Then

$$\iiint_{\vec{B}} P_x\, dx\, dy\, dz = \iint_{\vec{K}} \left(\int_{-\sqrt{1-y^2-z^2}}^{\sqrt{1-y^2-z^2}} P_x(x,y,z)\, dx \right) dy\, dz$$

$$= \iint_{\vec{K}} P(x,y,z) \Big|_{x=-\sqrt{1-y^2-z^2}}^{x=\sqrt{1-y^2-z^2}} dy\, dz$$

$$= \iint_{\vec{K}} \left(P(\sqrt{1-y^2-z^2},y,z) - P(-\sqrt{1-y^2-z^2},y,z) \right) dy\, dz,$$

so the triple integral is equal to the surface integral. By what has been said above, this proves the theorem. $\qquad\square$

Divergence theorem for a unit tetrahedron

Theorem 11.3. *Let B be the unit tetrahedron, $0 \le x,y,z$, $x+y+z \le 1$, and $\mathbf{n}$ the outward unit normal on ∂B. Let $\mathbb{V}$ be a continuously differentiable vector field defined on an open set containing B; then*

$$\iiint_B \operatorname{div} \mathbb{V}\, dV = \iint_{\partial B} \mathbb{V} \cdot \mathbf{n}\, dA.$$

Proof. In Exercise 11.8, you are asked to prove this theorem using differential forms, following the pattern of the last two proofs. For the sake of illustration, we take an alternate approach, integrating the scalar functions $\operatorname{div}\mathbb{V}$ and $\mathbb{V} \cdot \mathbf{n}$ directly over the unoriented domains B and ∂B, respectively. If $\mathbb{V} = (P,Q,R)$, then

$$\iiint_B \operatorname{div}\mathbb{V}\, dV = \iiint_B (P_x + Q_y + R_z)\, dV$$

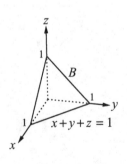

We convert each term into an iterated integral, describing B by inequalities in three different ways, each suited to the term being integrated. To integrate P_x, let

$$B: \quad \begin{array}{c} 0 \le y \le 1, \\ 0 \le z \le 1-y, \\ 0 \le x \le 1-y-z; \end{array}$$

then

$$\iiint_B P_x \, dV = \int_0^1 \int_0^{1-y} \left(\int_0^{1-y-z} P_x(x,y,z) \, dx \right) dz \, dy$$

$$= \int_0^1 \int_0^{1-y} P(x,y,z) \Big|_0^{1-y-z} dz \, dy$$

$$= \int_0^1 \int_0^{1-y} \big(P(1-y-z,y,z) - P(0,y,z) \big) \, dz \, dy.$$

By changing the description of B appropriately, we obtain similar expressions for the integrals of Q_y and R_z:

$$\iiint_B Q_y \, dV = \int_0^1 \int_0^{1-z} \big(Q(x,1-x-z,z) - Q(x,0,z) \big) \, dx \, dz,$$

$$\iiint_B R_z \, dV = \int_0^1 \int_0^{1-x} \big(R(x,y,1-x-y) - R(x,y,0) \big) \, dy \, dx.$$

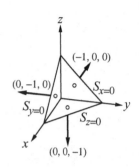

(Recall that the order of the differentials here indicates merely the order in which the integrations are to be carried out, not the orientation of the domain of integration.)

In the tetrahedral surface ∂B, three faces $S_{x=0}$, $S_{y=0}$, and $S_{z=0}$ lie in coordinate planes; the fourth, S_1, lies in the plane $x+y+z=1$. Because the outward unit normal on $S_{x=0}$ is $\mathbf{n} = (-1,0,0)$, we have

$$\iint_{S_{x=0}} \mathbb{V} \cdot \mathbf{n} \, dA = \iint_{S_{x=0}} (P,Q,R) \cdot (-1,0,0) \, dA = \int_0^1 \int_0^{1-y} -P(0,y,z) \, dz \, dy.$$

In a similar way, $\mathbf{n} = (0,-1,0)$ on $S_{y=0}$ and $\mathbf{n} = (0,0,-1)$ on $S_{z=0}$, so

$$\iint_{S_{y=0}} \mathbb{V} \cdot \mathbf{n} \, dA = \int_0^1 \int_0^{1-z} -Q(x,0,z) \, dx \, dz$$

$$\iint_{S_{z=0}} \mathbb{V} \cdot \mathbf{n} \, dA = \int_0^1 \int_0^{1-x} -R(x,y,0) \, dy \, dx$$

On the fourth face, S_1, the outward unit normal is $\mathbf{n} = \left(\frac{1}{\sqrt{3}}, \frac{1}{\sqrt{3}}, \frac{1}{\sqrt{3}} \right)$, so

$$\iint_{S_1} \mathbb{V} \cdot \mathbf{n} \, dA = \iint_{S_1} \frac{P+Q+R}{\sqrt{3}} \, dA.$$

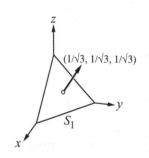

To integrate the first term, $P/\sqrt{3}$, treat S_1 as the graph of

$$x = 1-y-z \text{ on } S_{x=0}: \quad \begin{array}{c} 0 \le y \le 1, \\ 0 \le z \le 1-y. \end{array}$$

To get an expression of dA, we must calculate (cf. Definition 10.6, p. 409)

$$\sqrt{\left[\frac{\partial(y,z)}{\partial(y,z)}\right]^2 + \left[\frac{\partial(z,x)}{\partial(y,z)}\right]^2 + \left[\frac{\partial(x,y)}{\partial(y,z)}\right]^2} = \sqrt{\left|\begin{matrix}1 & 0\\0 & 1\end{matrix}\right|^2 + \left|\begin{matrix}0 & -1\\1 & -1\end{matrix}\right|^2 + \left|\begin{matrix}-1 & 1\\-1 & 0\end{matrix}\right|^2} = \sqrt{3}.$$

Thus $dA = \sqrt{3}\,dy\,dz$, and

$$\iint_{S_1} \frac{P}{\sqrt{3}}\,dA = \int_0^1 \int_0^{1-y} P(1-y-z,y,z)\,dz\,dy.$$

For $Q/\sqrt{3}$, treat S_1 as the graph of $y = 1 - x - z$ on $S_{y=0}$, and for $R/\sqrt{3}$, treat it as $z = 1 - x - y$ on $S_{z=0}$; then

$$\iint_{S_1} \frac{Q}{\sqrt{3}}\,dA = \int_0^1 \int_0^{1-z} Q(x,1-x-z,z)\,dx\,dz,$$

$$\iint_{S_1} \frac{R}{\sqrt{3}}\,dA = \int_0^1 \int_0^{1-x} R(x,y,1-x-y)\,dy\,dx.$$

The triple integral and the surface integral reduce to six iterated double integrals each, and these are equal in pairs.   □

Symbolic form of the divergence theorem

We must still show that the divergence theorem applies to other regions. To do this, it is helpful if we think of the integrals in the theorem as *symbolic pairings* (cf. p. 427). Thus if $\mathbb{V} = (P,Q,R)$ and

$$\mathbb{V} \cdot \mathbf{n}\,dA = P\,dy\,dz + Q\,dz\,dx + R\,dx\,dy = \alpha,$$
$$\operatorname{div}\mathbb{V}\,dV = (P_x + Q_y + R_z)\,dx\,dy\,dz = d\alpha,$$

we write

$$\iiint_B \operatorname{div}V\,dV = \iiint_{\vec{B}} d\alpha = \langle\vec{B},d\alpha\rangle, \qquad \iint_{\partial B} \mathbb{V} \cdot \mathbf{n}\,dA = \iint_{\partial\vec{B}} \alpha = \langle\partial\vec{B},\alpha\rangle.$$

Here $\vec{B}$ is the positively oriented unit cube, unit ball, or unit tetrahedron, and $\partial\vec{B}$ is its boundary with the induced orientation. In terms of symbolic pairings, the divergence theorem thus has the form

$$\langle\vec{B},d\alpha\rangle = \langle\partial\vec{B},\alpha\rangle.$$

Note that when Green's theorem and the fundamental theorem of calculus are expressed in terms of symbolic pairings (pp. 427–429), they have exactly the same form. The essential point of each theorem is that the exterior derivative d and the boundary operator ∂ are adjoints in the symbolic pairing.

Images under coordinate changes

We can now extend the divergence theorem to any region D that is the image, under a coordinate change, of a region (such as a unit cube) to which the divergence theorem already applies.

Theorem 11.4. *Let* $\boldsymbol{\varphi} : \Omega \to \mathbb{R}^3$ *be a coordinate change, and let* $B \subset \Omega$ *be a region for which the divergence theorem is known to hold. Suppose* $D = \boldsymbol{\varphi}(B)$, $\mathbf{n}$ *is the outward unit normal on* ∂D, *and* $\mathbb{V}$ *is a continuously differentiable vector field defined on* $\boldsymbol{\varphi}(\Omega)$; *then*

$$\iiint_D \operatorname{div} \mathbb{V} \, dV = \iint_{\partial D} \mathbb{V} \cdot \mathbf{n} \, dA.$$

Proof. We convert to differential forms, as in the earlier proofs. Thus, let $\alpha = \mathbb{V} \cdot \mathbf{n} \, dA$, $d\alpha = \operatorname{div} \mathbb{V} \, dV$. Let $\vec{B}$ be B with its positive orientation, let $\partial \vec{B}$, $\vec{D} = \boldsymbol{\varphi}(\vec{B})$, and $\partial \vec{D} = \boldsymbol{\varphi}(\partial \vec{B})$ receive the appropriate induced orientations (cf. p. 355), and let $\boldsymbol{\varphi}^*$ be the pullback of $\boldsymbol{\varphi}$ on forms (cf. pp. 433ff.). Then

$$
\begin{aligned}
\iiint_D \operatorname{div} \mathbb{V} \, dV = \iiint_{\vec{D}} d\alpha &= \langle \vec{D}, d\alpha \rangle \\
&= \langle \boldsymbol{\varphi}(\vec{B}), d\alpha \rangle && \text{definition of } \vec{D} \\
&= \langle \vec{B}, \boldsymbol{\varphi}^*(d\alpha) \rangle && \boldsymbol{\varphi} \text{ and } \boldsymbol{\varphi}^* \text{ are adjoints} \\
&= \langle \vec{B}, d(\boldsymbol{\varphi}^*\alpha) \rangle && d \text{ and } \boldsymbol{\varphi}^* \text{ commute} \\
&= \langle \partial \vec{B}, \boldsymbol{\varphi}^*\alpha \rangle && \text{divergence theorem for } \vec{B} \\
&= \langle \boldsymbol{\varphi}(\partial \vec{B}), \alpha \rangle && \boldsymbol{\varphi} \text{ and } \boldsymbol{\varphi}^* \text{ are adjoints} \\
&= \langle \partial \vec{D}, \alpha \rangle && \text{definition of } \partial \vec{D} \\
&= \iint_{\partial \vec{D}} \alpha = \iint_{\partial D} \mathbb{V} \cdot \mathbf{n} \, dA. && \square
\end{aligned}
$$

At two key points in the proof, we use the fact that a coordinate change $\boldsymbol{\varphi}$ and its pullback $\boldsymbol{\varphi}^*$ on differential forms are adjoints (Theorem 10.15, p. 436). This is just the change of variables formula for integrals expressed in terms of symbolic pairings.

A good example of the image of a cube under a coordinate change in $\mathbb{R}^3$ is the region D in (x,y,z)-space between two graphs $z = A(x,y)$ and $z = B(x,y)$, when A and B have a common domain of the form

$$D_0 : \begin{array}{l} a \le x \le b, \\ \alpha(x) \le y \le \beta(x). \end{array}$$

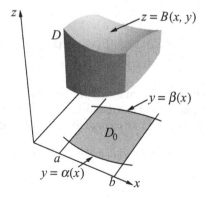

We assume that α and β are continuously differentiable on some open interval I containing $[a,b]$, and A and B are likewise continuously differentiable on some open set Ω_0 containing D_0 in the (x,y)-plane. The inequalites defining D allow us to transform a triple integral over D into a threefold iterated integral:

$$\iiint_D f(x,y,z)\,dV = \int_a^b \left(\int_{\alpha(x)}^{\beta(x)} \left(\int_{A(x,y)}^{B(x,y)} f(x,y,z)\,dz \right) dy \right) dx.$$

The image of a cube

Theorem 11.5. *Suppose there are continuously differentiable functions $\alpha(x)$, $\beta(x)$, $A(x,y)$, and $B(x,y)$ for which $\alpha_0 < \alpha(x) < \beta(x) < \beta_0$ for every x in an open interval I, and $A_0 < A(x,y) < B(x,y) < B_0$ for every (x,y) in an open set Ω_0. Suppose $[a,b] \subset I$, and suppose $D_0 \subset \Omega_0$ is given by*

$$D_0: \qquad a \le x \le b, \qquad \alpha(x) \le y \le \beta(x).$$

If D is the region in (x,y,z)-space defined by

$$D: \qquad (x,y) \in D_0, \qquad A(x,y) \le z \le B(x,y),$$

then D is the image of the unit cube B under a coordinate change.

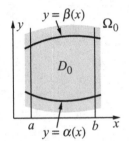

Proof. Let Ω be the open set in $\mathbb{R}^3$ constructed as the product of Ω_0 in the (x,y)-plane and the open interval (A_0,B_0) on the z-axis: $\Omega = \Omega_0 \times (A_0,B_0)$. Now define $\varphi : \Omega \to \mathbb{R}^3$ as

$$\varphi : \begin{cases} u = \dfrac{x-a}{b-a}, \\[2mm] v = \dfrac{y-\alpha(x)}{\beta(x)-\alpha(x)}, \\[2mm] w = \dfrac{z-A(x,y)}{B(x,y)-A(x,y)}. \end{cases}$$

Because the components are continuously differentiable and the denominators are never zero on Ω, φ is well-defined and continuously differentiable.

If $a \le x \le b$, then $0 \le u \le 1$. If, in addition, $\alpha(x) \le y \le \beta(x)$, then $0 \le v \le 1$. Finally, if $A(x,y) \le z \le B(x,y)$ as well, then $0 \le w \le 1$. Thus, $\varphi(D) = B$, the unit cube. We can solve for x, y, and z to get the inverse:

$$\varphi^{-1} : \begin{cases} x = a + (b-a)u, \\ y = \alpha(x) + \big(\beta(x)-\alpha(x)\big)v, \\ z = A(x,y) + \big(B(x,y)-A(x,y)\big)w, \end{cases}$$

understanding that x will be replaced by $a+(b-a)u$ in the formula for y, and these expressions will then replace x and y in the formula for z. Hence φ^{-1} is continuously differentiable, so φ^{-1} is a coordinate change for which $D = \varphi^{-1}(B)$. $\square$

Corollary 11.6 *The divergence theorem holds for the region D.* $\square$

The solid region D is the 3-dimensional analogue of the basic 2-dimensional region on which we established Green's theorem (cf. Theorem 9.18, p. 364, and pp. 367–368). In the final version of Green's theorem, we assumed the domain can be decomposed into a finite number of nonoverlapping regions on which Green's theorem is known to apply. Our final version of the divergence theorem is similar.

The divergence theorem on more general regions

Theorem 11.7 (Divergence theorem). *Suppose $\mathbb{V}$ is a continuously differentiable vector field defined on an open set Ω in $\mathbb{R}^3$, and D is a closed bounded subset of Ω. Suppose $D_1, \ldots, D_k$ are nonoverlapping regions in $\mathbb{R}^3$ on which the divergence theorem applies, and $\vec{D} = \vec{D}_1 + \cdots + \vec{D}_k$ when all regions are positively oriented; then*

$$\iiint_D \operatorname{div} \mathbb{V} \, dV = \iint_{\partial D} \mathbb{V} \cdot \mathbf{n} \, dA.$$

Proof. By the additivity of triple integrals, we know immediately that

$$\iiint_D \operatorname{div} \mathbb{V} \, dV = \iiint_{D_1} \operatorname{div} \mathbb{V} \, dV + \cdots + \iiint_{D_k} \operatorname{div} \mathbb{V} \, dV.$$

The surface integrals combine in a more interesting way. If two cells $\vec{D}_i$ and $\vec{D}_j$ meet along a face S, then, at any point $\mathbf{p}$ on S, the outward normal $\mathbf{n}_i(\mathbf{p})$ from $\vec{D}_i$ is opposite the outward normal $\mathbf{n}_j(\mathbf{p})$ from $\vec{D}_j$: $\mathbf{n}_j(\mathbf{p}) = -\mathbf{n}_i(\mathbf{p})$. Therefore,

$$\underbrace{\iint_S \mathbb{V} \cdot \mathbf{n}_j \, dA}_{S \text{ as part of } \partial D_j} = \iint_S \mathbb{V} \cdot -\mathbf{n}_i \, dA = -\underbrace{\iint_S \mathbb{V} \cdot \mathbf{n}_i \, dA}_{S \text{ as part of } \partial D_i},$$

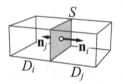

so the contributions that S makes to

$$\iint_{\partial D_i} \mathbb{V} \cdot \mathbf{n} \, dA \text{ and } \iint_{\partial D_j} \mathbb{V} \cdot \mathbf{n} \, dA$$

exactly cancel. The only contributions that do not cancel are from the faces S that $\partial \vec{D}_i$ shares with $\partial \vec{D}$ itself. In those circumstances (and because $\vec{D}_i$ lies in $\vec{D}$), the outward normal on S is the same for $\vec{D}_i$ and for $\vec{D}$, so

$$\underbrace{\iint_S \mathbb{V} \cdot \mathbf{n} \, dA}_{\text{as part of } \partial D_i} = \underbrace{\iint_S \mathbb{V} \cdot \mathbf{n}_i \, dA}_{\text{as part of } \partial D}.$$

Therefore, after all the cancellations are taken into account,

$$\iint_{\partial D} \mathbb{V} \cdot \mathbf{n} \, dA = \iint_{\partial D_1} \mathbb{V} \cdot \mathbf{n} \, dA + \cdots + \iint_{\partial D_k} \mathbb{V} \cdot \mathbf{n} \, dA.$$

By hypothesis,

$$\iiint_{D_i} \operatorname{div} \mathbb{V} \, dV = \iint_{\partial D_i} \mathbb{V} \cdot \mathbf{n} \, dA$$

for each $i = 1, \ldots, k$, so the proof is complete. $\square$

11.2 Circulation and vorticity

The differential forms corresponding to a field

In the previous section we used the connection between a vector field and its divergence, on the one hand, and a corresponding 2-form and its exterior derivative. Quite generally, there is a natural way to make a vector field correspond to either a 1-form or a 2-form, and a scalar field (i.e., a function) to either a 0-form or a 3-form:

$$f \leftrightarrow \omega_f^0 = f,$$

$$\mathbb{F} = (A, B, C) \leftrightarrow \omega_{\mathbb{F}}^1 = A\,dx + B\,dy + C\,dz,$$

$$\mathbb{V} = (P, Q, R) \leftrightarrow \omega_{\mathbb{V}}^2 = P\,dy\,dz + Q\,dz\,dx + R\,dx\,dy,$$

$$H \leftrightarrow \omega_H^3 = H\,dx\,dy\,dz.$$

$d(\omega_{\mathbb{V}}^2) = \omega_{\text{div}\,\mathbb{V}}^3.$

The connection we made between the divergence and the exterior derivative can now be viewed in the following light. For the 2-form corresponding to $\mathbb{V}$,

$$\omega_{\mathbb{V}}^2 = P\,dy\,dz + Q\,dz\,dx + R\,dx\,dy,$$

we have

$$d(\omega_{\mathbb{V}}^2) = (P_x + Q_y + R_z)\,dx\,dy\,dz = \omega_{\text{div}\,\mathbb{V}}^3.$$

Hence, we can reformulate the divergence theorem as

$$\iiint_{\vec{B}} \omega_{\text{div}\,\mathbb{V}}^3 = \iint_{\partial \vec{B}} \omega_{\mathbb{V}}^2, \quad \text{or} \quad \langle \vec{B}, \omega_{\text{div}\,\mathbb{V}}^3 \rangle = \langle \partial \vec{B}, \omega_{\mathbb{V}}^2 \rangle,$$

for every suitable oriented region $\vec{B}$ in $\mathbb{R}^3$.

$d(\omega_f^0) = \omega_{\text{grad}\,f}^1.$

Suppose instead that we begin with the 0-form ω_f^0 corresponding to a function f; then

$$d(\omega_f^0) = df = f_x\,dx + f_y\,dy + f_z\,dz = \omega_{\text{grad}\,f}^1,$$

where $\text{grad}\,f = (f_x, f_y, f_z)$ is the gradient vector field of f. For any piecewise-smooth oriented path $\vec{C}$, we have (cf. p. 425)

$$\int_{\vec{C}} \omega_{\text{grad}\,f}^1 = \int_{\vec{C}} \text{grad}\,f \cdot d\mathbf{x} = \int_{\vec{C}} df = f(\text{end of } \vec{C}) - f(\text{start of } \vec{C}).$$

The right-hand side of this equation is the "0-dimensional integral" of $f = \omega_f^0$ over $\partial \vec{C} = $ end of $\vec{C} - $ start of $\vec{C}$ (cf. pp. 428–429). Therefore, we can rewrite the equation itself as the symbolic pairing

$$\langle \vec{C}, \omega_{\text{grad}\,f}^1 \rangle = \langle \partial \vec{C}, \omega_f^0 \rangle.$$

This is, in essence, the fundamental theorem of calculus; compare it to our reformulation of the divergence theorem, above.

Corresponding
differential operators

The gradient and the divergence are differential operators. The gradient takes as input a scalar field and produces as output a vector field. The divergence does the reverse: the input is a vector field, the output a scalar. We have just noted that these two differential operators correspond to the exterior derivative operator on k-forms when $k = 0$ and 2, respectively. What differential operator corresponds to the exterior derivative when $k = 1$, and how is it defined? To answer these questions, we begin with the correspondence

$$\mathbb{F} = (A,B,C) \leftrightarrow \omega_{\mathbb{F}}^1 = A\,dx + B\,dy + C\,dz.$$

A straightforward calculation (cf. p. 425) then gives

$$d(\omega_{\mathbb{F}}^1) = (C_y - B_z)\,dy\,dz + (A_z - C_x)\,dz\,dx + (B_x - A_y)\,dx\,dy.$$

This is a 2-form; it corresponds to the new vector field

$$\mathbb{V} = (C_y - B_z, A_z - C_x, B_x - A_y),$$

whose components are particular combinations of the derivatives of the components of $\mathbb{F}$. For reasons that emerge later, we call $\mathbb{V}$ the *curl* of $\mathbb{F}$.

Definition 11.2 *The **curl** of the vector field $\mathbb{F} = (A,B,C)$ is the vector field*

curl $\mathbb{F}$

$$\mathrm{curl}\,\mathbb{F} = (C_y - B_z, A_z - C_x, B_x - A_y).$$

The curl is thus a differential operator whose input and output are both vector fields. It completes the trio of operators that correspond to the exterior derivative; we have

$d(\omega_{\mathbb{F}}^1) = \omega_{\mathrm{curl}\,\mathbb{F}}^2.$

$$d(\omega_{\mathbb{F}}^1) = \omega_{\mathrm{curl}\,\mathbb{F}}^2.$$

The gradient, divergence, and curl can all be expressed in terms of "nabla," the vector differential operator

nabla

$$\nabla = \left(\frac{\partial}{\partial x}, \frac{\partial}{\partial y}, \frac{\partial}{\partial z}\right)$$

introduced on page 93 for two variables and extended here to three. By treating nabla as if it were an ordinary vector, we can combine it with scalar and vector fields using scalar multiplication and the dot and cross-products. Scalar multiplication (by a function placed to the right of nabla) gives the gradient, a vector function:

$\nabla f = \mathrm{grad}\,f$
$\nabla \cdot \mathbb{V} = \mathrm{div}\,\mathbb{V}$
$\nabla \times \mathbb{F} = \mathrm{curl}\,\mathbb{F}$

$$\nabla f(x,y,z) = \left(\frac{\partial}{\partial x}, \frac{\partial}{\partial y}, \frac{\partial}{\partial z}\right) f = \left(\frac{\partial f}{\partial x}, \frac{\partial f}{\partial y}, \frac{\partial f}{\partial z}\right) = \mathrm{grad}\,f.$$

The dot (or scalar) product with a vector field gives the divergence, a scalar function:

$$\nabla \cdot \mathbb{V} = \left(\frac{\partial}{\partial x}, \frac{\partial}{\partial y}, \frac{\partial}{\partial z}\right) \cdot (P,Q,R) = \frac{\partial P}{\partial x} + \frac{\partial Q}{\partial y} + \frac{\partial R}{\partial z} = \mathrm{div}\,\mathbb{V}.$$

The cross (or vector) product gives the curl, a vector function:

$$\nabla \times \mathbb{F} = \left(\frac{\partial}{\partial x}, \frac{\partial}{\partial y}, \frac{\partial}{\partial z} \right) \times (A, B, C) = \left(\begin{vmatrix} \frac{\partial}{\partial y} & \frac{\partial}{\partial z} \\ B & C \end{vmatrix}, \begin{vmatrix} \frac{\partial}{\partial z} & \frac{\partial}{\partial x} \\ C & A \end{vmatrix}, \begin{vmatrix} \frac{\partial}{\partial x} & \frac{\partial}{\partial y} \\ A & B \end{vmatrix} \right)$$

$$= \left(\frac{\partial C}{\partial y} - \frac{\partial B}{\partial z}, \frac{\partial A}{\partial z} - \frac{\partial C}{\partial x}, \frac{\partial B}{\partial x} - \frac{\partial A}{\partial y}, \right) = \operatorname{curl} \mathbb{F}.$$

Physical meaning
of curl $\mathbb{F}$

What is the physical meaning of curl $\mathbb{F}$ when $\mathbb{F}$ describes a steady fluid flow? The answer to this question is complex and subtle, in part because curl $\mathbb{F}$ is itself a vector rather than a scalar. To explore the question, we begin by looking at some examples.

Example 1:
$\mathbb{F} = (-\omega y, \omega x, 0)$

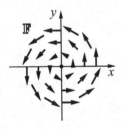

For the first example, take $\mathbb{F} = (-\omega y, \omega x, 0)$, where ω is a constant (not a differential form!). Because the z-component is zero, the field $\mathbb{F}$ is everywhere parallel to the (x, y)-plane, so fluid in the plane $z = $ constant stays in that plane. The figure shows $\mathbb{F}$ in one such plane; it suggests that the z-axis is a *vortex*. We now show, more exactly, that the fluid rotates around the z-axis with angular speed ω.

To begin, recall that $\mathbb{F}(x, y, z)$ represents the velocity of the fluid at the point (x, y, z). Thus, if $\mathbf{x}(t) = (x(t), y(t), z(t))$ represents the position of a particle of fluid at time t, then its velocity at that point is $\mathbb{F}(\mathbf{x}(t))$, so

$$\mathbf{x}'(t) = \mathbb{F}(\mathbf{x}(t)), \quad \text{or} \quad x' = -\omega y, \quad y' = \omega x, \quad z' = 0.$$

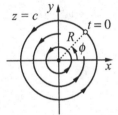

It follows that $x'' = (-\omega y)' = -\omega^2 x$, implying that the solutions are sines and cosines. The general solution is the three-parameter family

$$x(t) = R\cos(\omega t + \phi), \quad y(t) = R\sin(\omega t + \phi), \quad z(t) = c;$$

the parameters $R \geq 0$, ϕ, and c are the arbitrary constants of integration. These equations describe the motion of a fluid particle that is initially (i.e., when $t = 0$) at the point whose cylindrical coordinates are (R, ϕ, c) (cf. Exercise 5.11, p. 178). The particle remains in the plane $z = c$, moves on the circle of radius R centered on the z-axis, and makes an angle of $\theta = \omega t + \phi$ with the positive x-axis at time t. The angular speed is $\theta' = \omega$, as we wished to show.

Angular velocity vector

Any uniform rotation in space (such as we see in this example) is characterized by three elements:

1. Its axis of rotation

2. Its angular speed

3. The direction it rotates around its axis

We can use a vector, the **angular velocity vector**, $\boldsymbol{\omega}$, to represent these three elements. We take $\boldsymbol{\omega}$ to be parallel to the axis of rotation, to have magnitude $\omega = \|\boldsymbol{\omega}\|$ equal to the angular speed, and to have the direction that the thumb points when the fingers of the right hand curl in the direction of the rotation. Thus, a spinning disk with angular velocity $\boldsymbol{\omega}$ turns in the counterclockwise direction when viewed from the side toward which $\boldsymbol{\omega}$ points. (This is the definition to use when the coordinate frame itself is right-handed. If it is left-handed, then we would curl the fingers of the left hand to determine the direction of $\boldsymbol{\omega}$.) The angular velocity vector captures all aspects of a uniform rotation except the location—as distinct from the direction—of the axis of rotation in space.

According to our analysis, the rotation of the flow $\mathbb{F}$ at any point on the z-axis is given by the angular velocity vector $\boldsymbol{\omega} = (0, 0, \omega)$. A quick computation shows that

Angular velocity
and the curl

$$\operatorname{curl} \mathbb{F} = (0, 0, 2\omega) = 2\boldsymbol{\omega},$$

suggesting that $\operatorname{curl} \mathbb{F}$ essentially represents this uniform rotational motion (with $\| \operatorname{curl} \mathbb{F} \|/2$ equal to the angular speed). But there is a problem. Because $\operatorname{curl} \mathbb{F}$ is a field, it assigns a vector to each point (x, y, z) in space, namely, the constant vector $(0, 0, 2\omega)$. At any point $(0, 0, z)$ on the z-axis, this vector appears to explain the rotation we see in the flow. But at no other point is the flow a rotation around that point. What is $\operatorname{curl} \mathbb{F}$ telling us there?

In fact, the curl does gives us information about rotation at every point, but the rotation is not the rotation of the fluid itself. To see what is actually involved, it is helpful to study a second flow that lacks the obvious vortex of Example 1. For Example 2 we take $\mathbb{F} = (-ky, 0, 0)$, $k > 0$.

Example 2:
$\mathbb{F} = (-ky, 0, 0)$

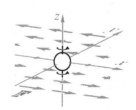

The figure shows how $\mathbb{F}$ looks in the (x, y)-plane; it looks the same in every parallel plane. The fluid flows in straight lines parallel to the x-axis, moving left when $y > 0$ and right when $y < 0$. Everywhere on the (x, z)-plane $(y = 0)$, the fluid is stationary. The fluid does not rotate. Nevertheless,

$$\operatorname{curl} \mathbb{F} = (0, 0, k)$$

at every point (x, y, z). If this were an angular velocity, it would represent a counterclockwise rotation with angular speed $k/2$ around the vertical axis. What, if anything, is rotating?

Place at the origin a little ball with a rough surface like a tennis ball; use one whose density is the same as the fluid's, so it will have no tendency to float or sink. Because the fluid at the origin is motionless, the ball will stay put, but it will not remain motionless. The shearing action of the nearby fluid will make it spin in place around a vertical axis. The fluid at higher and lower levels (i.e., where $z > 0$ and $z < 0$) flows the same way as in the (x, y)-plane; therefore that fluid will not alter the way the ball moves. The net effect is a counterclockwise spin around the vertical; only the magnitude of the spin (its angular speed ω) remains undetermined. The angular velocity vector of the little ball is thus a positive multiple of curl $\mathbb{F}$.

The ball spins the same way everywhere

A similar test ball placed anywhere on the x-axis or, for that matter, anywhere in the vertical (x,z)-plane should behave the same way. Off the (x,z)-plane, the flow is nonzero, and the ball will be carried along by the fluid. But fluid particles at points farther from the (x,z)-plane move even faster, so they drag that side of the ball forward; particles closer to the (x,z)-plane move more slowly, dragging that side of the ball back. The fluid thus has the same shearing effect on the moving ball that it does on the stationary one: it spins the ball counterclockwise as it carries it along. So curl $\mathbb{F}$ describes the rotation of the test ball everywhere, at least qualitatively. Only the quantitative link between angular speed and $\| \text{curl}\,\mathbb{F} \|$ remains undetermined.

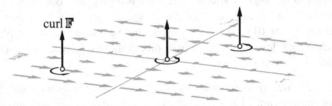

curl $\mathbb{F}$

Vorticity

Let us call this tendency of a moving fluid to spin an object that is carried along with it the **vorticity** of the flow. Example 2 suggests that the vorticity of $\mathbb{F}$ is caused by its shearing action and is decribed by curl $\mathbb{F}$. In fact, by associating the curl with a flow's vorticity instead of its rotation, *per se*, we can clear up the puzzle of Example 1. In that example, fluid farther from the z-axis moves faster than fluid that is closer, but here the flow at one level $z = $ constant is the same as at any other. Therefore, as a test ball moves with the fluid around the z-axis, it also spins because of the shearing action of nearby fluid. At every point, the spin is counterclockwise around a vertical axis, a motion described qualitatively by curl $\mathbb{F} = (0,0,2\omega)$ at that point.

Quantifying vorticity

To describe vorticity quantitatively and not just qualitatively, we need some way to specify the magnitude of the spin induced by the shearing action of a flow. Return to Example 1 and its simple rotational motion around the z-axis. As the fluid moves around the circle of radius r centered at the origin in the plane $z = c$, its (linear) speed at any point is ωr. If we think of speed as a measure of the "motion" of a fluid, then the quantity

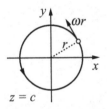

$$\text{speed} \times \text{length of path} = 2\omega\pi r^2$$

describes, in some sense, the total motion of the fluid as it travels around that circle. Note that this quantity, which we call the *circulation*, is proportional to the area of the circle. In fact,

$$\frac{\text{circulation}}{\text{area}} = \frac{2\omega\pi r^2}{\pi r^2} = 2\omega$$

is exactly the magnitude of the vector curl $\mathbb{F} = (0,0,2\omega)$.

Staying with Example 1, we ask: Can we determine the circulation of the fluid around a circle C in some plane $z = c$ but centered at a point away from the origin? Along C, the fluid flow is, in general, not longer tangent, so we need to decide how to measure the fluid's "motion." The figure in the margin suggests we replace the flow $\mathbb{F}$ by its tangential component $\mathbb{F}_t$. Here $\mathbf{t}$ is the unit tangent vector to C in the counterclockwise direction. With this choice of $\mathbf{t}$, C becomes the oriented path that we denote as $\vec{C}$. Because the speed of the flow given by the tangential component is

$$\|\mathbb{F}_t\| = \mathbb{F} \cdot \mathbf{t},$$

the "total motion" of this flow around $\vec{C}$ will be the integral of this scalar quantity with respect to arc length:

$$\text{circulation} = \oint_C \mathbb{F} \cdot \mathbf{t} \, ds.$$

(The domain of a scalar integral is an unoriented path; cf. Definition 1.6.) On page 19 we noted that this scalar integral has the same value as the vector integral

$$\oint_{\vec{C}} \mathbb{F} \cdot d\mathbf{x},$$

where $\vec{C}$ is C provided with the orientation given by the unit tangent $\mathbf{t}$. If we parametrize $\vec{C}$ as $(a + r\cos t, b + r\sin t, c)$ with $0 \le t \le 2\pi$, and recall that $\mathbb{F} = (-\omega y, \omega x, 0)$, then

$$\text{circulation} = \oint_{\vec{C}} \mathbb{F} \cdot d\mathbf{x} = \oint_{\vec{C}} -\omega y \, dx + \omega x \, dy$$

$$= \omega r \int_0^{2\pi} \left((b + r\sin t)\sin t + (a + r\cos t)\cos t \right) dt$$

$$= \omega r \int_0^{2\pi} \left(b\sin t + a\cos t + r \right) dt = 2\omega\pi r^2.$$

Thus, for every circle parallel to the (x, y)-plane, we have

$$\frac{\text{circulation}}{\text{area}} = \frac{2\omega\pi r^2}{\pi r^2} = 2\omega;$$

circulation per unit area equals the magnitude of the vorticity vector $\operatorname{curl} \mathbb{F}$ at every point.

Although we have not established that circulation per unit area measures vorticity in all cases, let us try it on the flow $\mathbb{F}$ of Example 2. This time we calculate the circulation around a square instead of a circle, because the calculation reduces to a simple product. Let $\vec{C}$ be the boundary of the square that lies in the plane $z = c$, has its lower-left corner at the point (a, b, c), and has sides of length s parallel to the x- and y-axes. Give $\vec{C}$ the counterclockwise orientation when seen from above (i.e., from where $z > c$).

Circulation as
a path integral

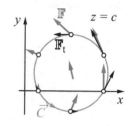

Circulation for
Example 2

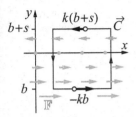

The contributions to the circulation from the left and right sides are zero because the tangential velocity is zero there. On the bottom side (where $y = b$), the tangential velocity is $-kb$, so the contribution is $-kbs$. (In the figure, b is chosen to be negative, so $-kb$ is positive, as shown.) On the top side (where $y = b + s$), the contribution is $+k(b+s)s$. Therefore,

$$\text{circulation} = k(b+s)s - kbs = ks^2.$$

The area of the square is s^2, and the vorticity vector is $\text{curl}\,\mathbb{F} = (0,0,k)$, so we find once again that the circulation per unit area equals the magnitude of the vorticity vector.

Staying with Example 2, let us determine circulation per unit area when the path is a circle instead of a square. We take the same oriented circle $\vec{C}$ we used for Example 1,

$$(x,y,z) = (a + r\cos t, b + r\sin t, c), \quad 0 \le t \le 2\pi.$$

With the flow field $\mathbb{F} = (-ky, 0, 0)$, we have

$$\text{circulation} = \oint_{\vec{C}} -ky\,dx = \int_0^{2\pi} (krb\sin t + kr^2 \sin^2 t)\,dt = k\pi r^2.$$

For these paths, at least, circulation per unit area also equals k.

Circulation of a flow

Definition 11.3 *The **circulation** of the flow $\mathbb{F}$ around the oriented closed loop $\vec{C}$ is the path integral*

$$\text{circulation of}\,\mathbb{F}\,\text{around}\,\vec{C} = \oint_{\vec{C}} \mathbb{F} \cdot d\mathbf{x}.$$

Circulation around other curves

In our two examples, vorticity was constant in both magnitude and direction, and we calculated the circulation only around curves lying in planes perpendicular to the vorticity vector $\text{curl}\,\mathbb{F}$. Suppose we take an arbitrary plane; how does the circulation around a curve in that plane depend on the orientation of the plane?

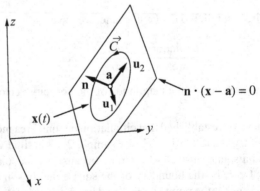

Circles with arbitrary orientation

We can parametrize the oriented circle $\vec{C}$ of radius r that is centered at the point $\mathbf{a}$ and lies in the plane with unit normal $\mathbf{n}$ by choosing two perpendicular unit vectors $\mathbf{u}_1$ and $\mathbf{u}_2$ for which $\mathbf{u}_1 \times \mathbf{u}_2 = \mathbf{n}$ and setting

$$\mathbf{x}(t) = \mathbf{a} + (r\cos t)\mathbf{u}_1 + (r\sin t)\mathbf{u}_2, \quad 0 \le t \le 2\pi.$$

Let $\mathbf{a} = (a, b, c)$, $\mathbf{u}_1 = (\alpha_1, \beta_1, \gamma_1)$, and $\mathbf{u}_2 = (\alpha_2, \beta_2, \gamma_2)$; then the circulation of $\mathbb{F} = (-ky, 0, 0)$ (Example 2) around $\vec{C}$ is

$$\oint_{\vec{C}} -ky\,dx = \int_0^{2\pi} -k(b + r\beta_1\cos t + r\beta_2\sin t)(-r\alpha_1\sin t + r\alpha_2\cos t)\,dt$$

$$= kr^2 \int_0^{2\pi} (\alpha_1\beta_2\sin^2 t - \beta_1\alpha_2\cos^2 t)\,dt$$

$$= kr^2(\alpha_1\beta_2\,\pi - \beta_1\alpha_2\,\pi) = \pi r^2 k \begin{vmatrix} \alpha_1 & \beta_1 \\ \alpha_2 & \beta_2 \end{vmatrix}.$$

(Of the six terms in the second integral, only the two that make a nonzero contribution have been carried over to the third integral.) Because

$$\mathbf{n} = (\alpha_1, \beta_1, \gamma_1) \times (\alpha_2, \beta_2, \gamma_2) = \left(\begin{vmatrix} \beta_1 & \gamma_1 \\ \beta_2 & \gamma_2 \end{vmatrix}, \begin{vmatrix} \gamma_1 & \alpha_1 \\ \gamma_2 & \alpha_2 \end{vmatrix}, \begin{vmatrix} \alpha_1 & \beta_1 \\ \alpha_2 & \beta_2 \end{vmatrix} \right)$$

and $\operatorname{curl}\mathbb{F} = (0, 0, k)$, we can express the circulation of $\mathbb{F}$ around $\vec{C}$ as

$$(\text{area inside } \vec{C}) \ \operatorname{curl}\mathbb{F} \cdot \mathbf{n}.$$

For this example, at least, we see how the orientation of the plane containing $\vec{C}$ affects the value of the circulation. It says that if the unit normal $\mathbf{n}$ makes an angle of θ radians with $\operatorname{curl}\mathbb{F}$, then

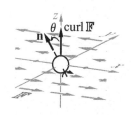

$$\frac{\text{circulation of } \mathbb{F} \text{ around } \vec{C}}{\text{area inside } \vec{C}} = \|\operatorname{curl}\mathbb{F}\| \cos\theta.$$

For example, suppose the test ball in Example 2 is constrained to rotate around an axis parallel to $\mathbf{n}$. If $\mathbf{n}$ is vertical (i.e., $\theta = 0$), the ball will spin as before. However, as we increase θ and tilt $\mathbf{n}$ away from the vertical, the shearing effect of the nearby fluid—and with it the ball's rate of spin—will decrease. As the axis becomes horizontal, the rate of spin will approach zero.

The connection between vorticity and circulation that we have noted in the examples holds, in fact, quite generally. To generalize, let us first replace the orthogonal vectors $r\mathbf{u}_1$ and $r\mathbf{u}_2$ in the parametrization of the circle $\vec{C}$ above by arbitrary linearly independent vectors $\mathbf{v}_1$ and $\mathbf{v}_2$:

Parametrizing an ellipse in any plane

$$\mathbf{x}(t) = \mathbf{a} + (\cos t)\mathbf{v}_1 + (\sin t)\mathbf{v}_2, \quad 0 \le t \le 2\pi.$$

The image is an oriented ellipse $\vec{E}$ whose area is $\pi\|\mathbf{v}_1 \times \mathbf{v}_2\|$. To see this, consider the linear map L of the plane containing $\vec{C}$ to the plane containing $\vec{E}$ given by $L(r\mathbf{u}_i) = \mathbf{v}_i$, $i = 1, 2$.

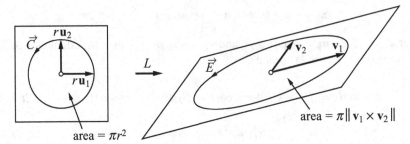

Then $\vec{E}$ is an ellipse because it is the image of a circle under a linear map. Furthermore, because L maps the square $r\mathbf{u}_1 \wedge r\mathbf{u}_2$ to the parallelogram $\mathbf{v}_1 \wedge \mathbf{v}_2$,

$$|\text{area magnification factor for } L| = \left| \frac{\text{area}(\mathbf{v}_1 \wedge \mathbf{v}_2)}{\text{area}(r\mathbf{u}_1 \wedge r\mathbf{u}_2)} \right| = \frac{\|\mathbf{v}_1 \times \mathbf{v}_2\|}{r^2}.$$

Finally, because $\vec{E} = L(\vec{C})$ and $\vec{C}$ has area πr^2, the area of $\vec{E}$ is $\pi \|\mathbf{v}_1 \times \mathbf{v}_2\|$.

Flows and maps

To continue with our generalization of the connection between vorticity and circulation, we note that a flow field $\mathbb{F} = (A, B, C)$ defined on an open region Ω in $\mathbb{R}^3$ can be considered a map $\mathbb{F} : \Omega \to \mathbb{R}^3$,

$$\mathbb{F} : \begin{cases} u = A(x, y, z), \\ v = B(x, y, z), \\ w = C(x, y, z). \end{cases}$$

In particular, if the map $\mathbb{F}$ is differentiable on Ω, then we have

$$\mathbb{F}(\mathbf{a} + \Delta\mathbf{x}) = \mathbb{F}(\mathbf{a}) + d\mathbb{F}_\mathbf{a}(\Delta\mathbf{x}) + o(\Delta\mathbf{x}) \quad \text{as } \Delta\mathbf{x} \to \mathbf{0},$$

for any point $\mathbf{a}$ in Ω (Definition 4.6, p. 129). The following theorem now considers the circulation of $\mathbb{F}$ around the family of ellipses

$$\vec{E}_\varepsilon : \mathbf{x}(t) = \mathbf{a} + (\varepsilon \cos t)\mathbf{v}_1 + (\varepsilon \sin t)\mathbf{v}_2, \quad 0 \leq t \leq 2\pi.$$

Note that area $\vec{E}_\varepsilon = \pi\varepsilon^2 \|\mathbf{v}_1 \times \mathbf{v}_2\|$. In our examples, the ratio of circulation to area had a constant value, namely, $\text{curl}\,\mathbb{F} \cdot \mathbf{n}$. Here the ratio is no longer constant; nevertheless, its limiting value equals $\text{curl}\,\mathbb{F} \cdot \mathbf{n}$ as the area approaches zero.

Limiting value of circulation/area

Theorem 11.8. *If* $\mathbb{F} : \Omega \to \mathbb{R}^3$ *is a differentiable flow field, then*

$$\lim_{\varepsilon \to 0} \frac{\text{circulation of } \mathbb{F} \text{ around } \vec{E}_\varepsilon}{\text{area inside } \vec{E}_\varepsilon} = \text{curl}\,\mathbb{F} \cdot \mathbf{n},$$

where $\mathbf{n}$ *is the unit normal in the plane containing the ellipses* $\vec{E}_\varepsilon$.

Proof. The proof follows from a sequence of lemmas.

Lemma 11.1. $\displaystyle\oint_{\vec{E}_\varepsilon} \mathbb{F} \cdot d\mathbf{x} = \pi\varepsilon^2 \left[d\mathbb{F}_\mathbf{a}(\mathbf{v}_1) \cdot \mathbf{v}_2 - d\mathbb{F}_\mathbf{a}(\mathbf{v}_2) \cdot \mathbf{v}_1 \right] + o(\varepsilon^2) \;\text{as } \varepsilon \to 0.$

Proof. The circulation of $\mathbb{F}$ around $\vec{E}_\varepsilon$ is the integral

$$\oint_{\vec{E}_\varepsilon} \mathbb{F} \cdot d\mathbf{x} = \int_0^{2\pi} \mathbb{F}(\mathbf{a} + \varepsilon \cos t \, \mathbf{v}_1 + \varepsilon \sin t \, \mathbf{v}_2) \cdot (-\varepsilon \sin t \, \mathbf{v}_1 + \varepsilon \cos t \, \mathbf{v}_2) \, dt.$$

Set $\Delta\mathbf{x} = \varepsilon \cos t \, \mathbf{v}_1 + \varepsilon \sin t \, \mathbf{v}_2$. Because $\Delta\mathbf{x} \to \mathbf{0}$ as $\varepsilon \to 0$, we have $o(\Delta\mathbf{x}) = o(\varepsilon)$. Therefore, using the differentiblity of $\mathbb{F}$, we can write the integral as

$$\int_0^{2\pi} (\mathbb{F}(\mathbf{a}) + \varepsilon \cos t \, d\mathbb{F}_\mathbf{a}(\mathbf{v}_1) + \varepsilon \sin t \, d\mathbb{F}_\mathbf{a}(\mathbf{v}_2) + o(\varepsilon)) \cdot (-\varepsilon \sin t \, \mathbf{v}_1 + \varepsilon \cos t \, \mathbf{v}_2) \, dt.$$

When we expand the dot product, we get eight terms. Four have average value zero and therefore vanish when integrated; the others combine to give

$$\oint_{\vec{E}_\varepsilon} \mathbb{F} \cdot d\mathbf{x} = \int_0^{2\pi} (\varepsilon^2 \cos^2 t \, d\mathbb{F}_\mathbf{a}(\mathbf{v}_1) \cdot \mathbf{v}_2 - \varepsilon^2 \sin^2 t \, d\mathbb{F}_\mathbf{a}(\mathbf{v}_2) \cdot \mathbf{v}_1 + o(\varepsilon^2)) \, dt$$

$$= \pi\varepsilon^2 [d\mathbb{F}_\mathbf{a}(\mathbf{v}_1) \cdot \mathbf{v}_2 - d\mathbb{F}_\mathbf{a}(\mathbf{v}_2) \cdot \mathbf{v}_1] + o(\varepsilon^2) \quad \text{as } \varepsilon \to 0. \qquad \square$$

We find that the expression $\psi(\mathbf{v}_1, \mathbf{v}_2) = d\mathbb{F}_\mathbf{a}(\mathbf{v}_1) \cdot \mathbf{v}_2 - d\mathbb{F}_\mathbf{a}(\mathbf{v}_2) \cdot \mathbf{v}_1$ is the key to relating the circulation of $\mathbb{F}$ to curl $\mathbb{F}$.

$\psi(\mathbf{v}_1, \mathbf{v}_2)$

Lemma 11.2. *There is a unique vector* $\mathbf{q}$ *for which*

$$\psi(\mathbf{v}_1, \mathbf{v}_2) = d\mathbb{F}_\mathbf{a}(\mathbf{v}_1) \cdot \mathbf{v}_2 - d\mathbb{F}_\mathbf{a}(\mathbf{v}_2) \cdot \mathbf{v}_1 = \mathbf{q} \cdot (\mathbf{v}_1 \times \mathbf{v}_2),$$

for all vectors $\mathbf{v}_1$, $\mathbf{v}_2$ *in* $\mathbb{R}^3$.

Proof. For any vector $\mathbf{q}$, define the function

$$\tau_\mathbf{q}(\mathbf{v}_1, \mathbf{v}_2) = \mathbf{q} \cdot (\mathbf{v}_1 \times \mathbf{v}_2).$$

Then ψ and $\tau_\mathbf{q}$ are both bilinear and antisymmetric (cf. p. 62); that is, they are linear functions of each of their inputs and, furthrmore,

$$\psi(\mathbf{v}_2, \mathbf{v}_1) = -\psi(\mathbf{v}_1, \mathbf{v}_2), \quad \tau_\mathbf{q}(\mathbf{v}_2, \mathbf{v}_1) = -\tau_\mathbf{q}(\mathbf{v}_1, \mathbf{v}_2),$$

for all pairs $\mathbf{v}_1, \mathbf{v}_2$. Therefore, if ψ and $\tau_\mathbf{q}$ agree on a basis for $\mathbb{R}^3$, they must agree everywhere (cf. Exercise 11.11).

Set $\mathbf{q} = (q_1, q_2, q_3)$ where

$$q_1 = \psi(\mathbf{e}_2, \mathbf{e}_3), \quad q_2 = \psi(\mathbf{e}_3, \mathbf{e}_1), \quad q_3 = \psi(\mathbf{e}_1, \mathbf{e}_2),$$

and $\{\mathbf{e}_1, \mathbf{e}_2, \mathbf{e}_3\}$ is the standard basis for $\mathbb{R}^3$. Then

$$\tau_\mathbf{q}(\mathbf{e}_2, \mathbf{e}_3) = \mathbf{q} \cdot (\mathbf{e}_2 \times \mathbf{e}_3) = \mathbf{q} \cdot \mathbf{e}_1 = q_1 = \psi(\mathbf{e}_2, \mathbf{e}_3),$$

$$\tau_\mathbf{q}(\mathbf{e}_3, \mathbf{e}_1) = \mathbf{q} \cdot (\mathbf{e}_3 \times \mathbf{e}_1) = \mathbf{q} \cdot \mathbf{e}_2 = q_2 = \psi(\mathbf{e}_3, \mathbf{e}_1),$$

$$\tau_\mathbf{q}(\mathbf{e}_1, \mathbf{e}_2) = \mathbf{q} \cdot (\mathbf{e}_1 \times \mathbf{e}_2) = \mathbf{q} \cdot \mathbf{e}_3 = q_3 = \psi(\mathbf{e}_1, \mathbf{e}_2);$$

by antisymmetry, we find $\tau_q(\mathbf{e}_i, \mathbf{e}_j) = \psi(\mathbf{e}_i, \mathbf{e}_j)$ for all $i, j = 1, 2, 3$.   □

Any square matrix M can be written uniquely as the sum of a symmetric and an antisymmetric matrix. Set

$$S = \tfrac{1}{2}(M + M^\dagger), \quad \mathcal{A} = \tfrac{1}{2}(M - M^\dagger),$$

where $M^\dagger$ is the transpose of M. Then $S^\dagger = S$, $\mathcal{A}^\dagger = -\mathcal{A}$, and $M = S + \mathcal{A}$.

$\psi_M = \psi_{\mathcal{A}}$

Lemma 11.3. *For any* 3×3 *matrix M,*

$$\psi_M = M\mathbf{v}_1 \cdot \mathbf{v}_2 - M\mathbf{v}_2 \cdot \mathbf{v}_1 = \mathcal{A}\mathbf{v}_1 \cdot \mathbf{v}_2 - \mathcal{A}\mathbf{v}_2 \cdot \mathbf{v}_1 = \psi_{\mathcal{A}},$$

where $\mathcal{A}$ *is the antisymmetric part of M:* $\mathcal{A} = \tfrac{1}{2}(M - M^\dagger)$.

Proof. Write $M = S + \mathcal{A}$, where $S = \tfrac{1}{2}(M + M^\dagger)$; then it is easy to see that $\psi_M = \psi_S + \psi_{\mathcal{A}}$. Now write the dot products as matrix multiplications using column vectors and their transposes:

$$\psi_S = (S\mathbf{v}_1)^\dagger \mathbf{v}_2 - (S\mathbf{v}_2)^\dagger \mathbf{v}_1 = \mathbf{v}_1^\dagger S^\dagger \mathbf{v}_2 - \mathbf{v}_2^\dagger S^\dagger \mathbf{v}_1 = \mathbf{v}_1^\dagger S^\dagger \mathbf{v}_2 - \mathbf{v}_2^\dagger S\mathbf{v}_1.$$

In the last term we used the symmetry of S. Each term is a scalar (i.e., a 1×1 matrix), so is equal to its own transpose. Thus we can write

$$\mathbf{v}_1^\dagger S^\dagger \mathbf{v}_2 = (\mathbf{v}_1^\dagger S^\dagger \mathbf{v}_2)^\dagger = \mathbf{v}_2^\dagger S\mathbf{v}_1,$$

from which it follows that $\psi_S \equiv 0$ and hence $\psi_M \equiv \psi_{\mathcal{A}}$.   □

Antisymmetric
part of $d\mathbb{F}_\mathbf{a}$

For the map $\mathbb{F}$ with components (A, B, C), the derivative and its antisymmetric part are

$$d\mathbb{F}_\mathbf{a} = \begin{pmatrix} A_x & A_y & A_z \\ B_x & B_y & B_z \\ C_x & C_y & C_z \end{pmatrix}, \quad \mathcal{A} = \frac{1}{2}\begin{pmatrix} 0 & -\gamma & \beta \\ \gamma & 0 & -\alpha \\ -\beta & \alpha & 0 \end{pmatrix},$$

where

$$\alpha = C_y - B_z, \quad \beta = A_z - C_x, \quad \gamma = B_x - A_y$$

and all components are evaluated at $\mathbf{x} = \mathbf{a}$. Labels are chosen for the components of $\mathcal{A}$ so that $(\alpha, \beta, \gamma) = \operatorname{curl} \mathbb{F}(\mathbf{a})$. With $\mathcal{A}$ we can now determine the vorticity vector $\mathbf{q}$ that is provided by Lemma 11.2.

Lemma 11.4. $d\mathbb{F}_\mathbf{a}(\mathbf{v}_1) \cdot \mathbf{v}_2 - d\mathbb{F}_\mathbf{a}(\mathbf{v}_2) \cdot \mathbf{v}_1 = \operatorname{curl} \mathbb{F}(\mathbf{a}) \cdot (\mathbf{v}_1 \times \mathbf{v}_2)$.

Proof. Because $\psi_{d\mathbb{F}_\mathbf{a}} = \psi_{\mathcal{A}}$, we have

$$q_1 = \psi_{\mathcal{A}}(\mathbf{e}_2, \mathbf{e}_3) = \mathcal{A}\mathbf{e}_2 \cdot \mathbf{e}_3 - \mathcal{A}\mathbf{e}_3 \cdot \mathbf{e}_2 = \tfrac{1}{2}\begin{pmatrix} -\gamma \\ 0 \\ \alpha \end{pmatrix} \cdot \begin{pmatrix} 0 \\ 0 \\ 1 \end{pmatrix} - \tfrac{1}{2}\begin{pmatrix} \beta \\ -\alpha \\ 0 \end{pmatrix} \cdot \begin{pmatrix} 0 \\ 1 \\ 0 \end{pmatrix} = \alpha.$$

In a similar way, you can show $q_2 = \beta$, $q_3 = \gamma$.   □

To complete the proof of Theorem 11.8, use Lemma 11.1 and Lemma 11.4 to write

$$\frac{\text{circulation of } \mathbb{F} \text{ around } \vec{E}_\varepsilon}{\text{area inside } \vec{E}_\varepsilon} = \frac{\pi\varepsilon^2 \left[d\mathbb{F}_\mathbf{a}(\mathbf{v}_1) \cdot \mathbf{v}_2 - d\mathbb{F}_\mathbf{a}(\mathbf{v}_2) \cdot \mathbf{v}_1 \right] + o(\varepsilon^2)}{\pi\varepsilon^2 \|\mathbf{v}_1 \times \mathbf{v}_2\|}$$

$$= \text{curl}\,\mathbb{F} \cdot \frac{\mathbf{v}_1 \times \mathbf{v}_2}{\|\mathbf{v}_1 \times \mathbf{v}_2\|} + \frac{o(\varepsilon^2)}{\varepsilon^2}$$

$$= \text{curl}\,\mathbb{F} \cdot \mathbf{n} + \frac{o(\varepsilon^2)}{\varepsilon^2}.$$

In the second term, the divisor $\pi\|\mathbf{v}_1 \times \mathbf{v}_2\|$ has been absorbed into $o(\varepsilon^2)$. The theorem then follows because, by definition, $o(\varepsilon^2)/\varepsilon^2 \to 0$ as $\varepsilon \to 0$. $\qquad\square$

In the proof of Lemma 11.2, there is no motivation (other than hindsight) for the formula

Reconstructing the
vorticity vector **q**

$$\mathbf{q} = (\psi(\mathbf{e}_2, \mathbf{e}_3), \psi(\mathbf{e}_3, \mathbf{e}_1), \psi(\mathbf{e}_1, \mathbf{e}_2))$$

that expresses the components of the vorticity vector **q** in terms of particular circulation calculations. According to this equation, the x-component of vorticity comes from circulation in a plane normal to the x-axis (i.e., a plane parallel to the vectors $\mathbf{e}_2$ and $\mathbf{e}_3$ that determine the (y,z)-plane). Similarly, the y-component of vorticity **q** uses a plane normal to the y-axis, and the z-component of **q** uses a plane normal to the z-axis. Let us now reconstruct **q** by calculating anew the circulation in those planes. The work will look similar to the initial work (pp. 449–451) that led to our identifying the divergence of a field.

It is convenient to calculate the circulation around the boundary of a rectangle—as we did in Example 2—rather than around an ellipse. On a rectangle whose sides are parallel to the axes, the tangential component of $\mathbb{F}$ on a side is just one of the Cartesian components of $\mathbb{F}$. We begin with the oriented rectangle $\vec{R}_x$ centered at the point $\mathbf{a} = (a,b,c)$ and lying in the plane $x = a$ (and thus parallel to the (y,z)-plane). Let the lengths of its sides be denoted Δy and Δz, and orient it counterclockwise when viewed from the side where $x > a$. Its orientation normal is then $\mathbf{n} = (1,0,0)$.

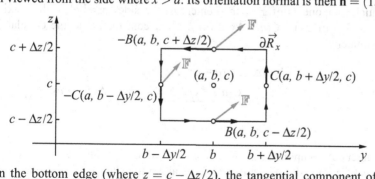

On the bottom edge (where $z = c - \Delta z/2$), the tangential component of $\mathbb{F} = (A,B,C)$ is $B(a,y,c-\Delta z/2)$. If we approximate B everywhere on this edge by its value at the center, $(a,b,c-\Delta z/2)$, then the contribution this edge makes to the circulation of $\mathbb{F}$ around $\partial \vec{R}_x$ is approximately

$$B(a,b,c - \Delta z/2)\, \Delta y.$$

Along the top edge, the tangential component of $\mathbb{F}$ is $-B(a,y,c+\Delta z/2)$. The minus sign is needed because the unit tangent to $\partial \vec{R}_x$ on this edge is $\mathbf{t} = (0,-1,0)$. If we approximate $-B$ everywhere by its value $-B(a,b,c+\Delta z/2)$ at the center, the approximate contribution this edge makes to the circulation is

$$-B(a,b,c + \Delta z/2)\, \Delta y.$$

If we put these together and use the microscope equation, we have

$$-B(a,b,c + \Delta z/2)\, \Delta y + B(a,b,c - \Delta z/2)\, \Delta y$$
$$= -\frac{B(a,b,c + \Delta z/2) - B(a,b,c - \Delta z/2)}{\Delta z}\, \Delta z \Delta y$$
$$\approx -B_z(a,b,c)\, \Delta y \Delta z.$$

For the right and left edges there is a similar result: together they contribute approximately

$$C(a,b + \Delta y/2,c)\, \Delta z - C(a,b - \Delta y/2)\, \Delta z$$
$$= \frac{C(a,b + \Delta y/2,c) - C(a,b - \Delta y/2,c)}{\Delta y}\, \Delta y \Delta z$$
$$\approx C_y(a,b,c)\, \Delta y \Delta z.$$

Thus we can write

$$\text{circulation of } \mathbb{F} \text{ around } \partial \vec{R}_x \approx (C_y(\mathbf{a}) - B_z(\mathbf{a}))\, \text{area}\, \vec{R}_x.$$

The factor $C_y - B_z$ is indeed the x-component of curl $\mathbb{F}$.

For a similar rectangle $\vec{R}_y$ in the plane $y = b$ parallel to the (z,x)-plane, the tangential components of $\mathbb{F}$ at the centers of the sides parallel to the z-axis are $C(a - \Delta x/2,b,c)$ and $-C(a + \Delta x/2,b,c)$. Their contribution to the circulation is approximately

$$-C(a + \Delta x/2,b,c)\, \Delta z + C(a - \Delta x/2,b,c)\, \Delta z$$
$$= -\frac{C(a + \Delta x/2,b,c) - C(a - \Delta x/2,b,c)}{\Delta x}\, \Delta x \Delta z$$
$$\approx -C_x(a,b,c)\, \Delta z \Delta x.$$

The tangential components of $\mathbb{F}$ at the centers of the remaining two sides are $A(a,b,c + \Delta z/2)$ and $-A(a,b,c - \Delta z/2)$, making together the approximate contribution

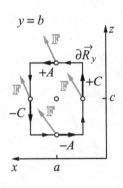

$$A(a,b,c+\Delta z/2)\,\Delta x - A(a,b,c-\Delta z/2)\,\Delta x$$
$$= \frac{A(a,b,c+\Delta z/2) - A(a,b,c-\Delta z/2)}{\Delta z}\,\Delta z\,\Delta x \approx A_z(a,b,c)\,\Delta z\,\Delta x$$

to the circulation around $\partial \vec{R}_y$. Thus

$$\text{circulation of } \mathbb{F} \text{ around } \partial \vec{R}_y \approx (A_z(\mathbf{a}) - C_x(\mathbf{a}))\ \text{area}\,\vec{R}_y;$$

the factor $A_z - C_x$ is the y-component of curl $\mathbb{F}$. A similar analysis of an oriented rectangle $\vec{R}_z$ in the plane $z = c$ leads to the result

$$\text{circulation of } \mathbb{F} \text{ around } \partial \vec{R}_z \approx (B_x(\mathbf{a}) - A_y(\mathbf{a}))\ \text{area}\,\vec{R}_z;$$

the factor $B_x - A_y$ gives the final component of curl $\mathbb{F}$.

Thus, the circulation around any one of the rectangles $\vec{R}$ is approximately curl $\mathbb{F}(\mathbf{a}) \cdot \mathbf{n}$ area $\vec{R}$. But this expression is itself an approximation to the flux of the vector field curl $\mathbb{F}$ through the small rectangle $\vec{R}$ (Definition 10.1, p. 388). Hence,

> Circulation and vorticity

$$\text{circulation of } \mathbb{F} \text{ around } \partial \vec{R} \approx \text{flux of curl}\,\mathbb{F} \text{ through } \vec{R}.$$

Even more is true. In the next section, we show that, for an oriented surface $\vec{S}$ with boundary $\partial \vec{S}$, the circulation of the vector field $\mathbb{F}$ around $\partial \vec{S}$ is equal to the flux of its vorticity field curl $\mathbb{F}$ through $\vec{S}$:

$$\text{circulation of } \mathbb{F} \text{ around } \partial \vec{S} = \text{flux of curl}\,\mathbb{F} \text{ through } \vec{S}$$

$$\oint_{\partial S} \mathbb{F} \cdot \mathbf{t}\, ds = \iint_S \text{curl}\,\mathbb{F} \cdot \mathbf{n}\, dA.$$

This is the physical content of Stokes' theorem. It also gives us a new way to look at vorticity. Instead of having to rely on the physical image of a little ball set spinning by the shear action of a fluid, we can use the mathematical notion of the circulation of that fluid in various planes. Note that the circulation/flux identity involves two distinct flows: $\mathbb{F}$ and its *vorticity* curl $\mathbb{F}$.

> The vorticity flow
> of a flow

Definition 11.4 *Let the continuously differentiable vector field $\mathbb{F}$ represent a steady fluid flow field; then the **vorticity flow field**, or the **vortex flow field**, of $\mathbb{F}$ is represented by the vector field $\mathbb{V} = \text{curl}\,\mathbb{F}$.*

We attribute the vorticity of a fluid flow—as given by the curl—to shearing forces induced by the flow. We attribute these forces, in turn, to the relative motions of nearby fluid particles. Let us now analyze the relative motions; as we show in Theorem 11.10, they explain the divergence as well as the curl.

> The relative motions of
> nearby fluid particles

The velocity field $\mathbb{F}$ determines the overall motion of the fluid; it must, therefore, determine the relative motion as well. To explore this connection and to clarify the distinction between the two kinds of motion, we work through a third example:

> Example 3:
> $\mathbb{F} = (ky^2, 0, 0)$

$$\mathbb{F} = (ky^2, 0, 0), \quad k > 0.$$

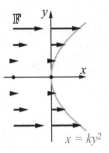

Think of this as a nonlinear modification of Example 2; the nonlinearity will help us see the relative motion more clearly. The fluid flows in straight lines parallel to the x-axis. In the (x,y)-plane, the speed of a particle is proportional to the square of its distance from the x-axis; in any parallel plane, the flow looks the same.

As in the previous examples, the position $\mathbf{x}(t) = (x(t), y(t), z(t))$ of a particle at time t is determined by the differential equations

$$\mathbf{x}'(t) = \mathbb{F}(\mathbf{x}(t)), \quad \text{that is,} \quad x' = ky^2, \quad y' = 0, \quad z' = 0.$$

The solutions are

$$x(t) = a + kb^2 t, \quad y(t) = b, \quad z(t) = c;$$

a, b, and c are arbitrary constants of integration. These equations parametrize the path of the particle that is initially (i.e., when $t = 0$) at the point $\mathbf{a} = (a, b, c)$. The particle does not move if $b = 0$; therefore we assume $b \neq 0$ for the rest of the discussion.

The flow maps $\mathbf{h}_t$

More generally, then, the equations say that the particle at the point $\mathbf{x} = (x, y, z)$ at time $t = 0$ is at the point $\mathbf{h}_t(\mathbf{x}) = (x + ky^2 t, y, z)$ at an arbitrary (earlier or later) time t. In other words, the flow defines, and is defined by, the family of maps $\mathbf{h}_t : \mathbb{R}^3 \to \mathbb{R}^3$:

$$\mathbf{h}_t : \begin{cases} u = x + ky^2 t, \\ v = y, \\ w = z; \end{cases} \qquad \mathrm{d}(\mathbf{h}_t)_{\mathbf{x}} = \begin{pmatrix} 1 & 2kyt & 0 \\ 0 & 1 & 0 \\ 0 & 0 & 1 \end{pmatrix}.$$

Note that $\mathbf{h}_s \circ \mathbf{h}_t = \mathbf{h}_{s+t}$ for every s, t. This implies each $\mathbf{h}_t$ is invertible, because $\mathbf{h}_{-t} \circ \mathbf{h}_t = \mathbf{h}_0 = $ identity.

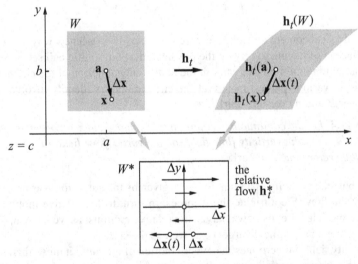

Describing the relative flow

To describe the relative motion of fluid particles near the particle that is intially at the point $\mathbf{a}$, choose a small cube W (a "window") centered at $\mathbf{a}$, and let $\mathbf{h}_t(W)$ denote the image of W under the flow $\mathbf{h}_t$ (t can take negative as well as positive values). If

$\mathbf{x}$ is a point in W, and $\Delta x = \mathbf{x} - \mathbf{a}$ indicates the position of $\mathbf{x}$ in relation to $\mathbf{a}$, then the vector

$$\Delta\mathbf{x}(t) = \mathbf{h}_t(\mathbf{x}) - \mathbf{h}_t(\mathbf{a})$$

in $\mathbf{h}_t(W)$ indicates how the relative position $\Delta\mathbf{x}$ varies over time. When the vectors $\Delta\mathbf{x}(t)$ (for t near 0) are translated to a common point (at the origin of a window W^* with local coordinates $\Delta\mathbf{x} = (\Delta x, \Delta y, \Delta z)$), they exhibit the relative flow we seek to describe. With these local coordinates in mind, we rewrite $\mathbf{h}_t(\mathbf{x})$ as

$$\mathbf{h}_t(\mathbf{a} + \Delta\mathbf{x}) = (a + \Delta x + k(b + \Delta y)^2 t,\ b + \Delta y, c + \Delta z),$$

which shows how $\Delta\mathbf{x}(t) = \mathbf{h}_t(\mathbf{x}) - \mathbf{h}_t(\mathbf{a})$ can be described by the equations

$$\mathbf{h}_t^* : \begin{cases} \Delta x(t) = \Delta x + t\,(2kb\,\Delta y + k(\Delta y)^2), \\ \Delta y(t) = \Delta y, \\ \Delta z(t) = \Delta z. \end{cases}$$

From one point of view, these equations parametrize a straight line that is parallel to the Δx-axis and passes through the arbitrary, but fixed, point $\Delta\mathbf{x} = (\Delta x, \Delta y, \Delta z)$ in W^*. This straight line is one of the relative flow lines we see in W^*, above. From a second point of view (in which $\Delta\mathbf{x}$ is variable, not fixed), these equations define the family of *relative-flow maps* $\mathbf{h}_t^* : W^* \to \mathbb{R}^3$. When the center of the original window W lies off the (z,x)-plane (i.e., when $b \neq 0$), the relative flow described by $\mathbf{h}_t^*$ is along paths that head in opposite directions on opposite sides of the Δx-axis.

The relative-flow
maps $\mathbf{h}_t^*$

The relative flow has its own velocity field; let us call it $\mathbb{F}^*$. The field vector $\mathbb{F}^*$ at the point $\Delta\mathbf{x} = (\Delta x, \Delta y, \Delta z)$ is, by definition, the velocity of the path $\Delta\mathbf{x}(t)$ at its intitial point:

The relative-flow
field $\mathbb{F}^*$

$$\mathbb{F}^*(\Delta x, \Delta y, \Delta z) = \left.\frac{d}{dt}\Delta\mathbf{x}(t)\right|_{t=0} = (2kb\Delta y + k(\Delta y)^2, 0, 0)$$

Alternatively, the relative flow field is the map $\mathbb{F}^* : W^* \to \mathbb{R}^3$,

$$\mathbb{F}^* : \begin{cases} \Delta u = 2kb\,\Delta y + k(\Delta y)^2, \\ \Delta v = 0, \\ \Delta w = 0. \end{cases}$$

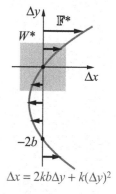

$$\Delta x = 2kb\Delta y + k(\Delta y)^2$$

Because we assume W^* is small, $k(\Delta y)^2$ is negligible in comparison to $2kb\,\Delta y$ (recall that $b \neq 0$), so $\mathbb{F}^*$ is well approximated by its linear part,

$$\begin{pmatrix} 0 & 2kb & 0 \\ 0 & 0 & 0 \\ 0 & 0 & 0 \end{pmatrix} \begin{pmatrix} \Delta x \\ \Delta y \\ \Delta z \end{pmatrix},$$

which is just $d\mathbb{F}_{\mathbf{a}}(\Delta\mathbf{x})$. This gives us the following links between the relative flow $\mathbb{F}^*$ in W^* and the overall flow $\mathbb{F}$.

$$\mathbb{F}^*(\Delta \mathbf{x}) = d\mathbb{F}_{\mathbf{a}}(\Delta \mathbf{x}) + \mathbf{o}(\Delta \mathbf{x}),$$
$$\mathbf{h}_t^*(\Delta \mathbf{x}) = \Delta \mathbf{x} + t\, d\mathbb{F}_{\mathbf{a}}(\Delta \mathbf{x}) + t\,\mathbf{o}(\Delta \mathbf{x}).$$

Because we can make W^* arbitrarily small, we can neglect the higher-order terms $\mathbf{o}(\Delta \mathbf{x})$ in analyzing the relative flow, so that

$$\mathbf{h}_t^* \approx d(\mathbf{h}_t^*)_0 = I + t\, d\mathbb{F}_{\mathbf{a}}.$$

Splitting $d\mathbb{F}_{\mathbf{a}}$ to describe its action

Hence, we look to the action of $d\mathbb{F}_{\mathbf{a}}$ to explain the relative flow. First split $d\mathbb{F}_{\mathbf{a}}$ into its symmetric and antisymmetric parts (cf. p. 470); thus, $d\mathbb{F}_{\mathbf{a}} = S + A$, where

$$S = \begin{pmatrix} 0 & kb & 0 \\ kb & 0 & 0 \\ 0 & 0 & 0 \end{pmatrix}, \quad A = \begin{pmatrix} 0 & kb & 0 \\ -kb & 0 & 0 \\ 0 & 0 & 0 \end{pmatrix}$$

Because S is symmetric, it is a pure strain (cf. Theorem 2.6, p. 40): it has three mutually perpendicular eigenvectors with real eigenvalues, as follows:

$$\lambda_1 = kb, \qquad \lambda_2 = -kb, \qquad \lambda_3 = 0,$$
$$\mathbf{b}_1 = \begin{pmatrix} \varepsilon/\sqrt{2} \\ \varepsilon/\sqrt{2} \\ 0 \end{pmatrix}, \quad \mathbf{b}_2 = \begin{pmatrix} -\varepsilon/\sqrt{2} \\ \varepsilon/\sqrt{2} \\ 0 \end{pmatrix}, \quad \mathbf{b}_3 = \begin{pmatrix} 0 \\ 0 \\ \varepsilon \end{pmatrix}.$$

We take $\varepsilon > 0$, so each eigenvector has length ε. As it happens, $\mathbf{b}_3$ is also an eigenvector for A (with the same eigenvalue, 0). Furthermore, A maps the plane determined by the other two eigenvectors to itself, because

$$A\mathbf{b}_1 = -kb\,\mathbf{b}_2 \quad \text{and} \quad A\mathbf{b}_2 = kb\,\mathbf{b}_1.$$

Therefore, the map $d(\mathbf{h}_t^*)_0 = I + t(S + A)$ acts in a simple way on the basis $\{\mathbf{b}_1, \mathbf{b}_2, \mathbf{b}_3\}$ for $\mathbb{R}^3$:

$$\mathbf{b}_1 \to \mathbf{b}_1 + kbt\,\mathbf{b}_1 - kbt\,\mathbf{b}_2,$$
$$\mathbf{b}_2 \to \mathbf{b}_2 - kbt\,\mathbf{b}_2 + kbt\,\mathbf{b}_1,$$
$$\mathbf{b}_3 \to \mathbf{b}_3.$$

Let us see how the cube $K = \mathbf{b}_1 \wedge \mathbf{b}_2 \wedge \mathbf{b}_3$ "flows" under this transformation.

The action of $I + tS$

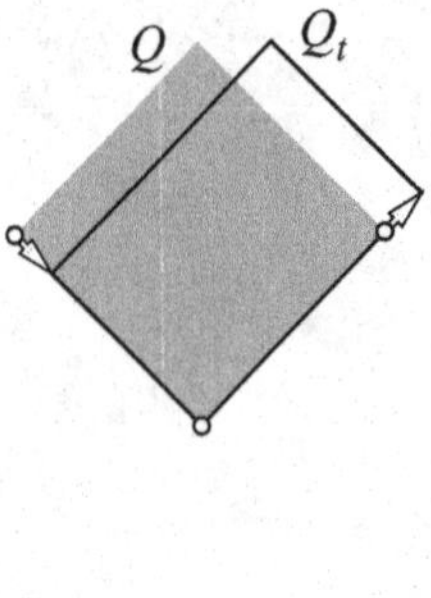

First remove A in order to isolate the action of the symmetric component S:

$$\mathbf{b}_1 \to \mathbf{b}_1 + kbt\,\mathbf{b}_1, \quad \mathbf{b}_2 \to \mathbf{b}_2 - kbt\,\mathbf{b}_2, \quad \mathbf{b}_3 \to \mathbf{b}_3.$$

The vertical edge $\mathbf{b}_3$ is left unchanged, so we concentrate on what happens to the base square $Q = \mathbf{b}_1 \wedge \mathbf{b}_2$ in the $(\Delta x, \Delta y)$-plane. When $t = 0$, we have the original square Q; as t changes, one edge of Q grows longer and the other grows shorter at the same rate, yielding a rectangle Q_t whose sides are parallel to the square. (The white arrows in the figure are $kbt\,\mathbf{b}_1$ and $-kbt\,\mathbf{b}_2$.) Because we are interested in $t \to 0$, we can assume $|kbt| \ll 1$, so the deformation $Q \to Q_t$ is small. The original cube K becomes a rectangular parallelepiped K_t (a "brick") whose base is the rectangle Q_t. In effect, $I + t\mathcal{S}$ is a *strain* that changes the shape of the cube, and also its volume:

$$\operatorname{vol} K_t = \varepsilon(1 + kbt) \times \varepsilon(1 - kbt) \times \varepsilon = \varepsilon^3(1 - (kbt)^2).$$

To first order, t has no effect on the volume. More precisely, the change in volume is $\Delta \operatorname{vol} K = \operatorname{vol} K_t - \operatorname{vol} K = -(kbt)^2 \varepsilon^3$, so the *relative* change (i.e., the change in volume as a fraction of the volume itself) is the function

$$V(t) = \frac{\Delta \operatorname{vol} K}{\operatorname{vol} K} = -(kbt)^2, \quad \text{for which} \quad V'(0) = 0.$$

It turns out that $V'(0) = 0$ is a consequence of the particular nature of $\mathcal{S}$ and thus, ultimately, of $\mathbb{F}$. We show that, for a general flow $\mathbb{F}$, the relative growth rate $V'(0)$ of volume is $\operatorname{div} \mathbb{F}$. Note that, in our example, $\operatorname{div} \mathbb{F} = 0$.

Now remove $\mathcal{S}$ from $I + t(\mathcal{S} + \mathcal{A})$ in order to isolate the action of the antisymmetric component $\mathcal{A}$:

the action of $I + t\mathcal{A}$

$$\mathbf{b}_1 \to \mathbf{b}_1 - kbt\,\mathbf{b}_2, \quad \mathbf{b}_2 \to \mathbf{b}_2 + kbt\,\mathbf{b}_1, \quad \mathbf{b}_3 \to \mathbf{b}_3.$$

Again it is sufficient to see what happens to the base square Q in the $(\Delta x, \Delta y)$-plane. This time the small changes (the black arrows in the figure) are perpendicular to the sides; the effect is to turn the square, rather than strain it. Nevertheless, Q_t is slightly larger than Q when $t \neq 0$, but t again has only a second-order effect on the volume of K_t:

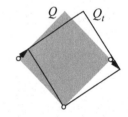

$$\operatorname{vol} K_t = \varepsilon\sqrt{1 + (kbt)^2} \times \varepsilon\sqrt{1 + (kbt)^2} \times \varepsilon = \varepsilon^3 + O(t^2) \text{ as } t \to 0.$$

We show that, in the general case, $I + t\mathcal{A}$ will continue to have only a second-order effect on volume. To first order, $I + t\mathcal{A}$ is a uniform rotation.

Lemma 11.5. *To first order in t, the flow*

$$I + t\mathcal{A} = \begin{pmatrix} 1 & kbt & 0 \\ -kbt & 1 & 0 \\ 0 & 0 & 1 \end{pmatrix}$$

is a uniform rotation with angular velocity $\boldsymbol{\omega} = (0, 0, -kb)$, *that is, a uniform rotation around the positive Δz-axis with angular speed* $-kb$.

Proof. In (x, y, z)-space, uniform rotation with angular speed ω around the positive z-axis is given by the matrix function

$$R_{\omega t} = \begin{pmatrix} \cos \omega t & -\sin \omega t & 0 \\ \sin \omega t & \cos \omega t & 0 \\ 0 & 0 & 1 \end{pmatrix}.$$

The Taylor approximations

$$\cos \omega t = 1 + O(t^2) \quad \text{and} \quad \sin \omega t = \omega t + O(t^3) \quad \text{as } t \to 0$$

imply

$$R_{\omega t} = \begin{pmatrix} 1 & -\omega t & 0 \\ \omega t & 1 & 0 \\ 0 & 0 & 1 \end{pmatrix} + O(t^2) \quad \text{as } t \to 0. \qquad \square$$

Combining the rotation and the strain

To first order in t, tS induces no rotation and $t\mathcal{A}$ induces no strain. Thus, to first order in t and in a sufficiently small window W^*, the relative flow

$$\mathbf{h}_t^* \approx I + tS + t\mathcal{A}$$

rotates the cube K around the positive Δz-axis with angular speed $-kb$ while altering the lengths of its sides at the rates kb, $-kb$, and 0.

The curl describes local rotation

For the flow $\mathbb{F} = (ky^2, 0, 0)$ of our example, $\operatorname{curl}\mathbb{F}(\mathbf{a}) = (0, 0, -2kb) = 2\boldsymbol{\omega}$. We originally interpreted the curl as describing the tendency of a flow to spin a small object carried along with it. Example 3 suggests a new interpretation, in which the curl directly describes the local rotation of a small blob of the fluid itself under the action of the relative flow $\mathbf{h}_t^*$.

With the decomposition $d\mathbb{F}_\mathbf{a} = S + \mathcal{A}$, we were able to focus separately on the two distinct aspects of the relative flow: strain and rotation. In fact, these aspects act together, and together they should produce the shear flow in W^* that we observed at the outset. The following figure confirms this. Each gray arrow, as the sum of a white and a black arrow, expresses the action of $I + tS + t\mathcal{A}$.

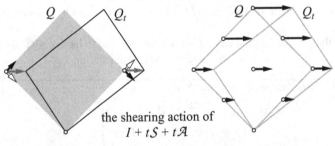

the shearing action of
$I + tS + t\mathcal{A}$

The relative flow for an arbitrary $\mathbb{F}$

Using Example 3 as a guide, we now take up the general case of an arbitrary continuously differentiable velocity field $\mathbb{F} = (A, B, C)$ defined on an open set Ω in $\mathbb{R}^3$. The corresponding flow $\mathbf{h}_t : \Omega \to \mathbb{R}^3$ is defined by $\mathbf{h}_t(\mathbf{x}) = \mathbf{x}(t)$, where $\mathbf{x}(t)$ is the unique solution of the initial-value problem

$$\mathbf{x}'(t) = \mathbb{F}(\mathbf{x}(t)), \quad \mathbf{x}(0) = \mathbf{x}.$$

To describe the relative flow of fluid particles near the particle that is initially at the point $\mathbf{a}$, let W be a small cube centered at $\mathbf{a}$, and let $\mathbf{x}$ be an arbitrary point in W. Then $\Delta \mathbf{x} = \mathbf{x} - \mathbf{a}$ gives the position of $\mathbf{x}$ in relation to $\mathbf{a}$, and the vector

$$\Delta \mathbf{x}(t) = \mathbf{h}_t(\mathbf{x}) - \mathbf{h}_t(\mathbf{a}) = \mathbf{h}_t(\mathbf{a} + \Delta \mathbf{x}) - \mathbf{h}_t(\mathbf{a}) = \mathbf{h}_t^*(\Delta \mathbf{x})$$

describes how that relative position varies over time. To get a formula that ties $\mathbf{h}_t^*$ back to $\mathbb{F}$, we first use Taylor's theorem to write

$$\mathbf{h}_t(\mathbf{x}) = \mathbf{x}(t) = \mathbf{x}(0) + t\mathbf{x}'(0) + \mathbf{O}(t^2) = \mathbf{x} + t\mathbb{F}(\mathbf{x}) + \mathbf{O}(t^2) \text{ as } t \to 0.$$

Note that $\mathbf{x}'(0) = \mathbb{F}(\mathbf{x}(0)) = \mathbb{F}(\mathbf{x})$ and that $\mathbf{x}(t)$ has the continuous second derivative required by Taylor's theorem, because $\mathbf{x}'(t) = \mathbb{F}(\mathbf{x}(t))$ and $\mathbb{F}$ is continuously differentiable. Because $\mathbf{x} = \mathbf{a} + \Delta \mathbf{x}$, it follows that

$$\mathbf{h}_t(\mathbf{a} + \Delta \mathbf{x}) = \mathbf{a} + \Delta \mathbf{x} + t\mathbb{F}(\mathbf{a} + \Delta \mathbf{x}) + \mathbf{O}(t^2)$$

$$\text{and } \mathbf{h}_t(\mathbf{a}) = \mathbf{a} + t\mathbb{F}(\mathbf{a}) + \mathbf{O}(t^2)$$

as $t \to 0$, and hence (using the differentiability of $\mathbb{F}$ at $\mathbf{a}$) that

$$\mathbf{h}_t^*(\Delta \mathbf{x}) = \mathbf{h}_t(\mathbf{a} + \Delta \mathbf{x}) - \mathbf{h}_t(\mathbf{a}) = \Delta \mathbf{x} + t(\mathbb{F}(\mathbf{a} + \Delta \mathbf{x}) - \mathbb{F}(\mathbf{a})) + \mathbf{O}(t^2)$$

$$= \Delta \mathbf{x} + t(d\mathbb{F}_{\mathbf{a}}(\Delta \mathbf{x}) + \mathbf{o}(\Delta \mathbf{x})) + \mathbf{O}(t^2)$$

as $\Delta \mathbf{x} \to \mathbf{0}$ and $t \to 0$.

As we did in Example 3, we consider that this formula for the relative flow defines a family of maps

$$\mathbf{h}_t^* : W^* \to \mathbb{R}^3$$

of a window W^* centered at the origin of local coordinates $\Delta \mathbf{x}$. Compare the general formula for $\mathbf{h}_t^*$ that we have just obtained with the earlier one in Example 3 (p. 475); the only new ingredient is the higher-order term $\mathbf{O}(t^2)$. The derviative $d\mathbb{F}_{\mathbf{a}}$ is still the key to the relative flow. The velocity field $\mathbb{F}^*$ of the relative flow is

The relative flow and d$\mathbb{F}_{\mathbf{a}}$

$$\mathbb{F}^*(\Delta \mathbf{x}) = \frac{d}{dt}\mathbf{h}_t^*(\Delta \mathbf{x})\bigg|_{t=0} = d\mathbb{F}_{\mathbf{a}}(\Delta \mathbf{x}) + \mathbf{o}(\Delta \mathbf{x}) \text{ as } \Delta \mathbf{x} \to \mathbf{0}.$$

Furthermore, when W^* is small enough for us to ignore higher-order terms in $\Delta \mathbf{x}$, we can approximate $\mathbf{h}_t^*$ by its derivative at the origin $\Delta \mathbf{x} = \mathbf{0}$:

$$\mathbf{h}_t^* \approx d(\mathbf{h}_t^*)_0 = I + t\,d\mathbb{F}_{\mathbf{a}} + \mathbf{O}(t^2) \text{ as } t \to 0.$$

Let us now use the approximation $d(\mathbf{h}_t^*)_0$ to get an idea how $\mathbf{h}_t^*$ affects volumes near $\Delta \mathbf{x} = \mathbf{0}$. We expect that the volume of a small cube of fluid may change as the fluid flows; we want to measure that change. As before, it is helpful to split $d\mathbb{F}_{\mathbf{a}}$ into its symmetric and antisymmetric parts: $d\mathbb{F}_{\mathbf{a}} = \mathcal{S} + \mathcal{A}$. The symmetric matrix $\mathcal{S}$ has mutually orthogonal eigenvectors $\mathbf{b}_1, \mathbf{b}_2, \mathbf{b}_3$ of length ε, with eigenvalues $\lambda_1, \lambda_2, \lambda_3$, respectively. Choose an order for the eigenvectors so that the cube $K = \mathbf{b}_1 \wedge \mathbf{b}_2 \wedge \mathbf{b}_3$ has positive volume ε^3. Its image $K_t = (I + t\mathcal{S})(K)$ is the rectangular parallelepiped with

How $d(\mathbf{h}_t^*)_0$ affects volumes

$$\text{vol}\,K_t = \varepsilon(1+t\lambda_1) \times \varepsilon(1+t\lambda_2) \times \varepsilon(1+t\lambda_3)$$
$$= \varepsilon^3(1+t(\lambda_1+\lambda_2+\lambda_3)+O(t^2)) \;\text{ as } t \to 0.$$

Because the sum of the eigenvalues of a matrix equals the trace of that matrix, and because the diagonal elements of an antisymmetric matrix are all 0, we have

$$\lambda_1 + \lambda_2 + \lambda_3 = \text{tr}\,\mathcal{S} = \text{tr}(\mathcal{S}+\mathcal{A}) = \text{tr}\,d\mathbb{F}_{\mathbf{a}}.$$

Furthermore, because $\mathbb{F} = (A,B,C)$,

$$\text{tr}\,d\mathbb{F}_{\mathbf{a}} = \frac{\partial A}{\partial x}(\mathbf{a}) + \frac{\partial B}{\partial y}(\mathbf{a}) + \frac{\partial C}{\partial z}(\mathbf{a}) = \text{div}\,F(\mathbf{a}),$$

finally linking volume change to divergence:

$$\text{vol}\,K_t = \text{vol}\,K(1 + t\,\text{div}\,\mathbb{F}(\mathbf{a}) + O(t^2)) \;\text{ as } t \to 0.$$

Thus, the relative (or percentage) change in volume (p. 477) is

$$V(t) = \frac{\text{vol}\,K_t - \text{vol}\,K}{\text{vol}\,K} = t\,\text{div}\,\mathbb{F}(\mathbf{a}) + O(t^2) \;\text{ as } t \to 0;$$

consequently, $V'(0) = \text{div}\,\mathbb{F}(\mathbf{a})$. Note that we obtained this result under the assumption we could replace the relative flow by its linear approximation and we could restrict ourselves to cubes whose edges were the eigenvectors of the symmetric part of the linear approximation.

How $\mathbf{h}_t^*$ itself affects volume   We now remove the restrictive assumptions and determine $V'(0)$ using the original nonlinear map $\mathbf{h}_t^*$ itself. First, let K range over closed sets with volume that contain the point $\mathbf{a}$ (so $\Delta \mathbf{x} = \mathbf{0}$). Let $K_t = \mathbf{h}_t^*(K)$; its volume is

$$\text{vol}\,\mathbf{h}_t^*(K) = \iiint_K J_{\mathbf{h}_t^*} \, dV$$

($J_{\mathbf{h}_t^*}$ is the Jacobian of $\mathbf{h}_t^*$). Because $\mathbf{h}_t^*$ is nonlinear, the ratio

$$\frac{\text{vol}\,\mathbf{h}_t^*(K) - \text{vol}\,K}{\text{vol}\,K}$$

is no longer independent of the diameter δK of K (Definition 8.14, p. 291). However, to determine percentage growth of volume, it is sufficient to see what happens to this ratio as $\delta K \to 0$. Because $\text{vol}\,\mathbf{h}_t^*(K)$ is a set function of integral type (cf. pp. 310–312), we can calculate its derivative as K shrinks down to the point $\mathbf{a}$; by Theorem 8.39, (p. 312), we find

$$\lim_{\delta K \to 0} \frac{\text{vol}\,\mathbf{h}_t^*(K)}{\text{vol}\,K} = J_{\mathbf{h}_t^*}(\mathbf{0}).$$

This allows us to define the percentage change of volume as the limit

$$V(t) = \lim_{\delta K \to 0} \frac{\operatorname{vol} \mathbf{h}_t^*(K) - \operatorname{vol} K}{\operatorname{vol} K} = J_{\mathbf{h}_t^*}(\mathbf{0}) - 1.$$

Because

$$J_{\mathbf{h}_t^*}(\mathbf{0}) = \det d(\mathbf{h}_t^*)_0 = 1 + t \operatorname{div} \mathbb{F}(\mathbf{a}) + O(t^2) \text{ as } t \to 0,$$

we again have

$$V(t) = t \operatorname{div} \mathbb{F}(\mathbf{a}) + O(t^2) \text{ as } t \to 0,$$

and $V'(0) = \operatorname{div} \mathbb{F}(\mathbf{a})$.

We now consider the action of the flow $I + t\mathcal{A}$, where $\mathcal{A}$ is the antisymmetric part of $d\mathbb{F}_{\mathbf{a}}$. As we noted on page 470, when $\mathbb{F} = (A, B, C)$, then

Action of $I + t\mathcal{A}$

$$\mathcal{A} = \frac{1}{2}\begin{pmatrix} 0 & -\gamma & \beta \\ \gamma & 0 & -\alpha \\ -\beta & \alpha & 0 \end{pmatrix}, \text{ where } Z = \begin{pmatrix} \alpha \\ \beta \\ \gamma \end{pmatrix} = \begin{pmatrix} C_y - B_z \\ A_z - C_x \\ B_x - A_y \end{pmatrix},$$

and all functions are evaluated at $\mathbf{x} = \mathbf{a}$.

Lemma 11.6. *The vector Z is an eigenvector of $\mathcal{A}$ with eigenvalue 0.*   □

The following theorem is the analogue, for a general flow, of Lemma 11.5 for the flow of Example 3. Note that $Z = \operatorname{curl} \mathbb{F}(\mathbf{a})$.

Theorem 11.9. *To first order in t, the flow $I + t\mathcal{A}$ is a uniform rotation with angular velocity $\frac{1}{2}Z$.*

Proof. By definition, a rotation matrix R is an orthogonal matrix (i.e., one whose transpose equals its inverse: $R^\dagger R = I$) with positive determinant. Let $R_t = I + t\mathcal{A}$. Then $R_t^\dagger = I - t\mathcal{A}$ and $(R_t)^\dagger R_t = I - t^2 \mathcal{A}$; thus, to first order in t, R_t is orthogonal. Becaue Z is an eigenvector of $\mathcal{A}$ with eigenvalue 0, $R_t Z = Z$. This implies Z is the rotation axis of each R_t.

To determine the angular speed and show that it is constant, we need more detailed information about $\mathcal{A}$. The vectors

$$X_1 = \frac{1}{\sqrt{\alpha^2 + \beta^2}}\begin{pmatrix} -\beta \\ \alpha \\ 0 \end{pmatrix}, \quad X_2 = \frac{Z \times X_1}{\|Z\|} = \frac{1}{\|Z\|\sqrt{\alpha^2 + \beta^2}}\begin{pmatrix} -\alpha\gamma \\ -\beta\gamma \\ \alpha^2 + \beta^2 \end{pmatrix}$$

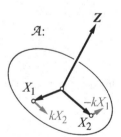

will provide this information. They are orthogonal unit vectors that span a plane orthogonal to Z; we claim the rotation leaves that plane invariant. This follows from quick calculations that show

$$\mathcal{A}X_1 = kX_2 \text{ and } \mathcal{A}X_2 = -kX_1,$$

where $k = \frac{1}{2}\|Z\|$. In terms of the basis $\{X_1, X_2, Z\}$, the matrix for $\mathcal{A}$ is

$$\begin{pmatrix} 0 & -k & 0 \\ k & 0 & 0 \\ 0 & 0 & 0 \end{pmatrix},$$

and the matrix for the flow $I + t\mathcal{A}$ is

$$\begin{pmatrix} 1 & -kt & 0 \\ kt & 1 & 0 \\ 0 & 0 & 1 \end{pmatrix}.$$

It follows from Lemma 11.5 that, to first order in t, $I + t\mathcal{A}$ is a uniform rotation with angular velocity $k = \frac{1}{2}Z$. $\qquad\qquad\square$

Theorem 11.10. *Suppose a steady fluid flow is governed by the velocity field* $\mathbb{F}$, *and* W^* *is a frame that is translated in space so as to remain centered on the fluid particle initially at the point* **a**. *Then a vanishingly small ball of fluid centered at the origin of* W^* *does the following.*

- *Rotates with instantaneous angular velocity* $\frac{1}{2}\operatorname{curl}\mathbb{F}(\mathbf{a})$

- *Changes its relative volume at the instantaneous rate* $\operatorname{div}\mathbb{F}(\mathbf{a})$ $\qquad\square$

11.3 Stokes' theorem

Stokes' theorem is our final setting for the assertion that the boundary operator and the exterior derivative are adjoints in the symbolic integral pairing: $\langle \partial\vec{S}, \omega \rangle = \langle \vec{S}, d\omega \rangle$. In this setting, $\vec{S}$ is a piecewise-smooth oriented surface in (x,y,z)-space, and $\omega = \omega(x,y,z)$ is a differential 1-form. In physical terms, Stokes' theorem equates the circulation of a flow around the boundary of a surface to the flux of the vorticity field (Definition 11.4, p. 473) of that flow through the surface.

Linking Stokes' and Green's theorems

To begin the process of proving Stokes' theorem, we first note how similar it is to Green's theorem. Both assert that

$$\oint_{\partial\vec{S}} \omega = \iint_{\vec{S}} d\omega,$$

where ω is a 1-form and $\vec{S}$ is an oriented 2-dimensional region. The only difference is the dimension of the ambient space: Green's theorem is set in $\mathbb{R}^2$ (forcing $\vec{S}$ to be planar), whereas Stokes' theorem is set in $\mathbb{R}^3$ (allowing $\vec{S}$ to be curved.) But because a surface patch in space is parametrized by a plane region, we are able to use Green's theorem to prove Stokes'.

Ingredients of the proof

The proof follows quickly once the ingredients are assembled. Recall that a *piecewise-smooth oriented surface* (Definitions 10.7, p. 412, and 10.12, p. 420) is a finite sum of oriented surface patches whose common boundary segments have opposite orientations. An *oriented surface patch* $\vec{S}$ (Definition 10.2, p. 392) is the image $\vec{S} = \mathbf{f}(\vec{U})$ of a closed bounded, positively oriented set $\vec{U} \subset \Omega$ with area, where the *parametrization*

$$\mathbf{f} : \Omega \to \mathbb{R}^3$$

is a continuously differentiable 1–1 immersion on the open set $\Omega \subseteq \mathbb{R}^2$. The map $\mathbf{f}$ is an *immersion* at the point **a** if the derivative $d\mathbf{f}_{\mathbf{a}} : \mathbb{R}^2 \to \mathbb{R}^3$ is 1–1.

The *surface integral* of a 2-form α defined everywhere on the surface patch $\vec{S}$ is given by the pullback $\mathbf{f}^*$:

$$\iint_{\vec{S}} \alpha = \iint_{\vec{U}} \mathbf{f}^* \alpha$$

(Definition 10.4, as reformulated on p. 432). The value of the surface integral is independent of the parametrization used to represent $\vec{S}$ (Corollary 10.4, p. 401). The surface integral of a 2-form over a piecewise-smooth oriented surface is the sum of the integrals over its smooth oriented pieces (Definition 10.13, p. 420). We also use the facts that the pullback $\mathbf{f}^*$ commutes with the exterior derivative (Theorem 10.16, p. 437), and that $\mathbf{f}$ and $\mathbf{f}^*$ are adjoints on piecewise-smooth curves $\vec{C}$ (Exercise 4.37, p. 149). Because these facts were first established in slightly different circumstances, we reconstruct them here, taking the target of $\mathbf{f}$ to be $\mathbb{R}^3$ instead of $\mathbb{R}^2$.

Lemma 11.7. *For any continuously differentiable map $\mathbf{f} : \Omega \to \mathbb{R}^3$ and k-form $\alpha(x,y,z)$, $\mathbf{f}^*(d\alpha) = d(\mathbf{f}^*\alpha)$.*

$\mathbf{f}^*$ and d commute

Proof. All aspects of the proof of Theorem 10.16 for $k \neq 2$ carry over, *mutatis mutandis*. The only difference occurs for $k = 2$, where the 3-form $d\alpha$ may be nonzero. However, the pullbacks $\mathbf{f}^*(d\alpha)$ and $d(\mathbf{f}^*\alpha)$ are 3-forms in two variables, so they are both zero and $\mathbf{f}^*(d\alpha) = d(\mathbf{f}^*\alpha)$ for all k. $\square$

The proof of the second lemma exploits, in addition, the fact that $\mathbf{f}$ is a 1–1 immersion.

Lemma 11.8. *For any piecewise-smooth oriented curve $\vec{C}$ and 1-form β defined everywhere on $\mathbf{f}(\vec{C})$,*

$\mathbf{f}$ and $\mathbf{f}^*$ are adjoints

$$\langle \mathbf{f}(\vec{C}), \beta \rangle = \int_{\mathbf{f}(\vec{C})} \beta = \int_{\vec{C}} \mathbf{f}^*\beta = \langle \vec{C}, \mathbf{f}^*(\beta) \rangle.$$

Proof. Let $\vec{C} = \vec{C}_1 + \cdots + \vec{C}_m$ be a decomposition into smooth oriented curves, each of which is either simple or is a simple closed curve. Because $\mathbf{f}$ is a 1–1 immersion, each $\mathbf{f}(\vec{C}_i)$ is likewise either simple, or a simple closed, smooth oriented curve, providing a decomposition

$$\mathbf{f}(\vec{C}) = \mathbf{f}(\vec{C}_1) + \cdots + \mathbf{f}(\vec{C}_m).$$

Let $\mathbf{u}_i(t) = (u_i(t), v_i(t))$, $a_i \leq t \leq b_i$ parametrize $\vec{C}_i$; then

$$\mathbf{x}_i(t) = \mathbf{f}(\mathbf{u}_i(t)),$$
$$(x_i(t), y_i(t), z_i(t)) = (x(u_i(t), v_i(t)), y(u_i(t), v_i(t)), z(u_i(t), v_i(t))),$$

parametrizes $\mathbf{f}(\vec{C}_i)$ with the same domain $a_i \leq t \leq b_i$. Suppose, for simplicity, that $\beta = P\,dx$; then

$$\mathbf{f}^*\beta = P(\mathbf{f}(\mathbf{u}))\,\mathbf{f}^*(dx) = P(\mathbf{f}(\mathbf{u}))(x_u\,du + x_v\,dv)$$

and

$$\int_{\vec{C}_i} \mathbf{f}^* \beta = \int_{a_i}^{b_i} P(\mathbf{f}(\mathbf{u}_i(t)))\,(x_u u_i' + x_v v_i')\,dt.$$

On the other hand,

$$\int_{\mathbf{f}(\vec{C}_i)} \beta = \int_{a_i}^{b_i} P(\mathbf{f}(\mathbf{u}_i(t)))\,x_i'\,dt = \int_{a_i}^{b_i} P(\mathbf{f}(\mathbf{u}_i(t)))\,(x_u u_i' + x_v v_i')\,dt = \int_{\vec{C}_i} \mathbf{f}^* \beta.$$

You can treat the 1-forms $\beta = Q\,dy$ and $\beta = R\,dz$ the same way. $\qquad\qquad\square$

Stokes' theorem for a surface patch

Theorem 11.11. *Let* $\mathbf{f}: \Omega \to \mathbb{R}^3$ *be a continuously differentiable 1–1 immersion on an open set* $\Omega \subseteq \mathbb{R}^2$. *Let* $\vec{U} \subset \Omega$ *be a closed, bounded, positively oriented set with area on which Green's theorem holds. If* ω *is a continuously differentiable 1-form defined on the oriented surface patch* $\vec{S} = \mathbf{f}(\vec{U})$, *then*

$$\oint_{\partial \vec{S}} \omega = \iint_{\vec{S}} d\omega.$$

Proof. We have the following sequence of equalities:

$$\iint_{\vec{S}} d\omega = \iint_{\mathbf{f}(\vec{U})} d\omega = \iint_{\vec{U}} \mathbf{f}^* d\omega \qquad\qquad \text{definition of surface integral}$$

$$= \iint_{\vec{U}} d(\mathbf{f}^* \omega) \qquad\qquad d \text{ and } \mathbf{f}^* \text{ commute (Lemma 11.7)}$$

$$= \oint_{\partial \vec{U}} \mathbf{f}^* \omega \qquad\qquad \text{Green's theorem for } \vec{U}$$

$$= \oint_{\mathbf{f}(\partial \vec{U})} \omega \qquad\qquad \mathbf{f} \text{ and } \mathbf{f}^* \text{ are adjoints (Lemma 11.8)}$$

$$= \oint_{\partial \vec{S}} \omega. \qquad\qquad\qquad\qquad \square$$

Stokes' theorem

Corollary 11.12 (Stokes' theorem) *Suppose* $\vec{S} = \vec{S}_1 + \cdots + \vec{S}_m$ *is a piecewise-smooth oriented surface, and suppose the theorem holds on each of the surface patches* $\vec{S}_i$, $i = 1, \ldots, m$; *then*

$$\oint_{\partial \vec{S}} \omega = \iint_{\vec{S}} d\omega.$$

Proof. Because the common segments of the various $\partial \vec{S}_i$ have opposite orientation, path integrals over those segments cancel in pairs; only the segments of $\partial \vec{S}_i$ that lie in $\partial \vec{S}$ make a nonzero contribution to the path integral. Therefore,

$$\oint_{\partial \vec{S}} \omega = \sum_{i=1}^m \oint_{\partial \vec{S}_i} \omega = \sum_{i=1}^m \iint_{\vec{S}_i} d\omega = \iint_{\vec{S}} d\omega.$$

The final equality is just the definition of a surface integral over $\vec{S}$. $\qquad\qquad\square$

From differential forms to vector fields

If $\omega = A\,dx + B\,dy + C\,dz$, then

$$d\omega = (C_y - B_z)\,dy\,dz + (A_z - C_x)\,dz\,dx + (B_x - A_y)\,dx\,dy,$$

and Stokes' theorem states that

$$\oint_{\partial\vec{S}} A\,dx + B\,dy + C\,dz = \iint_{\vec{S}}(C_y - B_z)\,dy\,dz + (A_z - C_x)\,dz\,dx + (B_x - A_y)\,dx\,dy.$$

Our discussion of the connection between differential forms and scalar and vector fields (pp. 460–462) makes it easy to convert Stokes' theorem into a statement about the integrals of vector fields. If $\omega = A\,dx + B\,dy + C\,dz$, then

$$\omega = \omega_{\mathbb{F}}^1 \leftrightarrow \mathbb{F} = (A,B,C)$$

and

$$d\omega = d\left(\omega_{\mathbb{F}}^1\right) = \omega_{\mathrm{curl}\,\mathbb{F}}^2 \leftrightarrow \mathrm{curl}\,\mathbb{F} = (C_y - B_z, A_z - C_x, B_x - A_y).$$

If $\mathbf{t}$ is the positively oriented unit tangent vector on $\partial\vec{S}$, then (cf. p. 19)

$$\oint_{\partial\vec{S}} A\,dx + B\,dy + C\,dz = \oint_{\partial S} \mathbb{F}\cdot\mathbf{t}\,ds = \text{circulation of }\mathbb{F}\text{ around }\partial\vec{S}.$$

If $\mathbf{n}$ is the unit normal that determines the orientation of $\vec{S}$, then (cf. pp. 403–404)

$$\iint_{\vec{S}}(C_y - B_z)\,dy\,dz + (A_z - C_x)\,dz\,dx + (B_x - A_y)\,dx\,dy$$

$$= \iint_S \mathrm{curl}\,\mathbb{F}\cdot\mathbf{n}\,dA = \text{total flux of curl}\,\mathbb{F}\text{ through }\vec{S}.$$

With these connections, we can restate Stokes' theorem for vector fields.

Theorem 11.13 (Physical form of Stokes' theorem). *If $\mathbb{F}$ is a continuously differentiable flow field defined on a piecewise-smooth oriented surface $\vec{S}$, and $\mathrm{curl}\,\mathbb{F}$ is its vorticity field, then*

$$\oint_{\partial S} \mathbb{F}\cdot\mathbf{t}\,ds = \iint_S \mathrm{curl}\,\mathbb{F}\cdot\mathbf{n}\,dA,$$

circulation of $\mathbb{F}$ around $\partial\vec{S}$ = total flux of $\mathrm{curl}\,\mathbb{F}$ through $\vec{S}$. $\qquad\square$

To illustrate the theorem in its physical form, let us add an extended example to the two we considered in the last section. We take the flow field and its vorticity field (Definition 11.4, p. 473) to be

Example 3:
$\mathbb{F} = (yz, -xz, 0)$

$$\mathbb{F} = (yz, -xz, 0), \quad \mathrm{curl}\,\mathbb{F} = (x, y, -2z).$$

This flow is similar to the flow of Example 1, where $\mathbb{F} = (-\omega y, \omega x, 0)$ (pp. 462–463). In that case, the entire fluid rotated rigidly (i.e., without the particles changing their relative positions over time) with constant angular speed ω around the z-axis. The flow in Example 3 is only slightly more complicated: the variable $-z$ simply replaces the constant ω. Thus the fluid at each level $z = c$ rotates rigidly around the z-axis with its own constant angular speed $\omega = -c$.

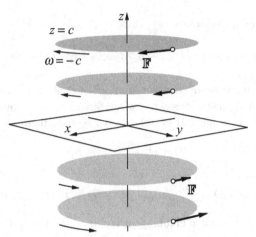

Imagine the fluid is separated into parallel disks. The disks above the (x,y)-plane rotate clockwise (when viewed from above); those below, counterclockwise. The farther a disk is from the (x,y)-plane, the faster it spins. This difference introduces a new shearing action within the fluid that we did not see in Example 1. There, the only shearing action was caused by the greater speed of particles farther from the z-axis. That is present here as well, and accounts for the component $-2z$ in the vorticity vector $\operatorname{curl}\mathbb{F} = (x,y,-2z)$. The new shearing action, between disks, accounts for the other components.

Vortex lines of $\mathbb{F}$

Let us examine all this in more detail. To connect the circulation of a field $\mathbb{F}$ to the flux of its vorticity field $\operatorname{curl}\mathbb{F}$, we naturally think of the second field, $\operatorname{curl}\mathbb{F}$, as a flow of "particles." The paths of these particles are called the **vortex lines** of $\mathbb{F}$. In the present case, the vortex lines are paths $\mathbf{x}(t) = (x(t),y(t),z(t))$ that satisfy the differential equations (cf. p. 462)

$$\mathbf{x}'(t) = \operatorname{curl}\mathbb{F}(\mathbf{x}(t)), \quad \text{or} \quad x' = x, \quad y' = y, \quad z' = -2z.$$

The general solution here is the three-parameter family

$$\mathbf{x}(t) = (ae^t, be^t, ce^{-2t}).$$

This describes the motion of the particle that is initially at the point $\mathbf{x}(0) = (a,b,c)$, which can thus be anywhere in space. In particular, if the initial point lies in the vertical plane $y = mx$ (so that $b = ma$), then the entire path is in the same plane, because

$$y(t) = be^t = ma\,e^t = mx(t).$$

In fact, this equation shows that we obtain all paths by rotating the paths that lie in a single vertical plane (e.g., in the plane $y = 0$) around the z-axis. The flow of $\operatorname{curl}\mathbb{F}$ has rotational symmetry around the z-axis.

The solutions on the plane $y = 0$ (i.e., where $b = 0$) are

$$x(t) = ae^t, \quad y(t) = 0, \quad z(t) = ce^{-2t}.$$

For a given $a \neq 0$ and c, this vortex line lies on the graph of the function

$$z = k/x^2, \quad k = ca^2,$$

in the (x,z)-plane. Particles on these trajectories flow simultaneously away from the z-axis and toward the (x,y)-plane. Particles on the z-axis (where $a = 0$) flow directly toward the origin. Particles in the (x,y)-plane ($c = 0$) flow radially away from the origin on straight lines. The origin is said to be a *saddle point* of the flow.

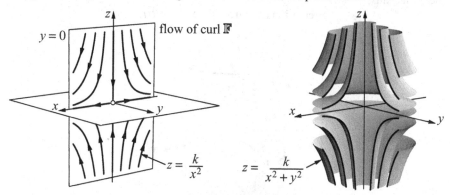

When the graph $z = k/x^2$ is rotated around the z-axis, it sweeps out the horn-shaped surface that is the graph of $z = k/(x^2 + y^2)$. (In the figure on the right, above, the portion of each surface that lies in the first quadrant has been cut away for better visibility.) The horn-shaped surfaces make it is easy to visualize the two flows and the way they are related: the flow lines of curl $\mathbb{F}$ (the vortex lines) are intersections of those surfaces with the vertical planes $y = mx$ that are "hinged" on the z-axis; the flow lines of $\mathbb{F}$ itself are their intersections with the horizontal planes $z = c$.

How the flows intersect
$z = k/(x^2 + y^2)$

To examine the link between the two fields, we need an oriented surface $\vec{S}$. We take $\vec{S}$ to be a cylinder centered at the origin; let its axis be the z-axis, and let its orientation normal $\mathbf{n}$ be outward-pointing. Let the radius be R and the height $2H$. The boundary of $\vec{S}$ is a pair of circles. The orientation induced by $\vec{S}$ on the upper one, $\partial \vec{S}_1$, is clockwise when viewed from above; on the lower one, $\partial \vec{S}_2$, it is counterclockwise. We want to compare the circulation of $\mathbb{F}$ around $\partial \vec{S} = \partial \vec{S}_1 + \partial \vec{S}_2$ with the total flux of curl $\mathbb{F}$ through $\vec{S}$.

The surface $\vec{S}$

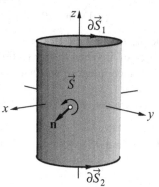

The flow $\mathbb{F}$ is everywhere tangent to $\partial\vec{S}$, so the circulation is quite simple to calculate. The upper circle, $\partial\vec{S}_1$, is in the horizontal plane $z = H$, wherein the fluid governed by $\mathbb{F}$ rotates with constant angular speed $\omega = H$ in the clockwise direction, the direction of $\partial\vec{S}_1$. Because $\partial\vec{S}_1$ is a circle of radius R, the fluid on it moves with speed HR. The circulation of $\mathbb{F}$ on $\partial\vec{S}_1$ is therefore just the product of this speed and the length of $\partial\vec{S}_1$, namely the positive quantity $2\pi R^2 H$. Taking orientations properly into account, we see that the circulation around the bottom of the cylinder, $\partial\vec{S}_2$, is the same; thus,

$$\text{circulation of } \mathbb{F} \text{ around } \partial\vec{S} = 4\pi R^2 H.$$

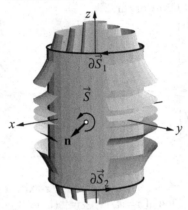

Now consider the vortex field $\text{curl}\,\mathbb{F}$ on $\vec{S}$. Although $\text{curl}\,\mathbb{F}$ is neither constant in magnitude nor perpendicular to $\vec{S}$, we now show that its projection onto the orienting normal $\mathbf{n}$ is constant, so total flux Φ is also simple to calculate. First, write the coordinates of a point on $\vec{S}$ in the form

$$(x,y,z) = (R\cos\theta, R\sin\theta, z);$$

(R, θ, z) are the *cylindrical coordinates* of the point. At this point, the vectors $\mathbf{n}$ and $\text{curl}\,\mathbb{F}$ have the form

$$\mathbf{n} = (\cos\theta, \sin\theta, 0) \quad \text{and} \quad \text{curl}\,\mathbb{F} = (R\cos\theta, R\sin\theta, -2z),$$

from which it follows that

$$\text{curl}\,\mathbb{F} \cdot \mathbf{n} = R$$

everywhere on $\vec{S}$. In principle (Definition 10.1, p. 389), Φ is the product of this projection length and the area of $\vec{S}$. When the projection length varies, the product needs to be rendered as a surface integral, but that is unnecessary here. Thus we have

$$\Phi = \text{curl}\,\mathbb{F} \cdot \mathbf{n} \; \text{area}\,\vec{S} = R \times (2\pi R \times 2H) = 4\pi R^2 H;$$

hence

$$\text{circulation of } \mathbb{F} \text{ around } \partial\vec{S} = \text{total flux of curl}\,\mathbb{F} \text{ through } \vec{S},$$

as we wished to show.

We now confirm this relation between $\mathbb{F}$ and $\operatorname{curl}\mathbb{F}$ on a second surface. Let $\vec{\Sigma}_1$ be the flat disk whose boundary is $\partial\vec{S}_1$. To make $\partial\vec{\Sigma}_1 = \partial\vec{S}_1$, that is, to make the orientations match, the orienting normal for $\vec{\Sigma}_1$ must point downward: $\mathbf{n}_1 = (0,0,-1)$. Let $\vec{\Sigma}_2$ be the disk whose boundary is $\partial\vec{S}_2$; here the orienting normal is $\mathbf{n}_2 = (0,0,+1)$. Finally, let $\vec{\Sigma} = \vec{\Sigma}_1 + \vec{\Sigma}_2$. Then $\partial\vec{\Sigma} = \partial\vec{S}$, so

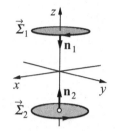

The surface Σ

$$\text{circulation of } \mathbb{F} \text{ around } \partial\vec{\Sigma} = 4\pi R^2 H,$$

as before. The total flux of $\operatorname{curl}\mathbb{F}$ through $\vec{\Sigma}$ is once again a simple calculation. On $\vec{\Sigma}_1$, a point has coordinates (x,y,H), so

$$\operatorname{curl}\mathbb{F}\cdot\mathbf{n}_1 = (x,y,-2H)\cdot(0,0,-1) = +2H$$

there. Because this is a constant, the flux Φ_1 through $\vec{\Sigma}_1$ is just

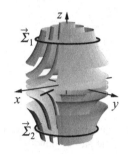

$$\Phi_1 = \operatorname{curl}\mathbb{F}\cdot\mathbf{n}_1 \ \text{area}\,\vec{\Sigma}_1 = 2H \times \pi R^2 = 2\pi R^2 H.$$

On $\vec{\Sigma}_2$, $\operatorname{curl}\mathbb{F}\cdot\mathbf{n}_2 = (x,y,2H)\cdot(0,0,1) = 2H$ once again, and the flux is $\Phi_2 = 2\pi R^2 H$. Hence,

$$\text{circulation of } \mathbb{F} \text{ around } \partial\vec{\Sigma} = \text{total flux of } \operatorname{curl}\mathbb{F} \text{ through } \vec{\Sigma}.$$

With $\vec{S}$ and $\vec{\Sigma}$, we were able to determine total flux without calculating an integral. Here is a third surface, $\vec{T}$, for which the integral is necessary. We define $\vec{T}$ by a parametrization $\mathbf{f}: \vec{U} \to \vec{T}$ (with α to be determined):

The surface $\vec{T}$

$$\mathbf{f}: \begin{cases} x = \alpha\cos u\cosh v, \\ y = \alpha\sin u\cosh v, \\ z = v, \end{cases} \qquad \vec{U}: \begin{array}{l} -\pi \le u \le \pi, \\ -H \le v \le H. \end{array}$$

This is a surface of revolution around the z-axis; if $\alpha = 1$, it is called a *catenoid*. However, we want to choose α so that $\partial\vec{T} = \partial\vec{S}_1 + \partial\vec{S}_2$. Thus, in the first quadrant in the (x,z)-plane (where $\cos u = 1$), we want $x = R$ when $z = H$. Consequently $R = \alpha\cosh H$, so $\alpha = R/\cosh H$. Note: $\mathbf{f}$ is not 1–1 on $\vec{U}$, so $\vec{T}$ is not a surface patch, strictly speaking. We should break up $\vec{T}$ into two separate pieces. We can accomplish that by breaking up $\vec{U}$ into two pieces (e.g., with $-\pi \le u \le 0$ and $0 \le u \le \pi$) and using the same formula $\mathbf{f}$ for each. But nothing essential is lost by treating these two pieces together, as we do; see a similar comment (p. 396) about parametrizing a sphere.

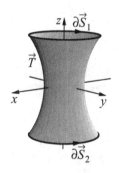

To determine the total flux of $\operatorname{curl}\mathbb{F}$ through $\vec{T}$, we must pull back

$$d\omega = x\,dy\,dz + y\,dz\,dx - 2z\,dx\,dy$$

to $\vec{U}$ using $\mathbf{f}^*$. We have

$$\mathbf{f}^*(dy\,dz) = \alpha\cos u\cosh v\,du\,dv, \qquad \mathbf{f}^*(dz\,dx) = \alpha\sin u\cosh v\,du\,dv,$$
$$\mathbf{f}^*(dx\,dy) = -\alpha^2\sinh v\cosh v\,du\,dv,$$

from which it follows that

$$\mathbf{f}^*(d\omega) = \alpha^2(\cosh^2 v + 2v\sinh v\cosh v)\,du\,dv.$$

Therefore, the total flux is

$$\iint_{\vec{T}} d\omega = \iint_{\vec{U}} \mathbf{f}^*(d\omega) = \alpha^2 \int_{-\pi}^{\pi} du \int_{-H}^{H} (\cosh^2 v + 2v\sinh v\cosh v)\,dv$$
$$= \alpha^2 \times 2\pi \times 2H\cosh^2 H.$$

(Note that $\cosh^2 v + 2v\sinh v\cosh v = (v\cosh^2 v)'$.) Because $\alpha^2 = R^2/\cosh^2 H$, we find

$$\text{total flux of curl}\,\mathbb{F} \text{ through } \vec{T} = 4\pi R^2 H,$$

agreeing with the values for $\vec{S}$ and $\vec{\Sigma}$.

Oriented surfaces with the same boundary

Of course it is not an accident that the total flux of curl $\mathbb{F}$ through one of these surfaces has the same value as through any other. It is a simple consequence of Stokes' theorem and the fact that the three surfaces have the same boundary, including orientation.

Theorem 11.14. *Suppose Stokes' theorem holds for the piecewise-smooth oriented surfaces $\vec{S}$ and $\vec{\Sigma}$, and $\partial\vec{S} = \partial\vec{\Sigma}$. If ω is any 1-form defined on a region containing $\vec{S}$ and $\vec{\Sigma}$, then*

$$\iint_{\vec{S}} d\omega = \iint_{\vec{\Sigma}} d\omega.$$

Proof. The proof involves two applications of Stokes' theorem:

$$\iint_{\vec{S}} d\omega = \oint_{\partial\vec{S}} \omega = \oint_{\partial\vec{\Sigma}} \omega = \iint_{\vec{\Sigma}} d\omega. \qquad \square$$

Implications of the divergence theorem

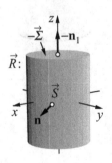

The divergence theorem gives us another way to show that the total flux of curl $\mathbb{F}$ through any two of the surfaces in our Example 3 must be equal. The key is that any two of the surfaces, properly reoriented, form the total boundary of a 3-dimensional region. For example, $\vec{S}$ and $-\vec{\Sigma}$ make up the boundary of the positively oriented cylindrical region

$$\vec{R}: \quad \begin{matrix} x^2 + y^2 \le R^2, \\ -H \le z \le H. \end{matrix}$$

Now apply the divergence theorem to the region $\vec{R}$ and the 2-form $d\omega$ on $\partial\vec{R}$; because $d(d\omega) = 0$ on $\vec{R}$ (because $d^2 = 0$ always), we find

$$0 = \iiint_{\vec{R}} d(d\omega) = \iint_{\partial\vec{R}} d\omega = \iint_{\vec{S} - \vec{\Sigma}} d\omega = \iint_{\vec{S}} d\omega - \iint_{\vec{\Sigma}} d\omega.$$

Of course, this argument using the divergence theorem works equally well for any pair of piecewise-smooth oriented surfaces that have a common boundary and that together form the complete boundary (when properly reoriented) of a 3-dimensional region.

In Theorem 11.8 and also in the discussion on pages 471–473, we established that the fundamental link between circulation and curl has the form

$$\lim_{\delta(\vec{S}) \to 0} \frac{\text{circulation of } \mathbb{F} \text{ around } \partial\vec{S}}{\text{area of } \vec{S}} = \operatorname{curl}\mathbb{F}(\mathbf{a}) \cdot \mathbf{n},$$

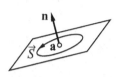

where $\vec{S}$ is centered at $\mathbf{a}$ and lies in the plane with normal $\mathbf{n}$ passing through $\mathbf{a}$; $\delta(\vec{S})$ is the *diameter* of $\vec{S}$ (Definition 8.14, p. 291). However, to carry out the computations, we needed $\vec{S}$ to be either an ellipse or a rectangle. Stokes' theorem is a more powerful computational tool; with it, we can now remove this restriction.

Theorem 11.15. *Suppose the flow field* $\mathbb{F}$ *is defined on an open set* $\Omega \subseteq \mathbb{R}^3$. *Let* $S_k \subset \Omega$ *be a sequence of closed surfaces with area that pass through a common point* $\mathbf{a}$ *and lie in a common plane with unit normal* $\mathbf{n}$. *Suppose that the diameter* $\delta(S_k) \to 0$ *as* $k \to \infty$ *and the boundary of each* S_k *is a piecewise-smooth simple closed curve. Then*

$$\lim_{k \to \infty} \frac{\text{circulation of } \mathbb{F} \text{ around } \partial S_k}{\text{area of } S_k} = \operatorname{curl}\mathbb{F}(\mathbf{a}) \cdot \mathbf{n},$$

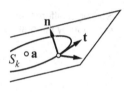

where circulation is computed in the direction $\mathbf{t}$ *along* ∂S_k *for which the ordered triple of vectors* $\{$*outward normal to* S_k *in the plane*, $\mathbf{t}$, $\mathbf{n}\}$ *has the same orientation as the coordinate axes.*

Proof. By Stokes' theorem, the circulation of $\mathbb{F}$ around ∂S_k is

$$\oint_{\partial S_k} \mathbb{F} \cdot \mathbf{t}\, ds = \iint_{S_k} \operatorname{curl}\mathbb{F} \cdot \mathbf{n}\, dA.$$

By an adaptation of the law of the mean for double integrals (Theorem 3.7, p. 76),

$$\iint_{S_k} \operatorname{curl}\mathbb{F} \cdot \mathbf{n}\, dA = \operatorname{curl}\mathbb{F}(\mathbf{a}_k) \cdot \mathbf{n} \,\operatorname{area} S_k,$$

where $\mathbf{a}_k$ is a point in S_k; note that $\mathbf{n}$ is constant in the integral. Now let $k \to \infty$; then $\delta(S_k) \to 0$, so $\mathbf{a}_k \to \mathbf{a}$. $\qquad\square$

The theorem calls our attention to the quantity

Circulation per unit area

$$q(\mathbf{a}, \mathbf{n}) = \lim_{k \to \infty} \frac{\text{circulation of } \mathbb{F} \text{ around } \partial S_k}{\text{area of } S_k}$$

that depends on the point $\mathbf{a}$ and the unit normal $\mathbf{n}$, but not on the particular sets S_k used to define it; let us call it the **circulation of** $\mathbb{F}$ **per unit area** at the point $\mathbf{a}$ in the direction $\mathbf{n}$. Now let

$$q_i(\mathbf{a}) = q(\mathbf{a}, \mathbf{e}_i), \quad i = 1, 2, 3,$$

where $\{\mathbf{e}_1, \mathbf{e}_2, \mathbf{e}_3\}$ is the standard basis in $\mathbb{R}^3$.

Corollary 11.16 $\operatorname{curl} \mathbb{F}(\mathbf{a}) = (q_1(\mathbf{a}), q_2(\mathbf{a}), q_3(\mathbf{a}))$. $\qquad\qquad\qquad\qquad$ □

This corollary serves, as Lemmas 11.2–11.4 did, to give us insight into the way the vorticity vector $\operatorname{curl}\mathbb{F}$ is linked to $\mathbb{F}$ itself. It provides us with an alternative to the physical imagery of a little ball set spinning by the shearing action of a fluid flow represented by $\mathbb{F}$.

11.4 Closed and exact forms

Using differential forms

In this section, we use differential forms to define and to solve ordinary and partial differential equations. When the differential forms have certain special properties (e.g., when they are *closed* or *exact*), the solutions can be particularly simple and elegant. In a different vein, we also show how those special forms give us information about the geometry of the domains on which they are defined.

Solving a differential equation by integrating

Analytic methods for solving differential equations frequently involve integration, or *quadrature*, as it is traditionally called in this context. For example, the solution of the basic equation $dy/dx = f(x)$ is a quadrature:

$$y = \int f(x)\,dx.$$

The integral represents the infinite collection of functions $F(x) + c$, where F is a specific antiderivative of f (i.e., $F'(x) = f(x)$), and c is an arbitrary constant. We say the solutions form a one-parameter family; c is the parameter.

For the more complicated equation $dy/dx = f(x)/g(y)$, a solution is a function $y = \varphi(x)$ for which

$$\varphi'(x) = \frac{f(x)}{g(\varphi(x))} \quad \text{for all } x \text{ in some nonempty interval.}$$

To find φ, first rewrite the differential equation as an "equation with differentials" in such a way that the two variables are separated:

$$g(y)\,dy - f(x)\,dx = 0.$$

This implies $G(y) - F(x) = c$, where c is an arbitrary constant, and G and F are specific antiderivatives of g and f, respectively. For each c that we specify, the equation $G(y) - F(x) = c$ is a relation between x and y that defines y implicitly as a function of x (cf. Chapter 6.1) if there is a "seed point" $(x, y) = (a, b)$ for which

$$G(b) - F(a) = c \quad \text{and} \quad G'(b) \neq 0.$$

The implicit function $y = \varphi_c(x)$ that is supplied by Theorem 6.1, page 189 (and by the specified c), has $\varphi_c(a) = b$ and satisfies the conditions

$$G(\varphi_c(x)) - F(x) = c \ \text{ and } \ \varphi_c'(x) = -\frac{-F'(x)}{G'(\varphi_c(x))} = \frac{f(x)}{g(\varphi_c(x))}$$

at all points x on some open interval including $x = a$. Thus $\varphi_c(x)$ is indeed a solution to the differential equation. As in the simpler case, c serves to parametrize an infinite family of such solutions.

The function $G(y) - F(x)$ is called a **primitive**, or **first integral**, of the differential equation. The first name is suggested by the fact that solutions emerge from it (as implicitly defined functions). The second name is suggested by the fact that we can write it as an integral:

Primitives and
first integrals

$$G(y) - F(x) = \int g(y) \, dy - \int f(x) \, dx.$$

Simply put, we solve the differential equation by integrating it to obtain a first integral/primitive that defines solutions implicitly.

There is a larger class of differential equations, called *exact*, that can be integrated the same way. An **exact differential equation** has the form

Exact differential
equations

$$\Phi_x(x,y) \, dx + \Phi_y(x,y) \, dy = 0;$$

it has this name because the left-hand side is exactly equal to the differential (i.e., the exterior derivative) of the function $\Phi(x,y)$: $d\Phi = \Phi_x \, dx + \Phi_y \, dy$. A solution is a function $y = \varphi(x)$ for which

$$\Phi_x(x, \varphi(x)) + \Phi_y(x, \varphi(x)) \, \varphi'(x) = 0$$

for all x in some nonempty interval. (In other words, the given differential equation is satisfied when we substitute $\varphi(x)$ for y and $\varphi'(x) \, dx$ for dy.) Because we can write the differential equation as $d\Phi = 0$, we have

$$\Phi(x,y) = \int d\Phi = c,$$

implying that Φ is a *first integral* for the differential equation. In other words, if we fix c and find a "seed point" $(x,y) = (a,b)$ for which

$$\Phi(a,b) = c \ \text{ and } \ \Phi_y(a,b) \neq 0,$$

then the implicit function theorem provides a function $y = \varphi_c(x)$ with $\varphi_c(a) = b$ and for which

$$\Phi(x, \varphi_c(x)) = c \ \text{ and } \ \varphi_c'(x) = \frac{-\Phi_x(x, \varphi_c(x))}{\Phi_y(x, \varphi_c(x))}.$$

for all x on some open interval containing $x = a$. Thus φ_c is a solution to the differential equation.

The differential equation $y\,dx + x\,dy = 0$ is exact. One first integral is $\Phi = xy$, and the solutions are the functions $\varphi_c(x) = c/x$. In other cases, it may be more difficult to see whether the differential equation is exact. For example,

$$\frac{-y}{x^2+y^2}\,dx + \frac{x}{x^2+y^2}\,dy = 0, \quad (x,y) \neq (0,0),$$

is exact; its first integral is

$$\theta(x,y) = \arctan\left(\frac{y}{x}\right),$$

as can be verified immediately (cf. pp. 429–431). The solution implicitly defined by the equation $\theta = c$ is the linear function $\varphi_c(x) = (\tan c)x$, $x \neq 0$. These solutions form a one-parameter family whose graphs are the straight lines that radiate from the origin. We cannot expect the solutions to be defined at the origin because the differential equation itself is not defined there. The example has an even more remarkable feature: the first integral $\theta(x,y)$ must be multiple-valued if it is to avoid discontinuities on the punctured plane. See the graph of $z = \theta(x,y)$ on page 430.

Each differential equation we have been considering can be written in the form $\omega = 0$ when ω is the general 1-form $P(x,y)\,dx + Q(x,y)\,dy$. In terms of ω, an *exact* differential equation is one for which $\omega = d\Phi$ for some 0-form Φ. In this case, we now say ω itself is exact, and then extend this definition to general k-forms.

Definition 11.5 *A differential k-form ω in n variables is said to be **exact** if there is a $(k-1)$-form α for which $\omega = d\alpha$.*

When $\omega = d\alpha$ is exact, then $d\omega = d^2\alpha = 0$ (because $d^2 = 0$ by Theorem 10.17, p. 439). Recall that the exterior derivative "d" and the boundary operator "∂" are paired as *adjoints* (cf. p. 428). We use the term *closed* for a curve or surface S that has zero boundary, $\partial S = \emptyset$; therefore the pairing suggests the same term, *closed*, for a differential form ω that has zero exterior derivative, $d\omega = 0$.

Definition 11.6 *We say the k-form ω is **closed** if $d\omega = 0$.*

These definitions lead to the following conclusion.

Corollary 11.17 *Every exact form is closed.* □

The corollary gives us a necessary condition for a differential equation of the more general form

$$\omega = P(x,y)\,dx + Q(x,y)\,dy = 0$$

to be exact: we must have $Q_x = P_y$ (because $d\omega = (Q_x - P_y)\,dx\,dy$). This also follows from the equality of mixed partial derivatives, for if ω is an exact 1-form, with $\omega = d\Phi = \Phi_x\,dx + \Phi_y\,dy$, then $P = \Phi_x$, $Q = \Phi_y$, and

$$Q_x = (\Phi_y)_x = (\Phi_x)_y = P_y.$$

Because an exact differential equation is "integrable" (i.e., solvable by integrations), we think of $Q_x = P_y$ as an **integrability condition**, in this case, a *necessary* condition. As we show below, $Q_x = P_y$ is also a *sufficient* condition for the integrability of $\omega = 0$, at least locally.

The integrability condition shows that, for example, $x\,dy - y\,dx = 0$ cannot be exact ($Q_x - P_y = 2$). Nevertheless, when we rewrite this differential equation as

Integrating factors

$$\frac{x\,dy - y\,dx}{x^2 + y^2} = 0,$$

it becomes exact. It has the first integral $\arctan(y/x)$, as noted above. Because the added factor $1/(x^2 + y^2)$ makes the differential equation integrable, we call it an **integrating factor**. Another integrating factor is $1/x^2$, because

$$\frac{x\,dy - y\,dx}{x^2} = \frac{-y}{x^2}\,dx + \frac{1}{x}\,dy = d\left(\frac{y}{x}\right).$$

In this case the first integral is y/x, not $\arctan(y/x)$, but the solution graphs are unchanged; they are the same straight lines that radiate from the origin. With yet another integrating factor, $1/xy$, the differential equation even becomes separable:

$$\frac{x\,dy - y\,dx}{xy} = \frac{dy}{y} - \frac{dx}{x} = d(\ln|y| - \ln|x|).$$

Here the first integral is $\ln|y/x|$; it leads once again to the same solution graphs. Of course most differential equations fail to be exact and fail to have integrating factors that make them exact. We leave further discussion of the art of finding integrating factors to texts on differential equations.

Every exact form is closed; is every closed form exact? The closed form

Is a closed form exact?

$$\omega = \frac{-y\,dx + x\,dy}{x^2 + y^2}$$

reveals a difficulty. The form is defined everywhere in the *punctured plane* $\mathcal{P} = \mathbb{R}^2 \setminus (0,0)$, but there is no continuously differentiable, single-valued function $\Phi(x,y)$ on $\mathcal{P}$ for which $d\Phi = \omega$ (see Exercise 11.12). As pointed out above (and on pp. 429–431), the angle function $\theta(x,y) = \arctan(y/x)$ does have $d\theta = \omega$ everywhere on $\mathcal{P}$, but it is multiple-valued. There is no way to assign a unique angle to every point in $\mathcal{P}$ without having discontinuities. (There is a similar difficulty with the earth's time zones: time increases steadily in the eastward direction, until the International Date Line, where it drops back 24 hours.) The obstruction to continuity disappears if we restrict ω to a domain that has no closed path encircling the origin. For example, we can use a disk or a rectangle that excludes the origin. More generally, we have the following result for closed 1-forms in two variables.

Theorem 11.18. *Suppose the 1-form* $\omega = P(x,y)\,dx + Q(x,y)\,dy$ *is defined and closed on a rectangular window W centered at a point* (a,b) *in* $\mathbb{R}^2$. *Then* $\omega(x,y) = d\Phi(x,y)$ *for every* (x,y) *in W, where*

$$\Phi(x,y) = \int_a^x P(t,b)\,dt + \int_b^y Q(x,t)\,dt.$$

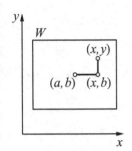

Proof. It suffices to show that $\Phi_x = P$, $\Phi_y = Q$ in W. Because y does not appear in the first integral, and appears only in the upper limit of integration in the second, the equality $\Phi_y = Q$ is immediate. To verify the other equality, we use the integrability condition $Q_x = P_y$ to write

$$\Phi_x(x,y) = P(x,b) + \int_b^y Q_x(x,t)\,dt = P(x,b) + \int_b^y P_y(x,t)\,dt$$

$$= P(x,b) + P(x,t)\Big|_b^y = P(x,b) + P(x,y) - P(x,b) = P(x,y). \qquad \square$$

Local exactness

Our goal is to prove this result in full generality: to show that a k-form in n variables that is closed inside a rectangular parallelepiped (a "window") is exact there. We say that such a closed form is **locally exact**.

From *partial*
to *ordinary*
differential equations

Consider how Theorem 11.18 accomplished the goal. The function $\Phi(x,y)$ for which

$$d\Phi = \Phi_x\,dx + \Phi_y\,dy = P\,dx + Q\,dy = \omega,$$

is a solution to the pair of partial differential equations

$$\Phi_x = P, \quad \Phi_y = Q,$$

where $P(x,y)$ and $Q(x,y)$ are given functions that satisfy the integrability condition $Q_x = P_y$. But the theorem presents $\Phi(x,y)$ as $F_b(x) + G_x(y)$, where F_b and G_x are the particular solutions of the ordinary differential equations

$$F_b'(t) = P(t,b), \quad G_x'(t) = Q(x,t),$$

that satisfy the initial conditions

$$F_b(a) = 0, \quad G_x(b) = 0.$$

In the first function, b is a parameter; in the second, x is.

In general, showing that a closed k-form in n variables is locally exact reduces to solving a set of partial differential equations in the presence of certain integrability conditions. For example, take $k = 2$, $n = 3$, and suppose

$$\omega = P(x,y,z)\,dy\,dz + Q(x,y,z)\,dz\,dx + R(x,y,z)\,dx\,dy$$

is closed; this implies $P_x + Q_y + R_z = 0$. If ω is to be exact, we need a 1-form

$$\alpha = A(x,y,z)\,dx + B(x,y,z)\,dy + C(x,y,z)\,dz$$

with $d\alpha = \omega$; this implies

$$C_y - B_z = P, \quad A_z - C_x = Q, \quad B_x - A_y = R.$$

These are three partial differential equations for the three unknown functions A, B, and C, together with one integrability condition $P_x + Q_y + R_z = 0$ imposed on the known functions P, Q, and R. More generally, local exactness of a closed k-form in n variables involves $\binom{n}{k}$ partial differential equations for $\binom{n}{k-1}$ unknown functions together with $\binom{n}{k+1}$ integrability conditions (see Exercise 11.25). To prove local exactness for a general k-form by the approach of Theorem 11.18, we must first reduce the partial differential equations with their integrability conditions into ordinary differential equations whose solutions (expressed as ordinary integrals) supply the coefficients of the needed $(k-1)$-form.

In "The Poincaré Lemma and an Elementary Construction of Vector Potentials" [22], Shirley Llamado Yap introduces an algorithm for carrying out this approach. The algorithm constructs a solution for every k and $n \geq k$, using induction on n. Because the argument involves a flurry of subscripts, we first step through the (subscript-free) example with $k = 2$ and $n = 3$ we have just introduced. Thus we are given three functions $P(x,y,z)$, $Q(x,y,z)$, and $R(x,y,z)$ that are defined in a window centered at $(x,y,z) = (a,b,c)$. They satisfy the integrability condition $P_x + Q_y + R_z = 0$. We seek three functions $A(x,y,z)$, $B(x,y,z)$, and $C(x,y,z)$ that satisfy the three partial differential equations

$$C_y - B_z = P, \quad A_z - C_x = Q, \quad B_x - A_y = R,$$

in that window.

A system of partial differential equations typically has many solutions. We seek a solution in which $C(x,y,z) \equiv 0$. In that case the first two equations reduce to differentiation with respect to z alone: $A_z = Q$, $B_z = -P$. By treating x and y as parameters, we can think of these as *ordinary* differential equations in z, whose solutions are then given by integration:

$$A(x,y,z) = A(x,y,c) + \int_c^z Q(x,y,t)\,dt,$$

$$B(x,y,z) = B(x,y,c) - \int_c^z P(x,y,t)\,dt.$$

The first equation expresses the values of A off the plane $z = c$ in terms of its values on that plane (and on the values of Q in the window). But we have not yet determined the values of A on the plane.

The situation is similar for B, but, following the approach we took with C, we seek a solution in which $B(x,y,c) \equiv 0$. In that case, the equation for B reduces to

$$B(x,y,z) = -\int_c^z P(x,y,t)\,dt.$$

Now consider the third partial differential equation, $B_x - A_y = R$. For the moment, we look for a solution only on the plane $z = c$; in Step 3, we remove this restriction. (The move from the plane to 3-space becomes the induction step in the general algorithm.) On $z = c$, we have $B = 0$ by Step 1, so $B_x - A_y = R$ reduces to

Solving the case
$k = 2$, $n = 3$

Step 1

Step 2

$$A_y(x,y,c) = -R(x,y,c).$$

We can treat this as an ordinary differential equation in y (with x and c as parameters); its solution is the integral

$$A(x,y,c) = A(x,b,c) - \int_b^y R(x,t,c)\,dt.$$

By analogy with what we have done with C and B, we seek a solution for which $A(x,b,c) = 0$; then

$$A(x,y,c) = -\int_b^y R(x,t,c)\,dt.$$

Step 3   Combining the results from Steps 1 and 2, we obtain the following formulas for A and B that are defined in the entire window:

$$A(x,y,z) = -\int_b^y R(x,t,c)\,dt + \int_c^z Q(x,y,t)\,dt,$$

$$B(x,y,z) = -\int_c^z P(x,y,t)\,dt.$$

But we have not yet verified that A and B, as defined by these formulas, satisfy the third partial differential equation everywhere in the window. In Step 2, we constructed A and B to satisfy that third equation *only on the plane $z = c$*.

To show that A and B also satisfy the third equation when $z \neq c$, we take the following approach. We can write the third equation as $E = 0$, where

$$E(x,y,z) = B_x(x,y,z) - A_y(x,y,z) - R(x,y,z).$$

By Step 2, E equals zero when $z = c$ (and (x,y,z) lies in the window); we must show that E remains equal to zero for all z in some open neighborhood of $z = c$. We claim that the derivative of E with respect to z is zero; it will then follow that the value of E does not change—and will thus remain equal to zero—as z moves away from c. To prove the claim, we invoke the integrability condition $P_x + Q_y + R_z = 0$:

$$\frac{\partial E}{\partial z} = B_{xz} - A_{yz} - R_z = (B_z)_x - (A_z)_y - R_z = -P_x - Q_y - R_z = 0.$$

This completes the construction of the 1-form α for which $d\alpha = \omega$, and thus proves the Poincaré lemma in this case.

Theorem 11.19. *If $\omega = P\,dy\,dz + Q\,dz\,dx + R\,dx\,dy$ is closed in a window centered at $(x,y,z) = (a,b,c)$, then $\omega = d\alpha$, where $\alpha = A\,dx + B\,dy + C\,dz$ and*

$$A(x,y,z) = -\int_b^y R(x,t,c)\,dt + \int_c^z Q(x,y,t)\,dt,$$

$$B(x,y,z) = -\int_c^z P(x,y,t)\,dt,$$

$$C(x,y,z) = 0. \qquad\qquad\qquad\qquad\qquad\qquad\qquad\qquad\qquad\qquad\quad \square$$

The 1-form α that makes ω locally exact is not unique. If β also makes ω locally exact (i.e., $d\beta = \omega$), then

$$d(\beta - \alpha) = \omega - \omega = 0.$$

Hence $\beta - \alpha$ is a *closed* 1-form, so by the Poincaré lemma for 1-forms (Theorem 11.18, as extended to higher dimensions in Exercises 11.23 and 11.24), $\beta - \alpha$ is itself locally exact: $\beta - \alpha = d\Phi$ for some 0-form Φ. The most general 1-form that makes ω locally exact is thus $\alpha + d\Phi$, where Φ is an arbitrary 0-form.

We move on now to Yap's general algorithm for constructing a $(k-1)$-form α that makes a *closed* k-form ω locally exact: $\omega = d\alpha$. The construction proceeds inductively on the dimension n of the space on which the forms are defined. For simplicity, we assume that ω is defined on a window (rectangular parallelepiped) centered at the origin in $\mathbb{R}^n$.

Throughout the argument, k is fixed. The induction on n begins with $n = k$. In this case, a k-form ω has only a single term,

$$\omega = P(x_1, \ldots, x_k)\, dx_1 \cdots dx_k,$$

and ω is automatically closed. Let us take

$$\alpha = A(x_1, \ldots, x_k)\, dx_1 \cdots dx_{k-1};$$

then

$$d\alpha = \frac{\partial A}{\partial x_k}\, dx_k\, dx_1 \cdots dx_{k-1} = (-1)^{k-1} \frac{\partial A}{\partial x_k}\, dx_1 \cdots dx_{k-1}\, dx_k.$$

Thus $\omega = d\alpha$ if A satisfies the partial differential equation

$$P = (-1)^{k-1} \frac{\partial A}{\partial x_k}.$$

We can solve this differential equation immediately by integration:

$$A(x_1, \ldots, x_k) = (-1)^{k-1} \int_0^{x_k} P(x_1, \ldots, x_{k-1}, t)\, dt.$$

This completes the construction of α when $n = k$.

Now take $n > k$ and use induction. That is, assume the algorithm works for k-forms in $\mathbb{R}^{n-1}$ and then show that it works for k-forms in $\mathbb{R}^n$. The arguments make extensive use of multi-indices; see pages 439–443.

We are given a closed k-form

$$\omega = \sum_I P_I(x_1, \ldots, x_n)\, dx_I, \quad I = (i_1, \ldots, i_k),$$

with $1 \le i_1 < \cdots < i_k \le n$. We want to find a $(k-1)$-form

$\omega = d(\alpha + d\Phi)$
for any Φ

The general algorithm

The base, with $n = k$

The induction,
with $n > k$

$$\alpha = \sum_{\widehat{I_s}} A_{\widehat{I_s}}(x_1,\ldots,x_n)\,d\mathbf{x}_{\widehat{I_s}}, \quad \widehat{I_s} = (1_i,\ldots,\widehat{i_s},\ldots,i_k),$$

for which $\omega = d\alpha$ in an open neighborhood of $\mathbf{x} = \mathbf{0}$. Because

$$d\alpha = \sum_I \left(\sum_{s=1}^{k} (-1)^{s-1} \frac{\partial A_{\widehat{I_s}}}{\partial x_{i_s}} \right) d\mathbf{x}_I$$

(Theorem 10.19, p. 441), the condition $\omega = d\alpha$ yields the following $\binom{n}{k}$ partial differential equations

$$P_I = \sum_{s=1}^{k} (-1)^{s-1} \frac{\partial A_{\widehat{I_s}}}{\partial x_{i_s}}$$

for the $\binom{n}{k-1}$ unknown functions $A_{\widehat{I_s}}$. To obtain these functions, we follow the same steps as in the example above (pp. 497–499).

Step 1

First, restrict the multi-index I to the case where $i_k = n$. Consider the new multi-index $J = (i_1,\ldots,i_{k-1})$ with $i_{k-1} \le n-1$. If we define $\widehat{J_s}$ by analogy with $\widehat{I_s}$, then the restriction $i_k = n$ means that $I = J, n$ and

$$\widehat{I_s} = \widehat{J_s}, n \ \text{ if } s < k, \quad \widehat{I_k} = J.$$

Now consider the partial differential equation for which $I = J, n$:

$$P_{J,n} = \frac{\partial A_{\widehat{J_1},n}}{\partial x_{i_1}} - \frac{\partial A_{\widehat{J_2},n}}{\partial x_{i_2}} + \cdots + (-1)^{k-1} \frac{\partial A_J}{\partial x_n}.$$

Following the example, we begin the process of determining the functions $A_{\widehat{I_s}}$ by setting

$$A_{\widehat{J_1},n}(x_1,\ldots,x_n) = \cdots = A_{\widehat{J_{k-1}},n}(x_1,\ldots,x_n) \equiv 0.$$

Because the multi-index $\widehat{J_s}$ selects $k-2$ distinct elements from the first $n-1$ positive integers, these equations for $A_{\widehat{J_s},n}$ determine $\binom{n-1}{k-2}$ of the functions we seek.

Each partial differential equation for which $I = J, n$ thus reduces to a single term on the right and involves only one unknown function, A_J. We write this equation in the form

$$\frac{\partial A_J}{\partial x_n} = (-1)^{k-1} P_{J,n}.$$

Integration yields

$$A_J(x_1,\ldots,x_n) = A_J(x_1,\ldots,x_{n-1},0) + (-1)^{k-1} \int_0^{x_n} P_{J,n}(x_1,\ldots,x_{n-1},t)\,dt.$$

This determines A_J in terms of its values on the hyperplane $x_n = 0$ (and the values of the known function $P_{J,n}$ in the window), but we have not yet determined the values of A_J on that hyperplane.

There are $\binom{n-1}{k-1}$ functions A_J defined by these integrals; together with the $\binom{n-1}{k-2}$ functions already set equal to zero, we have identified all the $\binom{n}{k-1}$ unknown functions, because (cf. Exercise 11.26)

$$\binom{n-1}{k-1} + \binom{n-1}{k-2} = \binom{n}{k-1}.$$

We now determine the functions A_J on the hyperplane $x_n = 0$. In Step 1 we exhausted the possibility that $i_k = n$, so from this point on we take $i_k < n$. This means that dx_n no longer appears in any basic differential $d\mathbf{x}_I$. Because x_n itself no longer appears as a variable on the hyperplane $x_n = 0$, we have reduced the setting to differential forms on $\mathbb{R}^{n-1}$. Therefore, if we can pull back ω and α to forms ω^* and α^* on $\mathbb{R}^{n-1}$ in such a way that $d\omega^* = 0$ and $d\alpha^* = \omega^*$ on $\mathbb{R}^{n-1}$, we can use the induction hypothesis to obtain the functions A_J.

Step 2

Consider the map $\mathbf{f} \colon \mathbb{R}^{n-1} \to \mathbb{R}^n : \mathbf{y} \to \mathbf{x}$ into the hyperplane $x_n = 0$:

$$\mathbf{f} : \begin{cases} x_1 = y_1, \\ \quad \vdots \\ x_{n-1} = y_{n-1}, \\ x_n = 0. \end{cases}$$

By Theorem 10.20, page 443, and the discussion preceding it, the pullback of $\mathbf{f}$ on a basic k-form is

$$\mathbf{f}^* dx_I = \begin{cases} d\mathbf{y}_I & \text{if } i_k < n, \\ 0 & \text{if } i_k = n. \end{cases}$$

This suggests we define a new multi-index $I^* = I$ with the restriction that $i_k \leq n - 1$. The pullback of ω is then

$$\omega^*(y_1, \dots, y_{n-1}) = \mathbf{f}^* \omega = \sum_{I^*} P_{I^*}(y_1, \dots, y_{n-1}, 0)\, d\mathbf{y}_{I^*}.$$

Because d and $\mathbf{f}^*$ commute, ω^* is closed:

$$d\omega^* = d\mathbf{f}^* \omega = \mathbf{f}^* d\omega = \mathbf{f}^* 0 = 0.$$

For α we have

$$\alpha^*(y_1, \dots, y_{n-1}) = \mathbf{f}^* \alpha = \sum_{\widehat{I}^*_s} A_{\widehat{I}^*_s}(y_1, \dots, y_{n-1}, 0)\, d\mathbf{y}_{\widehat{I}^*_s},$$

$$d\alpha^*(y_1, \dots, y_{n-1}) = d\mathbf{f}^* \alpha = \mathbf{f}^* d\alpha = \sum_{I^*} \left(\sum_{s=1}^{k} (-1)^{s-1} \frac{\partial A_{\widehat{I}^*_s}}{\partial x_{i_s}} \right) d\mathbf{y}_{I^*},$$

and the condition $\omega = d\alpha$ implies

$$\omega^* = \mathbf{f}^*\omega = \mathbf{f}^* d\alpha = d\mathbf{f}^*\alpha = d\alpha^*.$$

The equation $\omega^* = d\alpha^*$ then implies the partial differential equations

$$P_{I^*}(y_1,\ldots,y_{n-1},0) = \sum_{s=1}^{k}(-1)^{s-1}\frac{\partial A_{\widehat{I^*_s}}}{\partial x_{i_s}}(y_1,\ldots,y_{n-1},0).$$

By the induction hypothesis, the algorithm supplies solutions

$$A_{\widehat{I^*_s}}(y_1,\ldots,y_{n-1},0)$$

to these differential equations. The multi-index $\widehat{I^*}_s$ is an increasing sequence of $k-1$ positive integers between 1 and $n-1$, so it equals a unique multi-index heretofore written as J. Conversely, suppose J is an arbitrary $(k-1)$-multi-index. Because $k-1 \leq n-2$, at least one integer j in the range from 1 to $n-1$ is missing from J. Let I^* be the k-multi-index constructed by augmenting J by inserting j in the proper place. If j is in the ℓth place in I^*, then $\widehat{I^*}_\ell = J$. Thus, every J equals some $\widehat{I^*}_s$, and we have determined all the functions

$$A_J(x_1,\ldots,x_{n-1},0)$$

that remained to be found at the end of Step 1.

Step 3   Steps 1 and 2 together give us all the functions $A_J(x_1,\ldots,x_n)$ defined in a window centered at $\mathbf{0}$ in $\mathbb{R}^n$, but as yet we know only that those functions satisfy the partial differential equations when $x_n = 0$. We can put the matter this way (cf. Step 3 of the example). The partial differential equation indexed by I^* can be written as $E_{I^*} = 0$, where

$$E_{I^*}(x_1,\ldots,x_n) = \sum_{s=1}^{k}(-1)^{s-1}\frac{\partial A_{\widehat{I^*_s}}}{\partial x_{i_s}}(x_1,\ldots,x_n) - P_{I^*}(x_1,\ldots,x_n).$$

By Step 2, E_{I^*} equals zero when $x_n = 0$ (and when $(x_1,\ldots,x_n)$ is in the window). We claim E_{I^*} remains equal to zero for all x_n in an open interval centered at 0. To prove the claim, it is enough to show $\partial E_{I^*}/\partial x_n = 0$.

In the example, we invoked the integrability conditions to prove the claim. For the same reason, we invoke them here. The integrability conditions are the coefficients of the $(k+1)$-form $d\omega$ set equal to zero; therefore we begin by expressing ω in a way that allows us to read off those coefficients of $d\omega$. To index $(k+1)$-forms, let $L = (i_1,\ldots,i_{k+1})$, $1 \leq i_i < \cdots < i_{k+1} \leq n$, and write

$$\omega = \sum_{\widehat{L}_s} P_{\widehat{L}_s}(x_1,\ldots,x_n)\,d\mathbf{x}_{\widehat{L}_s},$$

where $s = 1,\ldots,k+1$. Then

$$dw = \sum_L \left(\sum_{s=1}^{k+1} (-1)^{s-1} \frac{\partial P_{\widehat{L}_s}}{\partial x_{i_s}} \right) d\mathbf{x}_L,$$

so the integrability conditions are

$$\sum_{s=1}^{k+1} (-1)^{s-1} \frac{\partial P_{\widehat{L}_s}}{\partial x_{i_s}} = 0.$$

There are $\binom{n}{k+1}$ such conditions, one for each multi-index $L = (i_1, \ldots, i_{k+1})$. We focus on those multi-indices L for which $i_{k+1} = n$. Then $L = I^*, n$ and

$$\widehat{L}_s = \widehat{I^*}_s, n \text{ if } s < k+1, \text{ and } \widehat{L}_{k+1} = I^*.$$

The integrability condition indexed by this L is

$$\sum_{s=1}^{k} (-1)^{s-1} \frac{\partial P_{\widehat{I^*}_s, n}}{\partial x_{i_s}} + (-1)^k \frac{\partial P_{I^*}}{\partial x_n} = 0.$$

Now let us determine $\partial E_{I^*} / \partial x_n$. For each multi-index I^*, we have

$$\frac{\partial E_{I^*}}{\partial x_n} = \sum_{s=1}^{k} (-1)^{s-1} \frac{\partial}{\partial x_n} \left(\frac{\partial A_{\widehat{I^*}_s}}{\partial x_{i_s}} \right) - \frac{\partial P_{I^*}}{\partial x_n} = \sum_{s=1}^{k} (-1)^{s-1} \frac{\partial}{\partial x_{i_s}} \left(\frac{\partial A_{\widehat{I^*}_s}}{\partial x_n} \right) - \frac{\partial P_{I^*}}{\partial x_n}$$

We know that each $\widehat{I^*}_s = J$ for a suitable multi-index J; thus we can write, using the partial differential equation for A_J from Step 1,

$$\frac{\partial A_{\widehat{I^*}_s}}{\partial x_n} = \frac{\partial A_J}{\partial x_n} = (-1)^{k-1} P_{J,n} = (-1)^{k-1} P_{\widehat{I^*}_s, n}.$$

Hence

$$\frac{\partial E_{I^*}}{\partial x_n} = (-1)^{k-1} \sum_{s=1}^{k} (-1)^{s-1} \frac{\partial P_{\widehat{I^*}_s, n}}{\partial x_{i_s}} - \frac{\partial P_{I^*}}{\partial x_n},$$

so

$$(-1)^{k-1} \frac{\partial E_{I^*}}{\partial x_n} = \sum_{s=1}^{k} (-1)^{s-1} \frac{\partial P_{\widehat{I^*}_s, n}}{\partial x_{i_s}} + (-1)^k \frac{\partial P_{I^*}}{\partial x_n} = 0$$

by the integrability condition. This proves the claim, and thus establishes the algorithm.

Theorem 11.20 (Poincaré lemma). *Suppose ω is a closed k-form defined in a window centered at the origin in $\mathbb{R}^n$. Then there is a $(k-1)$-form α for which $\omega = d\alpha$ in that window. The coefficients of α can be obtained from the coefficients of ω by integration (quadrature).* $\square$

The $(k-1)$-form α in the Poincaré lemma is not unique: if γ is any $(k-2)$-form, then $\omega = d(\alpha + d\gamma)$. In fact, it follows from the Poincaré lemma that *all* $(k-1)$-forms β with $\omega = d\beta$ can be expressed this way.

Corollary 11.21 *If $\omega = d\alpha = d\beta$, then locally $\beta = \alpha + d\gamma$ for some properly chosen $(k-2)$-form γ.*

Proof. Note that $\alpha - \beta$ is a closed $(k-1)$-form; therefore, by the Poincaré lemma it is locally exact. $\square$

The effect
of the domain

The Poincaré lemma says that a closed form will be exact if its domain is sufficiently simple. The closed 1-form

$$\omega = \frac{-y\,dx + x\,dy}{x^2 + y^2},$$

whose domain is the punctured plane $\mathcal{P} = \mathbb{R}^2 \setminus (0,0)$, shows that exactness may be lost if the domain is even slightly complicated. This example is not isolated; there is an analogue of ω in every dimension. We explore them now to get a better idea how the shape of a domain can become an obstruction to the exactness of a closed form.

We take the domain to be punctured 3-space $Q = \mathbb{R}^3 \setminus (0,0,0)$, and define the 2-form

$$\beta = X\,dy\,dz + Y\,dz\,dx + Z\,dx\,dy$$

by

$$X = \frac{x}{(x^2 + y^2 + z^2)^{3/2}}, \quad Y = \frac{y}{(x^2 + y^2 + z^2)^{3/2}}, \quad Z = \frac{z}{(x^2 + y^2 + z^2)^{3/2}}.$$

Because

$$d\beta = \left(\frac{\partial X}{\partial x} + \frac{\partial Y}{\partial y} + \frac{\partial Z}{\partial z} \right) dx\,dy\,dz$$

and

$$\frac{\partial X}{\partial x} = \frac{(x^2 + y^2 + z^2)^{3/2} - x \cdot 3x(x^2 + y^2 + z^2)^{1/2}}{(x^2 + y^2 + z^2)^3} = \frac{x^2 + y^2 + z^2 - 3x^2}{(x^2 + y^2 + z^2)^{5/2}},$$

$$\frac{\partial Y}{\partial y} = \frac{x^2 + y^2 + z^2 - 3y^2}{(x^2 + y^2 + z^2)^{5/2}}, \quad \frac{\partial Z}{\partial z} = \frac{x^2 + y^2 + z^2 - 3z^2}{(x^2 + y^2 + z^2)^{5/2}},$$

β is closed, but...

we see $d\beta = 0$ everywhere on Q.

If β were exact, so that $\beta = d\alpha$ for some 1-form α defined on Q, then we would have

$$\iint_{\vec{S}^2} \beta = \iint_{\vec{S}^2} d\alpha = \int_{\partial \vec{S}^2} \alpha = \int_{\emptyset} \alpha = 0,$$

where $\vec{S}^2$ is the outwardly oriented unit sphere in $\mathbb{R}^3$. The path integral equals zero because $\partial \vec{S}^2$ is empty. However, because $x^2 + y^2 + z^2 = 1$ on S^2, β reduces to radial flow out of the sphere (cf. p. 396), so

$$\iint_{\vec{S}2} \beta = \iint_{\vec{S}2} x\,dy\,dz + y\,dz\,dx + z\,dx\,dy = 4\pi \neq 0.$$

Consequently, β is not exact on Q.

... β is not exact

The 2-form β is the prototype for the following sequence of examples, one in each dimension. Let $Q_n = \mathbb{R}^n \setminus \mathbf{0}$, and let

An analogue of β
in each dimension

$$\beta_{n-1}(\mathbf{x}) = \sum_{s=1}^{n} (-1)^{s-1} \frac{x_s}{r^n} \, d\mathbf{x}_{\widehat{N}_s},$$

where $\mathbf{x} = (x_i, \dots, x_n)$, $r = \|\mathbf{x}\|$, $N = (1, \dots, n)$, and $\widehat{N}_s = (1, \dots, \widehat{s}, \dots, n)$. Note that

$$\beta_1 = \frac{x_2\,dx_1 - x_1\,dx_2}{x_1^2 + x_2^2} \quad \text{and} \quad \beta_2 = \frac{x_1\,dx_2\,dx_3 - x_2\,dx_1\,dx_3 + x_3\,dx_1\,dx_2}{(x_1^2 + x_2^2 + x_3^2)^{3/2}},$$

so the $(n-1)$-form β_{n-1} generalizes ω and β. We have

$$d\beta_{n-1} = \sum_{s=1}^{n} \frac{\partial}{\partial x_s}\left(\frac{x_s}{r^n}\right)(-1)^{s-1}\,dx_s\,dx_{\widehat{N}_s} = \sum_{s=1}^{n} \frac{r^2 - nx_s^2}{r^{n+2}}\,d\mathbf{x}_N = 0,$$

so β_{n-1} is closed. If β_{n-1} were exact, then we would have $\beta_{n-1} = d\alpha_{n-2}$ for some $(n-2)$-form defined on Q_n. Let $\vec{S}^{n-1}$ be the unit $(n-1)$-sphere in $\mathbb{R}^n$, oriented by its outward normal; as a set, S^{n-1} consists of all points $\mathbf{x}$ in $\mathbb{R}^n$ for which $r = 1$. Then we would have

$$\iint \cdots \int_{\vec{S}^{n-1}} \beta_{n-1} = \iint \cdots \int_{\vec{S}^{n-1}} d\alpha_{n-2} = \int \cdots \int_{\partial \vec{S}^{n-1}} \alpha_{n-2} = 0,$$

because $\partial \vec{S}^{n-1} = \emptyset$. Nevertheless, β is not exact, because

$$\iint \cdots \int_{\vec{S}^{n-1}} \beta_{n-1} \neq 0.$$

This follows immediately from the general n-dimensional Stokes' theorem (which we do not prove). In dimension $n = 4$, however, you can prove by a direct computation (cf. Exercise 11.13) that

$$\iiint_{\vec{S}^3} \beta_3 = 2\pi^2.$$

Let us resume our analysis of $Q = \mathbb{R}^3 \setminus (0,0,0)$. We distinguish between two kinds of 2-spheres in Q: those that enclose the origin, and those that do not. Each oriented sphere $\vec{S}_I$ of the first kind is the boundary of an oriented ball $\vec{B}_I$ that contains the origin. But the origin is not in Q, so, although $\vec{S}_I$ is the boundary of a ball in $\mathbb{R}^3$, it is not the boundary of a ball in Q. By contrast, each oriented sphere $\vec{S}_{II}$ of the second kind is the boundary of an oriented ball $\vec{B}_{II}$ that lies entirely in Q. For spheres of the first kind, we have the following result.

The two kinds
of spheres in Q

Theorem 11.22. *Suppose the origin lies in the interior of the positively oriented ball* $\vec{B}_I$ *in* $\mathbb{R}^3$. *Let* $\vec{S}_I = \partial \vec{B}_I$ *in* $\mathbb{R}^3$; *then*

$$E(\vec{S}_I) = \frac{1}{4\pi} \iint_{\vec{S}_I} \beta = +1.$$

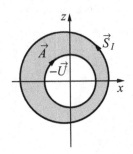

Proof. We show that $E(\vec{S}_I) = E(\vec{U})$, where $\vec{U}$ is the unit sphere (centered at the origin) with its outward orientation; we have already established (p. 504) that $E(\vec{U}) = +1$.

Suppose the given sphere, $\vec{S}_I$, lies everywhere outside $\vec{U}$, and suppose that $\vec{A}$ is the 3-dimensional positively oriented shell that lies between the two spheres; then $\partial \vec{A} = \vec{S}_I - \vec{U}$. The divergence theorem applies to β on $\vec{A}$; we thus find

$$E(\vec{S}_I) - E(\vec{U}) = \frac{1}{4\pi} \iint_{\vec{S}_I} \beta - \frac{1}{4\pi} \iint_{\vec{U}} \beta = \frac{1}{4\pi} \iint_{\partial \vec{A}} \beta = \frac{1}{4\pi} \iiint_{\vec{A}} d\beta = 0,$$

because β is closed on Q.

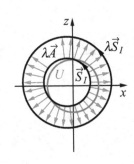

Even if $\vec{S}_I$ does not lie entirely outside the unit sphere U, some concentric enlargement $\lambda \vec{S}_I$ does. If $\lambda \vec{A}$ is the positively oriented 3-dimensional shell that lies between these concentric spheres, then $\partial(\lambda \vec{A}) = \lambda \vec{S}_I - \vec{S}_I$. Just as in the previous case, the divergence theorem applies, giving $E(\lambda \vec{S}_I) - E(\vec{S}_I) = 0$. Thus, in all cases, $E(\vec{S}_I) = E(\vec{U}) = +1$. $\quad\square$

Corollary 11.23 *If the sphere* $\vec{S}_I$ *encloses the origin and has inward orientation, then* $E(\vec{S}_I) = -1$. $\quad\square$

Theorem 11.24. *If* $\vec{B}_{II}$ *is an oriented ball that lies entirely in* Q *and* $\vec{S}_{II} = \partial \vec{B}_{II}$, *then* $E(\vec{S}_{II}) = 0$.

Proof. Because β is defined everywhere in an open neighborhood of $\vec{B}_{II}$, the divergence theorem applies, and

$$E(\vec{S}_{II}) = \frac{1}{4\pi} \iint_{\vec{S}_{II}} \beta = \frac{1}{4\pi} \iint_{\partial \vec{B}_{II}} \beta = \frac{1}{4\pi} \iiint_{\vec{B}_{II}} d\beta = 0. \quad\square$$

Differential forms that detect holes

We see that the function $E(\vec{S})$ plays the same role for oriented spheres in Q that the winding number

$$W(\vec{C}) = \frac{1}{2\pi} \oint_{\vec{C}} \beta_1, \quad \beta_1 = \frac{x\,dy - y\,dx}{x^2 + y^2},$$

(p. 430) plays for oriented circles $\vec{C}$ in the punctured plane. That is, each domain has a hole, and a nonzero value of the function indicates that the sphere or circle encloses that hole. The function is, in each case, determined by the differential form; thus we can just as well say it is the differential form that detects the hole.

Every closed 1-form on Q is exact

In the case of the 3-dimensional region Q, it is a 2-form that detects the hole. Is there a 1-form on Q that does the same thing? In other words, is there a 1-form α that is closed on Q but fails to be exact on Q?

Theorem 11.25. *Every closed* 1*-form* α *on* Q *is exact.*

Proof. We construct a function $\Phi(x,y,z)$ for which $d\Phi = \alpha$ on Q. Select a smooth path $\vec{C}$ in Q that starts at a fixed point (a,b,c) in Q and ends at an arbitrary point (x,y,z) in Q. Define

$$\Phi(x,y,z) = \int_{\vec{C}} \alpha;$$

for this to be meaningful, we must show the integral is path-independent.

Let $\vec{C}_1$ be any other smooth path in Q with the same starting and ending points as $\vec{C}$. Then $\vec{C}_1 - \vec{C}$ is a closed piecewise-smooth oriented path in Q, and is therefore the boundary of an oriented surface $\vec{\Sigma}$ that can be chosen to avoid the origin. In other words, $\vec{\Sigma}$ lies entirely in Q, so Stokes' theorem applies. Thus

$$\int_{\vec{C}_1} \alpha - \int_{\vec{C}} \alpha = \int_{\vec{C}_1 - \vec{C}} \alpha = \int_{\partial\vec{\Sigma}} \alpha = \iint_{\vec{\Sigma}} d\alpha = 0,$$

because $d\alpha = 0$ by hypothesis, so the integral is path-independent. $\square$

Because the exterior derivative on forms corresponds to the classical operations of the gradient, divergence, and curl on scalar and vector fields, we can translate the relations between closed and exact forms into relations between these operations. On pages 460–462, we made the following correspondence between fields and forms in $\mathbb{R}^3$.

$$f \leftrightarrow \omega_f^0 = f,$$

$$\mathbb{F} = (A,B,C) \leftrightarrow \omega_{\mathbb{F}}^1 = A\,dx + B\,dy + C\,dz,$$

$$\mathbb{V} = (P,Q,R) \leftrightarrow \omega_{\mathbb{V}}^2 = P\,dy\,dz + Q\,dz\,dx + R\,dx\,dy,$$

$$H \leftrightarrow \omega_H^3 = H\,dx\,dy\,dz.$$

Using the differential operator ∇ (i.e., *nabla*, Definition 3.3, p. 93) to express the classical operators,

$$\operatorname{grad} f = \nabla f, \quad \operatorname{curl} \mathbb{F} = \nabla \times \mathbb{F}, \quad \operatorname{div} \mathbb{V} = \nabla \cdot \mathbb{V},$$

we can express the correspondences between those operators and the exterior derivative in the following way.

$\operatorname{grad} f:\quad \nabla f = (f_x, f_y, f_z) \leftrightarrow d(\omega_f^0) = f_x\,dx + f_y\,dy + f_z\,dz$

$\operatorname{curl} \mathbb{F}:\quad \nabla \times (A,B,C) \leftrightarrow d(\omega_{\mathbb{F}}^1) = (C_y - B_z)\,dy\,dz + (A_z - C_x)\,dz\,dx$
$$+ (B_x - A_y)\,dx\,dy$$

$\operatorname{div} \mathbb{V}:\quad \nabla \cdot (P,Q,R) \leftrightarrow d(\omega_{\mathbb{V}}^2) = (P_x + Q_y + R_z)\,dx\,dy\,dz$

As already noted (pp. 460–462), these correspondences define the forms

$$\omega^1_{\operatorname{grad} f} = d\left(\omega^0_f\right) = f_x\, dx + f_y\, dy + f_z\, dz$$

$$\omega^2_{\operatorname{curl}\mathbb{F}} = d\left(\omega^1_{\mathbb{F}}\right) = (C_y - B_z)\, dy\, dz + (A_z - C_x)\, dz\, dx + (B_x - A_y)\, dx\, dy$$

$$\omega^3_{\operatorname{div}\mathbb{V}} = d\left(\omega^2_{\mathbb{V}}\right) = (P_x + Q_y + R_z)\, dx\, dy\, dz.$$

Theorem 11.26. *Suppose f is a scalar field, and $\mathbb{F}$ a vector field, and each is defined on an open set in $\mathbb{R}^3$; then*

$$\nabla \times \nabla f = \operatorname{curl}\operatorname{grad} f = \mathbf{0} \quad and \quad \nabla \cdot \nabla \times \mathbb{F} = \operatorname{div}\operatorname{curl}\mathbb{F} = 0.$$

Proof. These are just translations of $d^2 = 0$. □

The Poincaré lemma itself translates into the following pair of theorems.

Theorem 11.27. *Suppose $\mathbb{F}$ is a vector field and $\operatorname{curl}\mathbb{F} = \mathbf{0}$ in a neighborhood of some point in $\mathbb{R}^3$; then there is a scalar field Φ for which $\mathbb{F} = \nabla\Phi = \operatorname{grad}\Phi$ in a window centered at that point.* □

Theorem 11.28. *Suppose $\mathbb{V}$ is a vector field and $\operatorname{div}\mathbb{V} = 0$ in a neighborhood of some point in $\mathbb{R}^3$; then there is another vector field $\mathbb{P}$ for which*

$$\mathbb{V} = \nabla \times \mathbb{P} = \operatorname{curl}\mathbb{P}$$

in a window centered at that point. □

Vector and scalar potentials

The function Φ in Theorem 11.27 is a *potential function* of the vector field $\mathbb{F}$ (Definition 1.3, p. 25). Because the vector field $\mathbb{P}$ in Theorem 11.28 stands in the same relation to the field $\mathbb{V}$, we call $\mathbb{P}$ a **vector potential** for $\mathbb{V}$. (For the sake of clarity, we now refer to the function f as a **scalar potential** for the field $\mathbb{F}$ of Theorem 11.27.) By extension, we call α a *vector potential* for any k-form ω whenever $d\alpha = \omega$. The Poincaré lemma thus asserts the existence of a local vector potential for any closed k-form; indeed, Yap's result [22] is expressed in this language. By Corollary 11.21, a vector potential is not unique (when it exists).

Irrotational and incompressible flows

Suppose a fluid flow is represented by a continuously differentiable vector field $\mathbb{V}$. We say the flow is **irrotational** if $\operatorname{curl}\mathbb{V} = \nabla \times \mathbb{V} = \mathbf{0}$; we say it is **incompressible** if $\operatorname{div}\mathbb{V} = \nabla \cdot \mathbb{V} = 0$. In these terms the previous theorems say the following.

- A gradient flow is irrotational.

- A vortex flow (Definition 11.4, p. 473) is incompressible.

- An irrotational flow is locally a gradient.

- An incompressible flow is locally a vortex flow.

The Laplacian

There are meaningful ways to compose a pair of classical operators that do not correspond to the composition d^2. One is $\operatorname{div}\operatorname{grad} f = \nabla \cdot \nabla f$, where $f = f(x,y,z)$ is a scalar field. We have

$$\nabla f = \left(\frac{\partial f}{\partial x}, \frac{\partial f}{\partial y}, \frac{\partial f}{\partial z} \right) \quad \text{and} \quad \nabla \cdot \nabla f = \frac{\partial^2 f}{\partial x^2} + \frac{\partial^2 f}{\partial y^2} + \frac{\partial^2 f}{\partial z^2}.$$

The second-order differential operator

$$\nabla \cdot \nabla = \frac{\partial^2}{\partial x^2} + \frac{\partial^2}{\partial y^2} + \frac{\partial^2}{\partial z^2}$$

that appears here is called the **Laplacian**; it is also denoted as Δ:

$$\Delta f = \frac{\partial^2 f}{\partial x^2} + \frac{\partial^2 f}{\partial y^2} + \frac{\partial^2 f}{\partial z^2}.$$

The Laplacian operates on a scalar field (i.e., a function) to produce another scalar field. We can extend the Laplacian to vector fields by operating on each component. If $\mathbb{F} = (A, B, C)$, just set

$$\Delta \mathbb{F} = (\Delta A, \Delta B, \Delta C),$$

another vector field.

Two more classical composites that are similarly unrelated to d^2 are

$$\text{grad div}\, \mathbb{V} = \nabla(\nabla \cdot \mathbb{V}) \quad \text{and} \quad \text{curl}(\text{curl}\,\mathbb{F}) = \nabla \times (\nabla \times \mathbb{F}).$$

Each composite operates on a vector field to produce another vector field; the two are connected by the following identity (see Exercise 11.28):

$$\text{curl}(\text{curl}\,\mathbb{F}) = \text{grad}(\text{div}\,\mathbb{F}) - \text{div}(\text{grad}\,\mathbb{F})$$
$$\nabla \times (\nabla \times \mathbb{F}) = \nabla(\nabla \cdot \mathbb{F}) - \Delta \mathbb{F}.$$

Exercises

11.1. Calculate $\text{div}\, \mathbb{V} = \nabla \cdot \mathbb{V}$ when:

    a. $\mathbb{V} = (x \cos y, x \sin y, 0)$.

    b. $\mathbb{V} = (y + z, z + x, x + y)$.

    c. $\mathbb{V} = (x/yz, y/zx, z/xy)$.

    d. $\mathbb{V} = \text{grad}\, f$, $f(x, y, z) = ax + by + cz$.

    e. $\mathbb{V} = \text{grad}\, f$, $f(x, y, z)$ arbitrary.

11.2. Calculate $\nabla \times \mathbb{F} = \text{curl}\,\mathbb{F}$ when

    a. $\mathbb{F} = (yz, zx, xy)$.

    b. $\mathbb{F} = (y + z, z + x, x + y)$.

    c. $\mathbb{F} = \text{grad}\, f$, $f(x, y, z) = ax + by + cz$.

    d. $\mathbb{F} = \text{grad}\, \varphi$, $\varphi(x, y, z) = ax^2 + 2bxy + cy^2 + 2dyz + ez^2 + 2fzx$.

e. $\mathbb{F} = (y^2, z^2, x^2)$.

f. $\mathbb{F} = (z, x, y)$.

g. $\mathbb{F} = \operatorname{grad} H$, $H(x, y, z)$ arbitrary.

11.3. Calculate the **circulation** $\oint_C \mathbb{F} \cdot d\mathbf{s}$ of the field $\mathbb{F} = (z, 0, 0)$ around the closed oriented path $\vec{C}$ where:

a. $\vec{C} : (x, y, z) = (\cos t, \sin t, \frac{1}{3} \cos t)$, $0 \le t \le 2\pi$.

b. $\vec{C} : (x, y, z) = (0, 3\sin t, 3\cos t)$, $0 \le t \le 2\pi$.

c. $\vec{C}$ is the unit circle in the (x, y)-plane, traversed counterclockwise when viewed from the positive z-axis.

d. $\vec{C}$ is the circle in plane $z = 2$ of radius r centered at the point $(p, q, 2)$, traversed clockwise when viewed from a position where $z > 2$.

e. $\vec{C}$ is the circle in plane $y = 2$ of radius r centered at the point $(p, 2, q)$, traversed clockwise when viewed from a position where $y < 2$.

f. $\vec{C}$ is the rectangle with vertices $(0, 0, 0)$, $(1, 1, 0)$, $(0, 2, 1)$, $(-1, 1, 1)$, traversed in that order.

11.4. For each curve $\vec{C}$ in Exercise 11.3, let $\vec{R}$ be the plane region whose boundary is $\vec{C}$: $\partial \vec{R} = \vec{C}$. Now compute the surface integral

$$\iint_{\vec{R}} (\operatorname{curl} \mathbb{F} \cdot \mathbf{n}) \, dA.$$

(You might wish to rewrite this in a form that is more amenable to computation.)

By (the original version of) Stokes' theorem, this surface integral should have the same value as the path integral that gave the circulation in the corresponding problem. Do your values agree?

11.5. Calculate the flux of $\mathbb{V} = (x, y, z)$ out of the positively oriented sphere $\vec{S}$ of radius R centered at the origin $(x, y, z) = (0, 0, 0)$ in two ways:

a. First, directly as $\iint_{\vec{S}} \mathbb{V} \cdot \mathbf{n} \, dA$.

b. Second, by calculating the divergence of $\mathbb{V}$ over the interior of $\vec{S}$ and using the divergence theorem. State the theorem and indicate how you are using it.

Do the two values agree?

11.6. Let $\vec{P}$ be the positively oriented rectangular parallelepiped $\vec{P}$ in (x, y, z)-space with $0 \le x \le 5$, $0 \le y \le 3$, $0 \le z \le 2$. Calculate the flux of $\mathbb{V} = (-y, x, 0)$ out of $\vec{P}$ in two ways:

a. First, directly as $\iint_{\partial \vec{P}} \mathbb{V} \cdot \mathbf{n} \, dA$.

b. Second, by calculating the divergence of $\vec{V}$ over $\vec{P}$ and using the divergence theorem.

Do the two values agree?

11.7. Suppose S is a closed surface in $\mathbb{R}^3$, and the vector $\mathbf{x}$ points from the origin to an arbitrary point on S. Show that

$$\tfrac{1}{3} \iint_S (\mathbf{x} \cdot \mathbf{n}) \, dA = \text{vol}(S),$$

where $\mathbf{n}$ is the outward unit normal to S at $\mathbf{x}$.

11.8. Prove the divergence theorem for the positively oriented unit tetrahedron $\vec{B}$ (Theorem 11.3) in the form

$$\iiint_{\vec{B}} d\alpha = \iint_{\partial \vec{B}} \alpha,$$

where $\alpha = P\,dy\,dz + Q\,dz\,dx + R\,dx\,dy$.

11.9. Use the divergence theorem to calculate the flux of $\mathbf{V} = (x, y, z)$ out of the sphere S of radius R in (x, y, z)-space. Compare your answer with an earlier calculation of the same quantity (Exercise 10.4, p. 444).

11.10. Use the divergence theorem to calculate the flux of $\mathbf{V} = (-y, x, 0)$ out of the rectangular parallelepiped P in (x, y, z)-space given by

$$0 \le x \le 5, \quad 0 \le y \le 3, \quad 0 \le z \le 2.$$

Compare your answer with an earlier calculation of the same quantity (Exercise 10.5, p. 444).

11.11. Show that if two bilinear forms $A(\mathbf{v}, \mathbf{w})$ and $B(\mathbf{v}, \mathbf{w})$ defined on $\mathbb{R}^n$ agree on a basis for $\mathbb{R}^n$, they agree everywhere.

11.12. Suppose there is a continuously differentiable, single-valued function $\Phi(x, y)$ defined everywhere on the "punctured plane" $\mathcal{P} : \mathbb{R}^2 \setminus (0, 0)$ for which

$$d\Phi = \omega = (x\,dy - y\,dx)/(x^2 + y^2).$$

Show, in the following steps, that this assumption leads to a contradiction.

a. Let $\varphi(t) = \Phi(\cos t, \sin t)$. Show that $\varphi'(t) = 1$ for all t.
b. Deduce that $\varphi(t) = t + \Phi(1, 0)$ and then that $\Phi(1, 0) = 2\pi + \Phi(1, 0)$. The contradiction is then $0 = 2\pi$.

11.13. Show that $\iiint_{\vec{S}^3} \beta_3 = 2\pi^2$ (cf. p. 505). The following provides one approach.

a. Parametrize S^3 with the "spherical coordinates" map $\mathbf{x} = \mathbf{s}(1, t_1, t_2, t_3)$ (Exercise 5.25, p. 183) and show that $\mathbf{s}$ defines the positive (outward normal) orientation on S^3.

b. Show that $\mathbf{s}^*(\beta_3) = \cos t_2 \cos^2 t_3$; then integrate $\mathbf{s}^*(\beta_3)$ over the appropriat domain to obtain the result.

11.14. Show that $\mathbb{F} = (2x + y\cos xy, x\cos xy, 2z^2)$ has a scalar potential and find it.

11.15. Does $\mathbb{F} = (2x + y\cos xy, x\cos xy, 2z^2)$ have a vector potential? If so, find it; if not, explain why not.

11.16. a. Does $\mathbb{V} = (y+z, z+x, x+y)$ have a scalar potential? If so, find it; if not, explain why not.

b. Show that $\mathbb{V}$ has a vector potential and find it.

c. Find another vector potential for $\mathbb{V}$ that has the form $(0, B, C)$, for suitable functions $B(x,y,z)$ and $C(x,y,z)$. Suggestion: Take your solution $\mathbb{F} = (P, Q, R)$ to part (a) and construct a function f for which $\partial f / \partial x = -P$. Then $\mathbb{F} + \text{grad} f$ will solve the problem. Explain why.

11.17. Show that the following system of partial differential equations has a solution (consisting of three functions $P(x,y,z)$, $Q(x,y,z)$, and $R(x,y,z)$), and find the solution.

$$\frac{\partial R}{\partial y} - \frac{\partial Q}{\partial z} = yz, \quad \frac{\partial P}{\partial z} - \frac{\partial R}{\partial x} = zx, \quad \frac{\partial Q}{\partial x} - \frac{\partial P}{\partial y} = xy.$$

Explain how this question is connected with divergence, gradient, and curl.

11.18. Explain why the following system of partial differential equations has no solution $f(x,y,z)$.

$$\frac{\partial f}{\partial x} = -y+x, \quad \frac{\partial f}{\partial y} = x+y, \quad \frac{\partial f}{\partial z} = z.$$

Explain how this question is connected with divergence, gradient, and curl.

11.19. Explain why the following system of partial differential equations has no solutions $P(x,y,z)$, $Q(x,y,z)$, and $R(x,y,z)$.

$$\frac{\partial R}{\partial y} - \frac{\partial Q}{\partial z} = x, \quad \frac{\partial P}{\partial z} - \frac{\partial R}{\partial x} = y, \quad \frac{\partial Q}{\partial x} - \frac{\partial P}{\partial y} = z.$$

Explain how this question is connected with divergence, gradient, and curl.

11.20. Let $\omega(x,y,u,v)$ be the 2-form in $\mathbb{R}^4$ defined by

$$\omega = A\,dx\,dy + B\,dy\,du + C\,du\,dv + D\,dv\,dx + E\,dx\,du + F\,dy\,dv.$$

a. Show that if ω is closed (i.e., $d\omega = 0$), then the coefficients of ω must satisfy the following partial differential equations:

$$A_u + B_x - E_y = 0 \qquad C_x + D_u + E_v = 0$$
$$B_v + C_y - F_u = 0 \qquad D_y + A_v + F_x = 0$$

b. Suppose ω is exact; then $\omega = d\alpha$ for some 1-form $\alpha = P\,dx + Q\,dy + R\,du + S\,dv$. Show that the coefficients P, Q, R, and S of α must satisfy the following partial differential equations in terms of the known coefficients $A, \ldots, F$ of ω.

$$Q_x - P_y = A \qquad R_y - Q_u = B \qquad S_u - R_v = C$$
$$P_v - S_x = D \qquad R_x - P_u = E \qquad S_y - Q_v = F$$

11.21. Let $\omega = A\,dx\,dy + B\,dy\,dz + C\,dz\,dx$ be a 2-form in $\mathbb{R}^3$.

    a. Assume that ω is closed. Determine the partial differential equation (there is only one) that the coefficients A, B, and C must satisfy.

    b. Assume that ω is exact, with $\omega = d\alpha$, where $\alpha = P\,dx + Q\,dy + R\,dz$. Determine the conditions that the coefficients P, Q, and R of α must satisfy. (The conditions are three partial differential equations for P, Q, and R in terms of the given A, B, and C.)

11.22. Let $\beta = P\,dx + Q\,dy + R\,dz$ be a 1-form in $\mathbb{R}^3$.

    a. Assume that β is closed. Determine the conditions that the coefficients P, Q, and R must satisfy.

    b. Assume that β is exact, so $\beta = df$ for some function $f(x,y,z)$ on $\mathbb{R}^3$. Determine the conditions that f must satisfy.

11.23. Suppose the 1-form $\omega = P\,dx + Q\,dy + dz$ is closed in a window W centered at the point $(x,y,z) = (a,b,c)$. By analogy with Theorem 11.18, the exterior derivative of 0-form

$$\Phi(x,y,z) = \int_a^x P(t,b,c)\,dt + \int_b^y Q(x,t,c)\,dt + \int_c^z R(x,y,t)\,dt$$

should equal ω in W. Write down the integrability conditions defined by $d\omega = 0$; then use those conditions to establish $\Phi_x = P$, $\Phi_y = Q$, $\Phi_z = R$, and hence to prove that $d\Phi = \omega$.

11.24. Extend the result of the previous exercise to n dimensions, as follows. Suppose the 1-form $\omega = P_1\,dx_1 + \cdots P_n\,dx_n$ is closed in a window W centered at $(a_1, \ldots, a_n)$.

    a. Write down the integrability conditions described by $d\omega = 0$.

    b. Express the 0-form Φ as a sum of integrals of the various coefficients P_i of ω. The expression must reduce to the one in the previous exercise (*mutatis mutandis*) if $n = 3$.

    c. Use the integrability conditions to establish $\partial\Phi/\partial x_i = P_i$, $i = 1, \ldots, n$, and hence to prove that $d\Phi = \omega$.

11.25. Verify the statements in the text (p. 497) about the number of partial differential equations and integrability conditions that are involved in establishing local exactness of a closed form.

11.26. Show that
$$\binom{n-1}{k-1} + \binom{n-1}{k-2} = \binom{n}{k-1} \quad \text{if } n \geq k.$$

(Note: This yields "Pascal's triangle.")

11.27. The algorithm for the general Poincaré lemma (i.e., for a k-form in n variables) involves $\binom{n}{k+1}$ integrability conditions.

   a. Show that, in carrying out the induction step from $m-1$ variables to m variables, the algorithm uses only $\binom{m-1}{k}$ of those integrability conditions. Show, moreover, that different integrability conditions are used for each distinct value of m. Thus, as m successively takes the values $k+1, \ldots, n$, a total of
$$\sum_{m=k+1}^{n} \binom{m-1}{k}$$
   integrability conditions are used.

   b. Show that all integrability conditions are used in the algorithm by showing that
$$\sum_{m=k+1}^{n} \binom{m-1}{k} = \binom{n}{k+1}.$$

11.28. a. Express $\operatorname{curl}(\operatorname{curl}\mathbb{F}) = \nabla \times (\nabla \times \mathbb{F})$ in terms of the components of $\mathbb{F} = (A, B, C)$ and their derivatives.

   b. Interpret the identity $\mathbf{U} \times (\mathbf{V} \times \mathbf{W}) = (\mathbf{U} \cdot \mathbf{W})\mathbf{V} - (\mathbf{U} \cdot \mathbf{V})\mathbf{W}$ in a suitable way for $\nabla \times (\nabla \times \mathbb{F})$ to show that
$$\operatorname{curl}(\operatorname{curl}\mathbb{F}) = \operatorname{grad}(\operatorname{div}\mathbb{F}) - \operatorname{div}(\operatorname{grad}\mathbb{F}) = \nabla(\nabla \cdot \mathbb{F}) - (\nabla \cdot \nabla)\mathbb{F}.$$

11.29. Use the divergence theorem to prove that
$$\iiint_{R} \{\nabla f \cdot \nabla g + f(\nabla \cdot \nabla g)\} \, dV = \iint_{\partial R} (\mathbf{n} \cdot f\nabla g) \, dA,$$

where R is a region in $\mathbb{R}^3$, $\mathbf{n}$ is the unit normal outward on the surface ∂R, and f and g are smooth functions defined on R.

References

1. Buck, R.C.: Advanced Calculus. McGraw–Hill, New York (1956)
2. Callahan, J.: Singularities and plane maps. Amer. Math. Monthly **81**(3), 211–240 (1974)
3. Courant, R.C., John, F.: Introduction to Calculus and Analysis. Springer-Verlag, New York (1999)
4. Dieudonné, J.: Foundations of Modern Analysis. Academic Press, New York (1960)
5. Edwards, H.M.: Advanced Calculus: A Differential Forms Approach. Birkhäuser, Boston (1994)
6. Feynman, R.P., Leighton, R.B., Sands, M.: The Feynman Lectures on Physics. Addison–Wesley, Reading, MA (1963)
7. Flanders, H.: Differential Forms with Applications to the Physical Sciences. Academic Press, New York (1963)
8. Gleason, A.M.: The geometric content of advanced calculus. In: [12], CUPM Geometry Conference, Part II. Mathematical Association of America (1967)
9. Kaplan, W.: Advanced Calculus. Addison–Wesley, Reading, MA (1952)
10. Lang, S.: Differential Manifolds. Addison–Wesley, Reading, MA (1972)
11. Lang, S.: Differential and Riemannian Manifolds. Springer-Verlag, New York (1995)
12. Mathematical Association of America: CUPM Geometry Conference, Part II: Geometry in Other Subjects. Committee on the Undergraduate Program in Mathematics (CUPM), Berkeley, CA (1967)
13. Morse, M.: Relations between the critical points of a real function of n independent variables. Trans. Amer. Math. Soc. **27**(3), 345–396 (1925)
14. Protter, M.H., Morrey, C.B.: A First Course in Real Analysis, second edn. Springer-Verlag, New York (1991)
15. Rudin, W.: Principles of Mathematical Analysis, second edn. McGraw–Hill, New York (1976)
16. Schwartz, J.: The formula for change in variables in a multiple integral. Amer. Math. Monthly **61**(2), 81–85 (1954)
17. Steenrod, N.: The geometric content of freshman and sophomore mathematics courses. In: [12], CUPM Geometry Conference, Part II. Mathematical Association of America (1967)
18. Sylvester, J.J.: A demonstration of the theorem that every homogeneous quadratic polynomial is reducible by real orthogonal substitutions to the form of a sum of positive and negative squares. Philos. Mag. **IV**, 138–142 (1852)
19. Taylor, A.: Advanced Calculus. Ginn, Boston (1955)
20. Whitney, H.: On singularities of mappings of euclidean spaces, I. Mappings of the plane to the plane. Ann. of Math. **62**(3), 374–410 (1955)
21. Widder, D.V.: Advanced Calculus, second edn. Prentice–Hall, Englewood clifs, NJ (1961)
22. Yap, S.L.: The Poincaré lemma and an elementary construction of vector potentials. Amer. Math. Monthly **116**(3), 261–267 (2009)

Index

Printed in the United States
By Bookmasters